FORTSCHRITTE DER CHEMIE ORGANISCHER NATURSTOFFE

PROGRESS IN THE CHEMISTRY OF ORGANIC NATURAL PRODUCTS

BEGRÜNDET VON · FOUNDED BY

L. ZECHMEISTER

HERAUSGEGEBEN VON · EDITED BY

W. HERZ H. GRISEBACH G. W. KIRBY
TALLAHASSEE, FLA. FREIBURG i. BR. GLASGOW

VOL. 32

VERFASSER · AUTHORS

R. J. HIGHET · H. KÖSSEL · P. G. SAMMES
P. M. SCOPES · W. K. SEIFERT · H. SELIGER
E. A. SOKOLOSKI · H. C. VAN HUMMEL

1975

WIEN · SPRINGER-VERLAG · NEW YORK

Mit 31 Abbildungen · With 31 Figures

Library of Congress Catalog Card Number AC 39-1015

ISBN-13: 978-3-7091-7085-4   e-ISBN-13: 978-3-7091-7083-0
DOI: 10.1007/978-3-7091-7083-0

Inhaltsverzeichnis. Contents

Contents

Inhaltsverzeichnis. Contents VII

Mitarbeiterverzeichnis. List of Contributors

Highet, Dr. R. J., Department of Health, Education, and Welfare, Public Health Service, National Institutes of Health, National Heart and Lung Institute, Bethesda, MD 20014, U.S.A.

Kössel, Prof. Dr. H., Institut für Biologie III, Universität Freiburg, Schänzlestraße 9—11, D-7800 Freiburg i. Br., Bundesrepublik Deutschland.

Sammes, Dr. P. G., Department of Chemistry, Imperial College of Science and Technology, South Kensington, London SW7 2AY, United Kingdom.

Scopes, Dr. P. M., Department of Chemistry, Westfield College (University of London), Hampstead, London NW3 7ST, United Kingdom.

Seifert, Dr. W. K., Senior Research Associate, Chevron Oil Field Research Company, P.O. Box 1627, Richmond, CA 94802, U.S.A.

Seliger, Dr. H., Institut für makromolekulare Chemie, Universität Freiburg, Stefan-Meier-Straße 31, D-7800 Freiburg i. Br., Bundesrepublik Deutschland

Sokoloski, E. A., Department of Health, Education, and Welfare, Public Health Service, National Institutes of Health, National Heart and Lung Institute, Bethesda, MD 20014, U.S.A.

van Hummel, Dr. H. C., Botanisch Laboratorium, Faculteit der Wiskunde en Natuurwetenschappen, Katholieke Universiteit Nijmegen, Toernooiveld, Nijmegen, Netherlands.

Carboxylic Acids in Petroleum and Sediments

By W. K. SEIFERT, Chevron Oil Field Research Company,
Richmond, California, U.S.A.

With 12 Figures

Contents

I. General

1. Introduction

Historically, petroleum and bitumen in sediments were the step-children of natural product chemistry. One of the reasons for this phenomenon is the stigma that petroleum is a product of industry and not of nature. This concept was changed drastically by TREIBS' (*1*) discovery of porphyrins in petroleum in the 1930's which signaled its biological origin and represented the birth of modern organic geochemistry. Hundreds of papers have subsequently appeared documenting the biological fossil nature of petroleum.

The diagenesis of petroleum is presently understood as a complex process of conversion of the original lipid fraction of biological systems to hydrocarbons and to a lesser extent to polar compounds of varying degree of thermodynamic stability during sedimentation and maturation. Expulsion of mobile liquid reaction products from the sediment at elevated temperature and pressure (due to increased depths of burial) and subsequent migration from the source rock into porous reservoir rocks leads to formation of petroleum reservoirs. Consequently, compounds of similar structures are found in bitumen extracted from sediments and in petroleum; and, therefore, discussions of chemical composition of bitumen in sediments and petroleum from a natural products point of view belong together. Of the compounds structurally elucidated by organic petroleum chemists, those resembling the structures occurring in the living organisms are the most intriguing. They have been called petroleum biological markers and include such classes as porphyrins, isoprenoid, terpenoid, and steroid hydrocarbons, as well as an assembly of carboxylic acids (fatty, isoprenoid, steroid, amino, etc.), and most recently sterols. The amount of information regarding the origin, maturation, migration, and accumulation of petroleum which one can derive from such markers is directly proportional to the degree of molecular complexity. With the advent of modern instrumentation, rapid progress was achieved mainly elucidating the structures in the nonpolar, high molecular weight ($C_{15}-C_{40}$) hydrocarbon portion of petroleum. The progress in obtaining information on the structure of petroleum carboxylic acids which had always been retarded over that of petroleum hydrocarbons because of the enormous complexity of the mixtures and the difficulty of isolation and separation has received a boost to the extent that it appears safe to say that more information on the structure of petroleum carboxylic acids has been developed in the last decade than since the appearance of the first paper (*2*) on the subject 90 years ago.

References, pp. 44—49

It is the purpose of this review to summarize the progress on elucidating the structure of the monofunctional carboxylic acids and on the biogeochemical interpretations of the results. The last review on the subject dealing with petroleum alone and not with sediments dates back to 1955 (3). This book (3) is an illustration of the difficulties encountered when one was limited to elemental analysis plus a few physical measurements (e.g., refractive index, melting point, boiling point, and optical rotation) of the acids and some synthetically obtained derivatives. These difficulties have deterred researchers from getting involved in carboxylic acid research. The review by LOCHTE reveals that up to that time with the exception of fatty acids, only two acids had been identified with as many as 10 carbon atoms. The structures identified were mainly of the straight or branched paraffinic and of the cyclopentane and cyclohexane types, thus the name "Naphthenic Acids" was given to this mixture. It was, however, realized that bicyclic acids ($C_{12} - C_{18}$) were present.

The commercial isolation of these "Naphthenic Acids" consists of extraction not of the whole crude oil but of petroleum distillates boiling in the range $200 - 370°$ C, representing a molecular weight range of $250 - 400$ with dilute sodium hydroxide, subsequent acidification, water washing of the separated acids, and removal of such impurities as neutral oils, sulfur compounds, phenols, basic nitrogen compounds, and iron salts. Extraction with petroleum ether of a sodium carbonate solution of the acids will remove phenols and neutral oils. Esterification of the purified acids (tertiary acids don't react) with methanol and anhydrous HCl, subsequent distillation of the esters, saponification, and conversion to solid derivatives such as amides and various salts lead to the results obtained between 1925 and 1949 on these low molecular weight Naphthenic Acid constituents (Table 1, p. 7).

From a natural products point of view, the question arises whether the acids thus identified are indigenous to the petroleum or generated during the refinery operation. The present evidence suggests that most of the acids identified in petroleum distillates are also present in the original crude, and only a small quantity of acids appears to be formed during refinery processes.

More recently, a whole virgin crude (4) and several shale bitumens were used as starting material for elucidation of structures of carboxylic acids, thus removing doubts regarding their authenticity.

2. Occurrence

The main portion of the oxygen in petroleum is accounted for by carboxylic acids; inversely, the acidity of petroleum is a direct indication of its oxygen content. Thus, two approaches to analytical determination are practiced: (a) direct potentiometric titrations, (b) oxygen determination by neutron activation:

$$^{16}_{8}O + ^{1}_{0}n \xrightarrow[\text{Neutrons}]{14\ \text{MeV}} ^{16}_{7}N + ^{0}_{1}H$$

$$\Big| \beta\text{-decay, } t_{\frac{1}{2}} = 7.4\ \text{sec.}$$

$$\Big\downarrow \gamma = 6-7\ \text{MeV}$$

Gamma rays of >3.5 MeV energy are counted with a scintillation crystal.

The acid content of crudes varies from $0-3\%$ (5). Examples for 0% or very low contents are: Rangely, Colorado, U.S.A.; Pennsylvania, U.S.A.; Iraq; medium $(0.3-0.7\%)$: Lobitos (Peru), Balachany (U.S.S.R.), Gulf Coast (U.S.A.); high $(1.2-3.0\%)$: Lagunillas (Venezuela), Midway Sunset (California, U.S.A.); southeastern Europe (Rumania).

Geologically speaking, the heavy crudes from geologically young formations have the highest acid contents; paraffinic crudes usually have low acid contents. The internal acidity distribution in a crude is such that the acidity is low in the gasoline fraction, high in the kerosene fraction, and relatively low in heavy lubrication oils and distillation residues.

3. Derivatization

During the last decade, gas chromatography of the methyl esters and more recently gas chromatography in combination with mass spectrometry of acid derivatives and of hydrocarbons prepared from the acids, combined with other physical methods, have become the most efficient approaches towards structural analysis of carboxylic acid mixtures in petroleum. This opens up the area of high molecular weight (up to C_{30}) structures to an extent hitherto unexplored. Present approaches to derivatization are summarized in Chart 1. The very mild old method of esterification with diazomethane (6) is often replaced by the use of 1-alkyl-3p-tolyl triazine (7). The perfluoryl alcohol ester method (8), an application to acid analysis (9), will be discussed later. The trimethyl silylation approach is summarized in a recent book (10). Trimethyl silylation and perfluoro alcohol esterification result in derivatives (1) and (2) with vapor pressures comparable to those of methyl esters but with functional moieties which make mass

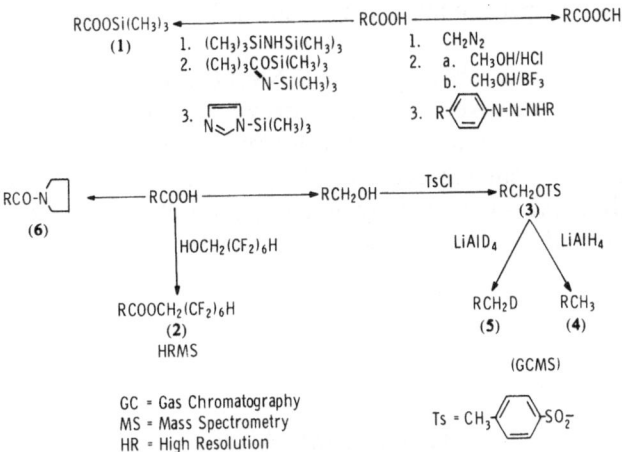

Chart 1. Modern methods of derivatization of petroleum acids

spectral interpretation considerably easier. Combined with fixation of a deuterium atom on the carbon originally carrying the carboxyl group (5) and mass spectral comparison with the unlabeled hydrocarbon (4) (Chart 1), derivatization is an extremely powerful tool for structural elucidation of complex petroleum acid derivatives (11). A recent approach applicable to unsaturated fatty acids deals with pyrrolidide (6) (Chart 1) [prepared from the acid chlorides with pyrrolidine (6)] and subsequent mass spectral investigation (12). Despite these advances, the purification and separation techniques for petroleum acids worked out during the first half of this century remain valuable assets.

The main driving force behind the earlier work on the structure of "Naphthenic Acids" was the industrial use of their various metal salts (5): Zn-, Co-, and Fe-salts serve as drying agents in the varnish industry, Ca- and Mg-salts as flotation agents in the paint industry, Cu-salts as fungicides and insecticides for the preservation of wood and textiles, sodium salts as emulsifying agents in the manufacture of cutting oils, Na-, Ca-, Zn-, or Al-salts are used in lubricating greases, and Ca- and Zn-salts as dispersants in motor oils.

The main impetus for recent work on the determination of carboxylic acid structure is interest in geochemistry. Carboxylic acids have been proposed as precursors of petroleum hydrocarbons (13). They have also been proved to be responsible for the large lowering of interfacial activity observed at alkaline pH between water/oil interfaces (4), an area of interest to organic geochemists (migration of petroleum) as well as to petroleum chemists in general because emulsions prepared from sodium hydroxide and asphaltic crudes are used to improve fluid flow in porous media as a process of secondary oil recovery (14). Most recently, steroid acids in petroleum have been linked to animal origin of petroleum (15).

A revival of research interest in "Naphthenic Acids" is presently occurring due to discovery of plant growth stimulation by potassium naphthenates (stimulation of glucose assimilation) (16, 17, 18). At this point, it should be remembered that "Naphthenic Acids" and petroleum acids are different in that they encompass different structures and ranges in molecular weight, although an overlap exists.

II. Paraffinic Acids

1. Linear Fatty Acids

n-Fatty acids in petroleum and sediments have been investigated intensively. They are also major components in most organisms and persist for geologically long times. Furthermore, they are structurally related to paraffin hydrocarbons common in petroleum and are at least in part responsible for their formation.

Normal fatty acids have been found in a variety of *petroleums* (Table 1): $C_1 - C_9$ n-fatty acids were first found in California, Caucasus, and Texas crudes, $C_{14} - C_{20}$ acids in Japan. While valeric acid occurs in gasoline, stearic acid appears in the gas oil fraction of petroleum. Recently Graham found $C_8 - C_{18}$ n-fatty acids with a very slight preference for even carbon numbers in a California petroleum (19).

Normal fatty acids occur in *sediments* ranging in age from pre-Cambrian to recent and in carbon number from $C_{10} - C_{36}$ (20). An excellent review on this last subject is available (21). In most sediments even-numbered fatty acids are much more abundant than those with odd carbon numbers. This fact is not surprising because normal fatty acid components of natural fats are predominantly even-numbered ranging from $C_4 - C_{26}$. Even carbon numbered normal fatty acids of $C_{26} - C_{38}$ are found principally in waxes of insect and plant origin. Odd carbon-numbered fatty acids have also been found in nature, but their concentration is smaller than that of the even ones. It is assumed (21) that natural fatty acids survive to some extent geochemical degradation and become incorporated in sediments. In recent sediments the distribution of normal fatty acids is most nearly like that of natural fatty acids. Generally speaking, as the age of the sediment increases, the even carbon number predominance decreases. This process can be quantitatively described by the carbon preference index (CPI).

Table 1. *Carboxylic Acids Identified in Crude Oils Prior to 1943**

Compound	Source
Aliphatic monocarboxylic acids	
Formic acid	California, Caucasus
Acetic acid	California, Caucasus
Propionic acid	California
Butyric acid	California, Texas
Isobutyric acid	
n-Valeric acid	California, Texas
Isovaleric acid	California
2-Methylpentanoic acid	California
3-Methylpentanoic acid	California
4-Methylpentanoic acid	Ploesti
3-Ethylpentanoic acid	Romania, Baku
Hexanoic acid	California
2-Methylhexanoic acid	California
3-Methylhexanoic acid	California, Romania
4-Methylhexanoic acid	California
5-Methylhexanoic acid	California, Ploesti, Baku
Heptanoic acid	California
Octanoic acid	California, Texas
Nonanoic acid	California
Palmitic Acid	Ishikari (Japan)
Stearic Acid	Ishikari
Myristic acid	Ishikari
Arachidic acid	Ishikari
Aliphatic dicarboxylic acids	
Dimethylmaleic anhydride	California
Dimethylmaleic acid	Texas
1,2,2-Trimethyl cyclopentane-1,3-dicarboxylic acid	
Cycloalkane monocarboxylic acids	
Cyclopentane carboxylic acid	California, Ploesti
2-Methylcyclopentane carboxylic acid	California
3-Methylcyclopentane carboxylic acid	California
Cyclopentylacetic acid	California, Ploesti
Cyclohexane carboxylic acid	California, Baku
3-Methyl cyclopentyl acetic acid	California, Ploesti
2,3-Dimethyl Cyclopentyl acetic acid	California
4-Methyl cyclohexane carboxylic acid	Texas
cis-2,2,6-Trimethylcyclohexane carboxylic acid	California
trans-2,2,6-Trimethyl cyclohexane carboxylic acid	California

* For literature references see ref. 5, pp. 169–175.

$$\text{CPI (Fatty acids)} = \frac{1}{2} \left(\frac{\Sigma \text{ concentrations of even n-fatty acids } C_{16} \text{ to } C_{30}}{\Sigma \text{ concentrations of odd } \text{ n-fatty acids } C_{15} \text{ to } C_{29}} + \frac{\Sigma \text{ concentrations of even n-fatty acids } C_{16} \text{ to } C_{30}}{\Sigma \text{ concentrations of odd } \text{ n-fatty acids } C_{17} \text{ to } C_{31}} \right)$$

In modern sediments the CPI is generally high (about $2-5$) reflecting the contribution from plants and lipids (*13, 22*). In ancient sediments high CPI values are sometimes observed, but generally they are much lower (around 1) than those in modern sediments.

Because palmitic (C_{16}) acid is the most abundant n-fatty acid in nature, investigators sometimes simply express the ratio $\frac{C_{16} + C_{18}}{2\,C_{17}}$ as a measure of even/odd relative concentration (*23*). Typical values of this ratio are: 15 for a recent sediment and 1.7 for an ancient sediment. For paraffin hydrocarbons this relationship is reversed in that odd-numbered paraffins distinctly dominate over even ones in recent sediments. In ancient sediments and crudes, the even/odd ratio approaches unity.

Cooper and Bray were the first to postulate a genetic relationship between fatty acidx and normal paraffins in petroleum and

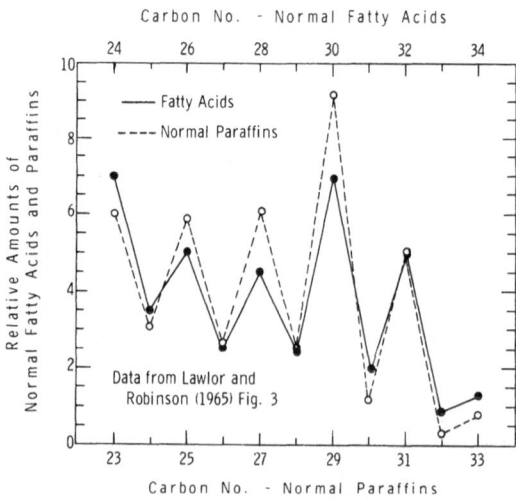

Fig. 1. Distribution of normal fatty acids and normal paraffin hydrocarbons in Green River formation oil shale

sediments (*13*). This relationship has been experimentally demonstrated (*24*) (Fig. 1): The carbon number distributions are almost identical when each normal fatty acid concentration is compared with the concentration of a normal paraffin having one less carbon atom than the acid (Fig. 1). The rationalization first advanced by COOPER and BRAY and later expanded by KVENVOLDEN and WEISER (*25*) (who designed a mathematical model to treat these processes) can be summarized as follows:

$$R_1CH_2COOH \rightarrow R_1CH_2{}^{\cdot} + CO_2 + H^{\cdot} \tag{1}$$

$$R_1CH_2{}^{\cdot} \xrightarrow{R_2H} R_1CH_3 + R_2{}^{\cdot} \tag{2}$$

$$R_1CH_2{}^{\cdot} \xrightarrow{\text{Oxidation}} R_1COOH \tag{3}$$

$$R_1CH_2COOH \xrightarrow{\text{Reduction}} R_1CH_2CH_3 \tag{4}$$

Equations (1) and (2) explain the observed phenomenon depicted in Fig. 1, namely, the correspondence of even-numbered fatty acids with odd-numbered paraffins. Superimposed is process (3) which results in carboxylic acids with one carbon atom less than the biological precursors and thus tends to decrease the relative abundance of even-numbered n-fatty acids over the odd-numbered ones with time, i.e., in going from recent to ancient sediments. Step (3) faces a problem in that petroleum maturation is generally considered to occur in a hydrogenating environment, allowing process (4) to become operative. The latter has been demonstrated (*26*) to occur in a recent sediment in the Persian Gulf: Even-numbered n-paraffin predominance, particularly in the range of $C_{16} - C_{18}$, indicating a strong superposition of step (4) on steps (1) and (2); however, this result has to be considered the unusual situation.

The attractive aspect of the combined concepts (1)—(4) is the involvement of hydrocarbons as hydrogen donors (step 2), thus randomizing not only even/odd fatty acid predominance with geological time but also the odd/even n-paraffin predominance. The net result is the disappearance of any carbon number predominance in both n-fatty acids and n-paraffins with time which is, in fact, observed in ancient sediments and crudes.

A by-product of the sequence of these reactions is the actually observed production of shorter chains, the increase in absolute concentration of n-alkanes, and the decrease in absolute concentration of fatty acids with geological time (27). Obviously, the mechanistic interpretation (1)−(4) is an oversimplification because fatty alcohols (23), which also occur in nature although in lesser quantity than acids and can also be converted either to acids or to hydrocarbons, have not been considered at all. In addition, the primary free radical formed by reaction (1) has other alternatives to react: (a) intramolecular hydrogen abstraction to form a more stable secondary radical, which may further rearrange to different secondary radicals. The net result would be the formation of rings and smaller paraffins and olefins (β-scission, Rice mechanism), which would, of course, be unstable in the hydrogenating petroleum medium. Indeed, some investigators (28) doubt that n-fatty acids are paraffin precursors for quantitative reasons, i.e., there are not enough fatty acids in recent sediments to be capable to account for all n-paraffins in petroleum, although there appears to be a genetic relationship in view of the similar distribution of acids in recent sediments and in marine plankton (23).

Some investigators actually tried to test the decarboxylation concept by simulating geological conditions in the laboratory. Behenic acid (n-C_{22}) was heated in a sealed tube with bentonite and water to 200° C (29). The main product was C_{21}-paraffin; in addition, lower cracking products and $C_{22} - C_{34}$ paraffins were found suggesting the formation and recombination of free radicals to form higher n-paraffins from n-fatty acids parallel with decarboxylation. Recent interpretations of this type of experimental data obtained under somewhat different conditions (30) suggests a radical decarboxylation oxidation sequence. While at first decarboxylation is the predominant process, prolonged reaction time signals catalytic cracking.

The origin of C_{36} n-fatty acids found in recent and ancient sediments is not known because plant waxes, which are assumed to be their potential precursors, are generally not abundant in marine environments (23).

In the face of all the pros and cons, the presently accepted concept (27) is that n-fatty acids are at least in part the precursors of geochemically formed n-alkanes.

For modern unconsolidated sediments, Kvenvolden reports (20) a bimodal distribution with n-C_{16} being the most abundant fatty acid (matching the abundance of palmitic acid in nature) followed by the $C_{24} - C_{26} - C_{28}$ n-fatty acids. $C_{22} - C_{32}$ n-fatty acids with even carbon number predominance isolated from Messel shale near Darmstadt, Germany, (Eocene) are definitely regarded as contributions from the organic material of higher plants (31).

The biological or recent bacterial origin of n-fatty acids was recently discussed, although not at all proved, during an investigation of old shales (up to 2.7 billion years) (32). An alternate proposal for the formation of n-fatty acids in such ancient shales was telomerization of ethylene terminated by water to produce an oxidizable functional end-group. The basis for these speculations was the absence of any parallelism between the distribution of the n-fatty acids and the n-alkanes having one less carbon such as that depicted in Fig. 1.

In one of the oldest sedimentary rocks known on earth from the Onverwacht, Swaziland, formation (3.4 – 3.7 billion years old), the n-C_{16} fatty acid was recently found as the most predominant n-fatty acid component, thus relating this sediment to microorganisms which are known to contain large quantities of palmitic acid (33).

2. Unsaturated Fatty Acids

Unsaturated fatty acids represent more than half of all fatty acids in the biosphere. However, in the geosphere, their concentration is comparatively small. This result is indicative of the relative instability of unsaturated fatty acids towards bacterial and other hydrogenative attack at an early and later stage of sedimentation. Evidence to this effect is a report (34) on a rapid decrease of unsaturated fatty acids in living and recently living layers of blue-green algae. More recently, a study was made (35) to find out whether in marine sediments bound acids (released by HCl treatment) correlate with the whole marine organism and whether free acids (extraction prior to acid treatment) correlate with bacterially produced acids. The work was directed towards the question of possible bacterial origin of fatty acids in general. The finding of a dominant diunsaturated acid which amounted to 90% of the total acid fraction in one recent lagoon sediment pointed against bacterial origin. In another recent sediment, the composition of saturated, unsaturated, and hydroxy acids pointed to the opposite conclusion, thus leaving some questions unanswered (35). However, the fact that unsaturated fatty acids disappear rapidly at the earliest stage of sedimentation was elegantly demonstrated by dual labeling of oleic acid with ^{14}C and ^{3}H and observing a small conversion to saturated fatty acids in a core of a modern estuarine sediment (36, 37). The labeled saturated fatty acids ($C_{12} - C_{18}$) which were isolated were proved to arise from C_2-unit degradation or complete resynthesis via β-oxidation by the anaerobes.

In spite of all these results, unsaturated fatty acids are not completely absent in ancient geological material. $C_{16:1}$ and $C_{18:1}$ were found in Green River shale (age 50 million years) (38). Esters of monounsaturated acids were postulated on the basis of mass spectral fragmentation patterns in Green River shale (*39*). In an even older shale (250–300 million years), oleic acid was suspected (*40*).

3. Iso and Anteiso Acids

The isomers of methyl mono-substituted pentanoic and hexanoic acids reported as occurring in petroleum in the older literature are summarized in Table 1 (p. 7). Isoacids (**7**) and anteiso acids (**8**) were found in recent marine sediments and in ancient Green River shale sediments (*41*) and petroleum (*42*).

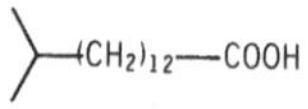

$$>\!\!-\!(CH_2)_{12}\!\!-\!COOH \qquad \wedge\!\!\vee\!(CH_2)_{10}\!\!-\!COOH$$

(**7**) 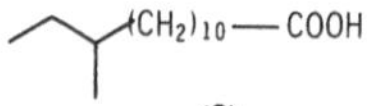   (**8**)

Bacterial lipids which are rich in branched-chain acids are believed to be their biological precursors; however, higher organisms and marine plankton are said' to contain insufficient quantities of these acids to account for the amounts observed in sediments (*23*). Furthermore, the ratio of n-fatty acids to branched fatty acids is considerably larger in the organisms as compared to the sediments, a disturbing fact considering the comparable stability of both types of acids. A genetic relationship between branched-chain paraffinic acids and branched-chain hydrocarbons such as that for the linear fatty acids discussed earlier in this review has been suggested (*41*).

4. Isoprenoid Acids

The first discovery of these very important compounds was made by Cason and collaborators at the University of California in Berkeley (*43*). Their starting material was a California refinery-processed commercial naphthenic acid sample. In a classical natural product-type approach, they confirmed by synthesis the presence of the following isoprenoid acids in the enormously complex mixture of polar compounds present in this petroleum: 2,6,10-trimethylhendecanoic acid (C_{14}) (**9**), 3,7,11-trimethyldodecanoic acid (C_{15}) (**10**), 2,6,10,14-tetramethylpentadecanoic acid (**11**) (pristanic acid), and 3,7,11,15-tetramethylhexadecanoic acid (C_{20}) (**12**) (phytanic acid).

The starting material for the synthesis of the C_{14} acid was farnesol, while phytol was the starting material for the C_{19} and C_{20}

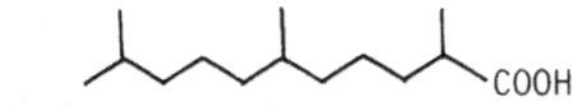

(9)

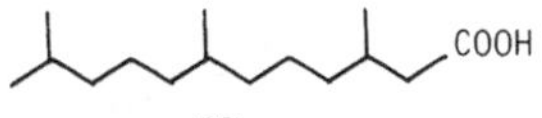

(10)

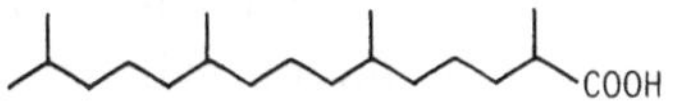

(11)

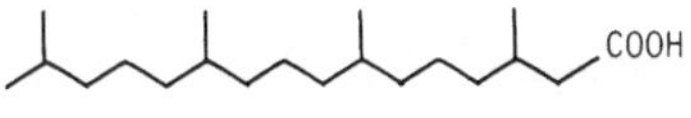

(12)

acids. One major tool for structural elucidation was mass spectro-metry of amides of these acids. For a full appreciation of the magnitude of the contribution of this team, the reader is referred to the original publications where also the presence of a C_{11} isoprenoid acid, namely, 4,8-dimethyl nonanoic acid in low concentration, is confirmed by mass spectrometry of its p-phthalimidophenacyl ester (44).

These isoprenoid acids are of great interest to organic geochemists because, on one hand, pristanic and phytanic acids occur in lipids of mammals (45) and direct preservation from marine organisms to sediments is possible; on the other hand, chlorophyll is abundant in recent sediments, and the phytol side chain of chlorophyll (13) could be a source of the isoprenoid acids.

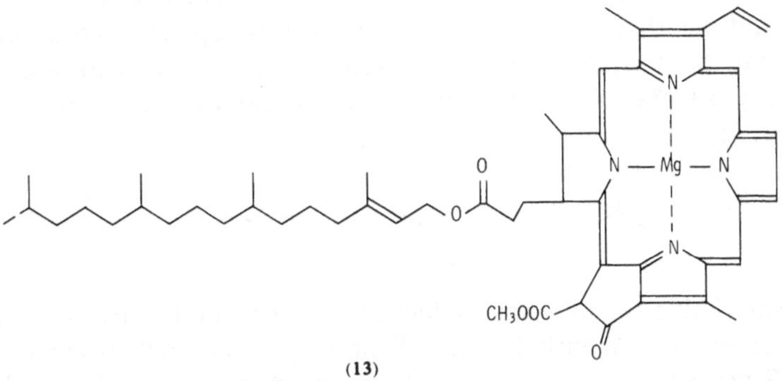

(13)

After the discovery of this class of biological marker compounds by the CASON team in a crude oil, a number of research workers found isoprenoid acids in sediments. Thus pristanic and phytanic acids were identified by gas chromatography combined with mass spectrometry in Green River shale (46), halophilic bacteria being discussed as another possible source. In the same sample and by the same technique, a different team (47) identified two lower isoprenoid acids, namely, 2,6-dimethylheptanoic and 3,7-dimethyloctanoic acids. Pristanic and phytanic acids were found to be the main constituents of the $C_8 - C_{22}$ branched-chain fraction of yet another shale (48). In a recent sediment (49), the ratio of palmitic to pristanic acid was found to be similar to that encountered in typical marine lipids, suggesting that the isoprenoid acids of recent sediments are derived directly from living organisms; this study of recent sediments also revealed the presence of a C_{16}-isoprenoid acid. The fact that finally (50) all isoprenoid acid homologues from C_{11} to C_{21} were found in ancient shales (age up to 210×10^6 years) suggests that those members missing in recent sediments (49) and present in ancient sediments are formed by a slow, postdepositional process.

DOUGLAS and collaborators prefer separation of normal and branched esters by urea clathration prior to GCMS; they also report detailed mass spectral fragmentation patterns of the methyl esters of the isoprenoid acids (50).

Regarding the possible genetic relationship of acids and hydrocarbons, it has been observed (51) that the abundances of C_{19} and C_{20} hydro-carbons are paralleled by the abundances of C_{19} and C_{20} isoprenoid acids, ruling out a simple decarboxylation relationship as demonstrated (Fig. 1, p. 8) for fatty acids/n-paraffins. This conclusion was further substantiated by the relative abundance of C_{15} isoprenoid acid and the corresponding absence of C_{14} isoprenoid hydrocarbon (51).

Thus, the problem of the origin of the isoprenoid acids was attacked by determining the relative and absolute stereochemistry of their methyl and menthyl esters. The net result of this sophisticated approach (52) is that the diastereomeric composition is compatible with a chloro-phyll (phytol) (14) derivation of the acids biosynthesized in Eocene times (52, 53).

(14)

This remarkable result was achieved experimentally by separating diastereomeric (−) menthyl esters of the isoprenoid acids isolated from Green River shale on a butanediol succinate-coated capillary column and relating the ratios of stereoisomers found to the known stereo-chemistry of natural phytol (54). Thus, the ratio of the 2(R), 6(S), 10(R) and 2(R), 6(R), 10(R) isomers of pristanic acid (11) was found to be about 1:1 which is the ratio expected if the acid had arisen from non-stereospecific maturation of natural phytol (52). Analogously, the stereo-chemistries of phytanic acid (12) and of C_{15} (10) and of C_{16} isoprenoid acids in Green River shale were determined (53). Other precursors were ruled out or made unlikely as follows. Nonstereospecific reduction of carotenoids and subsequent oxidation would yield all possible stereo-isomers (not just two as found), and stereospecific reduction would lead to the RRR and SSS isomers (55). The authors ruled out as major contributors vitamin K_1 and α-tocopherol, which have been shown to possess a phytyl side chain with the same stereochemistry as that of phytol from chlorophyll on the basis of relative abundances in nature; that is, the occurrence of these compounds in the biosphere is small compared to that of chlorophyll (*53*). In their most recent report (*56*), this research team elaborates further on the absolute stereochemistry of the most abundant isoprenoid acids present in Green River shale including the C_{17}-, C_{18}-, C_{21}-, and C_{22}-acids: It is concluded that pristane is *not* an intermediate in the genesis of pristanic acid because oxidation of the symmetrical pristane structure at both ends would be expected to yield four stereoisomers, but only two were found. This forces the conclusion that reaction takes place at the end of the phytyl chain carrying the hydroxyl group (**14**).

Speculation on the origin of the C_{21} and C_{22} isoprenoid acids includes that it may arise by degradation of higher biological iso-prenoids abundant in nature, e.g. the betulaphenols $(C_{30}-C_{45})$ in birchwood, solanesol (C_{45}) in plants or the bactoprenols (C_{55}) in lipids of lactobacillus. Possible alternate precursors of the C_{15} and C_{14} acids are farnesol (**15**) or farnesoic acid allowing for diagenetically early stereo-specific reduction by anaerobic bacteria to product the observed stereo-chemistry.

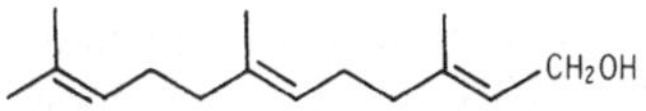

(**15**)

5. "Pseudo" Isoprenoid Acids

The first compound of this category was reported by BIEMANN *et al.* (*57*) and was a C_{18}-acid, namely, 6,10,14-trimethyl pentadecanoic acid (**16**) in Green River shale.

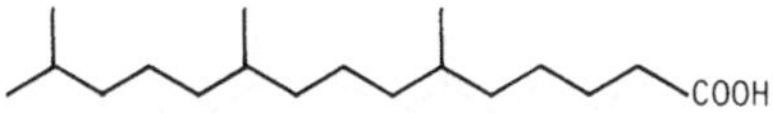

(**16**)

It was identified by GCMS and the structure proved by synthesis starting with farnesol (**15**). Hydrogenation of the latter to hexahydro-farnesol was followed by reaction with phosphorus tribromide to give hexahydrofarnesyl bromide, the Grignard reagent of which was coupled with methyl acrylate to give the methyl ester of (**16**). The acid (**16**) contains four unsubstituted methylene groups in sequence and, therefore, is not a true isoprenoid acid. The discoverers postulated its formation by biological oxidation prior to sedimentation and suggested two possible pathways: (a) oxidation of the irregular isoprenoid hydrocarbon squalane (17)

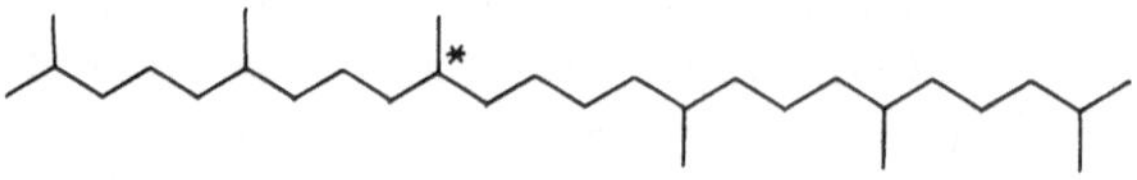

(17)

at the position indicated (*); (b) β-oxidation of pristanic acid (11) at the methyl group to give an alkyl malonic acid which could then decarboxylate to give the observed "pseudo" isoprenoid acid (16). It has been commented (50), relative to this proposition that in a biochemical system β-methylene oxidation of pristanic acid would be expected to give 4,8,12-trimethyltridecanoic acid rather than β-methyl oxidation to give the observed pseudoisoprenoid acid (16), and the question has been raised (50) why the C_{18}-acid (16) has not been found in the living system but only in sediments of considerable diagenetic change. The discoverers (57) themselves pointed out the unlikelyhood of the squalane oxidation route because of the confirmed absence of higher acid homologues which would also be expected to be formed along with (16). Consequently, the mechanism of formation of this interesting acid (16) is subject to dispute.

In the meantime, a whole suite of "pseudo" isoprenoid acids has been discovered (50) in a 210×10^6 year old shale, a bituminous dolomite from Switzerland which was laid down in a shallow coastal sea. This includes 3,7-dimethyldecanoic acid (18), 4,8-dimethyldecanoic acid (19), 2,6-dimethylundecanoic acid (20), and two C_{16} and two C_{17} acids

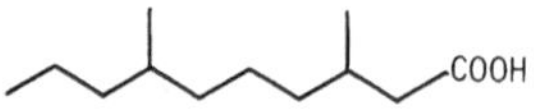

(18)

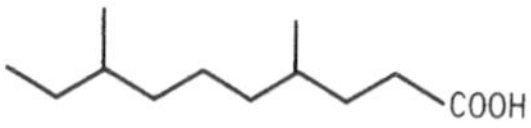

(19)

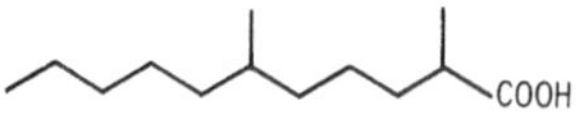

(20)

of not unequivocally ascertained structures. Such acids are rare in nature, although some have been found in feather waxes of birds, in the preen gland of the oyster catcher and in eider duck.

Thus, the origin of the "pseudo" isoprenoid acids appears unresolved. Biological remnants of products or sedimentary maturation of higher isoprenoid acids are the most likely alternatives.

III. Cyclic Saturated Acids

1. Monocyclic Acids

The name "Naphthenic Acids" implies a long-recognized pre-dominance of monocyclic classes (five- and six-membered rings) of acids in petroleum. The methyl-substituted cyclopentyl- and cyclohexylcarbo-xylic acids including some substituted cyclopentylacetic acid reported in the older literature are summarized in Table 1 (p. 7). These are all low molecular weight compounds. Interesting is the occurrence of cis- and trans-2,2,6-trimethylcyclohexanecarboxylic acids (21).

(21)   (22)

These acids could be isolated by virtue of the severe hindrance at the carboxyl group. It is speculated that they arise by oxidation of the β-ionone portion (22) of β-carotene. As mentioned earlier in this review, such processes would most likely have to occur at a very early stage of sedimentation or else by recent bacterial oxidation via oxygen-bearing aquifers. Observation of only one stereoisomer as is the case with trans-2,2,6-trimethylcyclohexylacetic acid (23), which can be con-

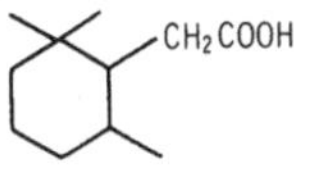

(23)

sidered a cyclic isoprenoid acid, makes a stereospecific bacterial oxidation mechanism more plausible. The structure of (23) was proved by synthesis (Chart 2) (58). α- or β-ionone (24) was converted via ketone (25) and

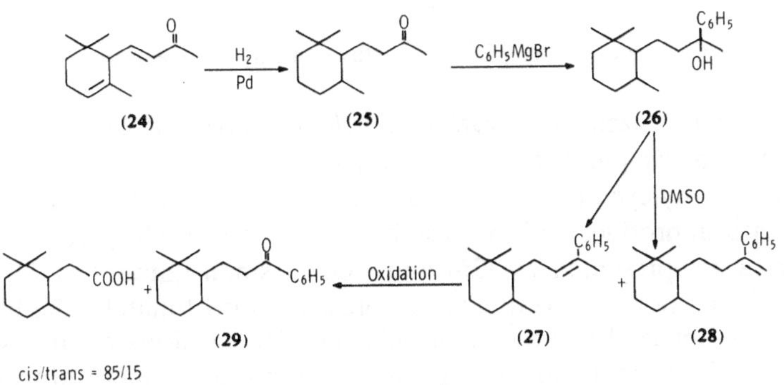

(24)   (25)   (26)

cis/trans = 85/15

Chart 2. Synthesis of 2,2,6-trimethylcyclohexyl acetic acid

carbinol (26) to a mixture of olefins (27) and (28) whose oxidation yielded
ketone (29) and a mixture of *cis* and *trans* acids (23); the latter were
separable in the form of the esters by gas liquid chromatography.

The Berkeley team proved the structure of yet another monocyclic
acid; 3-ethyl-4-methylcyclopentylacetic acid (30) (59) which is non-

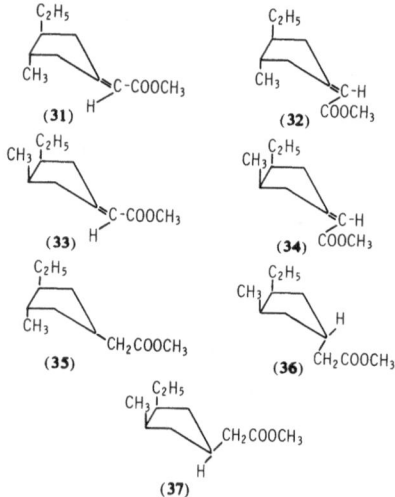

isoprenoid and in contrast to the isoprenoid trimethylcyclohexylacetic
acid shows no optical activity. The stereochemistry was elucidated by
conversion of the natural product acid to a mixture of 3-cyclopentylidene-
acetic acids *via* α-bromination and dehydrohalogenation. The esters
mixture was compared with the esters (31)—(34) of all four stereoisomeric 3-ethyl-4-methylcyclopentylideneacetic acids (Chart 3), which

Chart 3. Stereochemistry related to 3-ethyl-4-methylcyclopentylacetic acid

had been synthesized from *erythro* and *threo* isomers of 2-ethyl-3-methyl
succinic acid *via* the homologous adipic acids and the *cis-* and *trans-*
3-ethyl-4-methylcyclopentanones. The following stereochemical ranking
of the natural product acids was established (Chart 3). The predominating
isomers are represented by (35) and (36); a small amount of (37) was
observed, and none of the others was present. Unfortunately, this elegant
work does not lead to obvious geochemical conclusions because of the
low molecular weight and consequent loss of biogeochemical information
content of this acid.

2. Bi- and Polycyclic Naphthenic Acids

More recently, the low molecular weight barrier has been penetrated. Although the presence of bicyclic acids in the $C_{12}-C_{18}$ range of "Naphthenic Acids" has long been suspected (5) the first concrete proof for the abundance of high molecular weight $1-5$ ring polycyclic naphthenic acids was delivered only a few years ago by the author and his collaborators. A whole virgin California petroleum of Pliocene age was exhaustively extracted with sodium hydroxide (4). The extract was purified by a sequence of countercurrent extractions, ion exchange, and silica gel column separations. The complexity of the mixtures became apparent when the high resolution mass spectrum of a rather "pure" preparative thin-layer chromatographic subfraction was examined. The molecular weights ranged from 200 to about 700 with a maximum in the range of $300-400$. There was a peak at every mass across this wide range, and each peak showed multiplets from four to eight masses. Excluding odd masses which are minor constituents and must be due to nitrogen compounds and carbon-13, there were about 1500 compounds present, not counting the possibility of isomers (9). However, the low resolution mass spectrum showed a periodicity of 14 indicative of homologous series thus reducing the *types* of compounds by far.

The mass spectra of the 1,1,7-trihydroperfluoroheptyl esters (2) (Chart 1) are spectacularly different from those of the acids. An example is shown in Fig. 2 which illustrates the use of these derivatives

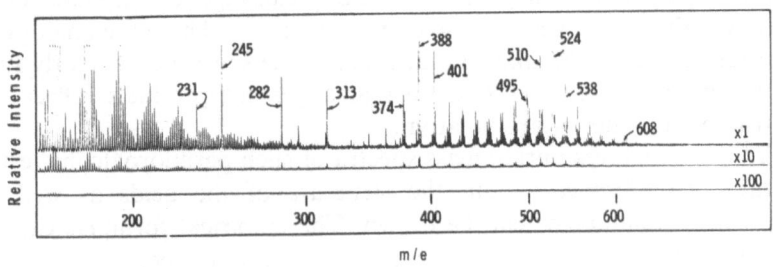

Fig. 2. Low resolution mass spectrum of 1,1,7-trihydroperfluoroheptyl esters of petroleum carboxylic acid mixture at a probe temperature of 57° C

for identification purpuses because the parent peaks are shifted by 300 mass units from the parent peaks of the acids. The even mass "parent peak region" starts at m/e 496 and continues with 510, 524, 538, etc., up to m/e 608. At the low mass end of the scale (up to about m/e 300), the spectrum consists of a mixture of ions derived from hydrocarbons, acyloxy, and related fragments which is referred to as the "hydrocarbon fragment region", and also contains a few large peaks from unreacted fluoroalcohol, e. g., at m/e 231, 245, 282, and 313. The middle range of the spectrum, the so-called "ester fragment region", contains, among the odd ester fragments, two predominant even mass peaks at m/e 374 (38) and 388 (39) commonly referred to as the McLafferty rearrangement fragments (Chart 4) (59). These result

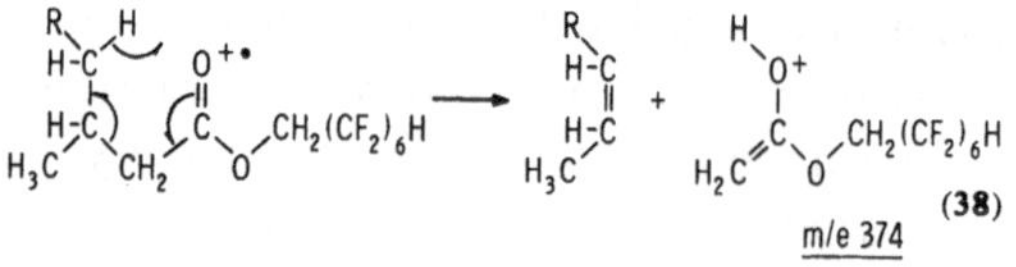

Chart 4. McLafferty rearrangement

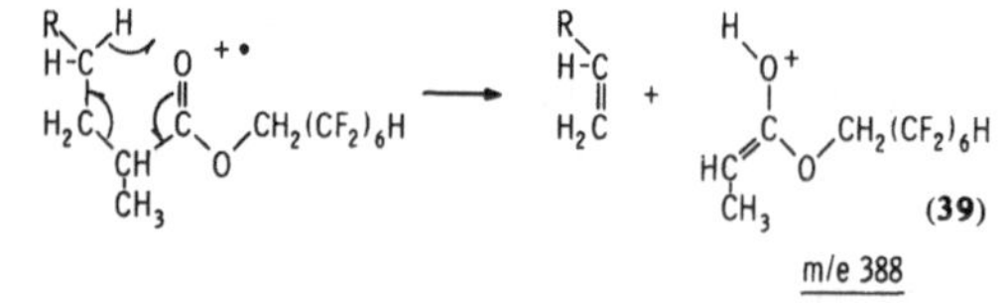

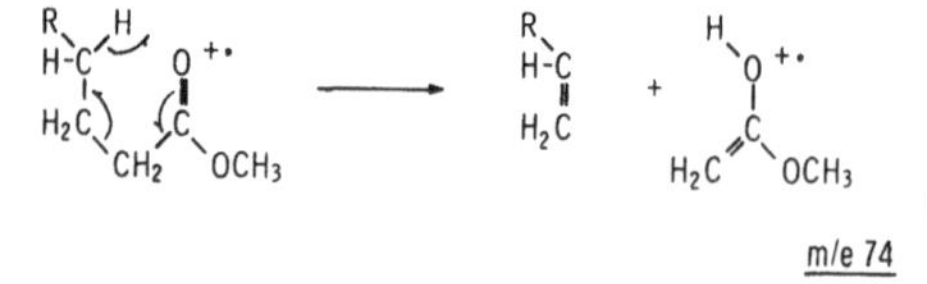

from the γ-hydrogen migration in and β-cleavage of aliphatic ester moieties without and with an α-methyl substituent. These are analogs of the peaks at m/e 74 (40) and 88 that occur in the spectra of similarly sub-stituted methyl esters. Thus, these rearrangement products reflect the substitution on the α-carbon atom.

From the high resolution mass spectra of such perfluoroalcohol ester mixtures (2) inferences about the structure of the acids as well as their relative abundance can be drawn. The empirical formulas can be derived from exact mass measurements of the molecular ions and subtraction of the moiety $C_7H_2F_{12}$ yields the empirical formula of the original acid. Hypothetical reduction to the related hydrocarbon leads to a Z-number calculated from the expression C_nH_{2n+z} and from the value of Z, the number of rings plus double bonds (called R) is calculated from the equation $R = (2-Z)/2$. Thus, compound classes of $C_{16} - C_{31}$ carboxylic acids ranging in Z-numbers from $+2$ to -26 were identified (9) representing $1-5$ ring polycyclic, naphthenic, aromatic, and naphthenoaromatic acids including some heterocyclic types. Their relative abundance in the original acid sample could be deduced semiquantitatively from a combination of three pieces of information: (a) the intensity of the parent peaks in the low resolution spectrum (Fig. 2); (b) the total ion monitor current of every low resolution scan; (c) the relative intensity of the multiplets obtained by high resolution mass measurement (9). The information derived in this fashion supported the quantitative data obtained by analysis of the hydrocarbons derived from the acids in a three-step reduction (4) (see Chart 1).

Separation of the hydrocarbons (discussed later in this review) and quantitative high resolution mass spectrometric group-type analysis (61) of saturated and aromatic fractions led to the conclusion depicted in Fig. 3: that 1−5 ring polycyclic naphthenic acids are the most abundant compound classes in this California petroleum.

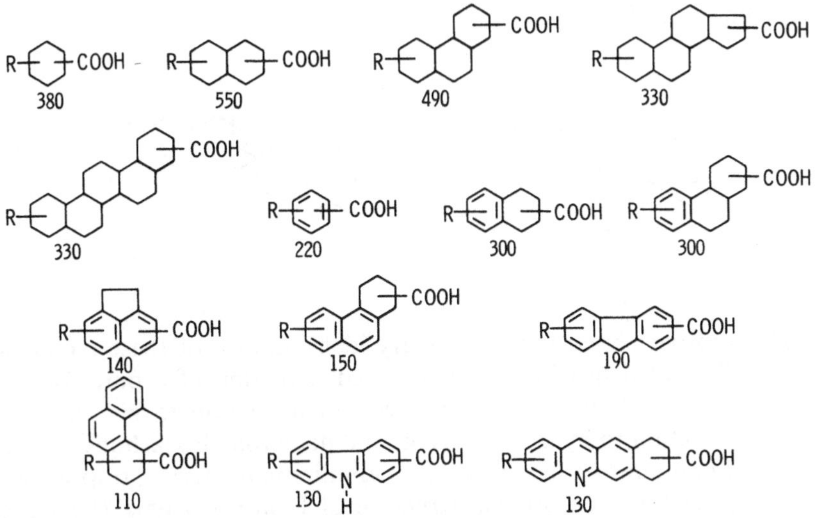

Fig. 3. Predominating (ppm)[a] compound classes [b] ($C_{16} - C_{31}$) of carboxylic acids in a California (Pliocene) petroleum

[a] Values are based on crude oil and represent minimum estimates.

[b] Position of substituents, rings, and ring sizes unknown.

The fluoroalcohol alcohol ester approach also led to the identification of a homologous series of bicyclic acids. The parent peak of the ester of the C_{11} member is seen in Fig. 2 (m/e 496). This $C_{10}H_{17}COOH$-compound could be 1-, 2-, or 9-decalin carboxylic acid or possibly a bridged bicyclic norbornane-type (41) acid which could account for the

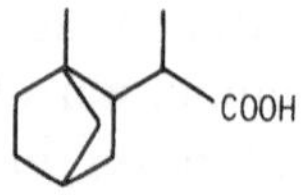

(41)

α-methyl group indicated from the McLafferty rearrangement fragment m/e 388.

3. Tricyclic Terpenoid Acids

Tricyclic acids are the second most abundant compound class in the California petroleum investigated (about 0.05% based on petroleum). The observation of a C_{21}- (42) and a C_{24}-tricyclic terpenoid carboxylic acid (43) in the above-described California petroleum occurred simultaneously with the first discovery of steroid carboxylic acids in this petroleum (62, 63).

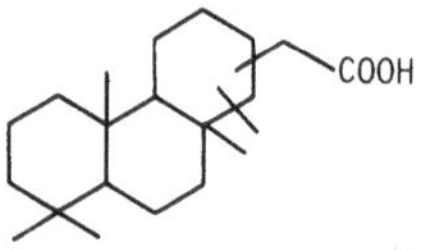

(42) 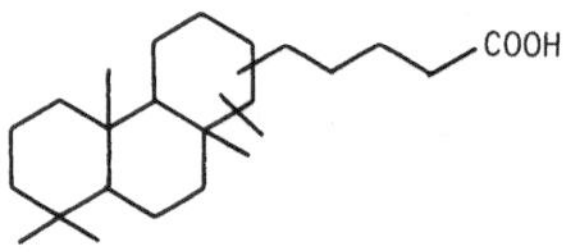   (43)

The approach used for preliminary identification of these terpenoid acids is illustrated in Chart 1. It involved reduction of a pure fraction of carboxylic acids to alcohols, subsequent tosylation to give (3), and parallel reduction to hydrocarbons (4) and deuterohydrocarbons (5) with lithium aluminum hydride and lithium aluminum deuteride, respectively. By this procedure, the original structure of R in (2) (Chart 1) remains largely unharmed (64), and the polar carboxyl group which inhibits further separations is transformed into a nonpolar deuteromethyl group (CH_2D) which allows further separation by silica gel and gel permeation chromatography (63). From a structural point of view no information is lost by this fixation of the carboxyl group with deuterium; the labeled and unlabeled hydrocarbons can now be subjected to gas chromatography combined with mass spectrometry, and their mass spectral fragmentation patterns can be compared.

Fig. 4, which depicts the reduced acids, illustrates the natural abundance of the C_{21}- and C_{24}-terpenoid acids (by the predominance of the terpane hydrocarbon peaks) over any single carboxylic acid in this

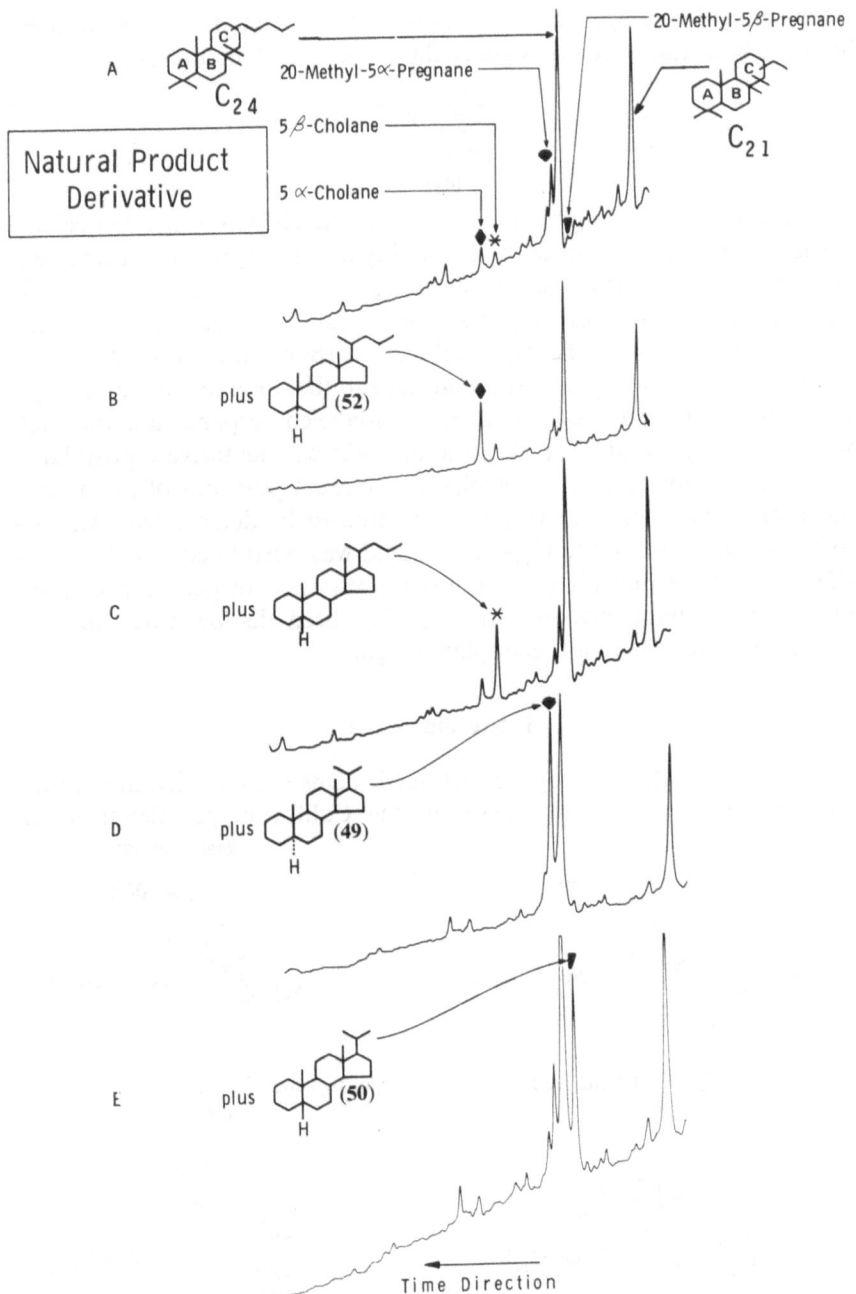

Fig. 4. Gas chromatogram[a] of carboxylic acid derived hydrocarbons coinjected with synthetic steroids

[a] Time is measured on a variable scale.

petroleum (15–30 ppm, based on crude). A fragment at m/e 191 in these hydrocarbon mass spectra signals the terpenoid structure (44) (65).

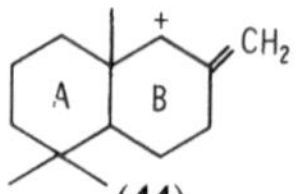

(44)

The absence of a fragment at m/e 192 in the GCMS of the deuterium-labeled compounds excludes the possibility of carboxyl group attachment to rings A and B. Because the empirical formula $C_{21}H_{38}$ of the acid (42) derived hydrocarbon requires three rings, only three carbon atoms, one of which has to be the carboxyl carbon, are structurally un-accounted for in the C_{21} terpane acid. Due to the observed loss of C_2H_4D in the mass spectrum of the deuterium-labeled terpane and the lack of loss of CH_2D and C_3H_6D, structure (42) was tentatively postulated for this first terpane acid in petroleum, the exact positions of the methyl and acetic acid groups in ring C remaining to be determined. Analog-ously, the structure of the C_{24}-terpane acid was postulated as (43). Acids (42) and (43) are presumably degradation products of pentacyclic triter-panes whose abundance in petroleum has been demonstrated (66, 67) and has been related to a higher plant origin.

4. Steroid Acids

Fig. 3 shows that tetracyclic carboxylic acids are included among the very abundant compound classes in the California petroleum which

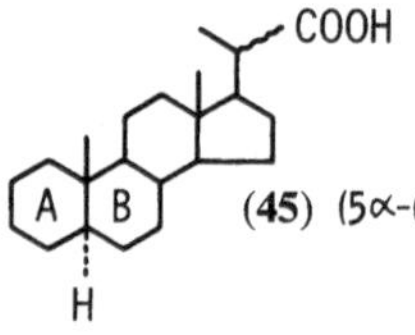

Nearly Absent

(45) (5α-C₂₂)

5α-Pregnane-20ξ-Carboxylic Acid
A/B trans

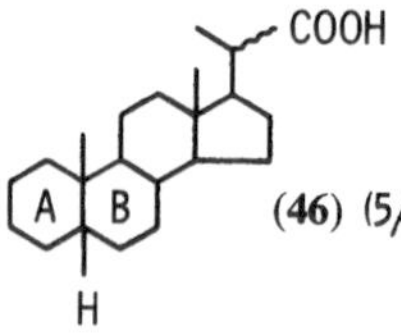

(46) (5β-C₂₂)

5β-Pregnane-20ξ-Carboxylic Acid
A/B cis

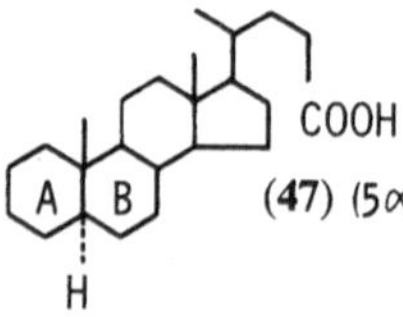

(47) (5α-C₂₄)

5α-Cholanic Acid
A/B trans

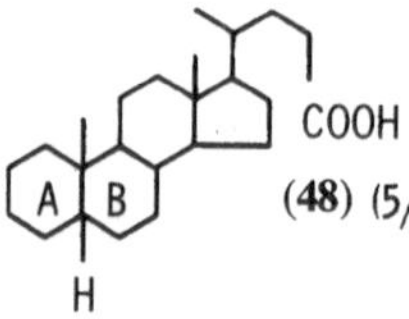

(48) (5β-C₂₄)

5β-Cholanic Acid
A/B cis

Chart 5. First discovery of steroid carboxylic acids in petroleum

constituted the subject of the investigation. Fig. 4 depicts a further refinement of the approach discussed above for the identification of the terpane acids which also led to the discovery (*62*) of steroid acids (Chart 5). The presence of 5α-pregnane-20ξ-carboxylic acid (**45**) and the absence of the stereoisomeric 5β-pregnane-20ξ-carboxylic acid (**46**) was proved (*63*) by synthesis of the corresponding hydrocarbons (Chart 6)

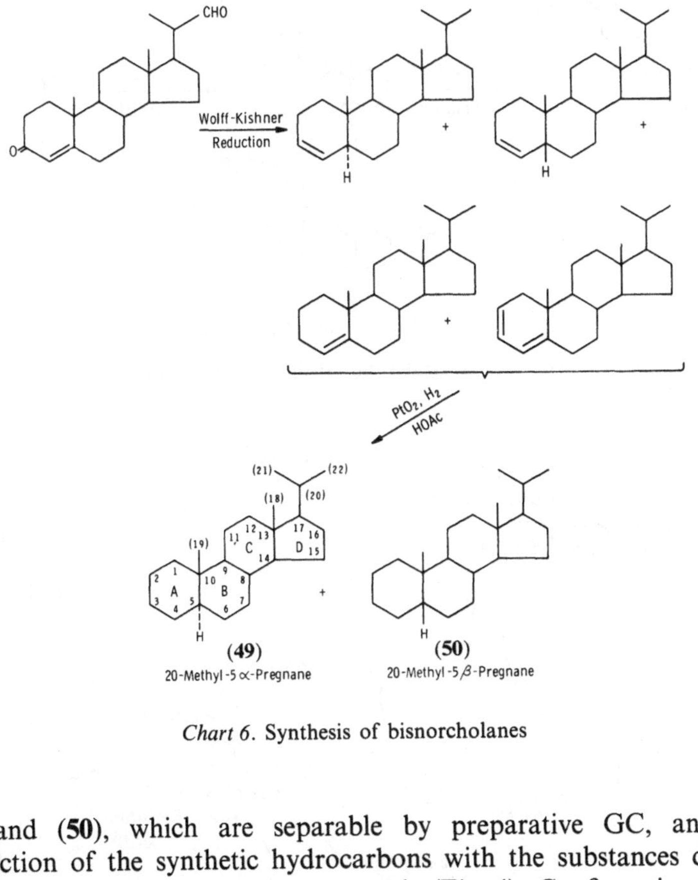

(**49**)
20-Methyl-5α-Pregnane

(**50**)
20-Methyl-5β-Pregnane

Chart 6. Synthesis of bisnorcholanes

(**49**) and (**50**), which are separable by preparative GC, and GC coinjection of the synthetic hydrocarbons with the substances derived from the naturally-occurring compounds (Fig. 4). Confirmation of the position of the carboxyl group in the C_{22}-steroid acid (**45**) was achieved by synthesis of the deuterium-labeled hydrocarbon (**51**) *via* the acid (**45**) (Chart 7).

The C_{22}-steroid acid (**45**) (Chart 5) was found to occur in two forms epimeric at C_{20}. This was shown by GCMS of the fluoroalcohol esters using the mass spectrometer parent mass position (m/e 646 for the C_{22} esters and m/e 674 for the C_{24} esters) as detectors (Fig. 5).

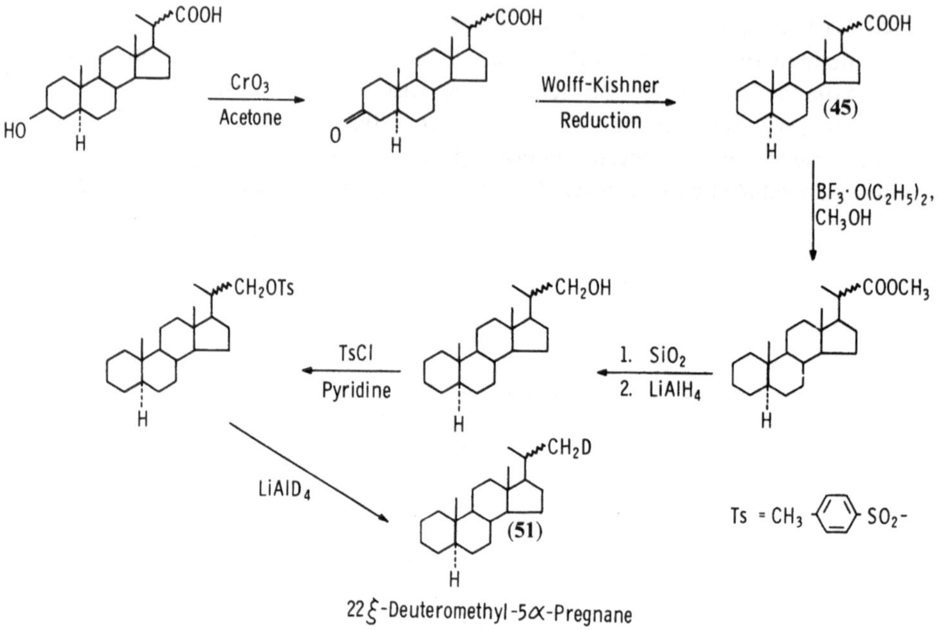

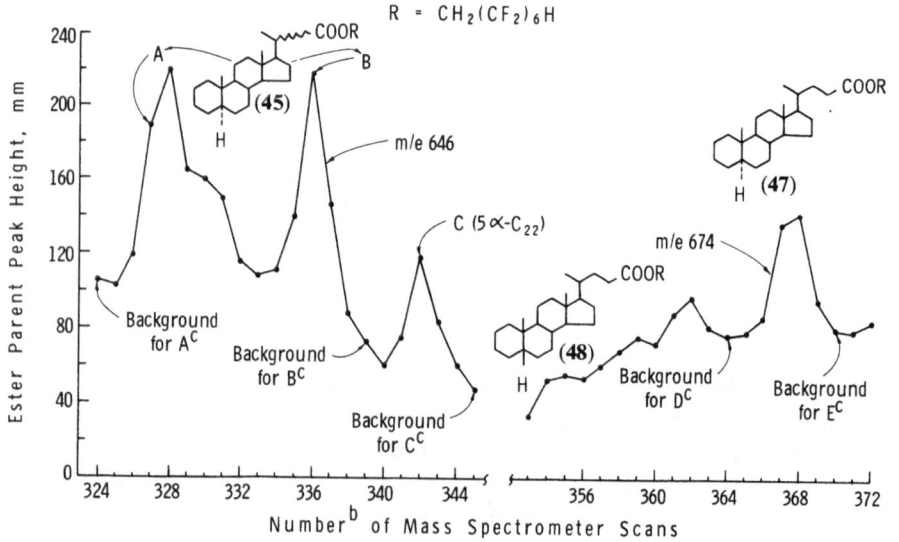

Chart 7. Synthesis of deuterium labelled C_{22}-steroid hydrocarbon

Fig. 5. Molecular ion mass chromatogram of esters[a] of petroleum steroid carboxylic acids

[a] With 1,1,7-Trihydroperfluoroheptanol.

[b] From sample injection. Scan cycle is 8.7 seconds.

[c] Height of steroid parent peaks shown.

References, pp. 44—49

This novel GCMS application to natural product chemistry has been called "mass chromatography". The two C_{20}-epimers A and B (Fig. 5) of the C_{22}-steroid acid (45) whose presence was further confirmed by GCMS coinjection with the synthetic esters prepared from synthetic acid (45) (Chart 7) are converted to one and the same hydrocarbon upon reduction (Chart 7). In addition, a third isomer C of the C_{22}-acid (45) was found. The mass spectrum indicates an A/B *trans*-5α-isomer. It may differ from A and B by having either a branched chain at C_{17} in the α-configuration or a straight chain in either the α- or β-configuration at C_{17}. From Fig. 5 the amount of the unidentified C_{22} steroid carboxylic acid (Isomer C) was estimated to be about one-tenth of the sum of the identified acids (Isomers A and B). 5α-Cholanic acid (47) and its derived hydrocarbon (52) and deuterium-labeled hydrocarbon (53) were synthesized as shown in Chart 8. Its presence in

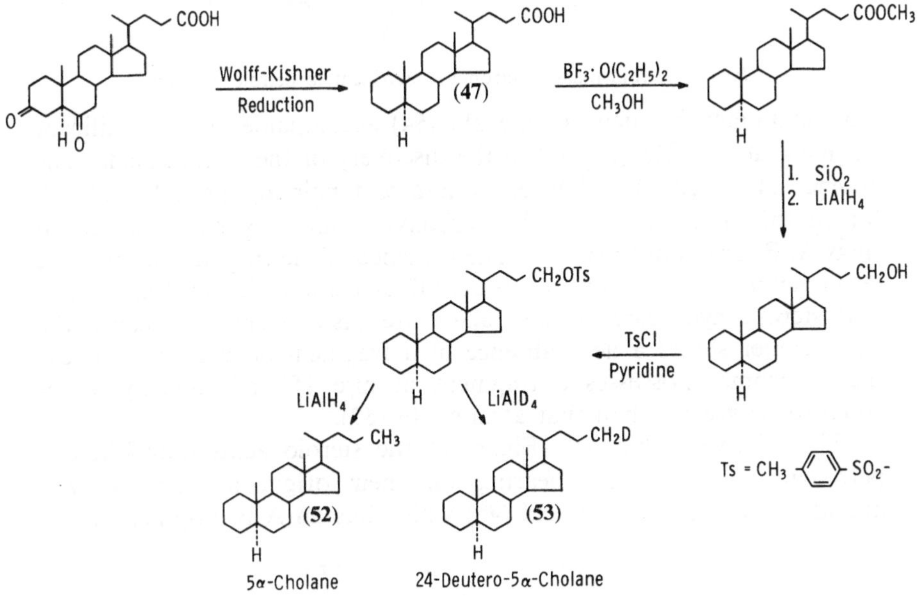

Chart 8. Synthesis of 5α-cholanic acid and the corresponding deuterium labelled hydrocarbon

petroleum as well as that of its 5β-isomer (48) (Figs. 4 and 5) was proved by the methods outlined above.

The results summarized here were in part achieved because of the excellent understanding of steroid mass spectrometry made available by previous work (68). A view of that portion of this work which is of interest to the steroid petroleum chemist is given in Chart 9: The

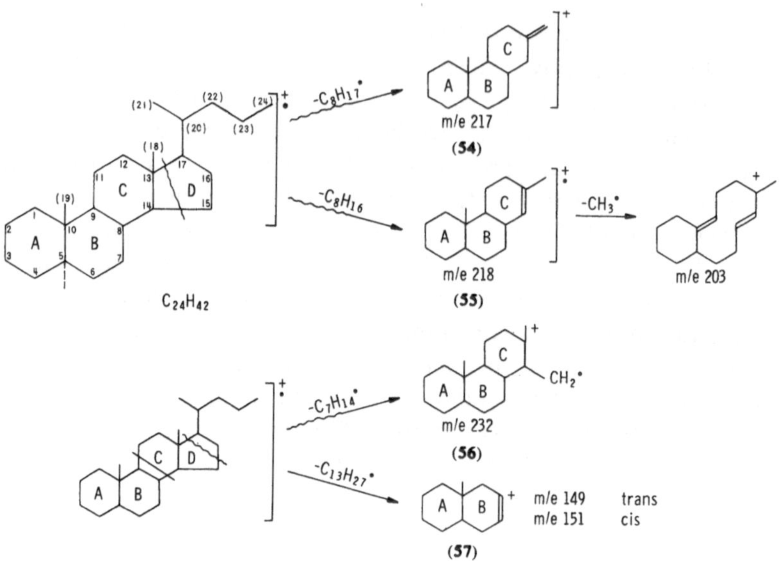

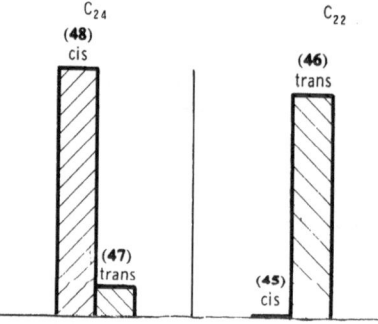

Chart 9. Steroid mass spectrometry pertinent to petroleum acids

most abundant fragment at m/e 217 (**54**) accompanied by a significant fragment at m/e 218 (**55**) led to the discovery of the steroid acids. The absence of a deuterium-labeled counterpart pair at m/e 218 and 219 forced the conclusion that the carboxyl group was not attached to rings A, B, or C but to ring D. The presence of the fragment at m/e 232 (**56**) pointed towards an unsubstituted C_{15}-position. *Cis*- (5β) and *trans*- (5α) steroid hydrocarbons and acids are distinguishable because the *trans*-isomer shows a near absence of a fragment at m/e 151 whereas the *cis* isomer possesses a fragment at m/e 151 of intensity about equal to or greater than that at m/e 149 (**57**).

The relative abundance (Fig. 6) of the steroid acids found in this petroleum has a direct bearing on their origin as follows. The abundance of 5β-cholanic acid (**48**) which has an A/B *cis*-junction and

Fig. 6. Relative ratios of stereoisomeric petroleum steroid acids

References, pp. 44—49

is thus thermodynamically less stable suggests that it is derived from animal bile acids, which are predominantly *cis*-C_{24}-hydroxycholanic acids (**58**) (Fig. 7), formed by a simple dehydration/hydrogenation process

	Acid	Representative Sources
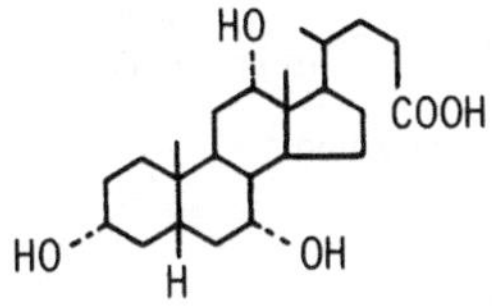	Cholic	Bird, Fish, Snake, and Others
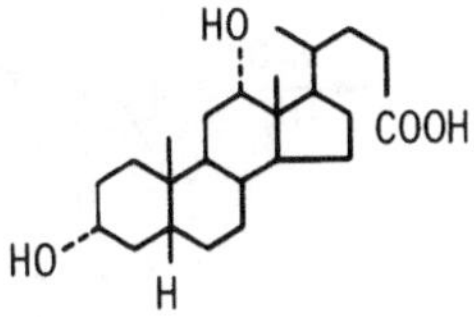	Desoxycholic	Ox, Rabbit, and Others
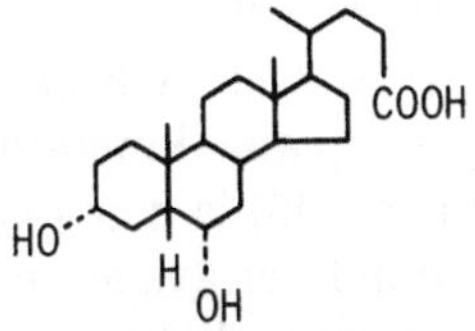	Hyodesoxycholic	Hog, Boar
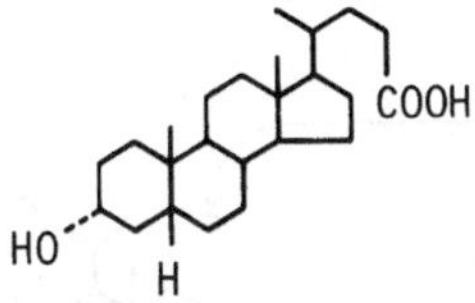	Lithocholic	Rabbit, Hog, Man
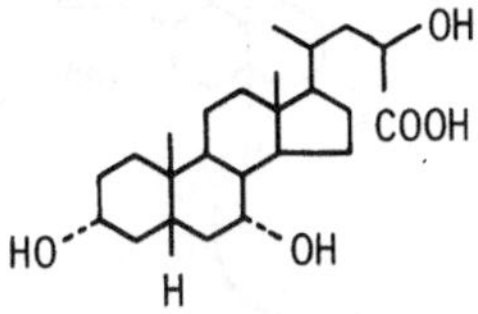	Phocaecholic	Sea Lion, Walrus
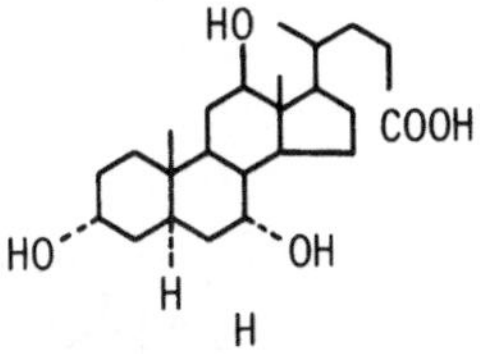	Allocholic	Lizard
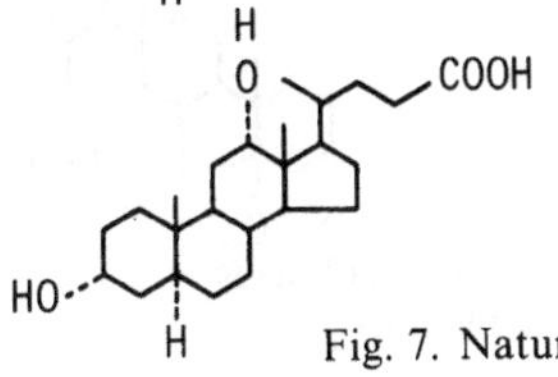	β-Lagodesoxycholic	Rabbit

Fig. 7. Naturally occurring C_{24}-bile acids

acceptable to petroleum chemists (Chart 10), petroleum being a

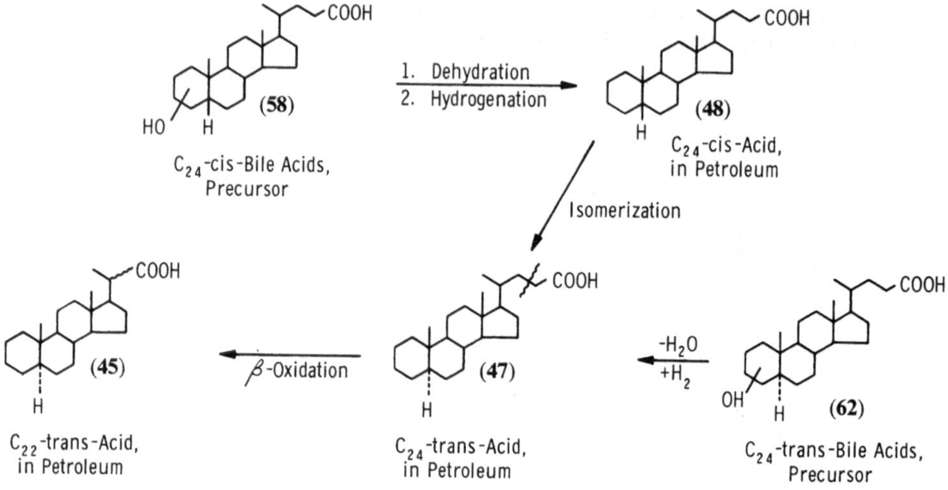

Chart 10. Possible animal origin of steroid acids

hydrogenating medium and clay a dehydration catalyst. A/B *cis*-steroid
stereochemistry has *not* been reported in the plant kingdom. The
presence of the two *trans*-C_{22}-steroid acid isomers (45) is best explained
by assuming oxidative $C_{22}-C_{23}$ double bond cleavage of such
naturally abundant sterols (in algae) as dihydroergosterol (59) (Chart 11).

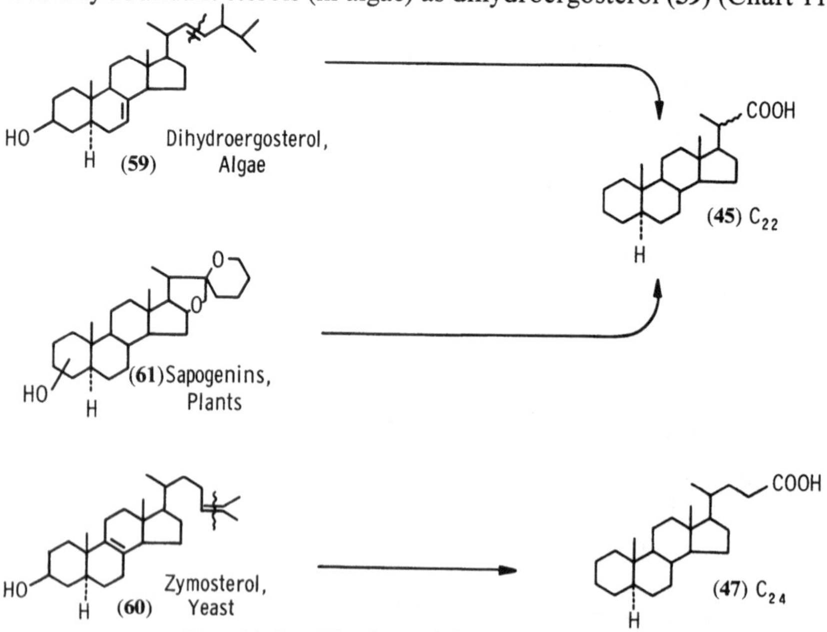

Chart 11. Possible plant origin of 5α-steroid acids
(oxidation, dehydration, reduction)

Analogously, 5α-cholanic acid (47) could come from plant zymosterol (60). An alternate but less preferred route for the 5α-C_{22}-acid (45) could start with steroidal sapogenins (61) (Chart 11).

The assumption that 5β-cholanic acid (48) could be formed by isomerization of plant-derived 5α-cholanic acid (47) fails because one would expect this to happen to a similar extent to the *trans*-C_{22}-acid (45), yet only traces of *cis*-C_{22}-acid (46) are found. Thus one is reduced to the hypothesis of an animal origin for 5β-cholanic acid (48). Although ascribing animal sources could in theory account for all the steroid acids found in California petroleum for e. g. if one invokes β-oxidation of *trans* bile acid (62) derived 5α-cholanic acid take out (47) (Chart 10) to give 5α-bisnorcholanic acid (45), the alternate plant route is more logical for the latter (15) because of the enormous predominance of sterols over bile acids in the biosphere. The finding of several C_{22}-*trans* isomers (Fig. 5) points to a variety of sources, as depicted in Chart 11.

Because the steroid acids found represent only trace amounts of the whole petroleum, the animal contribution to the origin of this California petroleum does not conflict with the predominant plant origin of petroleum. Concrete evidence for this was first obtained by A. TREIBS (1) who, with his discovery of desoxophylloerythroetioporphyrin, which is clearly linked to plant chlorophyll, laid the foundation for modern concepts of biological (plant-derived) petroleum diagenesis. Yet, based on spectroscopic evidence of a very small portion of mesoetioporphyrin which *can* be, but is not necessarily, derived from haemetype pigments (animal blood), TREIBS himself hinted at the possibility of some animal contribution to petroleum.

The above evidence regarding the intricate stereochemical details of steroid carboxylic acids in petroleum is another example, which, like the phytol/isoprenoid acid correlation discussed earlier in this review, sheds light on petroleum precursor/product relationships and thus becomes biogeochemically significant.

5. Pentacyclic Triterpenoid Acids

Most recently (69) a C_{32}-hopylacetic acid (63) was discovered in large quantities in Messel oil shale of Eocene age, constituting 50 ppm of dry rock! Its structure and membership of the 17βH, 21βH hopane series was proved unequivocally by synthesis (Chart 12). The biogeochemical significance of this acid is severalfold. Firstly it belongs to the thermodynamically less stable series of the hopanes which occurs in living organisms. Secondly it occurs as only one diastereomer at

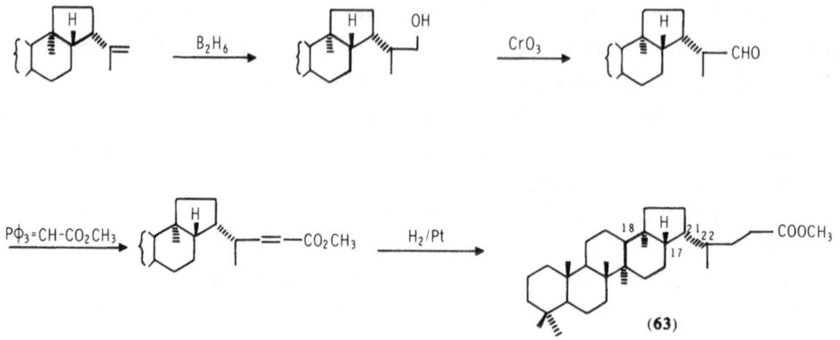

Chart 12. Synthesis of hopylacetic acid isolated from Messel shale (Eocene)

C_{22} (22R or 22S), which is again indicative of a relationship to living organisms. Thirdly it is accompanied by an abundance of a hydrocarbon with one less carbon atom, namely 17βH, 21βH-homohopane

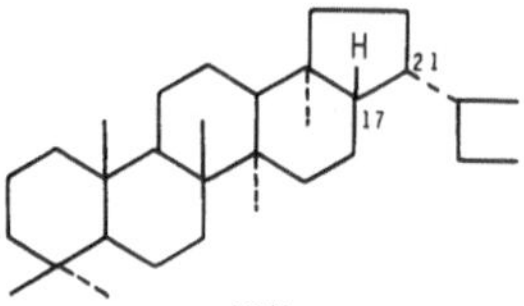

(C_{31}) (64) (70) which, in turn, also occurs as only one diastereomer at C_{22}. This circumstance provides additional evidence for the hypothesis discussed earlier in this review that petroleum n-paraffins are derived from n-fatty acids by decarboxylation. Fourth, the acid (63) may originate from C_{35}-precursors such as "bacteriohopanes" which contain four hydroxyl groups in the side chain. There is a dual significance to the occurrence of only one diastereomer in these terpanes: Members of the thermodynamically more stable 17αH, 21βH series, which are not found in living systems, generally occur in the form of both 22R and S diastereomers and are assumed to arise from the 17βH, 21βH series during maturation by isomerization. On the other hand, the presence of only one diastereomeric homohopane hydrocarbon has led the authors (69) to propose a possible bacterial synthesis of the C_{31} homohopane (64), by methylation of diploptene, a pentacyclic triterpene, which has recently been found in bacteria (71, 72). If true, such a mechanism would render doubtful, at least in part, the assumption of precursor/product relationships of petroleum pentacyclic triterpanes (67) from higher plants.

The same team identified another carboxylic acid, namely, bisnorhopanoic acid (65) in a 120 million-years-old marine cretaceous shale

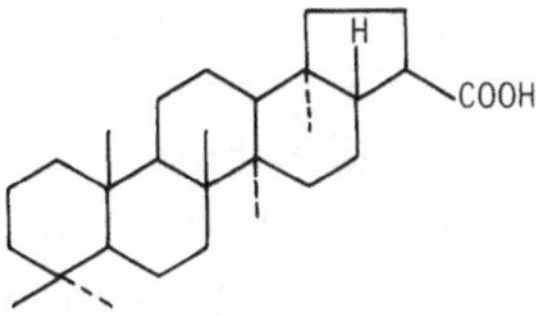

(65)

occurring near the coast of Gaboon.

IV. Aromatic Acids

1. Mono- and Diaromatic Acids

The first to realize that the so-called "Naphthenic Acids" also contain mono- and diaromatic acids as well as benzthiophenic acid was KNOTERUS (73), who examined the lubricating oil distillates of a Venezuelan crude oil. He proposed to replace the designation "Naphthenic Acids" by "Petroleum Acids". His findings were confirmed in the 250—450 molecular weight range and extended into the 700—1400 molecular weight range (74). The first results achieved with modern instrumentation (GCMS) deal with acids in the 150—250 molecular weight range of Green River shale (75). Besides compounds belonging to the classes of substituted benzoic, naphthoic, indanoic, and tetra-hydronaphthoic carboxylic acid, a phenyl alkanoic acid was found and tentatively identified as 2-methyl-4-(dimethylphenyl) butanoic acid (66).

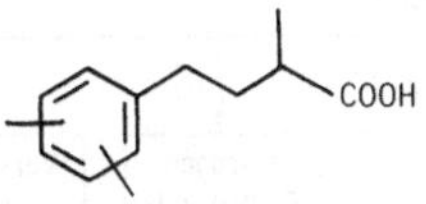

(66) 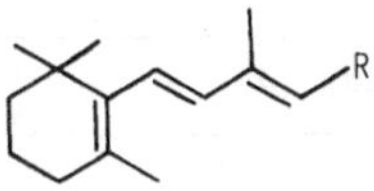   (67)

If both methyl groups were attached in the *ortho* position, a circumstance which has not yet been established, this acid could be rationalized as being derived from carotenes (67) or xanthophylls by oxidation and aromatization, the latter being accompanied by the loss of one methyl group.

The most comprehensive investigation of aromatic acids (including heterocyclic acids) to date was carried out in the author's laboratory (76, 77, 78) on the same California (Pliocene) petroleum in which the steroid acids had been discovered. The deuterium labeling approach which had been successful with the steroid acids succeeded with the aromatic acids only to the extent that it confirmed that the hydro-carbons were truly acid-derived and not artifacts of the isolation procedure (11); however, it failed with regard to detailed structural information because of the lacking predominance of individual components. Instead, a continuous distribution of very small amounts of many isomers and homologous series of many compound classes was encountered. The enormous complexity of these mixtures is illustrated in Fig. 8. After separation of the phenols from the acids by ion-exchange chromatography, 40% of all carboxylic acids present in this one California petroleum was converted to hydrocarbons under very mild conditions (64) and subsequently separated by double alumina chro-

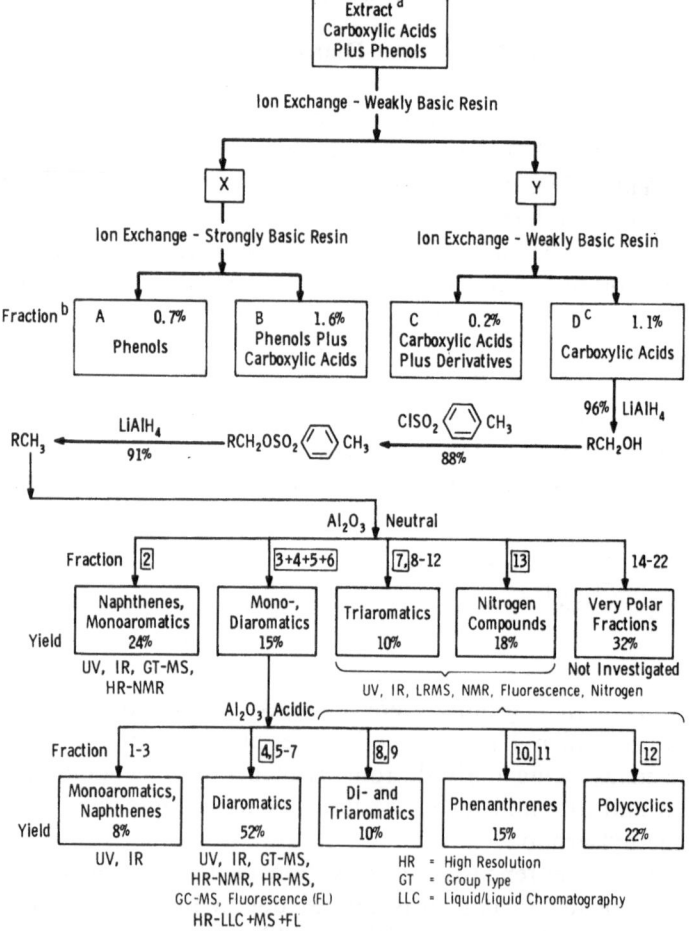

Fig. 8. Separation and transformation scheme leading to compound class identification of complex mixtures of polycyclic naphthenic, aromatic, and heterocyclic carboxylic acids
[a] This extract represents 3.5% of the total crude oil; 2.5% based on crude oil is RCOOH.
[b] Percentages are based on total crude oil.
[c] Represents 40% of all carboxylic acids present in crude oil.

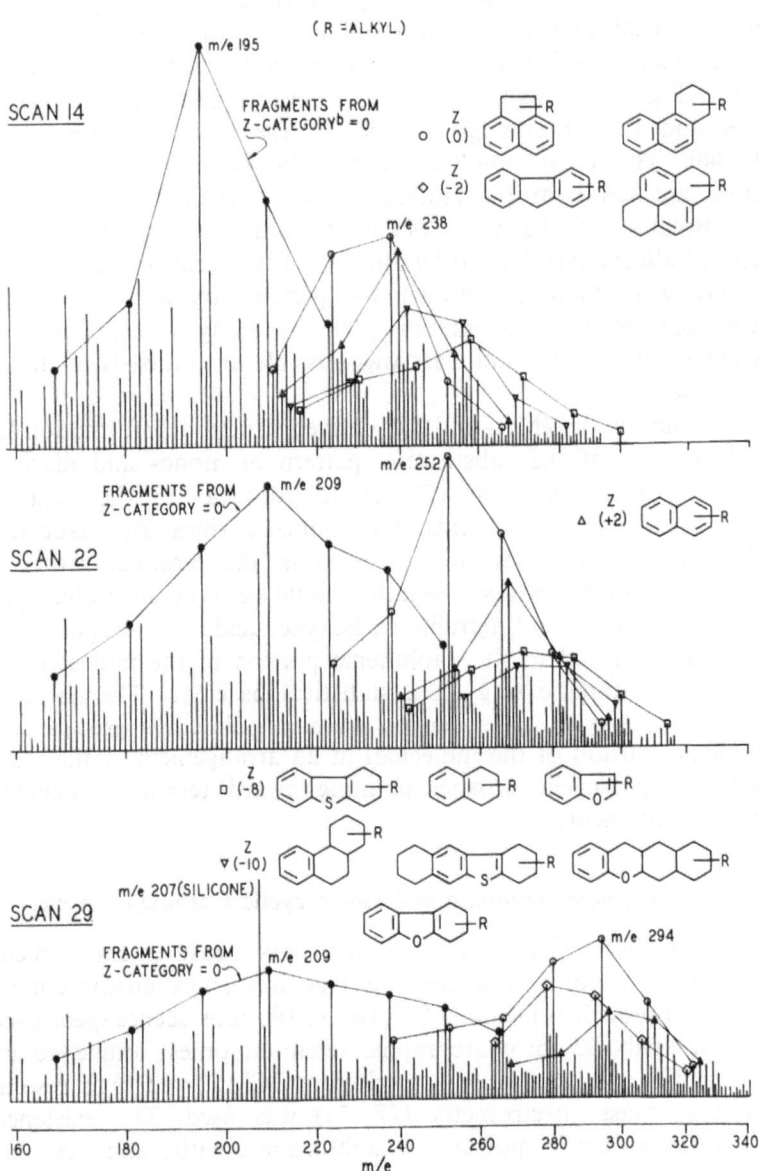

Fig. 9. Low resolution mass spectra (20 V) of selected GCMS fractions of diaromatic hydrocarbons[a] derived from carboxylic acids

(R = alkyl)

[a] Fraction 4 isolated by chromatography on acidic alumina.
[b] For definition, see original reference.

matography into fractions whose spectroscopic data indicated that
they belonged to chemically homogeneous compound types, e. g., di-
aromatics, triaromatics, nitrogen-containing types, etc.

The example of three GCMS scans performed on a diaromatic
fraction (Fig. 9) reveals the presence of several predominant compound
classes and their homologous series in the molecular weight range
160 — 340. The most abundant carboxylic acids in this petroleum
fraction are tetrahydrophenanthrene and/or acenaphthene carboxylic
acids. Because of the low resolution nature of the mass spectra,
structural alternatives had to be allowed. This situation was drastically
improved when high resolution mass spectrometry was applied to the
parent ions of these aromatics (77). In many cases, unequivocal
assignment of a compound class was possible on the basis of the exact
mass alone.

Combination of NMR- and MS-derived structural information led to
the elucidation of the substitution pattern of mono- and diaromatic
petroleum carboxylic acids (77) as follows: (a) In the octahydro-
phenanthrene carboxylic acids, the aromatic rings are disubstituted
predominantly, i.e., they are arranged in the terminal rather than
the center position, because the latter would be tetrasubstituted species.
(b) In the indane and tetralin carboxylic acids the carboxylic acid
groups are attached to the naphthenic portion of the molecule rather
than to the aromatic ring on a statistical basis. (c) The naphthenoaromatic carboxylic acids possess geminal dimethyl groups in the
naphthenic portion of the molecules in an arrangement similar to that
found in the polycyclic terpanes, pointing towards terpane precursors for
these aromatic acids.

2. Polynuclear Aromatic and Heterocyclic Carboxylic Acids

In those cases where high resolution mass measurements forced the
conclusion of two of several alternate structures, an exhaustive combined
analytical approach by means of UV, IR, NMR, fluorescence spectroscopy,
elemental analysis, chromatographic retention times, iodine complexing of sulfur compounds, low resolution GCMS, and high resolution
group-type mass spectrometry (77, 78) was used. The existence of
polynuclear aromatic, polycyclic naphthenoaromatic, and heterocyclic
petroleum carboxylic acids was virtually unknown prior to this work. The
conclusions about the structures of compound classes of the carboxylic
acids present in these complex mixtures are summarized in Fig. 10
(occurrence of 10 — 100 ppm based on petroleum) and Fig. 11 (< 10 ppm
occurrence). Some less likely structures which could not be ruled out
entirely are listed in the original paper (78).

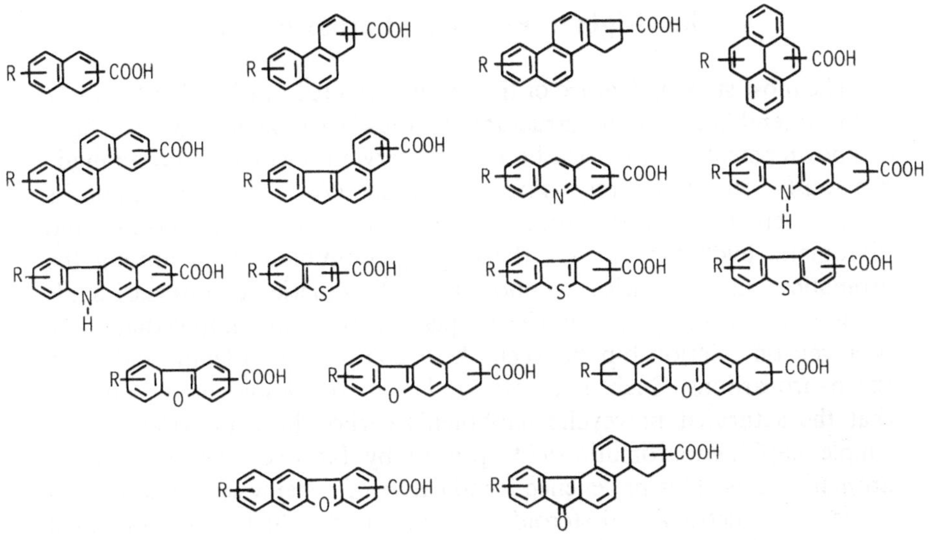

Fig. 10. Compound classes[a] $(C_{16}-C_{31})$ $(10-100$ ppm)[a] of aromatic and heterocyclic carboxylic acids in a California petroleum

[a] See footnotes Fig. 3.

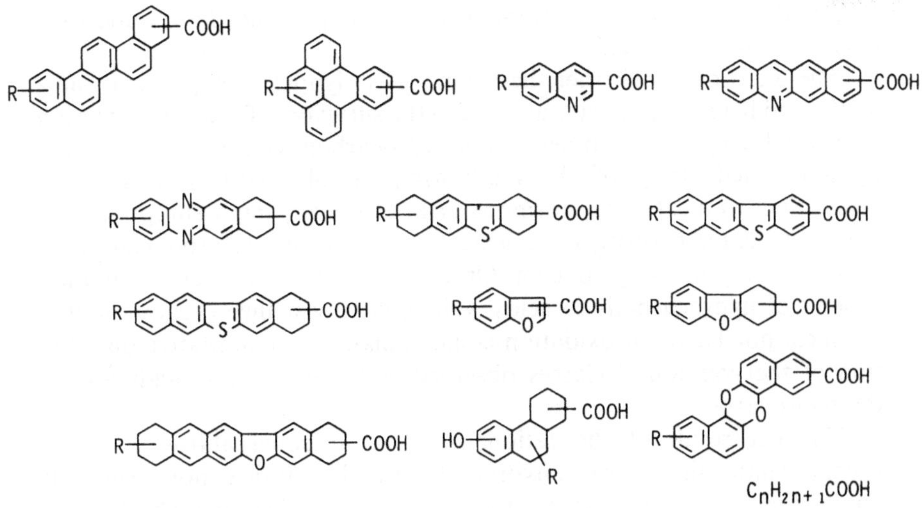

Fig. 11. Compound classes[a] $(C_{16}-C_{31})$ of aromatic and heterocyclic carboxylic acids present in very small quantities (<10 ppm) in a California petroleum

[a] See footnotes Fig. 3.

V. The Origin of Petroleum Carboxylic Acids

The most striking feature of the results depicted in Figs. 3, 10, and 11 is the resemblance of the structures of these carboxylic acids to those of known petroleum hydrocarbons, an observation which had been made previously on an entirely different sample (74). This conclusion is valid, even from a quantitative point of view; i. e., mono- and diaromatic hydrocarbons are the most predominant types among the aromatics and, likewise, mono- and diaromatic carboxylic acids predominate over the polynuclear types. These results imply that there is a genetic relationship between these complex petroleum acids and the hydrocarbons. Such a hypothesis has to be viewed in the context that the saturated polycyclic naphthenic carboxylic acids (Fig. 3) and simple naphthenoaromatic acids prevail by far over the polynuclear aromatic acids. This prevalence is indicative of a stepwise aromatization process of functionalized steroids and terpenoids which is accompanied either by the preservation of an originally present carboxylic acid group (e.g., steroid acid) or, at a very early stage of diagenesis, by the secondary formation of a carboxyl group from a functional group present in the original living material. [The conversion of a pentacyclic triterpene to a methylated picene has been experimentally demonstrated (79).] This whole process is accompanied by ring cleavage during maturation, which explains the predominance of mono- and bicyclic saturated over polycyclic saturated and of mono- and diaromatic over polycyclic aromatic acids.

The question of whether these complex polynuclear naphthenic, aromatic, and heterocyclic acids are indeed the survivors of a decarboxylation process which gave rise to petroleum hydrocarbons can only be answered by a detailed study of the relationship of all hydrocarbons to all acids. This would be an enormous task. However, the assumption that *most* petroleum hydrocarbons were formed by the decarboxylation of acids is a very unlikely concept. On the other hand, a reverse oxidation process of hydrocarbons to acids is difficult to visualize because neither chemical nor bacterial oxidation could explain the quantitative distribution of the compound classes observed in the carboxylic acids and in the hydrocarbons.

The net result of the rationalizations presented above is that a genetic relationship between hydrocarbons and acids does not necessarily imply that one arises from the other but rather that both have common precursors, such as functionalized steroids and terpenoids. After introduction of carboxyl groups during the early stage of sedimentation (or survival of originally present carboxyl groups), the hydrocarbon portions of the petroleum acids undergo transformational changes such as ring cleavage and partial or total aromatization during maturation similar to those suffered by the hydrocarbons themselves. This is sufficient to explain the observed resemblance of the structures of complex carboxylic acids (Figs. 3, 10, 11) and their internal quantitative distribution to that of petroleum hydrocarbons.

The predominance of carbazolic and acridinic acids (Fig. 3) (*80*) which could be of alkaloid origin is an interesting sidelight.

The evidence presented in this review points towards the existence of multiple diagenetic pathways of petroleum carboxylic acids and may be summarized as follows. Portions of the n-fatty acids (saturated and unsaturated) found in petroleum are biological remnants. Some branched-chain saturated acids and perhaps the "pseudo" isoprenoid acids are assumed to fall into the same category. At least one of the steroid acids (*cis*-C_{24}) is negligibly removed in the chain of degradation from its biological precursors, the bile acids. Isoprenoid acids are proven to arise predominantly from the phytyl side chain of chlorophyll. Polycyclic naphthenic and aromatic acids appear to arise from functionalized steroids and triterpenoids. Chemical or bacterial oxidation of hydrocarbons is the least likely general route to the acids and has been disproved in one instance, the possible genesis of phytanic acid from phytane. Oxidative paths involving functionalized species as starting material, preferably at the very early stage of sedimentation, are more plausible.

Finally, the most important determining factor regarding the origin of petroleum carboxylic acids and their hydrocarbon generation potential is the geological environment, which may vary so drastically from one case of sample to another that generalizations, at least at this stage of our knowledge, are premature.

VI. Bifunctional Acids

Although this review is intentionally limited to monofunctional carboxylic acids, some other acids existing in petroleum and sediments will be discussed briefly in the following sections.

1. Dicarboxylic Acids

The occurrence of dicarboxylic acids in petroleum was first recognized with the observation of dimethylmaleic acid and 1,2,2-trimethylcyclo-pentane-1,3-dicarboxylic acid in California and Texas crudes (Table I).

α, ω-Dicarboxylic acids ($C_9 - C_{21}$) (**68**) were found in Scottish torbanite (Carboniferous), an algal deposit (*51*), which is described as a boghead or cannel coal. α,ω-Dicarboxylic acids ($C_{12}-C_{18}$) were identified in Green River shale (*47*), and the range was recently extended (*50*) to C_{32}. The authors (*50*) also reported a series of another type of dicarboxylic acids (**69**)

$$HOOC(CH_2)_nCOOH \qquad\qquad CH_3(CH_2)_n\underset{\underset{\textstyle COOH}{|}}{C}HCH_2COOH$$

(**68**)   (**69**)

In addition, the branched-type (**70**) was observed (*39*).

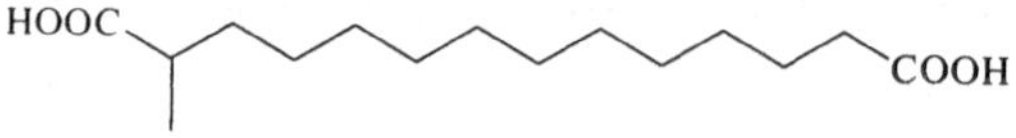

(**70**)

A C_4-diacid monoester (**71**) was observed (*9*) in the California petroleum investigated by this author.

$$ROOC(CH_2)_2COOH$$

(**71**)

A variety of dicarboxylic acids including aromatic types were identified in Alaskan tasmanite (*81*), a sedimentary rock rich in organic material of Late Jurassic to Early Cretaceous Age (130 — 190 million years old).

Such dicarboxylic acids are rare in nature. They occur as free acids or glycerides. Microbial oxidation has been suggested (*50*) as a possible route to their formation.

2. Hydroxy- and Ketoacids

The first hydroxy acid isolated from petroleum was a triterpene lactone, oxyallobetul-2-ene (**72**). Its discovery in 1954 in a U.S.

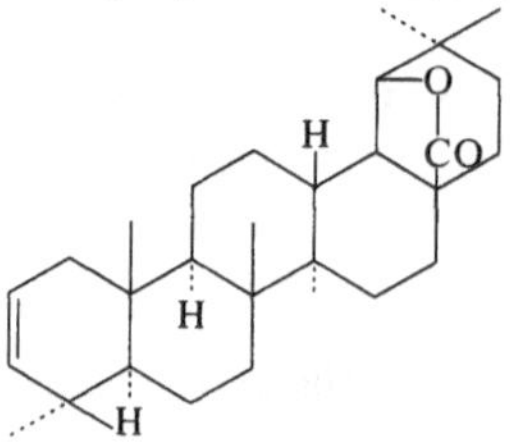

(**72**)

petroleum (*82*) and later in a Kuwait petroleum (*83*) was a milestone on the way to strengthening the proof of plant origin of petroleum. It may be an acid-catalyzed degradation product of betulin (**73**), a floral triterpene.

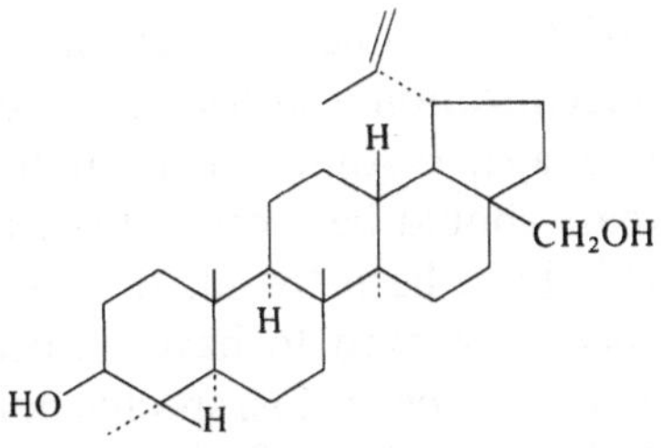

(73)

ω-Hydroxy acids ($C_{16}-C_{24}$), 10,16-dihydroxyhexadecanoic acid (both presumably derived from plant cutin and suberin), and $C_{10}-C_{24}$ α- and β-hydroxy acids (assumed to be formed by microbial oxidation of fatty acids) were all found in a 5000-year-old sediment from a freshwater lake in the English Lake District (*84*). Some α-ω-dicarboxylic acids of the type discussed above were also found in this sediment and explained as arising from ω-hydroxy acids by oxidation.

Most recently, (C_7-C_{15})-γ-lactones (**74**) were identified in a

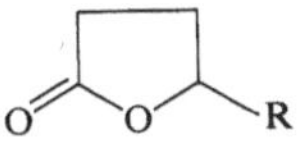

(74)

26 million year old shale in large concentrations (0.04% of rock). Because they are prevalent in biolipids and microbiological transformation products, they could either be biological remnants or, in the authors' (*85*) opinion, more likely cyclization products of Δ^3 and Δ^4 unsaturated fatty acids.

($\omega-1$)-Ketoacids ($C_{11}-C_{14}$) were identified in ancient Green River shale (*86*). These acids are uncommon in nature and believed to be formed by oxidation of the corresponding hydroxy acids.

3. Amino Acids

Amino acids have not yet been found in petroleum; however, they have caused considerable interest due to their occurrence in sediments. A review up to 1967 is available (*87*). An additional review on this subject including recent information is in preparation (*88*) and, therefore, only a few highlights are mentioned in the following paragraph.

Amino acids have been reported in recent and ancient sediments as old as Precambrian (*87*). Amino acids with asymmetric carbon atoms in modern geological materials are predominantly of the L-configuration. Therefore, they are assumed to be of biological origin.

With increasing age of the sediment, interconversion from L to D occurs. The degree of racemization is utilized for age dating. The method is most useful at young age. α-Amino acids indigenous to sediments older than 2 million years should be racemic (89). In modern materials, only protein amino acids have been found. In recent marine sediments, each amino acid has been observed to have its own rate of racemization (90). In ancient geological specimens, specific nonprotein amino acids have been reported (87, 91) such as β-alanine (75) (probable source aspartic acid), α- (76) and γ- (77) aminobutyric acid (probable precursor glutamic acid), alloisoleucene (78) (probable precursor L-isoleucine), and ornithine (79) (probable precursor arginine).

$$
\begin{array}{ccc}
& \overset{\displaystyle NH_2}{|} & \\
H_2NCH_2CH_2COOH & CH_3CH_2CHCOOH & H_2NCH_2CH_2CH_2COOH \\
(75) & (76) & (77)
\end{array}
$$

$$
\begin{array}{cc}
\overset{\displaystyle CH_3}{\underset{}{|}} & \\
H_5C_2CHCHNH_2COOH & H_2N(CH_2)_3CHNH_2COOH \\
(78) & (79)
\end{array}
$$

4. Miscellaneous

Another category of polyfunctional carboxylic acids in petroleum includes some petroporphyrins. An excellent review on that subject is available (92).

An interesting and important sidelight of carboxylic acids in petroleum is their profound effect on the interfacial activity of crude oils. Originally, porphyrins and related substances, polyfunctional "resins" and "asphaltenes" and later phenols had been claimed to be the major contributors to the interfacial activity in crude oils. Carboxyphenols have indeed been found in a California petroleum (4). The question of which acidic functional groups in petroleum are responsible for the interfacial activity of petroleum at alkaline pH has been settled for at least one California petroleum. Carboxylic acids were demonstrated experimentally and unequivocally to be the only responsible constituents as illustrated in Fig. 12, which depicts the effect of gradual esterification of "petroleum acids" [under such conditions that phenols cannot react (80)] on the change of interfacial acitivity at alkaline pH. The interfacial activity decreases parallel with the disappearance of carboxylic acids. Phenols, in fact, were proved (80) to have a detrimental effect on the activity of the carboxylic acids. The presence or contribution to inter-

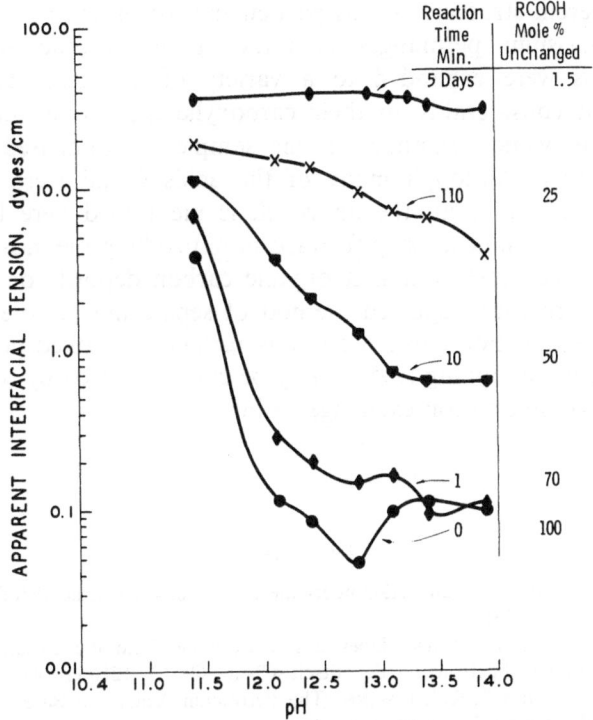

Fig. 12. Change of interfacial activity of "petroleum acids" by reaction with diazomethane at −50° C

facial activity of other acids such as sulfonic acids, sulfinic acids, carbazoles, mercaptans, etc., was ruled out independently.

In the course of this study (4), it was realized that derivatives of carboxylic acids exist in petroleum in considerable quantities. Examples of esters (93) and amides (94) of carboxylic acids had been reported earlier.

It had been pointed out that esters of fatty acids may be present in shales bound to the insoluble kerogen matrix (51). One step further in this direction was the isolation of $C_{14}-C_{17}$ and $C_{19}-C_{22}$ isoprenoid fatty acids by chromic acid oxidation of Green River shale kerogen (95), the interpretation being that these isoprenoid acids are indigenous, and were laid down at the time of sedimentation and are incorporated into the polymeric organic structures. More extensive oxidation led to predominant formation of dicarboxylic acids, an observation which supported the concept of the aliphatic nature of the Green River formation (96). In an organic-rich rock of Permian age (220−275 million years old), the absence of isoprenoid acids and production of polycyclic and aromatic acids by oxidation pointed toward a rather different structure of this particular kerogen matrix (97). Using alkaline potassium permanganate rather than chromic acid, these investigations were expanded to a variety of kerogens (98), all of which varied considerably in their carboxylic acid composition, e. g., from polymethylene structures in one sample to condensed aromatic ones in another. Although many of the acids found were obviously formed during oxidation, the approach is mentioned here because it appears to be a significant step forward in unravelling the most complex structures of the most abundant organic carbon deposits on earth.

Finally, a recently reported method of separating petroleum acids, namely, gel permeation chromatography using cross-linked polystyrene, appears worth mentioning (99). It may be complemented by separations on ferric chloride and ion-exchange resins (100).

References

1. Treibs, A.: Chlorophyll und Häminderivate in Organischen Mineralstoffen. Angew. Chem. **49**, 682 (1936).
2. Hell, C., und E. Medinger: Über das Vorkommen und die Zusammensetzung von Säuren im Rohpetroleum. Ber. dtsch. Chem. Ges. **7**, 1216 (1874).
3. Lochte, H. L., and E. R. Littmann: The Petroleum Acids and Bases. New York: Chemical Publishing Company, Inc. 1955.
4. Seifert, W. K., and W. G. Howells: Interfacially Active Acids in a California Crude Oil. Isolation of Carboxylic Acids and Phenols. Analyt. Chem. **41**, 554 (1969).
5. Costantinides, G., and G. Arich: Nonhydrocarbon Compounds in Petroleum: Fundamental Aspects of Petroleum Geochemistry. Ed. by B. Nagy and U. Colombo, pp. 143—151. New York: Elsevier. 1957.
6. De Boer, T. J., and H. J. Backer: A New Method for the Preparation of Diazomethane. Rec. Trav. Chim. **73**, 229 (1954).
7. White, E. H., A. A. Baum, and D. E. Eitel: 1-Methyl-3-p-tolyltriazene. Org. Syn. **48**, 102 (1968).
8. Teeter, R. M.: Fluoroalcohol Esters as Derivatives for Mass Spectrometry. Analyt. Chem. **39**, 1742 (1967).
9. Seifert, W. K., and R. M. Teeter: Preparative Thin-Layer Chromatography and High Resolution Mass Spectrometry of Crude Oil Carboxylic Acids. Analyt. Chem. **41**, 786 (1969).
10. Pierce, A. E.: Silylation of Organic Compounds; a Technique for Gas Phase Analysis. Pierce Chemical Co., Rockford, Ill., U.S.A. (1968).
11. Teeter, R. M., and W. K. Seifert: Observation of Derivatives of Petroleum Acids by Mass Spectrometry. ACS Meeting, Los Angeles, April 1971, Petroleum Division Preprints A7 (1971).
12. Vetter, W., W. Walther und M. Vecchi: Pyrrolidide als Derivate für die Strukturaufklärung aliphatischer und alicyclischer Carbonsäuren mittels Massenspektrometrie. Helv. Chim. Acta **54**, 1599 (1971).
13. Cooper, J. E., and E. E. Bray: A Postulated Role of Fatty Acids in Petroleum Formation. Geochimica et Cosmochimica Acta **27**, 1113 (1963).

14. McAuliffe, C. D.: Oil-in-Water Emulsions Improve Fluid Flow in Porous Media. J. Petroleum Technology, p. 729 (June 1973).

15. Seifert, W. K.: Steroid Acids in Petroleum. Animal Contribution to the Origin of Petroleum. Pure and Appl. Chem. **34**, 633 (1973).

16. Wort, D. J., J. G. Severson, Jr., and D. R. Peirson: Mechanism of Plant Growth Stimulation by Naphthenic Acid. Plant Physiol. **52**, 162 (1973).

17. Severson, J. G., Jr., and D. J. Wort: Phosphate Uptake and Distribution in Bush Bean Plants as Affected by Foliar Application of Naphthenate. Agronomy Journal **5**, 520 (1973).

18. Severson, J. G., Jr.: Stimulation of [^{14}C] Glucose Uptake and Metabolism in Bean Root Tips by Naphthenates. Phytochemistry **11**, 71 (1972).

19. Graham, D. W.: Ph. D. Dissertation, University of California, Berkeley; University Microfilms 65-8174, Ann Arbor, Michigan, 1965.

20. Kvenvolden, K. A.: Evidence for Transformations of Normal Fatty Acids in Sediments. Advances in Organic Geochemistry 1966 ed. by G. C. Hobson and G. C. Speers, p. 335. New York: Pergamon. 1970.

21. — Normal Fatty Acids in Sediments. J. Amer. Oil Chem. Soc. **44**, 628 (1967).

22. Douglas, A. G., K. Douraghi-Zadeh, G. Eglinton, J. R. Maxwell, and J. N. Ramsay: Fatty Acids in Sediments Including the Green River Shale (Eocene) and Scottish Torbanite (Carboniferous). Advances in Organic Geochemistry 1966 ed. G. D. Hobson and G. C. Speers, p. 315. Oxford: Pergamon Press. 1970.

23. Parker, P. L.: Fatty Acids and Alcohols. Organic Geochemistry ed. by G. Eglinton and M. T. J. Murphy, pp. 363—366. New York: Springer. 1969.

24. Lawlor, D. L., and W. E. Robinson: Fatty Acids in Green River Formation Oil Shale. Div. Petrol. Chem. Amer. Chem. Soc. Detroit Meeting 1965, 5.

25. Kvenvolden, K. A., and D. Weiser: A Mathematical Model of a Geochemical Process. Normal Paraffin Formation from Normal Fatty Acids. Geochimica et Cosmochimica Acta **31**, 1281 (1967).

26. Welte, D. H.: Distribution of Long Chain n-Paraffins and n-Fatty Acids in Sediments from the Persian Gulf. Geochimica et Cosmochimica Acta **32**, 465 (1968).

27. Maxwell, J. R., C. T. Pillinger, and G. Eglinton: Organic Geochemistry. Quart. Rev. (Chem. Soc. London) **25**, 585 (1971).

28. Abelson, D. H.: Organic Geochemistry and the Formation of Petroleum. Proc. Sixth World Petroleum Congress, Section I, 397 (1963).

29. Jurg, J. W., and E. Eisma: Petroleum Hydrocarbons: Generation From Fatty Acids. Science **144**, 1451 (1964); Advances in Organic Geochemistry 1966 ed. by G. D. Hobson and G. C. Speers, p. 367. Oxford-New York: Pergamon Press. 1970.

30. Waples, D. W.: Catalytic Formation of Hydrocarbons from Fatty Acids. Nature Physical Science **237**, 63 (1972).

31. Albrecht, P., and G. Ourisson: Biogenic Substances in Sediments and Fossils. Angew. Chem. Internat. Ed. **10**, 218 (1971).

32. Van Hoeven, W., J. R. Maxwell, and M. Calvin: Fatty Acids and Hydrocarbons as Evidence of Life Processes in Ancient Sediments and Crude Oils. Geochimica et Cosmochimica Acta **33**, 877 (1969).

33. Han, J., and M. Calvin: Occurrence of Fatty Acids and Aliphatic Hydrocarbons in a 3.4 Billion Year Old Sediment. Nature **224**, 576 (1969).

34. Parker, P. L., and R. F. Leo: Fatty Acids in Blue-Green Algal Mat Communities. Science **148**, 373 (1965).

35. Haug, P., and J. R. Sever: A Study of the Mechanism of Formation of the Acids in a Marine Sediment. Acids of the Excello Shale and Surtsey Lagoonal Sediment. Advances in Organic Geochemistry 1971, p. 293. Oxford-Braunschweig: Pergamon Press. 1972.

36. RHEAD, M. M., G. EGLINTON, G. H. DRAFFAN, and P. J. ENGLAND: Conversion of Oleic Acid to Saturated Fatty Acids in Severn Estuary Sediment. Nature **232,** 327 (1971).

37. — — — — Products of Short-Term Diagenesis of Oleic Acid in Estuary Sediment. Advances in Organic Geochemistry 1971, p. 323. Oxford-Braunschweig: Pergamon Press. 1972.

38. LEO, R. F.: The Geochemistry of Fatty Acids in Recent Marine Sediments. M. A. Thesis, University of Texas (1966).

39. HAUG, P., H. K. SCHNOES, and A. L. BURLINGAME: Studies of the Acidic Components of a Colorado Green River Formation Oil Shale: Mass Spectrometric Identification of the Methyl Esters of Extractable Acids. Chem. Geology **7,** 213 (1971).

40. RAMSAY, J. N.: Organic Geochemistry of Fatty Acids. M. S. Thesis, University of Glasgow (1966).

41. LEO, R. F., and P. L. PARKER: Branched Chain Fatty Acids in Sediments. Science **152,** 649 (1966).

42. BOCK, R., und K. BEHRENDS: Untersuchung eines Gemisches von Erdölsäuren. Ein Beitrag zum Naphthensäureproblem. Z. analyt. Chem. **208,** 338 (1965).

43. CASON, J., and D. W. GRAHAM: Isolation of Isoprenoid Acids from a California Petroleum. Tetrahedron **21,** 471 (1965).

44. CASON, J., and A. I. A. KHODAIR: Isolation of the C_{11}-Carbon Acyclic Isoprenoid Acid from Petroleum. J. Organ. Chem. (U.S.A.) **32,** 3430 (1967).

45. ACKMAN, R. G., and R. P. HANSEN: The Occurrence of Diastereomers of Phytanic and Pristanic Acids and Their Determination by Gas-Liquid Chromatography. Lipids **2,** 357 (1967).

46. EGLINTON, G., A. G. DOUGLAS, J. R. MAXWELL, J. N. RAMSAY, and S. STÄLLBERG-STENHAGEN: Occurrence of Isoprenoid Fatty Acids in Green River Shale. Science **153,** 1133 (1966).

47. HAUG, P., H. K. SCHNOES, and H. L. BURLINGAME: Isoprenoid and Dicarboxylic Acids Isolated from Colorado Green River Shale (Eocene). Science **158,** 772 (1967).

48. BURLINGAME, A. L., and B. R. SIMONEIT: High Resolution Mass Spectral Analysis of the Mineral Entrapped Fatty Acids Isolated from the Green River Formation (Eocene). Nature **218,** 252 (1968).

49. BLUMER, M., and W. J. COOPER: Isoprenoid Acids in Recent Sediments. Science **158,** 1463 (1967).

50. DOUGLAS, A. G., M. BLUMER, G. EGLINTON, and K. DOURAGHI-ZADEH: Mass Chromatographic-Mass Spectrometric Characterization of Naturally Occurring Acyclic Isoprenoid Carboxylic Acids. Tetrahedron **27,** 1071 (1971).

51. DOUGLAS, A. G., K. DOURAGHI-ZADEH, G. EGLINTON, J. R. MAXWELL, and J. N. RAMSAY: Fatty Acids in Sediments Including Green River Shale (Eocene) and Scottish Torbanite (Carboniferous). Advances in Organic Geochemistry 1966, ed. G. D. HOBSON and G. C. SPEERS, p. 315. Oxford: Pergamon Press. 1970.

52. MACLEAN, I., G. EGLINTON, K. DOURAGHI-ZADEH, R. J. ACKMAN, and S. N. HOOPER: Correlation of Stereoisomerism in Present-Day and Geologically Ancient Isoprenoid Fatty Acids. Nature **218,** 1019 (1968).

53. COX, R. E., J. R. MAXWELL, G. EGLINTON, and C. T. PILLINGER: The Geological Fate of Chlorophyll: The Absolute Stereochemistries of a Series of Acyclic Isoprenoid Acids in a 50 Million-Year-Old Lacustrine Sediment. Chem. Commun. **1970,** 1639.

54. ACKMAN, R. G., R. E. COX, G. EGLINTON, S. N. HOOPER, and J. R. MAXWELL: Stereochemical Studies of Acyclic Isoprenoid Components I – Gas Chromatographic Analysis of Stereoisomers of a Series of Standard Acyclic Isoprenoid Acids. J. Chromat. Science **10,** 392 (1972).

55. MAXWELL, J. R., C. T. PILLINGER, and G. EGLINTON: Organic Geochemistry. Quart. Rev. (Chem. Soc. London) **25**, 593 (1971).

56. MAXWELL, J. R., R. E. COX, G. EGLINTON, and C. T. PILLINGER: Stereochemical Studies of Acyclic Isoprenoid Components II – The Role of Chlorophyll in the Derivation of Isoprenoid-Type Acids in a Lacustrine Sediment. Geochimica et Cosmochimica Acta **37**, 297 (1973).

57. MURPHY, R. C., M. V. DJURICIC, S. P. MARKLEY, and K. BIEMAN: Acidic Components of Green River Shale Identified by a Gas Chromatographic Mass Spectrometry-Computer System. Science **165**, 695 (1969).

58. CASON, J., and K.-L. LIAUW: Characterization and Synthesis of a Monocyclic Eleven-Carbon Acid Isolated from a California Petroleum. J. Organ. Chem. **30**, 1763 (1965).

59. CASON, J., and A. I. A. KHODAIR: Separation from a California Petroleum and Characterization of Geometric Isomers of 3-Ethyl-4-Methylcyclopentylacetic Acid. J. Organ. Chem. **31**, 3618 (1966).

60. GILPIN, J. A., and F. W. MCLAFFERTY: Mass Spectrometric Analysis, Aliphatic Aldehydes. Analyt. Chem. **29**, 990 (1957).

61. GALLEGOS, E. J., J. W. GREEN, L. P. LINDEMAN, R. L. LETOURNEAU, and R. M. TEETER: Petroleum Group-Type Analysis by High Resolution Mass Spectrometry. Analyt. Chem. **39**, 1833 (1967).

62. SEIFERT, W. K., E. J. GALLEGOS, and R. M. TEETER: First Identification of a Steroid Carboxylic Acid in Petroleum. Angew. Chem. Internat. Ed. **10**, 747 (1971).

63. — — — Proof of Structure of Steroid Carboxylic Acids in a California Petroleum by Deuterium Labeling, Synthesis, and Mass Spectrometry. J. Am. Chem. Soc. **94**, 5880 (1972).

64. SEIFERT, W. K., R. M. TEETER, W. G. HOWELLS, and M. J. R. CANTOW: Analysis of Crude Oil Carboxylic Acids After Conversion to their Corresponding Hydrocarbons. Analyt. Chem. **41**, 1639 (1969).

65. BUDZIKIEWICZ, H., J. M. WILSON, and C. DJERASSI: Mass Spectrometry in Structural and Stereochemical Problems XXXII. Pentacyclic Triterpenes. J. Am. Chem. Soc. **85**, 3688 (1963).

66. HILLS, I. R., and E. V. WHITEHEAD: Triterpanes in Optically Active Petroleum Distillates. Nature **209**, 977 (1966).

67. WHITEHEAD, E. V.: The Structure of Petroleum Pentacyclanes. Sixth International Meeting on Organic Geochemistry, Paris, France, September 18–21, 1973, in press.

68. TÖKES, L., G. JONES, and C. DJERASSI: Mass Spectrometry in Structural and Stereochemical Problems CLXI: Elucidation of the Course of Characteristic Ring D Fragmentation of Steroids. J. Am. Chem. Soc. **90**, 5465 (1968).

69. ENSMINGER, A., A. VAN DORSSELAER, CH. SPYCKERELLE, P. ALBRECHT, and G. OURISSON: Pentacyclic Triterpenes of the Hopane Type as Ubiquitous Geochemical Markers: Origin and Significance. Sixth International Meeting of Organic Geochemistry, Paris, France, September 18–21, 1973, in press.

70. ENSMINGER, A., P. ALBRECHT, G. OURISSON, P. J. KIMBLE, J. R. MAXWELL, and G. EGLINTON: Homohopane in Messel Oil Shale: First Identification of a C_{31} Pentacyclic Triterpane in Nature. Bacterial Origin of Some Triterpanes in Ancient Sediments? Tetrahedron Letters **1972**, 3861.

71. BIRD, C. W., J. M. LYNCH, F. J. PIRT, and W. W. REID: Steroids and Squalene in Methylococcus Capsulatus Grown on Methane. Nature **230**, 473 (1971).

72. DE ROSA, M., A. CAMBACORTA, L. MINALE, and J. D. BU'LOCK: Bacterial Triterpenes. Chem. Commun. **1971**, 620.

73. KNOTERUS, J.: The Chemical Constitution of the Higher "Naphthenic Acids". J. Inst. Petroleum **43**, 307 (1957).

74. Caro, J. H.: High Molecular Weight Acid Compounds in Petroleum. Erdöl Zeitschrift für Bohr- und Fördertechnik **78**, 435 (1962).
75. Haug, P., H. K. Schnoes, and A. L. Burlingame: Aromatic Carboxylic Acids Isolated from the Colorado Green River Formation (Eocene). Geochimica et Cosmochimica Acta **32**, 358 (1968).
76. Seifert, W. K., and R. M. Teeter: Carboxylic Acids in a California Petroleum: Identification of Structural Types. Chem. Ind. (London) **1969**, 1464.
77. — — Identification of Polycyclic Naphthenic, Mono-, and Diaromatic Crude Oil Carboxylic Acids. Analyt. Chem. **42**, 180 (1970).
78. — — Identification of Polycyclic Aromatic and Heterocyclic Crude Oil Carboxylic Acids. Analyt. Chem. **42**, 750 (1970).
79. Streibl, M., and V. Herout: Terpenoids – Especially Oxygenated Mono-, Sesqui-, Di-, and Triterpenes. "Organic Geochemistry", ed. by G. Eglinton and M. T. J. Murphy, p. 411. New York: Springer. 1969.
80. Seifert, W. K.: Effect of Phenols on the Interfacial Activity of Crude Oil (California) Carboxylic Acids and the Identification of Carbazoles and Indoles. Analyt. Chem. **41**, 562 (1969).
81. Burlingame, A. L., P. C. Wszolek, and B. R. Simoneit: The Fatty Acid Content of Tasmanites. Advances in Organic Geochemistry 1968 ed. by P. A. Schenk and I. Havenaar, p. 131. Oxford: Pergamon. 1969.
82. Carruthers, W., and J. W. Cook: The Constituents of High Boiling Petroleum Distillates I. Preliminary Studies. J. Chem. Soc. (London) **1954**, 2047.
83. Barton, D. H. R., W. Carruthers, and K. H. Overton: Triterpenoids Part XXI. A Triterpenoid Lactone from Petroleum. J. Chem. Soc. (London) **1956**, 788.
84. Eglinton, G., D. H. Hunneman, and K. Douraghi-Zadeh: Gas Chromatographic-Mass Spectrometric Studies of Long-Chain Hydroxy Acids II. Tetrahedron **24**, 5929 (1968).
85. Hertz, H. S., D. D. Andresen, M. V. Duricic, K. Bieman, M. Saban, and D. Vitorovic: The Isolation and Identification of Gamma-Lactones in the Acidic Fraction of Aleksinac (Yugoslavia) Shale Bitumen. Geochimica et Cosmochimica Acta **37**, 1687 (1973).
86. Haug, P., H. K. Schnoes, and A. L. Burlingame: Ketocarboxylic Acids Isolated from the Colorado Green River Shale (Eocene). Chem. Commun. **1967**, 1130.
87. Hare, P. E.: Geochemistry of Proteins, Peptides, and Amino Acids. "Organic Geochemistry", ed. by G. Eglinton and M. T. J. Murphy, pp. 438—463. New York: Springer. 1969.
88. Kvenvolden, K. A.: Amino Acid Geochemistry. Annual Review of Earth and Planetary Sciences, Vol. III, 1974, ed. by F. A. Donath, Annual Reviews Inc., Palo Alto, California, in preparation.
89. Kvenvolden, K. A., and E. Peterson: Racemization of Amino Acids in Sediments from Saanich Inlet, British Colombia. Science **169**, 1079 (1970).
90. Kvenvolden, K. A., E. Peterson, J. Wehmiller, and P. E. Haere: Racemization of Amino Acids in Marine Sediments Determined by Gas Chromatography. Geochimica et Cosmochimica Acta **37**, 2215 (1973).
91. Kvenvolden, K. A.: Criteria for Distinguishing Biogenic and Abiogenic Amino Acids — Preliminary Considerations. Space Life Sciences **4**, 60 (1973).
92. Baker, E. W.: Porphyrins. Organic Geochemistry, ed. by G. Eglinton and M. T. J. Murphy, pp. 464—497. New York: Springer. 1969.
93. Jenkins, G. I.: The Occurrence and Determination of Carboxylic Acids and Esters in Petroleum. J. Inst. Petroleum **51**, 313 (1965).
94. Copelin, E. C.: Identification of 2-Quinolones in a California Crude Oil. Analyt. Chem. **36**, 2274 (1964).
95. Burlingame, A. L., and B. R. Simoneit: Isoprenoid Fatty Acids Isolated from the

Kerogen Matrix of the Green River Formation (Eocene). Science **160**, 531 (1968).

96. BURLINGAME, A. L., and B. R. SIMONEIT: High Resolution Mass Spectrometry of Green River Formation Kerogen Oxidations. Nature **222**, 741 (1969).

97. SIMONEIT, B. R., and A. L. BURLINGAME: Carboxylic Acids Derived from Tasmanian Tasmanite by Extractions and Kerogen Oxidations. Geochimica et Cosmochimica Acta **37**, 595 (1973).

98. DJURICIC, M. V., D. VITOROVIC, B. D. ANDRESEN, H. S. HERTZ, R. C. MURPHY, G. PRETI, and K. BIEMANN: Acids Obtained by Oxidation of Kerogens of Ancient Sediments of Different Geographic Origin. Advances in Organic Geochemistry 1971, p. 305. Oxford-Braunschweig: Pergamon. 1972.

99. COGSWELL, T. E., J. F. McKAY, and D. R. LATHAM: Gel Permeation Chromatographic Separation of Petroleum Acids. Analyt. Chem. **43**, 645 (1971).

100. McKAY, J. F., D. M. JEWELL, and D. R. LATHAM: The Separation of Acidic Compound Types Isolated from High Boiling Distillates. Separation Science **7**, 361 (1972).

(Received January 2, 1974)

Acknowledgments as to Sources of Figures

Figs. 3, 6, 7; Charts 5, 10, 11: From: W. K. SEIFERT: Steroid Acids in Petroleum — Animal Contribution to the Origin of Petroleum. Pure and Applied Chemistry **34**, 633—640 (1973). London: Butterworth's and Co. (Publishers Ltd.)

Figs. 4, 5; Charts 6, 7, 8: From: W. K. SEIFERT, E. J. GALLEGOS, and R. M. TEETER: Proof of Structure of Steroid Carboxylic Acids in a California Petroleum by Deuterium Labeling, Synthesis, and Mass Spectrometry. J. Amer. Chem. Soc. **94**, 5880 (1972).

Chart 12: From: A. ENSMINGER, A. VAN DORSSELAER, CH. SPYCKERELLE, P. ALBRECHT, and G. OURISSON: Pentacyclic Triterpenes of the Hopane Type as Ubiquitous Geochemical Markers: Origin and Significance. Presented at Sixth International Meeting of Organic Geochemistry. Paris, France, September 18—21, 1973, in press. Permission by Dr. P. ALBRECHT, Institut de Chimie, Université Louis Pasteur, Strasbourg, 1 Rue Blaise Pascal, Strasbourg, France.

Fig. 2.: From: W. K. SEIFERT, and R. M. TEETER: Preparative Thin-Layer Chromatography and High Resolution Mass Spectrometry of Crude Oil Carboxylic Acids. Analyt. Chemistry **41**, 786 (1969).

Figs. 8, 10, 11: From: W. K. SEIFERT, and R. M. TEETER: Identification of Polycyclic Aromatic, and Heterocyclic Crude Oil Carboxylic Acids. Analyt. Chemistry **42**, 750 (1970).

Fig. 9.: From: W. K. SEIFERT, and R. M. TEETER: Identification of Polycyclic, Naphthenic, Mono-, and Diaromatic Crude Oil Carboxylic Acids. Analyt. Chemistry **42**, 180 (1970).

Fig. 12: From: W. K. SEIFERT: Effect of Phenols on the Interfacial Activity of Crude Oil (California) Carboxylic Acids and the Identification of Carbazoles and Indoles. Analyt. Chemistry **41**, 562 (1969).

Fig. 1.: From: D. L. LAWLOR, and W. E. ROBINSON: Fatty Acids in Green River Formation Oil Shale. Div. of Petrol, Chem. Amer. Chem. Soc. Detroit Meeting, 1965.

Table 1: From: G. COSTANTINIDES, and G. ARICH: Nonhydrocarbon Compounds in Petroleum; in Fundamental Aspects of Petroleum Geochemistry. Ed. by B. NAGY and U. COLOMBO, pp. 143—151. New York: Elsevier. 1957.

Naturally Occurring 2,5-Dioxopiperazines and Related Compounds

By P. G. Sammes, Department of Chemistry, Imperial College of Science and Technology, London, U.K.

Contents

I. Introduction

Being relatively simple compounds, 2,5-dioxopiperazines are amongst the most ubiquitous peptide derivatives found in nature. Hydrolysates of proteins and polypeptides often contain these anhydro-dimers of amino-acids and they are commonly isolated from cultures of yeast, lichens and fungi. Their existence as a special group of compounds was first recognized around 1900 (1). The great Emil Fischer managed to synthesize many of the simpler members of this family in the early 1900's (2). The parent compound, 2,5-dioxopiperazine, often referred to as *cyclo*-gly-gly*, was made in 1888 (3).

---

* A variety of names are used in the literature to describe 2,5-dioxopiperazines. In this review such compounds are referred to either as derivatives of 2,5-dioxopiperazines (generally with the substitution 2,5-omitted), or as the *cyclo*-dipeptide, the nomenclature conforming to the IUPAC-IUB rules [*Biochemistry*, **5**, 1445 (1966)].

2,5-Dioxopiperazines differ in their properties from ordinary peptides. Although the simpler members are often water soluble, many — for example derivatives of phenylalanine — are only sparingly soluble. Because they do not exist as zwitterions, the simple members of this group of peptides are often neutral compounds. They do not give a test with ninhydrin and, in contrast with polypeptides, they form a reddish brown colour with picric acid or m-dinitrobenzene and sodium hydroxide (4). When the nitrogen atoms of the piperazine ring are unsubstituted they give a white precipitate with potassium mercuric iodide and they give a negative biuret test (5). The limitations and scope of the tests for these compounds have been discussed (6).

The spectroscopic properties of the 2,5-dioxopiperazines are of interest as models for more complex peptides (7). The o.r.d. spectra of these compounds show Cotton effects, the position of which is solvent dependent (8, 9), attributed to conformational changes in different media. The Cotton effects observed in both the o.r.d. and c.d. measurements are due to $n - \pi^*$ transitions with exciton interactions causing splitting of the associated $\pi - \pi^*$ transition of the cis-amide chromophore (8, 9). The observed exciton splitting can, in theory, only occur by overlap of the two π-bonds associated with the amide groups and this is only possible if the dioxopiperazine ring is puckered into a type of boat form. The Cotton effects are therefore attributed to non-planarity of the ring system (10). Interactions are also observed between the aromatic chromophores and amide bonds in derivatives of tyrosine and tryptophan. The observed c.d. molecular ellipticities are several times larger than those observed for linear analogues which suggests the adoption of rigid conformations in these dioxopiperazines (11).

The cis-amide bonds of the dioxopiperazines show modified infrared absorption properties from those associated with trans-amide bonds. The N—H stretching frequency of the cis-bond occurs at 3180—$3195\ cm^{-1}$, changed to 2250—$2350\ cm^{-1}$ upon deuteration; the trans-amide bond shows the same mode at about $3350\ cm^{-1}$, changing to $2450\ cm^{-1}$ upon deuteration. The bands associated with the carbonyl group also occur at different positions, cis-amide bonds showing the amide I mode (CO stretch) at 1670—$1690\ cm^{-1}$; trans-amide bonds have this at $1650\ cm^{-1}$. The amide II band (NH in-plane vibration) of the cis-amide bond is usually hidden under skeletal vibrations at 1440—$1450\ cm^{-1}$, whilst trans-amide bonds show the same mode at $1550\ cm^{-1}$ and the existence of this band is regarded as characteristic of a trans-amide bond (12). The cis-amide bond also shows a further (amide III) absorption at 1300—$1350\ cm^{-1}$ not shown by trans-amide bonds (13). The amide I and III bands are unaffected by deuteration, whilst the amide II band is moved to lower frequency in both cases. Whilst the presence of the amide

II band is characteristic of the *trans*-amide bond (*12*) its existence in a spectrum does not imply the absence of all *cis*-amide bonds. A categoric assignment can only be confirmed by deuteration. The situation is more complex in the case of N-substituted amides (*13*). The infrared results are summarized in Table 1.

Table 1. *Infrared Absorption Bands for Cis and Trans Secondary Amide Bands (12, 13)*

	NH-stretch		Amide I	Amide II	Amide III
H-*cis*	3180—3195 and 3040—3060		1670—1690	1440—1455	1305—1345
H-*trans*	∼3300	∼3100	∼1650	∼1550	none
D-*cis*	2250—2350	—	∼1650	1230—1250	1305—1345
D-*trans*	∼2450	—	∼1650	∼1450	none

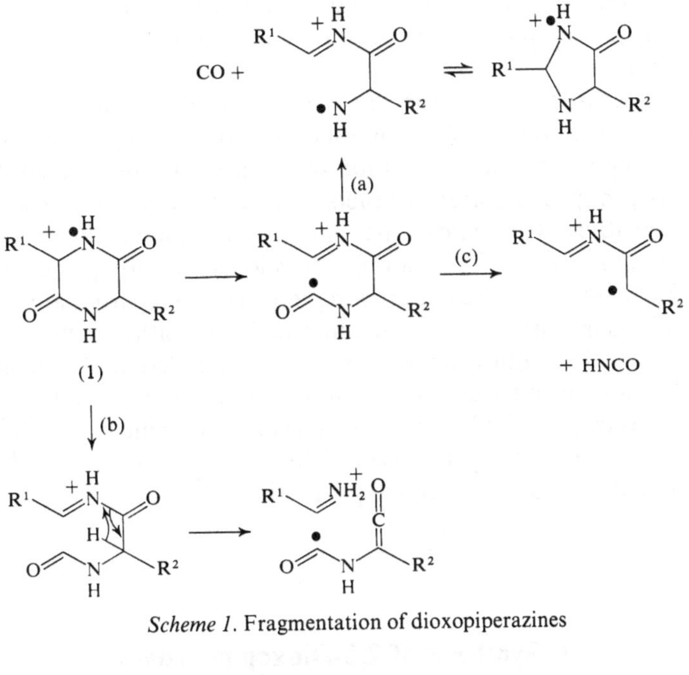

Scheme 1. Fragmentation of dioxopiperazines

Simple dioxopiperazines give rise to very characteristic mass spectral fragmentation patterns (*14, 15*). The parent ion (*e.g.* **1**) is generally prominent and is followed by the principal fragmentations depicted in Scheme 1. These include, a) loss of CO or CHO, b) amine fragmentation — a process of diagnostic value for the structure of dioxopiperazines, and c) elimination of HNCO. More complex derivatives, such as proline anhydride (2), also show the fragmentations depicted which were of value in elucidation of the structure of aranotin and related compounds (*16*).

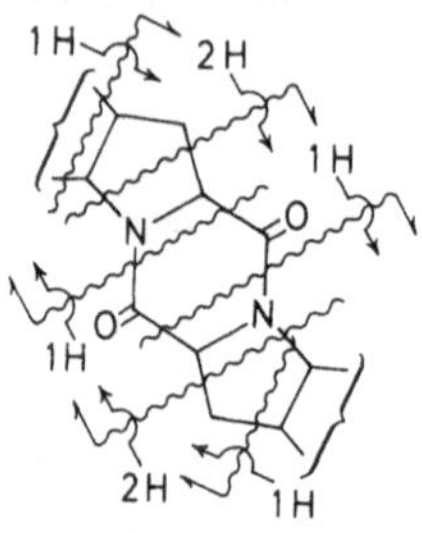

(2)

Proton magnetic resonance studies (*7, 17*) of dioxopiperazines have been made and recently these have been extended to include ^{15}N coupling studies, of value in elucidating the conformation of the cyclic system in solution (*18*). Europium induced shifts have also been used for this purpose (*19*).

Largely because of the rather inert nature of the simpler members of this group little interest in them was apparent five years ago. More recently there has been a growing recognition of the diversity of structural types within this group of natural products, currently numbering nearly 100. Many derivatives (the gliotoxins and sporidesmins) show antiviral properties and others (for example bicyclomycin) are powerful antibiotics. Furthermore, there is a growing tendency to regard many of these compounds as important metabolic intermediates, rather than as protein artefacts (*20*). The following review is not intended to be exhaustive but tries to summarize recent trends in the study of these materials. The group is arbitrarily divided into structural types which are discussed separately. This discussion is preceded by short sections on methods of preparation and on conformational properties.

II. Synthesis of 2,5-Dioxopiperazines

The cyclic dipeptides are commonly prepared by the cyclization of free dipeptide esters, generally liberated from the corresponding amine salts by the action of ammonia (*21, 22*). However, in some cases long exposure to the reagent is required to effect cyclization and this can lead to extensive racemization when employing optically pure dipeptide precursors (23). A convenient method for the preparation of symmetrically substituted dioxopiperazines is to heat the corresponding amino-acid in refluxing ethylene glycol. Dipeptides can also be used and yields are often good (24). Dimerization, rather than polymerization, of N-carboxylaminoacid anhydrides (Leuch's anhydrides) can be catalyzed by aziridine (25). In order to avoid racemization the use of the t-butoxycarbonyl-protected dipeptide methyl ester is recommended, removing the protecting group with formic acid and cyclizing by heating the derived formate salts in neutral solvents, such as refluxing toluene-sec-butanol mixtures. KOPPLE and GHAZARIAN have described the preparation of optically pure dioxopiperazines by heating the unprotected dipeptide or its hydrobromide salt in phenol (26), a modification of an earlier method in which β-naphthol was recommended (27).

A slightly different approach to the preparation of dioxopiperazines is by hydrogenolysis of benzyloxycarbonyl dipeptide methyl esters in methanol over a palladium on charcoal catalyst (28, 29). Using this method, optically homogeneous isomers of cyclo-ala-try were prepared (30). Activated esters have also been employed (30), but with these epimerization also occurred.

A g.l.c. method for estimating the optical purity of dioxopiperazines has been described (31) and the behaviour of diastereoisomeric derivatives on thin layer chromatography has also been studied (32).

In order to form dioxopiperazines, the dipeptide precursors have to adopt a folded conformation (3), rather than the more stable, extended form (4), in which the amide bond is in the favoured trans-configuration (33). Because proline containing peptides are forced to adopt the folded form by the presence of the pyrrolidine ring they are much more prone to cyclize with formation of the dioxopiperazine derivative (34). It is not too surprising, therefore, to find quite a large number of proline derivatives in this group of natural products.

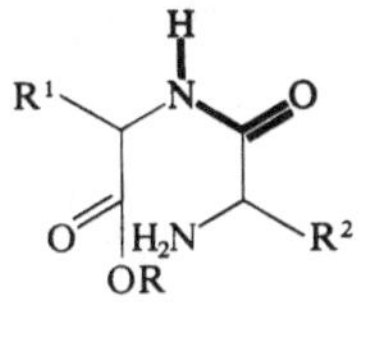

"folded" form   "open form"

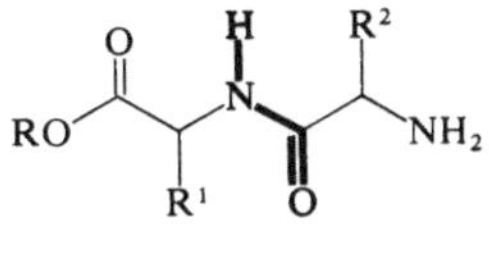

(3)   (4)

Steric accessibility in the cyclization step is also important. An example is in the spontaneous formation of the dioxopiperazine (6) from the cephalosporin derivative (5), (34). In contrast, the penicillin derivative (7) is stable because the side-chain amino group is inhibited from attacking the β-lactam function by the methyl group at position 2. As would be expected, the epimer (**8**) readily gives a dioxopiperazine on standing (*35*).

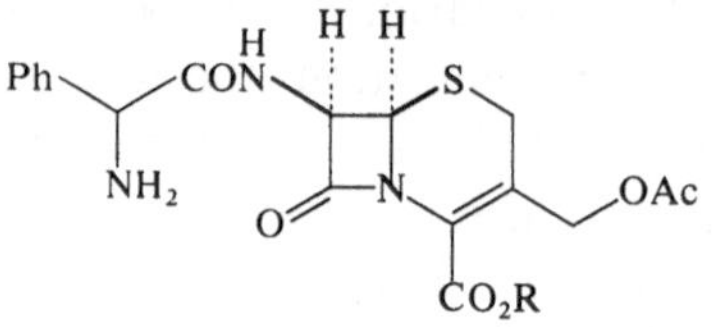

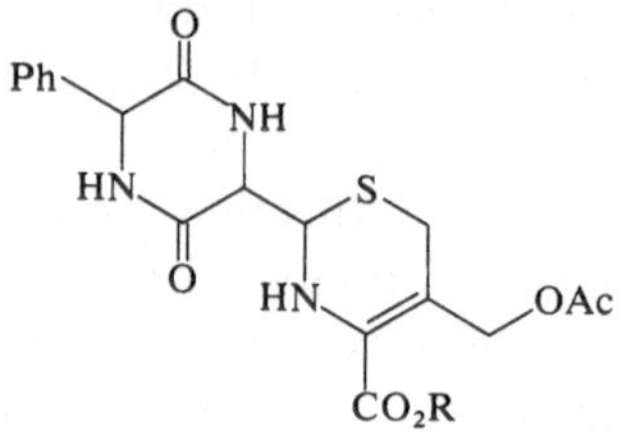

A careful study has recently been made on the effect of conformational preferences on the mode of cyclization of a variety of tetrapeptide esters containing combinations of sarcosine with glycine or alanine (*36*). These studies have allowed predictions to be made on the preferred reaction, *viz.* either formation of a cyclic tetrapeptide or the formation of two

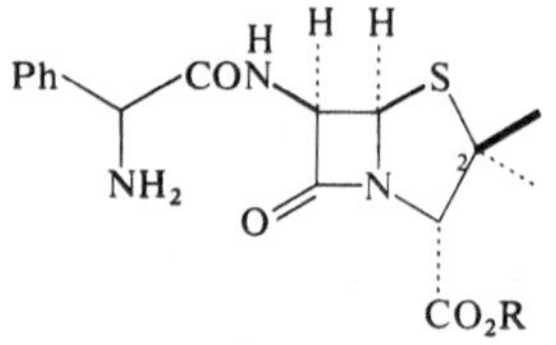

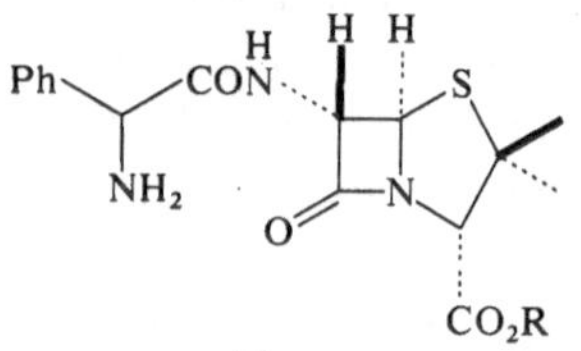

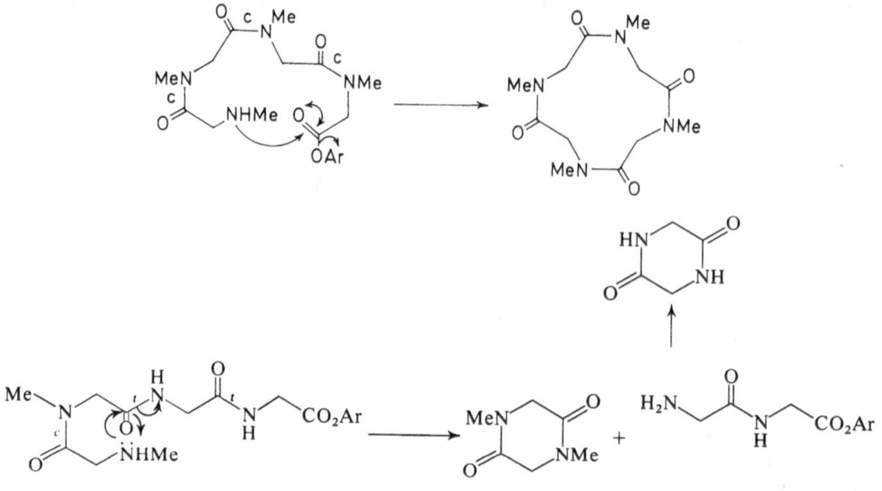

Scheme 2

dioxopiperazines. Amide bonds involving the sarcosyl nitrogen prefer the folded, *cis*-conformation compared to the normal peptide bond preference for *trans*-conformation. Essentially, the nature of the first and last peptide bonds of the tetrapeptides was found to be important. For *cis-cis* and *trans-trans* arrangements models showed that the termini of the tetrapeptide chain could easily be reached by one another and a cyclic tetrapeptide was formed. On the other hand, the *cis-trans* and *trans-cis* combinations tended to give the dioxopiperazines because the termini were then not capable of easy access to one another. An example is given in Scheme 2.

Dioxopiperazines are often formed during the manipulation of higher peptides. For example, the tripeptide Tos-L-ala-L-phe-D-pro-OC$_6$H$_4$NO$_2$ (9) gives the acyldioxopiperazine (**10**) on treatment with mild base (*31*). The formation of dioxopiperazines during the pyrolysis of actinomycins has been used to help determine the peptide sequences of these antibiotics (*38*).

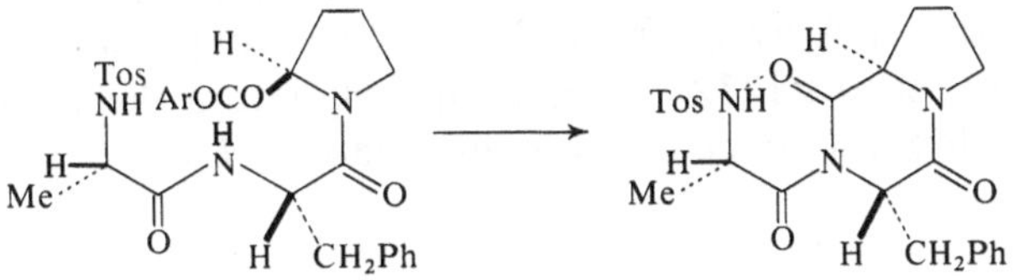

(9)   (10)

Once formed 2,5-dioxopiperazines generally behave as relatively stable dipeptides. The diastereoisomeric isomers of substituted derivatives do differ slightly in their chemical, as well as their physical, properties. For example, hydrolysis of the LL-isomer of *cyclo*-leu-leu is hydrolysed 3.5 times as fast as the DL isomer in 0.5 N hydrochloric acid. A similar result was also obtained in the alanine series. The result was explained in terms of both steric shielding (in the *trans*-isomer) and steric strain (in the LL-*cis* series), the latter effect being based on the known conformational preference of each isomer (*29*).

An interesting stereoselective method for the preparation of certain *cis*-substituted dioxopiperazines from arylidene substituted precursors has recently been encountered. Thus, arylidene derivatives of *cyclo*-gly-L-pro can be hydrogenated over palladium to give the *cis*-cyclodipeptides, but the specifity was lost when extended to the reduction of alkylidene analogues (*39*). The specific reduction of the arylidene derivatives has been employed as a method for preparing substituted phenylalanines, such as L-DOPA.

III. Conformations of Dioxopiperazines

Although dioxopiperazines possess the atypical *cis*-amide bonds they have, nevertheless, attracted attention as useful models for the study of certain subtle interactions between amide bonds and the amino-acid side chains. These effects are of importance in an understanding of the conformational constraints present in proteins. Dreiding models show that the six-membered ring in dioxopiperazines is generally slightly flexible and that it can exist in either a flat or slightly puckered (boat) form. Indeed, recent X-ray studies have shown that both forms exist. 2,5-Dioxopiperazine (*cyclo*-gly-gly) is planar (*40*) and so is (*trans*) *cyclo*-D-ala-L-ala (*41, 42*). In contrast, (*cis*) *cyclo*-L-ala-L-ala adopts a slightly skewed boat conformation with the methyl groups in the quasi-equatorial position (*43*). In order to explain the differences between the *cis*- and *trans-cyclo*-ala-ala compounds it has been assumed that small bond angle deformations can be magnified to produce large variations in the internal torsion angles of the ring, thus leading to large changes in its conformation (*44*). C. d. studies on *cyclo*-L-ala-L-ala also indicate a folding of the ring away from the methyl groups, as in (**11**), (*10*). *Cyclo*-sarcosyl-sarcosyl (**12**) is not completely flat in the solid state, but it tends to adopt a flattened chair conformation (*45*).

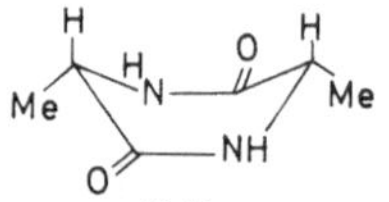

(11) 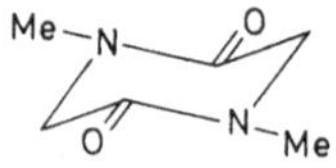   (12)

N.m.r. studies are of great value in complementing the X-ray diffraction analyses. For example, proline derivatives adopt a boat conformation in the solid state (*46*) and in solution (*47, 48*), reflecting the constraint imposed on the dioxopiperazine ring by fusion to the pyrrolidine ring. The degree of buckling is even more pronounced in the lower homologues, such as the azetidine derivative (**13**), (*49*). The nature of the solvent is important in n.m.r. studies since, in contrast to the results obtained in deuteriochloroform and dimethyl sulphoxide, the use of trifluoroacetic acid tends to cause the boat conformations to flatten (*47*). This is presumed to be due to protonation of the amide functions, which leads to greater double bond character across the carbonyl-nitrogen bonds.

References, pp. 107—118

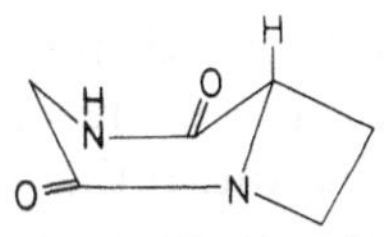

(13)

Interesting results have been accumulated for systems which contain an aromatic side chain. For example, *cyclo*-gly-tyr **(14)** shows a remarkable shielding of the *cis*-disposed glycyl hydrogen atom, moved upfield by 1.0 to 1.5 ppm from its expected position (*7*). Derivatives containing phenylalanine (*50*), histidine (*18, 51*) or tryptophan (*18*) behave similarly. The observed shielding can only be explained by the aromatic ring of the side chain adopting a folded conformation over the dioxopiperazine ring, thus subjecting the cisoidal hydrogen atom to the anisotropic shielding zone due to the aromatic ring current. Calculations for the phenylalanine derivative showed that the boat form of the dioxopiperazine ring is necessary, the planes containing the *cis*-amide bonds being buckled away from the side chain by about 5 to 10°, as in the conformation **(15)**, (*50, 52*). A study of the temperature effects on the chemical shifts allowed an equilibrium constant to be calculated for the folded and unfolded rotamers **(16 a, b,** and **c)** of the tyrosine derivative. In this way an interaction energy of some 12 kJ mol^{-1} was calculated to exist between the aromatic ring and the amide bonds (*7*). The interaction has been ascribed to direct dipole-induced dipole effects (*53*). Molecular orbital calculations predict a preference for the folded conformations in these cases (*54*).

The tendency for the aromatic side chain to adopt a folded conformation has been used to explain the facile acetyl transfer from the imidazole group to the tyrosyl ring in *cyclo*-L-histidyl-L-tyrosyl (**17** to **18**), which competes efficiently with solvolysis (*55*). For *cis*-disubstituted

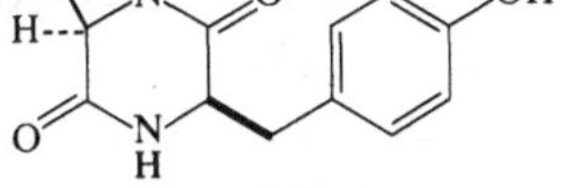

(14)

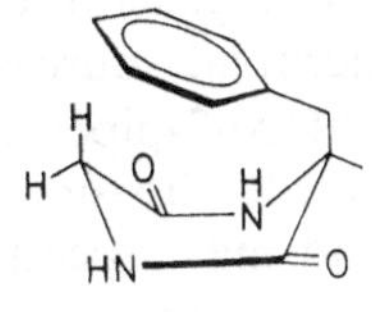

(15) 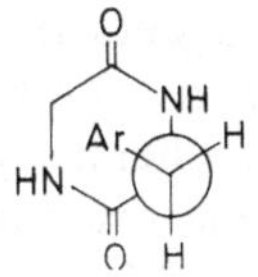   (16a) 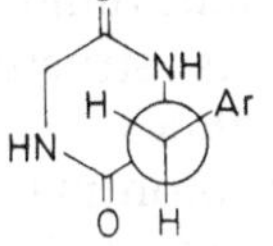   (16b) 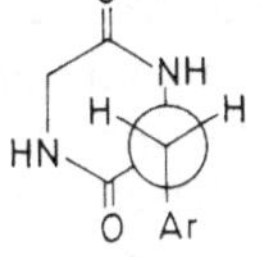   (16c)

systems such as *cyclo*-L-phe-L-phe there is competition for the space above the dioxopiperazine ring with the result that the aromatic groups do not sit centrally over the ring but rather that each phenyl ring is associated with just one of the amide groups (*18*). Interactions between the bulky *cis*-substituents also force the heterocyclic ring into a more planar arrangement in these cases. Low temperature c. d. studies on tyrosyl and tryptophanyl containing dioxopiperazines also show enhanced molecular ellipticities, consistent with the rigid conformations of these compounds (*11*). Studies on N-methylated dioxopiperazines bearing aromatic side chains have also shown that the folded conformation is maintained, even in non-polar solvents (*50*), which is consistent with the direct dipole-induced dipole nature of the interaction.

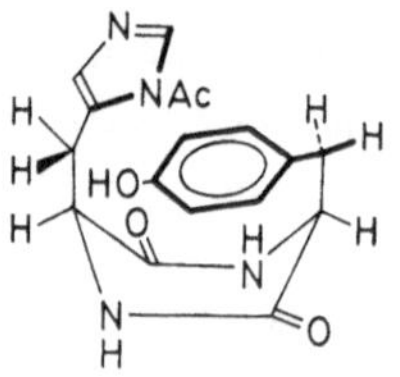

(17) 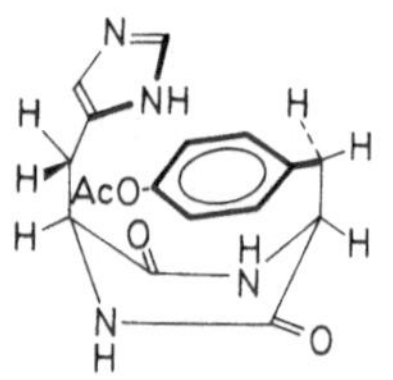   (18)

IV. Simple Dioxopiperazines

Dioxopiperazines are often isolated as hydrolytic products from peptides and proteins. Nevertheless, their widespread isolation from extracts and culture broths indicates their genuine metabolite character. Some examples are given in Table 2. Zizyphine (**19**), (*56*) and zizyphinine (**20**), (*57*) are derivatives of 4-hydroxy-prolylproline and are members of the peptide alkaloid family (*58*). Since common permutations are often found in this family it is expected that further derivatives of this dioxopiperazine system will be forthcoming. The antibiotic, amphomycin (*59*) contains the unit (**21**).

One of the most unusual members of the family of simple dioxopiperazines is the anti-tumour agent, compound 593 A (**190**). This material was isolated from *Streptomyces griseoluteus* as part of a general screening programme for new anti-tumour agents from microbial sources (*245*). The assignment of its structure as 3,6-bis(5-chloro-2-piperidyl)-2,5-dioxopiperazine (of unknown stereochemistry about the dioxopiperazine ring) was based on proton magnetic resonance and mass spectral data (*60*). The piperidyl ring was shown to bear equatorial substituents. The

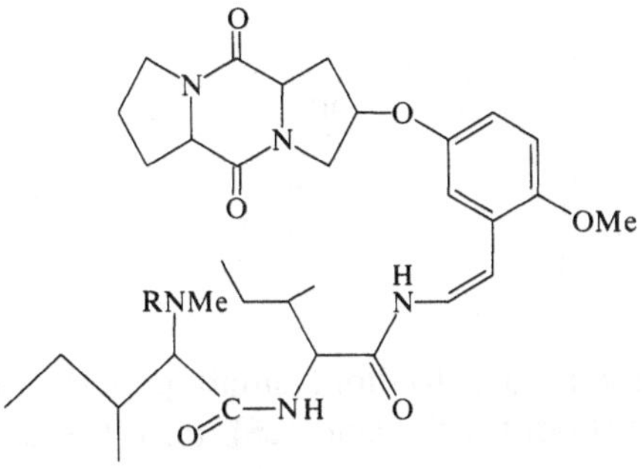

(19) R = Me
(20) R = H

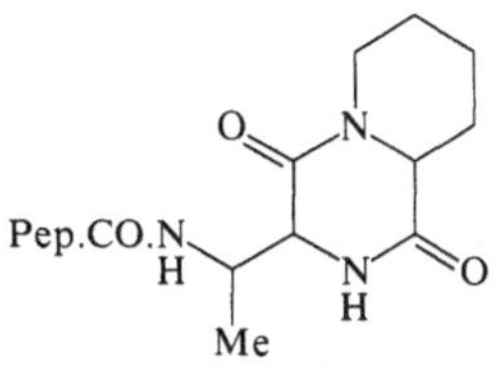

(21)

compound is remarkable in possessing the β-chloramine function typical of the nitrogen mustards, a group of compounds whose derivatives have been incorporated into many synthetic drugs of use in cancer chemotherapy. Once again, nature has precedented human efforts.

The compound picroroccellin (22), isolated from the lichen *Roccella fuciformis*, is soluble in base and reprecipitated by acid. The structure was assigned on the basis of chemical evidence only and it has not been verified by spectroscopic methods (*61*). Because there was some doubt about the source of this material its isolation has not been repeated since its original discovery. On pyrolysis it loses the elements of water and methanol to form the bis-arylidene derivative (23) of characteristic pale yellow colour. Methylation followed by heating gives the N-methylated derivative (24).

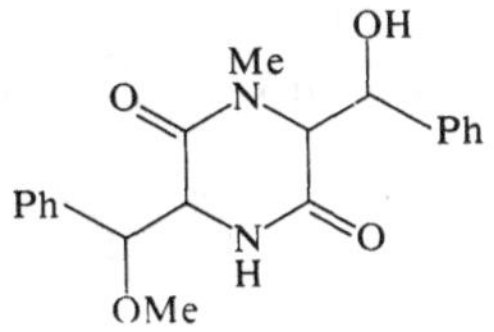

(22)

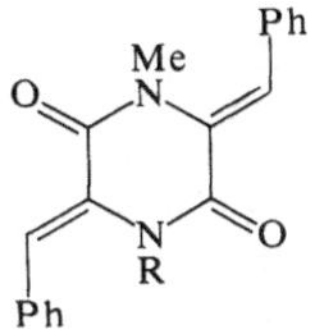

(23) R = H

(24) R = Me

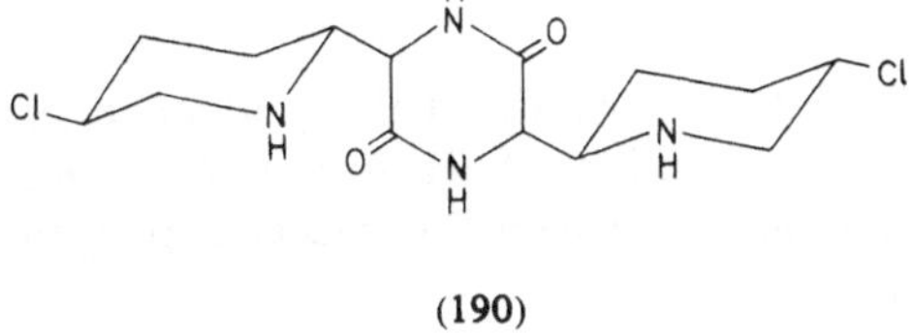

(190)

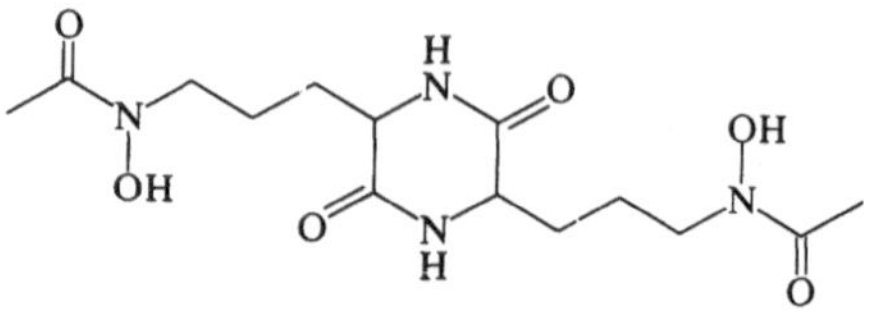

(25)

Dioxopiperazines are used by nature to hold small peptide links together, as in the growth factor rhodotorulic acid (**25**), (*62*), the dimer from δ-N-acetyl-L-δ-N-hydroxyornithine, which is also present in the ferrichrome, albomycin, and fusarimine group of siderochromes (*63*). Rhodotorulic acid is isolated from iron-deficient cultures of *Rhodotorula pilimanae* and is metabolized by the organism for its strong iron binding properties. Careful biosynthetic studies have shown that the acid (**25**) is metabolized from ornithine, which undergoes N-hydroxylation *before* acetylation and dimerization to the *cyclo*-dipeptide (*64*). A total synthesis of rhodotorulic acid has recently been completed (*65*). A key reaction in the synthesis (Scheme 3) was the discovery that hydrogen bromide in acetic acid cleaves the N-tosyl groups of the intermediate peptide (**26**) but leaves the O-benzyl residues of the protected hydroxylamine unaffected.

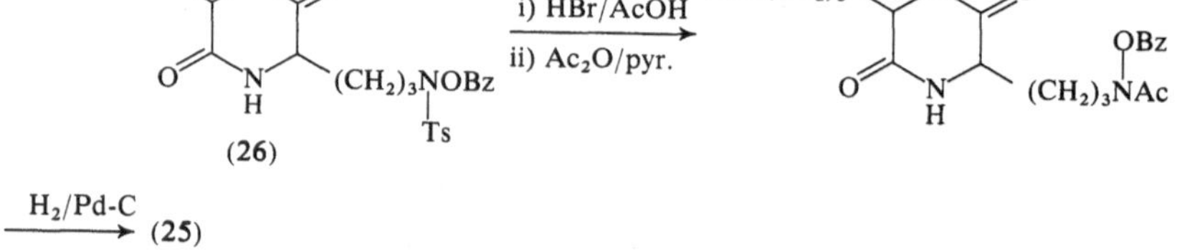

Scheme 3. Synthesis of rhodotorulic acid

Siderochromes related to rhodotorulic acid have recently been obtained from the culture fluids of *Fusarium dimerum*. These include dimerumic acid (**27**), (*66*), coprogen B (deacetylcoprogen) (**28**), (*66*), and coprogen (**29**), (*67*).

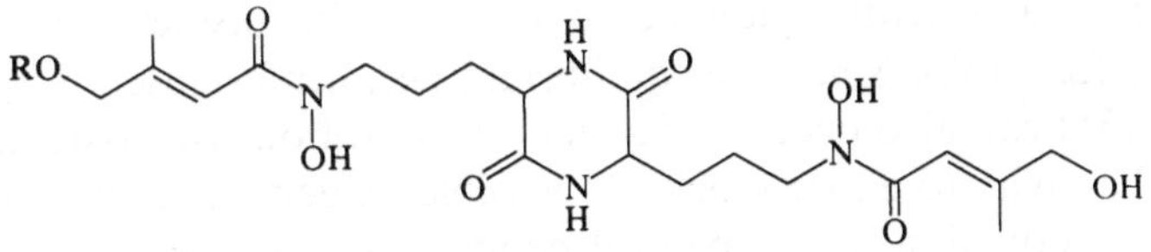

(27) R = H

(28) R = CO

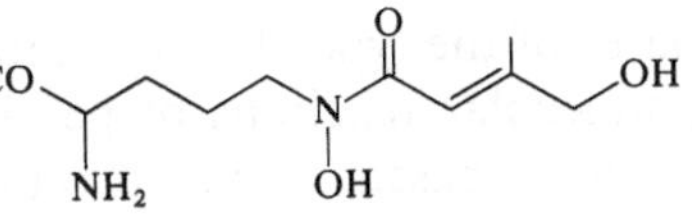

(29) R = CO

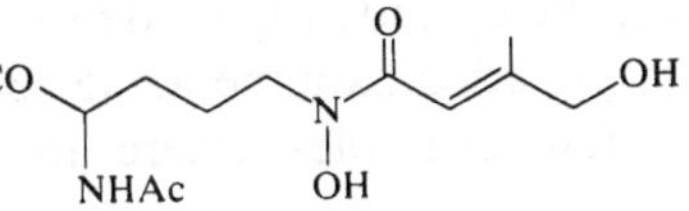

Cycloserine (**30**) is a broad spectrum antibiotic (*68*) which readily dimerizes, both in solution and in the solid state, (*69*) to give the dioxopiperazine, cycloserine dimer (**31**). The reaction is an equilibrium which is readily established at pH 1—2. Cyclol intermediates, such as (**32**), must be involved in these reactions (*70, 71*). A similar reaction has been found to occur with the α-amino-pyrrolidone (**33**), which dimerizes to the dioxopiperazine (**34**), (*72*). Under high pH conditions the dimer (**31**)

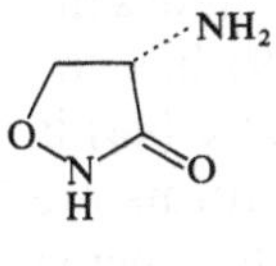

(30)

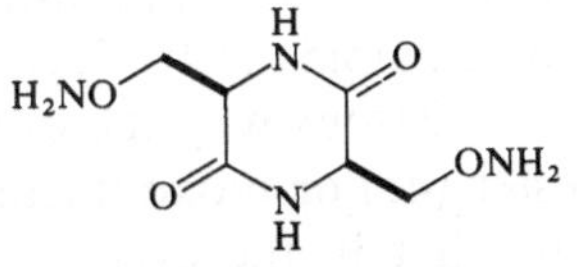

(31)

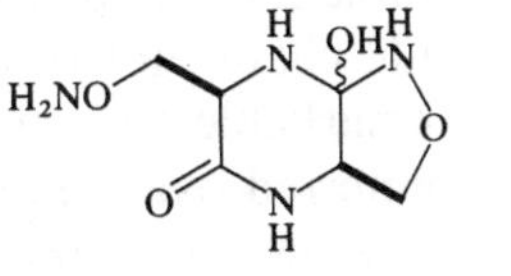

(32)

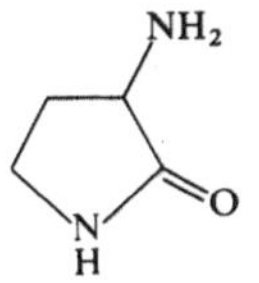

(33)

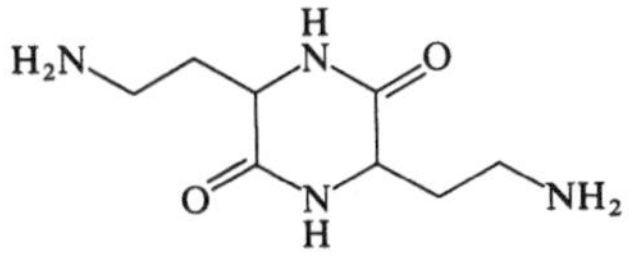

(34)

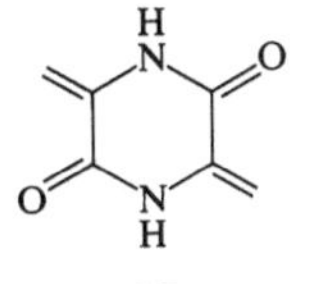

(35)

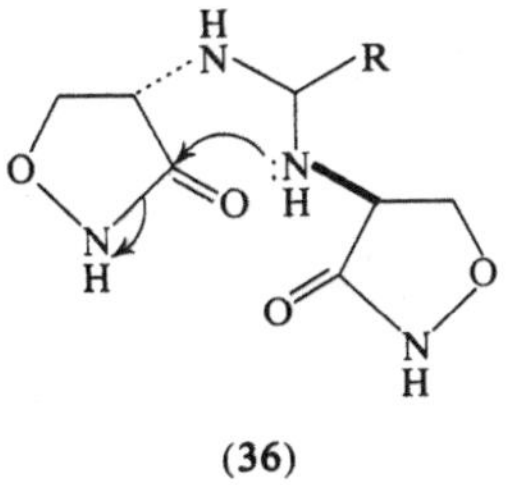

(36)

readily eliminates hydroxylamine to produce the bismethylene-dioxo-piperazine (35), (68). Since the equilibrium between cycloserine (30) and its dimer (31) can also occur under neutral conditions the assignment of biological activity to either species is possible. Aldehydes appear to cata-lyze the dimerization under neutral conditions, presumably by the in-volvement of species such as (36) (see arrows) (73).

Cyclol intermediates of the type (32) have often been postulated as of importance in intramolecular reactions of peptides (74, 75). One route to the cyclodepsipeptides is based on such functional group-amide group interaction (76). Thus, the cyclodepsipeptide serratamolide (37) has been prepared from the acylated dioxopiperazine (38) in this manner (77). However, there are few examples where the intermediate cyclol is

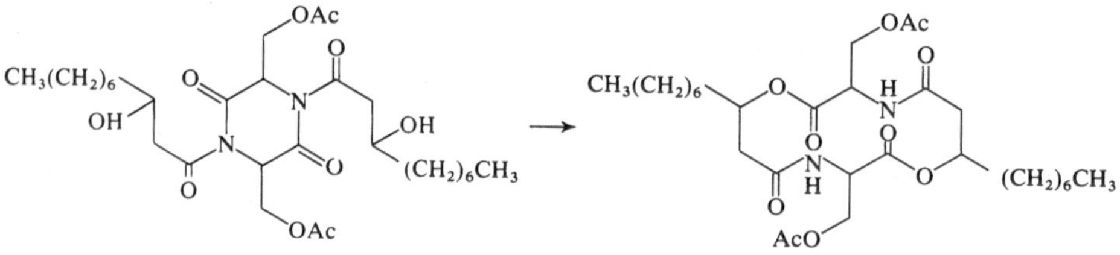

(38)   (37)

stable enough to be isolated (76, 78—81). One interesting class includes the side chain moiety of the ergot alkaloids (82). Ergotamine (39) has the remarkable structure indicated. Reasons why the cyclol structure (40) is preferred over the isomeric ester (41) or dioxopiperazine (42) forms are only just emerging. It appears that in the cyclic ester (41) the phenyl group of the phenylalanyl residue adopts its favoured conformation (50), viz. folded over the adjacent amide and ester groups (a) and (b) (see 41).

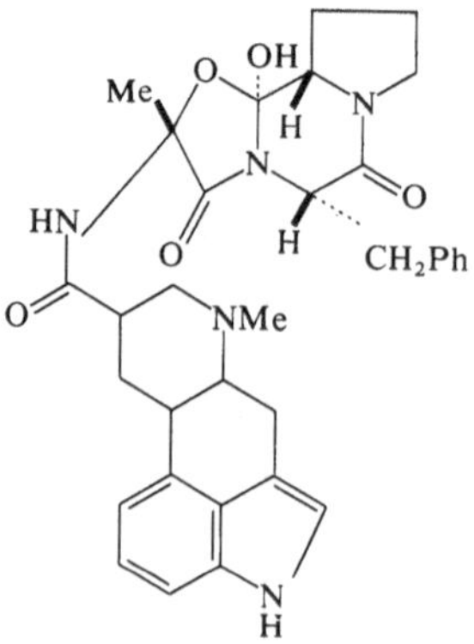

(39)

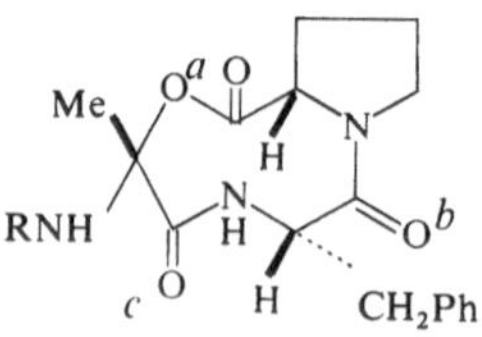

(41)

This conformation results in the other amide nitrogen (c) being held tightly against the ester carbonyl group, a situation known to result in cyclol formation (83). The stable cyclol (43), the first example of an amide-amide type cyclol (84), probably gains its stability from a similar intramolecular interaction (85), as well as from the hydrogen bonding indicated.

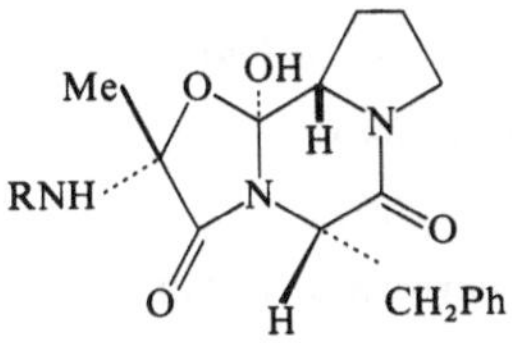

(40) 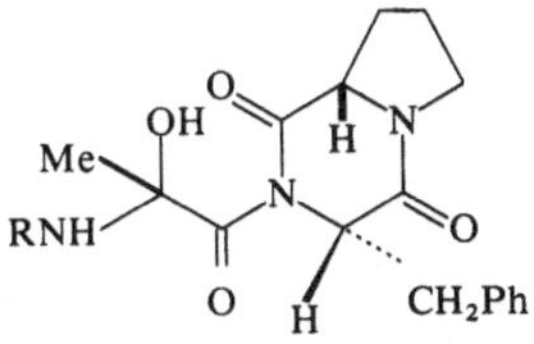   (42) 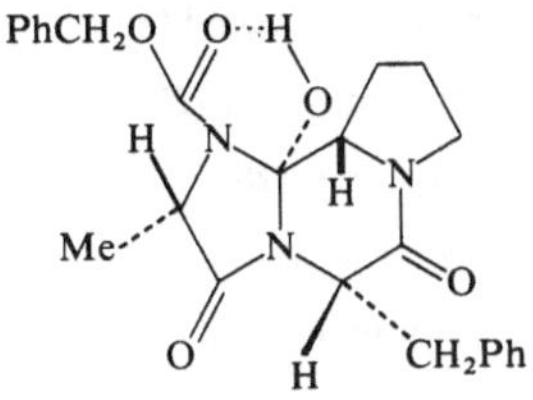   (43)

Dehydrodioxopiperazines often accompany the saturated analogues (see Table 2), suggesting their formation from the anhydride either by an *in vivo* dehydrogenation or by a hydroxylation-dehydration process, although they could also form by condensation of the amino-acid precursors at the pyruvoyl level of oxidation. Yet another possible pathway to the dehydro-compounds is via elimination from an appropriate precursor, for example dehydration from seryl derivatives. An *in vitro* example of dehydrogenation is the preparation of pyrocoll (44) by treatment of *cyclo*-L-pro-L-pro with dichlorodicyanobenzoquinone (86). Controlled potential oxidation of *cyclo*-leu-leu in acetonitrile afforded the interesting oxidation product (45) (87).

Table 2. *Some Simple Dioxopiperazines which Occur Naturally*

cyclo-L-pro-L-leu (185)	*Roselinia necatrix (239)* *Aspergillus fumigatus (240)*
cyclo-L-pro-L-val (186)	*Rosellinia necatrix (239)* *Aspergillus ochraceus (241)* *Metarrhizum ansiophae (241)*
cyclo-L-pro-L-phe (187)	*Rosellinia necatrix (239)*
cyclo-L-phe-L-phe (188)	*Penicillium nigricans (242)* *Streptomyces noursei (243)*
cyclo-L-ala-L-leu (189)	*Aspergillus niger (244)*
3,6-bis-2-(5-chloropiperazine)- 2,5-dioxopiperazine (190)	*Streptomyces griseoluteus (245)*
Cycloserine dimer. (31)	*Streptomyces orchidaceus (69)*

Table 2 (continued)

cyclo-L-pro-L-try (191)	Penicillium brevicompactum (112)
Picroroccellin (22)	Rocella fuciformis (61)
Zizyphine (19)	Zyzyphus oenoplia (56)
Zizyphinine (20)	Z. oenoplia (57)
Rhodotorulic acid (25)	Rhodotorula pilimanae (62)
Dimerumic acid (27)	Fusarium dimerum (66)
Coprogen B (28)	F. dimerum (66)
Coprogen (29)	F. dimerum (67)
Albonoursin (51)	Streptomyces noursei (92)

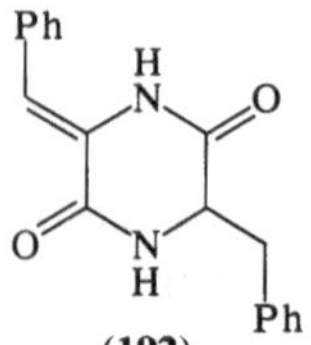

(192)

Streptomyces noursei (243)

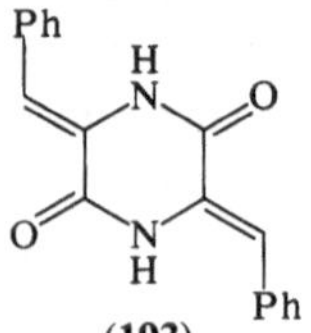

(193)

Streptomyces noursei (247)

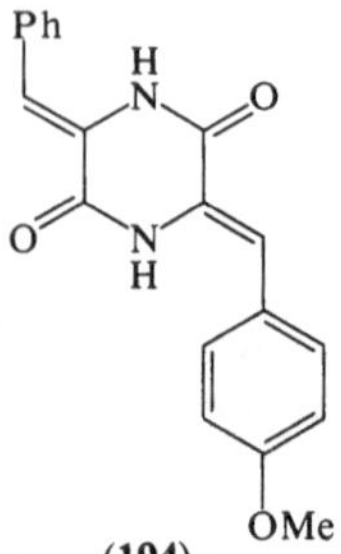

(194)

Streptomyces thioluteus (246)

Amphomycin (21)   Streptomyces canus (59, 247)

cyclo-gly-L-pro (195)   ex. commercial yeast extract (258)

cyclo-L-pro-L-tyr (196)   ex. commercial yeast extract (258)

References, pp. 107—118

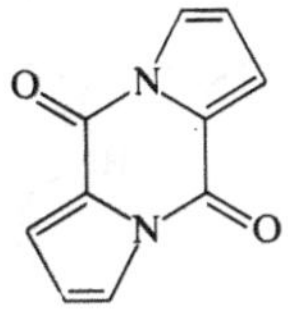

(44) 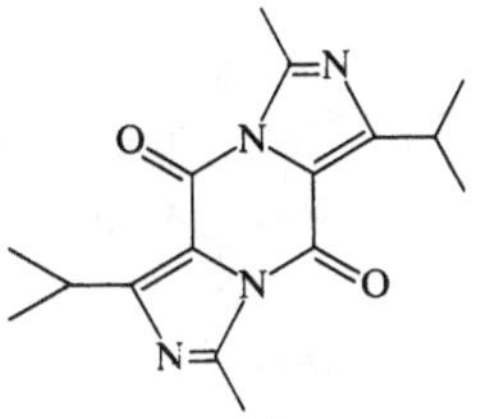　　　　　　　　　(45) 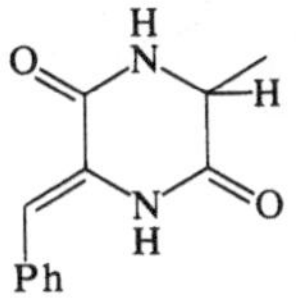　　　　　　　　　(46)

Arylidene derivatives of dioxopiperazines have been shown to possess the (Z)-configuration (88). Treatment of these compounds, for example (46), with base produces the isomeric pyrazine (47), (88). The isomerization is reversible (89) and can be used to explain the formation of compound (46) by treatment of the serine derivative (48) with base (90). Presumably dehydration is rapidly followed by isomerization into the pyrazine (47) and eventual isolation of the arylidene derivative. A similar type of tautomerism is exhibited by the synthetic pyrazine derivative (49), which exists mainly in the form (50), (91).

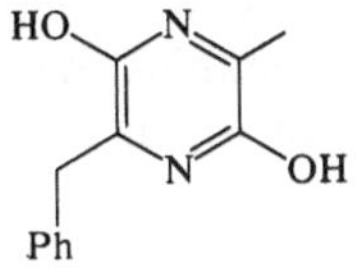

(47) 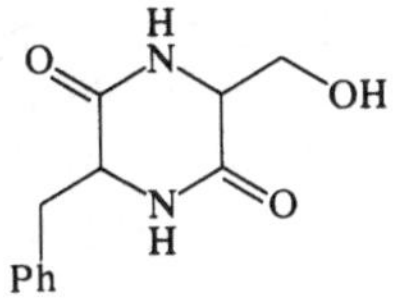　　　　　　　　　(48)

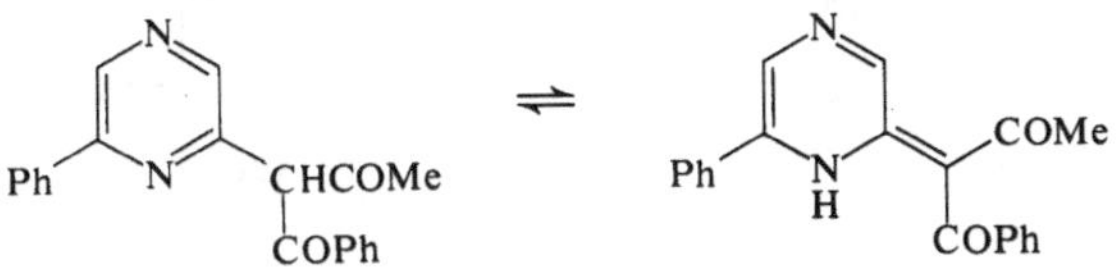

(49)　　　　　　　　　(50)

Albonoursin (51) was isolated from culture preparations of the antibiotics albofungin and nystatin (92). The synthesis of this unsummetrically substituted dioxopiperazine has been achieved (Scheme 4) by prior condensation of the α-ketoester (52) with chloroacetamide (93). The condensate (53) was treated with ammonia to give the piperazine (54), which was then condensed with benzaldehyde to give the natural product.

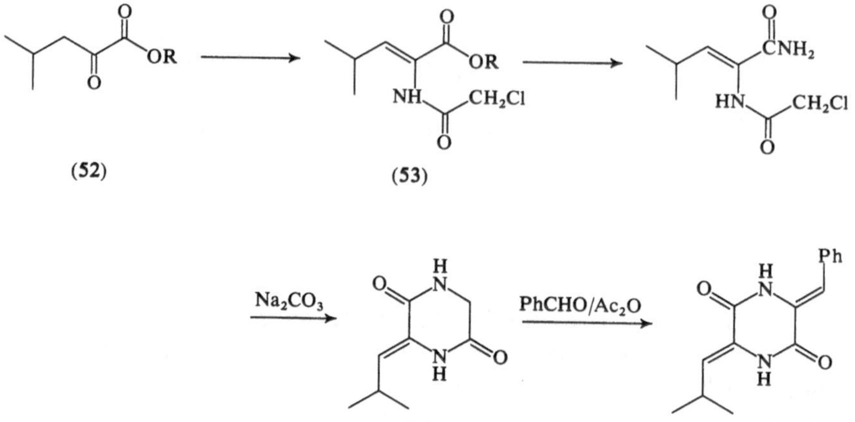

Scheme 4. Synthesis of albonoursin

Condensations of free methylene groups of dioxopiperazines with aldehydes are well exemplified and can be aided by prior acetylation of the amide groups (94). An alternative route to the mono-dehydroderivatives of the type (54) involves dehydration of the dichloroacetyl derivative (55) of an amino-acid (95). Dehydrohalogenation accompanies formation of the azlactone (56), which can be hydrolyzed and reacted with an amine to give the dioxopiperazine. This latter method enables the preparation of monoarylidene derivatives with one of the nitrogen atoms specifically substituted, e. g. with alkyl groups (96).

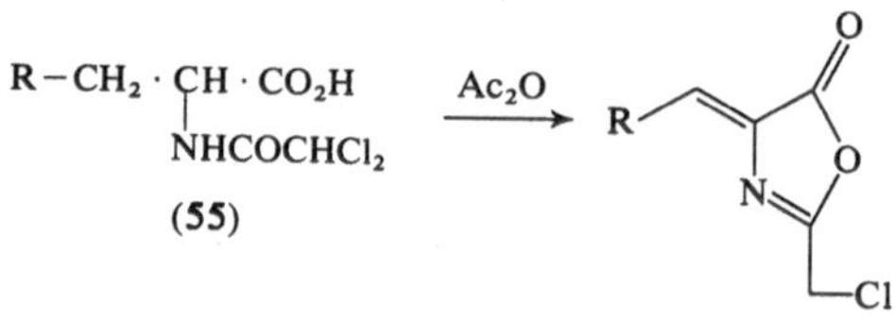

The tryptophan derivative (57) was obtained as an artefact by alkaline hydrolysis of the peptide antibiotic telomycin (97). Ultraviolet spectral studies on this compound and related systems indicated the presence of a dehydrotryptophan unit in the parent peptide.

The bisdehydrodioxopiperazine (58) has been synthesized by dimerisation of the enamine (59), prepared by reduction of the nitro-compound (60), (98). The enamine (59) did not convert with the isomeric imine (61); on heating, the imine afforded the imidazoline (62) rather than the piperazine (58).

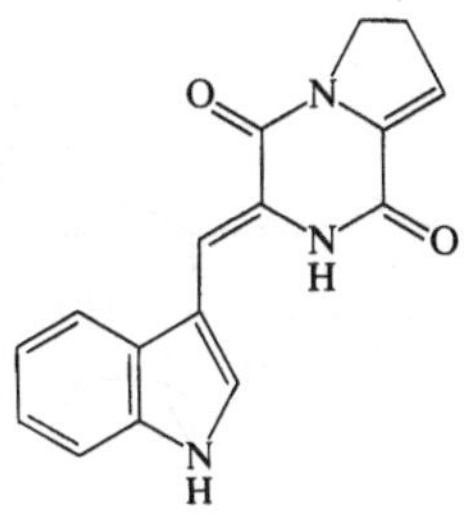

(57)

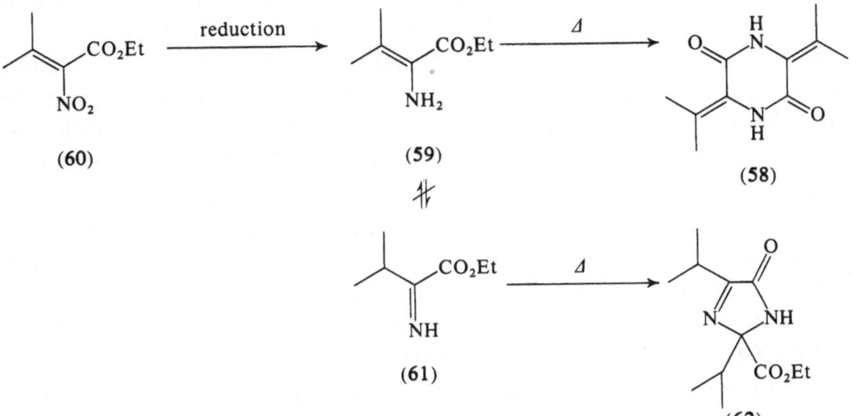

(60)   (59)   (58)

(61)   (62)

V. The Echinulins and Related Derivatives

Echinulin (**63**) was the first of a new group of dioxopiperazines isolated which contain an isoprenylated tryptophan unit (*99*). Isolated from *Aspergillus echinulatus,* it is a neutral, stable compound. The basic dioxopiperazine structure was recognized from the extensive work carried out by QUILICO *et al.* (*100*), but the structure was also proposed by BIRCH (*101*) on the basis of structure-biosynthesis relationships. Labelled-tryptophan (*102*), L-alanine (*101*), and mevalonic acid (*101*) are all efficiently incorporated into echinulin, giving a clear indication of its origin. *Cyclo*-L-ala-L-try is also incorporated, using *A. aristal* (*103*). Hydrogenation yields a hexahydro-derivative. Both echinulin and its reduction product show similar Cotton effects in the 200—300 nm region as *cyclo*-L-ala-L-try (*104, 105*), thus supporting the results obtained in the labelling experiments.

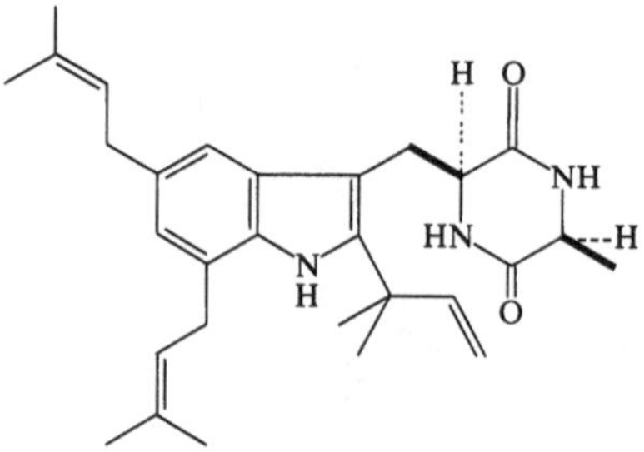

(63)

Although the alkylation pattern of the indole nucleus is quite normal
a novel feature of the molecule is the configuration of the isopentenyl
unit at position 2. Two possible explanations for the unusual orientation
of this group have been proposed. Either substitution occurs initially
at position 1 of the indole nucleus (*i.e.* at the nitrogen atom) followed
by rearrangement (*106*), or alkylation at position 3 is the first step, which
is then accompanied by rearrangement to position 2 (*107*). Recent work
appears to rule out both direct substitution (via an S_n2' process) (108)
and the pathway involving initial substitution at position 3 (*107*), which
leaves an initial alkylation at the nitrogen atoms as the most plausible
biosynthetic route. Model studies on the rearrangement of the alkylated
indole (**64**), which with acid gives mainly the 2-alkylated indoles (**65**) and
(**66**), (*109*) and the occurrence of the mould metabolite lanosulin (**67**),
(*110*) support an initial alkylation process at the indole nitrogen atom.
These results also appear to preclude a suggestion by Bycroft (*111*) in
which tryptophanyl groups are bound to the enzymes at the 2-position,
via sulphide bridges, prior to alkylation.

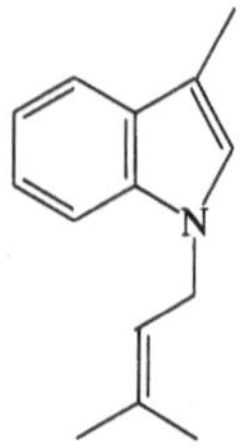

(64)

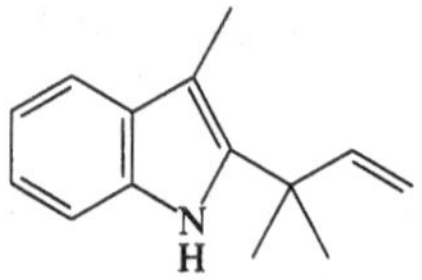

(65)

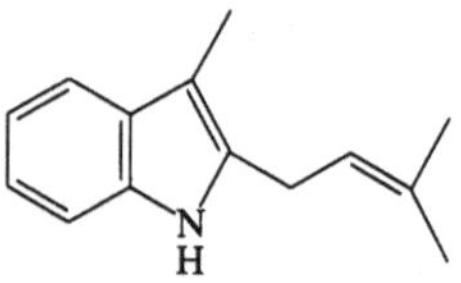

(66)

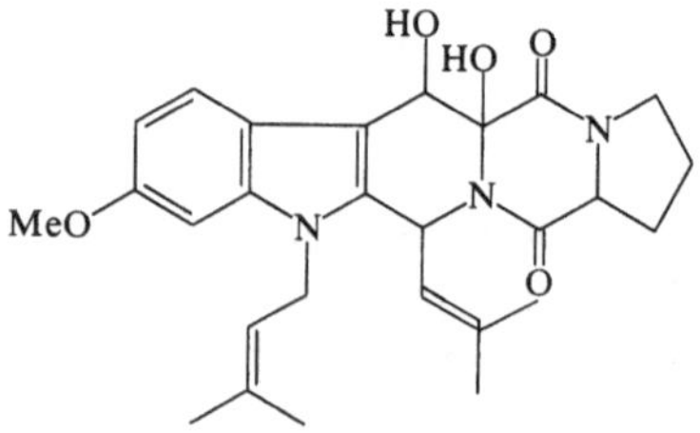

(67)

Approaches to the synthesis of echinulin have been described by several groups (*107, 112, 113*) and it has been achieved (*114*). The successful route involved prior formation of the indole (**68**). Aniline was alkylated to give the amine (**69**), which was heated with zinc chloride to give the rearranged aniline (**70**) in 70% yield (*115*). Reaction with ethyl 4-bromo-2,2-dimethyl-acetoacetate gave the indole (**71**) (Scheme 5). Reduction of the ester group followed by PFITZNER-MOFFATT oxidation and a WITTIG reaction gave the indole (**68**). Condensation of the corresponding gramine (**72**) with the dioxopiperazine (**73**), (*116*), followed by hydrolysis and decarboxylation gave a mixture of the *cis*- and *trans*-isomers of echinulin.

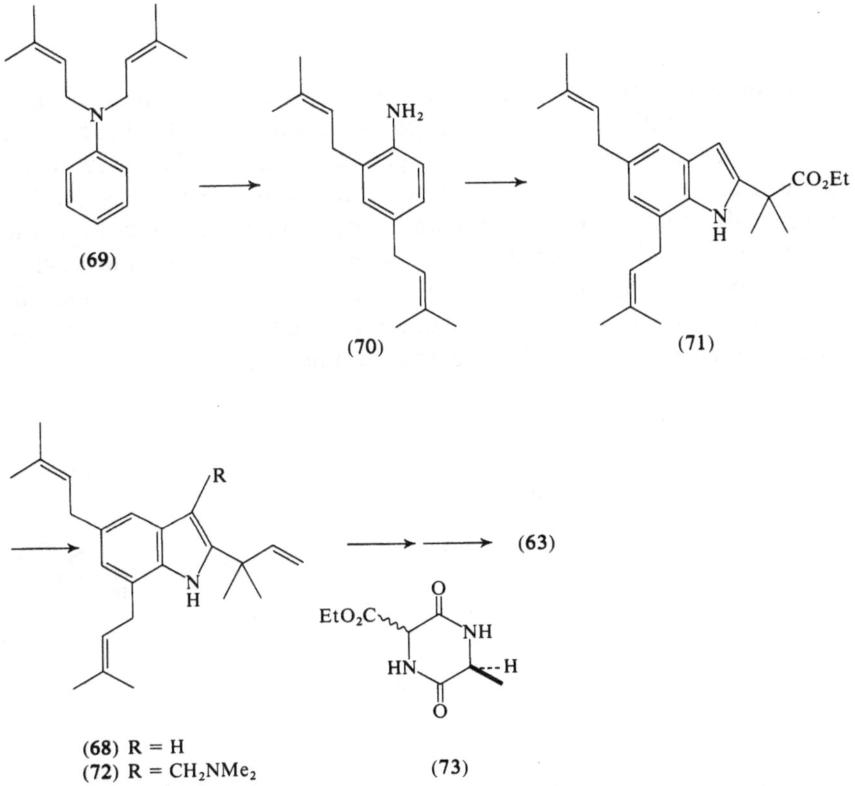

(69)

(70)

(71)

(63)

(68) R = H
(72) R = CH₂NMe₂

(73)

Scheme 5. Synthesis of echinulin

One further intriguing question on the biosynthesis of echinulin concerns the order in which the dimethylallyl groups are introduced into the indole nucleus. Some light has been shed on this problem by a study on the incorporation of *cyclo*-L-ala-L-try into a cell free extract of

Aspergillus amstelodami, using 3,3-dimethylallyl pyrophosphate as the isopentenyl source. Rapid formation of the known monoalkylated derivative (74) occurred (*117*), which was shown to be further incorporated into echinulin (*118*).

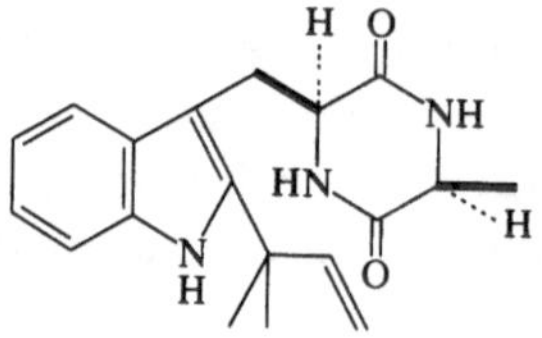

(74)

A compound related to echinulin is also found in *A. amstelodami*. Named neochinulin it occurs as orange-red crystals (*119*). On the basis of chemical degradations it has been assigned the structure (75), (*120*). Only two isopentenyl units are incorporated into the indole nucleus. Rather surprisingly, one of these was present at position 6; neoechinulin is the only indole alkaloid with an alkyl substituent at this position. Two chemical reactions used in the degradation studies deserve mention. Ozonolysis afforded the stable tetraoxopiperazine (76), whilst treatment with base initiated a retro-aldol hydrolysis to give the aldehyde (77).

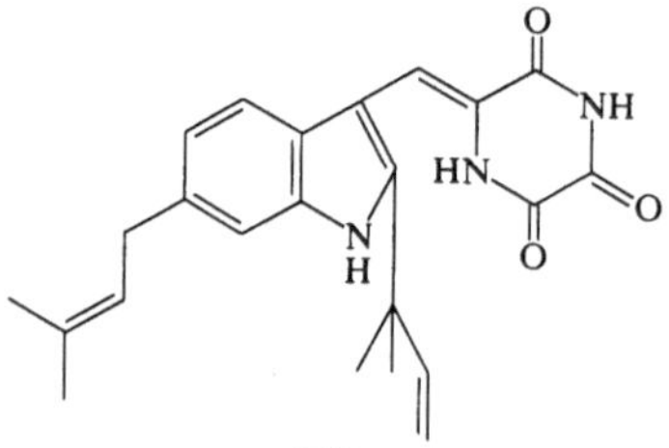

(75) 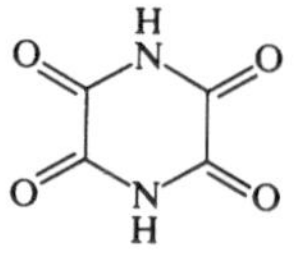   (76) 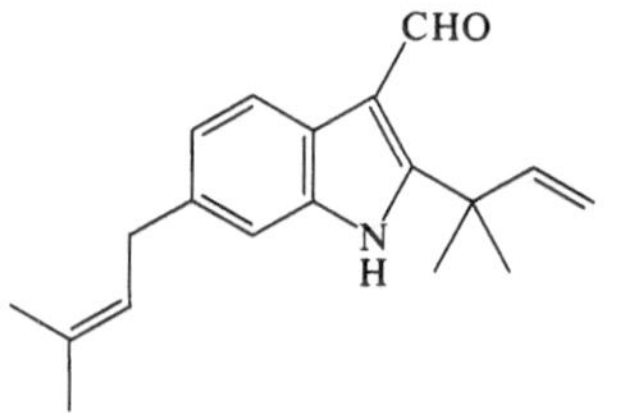   (77)

Parallel studies on both *Penicillium brevicompactum* (*121, 122, 123*) and *Aspergillus ustus* (*124, 125*) have lead to the isolation of ten isoprenylated *cyclo*-L-pro-L-try derivatives (see Table 3). These all contain the isoprenyl group at position 2 of the indole nucleus and each with the inverted (1′,1′-dimethylallyl) orientation, but differ in the site and level of oxidation. Studies on the fungus *A. ustus* were initiated because the fungus attacks maize, the metabolites causing acute toxicosis in animals such as ducklings.

Table 3. *The Echinulins and Related Compounds*

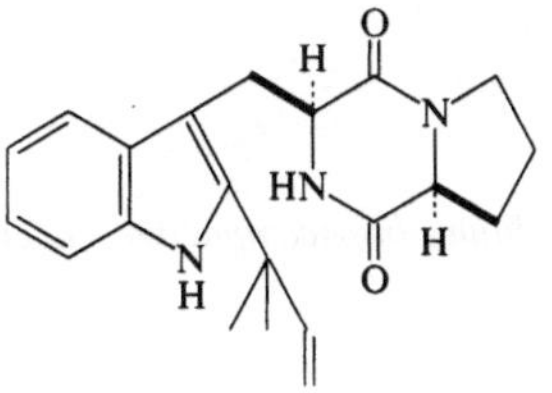

(78)

Aspergillus ustus (114)

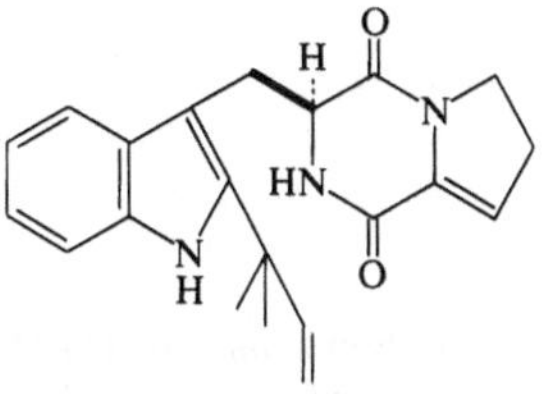

(79)

A. ustus (115)

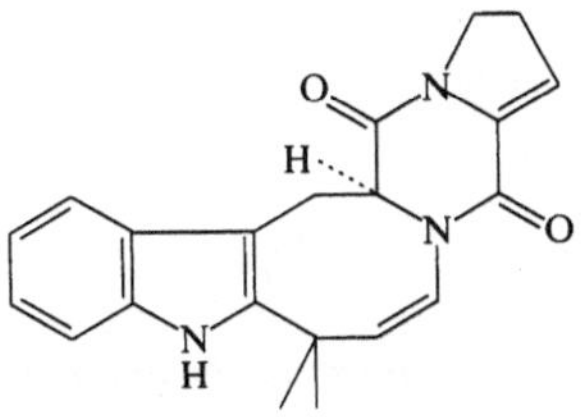

(80)

A. ustus (115)

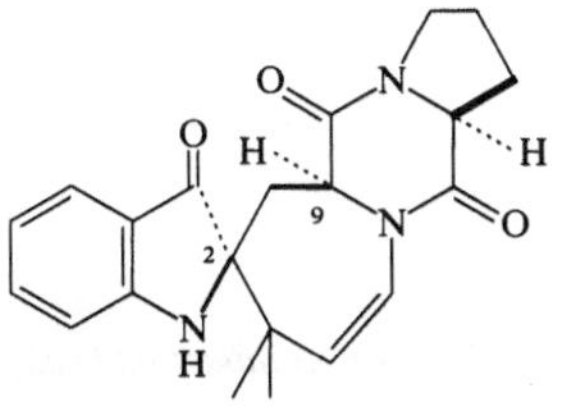

(81)

A. ustus (115)

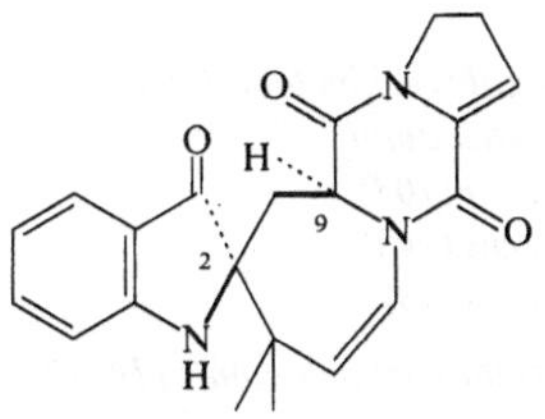

Austamide (82)

A. ustus (114)

Table 3 (continued)

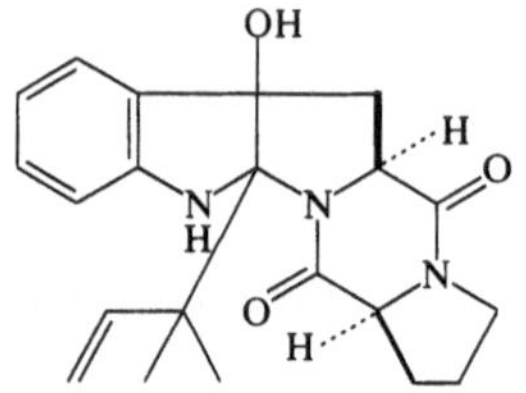

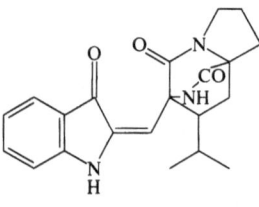

Brevianamide E (**84**)

Penicillium brevicompactum (122)

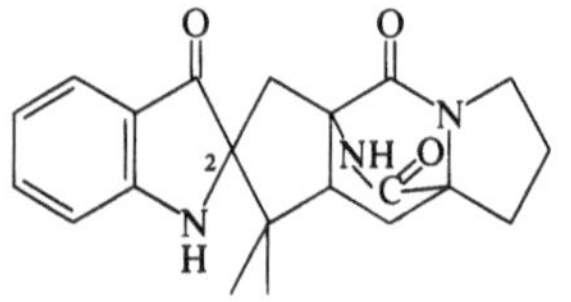

Brevianamide A (**86**)

Brevianamide B (epimer about position 2) (**87**)

P. brevicompactum (122)
P. brevicompactum (123)

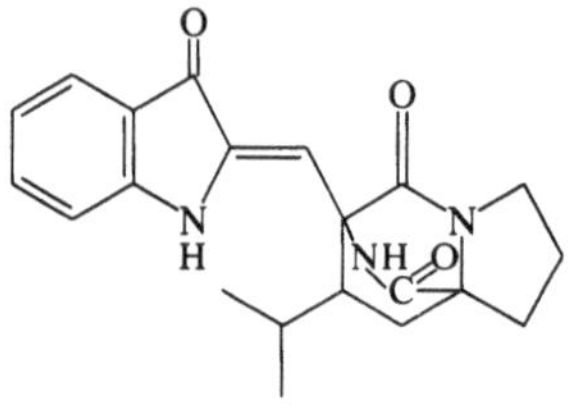

Brevianamide C (**89**)

P. brevicompactum (123)

Brevianamide D (**90**)

P. brevicompactum (123)

Echinulin (**63**)

Aspergillus echinulatus (99)
A. amstelodami
A. aristal (103)
A. repens (248)
A. ruber (248)

Neoechinulin (**75**)

Aspergillus amstelodami (119, 120)

Lanosulin (**67**)

Penicillium lanosum (110)

References, pp. 107—118

The simplest alkaloid from *A. ustus* (see Table 3 for formulas **78—82**) is *cyclo*-L-prolyl-[2-(1′,1′-dimethyl)allyl]-L-tryptophyl (**78**), (*125*) which is analogous to the alanyl derivative isolated by ALLEN from *A. amstelodami* (*117*). This was accompanied by the dehydrodipeptide (**79**). Substitution into the olefin bond of the dimethylallyl residue by the dioxopiperazine ring is also observed in products from this fungus. This results in the formation of a new nitrogen to carbon bond, as in compounds (**80**), (**81**), and (**82**). In the latter two cases further oxidation of the indole ring into the corresponding ψ-indoxyl system has also occurred. This oxidation, which results in a migration of the side chain from position 3 to position 2, is common in tryptophan derivatives (*126*) and can be performed *in vitro* by oxidation of 3-substituted indoles with oxygen over a platinum catalyst (*127*). Furthermore, this oxidation is reversible, treatment of the ψ-indoxyl system with sodium borohydride and treatment with acid can regenerate the parent indole system.

Of the oxidation products isolated the toxic principle has been identified as austamide (**82**), (*124*). The configuration about the dioxopiperazine nucleus retains the L-chirality in austamide (*i.e.* 9 S) and the related derivatives, but allows two possible orientations about the spiro-position 2 in the ψ-indoxyl series. On the basis of the relative deshielding of the 9-H proton by the ketone function, these groups are assumed to be syn-oriented, as depicted. The isolation of the compound (**80**) is considered to be of biosynthetic importance, since it is presumed to be the biosynthetic precursor of austamide (**82**), coupling of the dimethyllallyl group to the dioxopiperazine ring occurring before oxidation to the indoxyl series. The mechanism whereby the new carbon-nitrogen bond is formed is of interest. It represents an oxidation and is reminiscent of the case of gliotoxin (**83**) *(vide infra)*, in which an amide bond of the dioxopiperazine ring is also alkylated in an oxidative manner (*128*). The structure of austamide has also been confirmed by an X-ray analysis (*129*).

(**83**)

To date, a completely different series of compounds has been isolated from *P. brevicompactum* (see Table 3). The simplest of these is brevianamide E (**84**), (*121, 122*). Oxidation of the indole ring has only proceeded to the hydroxyindolenine stage in this compound, the oxidation being followed by an intramolecular addition of the amide group of the dioxopiperazine ring across the imine (85).

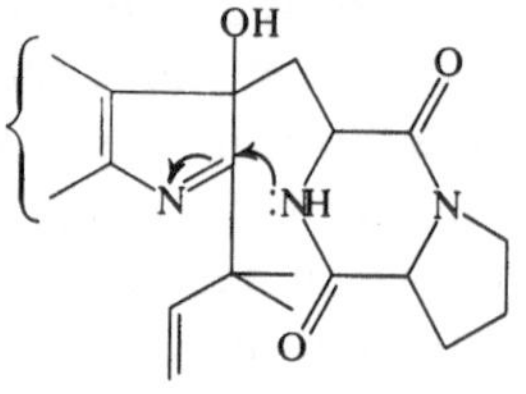

(85)

Four ψ-indoxyl type compounds accompany brevianamide E. Two of them are diastereoisomers, brevianamides A (86), (122) and B (87), (123). In these cyclo-pro-try derivatives the double bond of the incorporated dimethylallyl residue is missing. The structure of brevianamide A was elucidated partly by the use of biosynthetic arguments backed by a careful n.m.r. study. Thus, whereas the protons normally present at positions 3 and 6 in dioxopiperazines generally show characteristic resonances between τ 5.5 and 6.3 (17), no protons occurred between τ 4.0 to 6.7 in deuteriopyridine in the spectrum of brevianamide A, indicating substitution at these positions. The mass spectral fragmentation of this compound was also informative (Scheme 6), resulting in formulation of the structure (86) shown. The relative configurations about the chiral centres have not been established. Reduction of brevianamide A, followed by dehydration with acid afforded the indole (88), which could be reoxidized to give an isomeric indoxyl, shown to be identical to brevianamide B (87). The latter isomer must therefore be the epimer about position 2 of brevianamide A and, thermodynamically, the more stable of the pair.

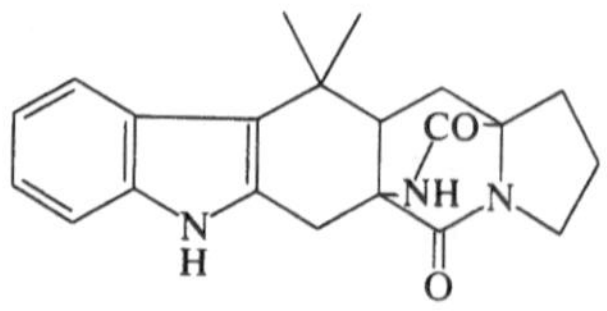

(88)

The remaining brevianamides, C (89) and D (90) (see Table 3) were non-crystalline orange to red compounds, shown to contain the 2-indoxyli-dene-alkane chromophore (91), (123). Spectral evidence confirmed the

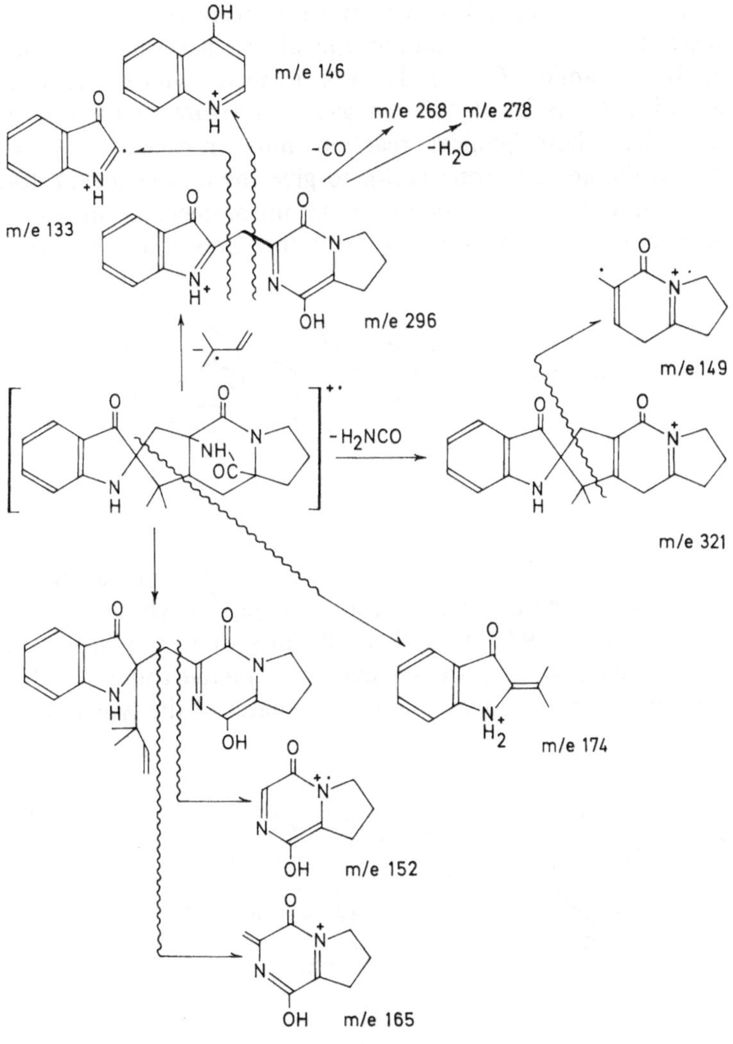

Scheme 6. Fragmentation of brevianamide A (**86**)

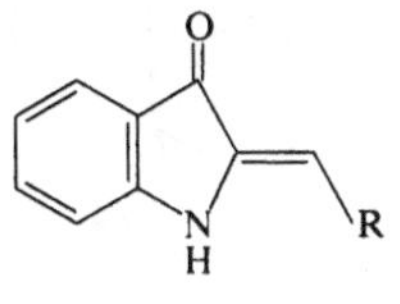

(91)

presence of the dioxopiperazine ring and also indicated the existence of an isopropyl group. The *cis*- and *trans*-relationship of these two isomers about the double bond was confirmed by irradiation, either isomer producing the same equilibrium mixture of both. It was further shown that these derivatives were photochemical artefacts since, in the dark, neither brevianamide C nor D was formed, whilst irradiation of brevianamide A, or B, at 430 nm gave a mixture of these two compounds. This photochemical reaction must proceed by α-cleavage adjacent to the ketone group (130), to give the diradical (92), followed by hydrogen migration, possibly in an intramolecular manner via a five-membered transition state, to give either brevianamide C or D.

(86) or (87) $\xrightarrow{h\nu}$

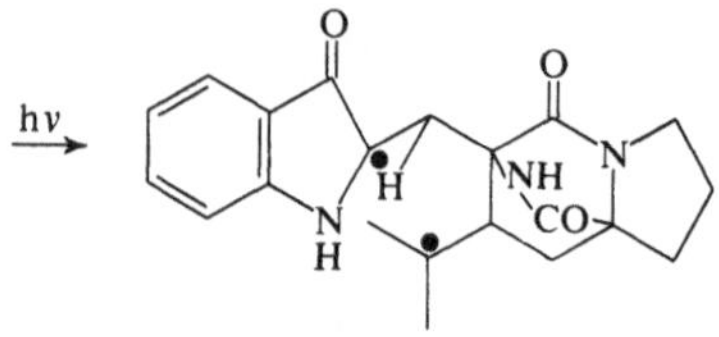

(92)

Formation of brevianamide A, and its isomer B, demands the creation of two new carbon-carbon bonds from its likely biosynthetic precursors, compounds (78) or (93), as well as the loss of two hydrogen atoms. Although radical processes are possible, an attractive route would involve a cycloaddition reaction involving intermediates of the type (94) or (95).

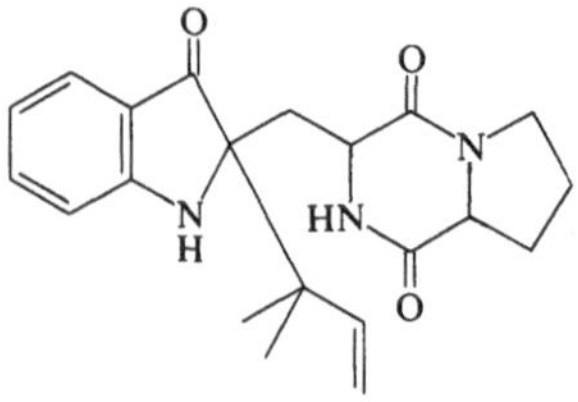

(93)

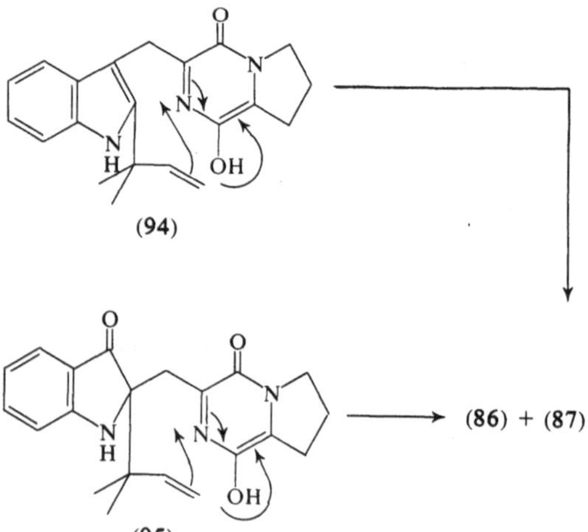

(94)

(95) $\longrightarrow$ (86) + (87)

Precedent for such cycloaddition reactions exists. Dihydroxypyrazines of the type (96) undergo cycloaddition with norbornadiene (131), whilst pyrazinones of the type (97) react with cyclopentene to give the adducts (98) and (99), (89).

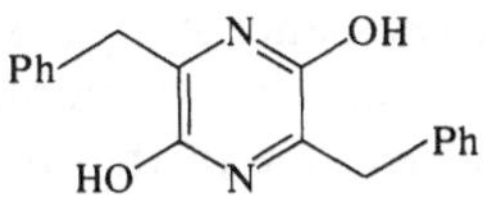

(96)

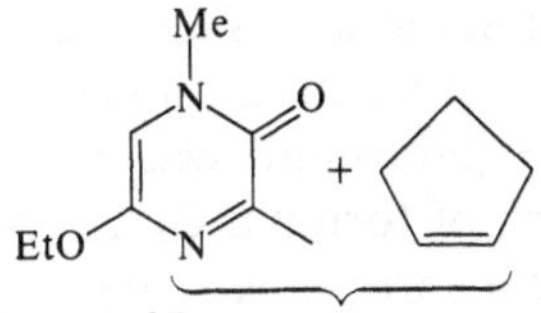

(97)

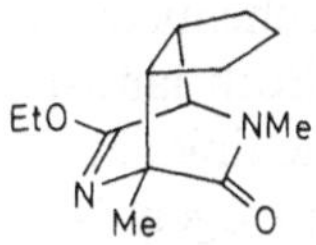

(98) 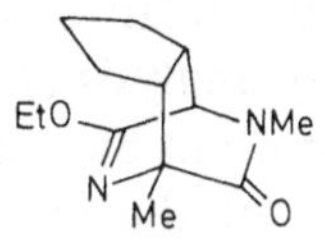   (99)

Brief mention has already been made of the interesting metabolite lanosulin (67), originally isolated from *Penicillium lanosum* (*110*). Structure elucidation of this compound was greatly aided by extensive n.m.r. and mass spectral studies. The presence of the dioxopiperazine ring was first indicated by the strong infrared bands at 1685 and 1645 cm^{-1} coupled with the absence of the amide II bands. Hydrolysis afforded proline. Its mass spectrum indicated the incorporation of two isoprene units. Comparison of its n.m.r. properties with the model compound (64) showed a remarkable resemblance, allowing one of the isoprene units to be placed as a 3,3-dimethylallyl substituent on the indole nitrogen. Placement of the remaining isoprene group on the dioxopiperazine ring nitrogen was also consistent with the spectral evidence. Lanosulin was originally assigned the structure (100), (*110*) but this has since been amended (*133*).

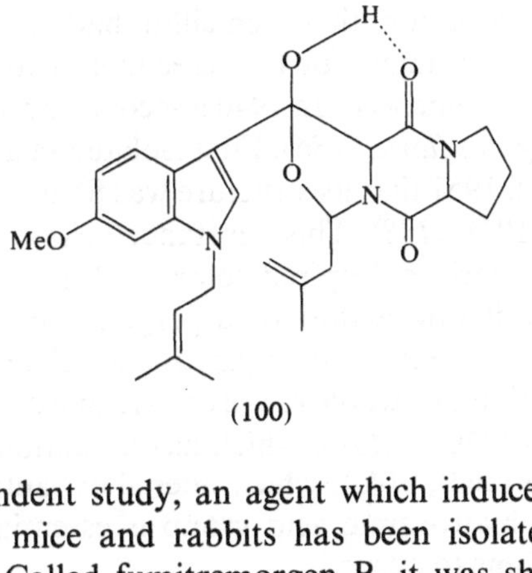

(100)

In an independent study, an agent which induces strong tremors in animals such as mice and rabbits has been isolated from *Aspergillus fumigatus* (*132*). Called fumitremorgen B, it was shown to be identical with lanosulin, but it was assigned, instead, the amended structure (67) on the basis of a very detailed nuclear magnetic resonance study (133). Fumitremorgen B (lanosulin) was accompanied by another tremorgen, fumitremorgen A, of formula $C_{32}H_{41}N_3O_7$, in which one further isopentenyl unit must be incorporated. The revised formulation of lanosulin (structure 67) bears a very strong resemblance to the naturally occurring peroxide, verruculogen (101), another tremorgen isolated from *Penicillium verruculosum*, the structure and relative stereochemistry of which have been confirmed by an X-ray analysis (134). Both lanosulin and verruculogen possibly represent important biosynthetic links in this group of alkaloids in that isoprenylation of the indole nitrogen is supposed to precede migration, with inversion, to the position 2. The second isopentenyl unit is probably also incorporated by prior alkylation of the dioxopiperazine nitrogen atom.

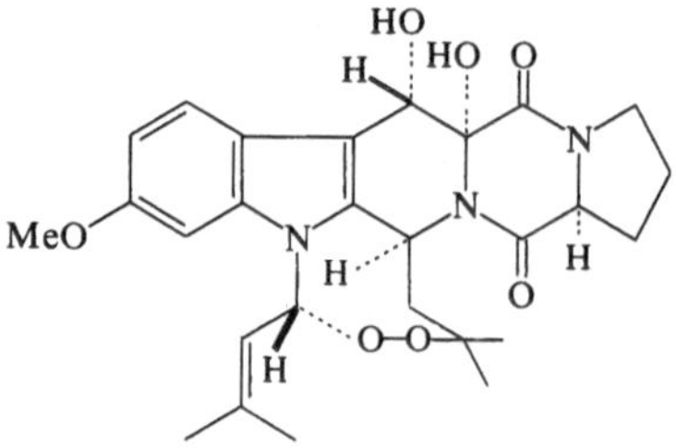

(101)

VI. Hydroxypyrazine Derivatives

Following the revelation that penicillin had antibiotic properties (135) a widespread programme for the screening of fungal culture fluids for other antibiotics commenced. In 1940 a second substance was detected in cultures of *Aspergillus flavus* (136). First isolated in a pure state in 1943 (137), it was not until 1951 that its structure was finally established as that of aspergillic acid (102), (138). This substance behaves as a hydroxamic acid and gives characteristic deep red colors with iron salts. Encouraged by the powerful antibiotic action of aspergillic acid a more detailed examination of the culture fluids of *A. flavus* and related fungi was made, resulting in the isolation of a whole range of pyrazine derivatives (Table 4), varying from flavacol (103), (139), which has no antibiotic properties, to mutaaspergillic acid (104), (140), which does show antibiotic behaviour. It appears that the hydroxamic acid group is essential for antagonism against microorganisms to occur.

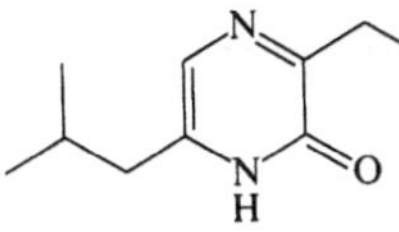

(103) 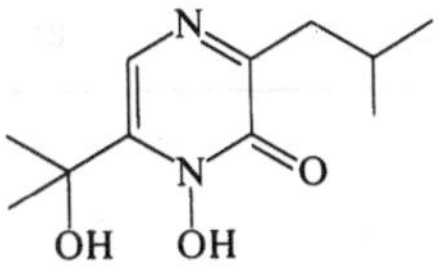   (104)

Table 4. *Some Natural Hydroxy-Pyrazine Derivatives**

Pulcherrimic acid (**109**)	*Candida pulcherrima* (*151*)
Aspergillic acid (**102**)	*Aspergillus flavus* (*136*)
Flavacol (**103**)	*A. flavus* (*139*)
Mutaaspergillic acid (**104**)	*A. oryzae* (*140*)
Hydroxyaspergillic acid (**106**)	*A. flavus* (*249*)
Deoxyaspergillic acid (**107**)	*A. sclerotiorum* (*147*)
Hydroxyneoaspergillic acid (**108**)	*A. sclerotiorum* (*147*)
Neoaspergillic acid (**105**)	*A. sclerotiorum* (*147*)

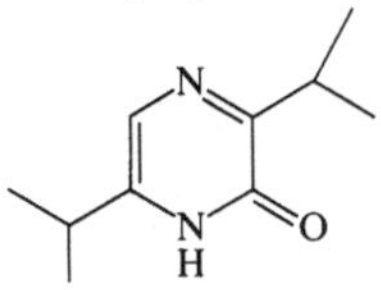

(**197**)   *A. sojae* (*250*)

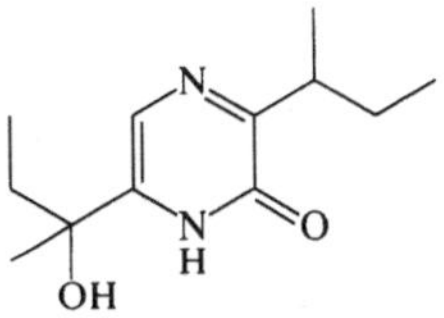

(**198**)   *A. sojae* (*250*)

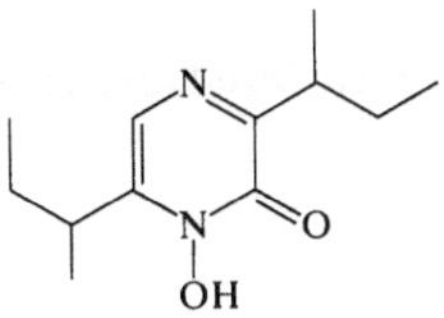

(**199**)   *A. sojae* (*250*)

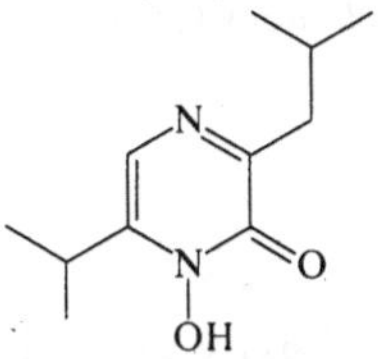

(**200**)   *A. sojae* (*250*)

* For a recent review on pyrazine chemistry see (*257*).

Table 4 (continued)

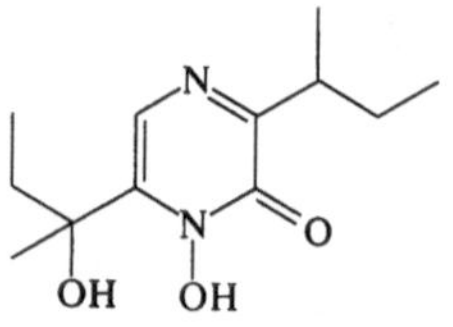 **(201)**	*A. sojae (250)*
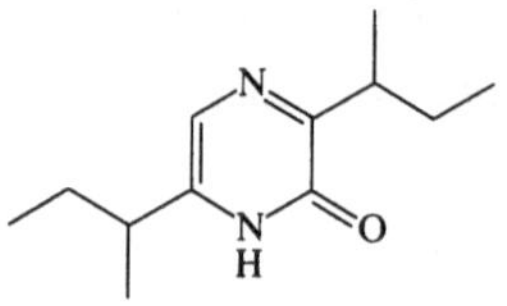 **(202)**	*A. M 4-1 (251)*
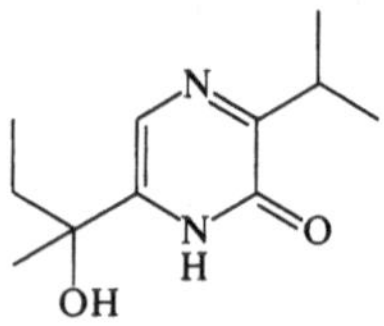 **(203)**	*A. M 4-1 (251)*
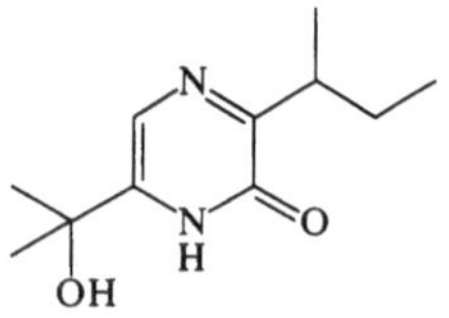 **(204)**	*A. M 4-1 (251)*
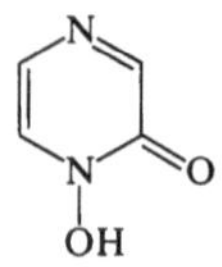 Enimycin **(205)**	*Streptomyces* No 2020-I *(252, 253)*
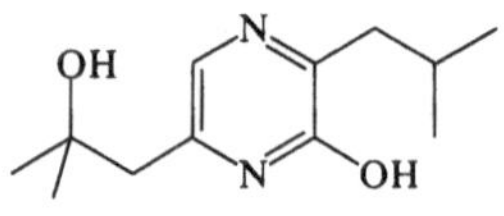 **(206)**	*Aspergillus ochraceus (254)*

A total synthesis of aspergillic acid has effected by OHTA and his collaborators *(141)* using an original and adaptable route (Scheme 7) starting from oximes *(142)*. This route has also been used to make neo-

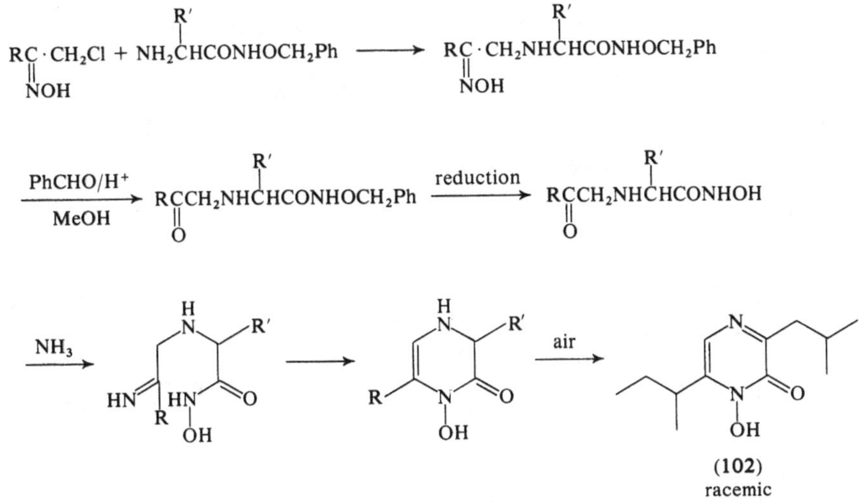

Scheme 7. Synthesis of aspergillic acid

aspergillic acid (**105**), (*141*) and mutaaspergillic acid (**104**), (*143*). However, in the latter case, the synthetic and natural compounds did not have identical ultraviolet properties and this point needs clarification. An alternative route to these cyclic hydroxamic acid derivatives has also been devised, which commences with the dioxopiperazines (*cf.* Scheme 9) and has been applied to the synthesis of aspergillic acid (*144*) and neo-aspergillic acid (*145*).

A considerable number of biosynthetic studies on these systems has been made by MacDonald and his collaborators. The systems are derived from amino-acids, combinations of valine, leucine, and isoleucine predominating. The question arises as to whether they are derived from the corresponding dioxopiperazine systems or not. The evidence suggests that, generally, they are *not* derived from these precursors (*147*). In a ·study on the biosynthesis of aspergillic acid (**102**) and hydroxyaspergillic acid (**106**) it has been shown that, whereas leucine and isoleucine are effective precursors in *A. flavus, cyclo*-isoleu-leu is less efficient (*148*). Furthermore, both D- and L-leucine were incorporated into these compounds, indicating that the amino acids are metabolized into the pyruvate pool before incorporation. It is interesting that *cyclo*-isoleu-leu is more efficiently incorporated into mycelial protein than into aspergillic acid, which indicates its faster hydrolysis into the starting amino-acids than its incorporation into the pyrazines. Other labelling experiments, however, confirmed that deoxyaspergillic acid (**107**) is converted into aspergillic acid (**102**) and then into hydroxyaspergillic acid (**106**), (*149*).

6*

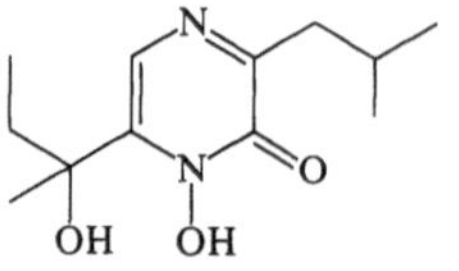

P. G. SAMMES:

(106) 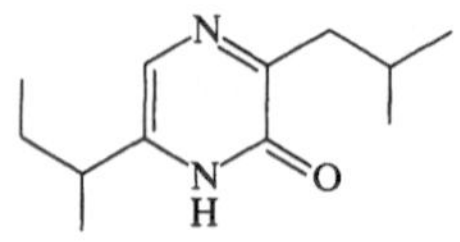   (107)

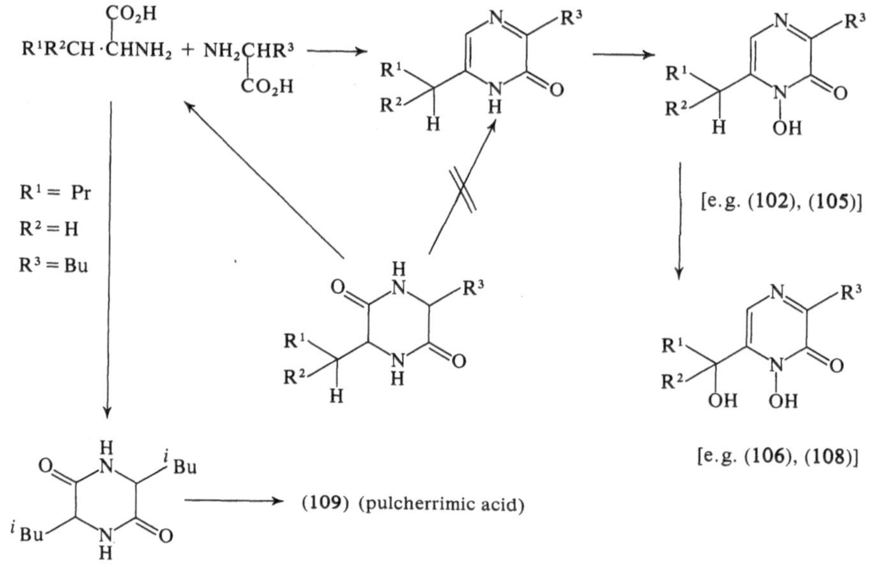

Scheme 8

A similar result was obtained in the flavacol (103) to hydroxyneo-aspergillic acid (108) series. Both neoaspergillic acid (105) and hydroxy-neoaspergillic acid (108) are synthesized by *A. sclerotiorum* entirely from leucine (150). These labelling experiments are summarized in Scheme 8. To summarize, although a categoric rejection of the dioxopiperazines as true precursors to these pyrazines cannot be given, they are, at best, only inefficient precursors.

An exception to the above results was found with studies on pulcher-rimic acid (109), (150), isolated from *Candida pulcherrima* (151). Labelling experiments have shown for this compound that L-leucine is converted into *cyclo*-L-leu-L-leu, which is then assimilated into the acid (98) without labelling mycelial protein (152).

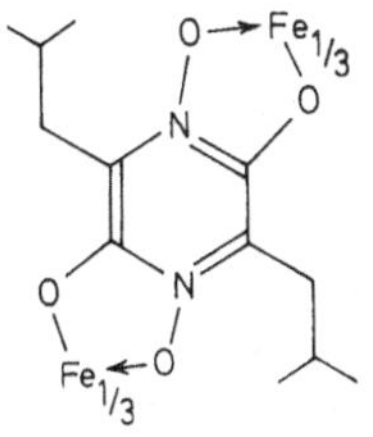

(110)

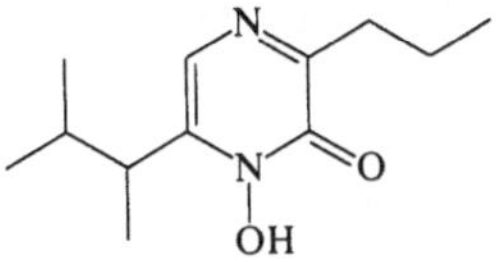

(105) R = H
(108) R = OH

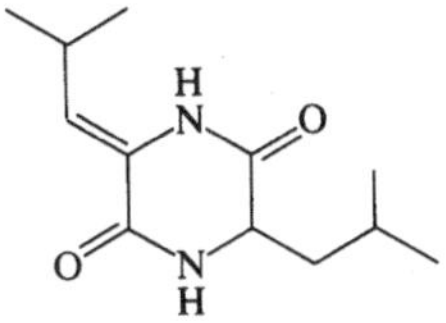

(111)

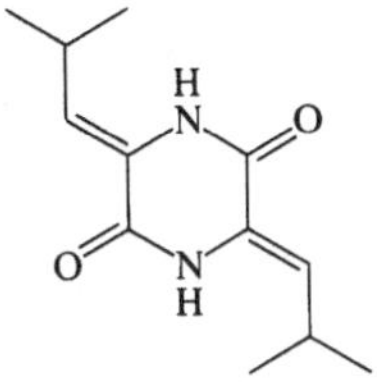

(112)

Pulcherrimic acid (109) was originally isolated as its ferric complex, pulcherrimin (110), from which the acid can be released by treatment with base, removal of ferric hydroxide, and reacidification (150). In attempting to repeat this work MACDONALD (153) isolated a sample of pulcherrimic acid which differed in its properties from that originally described. Reduction of the new material with iodine and red phosphorus, a reagent which had been used to reduce aspergillic acid (102) into its deoxy-analogue (107), (152), afforded three compounds, cyclo-leu-leu, and the two dehydro-compounds (111) and (112). On the basis of its n.m.r. spectrum the new acid was given the structure shown, viz. (109). In order to explain the differences between the new sample of pulcherrimic acid compared to the original sample, alternative structures such as (113) were proposed for the latter. More likely are formulations of the type (114), which is analogous to the structure found in mycelianamide (115), (156). A total synthesis of pulcherrimin has been claimed (Scheme 9 (157).

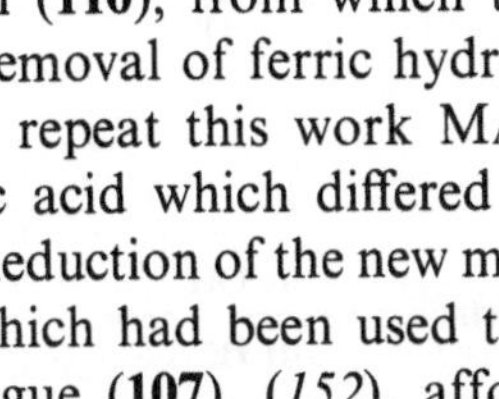

(113)

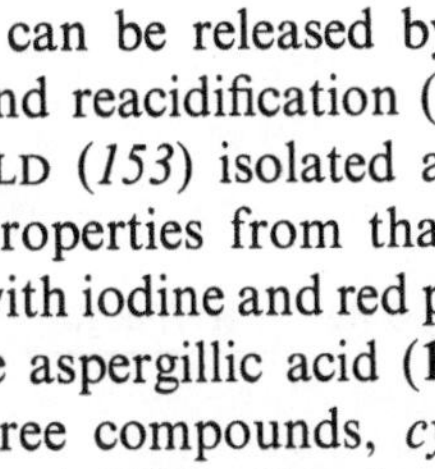

(114)

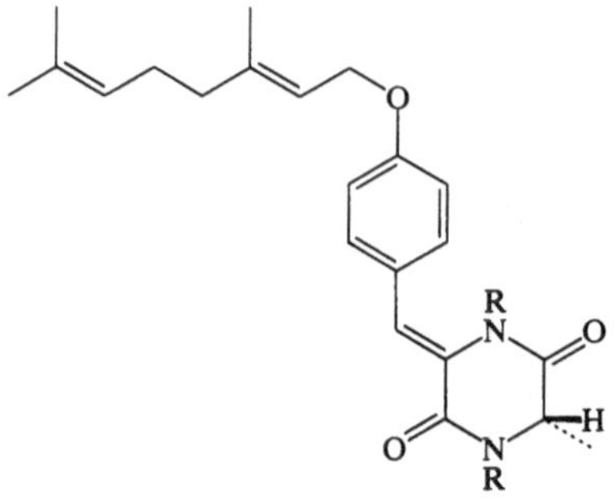

(115) R = OH
(116) R = H
(117) R = OMe

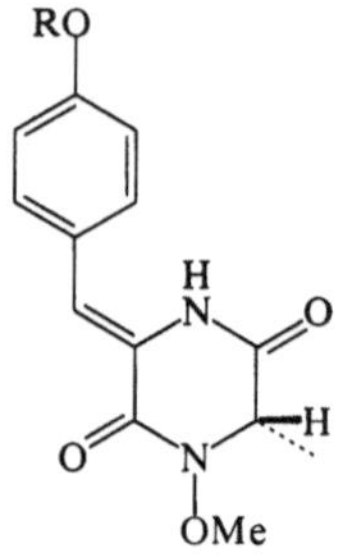

(118) R = geranyl

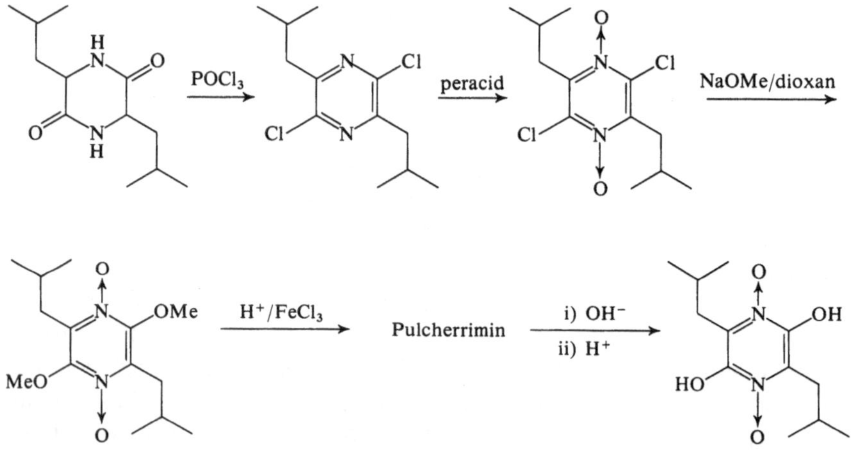

Scheme 9. Synthesis of pulcherrimin

Mycelianamide (115) has been isolated from *Penicillium griseofulvum* Dierckx (157) and its structure was determined largely on the basis of work by BIRCH and co-workers (158). The structure of the terpenoid side-chain was later revised (155) and confirmed by a synthesis of racemic deoxymycelianamide (116), (159). The chemistry of mycelianamide has been reviewed (160). It is noticeably unstable to both acid and base as well as being thermally unstable. It can be alkylated with diazomethane to give a dimethyl ether (117). This ether undergoes thermal decomposition at the remarkably low temperature of 65° within a few hours, losing the N-4 methoxy group as formaldehyde, to give the amide (118), (161). This reaction is probably enhanced by the relief of steric strain due to the cis-arylidene-N-methoxy interaction (Scheme 10).

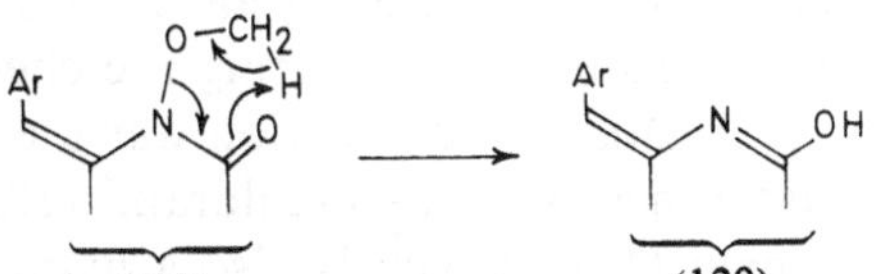

(119)   (120)

Scheme 10

Pyrazines related to the above compounds include methoxy-pyrazines (162). These are extremely widespread and have powerful flavouring properties-amongst the strongest known (163). Like some of the aspergillic-type pyrazines, it is believed that these derivatives are not produced from dioxopiperazine precursors (164).

VII. Sulphur-bridged Dioxopiperazines

This group of compounds can be divided into those related to gliotoxin (83) and those of the sporidesmin type (121).

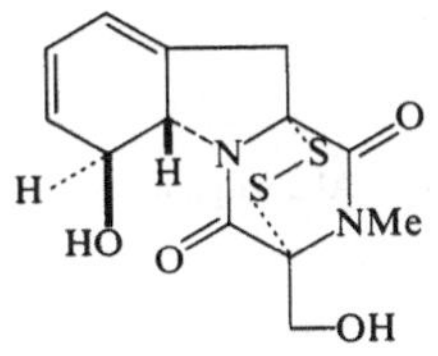

(83) 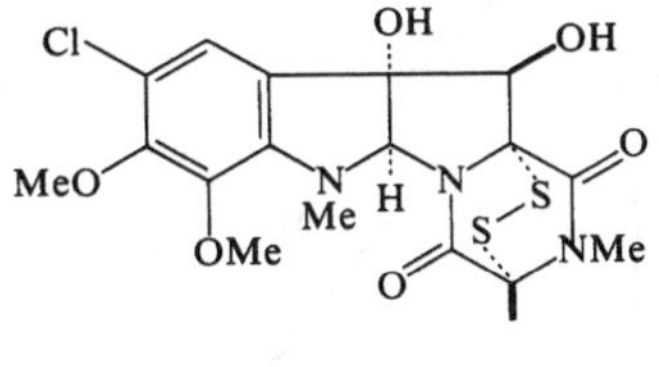   (121)

Gliotoxin (83) was first isolated by WEINDLING and EMERSON in 1936 from a species of *Trichoderma viride* (165). This toxic substance was later isolated from a variety of microorganisms, including *Gliocladium fimbriatum* (166), *Aspergillus fumigatus* (167), and *Penicillium terlikowskii* (168). The substance is highly antifungal — in some cases it is nearly as potent as mercuric chloride! — but it shows a much greater selectivity in

its action (*166*). It also possesses bacteriostatic and antiviral properties
(*169*) and it is these physiological properties, coupled with its unusual
chemical structure, which have created such an interest in gliotoxin and
related systems (*170*).

The structure of gliotoxin was first proposed on the basis of extensive
chemical studies (*171*) and was supported by an X-ray diffraction analysis
(*172*). Some of the reactions employed during the chemical studies are
depicted in Scheme 11. A most useful reaction was the dehydrogenation
with high potential quinones, such as o-chloranil (*173*). This reaction
afforded dehydrogliotoxin (**118**), which is also a natural product (*173*),
and in which the very unstable dienol function is converted into the much
more stable phenol without destruction of the labile disulphide bridge.
This enabled reactions to be carried out specifically at this unusual

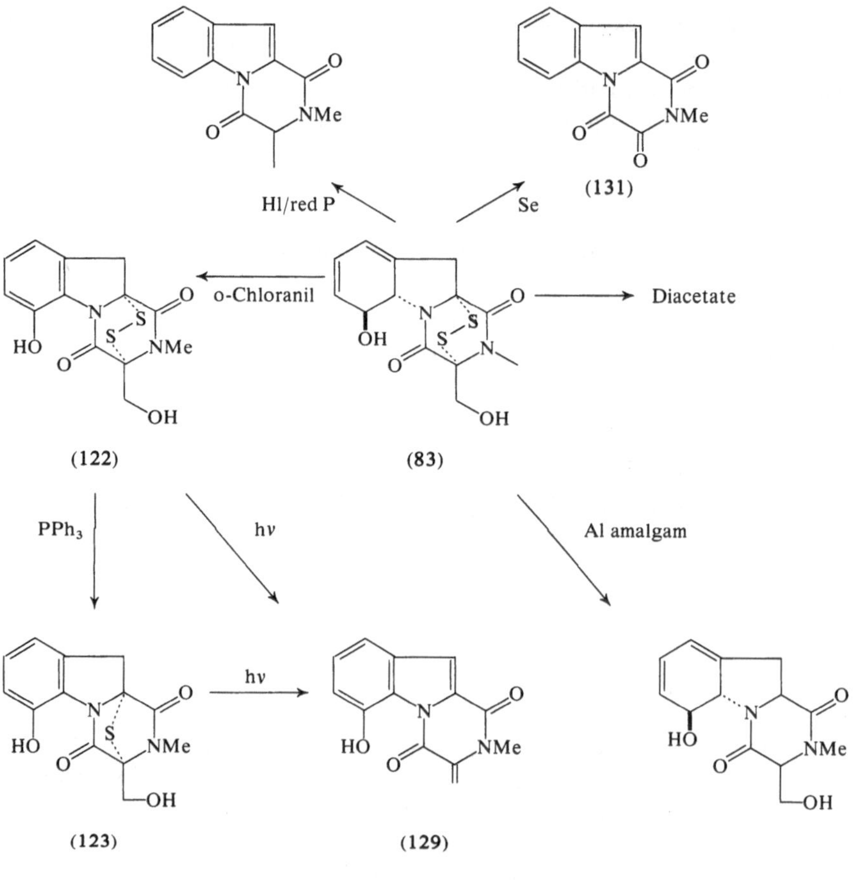

Scheme 11. Chemistry of gliotoxin

construction without conflicting reactions ensuing from the dienol group. For example, reduction with triphenyl-phosphine converts the disulphide (**122**) into the monosulphide (**123**), (*174*).

The absolute configuration of gliotoxin has also been established by X-ray crystallographic analysis (*172*) and these studies also revealed that the highly strained disulphide bridge does retain a twisted conformation, the dihedral angle about the $C-S-S-C$ system being 12°, compared to the 90—100° angle normally present in acyclic disulphides. The sulphur atoms are spatially nearer the carbonyl groups than they are to the nitrogen atoms of the dioxopiperazine ring. A similar disposition of the sulphur atoms has also been found for sporidesmin (**121**) and chaetocin (**124**), (*176*) as well as the tetrasulphide (**125**), (*177*) and is probably common to all the bridged di- or tetra-sulphides, implying an interaction between the sulphur atoms and the amide groups. This interaction is reminiscent of that known to occur in a transannular manner for certain keto-sulphides (*178*).

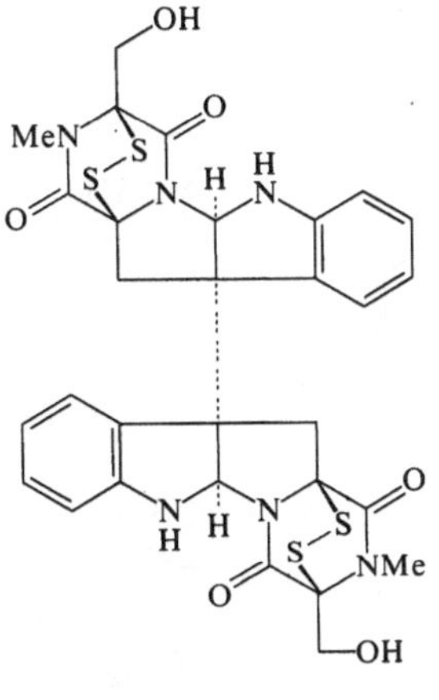

(124) 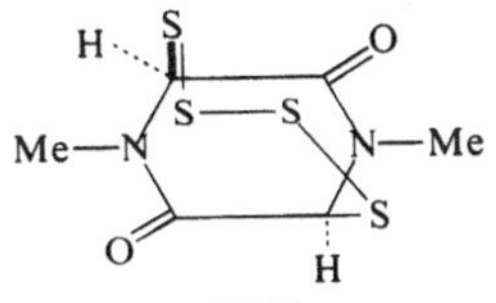   (125)

The strain in these disulphide systems is reflected by the ease and speed with which they can be opened. For example, reduction of the model disulphide (**126**) with sodium borohydride is very rapid (*179*). That the strain is more torsional than angular is reflected by the fact that reoxidation of the dithiol (**127**) back to the starting disulphide (**126**) is readily accomplished. Some reactions involving the disulphide bridge of acetylaranotin (**128**) have also been reported (see Scheme 12) (*180*). The reactions were carried out in pyridine at room temperature. It is interesting to note that hydrogen cyanide does not react with simple dialkyl-disulphides under these conditions (*181*). The presence of the

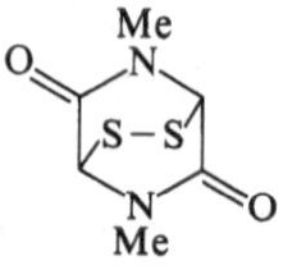

(126) 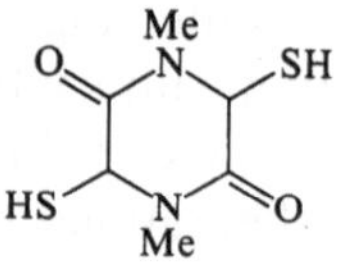   (127)

reactive disulphide bridge in these compounds appears to be essential for antiviral and antibacterial activity; related compounds in which the disulphide bridge is absent are notably inactive (*170, 171*).

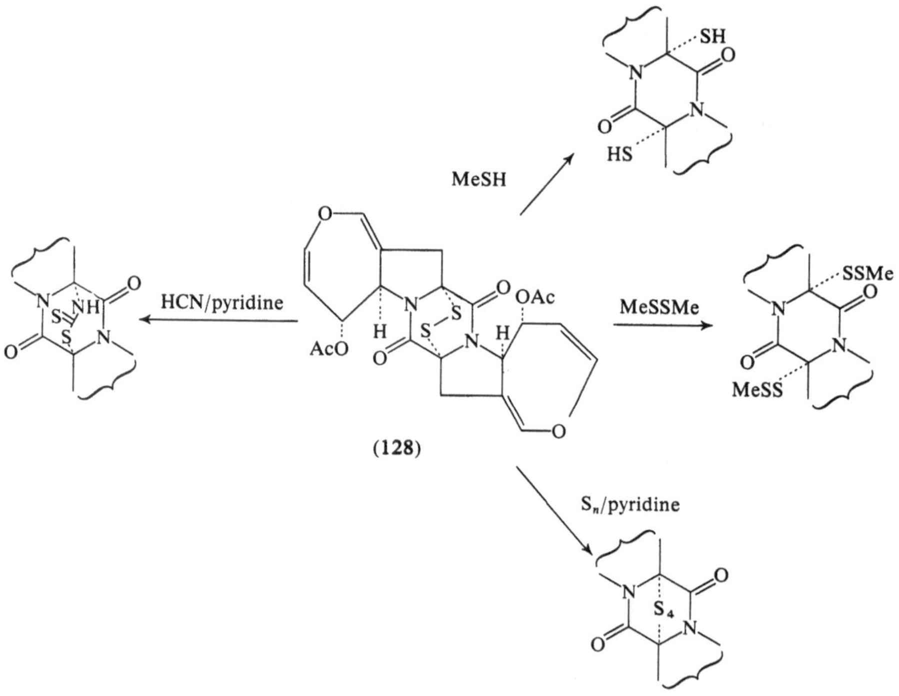

Scheme 12. Some reactions of acetylaranotin

Circular dichroism studies (*182, 183*) on the skew angle of the diene chromophore of gliotoxin were found to be anomalous, having an opposite sign to those found for simple diene systems (*184*). This has been explained in terms of an interaction between the disulphide and diene chromophores. On the basis of separate c. d. studies it has been proposed that reduction of the disulphide bridge with triphenylphosphine, for example, of dehydrogliotoxin (**122**) to produce the sulphide (**123**), proceeds with inversion of configuration at both the bridgehead carbons (*175*). This was suggested because the c. d. curves show opposite signs for the two compounds. This explanation is mechanistically unfeasible for the reduction process and it is more likely that a similar interaction of chromophores occurs in the disulphide, which is not present in the monosulphide and which again discredits a comparison of the c. d. curves.

In a detailed examination of the culture fluids from *P. terlikowski* for possible biosynthetic precursors of gliotoxin, three desthio-dioxopiperazines were isolated (**129**), (**130**), and (**131**), (*185*). Both compounds (**129**) and (**130**) were also obtained in small amounts from the oxidation of gliotoxin with chloranil.

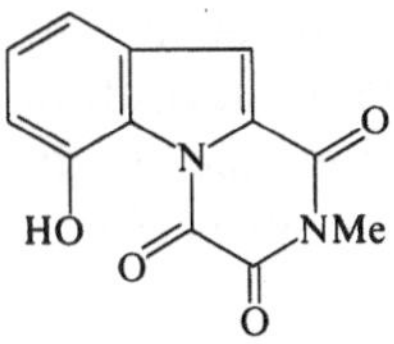

(130)

Compounds closely related to gliotoxin were isolated from the mould *Arachniotus aureus* which had attracted attention because extracts showed a very high antiviral activity (*186*). The principal component was shown to be aranotin (**132**). Aranotin acetate (**128**), (*186*), and the related compounds apoaranotin (**133**), (*128*), bisdethiodi(methylthio)acetylapoaranotin (**134**), and bisdethiodi(methylthio)acetylaranotin (**135**), (*186*) were also isolated from *A. aureus*. The antibiotics LL-S 88 α and LL-S 88 β were isolated independently (*187*) at about the same time from *Aspergillus terreus* and shown to be identical to acetylaranotin (**128**) and the thiomethyl derivative (**135**) respectively. The structure of acetyl-aranotin was con-

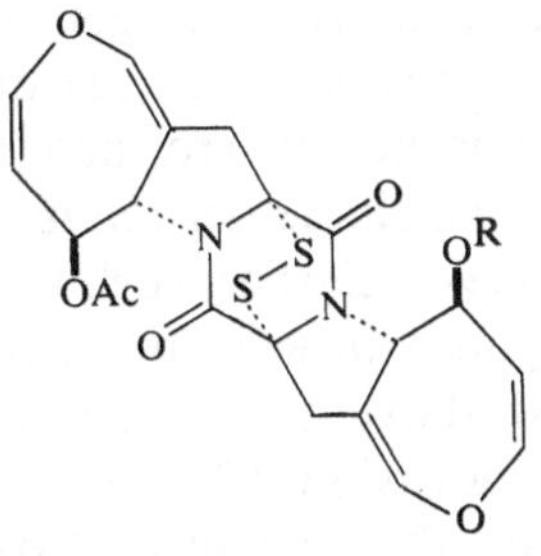

(132) R = H
(128) R = Ac

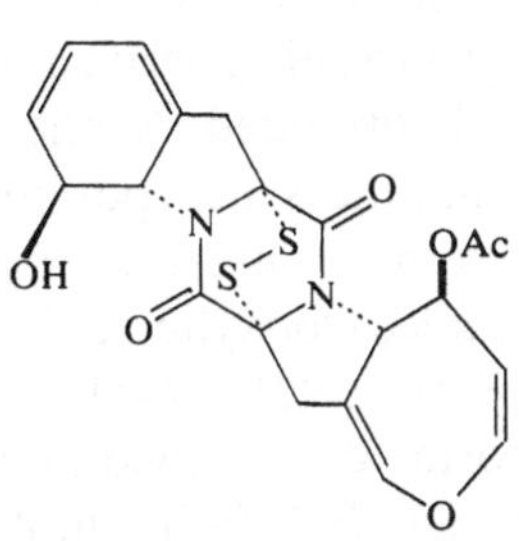

(133)

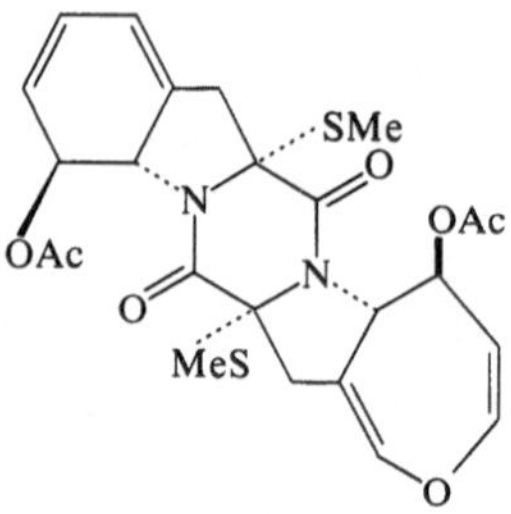

(134)

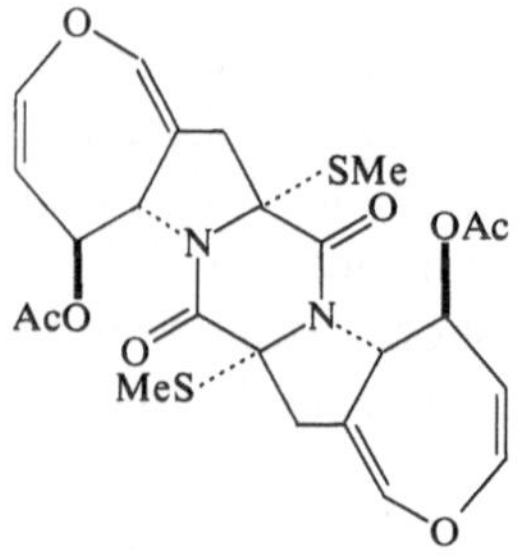

(135)

firmed by an X-ray analysis in this latter work and, subsequently, an X-ray analysis on the thiomethyl derivative (135) has also been published (*188*). C. d. studies confirmed the absolute configuration assigned to this group and also established their identity to the disulphide configuration in gliotoxin. Raney nickel desulphurisation of acetylaranotin proceeded with retention of configuration, the product (136) having a c. d. curve similar to that of *cyclo*-L-pro-L-pro (*189*).

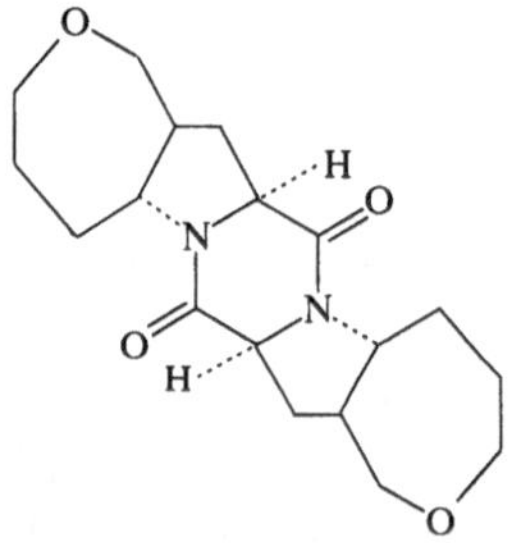

(136)

The striking structural resemblance of the aranotins to gliotoxin has led to considerable speculation on their biosynthesis. Labelling studies have demonstrated that serine or glycine (probably incorporated via serine) provides half of the dioxopiperazine ring in gliotoxin, the N-methyl group being derived from methionine (*190, 191*). The other part of the ring is provided by phenylalanine and not m-tyrosine (*192, 193*). D$_8$-Phenylalanine is incorporated into gliotoxin, in *Aspergillus terreus*, and into acetylaranotin in *Arachniotus aureus* (*194*). Earlier work had shown that the indolecarboxylic acid moiety of gliotoxin was not derived from tryptophan or acetate (*195*). As explained above, the earlier suggestion that m-tyrosine can be incorporated into gliotoxin (*182*) has been shown to be incorrect (*192, 193*). Similarly, a report that dehydrogliotoxin (122) can act as a precursor of gliotoxin has also been contradicted (*193*).

Three major questions arise on the further details of the biosynthetic processes involved. They are, how does the dihydro-aromatic ring form, how does the sulphur become incorporated, and what is the order of these processes? The most likely answer to the first question was provided by NEUSS et al. (128), who invoked the intermediacy of benzene oxides (196). This system is in equilibrium with the oxepin isomer (Scheme 13). Nucleophilic attack by the dioxopiperazine amide group would produce a substituted cyclohexadienol, of the type found in gliotoxin. Alternatively, further oxidation of the oxepin, followed by a similar nucleophilic process yields an aranotin-type system. Labelling experiments are fully consistent with this scheme. It is interesting to note that *Aspergillus terreus* produces both gliotoxin and acetylaranotin (97).

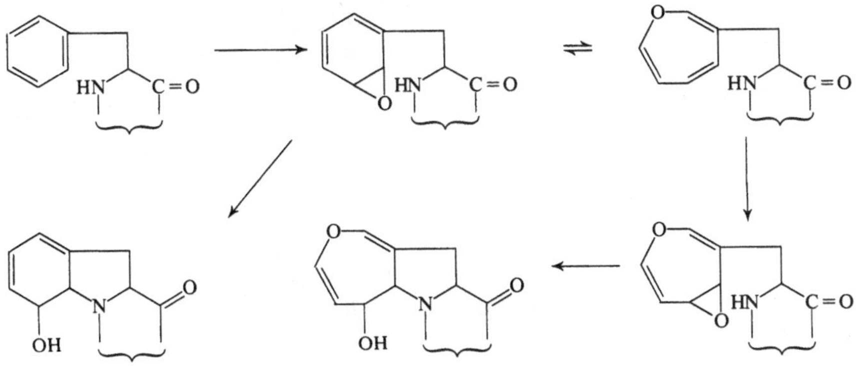

gliotoxin type   aranotin type

Scheme 13. Biogenesis of dihydroaromatic ring

The problem of sulphur incorporation into these compounds has also attracted attention. It is most probable that sulphur is added to the dioxopiperazine ring system after, or during, oxidation of the phenyl ring. The labelling experiments, involving perdeuterated phenylalanine, indicate that addition of sulphur to dehydropeptides of the type (129) does not occur (194), since the benzylic hydrogens are not exchanged during biosynthesis. In an oxidative environment, however, it is likely that both aromatic hydroxylation and dioxopiperazine oxidation could occur, the latter giving rise to ions (or the corresponding radicals) of the type (137), which can trap a sulphur nucleophile before loss of a proton to give a dehydrocyclodipeptide (198). Because of the potential usefulness of bridged epidithiodioxopiperazines much effort has recently been made to develop *in vitro* methods for the synthesis of these systems. Two strategies have become apparent in these synthetic attempts. In the first, attempts to preform the disulphide bond in a dipeptide precursor have

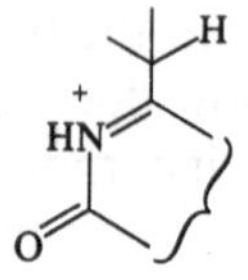

(137)

been followed by formation of the second amide bond. In the second route, introduction of sulphur into the preformed dioxopiperazine ring and its derivatives has been tried (Scheme 14).

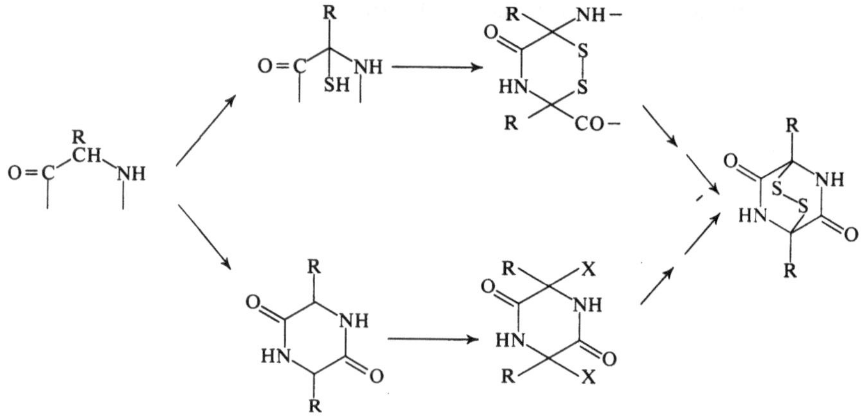

Scheme 14. Synthesis of epidithiodioxopiperazines

To date, the former route has been the most difficult and least success-ful. POJER and RAE prepared the precursor (138), (199), via chlorination of the azlactone with sulphuryl chloride, followed by displacement of the resulting chloride (139) with thioacetic acid to give the ester (140), then mild acid hydrolysis and oxidation with ferric chloride. Intermediates of the type (140) were also prepared by STEGLICH (200), for example by

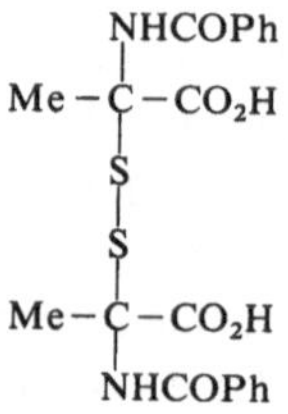

(138)

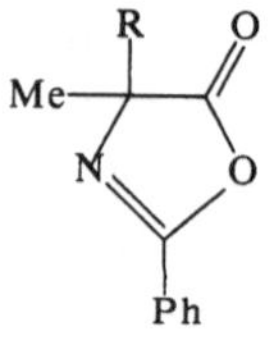

(139) R=Cl
(140) R=SAc

addition of mercaptans to the oxazolinone (141), and this method has been extended (201). α-Mercapto-α-amidoacids have been made by reaction of α-chloro-α-amido acids with hydrogen sulphide (202), but they are difficult compounds to work with. By analogy with the reported dimerisation of N-phenyl-pyruvamide into the dioxopiperazine (142), (203), an attempt to dimerize related 2-thiopyruvamides has been made. Treatment of the derived disulphide (143) with triethylamine, for example, did induce cyclisation to the monocyclic system (144), but further bridging did not occur (204).

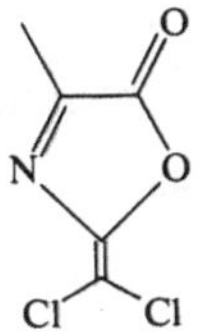

(141) 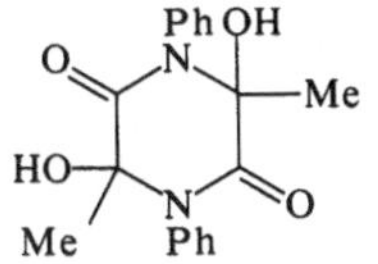   (142) 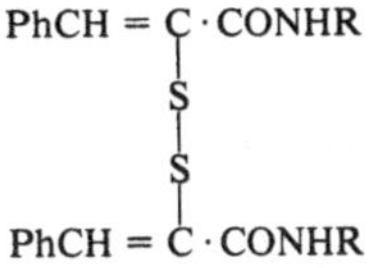   (143)

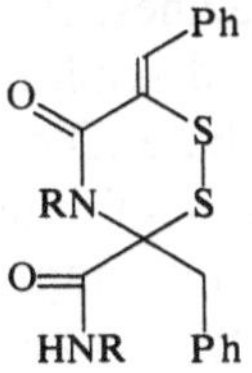

(144) 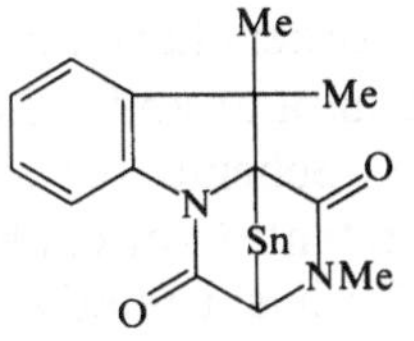   (145)

Partial success in the preparation of anhydrogliotoxin analogues cf. (145) has been achieved by WITKOP and his collaborators (205). The key reaction in this approach was that between the indolenine (146) and the chloride (147) (Scheme 15). The intermediate (148) arising from this route has yet to be cyclized to the cyclodipeptide.

More success has attended routes based on the introduction of sulphur after cyclodipeptide formation. TROWN brominated sarcosine anhydride to obtain the dibromide (149), (179). Subsequent reaction with potassium thiolacetate, followed by hydrolysis and oxidation, gave the disulphide (126). Unfortunately this method is not general, dehydrobromination

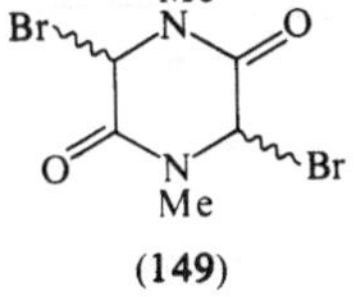

(149)

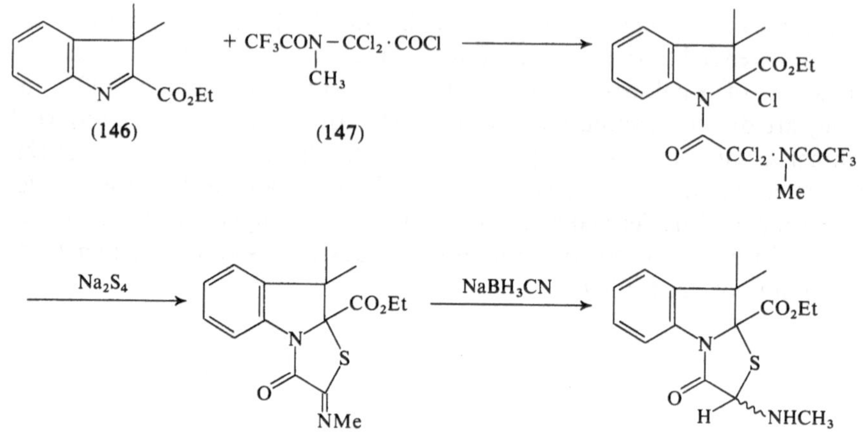

Scheme 15

occurring with 3,6-disubstituted dioxopiperazines. In an alternative method Schmidt and co-workers have introduced sulphur into dioxopiperazines by reaction with sulphur and a strong base (*206, 207*). The sulphur
is introduced stepwise (Scheme 16) and it is of interest that *cyclo*-L-
pro-L-pro reacts with retention of configuration, implying that the intermediate carbanion (**150**) has a relatively long lifetime. Strong bases are
needed in this method but these may be avoided by using activated
dioxopiperazines such as the ester (**151**), (*208*). In the latter case
formation of the bis-anion, with sodium hydride in dioxan, followed
by reaction with sulphur monochloride gave the disulphide (**152**) in
17% yield.

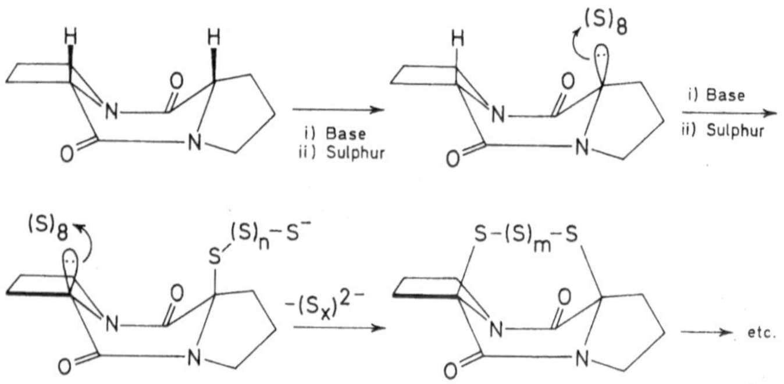

Scheme 16

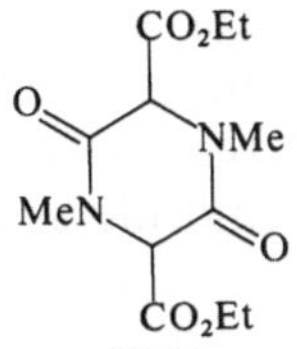

(151) 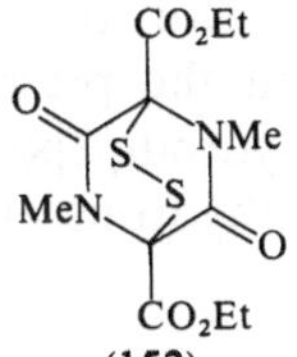   (152)

A versatile method for activating the dioxopiperazine system has recently been developed. Lead tetraacetate oxidation of cyclo-L-pro-L-pro affords the diacetoxy derivative (153), which reacts with thiols, in the presence of mild Lewis acids, to give the derivative (154), (209). One method for converting these derivatives into bridged disulphides is to hydrolyse the acetate (153), which gives the *cis*-diol (155), followed by Lewis acid-catalyzed addition of hydrogen sulphide and oxidation, to give the disulphide (156). In an alternative method (210) the derivative (157) was oxidized to the sulphone (158) followed by reaction with sodium tetrasulphide to give the bridged sulphide (159).

Although biosynthetic studies (194) rule out the possibility of dehydropeptides as intermediates in the synthesis of gliotoxin, protonation of the β-carbon in such systems would also produce ions of the type (137), and

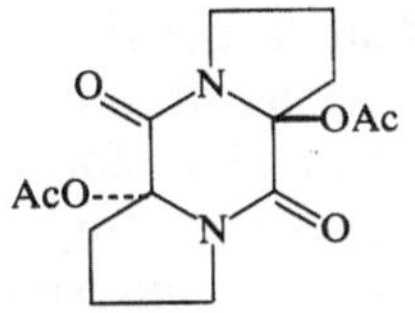

(153)

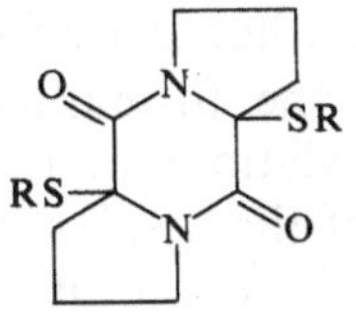

(154) R = H
(161) R = Et

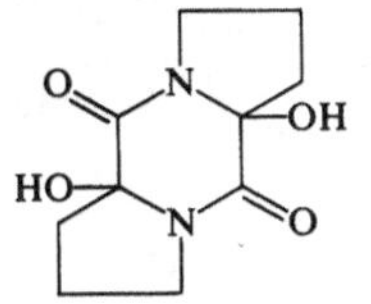

(155) 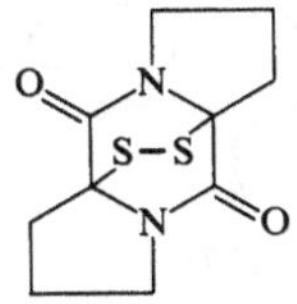   (156)

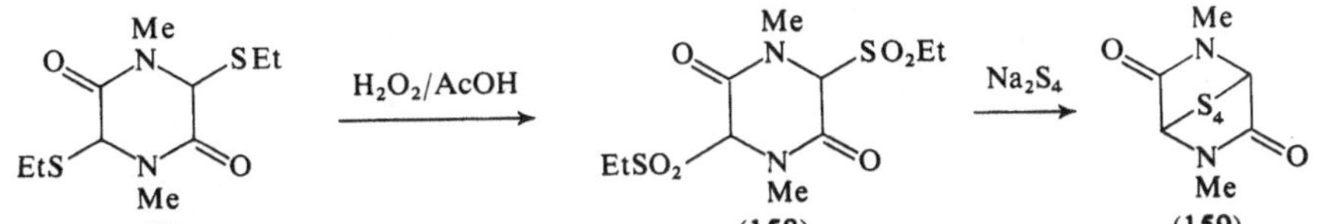

(157)   (158)   (159)

hence another route for the introduction of sulphur. This process has been achieved in the presence of catalytic amounts of acid. For example, the bisdehydrodipeptide (160) reacted with ethyl mercaptan, in the presence of anhydrous hydrogen chloride as catalyst, to give the adduct (161), (198).

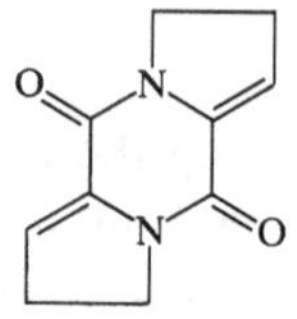

(160)

An ingenious method for the protection of the disulphide function of the bridged epidithiodioxopiperazines has been devised (211) and brilliantly exploited in a synthesis of both dehydrogliotoxin (122), (212) and sporidesmin (121), (213). Protection of the dithiol, (e. g. 127) with anisole (see Scheme 17) gives the acetal (162). This type of intermediate is useful because the bridgehead positions can form anions with butyl lithium and can subsequently be alkylated. Since the acetal bridge is achiral, one of the sulphur atoms being disposed near one of the carbonyl groups and the other away from the second carbonyl group, selectivity is observed in the formation of these anions at positions 3 and 6. Subsequent alkylations at these positions are therefore also selective. Re-formation of the disulphide bridge to produce compound (126) was readily effected by consecutive oxidation to the monosulphoxide (163), followed by treatment with a protic or Lewis acid.

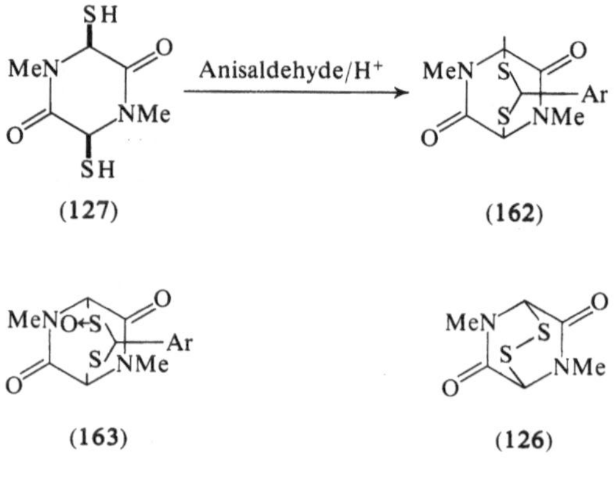

(127)　　　　　　　　　　　　　　(162)

(163)　　　　　　　　　　　　　　(126)

Scheme 17

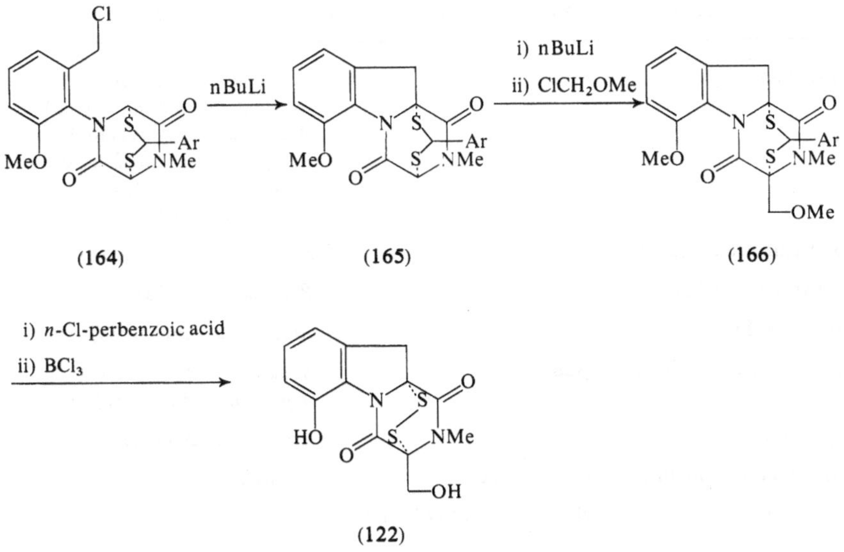

(164)   (165)   (166)

(122)

Scheme 18. Synthesis of dehydrogliotoxin

This scheme was immediately applied to the preparation of dehydro-gliotoxin (Scheme 18). The precursor (164) was treated with base to give the indoline derivative (165), that was further methoxymethylated to the compound (166). Subsequent treatment with m-chloroperbenzoic acid, followed by reaction with boron trichloride both regenerated the disulphide bridge and cleaved the methyl protecting groups to produce dehydrogliotoxin (122).

The sporidesmins are a group of dioxopiperazines based on *cyclo*-try-ala which are related in structure to the gliotoxin type (*171*). The first member of this group to be isolated, sporidesmin (121), (*214, 215*), was obtained as the active principle causing facial eczema in sheep and was obtained from the fungus *Pithomyces chartarum* which occurs in certain pasture areas of New Zealand. The related sporidesmins B to G (Table 5) have all been isolated from the same fungus.

Table 5. *Some Naturally Occurring Epipolythiodioxopiperazines**

Gliotoxin (**83**)	*Trichoderma viride (165)*
	Gliocladium fimbriatum (166)
	Aspergillus fumigatus (167)
	Penicillium terlikowski (168)
	Aspergillus terreus (197)
	Penicillium sp. (226)
Gliotoxin acetate (**207**)	*P. terlikowski (226, 168)*
Dehydrogliotoxin (**122**)	*P. terlikowski (173)*
Aranotin (**132**)	*Arachniotus aureus (186)*
Aranotin acetate (LLS 88 α) (**128**)	*A. aureus (186)*
	Aspergillus terreus (187)
Apoaranotin (**133**)	*Arachniotus aureus (128)*
Bisdethiodi(methylthio)acetylapoaranotin	*A. aureus (128)*
Bisdethiodi(methylthio)acetylaranotin (LLS 88 β)(**135**)	*A. aureus (128)*
	Aspergillus terreus (187)
Sporidesmin (**121**)	*Pithomyces chartarum (214)*

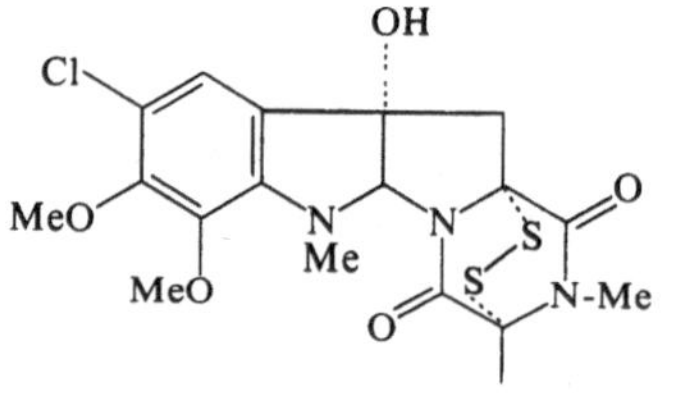

Sporidesmin B (**208**)

Sporidesmin C (**172**)	*P. chartarum (221)*
Sporidesmin D (**168**)	*P. chartarum (216)*
Sporidesmin E (**167**)	*P. chartarum (255)*

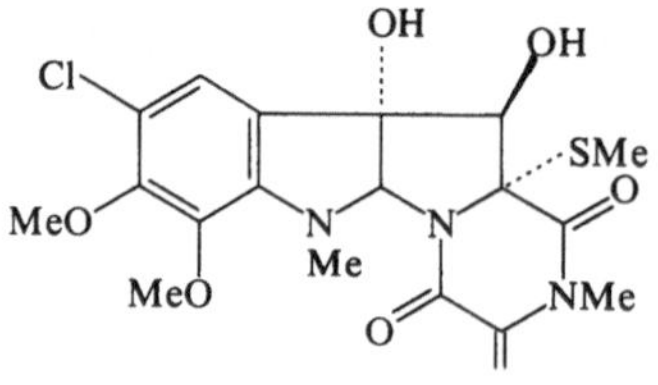

Sporidesmin F (**209**)

P. chartarum (216)

P. chartarum (214)

---

* For a detailed discussion on the occurrence of the sporidesmins and related compounds see (*256*).

Table 5 (continued)

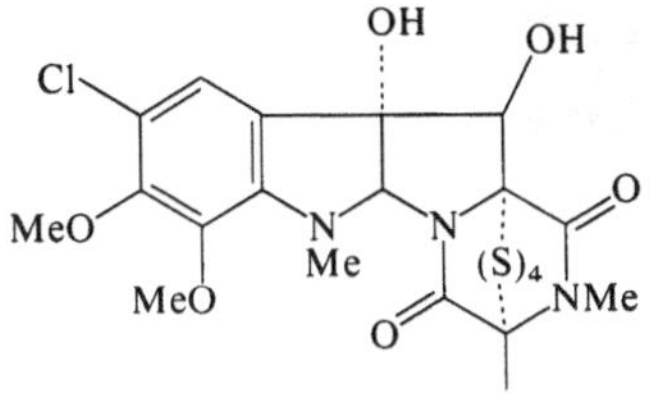

P. chartarum (218)

Sporidesmin G (169)

Chaetocin (124)	Chaetomium minutum (229)
Verticillin A (179)	Verticillium sp. (227, 228)
Chetomin (182)	Chaetomium cochliodes (231)
Dihydroxychaetocin (181)	Verticillium tenerum (230)
Verticillin B (180)	Verticillium sp. (228)
Verticillin C (210)	Verticillium sp. (228)

The structure of sporidesmin (121) has been largely deduced by spectroscopic methods, including mass spectrometry. An X-ray analysis on sporidesmin (175) confirmed the structure and subsequently revealed its absolute stereochemistry (173), identical to that predicted by circular dichroism studies, in which the disulphide function has the same configuration as that in gliotoxin (161, 175). Some chemical correlations in this series have been achieved. Thus sporidesmin E (167) can be reduced by one equivalent of triphenylphosphine, which abstracts only one sulphur atom, shown to be the central one, to give sporidesmin (174). Reduction of the disulphide bond and selective alkylation affords sporidesmin D (168, 216). The tetrasulphide, sporidesmin G (169), the structure of which has also been determined by an X-ray analysis (217), can be prepared from either sporidesmin or sporidesmin E by reaction with hydrogen polysulphide in chloroform (218). The use of dihydrogen disulphide has also been advocated (219). The trisulphide, sporidesmin E (167) also shows antiviral activity, whilst the tetrasulphide (169) has only low activity, and this possibly due to its in vivo conversion to the lower sulphides. From the n.m.r. spectrum of the trisulphide there is some evidence that the sulphur bridge can exist in two relatively stable conformations (170) and (171), (174). One of the two conformers reacts faster with the thiophile triphenylphosphine than the other. The chemistry of polysulphide systems has been reviewed (220).

The most unusual member of the sporidesmins is sporidesmin C (172), (221), in which the sulphur chain is attached to only one end of

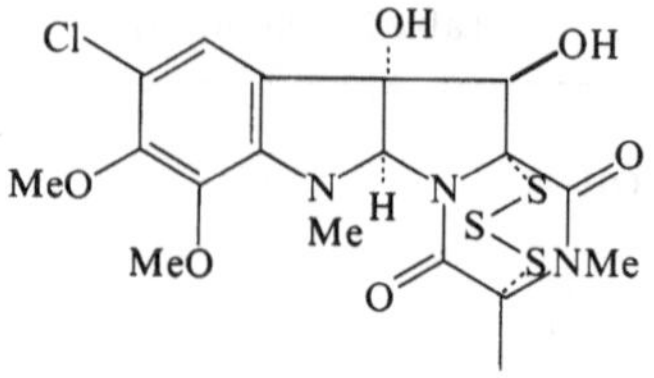

(167)

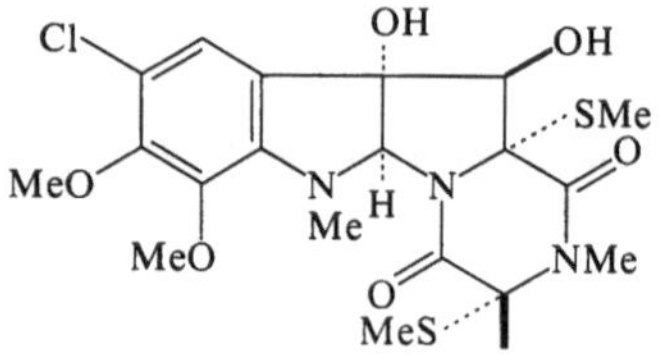

(168)

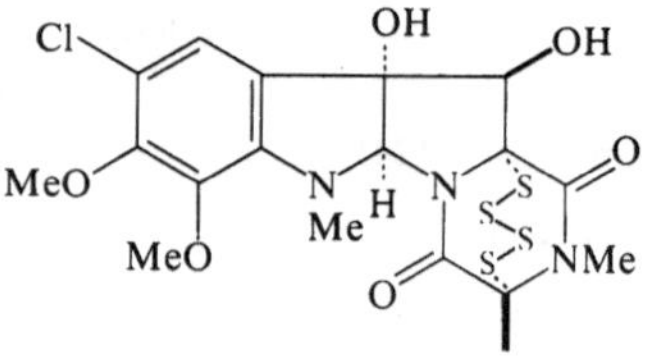

(169)

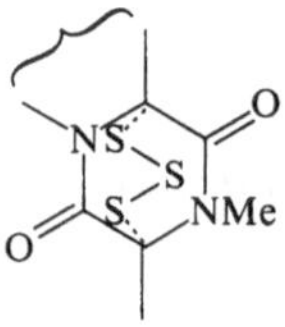

(170)

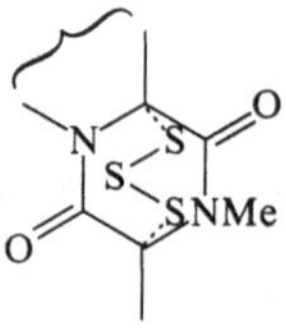

(171)

the dioxopiperazine system. The structure of sporidesmin C was largely deduced on the basis of mass spectral evidence on the derived acetate (173), produced by heating the natural alcohol in acetic anhydride-

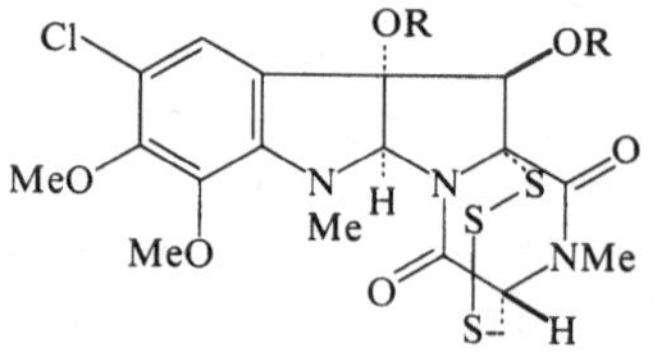

(172) R = H
(173) R = Ac

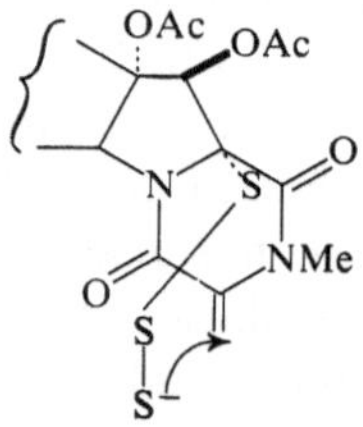

(174)

pyridine for 30 minutes. Since the proposed structure for sporidesmin C is exceptional, and since heating the normal sporidesmins with pyridine is known to effect the elimination of sulphur, it is possible that sporidesmin C diacetate is an artefact, produced by cleavage of a normal epipolythiodioxopiperazine system to give a species of the type (**174**). Re-addition across the dehydrodipeptide system, a known process (*198, 222*), would produce the observed structure for the diacetate (**173**).

Relatively little work has been carried out so far on the biosynthesis of the sporidesmins. The oxidative ring closure of tryptamine derivatives to produce 2,3-dihydropyrrolo[2,3-b]indoles, a system common to these compounds, is well exemplified (*223, 224*). One, of several possible biosynthetic routes to sporidesmin, is outlined in the scheme (Scheme 19). The addition of singlet oxygen to related diene systems has been described (*225*). Although initial details of one approach to the synthesis of these compounds have been described (*226*), a total synthesis of

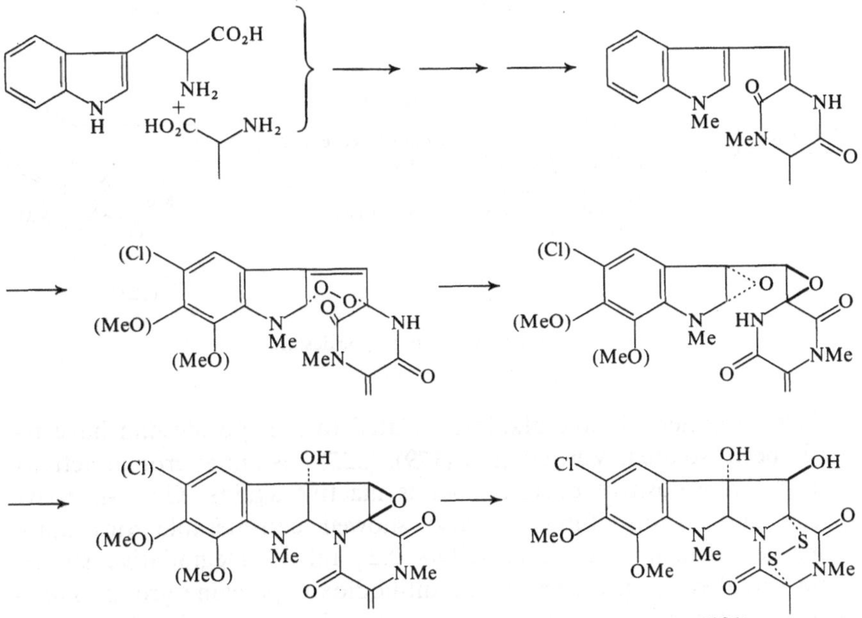

Scheme 19. Possible biosynthesis of sporidesmin

sporidesmin has also been reported (*213*). The method was similar to that used in the preparation of dehydrogliotoxin (*212*). The dithioacetal (**175**) (Scheme 20) was prepared from the dehydro-precursor (**176**) by a method involving an acid catalyzed addition of hydrogen sulphide. Ensuing alkylation, reduction and acetylation gave the indole (**177**) that was oxidized with iodosobenzene diacetate to give the cyclized derivative (**178**) and which was hydrolyzed to sporidesmin (**121**).

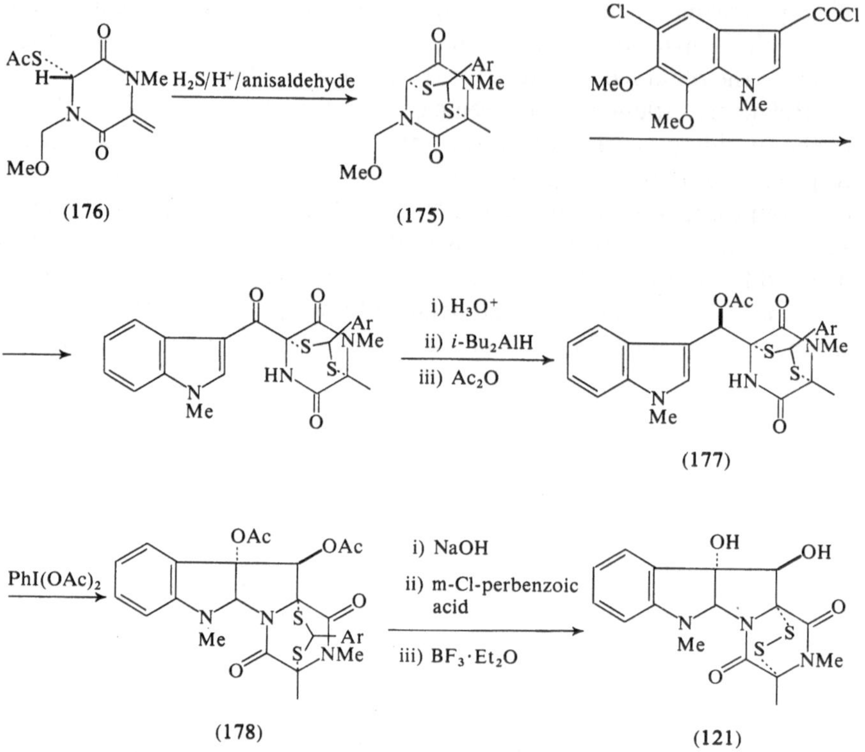

(176)   (175)

(177)

(178)   (121)

Scheme 20. Synthesis of sporidesmin

Several dimeric indole alkaloids related to the sporidesmin have recently been isolated. Verticillin A (**179**), (*227*) has antimicrobial activity against Gram-positive bacteria, but is inactive against Gram-negative species and against fungi. It also showed considerable anti-tumor properties against HeLa cells (ED_{50} 0.2γ ml^{-1}). Degradative studies again revealed the presence of the dithiodioxopiperazine group. Treatment with potassium hydroxide in aqueous dioxan at reflux for several hours gave over 50% of bi-indol-3-yl, indicative of a dimeric structure.

Verticillin A was accompanied by the related verticillins B (**180**) and C, the latter thought to be a trisulphide analogue of verticillin B (*228*). It is not known in what sequence the hydroxyl groups are incorporated into verticillin B.

A closely related compound, chaetocin (**124**) has been isolated from *Chaetomium minutum* and was assigned its structure on the basis of degradative studies (*229*). These were later confirmed by an X-ray analysis which also confirmed the absolute configuration of the molecule (*176*). C. d. studies had already shown that both this compound and verticillin A

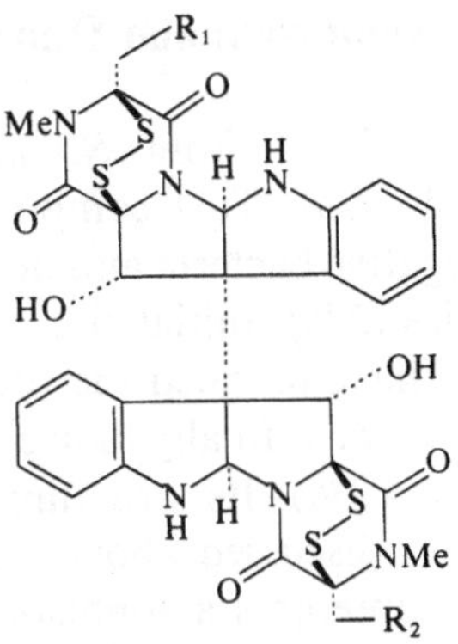

(179) $R_1 = R_2 = H$

(180) $R_1 = H, R_2 = OH$

(181) $R_1 = R_2 = OH$

are antidopal compared to the sporidesmins. It is interesting to note. therefore, that chaetocin does not show the antiviral activity associated with the sporidesmins and gliotoxins. Recently $11\alpha,11\alpha'$-dehydroxy-chaetocin (181) has also been discovered (230).

Chetomin was originally isolated in 1944 from a strain of *Chaetomium cochliodes* (231). A tentative structure (182) has been proposed for this compound (232). Since the material does possess antiviral activity it will be interesting to discover if its absolute configuration is of the expected gliotoxin type. Other compounds which appear to be related to the sporidesmins are oryzachloride (233), and the melinacidins (234), but the detailed structures of these compounds remain unravelled.

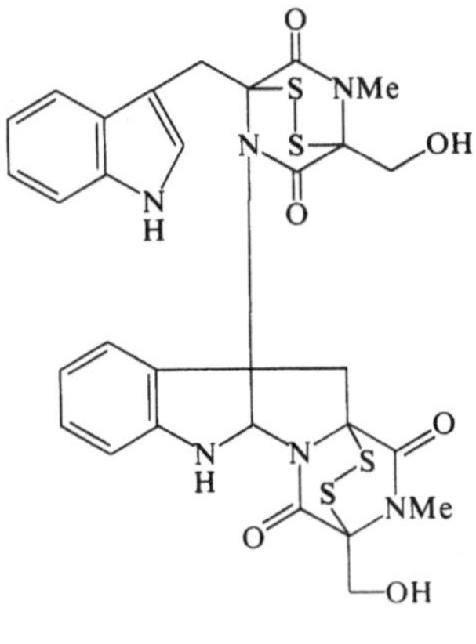

(182)

VIII. Bicyclomycin and Dibromophakellin

A new antibiotic isolated from *Streptomyces sapporonensis* has recently been described (*235*). This compound, bicyclomycin (**183**), is active against Gram-negative bacteria and does not cause cross resistance to the normal antibiotics (*236*). Initial clinical trials appear encouraging (*237*). Chemical studies have revealed the presence of a dioxopiperazine ring, but the structure was finally solved by means of an X-ray crystallographic analysis (*236*). Its structure is unrelated to any of the dioxopiperazines systems discussed above and much work needs to be carried out to see if it represents a sophisticated, highly functionalized derivative of *cyclo*-leu-isoleu, upon which the carbon skeleton appears to be based.

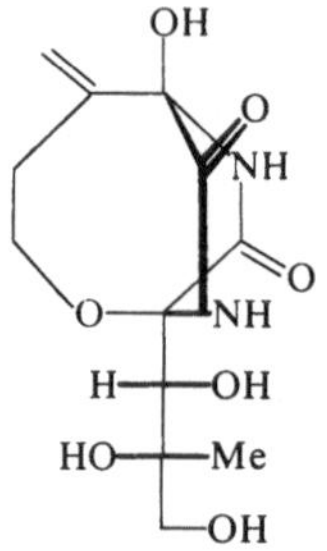

(183) 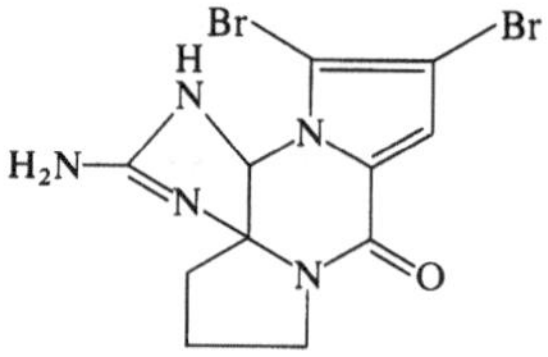   (184)

The dibrominated substance dibromophakellin (**184**) has been isolated from the marine sponge, *Phakellia flabellata* (*238*). It is interesting to speculate that this might have arisen from a *cyclo*-pro-pro precursor.

References:

1. FISCHER, E., and K. RASKE: Beitrag zur Stereochemie der 2,5-Diketopiperazine. Ber. **39**, 3981 (1906).
2. FISCHER, E.: Untersuchungen über Aminosäuren, Polypeptide und Proteine. Ber. **39**, 530 (1906).
3. CURTIUS, T., and GOEBEL: J. prakt. Chem. **37**, 173 (1888).
4. ABDERHALDEN, E., and E. KOMM: The Formation of Diketopiperazines from Polypeptides Under Various Conditions. Z. physiol. Chem. **139**, 147 (1924).
5. ABDERHALDEN, E., and R. HAAS: Further Studies on the Structure of Proteins: Studies on the Physical and Chemical Properties of 2,5-Diketopiperazines. Z. physiol. Chem. **151**, 114 (1926).
6. KATCHALSKI, E., I. GROSSFIELD, and F. M. FRANKEL: Synthesis of Lysine Anhydride. J. Am. Chem. Soc. **68**, 879 (1946).
7. KOPPLE, K. D., and D. H. MARR: Conformations of Cyclic Dipeptides: The Folding of Cyclic Dipeptides containing an Aromatic Side Chain. J. Am. Chem. Soc. **89**, 6193 (1967).
8. GREENFIELD, N. J., and G. D. FASMAN: Optical Activity of Simple Cyclic Amides in Solution. Biopolymers **7**, 595 (1969).
9. BALASUBRAMANIAN, D., and D. B. WETLAUFER: Optical Rotatory Properties of Diketopiperazines. J. Am. Chem. Soc. **88**, 3449 (1966).
10. SCHELLMAN, J. A., and B. E. NIELSON: In "Conformations of Biopolymers", ed. G. N. RAMACHANDRAN, vol. 1, p. 109. New York: Academic Press. 1967.
11. EDELHOCH, H., R. E. LIPPOLDT, and M. WILCHECK: The Circular Dichroism of Tyrosyl and Tryptophanyl Diketopiperazines. J. Biol. Chem. **243**, 4799 (1968).
12. BELLAMY, L. J.: The Infrared Spectra of Complex Molecules. Chapter 12. London: Methuen and Co. Ltd. 1957.
13. BLÁHA, K., J. SMOLÍKOVÁ, and A. VÍTEK: Aminoacids and Peptides, LXIV. Infrared Spectra of Substituted 2,5-Piperazinediones and the Detection of *cis*-Peptide Bonds in Diastereoisomeric Cyclohexapeptides. Coll. Czech. Chem. Commun. **31**, 4296 (1966).
14. JANKOWSKY, K., and L. VARFALVY: Mass Spectroscopy of 2,5-Dioxopiperazines. II. A Study of *cyclo*-Ala-ala. Bull. Acad. Pol. Sci. Ser. Sci. Chim. **20**, 493 (1966).
15. SVEC, H. J., and G. A. JUNE: The Mass Spectra of Dipeptides. J. Am. Chem. Soc. **86**, 2278 (1964).
16. NAGARAJAN, R., J. L. OCCOLOWITZ, N. NEUSS, and S. M. NASH: Mass Spectra of Diketopiperazines from Aranotin and Related Metabolites. Chem. Commun. **1969**, 359.
17. (a) ROMANET, R., A. CHEMIZARD, S. DUHOUX, and S. DAVID: Etudes par resonance magnetique nucleaire de l'echinuline, de certains derives et de modeles indoliques. Bull. soc. Chim. France 1048 (1963).
    (b) CHEMIZARD, A., and S. DAVID: Remarques sur les dioxo-2,5-piperazines. Bull. soc. Chim. France **1966**, 184.
18. KOPPLE, K. D., and M. OHNISHI: Conformations of Cyclic Peptides II: Side-chain Conformation and Ring Shape in Cyclic Dipeptides. J. Am. Chem. Soc. **91**, 962 (1969).
19. YOUNG, P. E., V. MADISON, and E. R. BLOUT: Cyclic Peptides VI. Europium Assisted N. M. R. Study of the Solution Conformation of *cyclo* (L-pro-L-pro) and *cyclo* (L-pro-D-pro). J. Am. Chem. Soc. **95**, 6142 (1973).
20. KREZCAREK, G. E., B. W. DOMINY, and R. G. LAWTON: The Interaction of Reactive Functional Groups along Peptide Chains: A Model for Alkaloid Biosynthesis. Chem. Commun. **1968**, 1450.
21. FISCHER, E.: Syntheses von Polypeptiden, XV. Ber. **39**, 2893 (1906).
22. BLÁHA, K.: Amino-acids and Peptides, XCV. Synthesis of Some Diastereoisomeric 2,5-Piperazine-diones. Coll. Czech. Chem. Commun. **34**, 4000 (1969).

23. Nitecki, D. E., B. Halpern, and J. W. Westley: A Simple Route to Sterically Pure Diketopiperazines. J. Org. Chem. **33**, 864 (1968).

24. Schott, H. F., J. B. Larkin, L. B. Rockland, and M. S. Dunn: The Synthesis of 1 (−) Leucylglycylglycine. J. Org. Chem. **12**, 490 (1947).

25. Rosenmund, P., and K. Kains: Diketopiperazines from Leuchs' Anhydrides. Angew. Chem. Internat. Edn. **9**, 162 (1970).

26. Kopple, K. D., and H. G. Ghazarian: A Convenient Synthesis of 2,5-Dioxopiperazines. J. Org. Chem. **33**, 862 (1968).

27. Lichtenstein, N.: The Behaviour of Dipeptides when Heated in β-Naphthol. J. Am. Chem. Soc. **60**, 560 (1938).

28. Zahn, H., and D. Brandenburg: Synthese einer geschützten Heptapeptidsequenz aus dem Tyrocidin B. Annalen **692**, 220 (1966).

29. Grahl-Nielson, O.: Acid Hydrolysis of Diastereoisomeric Dioxopiperazines. Tetrahedron Letters **1969**, 2827.

30. Slater, G. P.: Synthesis of Piperazine-2,5-diones. Chem. & Ind. (London) **1969**, 1092.

31. Goodman, M., and K. C. Steuben: Peptide Synthesis via Aminoacid Esters, II. Some Abnormal Reactions during Peptide Synthesis. J. Am. Chem. Soc. **84**, 1279 (1962).

32. Westley, J. W., V. A. Close, D. E. Nitecki, and B. Halpern: Determination of Steric Purity and Configuration of Diketopiperazines by Gas-liquid Chromatography, Thin-layer chromatography and Nuclear Magnetic Resonance Spectrometry. Anal. Chem. **40**, 1888 (1968).

33. Tsuboi, M., T. Shimanouchi, and S. Miyushima: Near Infrared Spectra of Compounds with Two Peptide Bonds and the Configuration of a Polypeptide Chain, VII. On the Extended Form of Polypeptide Chains. J. Am. Chem. Soc. **81**, 1406 (1959).

34. Indelicato, T. M., T. T. Norvilas, and W. J. Wheeler: Intramolecular Nucleophilic Attack in 7 (α-Amino) phenylcephalosporanic Esters. J. C. S. Chem. Commun. **1972**, 1162.

35. Roets, E. R., A. J. Vlietnieck, G. A. Janssen, and H. Vanderhaeghe: Intramolecular Nucleophilic Attack in 6-Epiampicillin. J. C. S. Chem. Commun. 484 (1973).

36. Titlestad, K.: Cleavage of Linear Tetrapeptides into Two Cyclic Dipeptides. Chem. Commun. **1971**, 1527.

37. Lucente, G., and P. Frattesi: Cyclisation of Activated Tosyl-peptides. Tetrahedron Letters **1972**, 4283.

38. Mauger, A. B.: Degradation of Peptides to Diketopiperazines: Applications of Pyrolysis-Gas Chromatography to Sequence Determination in Actinomycins. Chem. Commun. **1971**, 39.

39. Poisel, H., and U. Schmidt: Asymmetrische Induktion bis Reaktionen von Aminosäuren und Peptiden, I. Asymmetrische Synthese Aromatische α-Aminosäuren und N-Methyl α-Aminosäuren. Synthese von L-DOPA. Über die Katalytische Hydrierung ungesättigter Cyclopeptide. Chem. Ber. **106**, 3408 (1973).

40. Degeilh, R., and R. E. Marsh: A Refinement of the Crystal Structure of Diketopiperazine (2,5-Piperazine-dione). Acta Crystallog. **12**, 1007 (1959).

41. Benedetti, E., P. Corradini, and C. Pedone: Crystal and Molecular Structure of L-cis-3,6-Dimethyl-2,5-piperazinedione (L-alanyl-L-alanine-2,5-diketopiperazine). Biopolymers **7**, 751 (1969).

42. — — — The Crystal and Molecular Structure of trans-3,6-Dimethyl-2,5-piperazinedione. J. Phys. Chem. **73**, 2891 (1969).

43. Sletten, E.: Conformation of Cyclic Dipeptides: The Crystal and Molecular Structures of Cyclo-D-alanyl-L-alanyl and cyclo-L-Alanyl-L-alanyl. J. Am. Chem. Soc. **92**, 172 (1970).

44. Benedetti, E., P. Corradini, M. Goodman, and C. Pedone: Flexibility of Supposed "Rigid" Molecules: Substituted 2,5-Piperazinediones (Diketopiperazines). Proc. Natn. Acad. Sci. USA, **62**, 650 (1969).

45. GROTH, P.: Crystal Structure of N,N'-Dimethyldioxopiperazine. Acta Chem. Scand. 23, 3155 (1969).

46. KARLE, I. L.: Crystal Structure and Conformation of the Cyclic Dipeptide, cyclo-L-Prolyl-L-leucyl. J. Am. Chem. Soc. 94, 81 (1972).

47. SIEMION, I. Z.: Die Konformation des Prolin Ringes in Diketopiperazin Systemen. Annalen 748, 88 (1971).

48. — NMR Investigation of Proline Containing Dioxopiperazines. Org. Magn. Resonance 3, 545 (1971).

49. BLÁHA, K., M. BODESINSKY, I. FRIC, J. SMOLIKOVA, and J. VICAR: Cyclodipeptides. Conformational Analysis and Spectroscopic Studies. Tetrahedron Letters 1972, 4437.

50. GAWNE, G., G. W. KENNER, N. H. ROGERS, R. C. SHEPPARD, and K. TITLESTAD: In "Peptides", ed. E. BRICAS, p. 28. Amsterdam: North-Holland Publishing Co. 1968.

51. ZIAUDDIN, K. D. KOPPLE, and C. A. BUSH: Conformations of cyclo-L-His-L-Ser, cyclo-L-His-L-Asp, and cyclo-L-His-L-His. Tetrahedron Letters 1972, 483.

52. cf. JOHNSON, C. E., and F. A. BOVEY: Calculation of Nuclear Magnetic Resonance Spectra of Aromatic Hydrocarbons. J. Chem. Phys. 29, 1012 (1958).

53. ZIAUDDIN, and K. D. KOPPLE: Conformations of Folded Peptides: Stabilities of Folded Conformations of para-substituted 3-Benzylpiperazine-2,5-diones. J. Org. Chem. 35, 253 (1970).

54. CAILLET, J., B. PULLMANN, and B. MAIGRET: Molecular Orbital Calculations on the Folding of Cyclic Dipeptides with Aromatic and Aliphatic Side Chains. Biopolymers 10, 221 (1971).

55. KOPPLE, K. D., R. R. JARABAK, and P. L. MULLER: Reactivity of Cyclic Peptides. III. Reaction of Isomeric Histidine, Tyrosine Peptides with p-Nitrophenyl Acetate. Biochem. 2, 958 (1963).

56. ZBIRAL, E., E. L. MENARD, and J. M. MULLER: über die Inhaltsstoffe von Zizyphus oenoplia Mill. II. Zur Konstitutionsmittlung des Zizyphins. Helv. Chim. Acta 48, 1608 (1965).

57. PAILER, M., E. HASLINGER, and E. ZBIRAL: Zur Konstitution des Zizyphinins von Zizyphus oenoplia Mill. Monatsh. Chem. 100, 1608 (1968).

58. WARNHOFF, E. W.: Peptide Alakoids. Fortschr. Chem. Organ. Naturstoffe 28, 162 (1970).

59. BODANSKY, M., G. F. SINGLER, and A. BODANSKY: Structure of the Peptide Antibiotic Amphomycin. J. Am. Chem. Soc. 95, 2352 (1973).

60. ARISON, B. H., and J. L. BECK: The Structure of Compound 593 A, A New Anti-tumor Agent. Tetrahedron 29, 2743 (1973).

61. FORSTER. M. O., and W. B. SAVILLE: Isolation of Picroroccellin from Rocella fuciformis. J. Chem. Soc. 121, 816 (1922).

62. ATKINS, C. L., and J. B. NEILANDS: Rhodotorulic Acid. A Diketopiperazine Dihydroxamic Acid with Growth Regulatory Properties. I. Isolation and Characterization. Biochem. 7, 3734 (1968).

63. KELLER-SCHIERLEIN, W., V. PRELOG, and H. ZAHNER: Siderochrome. Fortschr. Chem. Organ. Naturstoffe 22, 279 (1964).

64. AKERS, H. A., M. LLINAS, and J. B. NEILANDS: Protonated Amino Acid Studies on Rhodotorulic Acid Biosynthesis in D_2O Media. Biochem. 11, 2283 (1972).

65. ISOWA, Y. T., TAKASHIMA, M. OHMORI, H. KURITA, M. SATO and K. MORI, Synthesis of Rhodotorulic Acid. Bull Chem. Soc. Japan 45, 1467 (1972).

66. DIEKMANN, H.: Metabolic Products of Microorganism. Part 81. Occurrence and Structure of Coprogen B and Dimerumic Acid. Arch. Mikrobiol. 73, 65 (1970).

67. KELLER-SCHIERLEIN, W., and H. DIEKMANN: Zur Konstitution des Coprogens. Helv. Chim. Acta 53, 2035 (1970).

68. HEDY, P. H., E. B. HODGE, V. V. YOUNG, R. L. HARRIED, G. A. BREWER, W. F.

Phillips, W. F. Runge, H. E. Stavely, A. Pohland, H. Boaz, and H. R. Sullivan: Structure and Reactions of Cycloserine. J. Am. Chem. Soc. **77**, 2345 (1955).

69. Karpeiskii, M. Yu., Yu. N. Breusov, R. M. Khomatov, E. S. Severin, and O. C. Polyanovskii: The Mechanism of Action of Cycloserine and Related Compounds with Aspartic-Glutamic Transaminase. Biokhimiya **28**, 342 (1963). Chem. Abs. **59**, 4219 f (1963).

70. Lassen, F. O., and C. H. Stammer: Cycloserine Dimer Hydrolysis and its Equilibration with Cycloserine. J. Org. Chem. **36**, 2631 (1971).

71. Miller, J. C., F. C. Neuhaus, F. O. Lassen, and C. H. Stammer: The Reactions of 3,6-Bis(aminoxymethyl)-2,5-piperazinedione with Acid and Alkali. A Kinetic Study. J. Org. Chem. **33**, 3908 (1968).

72. Poduska, K., G. S. Katrukha, A. B. Silaev, and J. Rudinger: Amino Acids and Peptides. LII. Intramolecular Aminolysis of Amide Bonds in Derivatives of $\alpha\gamma$-Diaminobutyric Acid, $\alpha\beta$-Diaminopropionic Acid, and Ornithine. Coll. Czech. Chem. Commun. **30**, 2410 (1965).

73. McKinney, J. D., and C. H. Stammer: Role of Azomethines in the Dimerisation of Cycloserine by Aldehydes. Tetrahedron **25**, 163 (1969).

74. Wrinch, D.: The Cyclol Theory in the "Globular" proteins. Nature **139**, 972 (1969).

75. — Chemical Aspects of Polypeptide Chain Structures and the Cyclol Theory. Copenhagen: Munksgaard. 1956.

76. Shemyakin, M. M., V. K. Antonov, A. M. Shkrob, V. I. Shchelekov, and Z. E. Agadzhanyan: Activation of the Amide Group by Acylation. Tetrahedron **21**, 3537 (1965).

77. Shemyakin, M. M., Y. A. Ovchimichov, V. K. Antonov, A. A. Kiryashkin, V. I. Ivanov, V. I. Shchelekov, A. M. Shkrob: Total Synthesis of Serratamolide, I. Synthesis of O,O'-Diacetyl Serratamolide. Tetrahedron Letters **1964**, 47.

78. Stoll, A.: Recent Investigations on Ergot Alkaloids. Fortschr. Chem. Organ. Naturstoffe **9**, 114 (1952).

79. Hofmann, A., A. J. Frey, and H. Ott: Die Totalsynthese des Ergotamins. Experientia **17**, 206 (1961).

80. Rothe, M., and R. Steinberger: Thiocyclols and Cyclothio-depsipeptides. Angew. Chem., Internat. Edn. **7**, 884 (1968).

81. Rothe, M., T. Tothe, and D. Jacob: Synthesis of an Azacyclol. Angew. Chem., Internat. Edn., **10**, 128 (1971).

82. Stoll, A., and Hofmann: The Ergot Alkaloids in "The Alkaloids", ed. R. F. Manske, Vol. **8**, p. 725. New York: Academic Press. 1965.

83. Leonard, N. J.: Transannular Nitrogen-Carbonyl Interactions. Record Chem. Progress **17**, 243 (1956).

84. Lucente, G., and A. Romeo: Synthesis of Cyclols from some small Peptides via Amide-Amide Reaction. Chem. Commun. **1971**, 1605.

85. Cerrini, S., W. Fedeli, and F. Mazza: X-Ray Crystallographic Proof of a Cyclol Structure in a Tripeptide. Chem. Commun. **1971**, 1607.

86. Machin, P. J., and P. G. Sammes: Unpublished work.

87. Simonson, L. A., and C. K. Mann: Anodically Induced 1,3-Cyclo addition of Acetonitrile to 3,6-Diisobutylpiperazine-2,5-dione. Tetrahedron Letters **1970**, 3303.

88. Blake, K. W., and P. G. Sammes: Geometrical Isomerism and Tautomerism of 3-Arylidene-6-methylpiperazine-2,5-diones. J. Chem. Soc. (C) 1970 (980).

89. Machin, P. J., A. E. A. Porter, and P. G. Sammes: Pyrazine Chemistry. Part V. Diels-Alder Reactions of Some 2,5-Dihydroxypyrazines. J. C. S., Perkin I, **1973**, 404.

90. Bergmann, M., and A. Miekeley: Neue Desmotrope Aminosäureanhydride von Piperazintypus. Zur Kenntnis des Abbau der Aminosäuren. Serine als Dehydrierungsmittel. Annalen **458**, 40 (1927).

91. Chakrabartty, S. K., and R. Levine: Chemistry of Pyrazine and its Derivatives.

XII. Reaction of Acetonylpyrazine with Phenyllithium in the Presence and Absence of Methyl Benzoate. J. Heterocyclic Chem. **4**, 109 (1967).

92. KHOKLOV, A. S., and G. B. LOSHKIN: The Structure of Albonoursin. Tetrahedron Letters **1963**, 1881.
93. SHIN, C., Y. CHIGERA, M. MASAKI, and A. OHTA: Total Synthesis of Albonoursin. Tetrahedron Letters **1967**, 4601.
94. GALLINA, C., and A. LIBERATORI: A New Synthesis of 1-Acetyl-3-Arylidene (alkylidene)piperazine-2,5-diones. Tetrahedron Letters **1973**, 1135.
95. SHIN, C., Y. CHIGERA, M. MASAKI, and A. OHTA: Synthesis of Albonoursin. Bull. Chem. Soc. Japan **42**, 191 (1969).
96. PORTER, A. E. A., and P. G. SAMMES: On the Synthesis of 3-Benzylidenepiperazine-2,5-diones. J. Chem. Soc. C **1970**, 2530.
97. SHEEHAN, J. C., D. MANIA, S. NAKAMURA, J. A. STOCK, and K. MAEDA: The Structure of Telomycin. J. Am. Chem. Soc. **90**, 462 (1968).
98. SHIN, C., M. MASAKI, and A. OHTA: The Independent Isolation of a Primary Enamine and the Tautomeric Imine. Bull. Chem. Soc. Japan **44**, 1657 (1971).
99. QUILICO, A., and L. PANIZZI: Chemische Untersuchungen über *Aspergillus echinulatus*. I. Mitteilung. Ber. **76**, 348 (1943).
100. QUILICO, A.: The Constitution of Echinulin. Res. Progr. org. biol. med. Chem. **1**, 225 (1964).
101. BIRCH, A. J., G. E. BLANCE, S. DAVID, and H. SMITH: Studies in relation to Biosynthesis. Part XXIV. Some Remarks on the Structure of Echinulin. J. Chem. Soc. **1961**, 3128.
102. MACDONALD, J. C., and G. P. SLATER: The Utilization of Tryptophan in the Biosynthesis of Echinulin. Canad. J. Microbiol. **12**, 455 (1966).
103. SLATER, G. P., J. C. MACDONALD, and R. NAKASHIMA: Biosynthesis of Echinulin by *Aspergillus amstelodami* from Cyclo-L-alanyl-L-tryptophanyl-^{14}C. Biochem. **9**, 2886 (1970).
104. NAKASHIMA, R., and G. P. SLATER: Configuration of Echinulin II. Optical Rotatory Dispersion of Echinulin, Hydroechinulin, and the Stereoisomeric 3-Methyl-6-(indolyl-3-methyl)piperazine-2,5-diones. Canad. J. Chem. **47**, 2069 (1969).
105. HOUGHTON, E., and J. E. SAXTON: The Echinulins: Preliminary Synthetic Studies and the Absolute Configuration of Echinulin. Tetrahedron Letters **1968**, 5475.
106. BIRCH, A. J., and K. R. FARRAR: Studies in Relation to Biosynthesis. Part XXXIII. Incorporation of Tryptophan into Echinulin. J. Chem. Soc. **1963**, 4277.
107. JACKSON, A. H., and A. E. SMITH: Electrophilic Substitution in Indoles I. Model Experiments Related to the Synthesis of Echinulin. Tetrahedron **21**, 989 (1965).
108. CASNATI, G., M. FRANCIONI, A. GUARESCHI, and A. POCHINI: Insertion of Isoprene Units into Indole Systems. Tetrahedron Letters **1969**, 2485.
109. CASNATI, G., and A. POCHINI: Rearrangement of 3-Alkyl-1-allylindoles; A Model Reaction for the Biogenesis of Echinulin-type Compounds. Chem. Commun. **1970**, 1328.
110. DIX, D. T., J. MARTIN, and C. E. MOPPETT: Molecular Structure of the Metabolite Lanosulin. J. C. S. Chem. Commun. **1972**, 1168.
111. BYCROFT, B. W., and W. LANDON: Thio-Claisen Rearrangements of Sulphonium Salts: Implications in Indole Alkaloid Biosynthesis. Chem. Commun. **1970**, 967.
112. PLIENINGER, H., and H. HERZOG: Synthesis of O- and C-Alkylated Indoxyl Derivatives. Preliminary work for the Synthesis of Echinulin. Monatsh. Chem. **98**, 807 (1967).
113. HOUGHTON, E., and J. E. SAXTON: Echinulin Series. Part II. Synthesis of ± Alanyltryptophan Anhydride and L-Alanyl-2-(1,1-dimethyl)allyltryptophan Anhydride. J. Chem. Soc. (C) **1969**, 1003.
114. TAKAMATSU, N., S. INOUE, and Y. KISHI: Synthetic Study on Echinulin and Related Compounds. Part II. A Stereoselective Total Synthesis of Optically Active Echinulin. Tetrahedron Letters **1971**, 4665.

115. TAKAMATSU, N., S. INOUE, and Y. KISHI: Synthetic Study on Echinulin and Related Compounds. Part I. Acid-catalyzed Amino-Claisen Rearrangement of allyl- and 3,3-Dimethylallyl Aniline Derivatives. Tetrahedron Letters **1971**, 4661.

116. KISHI, Y., S. NAKATSUKU, T. FUKUYAMA, and T. GOTO: A Stereoselective Decarboxylation of 1,6-Dimethyl-3(3′-indolyl)methyl-3-carboxy-2,5-piperazinedione. Tetrahedron Letters **1971**, 4657.

117. ALLEN, C. M.: Biosynthesis of Echinulin. Isoprenylation of Cyclo-L-alanyl-L-tryptophanyl. Biochem. **11**, 2154 (1972).

118. — Monoisoprenylated Cyclo-L-ala-L-try as a Biosynthetic Precursor of Echinulin. J. Am. Chem. Soc. **95**, 2386 (1973).

119. BARBETTA, M., G. CASNATI, A. POCHINI, and A. SILVA: Neoechinulin — a New Indole Metabolite from *Aspergillus echinulatus*. Tetrahedron Letters **1967**, 4457.

120. CASNATI, G., A. POCHINI, and R. UNGARO: Neoechinulin: A New Isoprenylindole Metabolite from *Aspergillus amstelodami*. Gazz. Chim. Ital., **103**, 141 (1973).

121. BIRCH, A. J., and J. J. WRIGHT: The Brevianamides: A New Class of Fungal Alkaloid. Chem. Commun. **1969**, 644.

122. — — Studies in Relation to Biosynthesis. Part XLII. The Structural Elucidation and some Aspects of the Biosynthesis of the Brevianamides A and E. Tetrahedron **26**, 2329 (1970).

123. BIRCH, A. J., and R. A. RUSSELL: Studies in Relation to Biosynthesis. Part XLIV. Structural Elucidations of Brevianamides B, C, D, and F. Tetrahedron **28**, 2999 (1972).

124. STEYN, P. S.: Austamide: A New Toxic Metabolite from *Aspergillus ustus*. Tetrahedron Letters **1971**, 3331.

125. — The Structures of Five Diketopiperazines from *Aspergillus ustus*. Tetrahedron **29**, 107 (1973).

126. cf. GILBERT, B.: The Alkaloids of *Aspidosperma* and Related Genera, in "The Alkaloids", ed. R. H. F. MANSKE, p. 335. New York: Academic Press. 1965.

127. WITKOP, B., and J. B. PATRICK: The Course and Kinetics of the Acid-Base Catalyzed Rearrangements of 11-Hydroxytetrahydrocarbazolenine. J. Am. Chem. Soc. **73**, 2188 (1951).

128. NEUSS, N., R. NAGARAJAN, B. B. MOLLOY, and L. L. HUCKSTEP: Aranotin and Related Metabolites II. Isolation, Characterization and Structure of Two New Metabolites. Tetrahedron Letters **1968**, 4467.

129. COETZER, J., and P. S. STEYN: The Crystal Structure of 5-Bromo-12S-tetrahydroaustamide. Acta Cryst. **B29**, 685 (1973).

130. SRINIVASAN, R.: Photochemistry of Cyclic Ketones. Adv. Photochem. **1**, 83 (1963).

131. PORTER, A. E. A., and P. G. SAMMES: A Diels-Alder Reaction of Possible Biosynthetic Importance. Chem. Commun. **1970**, 1103.

132. YAMAZAKI, A. S. SUZUKI, and K. MIZAKI: Tremorgenic Toxins from *Aspergillus fumigatus* Fres. Chem. Pharm. Bull. (Japan) **19**, 1739 (1971).

133. The author thanks Professor A. YAMAZAKI, Institute of Food Microbiology, Chiba University, Japan, for this information. Details of the revised structure for lanosulin were revealed at the IUPAC Congress, Hamburg, September, 1973.

134. The Author is indebted to Professor J. CLARDY, Ames Laboratory, Iowa State University, U.S.A., for the details of the X-ray analysis of verruculogen before its publication.

135. CLARKE, H. J., J. R. JOHNSON, and R. ROBINSON: The Chemistry of Penicillin, Princeton University Press, 1949.

136. WHITE, E. C.: Bactericidal Filtrates from a Mould Culture. Science **92**, 127 (1940).

137. WHITE, E. C., and J. H. HILL: Studies on Antibacterial Products Formed from Moulds. I. Aspergillic Acid. A Product of a Strain of *Aspergillus flavus*. J. Bacteriol. **45**, 433 (1943).

138. NEWBOLD, G. T., W. SHARP, and F. S. SPRING: Aspergillic Acid. Part III. The Synthesis of Racemic Deoxyaspergillic Acid. J. Chem. Soc. **1951**, 2679.

139. DUNN, G., G. T. NEWBOLD, and F. S. SPRING: Synthesis of Flavacol, a Metabolic Product of *Aspergillic flavus*. J. Chem. Soc. **1949**, 2586.

140. NAKAMURA, S.: Structure of Muta-aspergillic Acid. Agr. Biol. Chem. (Tokyo) **25**, 74 (1961).

141. MASAKI, M., Y. CHIGURA, and M. OHTA: Total Synthesis of Racemic Aspergillic Acid and Neoaspergillic Acid. J. Org. Chem. **31**, 4143 (1966).

142. MASAKI, M., and M. OHTA: Synthesis of a Homologue of Aspergillic Acid. J. Org. Chem. **29**, 3165 (1964).

143. SUGIYAMA, M., MASAKI, and M. OHTA: Synthesis of 1-Hydroxy-6-(1-hydroxy-1-methylethyl)-2-pyraxinone and the Structure of Muta-aspergillic Acid. Tetrahedron Letters **1967**, 845.

144. OHTA, A., and S. FUTII: Synthesis of DL-Aspergillic Acid and DL-Deoxyaspergillic Acid. Chem. Pharm. Bull. (Japan) **17**, 851 (1969).

145. OHTA, A.: Synthesis of Neoaspergillic Acid. Chem. Pharm. Bull. (Japan) **16**, 1160 (1968).

146. MacDONALD, J. C.: in: The Antibiotics, Vol. II. Biosynthesis, ed. D. GOTTLIEB and P. D. SHAW, p. 43. New York: Springer. 1967.

147. MICETICH, R. G., and J. C. MacDONALD: Metabolities from *Aspergillus sclerotiorum* Huber. J. Chem. Soc. **1964**, 1507.

148. MacDONALD, J. C.: Biosynthesis of Aspergillic Acid. J. Biol. Chem. **236**, 512 (1961).

149. — Biosynthesis of Hydroxyaspergillic Acid. J. Biol. Chem. **237**, 1977 (1962).

150. COOK, A. H., and C. A. SLATER: The Structure of Pulcherrimin. J. Chem. Soc. **1956**, 4133.

151. KLUYVER, A. J., J. P. VEN DER WALT, and A. J. CAN TRIET: Pulcherrimin, the Pigment of *Candida pulcherrimin*. Proc. Natl. Acad. Sci. U.S. **39**, 583 (1953).

152. MacDONALD, J. C.: The Biosynthesis of Pulcherriminic Acid. Biochem. J. **96**, 533 (1965).

153. — The Structure of Pulcherriminic Acid. Canad. J. Chem. **41**, 165 (1963).

154. DUTCHER, J. D.: Aspergillic Acid. An Antibiotic Substance from *Aspergillus flavus*. J. Biol. Chem. **171**, 321 (1947).

155. BATES, R. B., J. H. SCHAUBLE, and M. SOUCEK: The $C_{10}H_{17}$ Side Chain in Mycelianamide. The Stereochemistry of Bergamottin and Umbelliprenin. Tetrahedron Letters **1963**, 1683.

156. OHTA, A.: Synthesis of Pulcherrimin and Pulcherriminic Acid. Chem. Pharm. Bull. (Japan) **12**, 125 (1964).

157. OXFORD, A. E., and H. RAISTRICK: Studies on the Biochemistry of Microorganisms. Part 76. Mycelianamide. Biochem. J. **43**, 323 (1948).

158. BIRCH, A. J., L. A. MASSEY-WESTROPP, and R. W. RICKARDS: Studies Related to Biosynthesis. Part VIII. The Structure of Mycelianamide. J. Chem. Soc. **1956**, 3717.

159. GALLINA, C., A. ROMEO, V. TORTORELLA, and G. D'AGNELO: Synthesis of Racemic Deoxymycelianamide. Chem. & Ind. (London) **1966**, 1300.

160. BAPAT, J. B., D. ST. C. BLACK, and R. F. C. BROWN: Cyclic Hydroxamic Acids. Adv. Heterocyclic Chem. **10**, 199 (1969).

161. BROWN, R. F. C., and G. C. MEEHAN: Synthetic Approaches to Mycelianamide. Austral. J. Chem. **21**, 1581 (1968).

162. TERANISHI, R.: Odor and Molecular Structure, in "Gustation and Olfaction", ed. G. OHLOFF and A. F. THOMAS, p. 165. New York: Academic Press. 1971.

163. SEIFERT, R. M., R. G. BUTTERY, D. G. GUADAGIN, D. R. BLACK, and J. G. HARRIS: Synthesis of some 2-Methoxy-3-alkylpyrazines with strong Bell-pepper like Odours. J. Agr. Food Chem. **18**, 246 (1970).

164. MURRAY, K. E., J. SHIPTON, and F. B. WHITFIELD: 2-Methoxypyrazines and the Flavour of Green Peas (*Pisum sativum*). Chem. & Ind. (London) **1970**, 897.

165. WEINDLING, R., and O. EMERSON: Isolation of a Toxic Substance from the Culture Filtrate of *Trichoderma*. Phytopath. **26**, 1068 (1936).

166. JOHNSON, J. R., F. W. BRUCE, and J. D. DUTCHER: Gliotoxin, the Antibiotic Principle of *Gliocladium fimbriatin*. J. Am. Chem. Soc. **65**, 2005 (1943).
167. CROWFOOT, D., and B. W. ROGERS-LOW: X-Ray Crystallography of Gliotoxin. Nature **153**, 651 (1944).
168. JOHNSON, J. R., A. R. KIDWA, and J. S. WARNER: Gliotoxin XI. A Related Antibiotic from *Penicillium terlikowskii*. Gliotoxin Monoacetate. J. Am. Chem. Soc. **75**, 2110 (1953).
169. RICHTSEL, W. A., H. G. SCHNEIDER, B. J. SLOAN, P. R. GROF, F. A. MILLER, Q. R. BARTZ, J. EHRLICH, and G. J. DIXON: Antiviral Activity of Gliotoxin. Nature **204**, 1333 (1964).
170. TAYLOR, A.: In Biochemistry of Some Foodborne Microbial Toxins, ed. R. I. MATELES and G. N. WOGAN, p. 69. Cambridge, Massachusetts: The M. I. T. Press. 1967.
171. BELL, M. R., J. R. JOHNSON, B. S. WILDI, and R. B. WOODWARD: The Structure of Gliotoxin. J. Am. Chem. Soc. **80**, 1001 (1958).
172. BEECHAM, A. F., J. FRIDRICHSONS, and A. MC. L. MATHIESON: The Structure and Absolute Configuration of Gliotoxin and the Absolute Configuration of Sporidesmin. Tetrahedron Letters **1956**, 3131.
173. LOWE, G., A. TAYLOR, and L. C. VINING: Sporidesmins. VI. Isolation and Structure of Dehydrogliotoxin, a Metabolite of *Penicillium terlikowskii*. J. Chem. Soc. (C) **1966**, 1799.
174. SAFE, S., and A. TAYLOR: Sporidesmins. XI. The Reaction of Triphenylphosphine with Epipolythiodioxopiperazines. J. Chem. Soc. (C) **1971**, 1189.
175. FRIDRICHSONS, J., and A. MC. L. MATHIESON: The Structure of the Methylene Dibromide Adduct of Sporidesmin at $-150°$. Acta Cryst. **18**, 1043 (1965).
176. WEBER, H. P.: Molecular Structure and Absolute Configuration of Chaetocin. Acta Cryst. **B28**, 2945 (1972).
177. DAVIS, B. R., and I. BERNAL: The Crystal Structure of 2,5-Piperazinediones Having Epipolysulphide Bridges Between C3 and C6: The Structure of N,N'-Dimethyl-3,6-epitetrathio-2,5-piperazinedione. Proc. Nat. Acad. Sci. **70**, 279 (1973).
178. LEONARD, N. J., T. W. MILLIGAN, and T. L. BROWN: Transannular Interactions between Sulphide and Ketone Groups. J. Am. Chem. Soc. **82**, 4075 (1960).
179. TROWN, P. W.: Antiviral Activity of N,N'-Dimethyl-3,6-epidithio piperazine-2,5-dione. A Synthetic Compound Related to the Gliotoxins, LLS 88 α and β-Chetomin, and the Sporidesmins. Biochem. Biophys. Res. Commun. **33**, 402 (1968).
180. MURDOCK, K. C., and R. B. ANGIER: Acetylaranotin: Displacement Reactions at the Disulphide Linkage. Chem. Commun. **1970**, 55.
181. SCHOBERL, A., and E. LUDWIG: Die Aufspaltung der Disulfidbindung mit Natriumsulfit und Kaliumcyanid und über die Colorimetrische Bestimmung von Sulfhydrylverbindungen und Disulfiden. Ber. **70 B**, 1422 (1937).
182. BEECHAM, A. F., and A. MC. L. MATHIESON: The Circular Dichroism of Gliotoxin. Tetrahedron Letters **1966**, 3139.
183. ZIFFER, H., U. WEISS, and E. CHARNEY: Optical Activity of Non-planar Conjugated Dienes. IV. Interacting Chromophores in Gliotoxin. Tetrahedron **23**, 3881 (1967).
184. MOSCOWITZ, A., E. CHARNEY, U. WEISS, and H. ZIFFER: Optical Activity in Skewed Dienes. J. Am. Chem. Soc. **83**, 4661 (1961).
185. ALI, M. S., J. S. SHANNON, and A. TAYLOR: Isolation and Structures of 1,2,3,4-Tetrahydro-1,4-dioxopyrazino[1,2-a]indoles from Cultures of *Penicillium terlikowskii*. J. Chem. Soc. **1968**, 2044.
186. NAGARAJAN, N., L. L. HUCKSTEP, D. H. LIVELY, D. L. DeLONG, M. M. MARSH, and N. NEUSS: Aranotin and Related Metabolites from *Arachniotus aureus*. I. Determination of Structure. J. Am. Chem. Soc. **90**, 2980 (1968).
187. COSULICH, D. B., N. R. NELSON, and J. H. VAN DEN HENDE: Crystal and Molecular Structure of LLS88α, an Antiviral Epidithiapiperazinedione Derivative from *Aspergillus terreus*. J. Am. Chem. Soc. **90**, 6519 (1968).

188. MONCRIEF, J. W.: Molecular Structure of Bisdethiodi(thiomethyl)acetylaranotin including Absolute Configuration. J. Am. Chem. Soc. **90**, 6516 (1968).

189. NAGARAJAN, N., N. NEUSS, and M. M. MARSH: Aranotin and Related Metabolites. III. Configuration and Conformation of Acetylaranotin. J. Am. Chem. Soc. **90**, 6518 (1968).

190. SUHADOLNIK, R. J, and R. E. CHENOWETH: Biosynthesis of Gliotoxin. I. Incorporation of Phenylalanine-1- and 2-C^{14}. J. Am. Chem. Soc. **80**, 4391 (1958).

191. WINSTEAD, J. A., and R. J. SUHADOLNIK: Biosynthesis of Gliotoxin. II. Further Studies on the Incorporation of Carbon-14 and Tritium Labelled Precursors. J. Am. Chem. Soc. **82**, 1644 (1960).

192. JOHNS, N., and G. W. KIRBY: The Biosynthesis of Gliotoxin. Possible Involvement of a Phenylalanine Epoxide. Chem. Commun. **1971**, 163.

193. BU'LOCK, J. D., and A. P. RYLES: The Biosynthesis of the Fungal Toxin, Gliotoxin. Chem. Commun. **1970**, 1404.

194. BRANNON, D. R., J. A. MABE, B. B. MOLLOY, and W. A. DAY: Biosynthesis of Dithiadiketopiperazine Antibiotics. Comparison of Possible Aromatic Amino Acid Precursors. Biochem. Biophys. Res. Commun. **43**, 588 (1971).

195. BOSE, A. K., K. G. DAS, P. T. FUNKE, I. KUGAJERSKY, O. P. SHUKLA, K. S. KHANDANCHANI, and R. J. SUHADOLNIK: Biosynthetic Studies on Gliotoxin Using Stable Isotopes and Mass Spectral Methods. J. Am. Chem. Soc. **90**, 1038 (1968).

196. JERINA, D. M., J. W. DALY, B. WITKOP, P. ZALZMAN-NIRENBERG, and S. UDENFRIEND: The Role of Areneoxide-Oxepin Systems in the Metabolism of Aromatic Substrates. III. Formation of 1,2-Naphthalene Oxide from Naphthalene by Liver Microsomes. J. Am. Chem. Soc. **90**, 6525 (1968).

197. MILLER, P. A., P. W. TROWN, W. FULMAR, G. O. MORTON, and J. KARLINER: An Epidithiapiperazinedione Antiviral Agent from *Aspergillus terreus*. Biochem. Biophys. Chem. Commun. **33**, 219 (1968).

198. MACHIN, P. J., and P. G. SAMMES: Addition of Sulphur Nucleophiles Across Dehydrocyclodipeptides. J. C. S. Perkin I **1974**, 698.

199. POJER, P. M., and I. D. RAE: Synthesis of 2-Benzamido-2-mercaptopropionic Acid. Tetrahedron Letters **1971**, 3077.

200. STEGLICH, W., H. TANNER, and R. HURNAUS: 2-Dichlormethylenpseudooxazolon-(5). Chem. Ber. **100**, 1824 (1967).

201. KANEDA, A., and R. SUDO: The Preparation of α-Amino-α-benzylmercaptopropionic Acid Derivatives. Bull. Chem. Soc. (Japan) **43**, 2159 (1970).

202. PATEL, S. M., J. O. CURRIE, and R. K. OLSEN: The Synthesis of N-Acyl-α-mercaptoalanine Derivatives. J. Org. Chem. **38**, 126 (1973).

203. WOHL, A., and C. OESTERLIN: Überführung der Weinsäure in Oxalessigsäure durch Wasserspaltung bei niederer Temperatur. Ber. **34**, 1139 (1901).

204. YOSHIMURA, J., and Y. SUGIYAMA: An Attempted Synthesis of 3,6-Epidithio-2,5-piperazinediones by Cyclization of N,N'-dialkyl-2,2'-dithiocinnamamides. Bull. Chem. Soc. (Japan) **45**, 1554 (1972).

205. OTTENHEYM, H. C. J., T. F. SPANDE, and B. WITKOP: Approaches to Analogs of Anhydrogliotoxin. J. Am. Chem. Soc. **95**, 1989 (1973).

206. POISEL, H., and U. SCHMIDT: Über die elektrophile Einführung von Alkylgruppen und Schwefelfunktionen in den 2,5-Dioxopiperazin-Kern. Chem. Ber. **105**, 625 (1972).

207. OHLER, E., H. POISEL, F. TATARUCH, and U. SCHMIDT: Synthese des Epidithio-L-prolyl-L-prolin Anhydrids. Chem. Ber. **105**, 635 (1972).

208. HINO, T., and T. SAKO: Synthesis of 3,6-Diethoxycarbonyl-3,6-epipolythio-2,5-piperazinediones. Tetrahedron Letters **1971**, 3127.

209. OHLER, E., F. TATARUCH, and U. SCHMIDT: Über die Einführung von Säurestofffunktionen in Prolyl-prolinanhydrid mit Bleitetraacetat: Ein neuer Weg zum Epidisulfid des Prolyl-prolin-anhydrid. Chem. Ber. **106**, 396 (1973).

210. Ohler, E., F. Tataruch, and U. Schmidt: Nucleophile Einführung von Schwefel-funktionen über Sulfon und Hydroxyderivate Cyclisches Dipeptide (Dioxopiperazine). Chem. Ber. **106**, 165 (1973).

211. Kishi, Y., T. Fukuyama, and S. Nakatsuka: A New Method for the Synthesis of Epidithiodiketopiperazines. J. Am. Chem. Soc. **95**, 6490 (1973).

212. — — — A Total Synthesis of Dehydrogliotoxin. J. Am. Chem. Soc. **95**, 6492 (1973).

213. Kishi, Y., S. Nakatsuka, T. Fukuyama, and M. Havel: A Total Synthesis of Sporidesmin A. J. Am. Chem. Soc. **95**, 6493 (1973).

214. Ronaldson, J. W., A. Taylor, E. P. White, and R. J. Abraham: Sporidesmins. Part I. Isolation and Characterisation of Sporidesmin and Sporidesmin B. J. Chem. Soc. **1963**, 3172.

215. Hodges, R., J. W. Ronaldson, A. Taylor, and E. P. White: Sporidesmin and Sporidesmin B. Chem. & Ind. (London) **1963**, 42.

216. Jamieson, W. D., R. Rahman, and A. Taylor: Sporidesmins. Part VIII. Isolation and Structure of Sporidesmin D and Sporidesmin F. J. Chem. Soc. (C) **1969**, 1564.

217. Przybylska, M., E. M. Gopalkrishna, A. Taylor, and S. Safe: X-ray Crystallogra-phic Determination of the Stereochemistry of the Tetrathio-bridge in Sporidesmin G. J. C. S. Chem. Commun. **1973**, 554.

218. Francis, E., R. Rahman, S. Safe, and A. Taylor: Sporidesmins. Part XII. Isolation and Structure of Sporidesmin G, a Naturally Occurring 3,6-Epitetrathiopiperazine-2,5-dione. J. C. S. Perkin I **1972**, 470.

219. Safe, S., and A. Taylor: Sporidesmins. Part X. Synthesis of Polysulphides by Reaction of Dihydrogen Disulphide with Disulphides and Thiols. J. Chem. Soc. (C) **1970**, 432.

220. Rahman, R., S. Safe, and A. Taylor: The Stereochemistry of Polysulphides. Quart. Rev. **24**, 233 (1970).

221. Hodges, R., and J. S. Shannon: The Isolation and Structure of Sporidesmin C. Austral. J. Chem. **19**, 1059 (1966).

222. Horn, M. J., D. B. Jones, and S. J. Ringel: Isolation of a New Sulphur-containing Amino Acid (Lanthionine) from Sodium Carbonate Treated Wool. J. Biol. Chem. **138**, 141 (1941).

223. Nakagawa, M., T. Kaneko, and H. Yamaguchi: Photoinduced Oxidation of Tryptamine Derivatives. Formation of Pyrrolo[2,3-b]indole and N^b-(4-Cyanobuta-dienyl)tryptamine. J. C. S. Chem. Commun. **1972**, 603.

224. Ohno, M., T. F. Spande, and B. Witkop: Cyclisation of Tryptophan and Tryptamine Derivatives to 2,3-Dihydropyrrolo[2,3-b]indoles. J. Am. Chem. Soc. **92**, 343 (1970).

225. Foote, C. S., S. Mazur, P. A. Burns, and D. Lerdal: Chemistry of Singlet Oxygen. XVII. 1,4-Addition Products from Styrene Derivatives. J. Am. Chem. Soc. **95**, 586 (1973).

226. Amit, R. G., F. W. Eastwood, and I. D. Rae: Addition of a Highly Oxygenated Side Chain to an Indole Derivative. Chem. Commun. **1971**, 1614.

227. Minato, H., M. Matsumoto, and T. Katayama: Verticillin A, a New Antibiotic from *Verticillium sp*. Chem. Commun. **1971**, 44.

228. — — — Studies on the Metabolites of *Verticillium sp*. Structures of Verticillin A, B, and C. J. C. S. Perkin I **1973**, 1819.

229. Hauser, D., H. P. Weber, and H. P. Sigg: Isolierung und Strukturaufklärung von Chaetocin. Helv. chim. Acta **53**, 1061 (1970).

230. Hauser, D., H. R. Loosli, and P. Niklaus: Isolierung von 11α,11α′-Dihydroxy-chaetoxin am *Verticillium tenerum*. Helv. chim. Acta **55**, 2182 (1972).

231. Waksman, S. A., and E. Bugie: Chaetomin, a New Antibiotic Substance produced by *Chaetomium cochliodes*. J. Bacteriol. **48**, 527 (1944).

232. Safe, S., and A. Taylor: The Characterisation of Chetomin, a Toxic Metabolite of *Chaetomium cochliodes* and *Chaetomium globosum*. J. C. S. Perkin I **1972**, 472.

233. Kato, A., T. Saeki, S. Suzuki, K. Ando, G. Tamura, and K. Arima: Oryzachloride,

a New Antiviral Disulphide Dioxopiperazine Derivative. J. Antibiot. (Tokyo) **22**, 322 (1969).

234. ARGOUDELIS, A. D.: Melinacidins II, III, and IV. New 3,6-Epidithiadiketopiperazine Antibiotics. J. Antibiot. (Tokyo) **25**, 171 (1972).

235. KAMIYA, T., S. MAENO, M. HASHIMOTO, and Y. MINI: Bicyclomycin, a New Antibiotic. II. Structure Elucidation and Acyl Derivatives. J. Antibiot. (Tokyo) **25**, 576 (1972).

236. MIYOSHI, T., N. MIYAWA, H. AOBI, M. KOHSAKA, H. SAKAI, and H. IMANAKA: Bicyclomycin, a New Antibiotic. II. Taxonomy, Isolation, and Characterization. J. Antibiot. (Tokyo) **25**, 569 (1972).

237. NISHIDA, M., Y. MINI, and T. MATSUBARA: Bicyclomycin, a New Antibiotic. III. *In vitro* and *in vivo* Antimicrobial Activity. J. Antibiot. (Tokyo) **25**, 582 (1972).

238. SHARMA, G. M., and P. R. BURKHOLDER: Structure of Dibromophakellin, a New Bromine-containing Alkaloid from the Marine Sponge, *Phakellia flabellata*. Chem. Commun. **1971**, 151.

239. CHEN, Y.-S.: Studies on the Metabolic Products of *Roselinia necatrix*. I. Isolation and Characterization of Several Physiologically Active, Neutral Substances. Bull. Agric. Chem. Soc. Japan **24**, 372 (1960).

240. JOHNSON, J. L., W. G. JACKSON, and T. E. EBLE: Isolation of L-leucyl-L-proline Anhydride from Microbiological Formulations. J. Am. Chem. Soc. **73**, 2947 (1951).

241. KODAIRA, Y.: Toxic Substances to Insects Produced by *Aspergillus achraceus* and *Oospora destructor*. Agr. Biol. Chem. (Tokyo) **25** 261 (1961).

242. BIRKENSHAW, J. H., and Y. S. MOHAMMED: Studies in the Biosynthesis of Microorganisms. 111. The Production of L-Phenylalanine Anhydride (*cis*-L-3,6-dibenzyl-2,5-dioxopiperazine) by *Penicillium nigricans* (Bainier) Thom. Biochem. J. **85**, 523 (1962).

243. BROWN, R., C. KELLY, and S. E. WIBBERLEY: The Production of 3-Benzylidene-6-isobutylidene-2,5-dioxopiperazine, 3,6-Dibenzylidene-2,5-dioxopiperazine, and 3,6-Dibenzyl-2,5-dioxopiperazine by a Variant of *Streptomyces noursei*. J. Org. Chem. **30**, 277 (1965).

244. CAESAR, F., K. JANSSEN, and E. MUTSCHLER: Nigragillin, a New Alkaloid from the *Aspergillus niger* Group. 1. Isolation and Structure Elucidation of Nigragillin and a Dioxopiperazine. Pharm. Acta Helv. **44**, 676 (1969).

245. JENSEN, N. P., C. O. GITTERMAN, T. Y. CHEN, B. H. ARISON, and J. L. BECK: Isolation of a New Antitumour Antibiotic from *Streptomyces griseoluteus*. Chem. and Eng. News April 14-th, 1973, p. 24.

246. GERBER, N. N.: Phenazines, Phenoxazinones, and Dioxopiperazines from *Streptomyces thioluteus*. J. Org. Chem. **32**, 4055 (1967).

247. HEINEMANN, B., M. A. KAPLAN, R. D. MUIR, and I. R. HOOPER: Amphomycin, a New Antibiotic. Antibiot. and Chemother. **3**, 1239 (1953).

248. CASNATI, G., A. QUILICO, and A. RICCA: *Aspergillus glaucus* Group. XVIII. Echinulin. 12. Gazz. Chim. Ital. **92**, 129 (1962).

249. DUTCHER, J. D.: Aspergillic Acid, an Antibiotic Substance produced by *Aspergillus flavus*. III. The Structure of Hydroxyaspergillic Acid. J. Biol. Chem. **232**, 785 (1958).

250. YOKOTSUKA, T., M. SASAKI, T. KIKUCHI, Y. ASAO, and A. NOBUHARA: Compounds Produced by Moulds. I. Fluorescent Compounds Produced by Japanese Industrial Moulds. Bull. Agric. Chem. Soc. Japan **41**, 32 (1967).

251. YOKOTSUKA, T., T. KIKUCHI, M. SASAKI, and K. OSHITA: Aflatoxin G — like Compounds with Green Fluorescence Produced by Japanese Industrial Moulds. Bull. Agric. Chem. Soc. Japan **42**, 581 (1968).

252. TERAO, M., K. KARASAWA, N. TANAKA, H. YONEHARA, and H. UMEZAWA: A New Antibiotic, Emimycin. J. Antibiot. Ser. A **13**, 401 (1960).

253. TERAO, M.: Emimycin, a New Antibiotic. II. The Structure of Emimycin. J. Antibiot. Ser. A **16**, 182 (1963).

254. Yamazaki, M.: Deoxyneo-β-hydroxyaspergillic Acid. Chem. Pharm. Bull. (Japan) **20,** 2274 (1972).

255. Rahman, R., S. Safe, and A. Taylor: Sporidesmins. Part IX. Isolation and Structure of Sporidesmin E. J. Chem. Soc. (C) **1969,** 1665.

256. Taylor, A.: The Toxicology of Sporidesmins and Other Epipolythiodioxopiperazines, in "Microbial Toxins", ed. S. Kadis, A. Ciegler, and S. J. Ajl, Vol. VII, chapter 10. New York: Academic Press. 1971.

257. Cheeseman, G. W. H., and E. S. G. Werstuik: Recent Advances in Pyrazine Chemistry. Adv. Heterocyclic Chem. **14,** 99 (1972).

258. Tamura, S., A. Susuki, Y. Aoka, and N. Otaki: Isolation of Several Dioxo-piperazines from Peptone. Agr. Biol. Chem. (Japan) **28,** 650 (1964).

(Received December 3, 1973)

Structural Investigations of Natural Products by Newer Methods of NMR Spectroscopy

By R. J. HIGHET and E. A. SOKOLOSKI,
Department of Health, Education, and Welfare, Public Health Service,
National Institutes of Health, National Heart and Lung Institute,
Bethesda, Maryland, U.S.A.

With 18 Figures

Contents

I. Introduction

When the use of nuclear magnetic resonance spectroscopy in natural product studies was reviewed in an earlier volume of this series (*1*), the technique was already established as a primary tool of structural investigation. Straight-forward nmr techniques suitable for simple analytical instruments and mass spectroscopy remain the natural product chemist's first tools in the examination of natural materials. Often these methods together suffice either to demonstrate the structure of a compound, or to indicate the need for prompt recourse to the most definitive tool of structural investigation, x-ray crystallography. In recent years, methods have been developed to extend nmr methods to samples not previously amenable to this technique and to extract further structural information from the spectrum.

These methods have taken several forms: The use of solvent effects and shift reagents provide methods of simplifying and elucidating the spectra from simple spectrometers. Extensive theoretical and empirical studies of the relation of structure to nmr parameters allow a more exact interpretation of spectra so obtained: The double resonance techniques earlier described have been extended by INDOR and Nuclear Overhauser Effects, and the development of Fourier transform spectrometers has greatly extended the sensitivity of the instruments and allowed the study of the resonance of other nuclei. This last development opens up whole new areas of nmr applications.

II. Solvent Effects

One of the simplest methods of obtaining more information from an nmr spectrum proves to be the use of a different solvent with the same sample, particularly an aromatic solvent. The systems which have been studied most extensively have been those of steroidal ketones in benzene (*2*). Protons lying behind a plane normal to and bisecting the carbonyl bond suffer an upfield shift. The environment of methoxyl groups on aromatic systems can also be studied in benzene solution; methoxyl groups with an ortho proton will be shifted upfield at least 0.3 ppm from the position in deuteriochloroform, while o,o′ disubstituted methoxyls will not be affected (3). The method assisted QUIJANO et al. in establishing the substitution pattern of the Compositae flavone, eupalitin

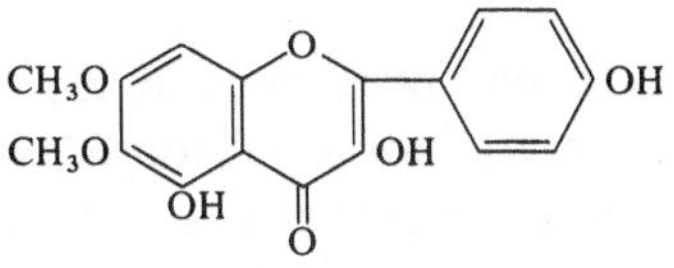

(1)

(1) (4). A benzene solution of the triethyl ether of (1) showed changes in the chemical shifts of the methoxyls corresponding to 0.01 and 0.64 ppm. In the context of other information available for the compound, only methoxyls at C-6 and C-7 could account for these changes.

MA and WARNHOFF have published a useful study of the chemical shifts of protons on carbon atoms bonded to nitrogen (5). In particular, N−CH₃ groups may be characterized and differentiated from acetyl groups or aromatic methyl groups by comparing the chemical shifts in such a neutral solvent as deuteriochloroform with those observed in acidic solvents.

III. Derivatives and *In Situ* Reactions

It is a common problem in the characterization of natural products to encounter aromatic rings bearing several oxygen atoms, one of which is a free phenol. The problem may be approached by comparing the chemical shifts of the aromatic protons of the free phenol in dimethylsulfoxide solution to those of the phenol anion (6). The upfield shift of the aromatic protons depends on their relation to the phenolic group,

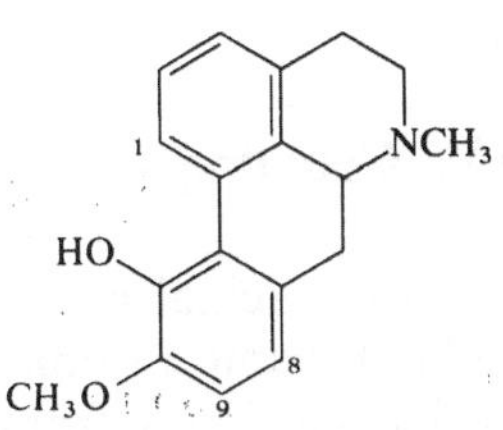

(2)

ortho shifting ca. 0.5 ppm, meta ca. 0.2, and para ca. 0.8. The method allowed PACHLER et al. to demonstrate the substitution pattern of the aporphine (2), as the two protons on C-8 and C-9 shifted upfield 0.8 and 0.4 ppm, respectively (7). The proton on C-1 was shifted 0.8 ppm downfield. The alternative method of comparing the chemical shifts of the parent phenol with those of its acetate is rather more popular, although less explicit.

An ingenious method has been devised for the characterization of alcohols by reaction with trichloracetylisocyanate in the nmr tube (8). This commercially available reagent reacts instantaneously with alcohols to form trichloroacetyl carbamates; the carbinyl proton of a primary alcohol is shifted 0.5 to 0.9 ppm, and that of a secondary, 1—1.5 ppm.

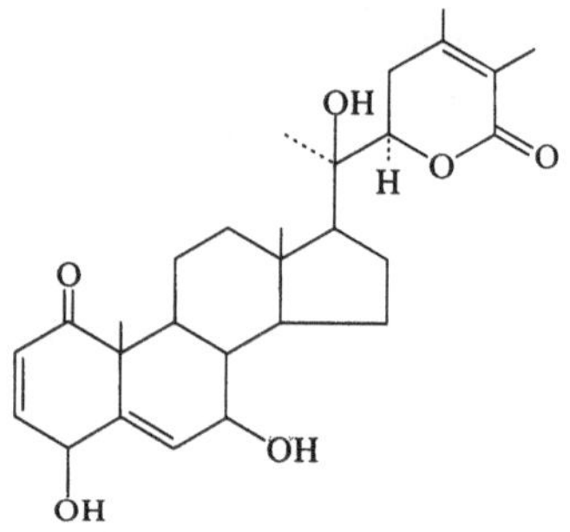

(3)

KIRSON et al. were able to establish the presence of three alcohol functions in the withanolide (3) from Acnistus australis. Addition of the isocyanate was followed by the appearance of three peaks corresponding to the NH resonances of three carbamates at δ 8.61, 8.68 and 8.81 (9). The method deserves wider use than it has received.

IV. Coupling Constants

While the interpretation of vicinal coupling constants by means of the KARPLUS relation quickly became a primary tool of nmr spectroscopy (1), inferences of comparable specificity from geminal couplings are not commonly made. However, the dependence of this coupling upon molecular environment is better understood (*10*), and an extensive compendium of coupling constant-structure relations is available (*11*). Indeed, the relationships revealed by this tabulation were sufficiently specific to show that several structures previously assigned to natural products might require revision. Unfortunately, the relations are only rarely quoted in support of proposed structures. RUEDI and EUGSTER, however, were able

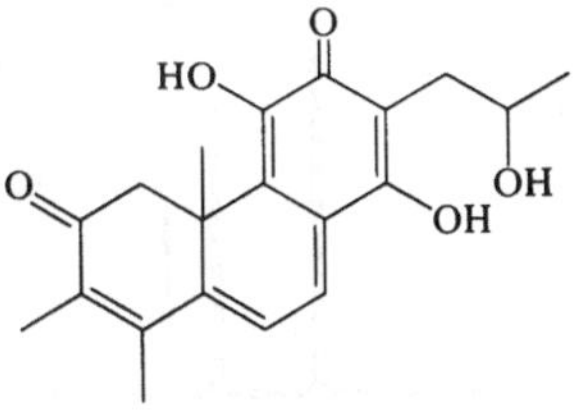

(4)

to use a coupling of (−) 17 Hz observed in the spectrum of Coleon E (**4**) in support of a structure with a methylene group α to a carbonyl in a six-membered ring (*12*).

Long range couplings (over more than three bonds) have been adequately investigated to provide sound structural information (*13*). In particular, rigid structures possessing protons connected by a zig-zag conformation such as that of (**5**) may be expected to show an appreciable

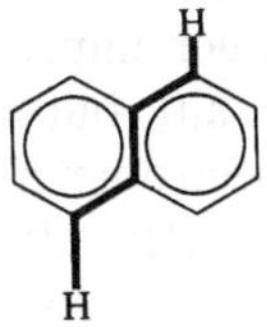

(5)

coupling. In Fig. 1, the downfield portion of the spectra of the two pigments fuerstione and pristimerin are compared (*14*). The absence of fine structure in the first spectrum shows that the three protons of the system must have the configuration shown, without two protons in the zig-zag configuration such as exists between Hb and Hx of pristimerin.

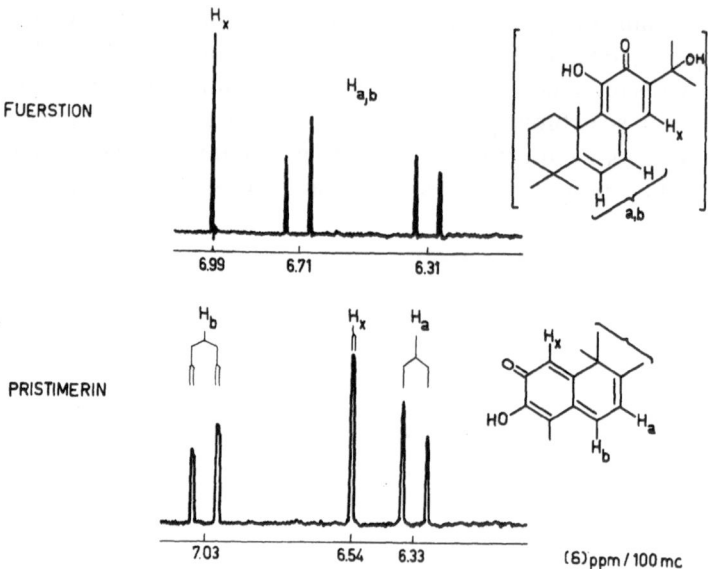

Fig. 1. Olefinic proton peaks of fuerstione and pristimerin. (From Helv. Chim. Acta **49**, 1151 (1966), courtesy of Prof. EUGSTER)

V. Superconducting Magnets

The powerful tools of coupling constants and exact knowledge of the chemical shift are often frustrated in the nmr spectra of proton-rich compounds by the complexity of their spectra. Most of the protons of a steroid lie in a blur of superposed absorptions between 1 and 3 ppm, while those of a carbohydrate are crowded in a confused huddle between 3 and 5 ppm. The situation is much improved if the spectra can be taken at higher field, as the chemical shift, but not the coupling constants, are dependent upon the field. Spectrometers are commercially available which operate from 200 to 360 MHz, although at a price of something like 1 dollar/1000 Hz. Operation of such an instrument involves special, but not formidable, cryogenic skills, and the supply of 75 to 100 l of liquid helium each month.

In spite of these obstacles, use of these spectrometers is rapidly becoming common. Because the higher fields cause a wider separation of the energy states of nuclei aligned for and against the field, theory predicts an increase of sensitivity at the higher field proportional to the square of the frequency (*15*). Unfortunately, at proton frequencies, this effect is largely cancelled by the more difficult electronic requirements. As a result, sensitivities at higher field are approximately equal to those of 100 MHz instruments. Furthermore, difficulties in obtaining homogeneous super-conducting fields result in a slight degradation of the resolution to produce peak widths at half height of approximately 0.5 Hz.

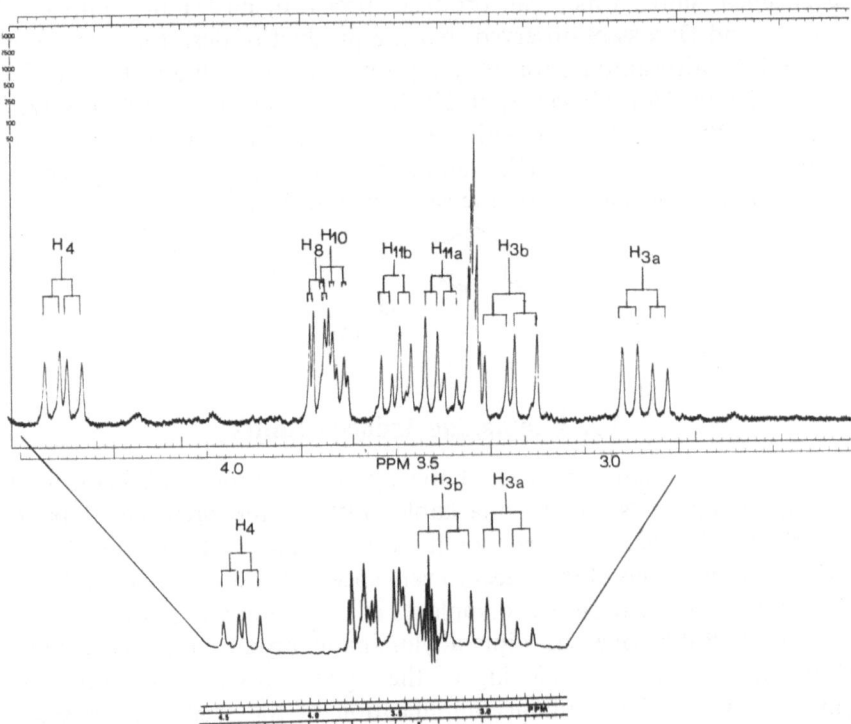

(6)

Too much should not be made of these problems, however, for the effect of the high field upon a spectrum can be quite helpful. Fig. 2 compares the midfield region of the spectrum of the plant phenol leucodrin (6)

Fig. 2. Midfield region of the ^{1}H spectrum of leucodrin (6) at 100 MHz (lower trace) and 220 MHz (upper trace)

at 100 and 220 MHz (*16*). The confused region of overlapping absorption at the lower frequency is well resolved at the higher frequency.

The utility of the greater resolution is nicely illustrated by a French group's study of the steroidal alkaloid holantosine A (*17*). A single spectrum was sufficient to establish all of the coupling constants of the sugar moiety (**7**) and its identity as 4-amino-3-O-methyl-2,4,6-tri-desoxyribohexose.

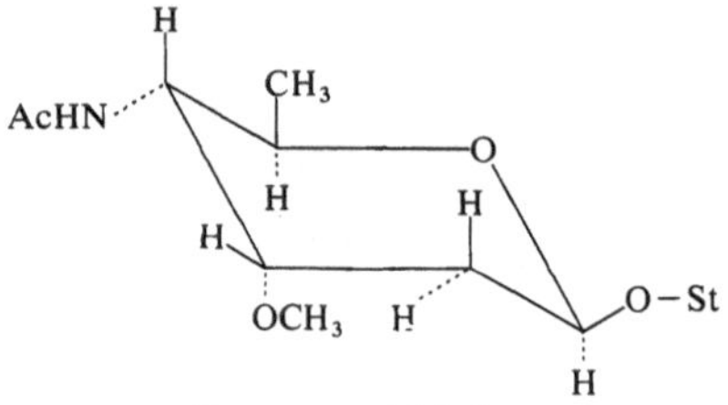

St = a steroidal ring
(7)

A potential of the high field spectra which is perhaps not immediately apparent is the examination of structures with accidental near-identity of chemical shifts, which can produce spectra of misleading simplicity. Perold and Ourisson observed that the product of desk-top photolysis of a 1,2-dihydrosantonin was a cyclopropyl ketone with a methyl singlet at δ 1.17 ppm (*18*). However, at 220 MHz the singlet could be resolved into a doublet, consistent with structure (**8**). The appearance of the methyl as a singlet at 60 MHz had arisen from the $-CHCH_3$ system in which the protons had nearly the same chemical shift.

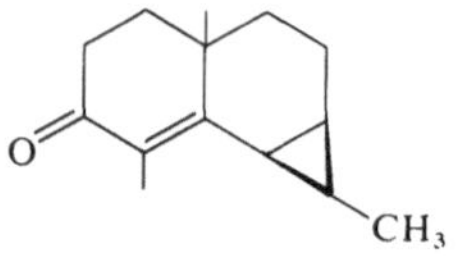

(8)

VI. Lanthanide Induced Shifts

Although it had long been known that the proton resonances of organic compounds suffer remarkable shifts in the presence of para-magnetic substances, the concomitant broadening of the signals re-stricted observations of this effect to very specialized studies of complexes. Hinckley's observation that complexes of europium and praseodymium with such β-diketones as dipivaloylmethane produced pseudocontact shifts with minimal broadening of the signals was greeted with great enthusiasm, for the resultant simplification of the spectra promised a multitude of applications (*19*). As the configuration of the substrates was not altered, the coupling constants were little changed, and meaning-ful interpretation of the simplified spectra as well as decoupling experiments were possible. Recovery of the samples from the mixture was possible by conventional chromatographic techniques.

In the intervening years a large literature has appeared on the applications and theoretical investigations of the method, including a useful review and a book (*20, 21*). This section will be restricted to the applications of the technique to the investigation of unknown structures.

An experimental difficulty was quickly recognized. Optimum separation of resonances with minimal blurring of the patterns occurs at a ratio of shift reagent to substrate of approximately 0.4. However, the molecular weight of the reagent is rather large (tris(dipivaloylmethanato)-europium, 704); as the interaction is that of a Lewis base with the reagent, the competition by the solvent is best avoided through the use of non-polar solvents, such as carbon tetrachloride, in which the reagent is only sparingly soluble. As a result, concentration of the substrate must be small and near the limits of the sensitivity of simple analytical instruments, 0.1 to 0.2 M. This problem has been much ameliorated by the use of more soluble derivatives, notably Eu(fod)$_3$, 1,1,1,2,2,3,3,-heptafluoro-7,7-dimethyloctane-4,6-dionatoeuropium.

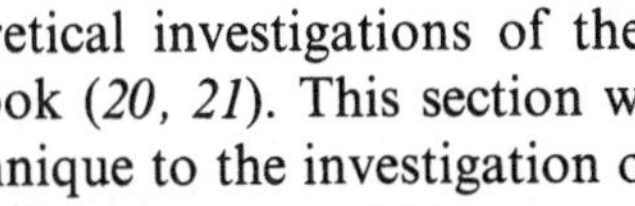

(9)

Fig. 3 shows the effect of the shift reagent Eu(fod)$_3$ upon the spectrum of citronellol (**9**). As increased amounts of the shift reagent are added, the peaks corresponding to the protons on C-2 and those on carbons farther from the hydroxyl are increasingly shifted away from the general absorption at 1—3 ppm. The difference in the chemical shifts of the C-2 protons arising from the adjacent asymmetric center is readily recognized.

When the technique is applied to a complicated molecule, the possibility arises of the apparent groupings changing order in the spectrum with increasing amount of the reagent. This requires that several spectra be run, to follow the shift of each group with the addition of more reagent. In polyfunctional molecules, it is not necessarily apparent which groups are responsible for complexing, but the general order of strength of interaction is that of the strength of the groups as Lewis bases, amines > hydroxyls > ketones > aldehydes > ethers > esters > nitriles.

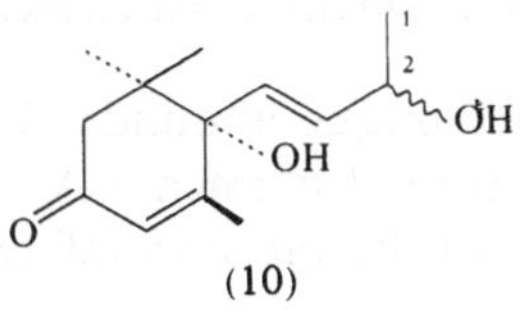

(10)

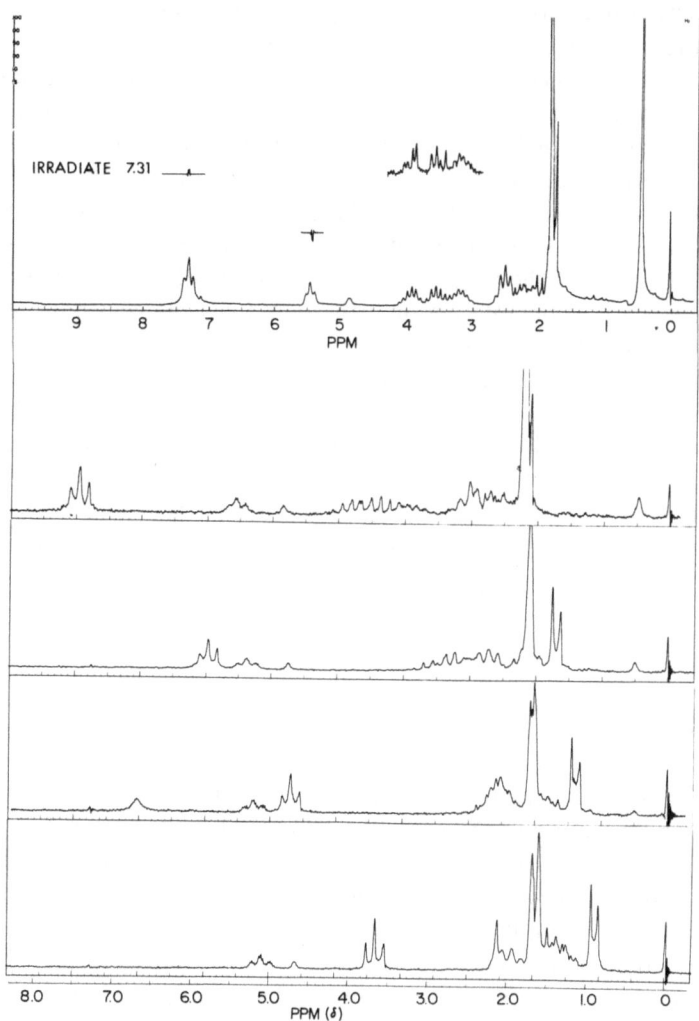

Fig. 3. Effect of the shift reagent Eu(fod)$_3$ upon the ^{1}H spectrum of citronellol (**9**). The bottom trace is that of the terpene at 60 MHz. With increasing amounts of the reagent (next three traces) the carbinyl triplet at δ 3.6 is shifted downfield passing that of the olefinic protons, which are more distant from the complexing site. The C-2 protons are shifted downfield to give discrete multiplets. The top trace repeats the spectrum of the highest concentration of reagent at 100 MHz. Irradiation of the carbinyl protons at δ 7.31 simplifies the absorption of the adjacent methylene

Applications of this technique abound. Galbraith and Horn were able to establish the structure of blumenol A from *Podocarpus blumei* as (**10**) by its pmr spectrum in the presence of DPM. The presence of the

References, pp. 162—166

grouping $-CH=CHCH(OH)CH_3$ was revealed after addition of the shift reagent, which caused the broad signal from $C-2-H$ at 4.42 to shift to δ 9.6, the methyl doublet of C-1 to δ 4.2, while the ethylenic protons now appeared cleanly separated at δ 8.8 and 9.0, with $J=16$ Hz; the lower field proton showed the further coupling to the carbinyl proton at δ 4.2, $J=5$ Hz (22). MUNTWYLER and KELLER-SCHIERLEIN were able to

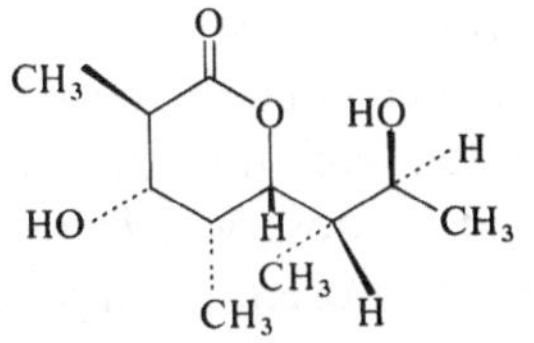

(11)

elucidate the structure of a lankamycin degradative product, the lactone (11), after the addition of Eu(fod)$_3$ (23). The signal at lowest field, δ 6.03, could be recognized as that of the methyl carbinol by decoupling the methyl doublet at δ 2.37; the coupling constants of each of the other protons could then be interpreted. Together with the knowledge of the original chemical shifts, this demonstrated the structure (11).

Such qualitative use of shift reagents to simplify complex spectra is relatively straight-forward, but quantitative interpretations offer even richer possibilities of structural investigations. The simplest statement of the relation of the change in chemical shift to the geometry of the complex is

$$\delta = k(3\cos^2\theta - 1)/r^3,$$

in which the angle θ is that defined by the site of complexing, the lanthanide atom, and the nucleus affected, and r is the distance between the proton and the lanthanide nucleus (24). This relation assumes that the phenomenon observed is a pseudocontact shift, with negligible contribution from contact shifts, which seems to be justified for protons, although sometimes violated in ^{13}C spectra. Distances and angles may be estimated from models, and speculative structures may be tested against this relation by least squares methods (25). In an extensive study of the interaction of chloroquin with Pr(fod)$_3$, it was possible to obtain the coordinates of each proton of the molecule (26).

VII. Computer-Aided Interpretation of Spectra

If such techniques for the simplification of spectra as those discussed above are not applicable or available, it is sometimes nevertheless necessary to extract values of nmr parameters from spectra too complex to allow safe first order analysis. Several computer programs are now available for this purpose (27, 28, 29). They afford a calculated spectrum based on parameters inferred from the observed spectrum or knowledge of the structure obtained from other sources. By comparison of the two spectra, the transitions of the calculated spectrum are assigned observed frequencies, and these data fed back into the computer program to allow an iterative fit of calculated to observed frequencies. For use with the spectrum of more than a few nuclei, a large computer is required. A successful solution provides nmr parameters of substantially better accuracy than those inferred from first order spectra. However, if the observed spectrum is at all complex, the process of assignment of the calculated transitions is tedious and often uncertain. There often results a time-consuming trial-and-error attempt to obtain assignments that will produce credible parameters, and chemists involved in repetitive attempts to complete these assignments must often reflect ruefully on the apt naming of the LAOCOON program. As a result, the iterative procedure is rarely used in structural investigations except to refine the knowledge obtained from more or less obvious first order analysis.

Examples of the profitable use of these methods in the study of natural materials do exist. Gordon, Stoessl and Stothers analysed the ABMNX system of altersolanol B acetonide (12) by the LAOCN3 program to obtain coupling constants of 4.28 and 2.42 Hz for the X proton (30). These values precluded an axial conformation for this proton and demonstrated that the ring existed in the conformation (13). Findlay et al. were similarly able to demonstrate the character of the side chain of the antibiotic flavipucine by analysis of the ABMQXY system of the hydrogenation product (14) (31).

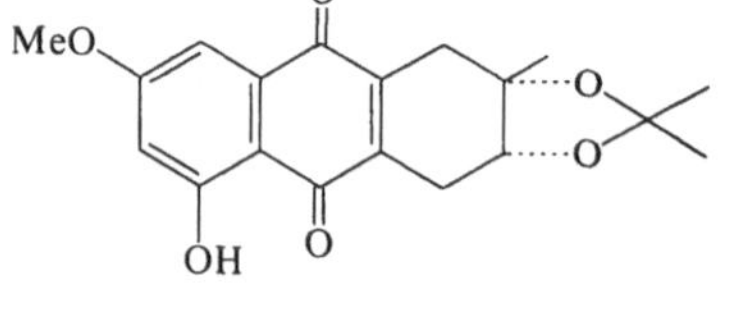

(12)

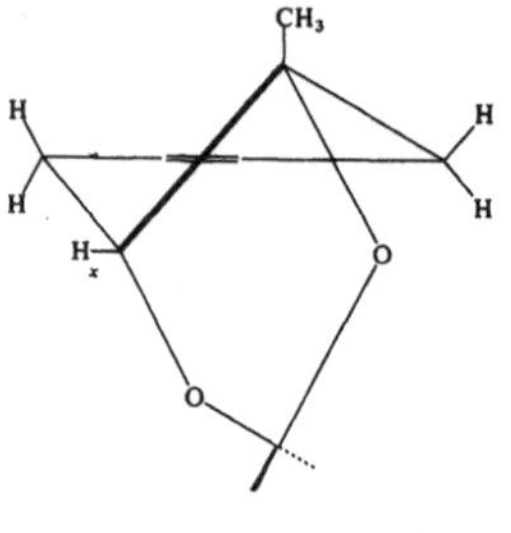

(13)

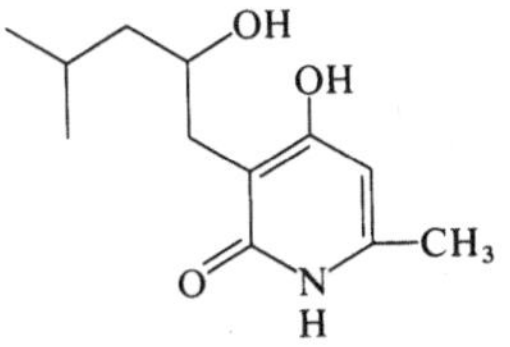

(14)

VIII. INDOR

The earlier review (*1*) described the use of double resonance techniques for the simplification of spectra, the demonstration of coupling between two nuclei in a complex spectrum and the determination of chemical shifts of protons obscured under overlapping resonances. These methods remain mainstays of nmr structural investigations. To them have been added two more sophisticated techniques, the rarely used INDOR method, and the very popular Nuclear Overhauser Effect.

These methods require the use of spectrometers more elaborate than the simplest analytical instruments, capable of frequency swept spectra. Except in the very stable superconducting fields, this calls for an internal lock, the relation of field to frequency being held constant by reference to a resonance frequency of the sample. Older instruments use homonuclear locks, with reference to a well-defined peak of the spectrum, such as that of tetramethylsilane. Newer instruments may use the resonance absorption of the deuterium or fluorine of the solvent.

The INDOR method (InterNuclear DOuble Resonance) differs from the more familiar mode of double resonance in that the irradiating frequency, H_2, is varied while the observing frequency, H_1, is fixed on the resonance to be observed. At irradiating fields of approximately one-tenth the strength used for complete decoupling the response of the observed nucleus is recorded as erect and inverted peaks corresponding to the spectral transitions of coupled nuclei. These result from the changes of the population of the energy states of the observed nuclei as the irradiating frequency encounters transitions with energy levels in common. As apparent in Fig. 4, the results provide the chemical shifts and the coupling constants of the nuclei.

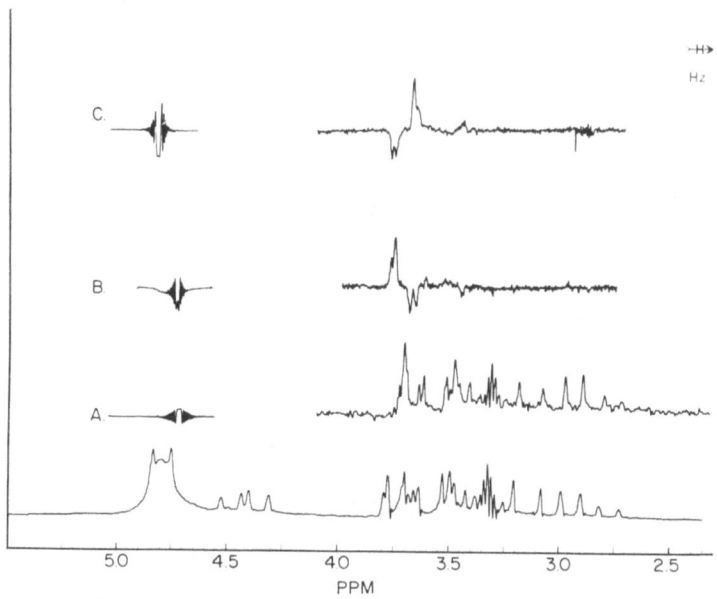

Fig. 4. Aliphatic portion of the ¹H spectrum of leucodrin (**6**).
In (A), decoupling H-9 by high power irradiation at δ 4.8 produces some simplification of the peaks near δ 3.8. INDOR experiment (B) and (C) were carried out by observing each of the H-9 peaks while sweeping a lower rf power through the region of 4 to 3 ppm

The practical limitations of the method are considerable. It is a somewhat insensitive technique, requiring several times the amount of material needed for a conventional spectrum. More important, the nature of the spectrum may preclude the use of INDOR, for the observed peak must be well separated from the coupled peak, and strong and discrete enough to allow effective monitoring. Furthermore, if the transitions to be detected are part of a very complex pattern, as that of a saturated alicyclic system, detection will be impossible. However, the method does seem to have more potential than indicated by its infrequent use.

A situation eminently suitable for the technique is that of the anomeric protons of sugars. Burton *et al.* have described studies in which the anomeric protons of fully mutarotated 2-deoxy-D-arabinohexapyranoses are monitored, and the C-2 protons fully characterized (*32*).

IX. The Nuclear Overhauser Effect

Magnetic nuclei placed in a strong magnetic field achieve an equilibrium distribution of populations of spin energy levels by a process, termed relaxation, largely dependent upon dipolar interactions with other nuclei. If two nuclei are so related by a mutual relaxation process, a disturbance of the population of the energy levels of one will be reflected by corresponding changes in those of the other. Experimentally this may be observed by irradiating the one with a resonant frequency field and observing a change in the integrated intensity of the peak corresponding to the other. The change may be a decrease in intensity, but more commonly is an increase. Because the change in intensity is dependent upon the inverse of the sixth power (v.i.) of the distance between nuclei, the effect is strongly dependent upon the geometry of the situation and, if the interaction can be restricted to intramolecular effects, it can provide substantial structural information.

Since this process was first demonstrated between protons by ANET and BOURN (33), the technique has become one of the most popular methods of structural examination. As an excellent review is available (34) and an even more extensive examination of the phenomenon has been published (35), this section will be restricted to brief comment upon the experimental procedure and some illustrations of the applications.

Because the effect can be observed only when the nuclei concerned are mutually dependent for relaxation, it is essential that paramagnetic substances be excluded from the sample, as they offer a much more efficient method of relaxation. Practically, this requires that the sample be degassed to remove oxygen. This is effected either by passing nitrogen through the dissolved sample, or, by alternate freezing and thawing of the solution with evacuation of the tube while the sample is frozen. The effectiveness of the process may be judged by the appearance of a sharp peak such as tetramethylsilane, working for a width at half height of less than 0.3 Hz.

The greater difficulty is one familiar to all who use nmr spectrometers, viz. that the integration is the least satisfactory capability of the system. This requires that somewhat more than minimal amount of sample be studied, such as 0.1 mmole, and that the integration of the observed peaks be repeated often enough with the irradiating frequency at the resonance frequency and then substantially removed to provide a reliable comparison. The minimum credible observation is probably 7%, although smaller values are sometimes quoted. If several nuclei are responsible for the relaxation of the proton being studied, the effect will be divided among them. As the sum of the enhancements cannot exceed 50%, dividing the effect among numerous nuclei may well render it impossible to observe; i.e., a proton in a proton-rich environment may not be suitable for study. Systems of paired nuclei separate from others will be optimum for this purpose.

BELL and SAUNDERS have examined the quantitative relation between the observed enhancement and the internuclear distance by determining the enhancement observed in a number of well defined and rigid systems

(36). The linear relation between the enhancement and the inverse sixth
power of the distance is so well obeyed that one can reasonably estimate
the distances between the nuclei of unknown structures by the observed
enhancement. When two isolated protons are observed, enhancements
range from 30% at 2.7 Å to 5% at 3.4 Å.

Commonly, use of the effect is limited to a qualitative demonstration
of the proximity of protons. Aromatic substitution patterns are eminently
suitable for study. McCorkindale was able to establish the substitution

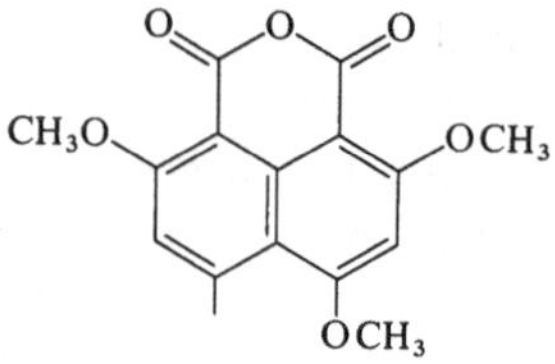

(15)

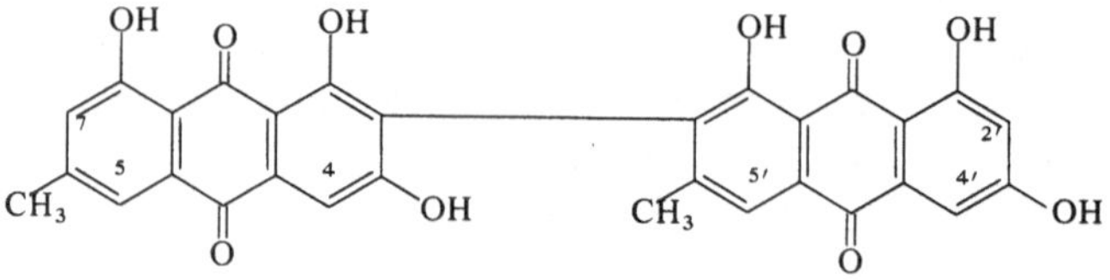

(16)

in limellicolic acid trimethyl ether (15) by demonstrating an 11%
enhancement of the absorption of the ring proton on irradiation of the
methoxyl group and a 21% enhancement by irradiation of the methyl
group (37). The methyl ether of the pigment madagascarin (16) provided
a particularly fruitful exercise for NOE investigation, with these
results (38):

Proton Observed	Methoxyl Group	Irradiated	%NOE
H 4	3	3.91 ppm	37
H 7	8	4.00	18
H 2'	1',3'	3.97	38
H 4'	1',3'	3.97	32

The effect is not limited to groups on a single ring. Murray and
Ballantyne showed the position of the methoxyl on the coumarin
nieshoutol from sneezewood to be that of (17) by irradiating the C-4
proton to cause a 12% increase in the absorption of the methoxyl, or by
irradiation of the methoxyl to cause an 11% increase in that of the C-4
proton (39).

(17)

X. Fourier Transform Techniques

1. Instrumentation

Of all the methods described here, the one which promises to affect nmr studies the most is Fourier transform spectroscopy. When ERNST and ANDERSON first described the use of this method for sensitivity enhancement, it was clear that they had introduced a new generation of spectrometers (*40*), but the proliferation of commercial instruments has awaited the development of the inexpensive minicomputer. At present, the combination of a suitable spectrometer and the Fourier transform accessory represents a capital investment approaching that of a superconducting-magnet spectrometer, but newer commercial instruments are being introduced whose cost will approximate that of the analytical spectrometers. Similarly, although at present the operation of the spectrometer-FT combination requires the expertise of the specialist, the new instruments promise to be suitable for operation by the chemical generalist. A valuable introduction to the technique has been published (*41*).

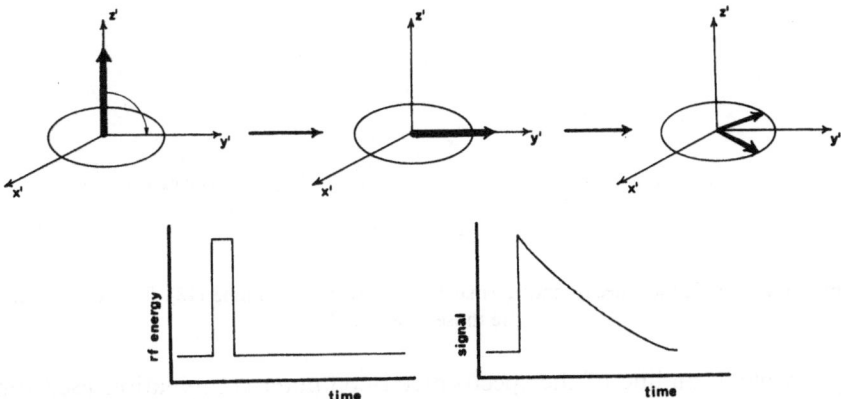

Fig. 5. Diagrammatic representation of the effect of a pulse of rf energy upon a magnetic nucleus in the rotating frame and the resulting free induction decay of the signal

The parameters of the system are best understood by reference to the rotating frame of the nmr experiment, Fig. 5. In conventional "cw" spectrometry, the weak observing field H_1 of the spectrometer is varied. At resonance, the nucleus is perturbed and precesses about the effective field. In the reference frame which rotates at the resonant frequency this corresponds to a slight displacement of the magnetic dipole of the nucleus out of alignment with the magnetic field, producing a component in the $x'y'$ plane detectable by a suitably placed receiver coil. However, if a much stronger H_1 is used, the dipole may be substantially displaced from alignment by a very short pulse. When this pulse is stopped, the nucleus will return to equilibrium by normal relaxation processes, resulting in a detectable signal termed a free induction decay (FID).

It can be shown that the free induction decay is the Fourier transform of the normal cw spectrum (40); i.e., the signal which can be collected in a second or two contains all of the information which is collected in 250 or 500 seconds by a cw spectrometer. Contemporary electronic technology allows the collection and coherent adding of a series of FID's to produce an enhancement of signal to noise proportional to the square root of the number collected. Thus it is practical to collect many thousand FID's to produce sensitivity enhancement of several orders of magnitude above that of a single cw run. Except in the case of very simple spectra, the FID does not yield spectral information on direct inspection, but must be converted by a small computer into the familiar absorption spectrum. A comparison of the two appears in Fig. 6.

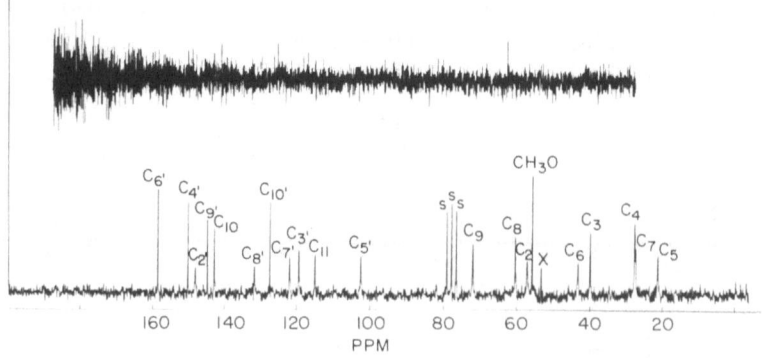

Fig. 6. Free induction decay and normal ^{13}C spectrum of quinine (18). The assignments are those of Ref. (42)

A block outline of the spectrometer-computer combination used for the experiment appears in Fig. 7. In operation, the receiver is turned off during the strong burst of rf power which the transmitter produces and

$CH_2{=}CH$ H

H

HO—$\overset{.}{C}$—H

CH_3O

(18)

very briefly thereafter to avoid swamping its circuitry. The analogue signal received is converted into digital information by the A/D converter, and stored either in the computer or an associated magnetic disk, sequential FID's being coherently added. Throughout the experiment, the field and frequency are locked by reference to a solvent resonance. After enough FID's are collected the computer effects a Fourier transform of the stored data, and the resultant absorption spectrum is plotted as a conventional spectrum.

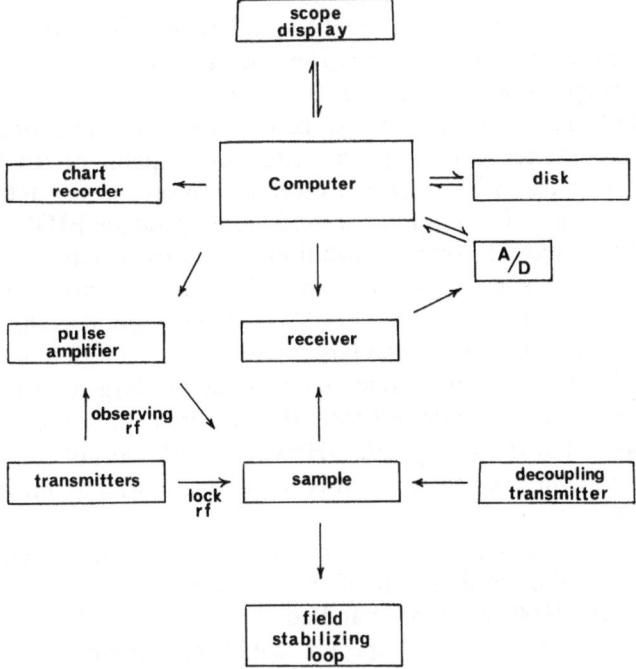

Fig. 7. Block diagram of a Fourier transform spectrometer

The experiment adds some new parameters to nmr spectrometry which are very satisfactorily handled by mathematical treatment and somewhat less satisfactorily by available equipment. It can be shown that all of the nuclei with resonant frequencies within the range Δ from a pulse will experience equivalent displacement if the strength of the pulse is large enough that

$$\gamma H_1 \gg 2\pi\Delta.$$

Since the angle of displacement is proportional to both pulse length and field strength,

$$\theta = \gamma H_1 t.$$

For a 90° pulse,

$$\gamma H_1 t = \pi/2,$$

and

$$t \ll 1/4\Delta.$$

To allow study of the full 6000 Hz spread of a ^{13}C spectrum at 23,500 gauss (at which field the proton frequency is 100 MHz), this relation implies that the 90° pulse should be less than 50 µsec. Spectrometers now sold meet or exceed this requirement, but many older instruments are successfully operated with much inferior power.

If the resultant absorption spectrum is not to be distorted, the A/D converter must sample the spectrum at a rate at least twice the frequency range to be encountered; this is readily feasible. However, the resolution of the spectrum proves to be the inverse of the period of data collection. As contemporary spectrometers are capable of resolution of better than 0.25 Hz in ^{13}C spectra as well as proton spectra, FT spectrometers collecting 12,000 data per second would produce FID's of 48,000 points to be stored. However, resolution of approximately 1 Hz retains nearly all of the spectral information obtainable in noise decoupled spectra, allowing the use of FID's of 8000 or 16000 points. This requires the use of a computer with much larger storage capacity than most minicomputers have, or an interactive magnetic disk of large capacity. Most spectrometers now in use do not meet this requirement, but compromise either resolution or frequency range. However, as the cost of core capacity in computers continues to drop, this requirement will be met with increasing ease.

Finally, the dynamic range of the spectra accumulated sets the requirements for the word length of the computer. A dynamic range of approximately 1000 can be handled by a 12-bit A/D computer ($2^{10} = 1024$) but only a few spectra could be accumulated without distortion. As a result, the use of computers with 16-bit words is now general, frequently with provision for the use of double precision words.

Commercial instruments are now available which satisfy all of these requirements without causing the chemist undue preoccupation with electronic or computer technology. Their use has permitted the facile investigation of micro samples by proton nmr, extended routine nmr investigations to carbon and phosphorus, and introduced the study of relaxation measurements to aid in the characterization of organic compounds and biochemical complexes.

2. Proton Spectra from Small Samples

It has long been possible to obtain useful proton spectra from very small quantities of material by the use of microcells and extensive collection of cw spectra. However, in the experience of most spectroscopists, micro cells are unsatisfactory, generally degrading resolution and frequently entailing the risk of losing the sample by inept handling or failure of the cell. FT spectroscopy allows the study of millimolar solutions in normal cells in much less time. Although it has been demonstrated that homonuclear decoupling is feasible, many commercial systems do not allow it, and it is not commonly used.

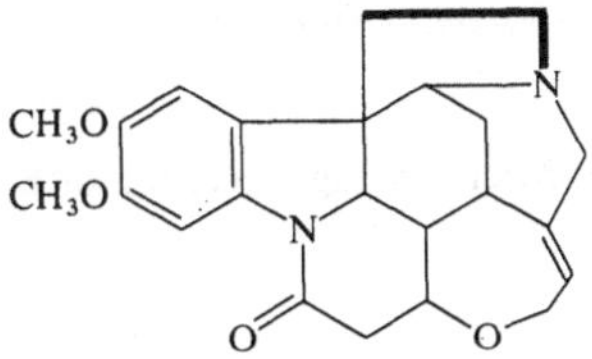

(19)

An important caveat must be entered in the study of spectra of small samples, as exemplified by Fig. 8, which is that of 50 µg of brucine (19) in .3 ml of deuteriochloroform, prepared with normal care. Such anticipated features of the spectrum as the methoxyl and aromatic protons are visible, even including the blurred olefinic signal. However, the most prominent peaks are artifacts. The proton remaining in the deuteriochloroform which dominates the downfield portion of the spectrum can be reduced by the use of "100%" deuteriochloroform. However, at high field strong peaks appear due to water and grease. If the artifactual character of such peaks were not recognized, the plot amplitude would be reduced, and the chemist would be forced to some such conclusion as that a methyl ketone and a long aliphatic chain were present in his sample.

At present no satisfactory solution to this problem is generally available. Special sample handling techniques do not much reduce the offending peaks. When the computer facilities are extensive enough, special programs allow the subtraction of a blank spectrum, but they are neither generally available nor adequately tested. Multiple pulse techniques (43) allow the reduction of the water peak, at the hazard of distorting other peaks. At the moment of writing, a commercial insert allowing the use of capillary sample tubes has been introduced, which promises to improve the problem notably by reducing the amount of solvent required a hundred-fold. An improvement of the signal to noise ratio is also obtained. Fig. 9 constrasts the spectrum thus provided with that of Fig. 8.

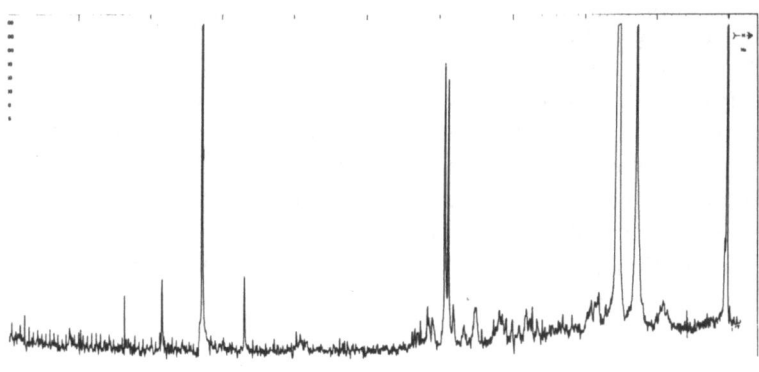

Fig. 8. Micro (50 μg) spectrum of brucine (19)

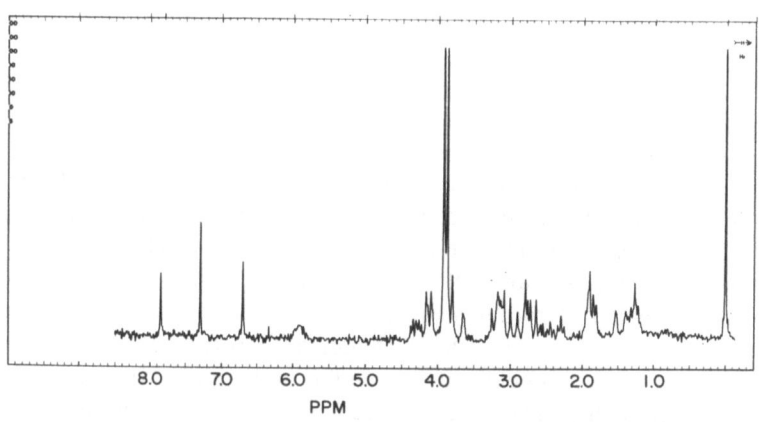

Fig. 9. Micro (50 μg) spectrum of brucine in 5 μl of CDCl₃.
The spectrum was obtained from 500 FID's, using an external ^{19}F lock. (Courtesy Dr. J. N. SHOOLERY)

References, pp. 162—166

XI. ^{13}C Satellites

Protons attached to a ^{13}C nucleus exhibit spin-spin coupling with very large coupling constants. Since the natural abundance of ^{13}C is 1.1%, satellite peaks of 0.55% relative intensity appear, flanking the main peak by J/2. The satellites of sharp peaks can frequently be recognized and provide the ^{13}C$-$H coupling constant, which can yield useful chemical information. The coupling constants reflect the amount of s character of the bond, sp$_3$ ^{13}C$-$H coupling being ca. 125 Hz, sp$_2$ ca. 160 Hz, and sp ca. 250 Hz. These values are modified by inductive effects and ring

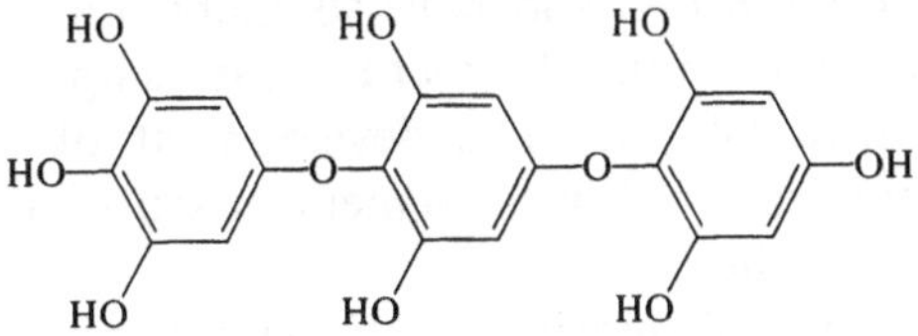

(20)

strain, and may be interpreted with the aid of extensive tables of the data from known structures (44). However, a variety of factors has prevented their being much used in natural product chemistry. Generally, the conventional proton spectrum provides much the same information from the chemical shifts and coupling constants. The necessity of studying the very weak peaks requires extensive collection of cw spectra or FID's with the attendant difficulties from artifacts and impurities.

However, the ^{13}C satellites can occasionally provide unique information. The proton on the ^{13}C atom is magnetically non-equivalent to other protons in the same molecule which may possess the same chemical shift by virtue of structural symmetry. The coupling can therefore be observed only in the satellite. GLOMBITZA and SATTLER were able to use ^{13}C satellites in this manner to support the structure of the polyphenol (20) isolated from *Halidrys siliquosa* (45). The octaacetate of the phenol showed three singlets at δ 6.31, 6.76 and 7.05 ppm; however, ^{13}C satellites of the three showed H$-$H couplings of 2.5 to 2.7 Hz, demonstrating that the singlets represented meta-situated protons in symmetrically substituted rings.

XII. Other Nuclei

With suitable modifications, modern nmr spectrometers are capable of observing the resonance of most nuclei with magnetic moments. However, the natural occurrence and nuclear properties of the isotopes limit the interest for natural product investigations to ^{13}C and ^{31}P.

At 23,500 gauss, the resonance frequency of ^{2}H is 15.4 MHz, with a sensitivity 1% that of ^{1}H. The natural abundance of .015% limits investigations to enriched samples. The relative sensitivity of ^{14}N, with a frequency of 7.2 MHz at 23,500 gauss is 0.1% that of ^{1}H, but the quadrupole results in very broad peaks except in symmetrical environments. The less abundant isotope ^{15}N possesses a spin of 1/2, and is suitable for high resolution nmr, but the relative sensitivity of 0.1% and natural abundance of 0.4% combine to put the spectrometer at a disadvantage forty times worse than that presented by ^{13}C. Concentrated solutions (5 to 9 M) of amino acids in 10 mm tubes have yielded useful spectra after 7 hours pulsing, and have so enabled an English group to make preliminary observations of structural effects upon the nucleus [46], which resemble those of ^{13}C. The development of spectrometers to produce an one-hundred fold improvement in sensitivity would be very interesting to alkaloid chemists.

The natural isotope of fluorine is ^{19}F, with a spin of 1/2 and a sensitivity 83% that of the proton. It is therefore eminently suitable for study by analytical instruments equipped to handle its frequency, 94MHz at 23,500 gauss. However, fluorine-containing natural products are so rare that there is no real application here.

Phosphorus occurs exclusively as ^{31}P, with 6% the response of protons and a spin of 1/2. The study of small samples of phosphorus-containing materials is therefore particularly facile by FT spectrometry, and is supported by extensive knowledge of the chemical shifts and coupling constants which characterize phosphorus [47]. Although the chemical shifts of ^{31}P extend over a range of nearly 700 ppm, natural materials, being mostly pentavalent, are limited to high field. No reference as satisfactory as tetramethylsilane is available, external 85% orthophosphoric acid being commonly used in spite of the broad peak it places in the middle of an interesting area. External P_4O_6 is a useful secondary standard, producing a sharp peak downfield -112.5 ppm from H_3PO_4. It is highly reactive, but sealed capillaries are commercially available. Because anionic phosphates have a high affinity for metal ions, it may be necessary to pass samples through an ion exchange column to remove paramagnetic impurities; alternatively, addition of micromolar amounts of ethylenediamine tetraacetic acid will remove the ion from the phosphate by formation of a complex and improve the resolution of the spectra. In any event, phosphate peaks are generally several hertz broad.

The chemical shifts of phosphorus are determined primarily by (1) the amount of $d\pi$ orbital involved in bonding; (2) the electronegativities of the substituents on the sigma bonds; and (3) deviations from geometric symmetry. The resulting complex situation makes it difficult to predict the shifts precisely. The problem for natural materials is simplified, since these factors are dominated by nearest-neighbor contributions and these materials are generally found in one of three classes: phosphates with shifts over a 6 ppm range centered on H_3PO_4; polyphosphates with shifts upfield, $+6$ to $+30$ ppm; or phosphonic acid derivatives which are more affected by substitution at remote carbons and are found downfield, -10 to -40 ppm.

[31]P nmr is therefore a particularly promising tool for biochemical investigations, which are vigorously under way. These are largely beyond the scope of this review, but a few applications are mentioned briefly. Because the chemical shifts of phosphates reflect pH, intracellular pH can be investigated (48). Complexes of enzymes with nucleotides can be investigated, LEE and CHAN having examined that between ribonuclease A and uridine-3'-monophosphate at 53,000 gauss (49).

Phosphonates are very effectively investigated by ^{31}P nmr because the chemical shift is sensitive to substitution. HENDERSON and his colleagues identified aminoethylphosphonic acid and phosphoserine in the extracts of the sea anemone (50).

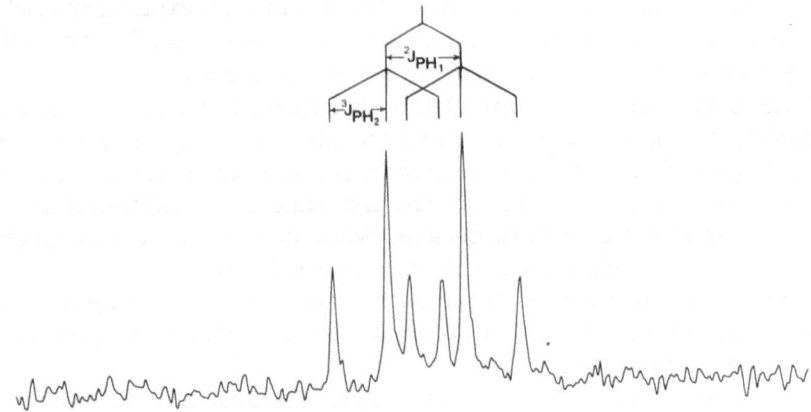

Fig. 10. [31]P spectrum of 2-amino-1-hydroxyethyl phosphonic acid from *Acanthamoeba*

Hydrolysis of a membrane polysaccharide from *Acanthamoeba castellanii* provided material of the constitution $C_2H_8NO_4P$, with the ^{31}P spectrum of Fig. 10 (51). The chemical shift of -15.6 (vs. H_3PO_4) showed it to be a phosphonic acid, and coupling to two protons with

$$CH_2-CH-PO_3H_2$$
$$NH_2 \quad OH$$

(21)

$J = 5.8$ Hz and to a single proton, $J = 8.2$ Hz could be seen. The ^{1}H spectrum showed the larger coupling to be to a proton of δ 3.49 ppm, while the smaller was to protons of δ 2.69 and 2.90 ppm, results consistent only with structure (21), 2-amino-1-hydroxyethyl phosphonic acid.

XIII. ^{13}C NMR Spectra

1. General Discussion

The most potent development from the introduction of Fourier transform techniques is the possibility for routine use of ^{13}C nmr spectra. Because of the greater paramagnetic contribution to the chemical shifts of nuclei beyond the first row of the periodic table, the range of chemical shifts of carbon is several hundred parts per million, in contrast to the 10 ppm range of the proton. Consequently, it has long been apparent that ^{13}C nmr spectra promised to be a valuable tool for the characterization of organic compounds. However, the low natural abundance of the isotope (1.1%), and its inherent low response in the nmr experiment (1.6% that of the proton at the same field) combined to make it totally unsuitable for the study of natural products in early spectrometers.

The development of multi-nuclear spectrometers made possible the complete decoupling of protons from ^{13}C nuclei by a high power noise-modulated rf field at the proton frequency, while the spectrometer observed resonances at ^{13}C frequencies, and achieved a field-to-frequency lock by the signal of a third nucleus. Noise decoupled ^{13}C resonances now appear as single peaks, with their intensities enhanced by hetero-nuclear Overhauser effects. When such a spectrometer is coupled to a Fourier transform accessory, it becomes feasible to obtain ^{13}C spectra on samples of familiar size.

These developments have permitted a great burgeoning of information on ^{13}C nmr spectra. Recently an excellent introductory book has appeared (52), as well as a definitive reference work (44), and a catalogue of spectra (42). These works describe current knowledge of the relation of structure

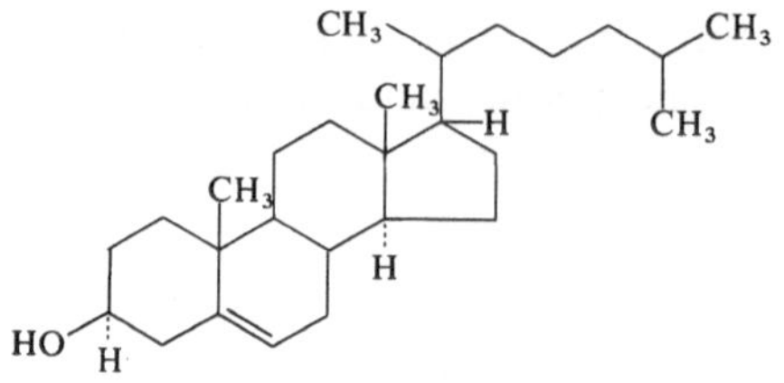

(22)

to the parameters of ^{13}C spectra in detail. This section will attempt only to sketch the primary characteristics of the method as a structural tool.

Fig. 11 contrasts the spectra obtainable from 10 mg of cholesterol (**22**) on a 23,500 gauss spectrometer*. Although extensive study has so developed knowledge of the spectra of steroids that the few sharp peaks of the proton spectrum can yield substantial chemical information, most of the proton resonances are huddled and unintelligible. The ^{13}C

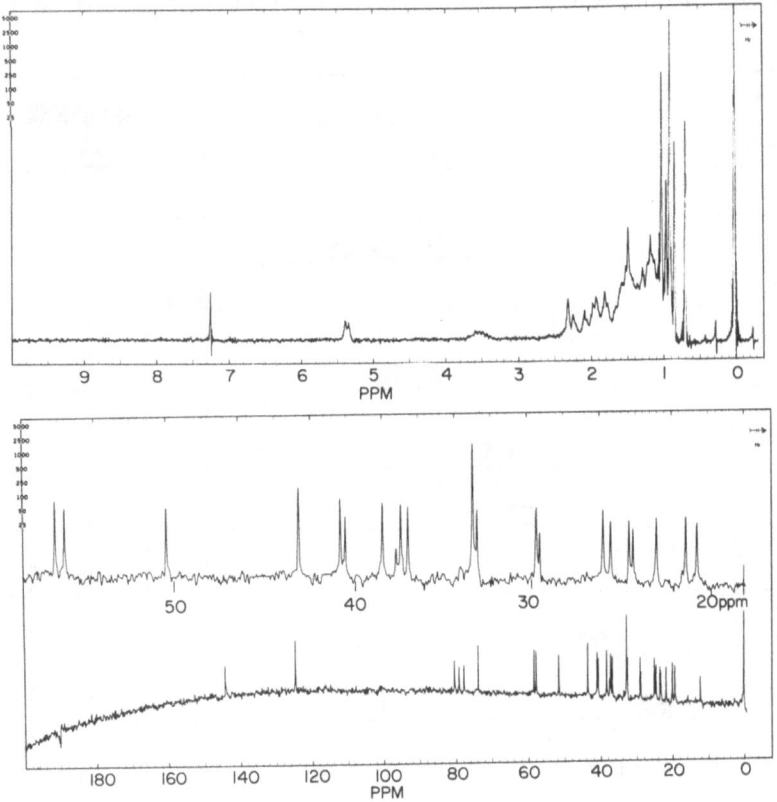

Fig. 11. ^{1}H (upper) and ^{13}C (lower) spectra of 10 mg. of cholesterol (**22**) at 23,500 gauss. The ^{1}H spectrum was taken in a single cw scan, while that of the cholesterol required the collection of 70,000 FID's in an overnight run

* Chemical shifts are shown in parts per million downfield from tetramethylsilane, as in proton spectra and will be quoted here as "δ 10". A variety of reference standards were used in earlier work, and some current work is still referred to carbon disulfide, δ 192.8 downfield. However, inasmuch as the standard works on the subject have adopted TMS as the standard, it is to be hoped that use of this reference will become general.

spectrum shows 25 peaks for the 27 carbon atoms. The peak at δ 42.4 can be expanded to show a doublet of 1.3 Hz separation, while that at δ 32.0 evidently corresponds to two carbons. Thus a datum for each carbon is available to assist characterization of the molecule.

Earlier work provided the chemical shift ranges characteristic of carbon in various functional groups, as summarized in Fig. 12. It is apparent that the two peaks at low field in Fig. 11 are those of the olefinic group, while that at δ 72 is that of the carbinyl carbon. The effect of various substitutions on carbon shifts has been investigated, and the

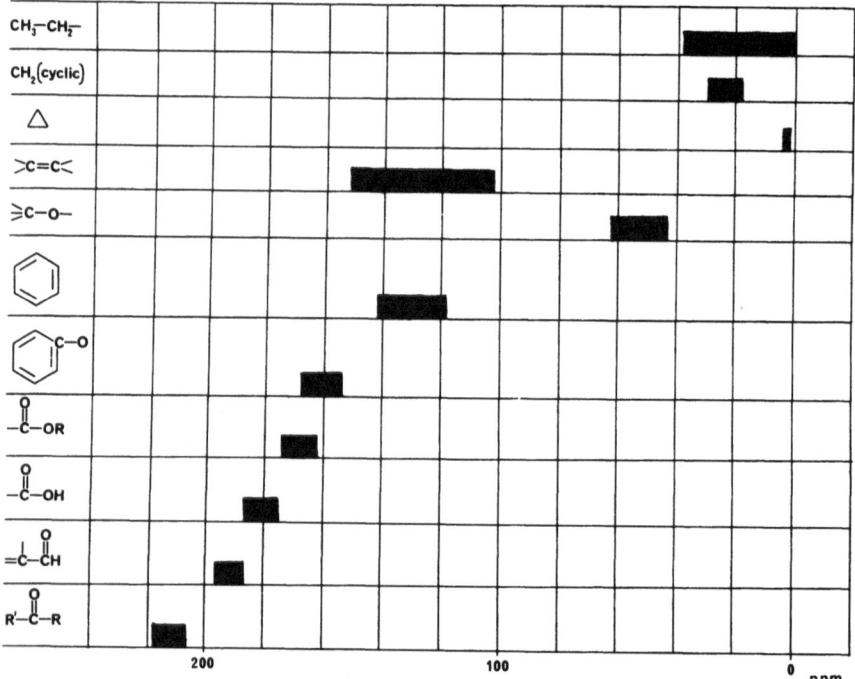

Fig. 12. Chart of the chemical shifts of variously situated ^{13}C atoms

general effects are well established, but exact structural inference from the chemical shifts of complicated cyclic molecules is not generally possible without the context of closely related compounds. Most current work on natural products consists of the compilation of the spectral characteristics of many closely related series of natural materials (53), but enough examples of structural investigations have appeared to indicate the manner in which ^{13}C nmr can be used.

The most obvious approach to the assignment of chemical shifts is the comparison of the spectra of a group of closely related compounds. As a change of substituent rarely affects chemical shifts of carbon atoms more than three bonds distant, carbon atoms close to the substituent can be identified and differentiated by the effects established in simpler systems (44, 52). Various techniques are available to aid this task. The most familiar of these, off-resonance decoupling, is achieved by a coherent (i. e., single-frequency) decoupling field several hundred hertz from the proton absorption of the sample. The effect is to reduce markedly all of the proton couplings. As a consequence, only one-bond couplings remain large enough to produce visible multiplets from the carbon atoms. The carbon absorption now takes on a multiplicity characteristic of the number of protons on the atom: Methyl group carbons appear as quartets, methylene carbons as triplets, methine carbons as doublets, and quaternary carbons as singlets. A spectrum with many carbons of similar chemical shifts results in a confusing overlap of multiplicities, which may well be difficult or impossible to unravel completely. Quartets and doublets may be distinguished from singlets and triplets by overlaying the partially and fully decoupled spectra. In the former, only singlets and triplets can have a peak corresponding to that in the latter. The outermost peaks needed to identify a group as a quartet or triplet may well be lost in noise or overlapping peaks. Finally, methylene groups bearing protons of markedly different chemical shifts may produce doubled doublets, rather than triplets.

A multiple pulse technique, termed partially relaxed Fourier transform spectroscopy ("PRFT") is available to supplement off-resonance decoupling within appropriate series (54, 55). As explained below, if the magnetic dipole of a nucleus is inverted by a 180° pulse, followed by a pause, and then displaced by a 90° pulse, the peak may appear erect or inverted or, if the pause was approximately $0.7 \, T_1$, may be lost in the baseline. Within the steroid series, the T_1's of the methyl and quarternary carbons are approximately 1 sec. (56), those of the methine approximately 0.5 and those of the methylene 0.3 sec. A pause of approximately 0.35 sec, then, will produce a spectrum such as Fig. 13. Methylene carbons are represented by erect peaks, methyl and quarternary carbons by inverted peaks, and methines by the absence of peaks. Some experimentation may be necessary to find appropriate pauses for other series. As knowledge of relaxation times is still somewhat limited, the technique should be used warily while experience is gained.

When a proton of the molecule can be replaced by a deuterium atom, as those of a methylene group α to a ketone, another tool is available. The carbon atom bearing a deuterium is not decoupled, does not receive Overhauser enhancement, and is generally, therefore, lost in the noise of the spectrum, while adjacent carbons may show small chemical shift differences.

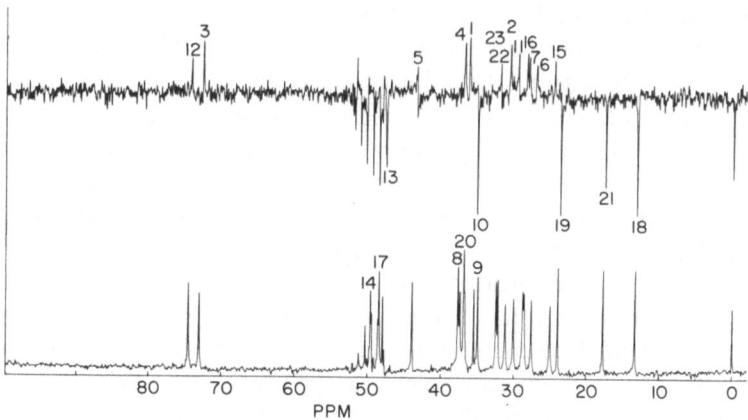

Fig. 13. Partially relaxed Fourier transform spectrum of desoxycholic acid (upper trace) and normal ^{13}C spectrum (lower trace) (cf. Fig. 18).

The 180° pulse was followed by a delay of 0.25 sec. In this period, ^{13}C atoms with a short relaxation time relaxed sufficiently to produce erect peaks, while those with T_1's near 1 sec remain inverted. Methines, with T_1's near .5 sec are mostly lost in the noisy baseline. The carbinyl methines, approximately 2500 Hz from the pulse, did not experience an effective 180° pulse, and appear distorted

Lanthanide-induced shifts can also be used to good effect. It appears that the carbon atoms bearing the complexing functional group may suffer a contact shift, but other carbon atoms should show effects interpretable in terms of their distances from the complex. The technique has been exemplified by studies by APSIMON et al. on hydroxy and keto steroids which required the reassignment of the chemical shifts of C-12 and C-16 of earlier studies of cholesterol (57). RANDALL et al. have made similar studies of borneol (58).

In appropriate circumstances, the use of single proton decoupling is particularly effective. Protons appearing as reasonably discrete peaks

(23)

can be selectively decoupled by a single frequency chosen from their chemical shift. As Fig. 14 shows, there results a spectrum in which a single carbon appears essentially as a singlet. Obviously, if the technique were applied to each carbon, the exercise would involve a great deal of instrument time.

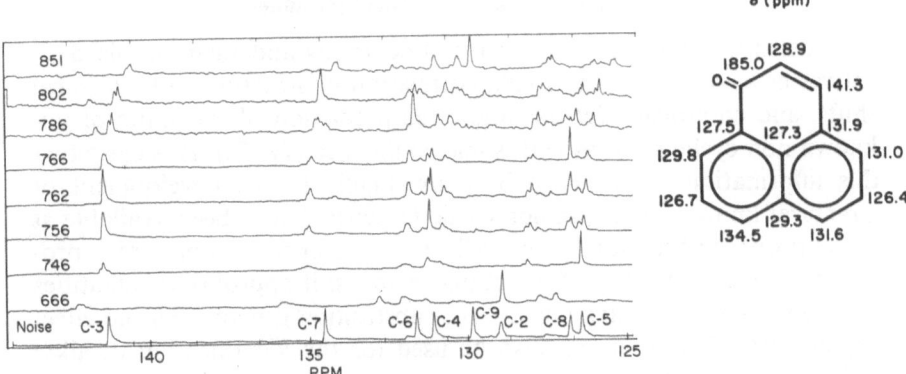

Fig. 14. Single proton decoupled ^{13}C spectra of phenalenone (**23**). The spectra were obtained by rapid 90° pulses which resulted in the saturation of ^{13}C atoms bearing no protons. They consequently do not appear. The bottom spectrum is a normal noise-decoupled spectrum. Those above were obtained with a coherent 100 MHz. rf irradiation set according to the proton frequencies previously determined (*59*)

2. Sample Size and Acquisition Times

Most modern spectrometers intended for ^{13}C nmr spectroscopy are capable of using large sample tubes, such as those of 12 mm diameter. When the supply of sample is generous, large tubes are very advantageous, for a single FID from 3 ml of a 2 M solution can provide a satisfactory spectrum. However, it is the common experience of natural product chemists to work with too little material, and they will generally prefer to use the familiar 5 mm tubes. If relaxation effects present no problem, 0.3 ml of 1 M solution can provide a satisfactory spectrum from 250 FID's, taken in 5 min. Samples of 0.1 mmole will provide satisfactory spectra in an hour or two, while 0.03 mmole will require an overnight run, and represents a practical lower limit.

If the molecule being studied has carbon atoms of long T_1, such as a carbonyl group or the substituted carbon atoms of aromatic rings, a rapid sequence of 90° pulses will quickly result in saturation and very little signal (*60*). The accepted approach to this problem is to use smaller flip angles from shorter pulses (*40, 61, 62, 63*). The consequence may well be that the total time will have to be as much as ten times as long and the minimum samples ca. .1 mmole. Off-resonance and single-proton decoupled spectra take approximately four times as long as noise-decoupled spectra.

3. Chemical Shifts of Alkyl Residues

Extensive studies of branched hydrocarbons and their simple derivatives have revealed a satisfying correlation of structure with chemical
shifts, and empirical relations allow the prediction of the shifts of unknown non-cyclic systems with some confidence (44, 52). It is clear that
this information would have been invaluable to the development of
structural knowledge of the non-cyclic terpenes, had it been available at
the appropriate time, and may well prove useful to contemporary problems, such as the study of insect pheromones, if appropriate quantities
of material were to be available. Since substitution generally has no effect
beyond three bonds, it can also be used for the identification of alkyl
residues attached to complex molecules.

Plattner and his colleagues were able to use such studies to support
the structure assigned to a chromanone from *Calophyllum brasiliense* oils
(64). Conventional spectra and context showed the structure to be (24),

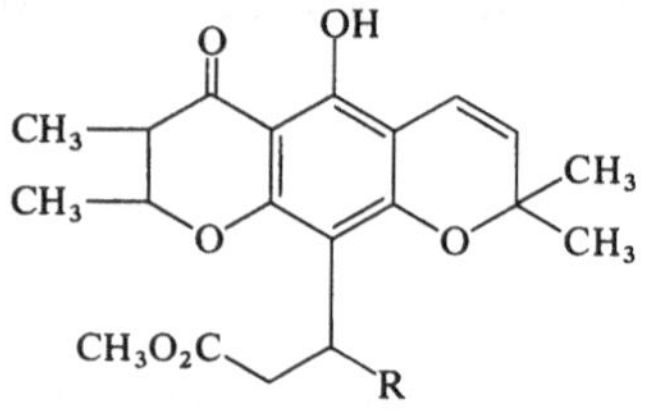

(24) R = Bu
(25) R = n−Pr

with the single point of uncertainty the nature of the butyl group. While
it was reasonably clear that only a n-butyl group could produce the
pattern observed in the proton spectrum at high field, ^{13}C nmr spectra
provided an unambiguous assignment of structure. Comparison of the
spectrum of the unknown with that of the known compound of
structure (25) revealed three peaks not duplicated in the spectrum of (25)
with δ 22.6, 30.0 and 33.0. The peak at δ 14.0 ppm, which produced a
clear quartet in the off-resonance spectrum, was readily identified as the
terminal methyl of the side chain. The chemical shifts could now be
compared with published data for model structures. Those for 3-methyl-
heptane provided a good fit, δ 14.1, 23.2, 29.7 and 36.5 for C-7, 6, 5 and 4
(65).

4. Functional Group Recognition

It is apparent from Fig. 12 that ^{13}C nmr spectra provide a means of
characterizing the functional groups of a molecule. This capability was
exploited in demonstrating the striking reversible rearrangement of

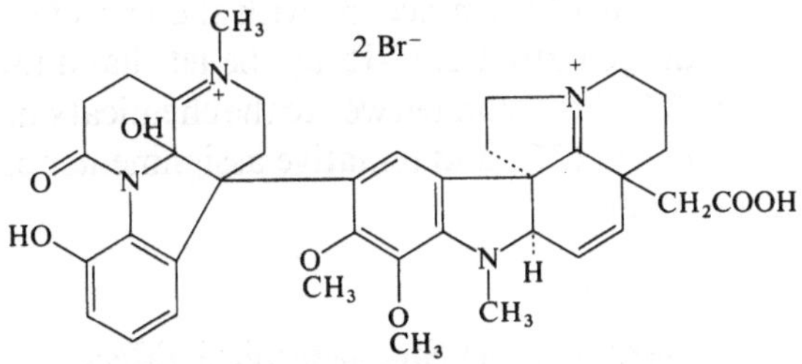

(26)

haplophytine and its dihydrobromide (66). Although x-ray studies of the hydrobromide had shown its structure to be (26), the ^{13}C nmr spectrum of the parent alkaloid revealed three carbonyl groups represented by singlets at δ 175.1, 175.6 and 197.2. The first two could be assigned to lactam and lactone carbonyl carbon atoms, and the third demonstrated the presence of an unsuspected ketonic carbonyl.

These results are accommodated by structure (27). Support for the nature of the lactone group and the entire right-hand half of the

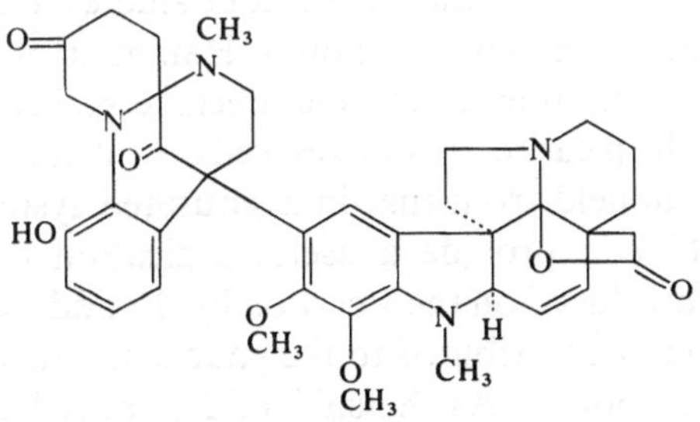

(27)

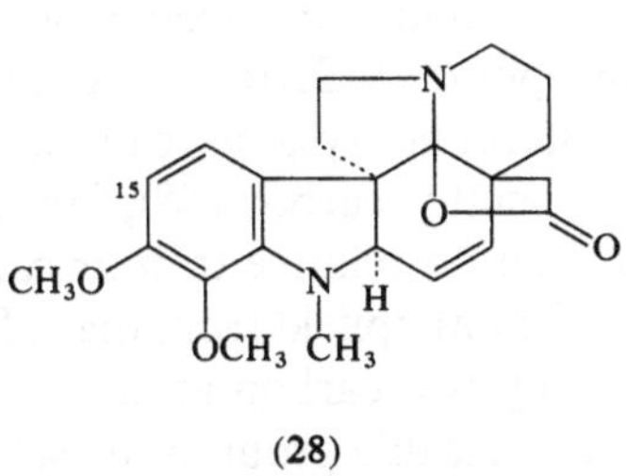

(28)

molecule was found in comparison of the ^{13}C nmr spectrum of aspidophytine (28). Peaks in the spectrum of (27) corresponded to those of (28) with the exception of a doublet of δ 102.3 assigned to C-15 of (28); it is replaced by a singlet of δ 124.1 in accord with the expected effect of the replacement of a hydrogen atom at C-15 by the left hand moiety of (27). The low field peak of (27) corresponded well to the chemical shift anticipated for a ketone situated as in (27), and tentative assignments for the remaining peaks could be made.

5. Substituent and Stereochemical Effects

As the discussion of substituent effects on the chemical shifts of relatively simple systems runs to hundreds of pages in STOTHERS' comprehensive work (44), no attempt to produce a general summary is in order here. In the study of many groups of natural products, many of the differences observed in the comparison of the spectra of closely related compounds can be understood by the generalization that the substitution of a methyl group into a saturated system commonly shifts the α and β carbon resonances downfield about 8 ppm, and the γ carbons upfield about 2 ppm. The substitution of a hydroxyl group into a saturated system has a more marked effect on the α carbon, a downfield shift of about 40 ppm, but effects on β and γ carbons are similar to those of the methyl group. Substitution effects on aromatic systems require closer attention to the tabulated results of previous studies, but changes on para carbon atoms can be shown to follow a Hammett plot (44). Structural changes often follow familiar electronic effects: e. g., in an α,β-unsaturated carbonyl system, the β carbon resonates well downfield, while the carbonyl carbon resonates upfield from that in a saturated system.

The "γ effect" can provide a useful indication of stereochemistry. The upfield shift produced on the γ carbon by the addition of a substituent to an aliphatic chain is attributed to the gauche interaction of the carbon atoms in a butane moiety. As the shift of 2 ppm reflects the fraction of the molecular population in that conformation, it is not surprising that an axial methyl group in a rigid cyclohexane system suffers an upfield shift of approximately 6 ppm (67). STOTHERS' group has recently shown that A/B cis and trans steroidal systems can be differentiated by the chemical shifts of the C-19 methyl carbon (68), for that of the trans isomer exists in a gauche conformation relative to four carbon atoms, and as a consequence occurs 6 to 12 ppm upfield from that of the cis isomer, which bears such a relation to only two carbon atoms.

Within the context of a sizeable group of closely related compounds, chemical shifts can provide fairly specific information on the substitution pattern and stereochemistry of an unknown compound. Recent studies on indole alkaloids illustrate this. Two new alkaloids from Aspidosperma, vandrikidine and hazuntinine, were shown by conventional spectrometry

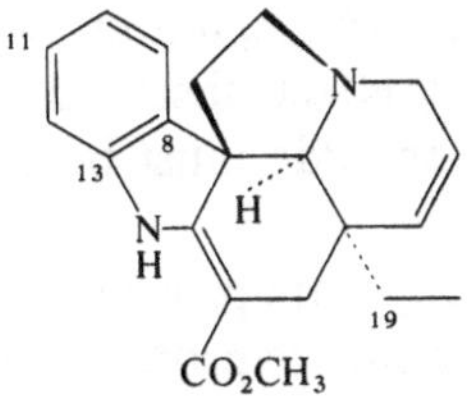

(29)

to be related to tabersonine (29) (69). ^{13}C nmr spectra, and presumably proton nmr sprectra as well, showed the presence of an additional aromatic methoxyl, δ 55.3; methoxyl substituent parameters (α, $+30.2$; o, -15.5; m, 0; p, -8.9) were applied to the aromatic shift data of tabersonine. The two downfield peaks of (29), δ 137.8 and 143.1, were known to correspond to C-8 and C-13 (70). The occurrence of peaks in the spectrum of vandrikidine at δ 130.4 and 144.0 required that the new methoxyl be at C-11. An additional hydroxyl was evidently to be placed at C-19 since only carbon atoms within three bonds of this position were altered. The methoxyl

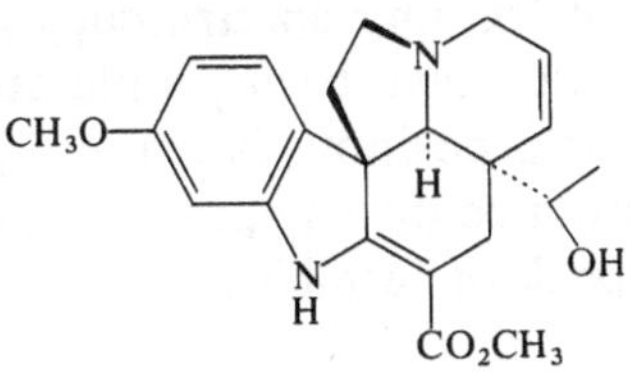

(30)

and hydroxyl groups of vandrikidine are thus placed as in (30). It would be instructive to be able to compare the information available from the proton nmr characteristics of this compound. The hydroxyl at C-19 was surely indicated by the appearance of the carbinyl proton and the doublet methyl group, but the pattern resulting from the 1,2,4-substituted aromatic ring might well not have permitted an unambiguous assignment.

The presence of an epoxide group in hazuntinine was inferred from the appearance of methine carbons at δ 51.8 and 57.0. Since carbons 3, 5

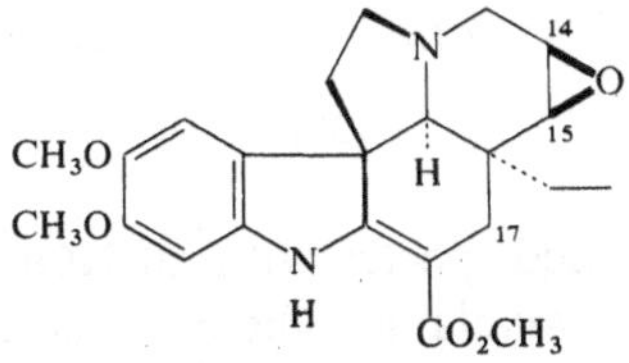

(31)

and 6 appeared as methylene groups, it must be at C-14 and C-15, and (31) must be the structure. The γ effect observed in the 2.3 ppm upfield shift of C-17 indicated that the epoxide must be *cis* to C-17 and *trans* to the ethyl side chain.

The complexity of carbohydrate proton nmr spectra has always placed a limit on their utility, the anomeric protons often being the only resonances readily interpretable. Proton nmr spectra of polysaccharides are further hampered by the line broadening common to high molecular weight materials. These difficulties are both much less troublesome in ^{13}C nmr spectra. The spectrum of the glucan from *Tremella mesenterica* shows multiplets for both the C-1 and C-6 of the glucose residues which can be identified by comparison with the corresponding atoms of disaccharides of known configuration (*71*).

6. Biogenetic Studies

The possibility of obtaining a measurable response from each carbon atom of a natural product immediately suggests that one of the most rewarding applications of ^{13}C nmr spectroscopy may be in the study of biogenesis, and such studies have been proliferating. When a substrate enriched with ^{13}C at known positions is fed to a growing organism, the products subsequently isolated display peaks of enhanced strength; once the identity of each peak is established, the biogenetic pattern is demonstrated.

There is a fundamental difficulty in the use of ^{13}C intensities, in that nuclear Overhauser enhancements differ from one carbon to another within a molecule. The problem may be avoided by the study of undecoupled spectra, but the resultant loss of sensitivity and discrete peaks is generally prohibitive. Appropriate gating of decoupler and receiver

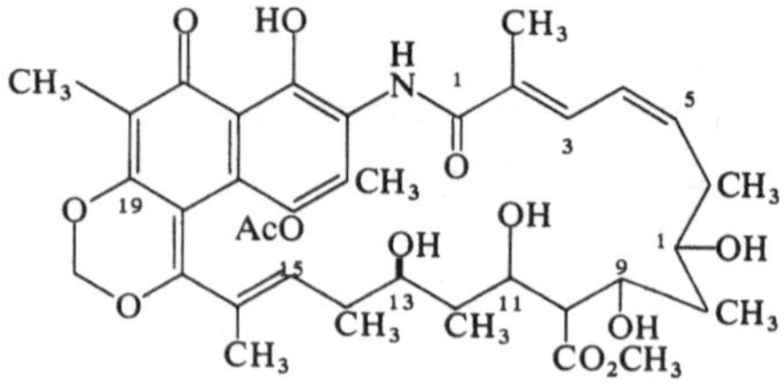

(32)

allows the use of decoupled spectra without NOE, but the programs and hardware are not widely available. The more straight-forward technique is simply to compare samples of normal ^{13}C content with those prepared by enrichment. Such spectra are said to provide reliable indication of carbon atoms with as little as 5% enrichment (72), the limitation being primarily the substantial noise level of most cmr spectra from limited amounts of material. A useful ancillary technique which does not seem to be widely employed is the simultaneous enrichment by both ^{13}C and ^{14}C. By this method, the degree of incorporation and resulting anticipated enhancement of the ^{13}C spectrum can be measured by the amount of radioactivity of the sample.

Feeding sodium (carboxy-^{13}C)propionate to *Streptomyces spectabilis* provided streptovaricin D (**32**) with an average of 15% enrichment at 1, 5, 7, 9, 11, 13, 15, and 19 carbons. Since the amide carbonyl is

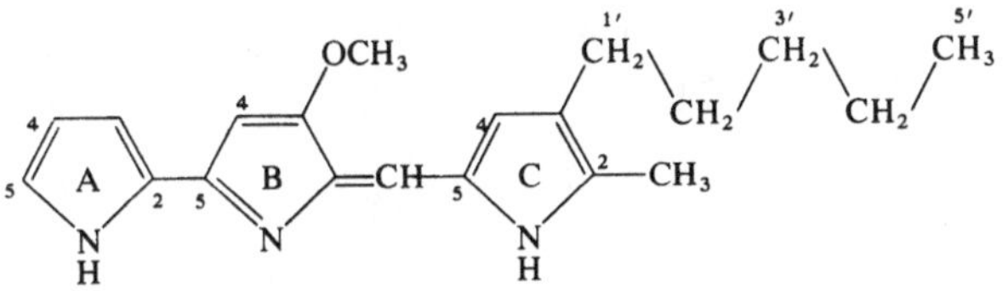

(33)

enriched, the results establish the "amide-head" pathway, rather than that in which the amide carbonyl is derived from a one-carbon fragment, with the propionate carbonyl appearing at C-3 (73). The complex pathways of primary and secondary routes leading to prodigiosin (**33**) were examined by the ^{13}C nmr spectra obtained on samples resulting from feeding acetate, alanine, glycine, proline, methionine and serine (74). It could be shown, for instance, that feeding (carboxy-^{13}C)proline resulted in the enhancement of only the signal from B-5 (28%), while feeding alanine produced enhancement in the amyl side chain at 1', 3' and 5', at B4, at C2, and the CH$_3$ at C2. As the first four of these were also enriched by acetate, it was concluded that only C2 and C2−CH$_3$ arise directly from alanine, the others resulting from the well established metabolism of alanine to acetate.

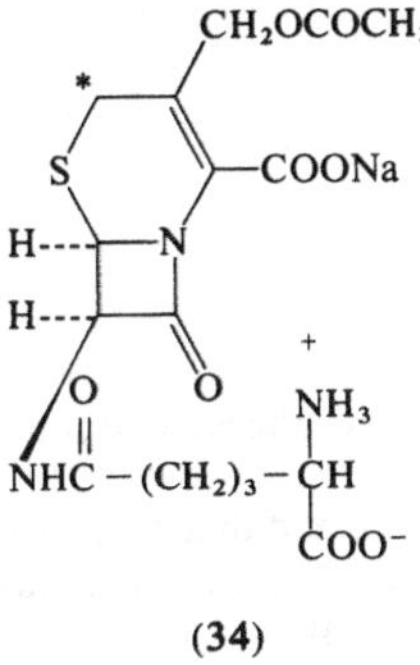

(34)    (35) 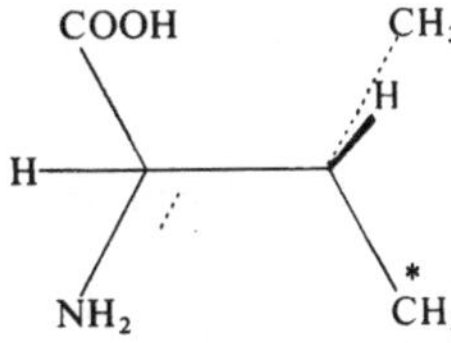   (36)

Studies on the biogenesis of cephalosporin C (**34**) have demonstrated
the value of the use of chirally labelled precursors in these studies (*72*).
Carboxylation of 2-phenylcyclopropyl magnesium bromide with $^{13}CO_2$
and resolution as the quinine salt provided (+)-*trans*-(1*S*,2*S*)-((1 – ^{13}C)-
carboxy)-2-phenylcyclopropane (**35**), which led by established steps to
(2*RS*,3*S*)-(4 – ^{13}C)valine (**36**) (*75*). When this was fed to *Cephalosporium
acremonium*, the cephalosporin C isolated showed an enhanced intensity
at C-2 exclusively.

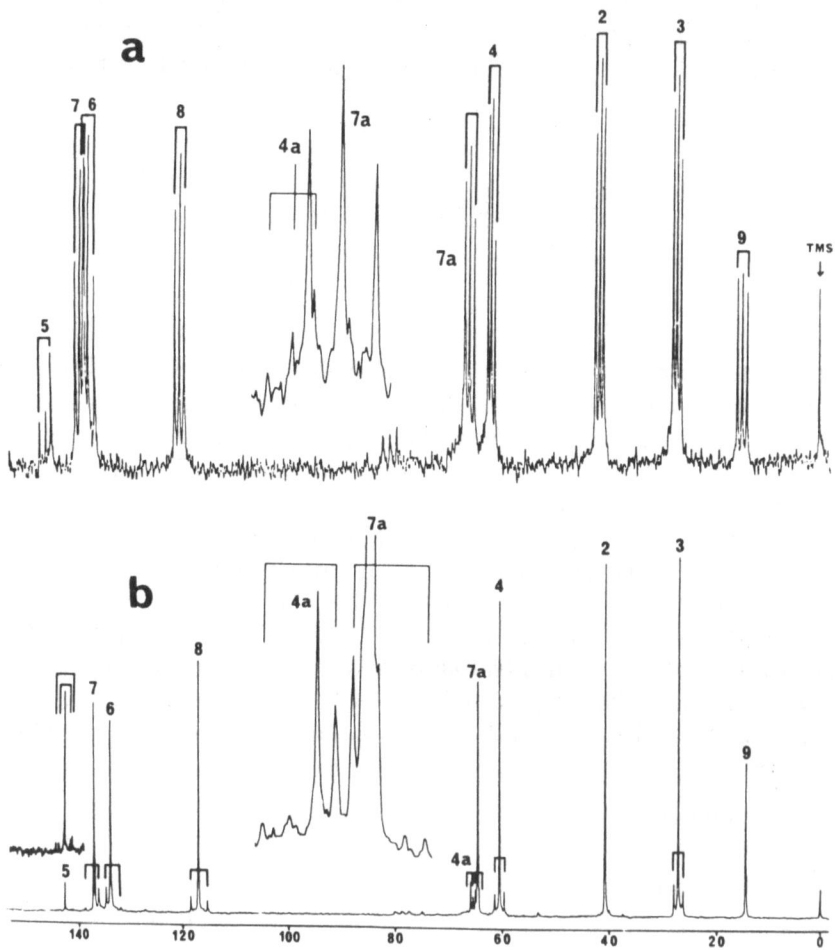

Fig. 15. ^{13}C nmr spectra from dihydrolatumcidin (**37**) prepared by alternate double
labelling.
The upper spectrum (a) is that of material obtained from $^{13}CH_3^{13}COOH$, and that below (b)
is that from the mixture of $^{13}CH_3COOH$ and $CH_3^{13}COOH$. Excess normal acetate was
added to avoid excess labelling. (Courtesy of Dr. H. Seto)

Japanese chemists have recently developed a method of striking ingenuity, "alternate double labelling" (76). $^{13}CH_3^{13}COOH$ is fed to a microorganism, with the result that acetate-derived carbons in metabolites appear as triplets, the center peak being that of the (preponderant) atoms of natural abundance (Fig. 15). Measurement of the ^{13}C-coupling constants demonstrates the pairing of carbons from the same acetate molecule. In a complementary experiment, a mixture of $^{13}CH_3COOH$

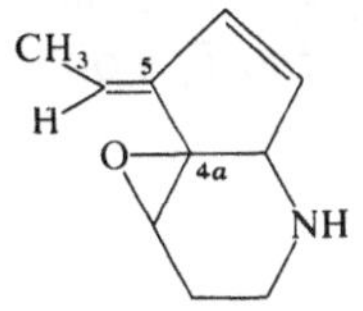

(37)

and $CH_3^{13}COOH$ is fed, resulting in triplets observable from the joining of acetate units in a head-to-tail manner. The authors illustrate the method by demonstrating that the spectra of dihydrolatumcidin (37) so prepared provide both a structure and biogenetic pattern consistent with previous results. $^{13}C - ^{13}C$ couplings observed varied between 35.4 and 78.7 Hz, without ambiguous indication of pairings. Coupling with two atoms was observed for C-4a and C-5.

7. Relaxation Studies

When the equilibrium population of a group of magnetic nuclei in a magnetic field has been perturbed by a pulse of rf energy, it returns to equilibrium by mechanisms characteristic of the molecular situation of the nuclei (41). Determination of these parameters, the spin-lattice relaxation time (T_1) and the spin-spin relaxation time (T_2) provides further characterization of the nuclei which may be useful in structural investigations. The spin-spin relaxation time, T_2, should provide useful information on polymeric and complexed materials, but only investigations of T_1 have provided results relevant to chemical structures (77).

Determination of T_1 is to be understood by reference again to the rotating frame (Fig. 16). In the most commonly used method, the "inversion-recovery" method, the nuclei are subjected to a 180° pulse to invert the magnetic dipole; the nuclei then begin to return to equilibrium by a first order process. If, after a pause of τ seconds, a 90° pulse is applied, the extent to which relaxation has proceeded can be measured by the signal received. It can be readily shown that the signal will follow the relation:

$$\ln(A_\infty - A_\tau) = \ln 2 A_\infty - \tau/T_1$$

where A_∞ is the intensity observed after a long pause ($>5T_1$). A semilog plot of intensities of peaks against the pauses used provides a line with the slope of $-1/T_1$. An example of such an experiment is shown in Fig. 17.

It is seen that after a pause of approximately $0.7\,T_1$ the signal observed will be 0; this is, indeed, the basis for PRFT, described above. For accurate determinations of T_1, however, each pulse sequence must be followed by a pause of five times the longest T_1 of the systems, to allow the nuclei to achieve equilibrium states. As the T_1's of commonly encountered systems

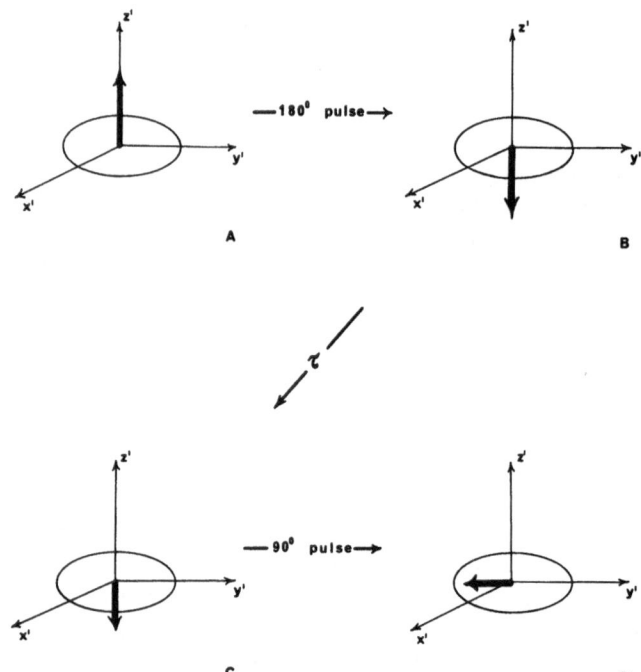

Fig. 16. Diagrammatic representation of the inversion-recovery experiment

may reach up to 60 sec, the experiments can be fairly lengthy. Since 250 FID's of a 1 M solution in a 5 mm tube provide a satisfactory spectrum, a 24 hr run should provide enough points for measurement of T_1 on .5 mmole of sample. If more generous quantities are available, a satisfactory signal can be obtained from a few FID's from 5 ml of a 1 M solution, and the determination can be effected much more quickly. Many molecules of interest to natural product chemists have no atoms of relaxation times greater than a few seconds, so that these experiments can be carried out relatively quickly; alternatively, the atoms of long relaxation times can be ignored. If long T_1's are to be measured (greater than perhaps 20 sec for ^{13}C) the samples must be degassed by freeze-evacuate-thaw cycles or by passing nitrogen through the sample. Because diffusion can affect the T_1's observed, it is well to standardize sample depth to 5 cm. With these precautions, observations reproducible within 10% can be obtained (78).

To return to the equilibrium population distribution, excited nuclei must interact with a fluctuating magnetic field, as that of other nuclei moving in the vicinity. Of the five mechanisms discussed for this

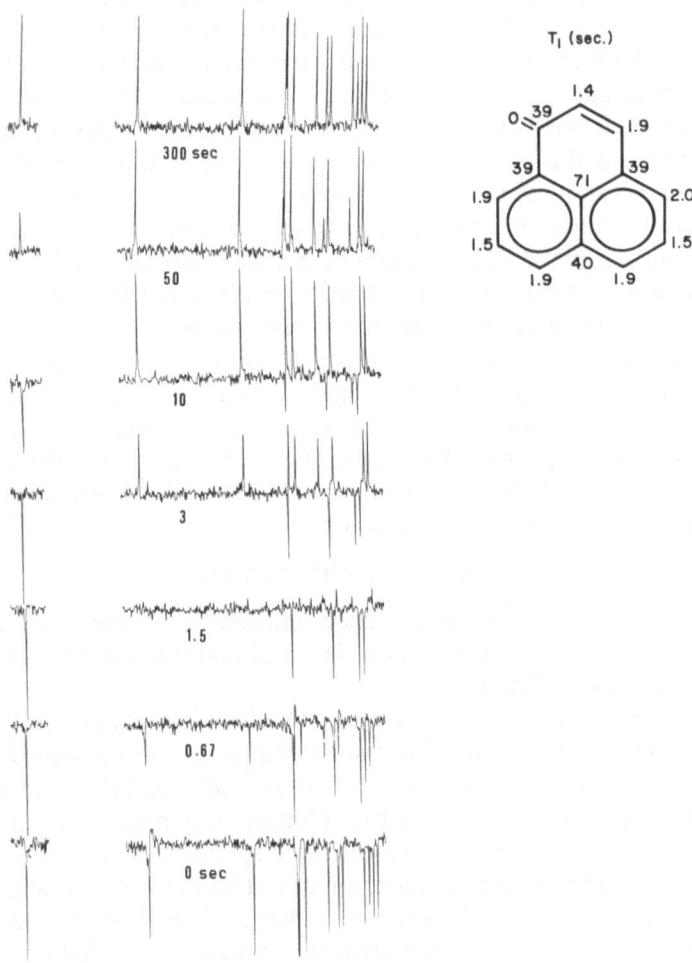

Fig. 17. Inversion-recovery spectra of phenalenone leading to the T_1's shown at right. Spectra were obtained from a single FID from a 5 ml. 2 M solution, with a 5 min. pause after each pulse sequence

process (41), one has been shown to provide useful structural information. Of the others, that based on chemical shift anisotropy has been demonstrated only for very small molecules, such as acetic acid. The T_1 of a nucleus can be affected by "scalar" relaxation if it is coupled to a nucleus of short relaxation time and very similar chemical shift. ^{13}C and ^{79}Br are such a pair. The quadrupole of a nucleus with spin $> 1/2$ can provide a relaxation mechanism if the quadrupole coupling is sufficiently great. ^{14}N has a spin of 1 and a quadrupole coupling which varies widely with the molecular situation. This mechanism should eventually interest alkaloid chemists, but at present the detection of ^{14}N resonances remains difficult and investigations have so far been restricted to very simple molecules. Spin rotation relaxation arises from fluctuating magnetic fields resulting from the electrons of spinning molecules. Although it is most significant in the study of small molecules, Levy, Cargioli and Anet have shown that it contributes substantially to the relaxation of such atoms as the substituted aromatic carbon of toluene (78).

The relaxation of atoms in molecules of interest to natural products chemists is dominated by dipole-dipole relaxation. It arises from the fluctuating magnetic field of interacting nuclei moving in relation to the main magnetic field. Molecular motion is measured by the correlation time, τ_c, which may be understood qualitatively as the average time required for a molecule to turn through one radian. Typical values for dilute solutions of non-polymeric molecules are near 10^{-12} sec. Slower motion produces more effective relaxation (and shorter T_1's) until correlation times approach the reciprocal of the resonance frequency, a possibility for polymers or viscous solutions. However, in solutions considered here, the rate of relaxation,

$$R_1 = 1/T_1 = h^2 \gamma_1 \gamma_2 \tau_c / r^6.$$

Since the nuclear properties are fixed, interest lies in the fact that T_1 is inversely related to the correlation time and that it is strongly dependent upon internuclear distances.

If molecular tumbling is isotropic, the relaxation rate $(1/T_1)$ of carbon atoms is simply proportional to the number of protons attached. This is clearly the case for the ring carbons of desoxycholic acid (Fig. 18). Internal motion in a molecule, such as the spinning of a methyl group, is also reflected. In Fig. 18 it is seen that the rapid spinning of the methyl groups has resulted in less effective relaxation of C-18 and C-19, and longer T_1's. The segmental motion of methylene carbons of an aliphatic chain has a similar effect; as a result, the terminal carbons of aliphatic chains in a wide variety of molecules show very similar T_1's (79, 80).

Although many large molecules behave approximately as though their molecular motion were isotropic, deviations from isotropic be-

haviour are to be anticipated, and may provide useful structural information. Studies on *p*-substituted benzenes have shown that the carbon
atoms para to the substituent have smaller T_1's than other protonated
carbon atoms, because rotation about the long axis of the molecule,
through the para substituent, is preferred. As a result of the more
rapid motion, the ortho and meta protons provide less effective means
of relaxation, and these carbon atoms have longer T_1's (*78*).

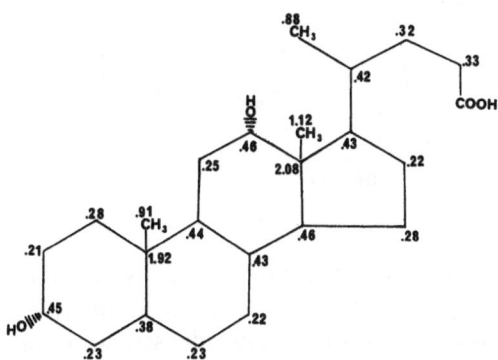

Fig. 18. Desoxycholic acid T_1's from Ref. (*56*)

The use of relaxation studies to aid structural investigations will
clearly be limited by the time required to obtain spectra on small quantities
of material. However, in appropriate circumstances, such determinations
will be valuable aids in assigning chemical shifts. Those of phytol (**38**)
could be assigned largely by comparison with a hydrocarbon model and
shifts calculated from empirical correlations (*81*). The two peaks at δ 39.5
and 40.0 could be assigned to C-4 and C-14, but not distinguished from
each other. However, it could be anticipated that C-4 would have a shorter
relaxation time, as the segmental motion of the portion of the chain near
the hydroxyl group would be slowed by hydrogen bonding of the hydroxyl

CH$_3$ CH$_3$ CH$_3$ CH$_3$

CH$_3$ CH$_2$OH

(**38**)

of the alcohol anchoring that end of the chain. As the peak at 40.0 ppm
had a T_1 one-third that of the peak at 39.5, the former was assigned to
C-4 and the latter to C-14.

XIV. Complementary Studies

We have been at some pains to indicate that the techniques described here complement rather than supplant older nmr methods. The interplay of these methods is nicely illustrated in a recent study of the structure of nicandrenone (39), an insecticide from *Nicandra physalodes* (82). The ^{13}C nmr spectrum showed 28 carbons: 4 methyls, 4 methylenes and 12 methinyls, and therewith 32 carbon-bound protons. The molecular

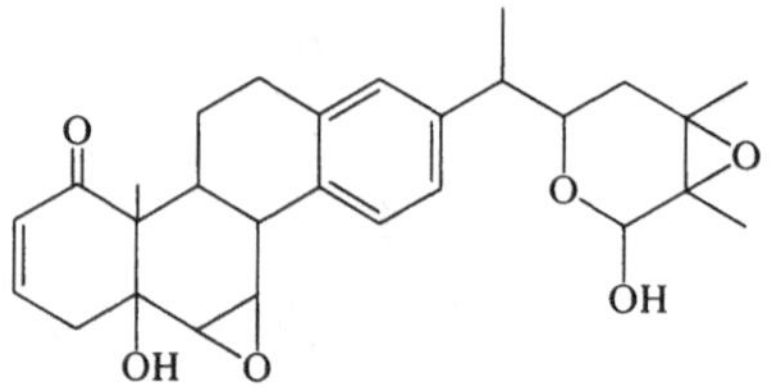

(39)

formula could therefore be revised from $C_{34}H_{42}O_7$ to $C_{28}H_{34}O_6$, allowing for two hydroxyl protons and consistent with the parent peak in the mass spectrum at m/e 466. A steroidal structure was thus indicated. The α,β-unsaturated ketone earlier demonstrated could only be accommodated by ring A of a steroid, and the carbonyl must be at C-1 to account for the downfield position of C-19 protons (δ 1.85). The aromatic ring indicated earlier was confirmed by the proton nmr spectrum showing aromatic protons in a 1,2,4 pattern and by peaks in the ^{13}C nmr spectrum at δ 124—129 (unsubstituted) and 135—142 (substituted). The absence of other sp^2 carbons required seven rings.

The three unassigned oxygens must be ethers, two of them epoxides to account for ^{13}C peaks at δ 56 and 57 (methinyl, ^{1}H δ 3.2 and 4.0, J = 4 Hz) and at δ 64 and 65 ppm (quaternary). The unusual chemical shift of the downfield proton of the methinyl group could be accounted for by its position in the plane of the aromatic ring, and the situation was demonstrated by an NOE of 20% of the aromatic proton at δ 7.4 upon irradiation at δ 4.0. A methinyl carbon at δ 92 and a proton at δ 5.0 coupled to a hydroxyl proton together required that a hemiacetal be present. A reasonable ring size for this group allowed only (39) to accommodate the remaining couplings and quaternary centers known.

References

1. JACKMAN, L. M.: Some Applications of Nuclear Magnetic Resonance Spectroscopy in Natural Product Chemistry. Fortschr. Chem. organ. Naturstoffe 23, 315 (1965).
2. WILLIAMS, D. H., and D. A. WILSON: Solvent Effects in Nuclear Magnetic Resonance Spectroscopy. Part V. Solvent Shifts in 11-Oxosteroids. The Geometry of a Benzene Ketone Complex. J. Chem. Soc. (B) (London) 144 (1966).

3. FALES, H. M., and K. S. WARREN: The Use of Benzene in Separating Aromatic Methoxyl Bands in Nuclear Magnetic Resonance Spectroscopy. J. Organ. Chem. (USA) 32, 501 (1967).

4. QUIJANO, L., F. MALANCO, and T. RIOS: The Structures and Eupatolin of Eupalin. The New Flavanol Rhamnosides Isolated from *Eupatorium ligustrinum* D.C. Tetrahedron 26, 2851 (1970).

5. MA, J. C. N., and E. W. WARNHOFF: On the Use of Nuclear Magnetic Resonance for the Detection, Estimation and Characterization of N-Methyl Groups. Canad. J. Chem. 43, 1849 (1965).

6. HIGHET, R. J., and P. F. HIGHET: The Characterization of Complex Phenols by Nuclear Magnetic Resonance Spectra. J. Organ. Chem. (USA) 30, 902 (1965).

7. PACHLER, K. G., R. R. ARNDT, W. H. BAARSCHERS: Nuclear Magnetic Resonance Study of Aporphine Alkaloids. II. The Structure of Rogersine. Tetrahedron 21, 2159 (1965).

8. GOODLETT, V. W.: Use of In Situ Reactions for Characterization of Alcohols and Glycols by Nuclear Magnetic Resonance. Anal. Chem. 37, 431 (1965).

9. KIRSON, I., D. LAVIE, S. M. ALBONICO, and H. R. JULIANI: The Withanolides of *Acnistus australis* (Griseb). Tetrahedron 26, 5062 (1970).

10. POPLE, J. A., and A. A. BOTHNER-BY: Nuclear Spin Coupling between Geminal Hydrogen Atoms. J. Chem. Phys. 42, 1339 (1965).

11. COOKSON, R. C., T. A. CRABB, J. J. FRANKEL, and J. HUDEC: Geminal Coupling Constants in Methylene Groups. Tetrahedron 22, suppl. 7, 355 (1966).

12. RÜEDI, P., and C. H. EUGSTER: Struktur von Coleon E, einem neuen diterpenoiden Methylenchinon aus der *Coleus barbatus*-Gruppe *(Labiatae)*. Helv. Chim. Acta 55, 1994 (1972).

13. STERNHELL, S.: Long-Range ^{1}H—^{1}H Spin-Spin Coupling in Nuclear Magnetic Resonance Spectroscopy. Rev. Pure and Appl. Chem. 14, 15 (1964).

14. KARANATSIOS, D., J. S. SCARPA, and C. H. EUGSTER: Struktur von Fuerstion. Helv. Chim. Acta 49, 1151 (1966).

15. POPLE, J. A., W. G. SCHNEIDER, and H. J. BERNSTEIN: High-Resolution Nuclear Magnetic Resonance. New York: McGraw-Hill. 1959.

16. PEROLD, G. W., and K. G. R. PACHLER: The Structure and Chemistry of Leucodrin. J. Chem. Soc. (C) (London) 1918 (1966).

17. JANOT, M. M., Q. KHUONG-HUU, C. MONNERET, I. KABORE, J. HILDESHEIM, S. D. GERO, and R. GOUTAREL: Alcaloides Steroidiques-C. Les Holantosines A et B. Nouveaux Aminoglyco Steroides isolés des Feuilles de L'*Holarrhena antidysenterica* (Roxb.) Wall *(Apocynacees)*. Tetrahedron 26, 1695 (1970).

18. PEROLD, G. W., and G. OURISSON: Photolyse des dihydro-1,2(6β, 11α et 11β)-Santonines. Tetrahedron Letters 1969, 3871.

19. HINCKLEY, C. C.: Paramagnetic Shifts in Solutions of Cholesterol and the Dipyridine Adduct of Tris-Dipivalomethanato Europium III. A Shift Reagent. J. Amer. Chem. Soc. 91, 5160 (1969).

20. MAYO, B. C.: Lanthanide Shift Reagents in Nuclear Magnetic Resonance Spectroscopy. Chem. Soc. Rev. (London) 2, 49 (1973).

21. SIEVERS, R. E.: Nuclear Magnetic Resonance Shift Reagents. New York: Academic Press. 1973.

22. GALBRAITH, M. N., and D. H. S. HORN: Structures of the Natural Products Blumenols A, B and C. Chem. Communs. (London) 1972, 113.

23. MUNTWYLER, R., and W. KELLER-SCHIERLEIN: Stoffwechsel-Produkte von Mikroorganismen. Zur Stereochemie des Lankamycins. Helv. Chim. Acta 55, 460 (1972).

24. McCONNELL, H. M., and R. E. ROBERTSON: Isotropic Nuclear Resonance Shifts. J. Chem. Phys. 29, 1361 (1958).

25. APSIMON, J. W., and H. BEIERBECK: Lanthanide Shift Reagents. A Novel Method for Fitting the Pseudocontact Shielding Equation to Experimental Induced Shifts. Tetrahedron Letters 1973, 581.

26. Angerman, N. S., S. S. Danyluk, and T. A. Victor: A Direct Determination of the Spatial Geometry of Molecules in Solution. I. Conformation of Chloroquine, an Antimalarial. J. Amer. Chem. Soc. **94**, 7137 (1972).

27. Bothner-By, A. A., and S. M. Castellano: LAOCN3. In Computer Programs for Chemistry, Vol. I. Edit. D. F. Detar, p. 10. New York: W. A. Benjamin, Inc. 1968.

28. Swalen, J. D.: NMRIT and NMREN. In Computer Programs for Chemistry, Vol. 1. Edit. D. F. Detar, p. 54. New York: W. A. Benjamin, Inc. 1968.

29. Johannesen, R. B., J. A. Ferretti, and R. K. Harris: UEAITR: A New Computer Program for Analysis of NMR Spectra. Analysis of the Proton Spectrum of Triisopropylphosphine. J. Magn. Reson. (London) **3**, 84 (1970).

30. Gordon, M., A. Stoessl, and J. B. Stothers: Stereochemistry of some Altersolanol B Derivatives and their Correlation with Bostrycin. Canad. J. Chem. **50**, 122 (1972).

31. Findlay, J. A., and L. Radics: Flavipucine (3'-Isovaleryl-6-Methyl Pyridine-3-spiro-2'-oxiran-2(1H)-4(3H)-dione), an Antibiotic from *Aspergillus flavipes*. J. Chem. Soc. (London). Perkin I **1972**, 2071.

32. Burton, R., L. D. Hall, and P. R. Steiner: Studies of Carbohydrate Derivatives by Nuclear Magnetic Double Resonance. Part V. Some ^{1}H—[^{1}H] INDOR Experiments. Canad. J. Chem. **48**, 2679 (1970).

33. Anet, F. A. L., and A. J. R. Bourn: Nuclear Magnetic Resonance Spectral Assignments from Nuclear Overhauser Effects. J. Amer. Chem. Soc. **87**, 5250 (1965).

34. Bell, R. A., and J. K. Saunders: Some Chemical Applications of the Nuclear Overhauser Effect. Topics in Stereochem. **7**, 1 (1973).

35. Noggle, J. H., and R. E. Schirmer: The Nuclear Overhauser Effect. New York: Academic Press. 1971.

36. Bell, R. A., and J. K. Saunders: Correlation of the Intramolecular Nuclear Overhauser Effect with Internuclear Distance. Canad. J. Chem. **48**, 1114 (1970).

37. McCorkindale, N. J., A. McRitchie, and S. A. Hutchinson: Lamellicolic Anhydride — A Heptaketide Naphthalic Anhydride. Chem. Communs. (London) **1973**, 108.

38. Buckley, D. G., E. Ritchie, W. C. Taylor, and L. M. Young: Madagascarin, A New Pigment from the Leaves of *Harungana madagascariensis*. Austral. J. Chem. **25**, 843 (1972).

39. Murray, R. D. H., and M. M. Ballantyne: Constituents of Sneezewood, *Ptaeroxylon obliquum*. II. Coumarins. The Structure of Nieshoutol. Tetrahedron **26**, 4473 (1970).

40. Ernst, R. R., and W. A. Anderson: Application of Fourier Transform Spectroscopy to Magnetic Resonance. Rev. Sci. Instr. **37**, 93 (1966).

41. Farrar, T. C., and E. D. Becker: Pulse and Fourier Transform NMR. New York: Academic Press. 1971.

42. Johnson, L. F., and W. C. Jankowski: Carbon-13 NMR Spectra. New York: Wiley-Interscience. 1972.

43. Jesson, J. P., P. Meakin, and G. Kneissel: Homonuclear Decoupling and Peak Elimination in Fourier Transform Nuclear Magnetic Resonance. J. Amer. Chem. Soc. **95**, 618 (1973).

44. Stothers, J. B.: Carbon-13 NMR Spectroscopy. New York: Academic Press. 1972.

45. Glombitza, K. W., and E. Sattler: Trifuhalol, Ein neuer Triphenyldiäther aus *Halidrys siliquosa*. Tetrahedron Letters **1973**, 4277.

46. Pregosin, P. S., E. W. Randall, and A. I. White: Natural Abundance Nitrogen-15 Nuclear Magnetic Resonance Spectroscopy — Amino Acid Derivatives. Chem. Communs. (London) **1971**, 1602.

47. Crutchfield, M. M., C. H. Dungan, J. H. Letcher, V. Mark, and J. R. van Wazer: Topics in Phosphorus Chemistry, vol. 5. P-31 Nuclear Magnetic Resonance. New York: Wiley-Interscience. 1967.

48. Moon, R. B., and J. H. Richards: Determination of Intracellular pH by ^{31}P Magnetic Resonance. J. Biol. Chem. **248**, 7276 (1973).

49. LEE, G. C. Y., and S. I. CHAN: A ^{31}P NMR Study of the Association of Uridine-3'-Monophosphate to Ribonuclease A. Biochem. Biophys. Res. Commun. **43**, 142 (1971).

50. HENDERSON, T. O., T. GLONEK, R. L. HILDEBRAND, and T. C. MYERS: Phosphorus-31 Nuclear Magnetic Resonance Studies of the Phosphonate and Phosphate Composition of the Sea Anemone, *Bunadosoma* sp. Archiv. Biochem. Biophys. **149**, 484 (1972).

51. KORN, E. D., D. G. DEARBORN, H. M. FALES, and E. A. SOKOLOSKI: Phosphonoglycan — A Major Polysaccharide Constituent of the Amoeba Plasma Membrane Contains 2-Aminoethyl Phosphonic Acid and 1-Hydroxy-2-Aminoethyl Phosphonic acid. J. Biol. Chem. **248**, 2257 (1973).

52. LEVY, G. C., and G. L. NELSON: Carbon-13 Nuclear Magnetic Resonance for Organic Chemists. New York: Wiley-Interscience. 1972.

53. GRAY, G. A.: Applications of ^{13}C NMR in Biochemistry. A Collection of Titled References. Palo Alto: Varian Associates (USA). 1973.

54. DODDRELL, D., and A. ALLERHAND: Assignments in the Carbon-13 Nuclear Magnetic Resonance Spectra of Vitamin B_{12}, Coenzyme B_{12} and other Corrinoids. Application of Partially-Relaxed Fourier Transform Spectroscopy. Proc. Nat. Acad. Sci. (USA) **68**, 1083 (1971).

55. NAKANISHI, K., V. P. GULLO, I. MIURA, T. R. GOVINDACHARI, and N. VISWANATHAN: Structure of Two Triterpenes. Application of Partially Relaxed Fourier Transform ^{13}C Nuclear Magnetic Resonance. J. Amer. Chem. Soc. **95**, 6473 (1973).

56. LEIBFRITZ, D., and J. D. ROBERTS: Nuclear Magnetic Resonance Spectroscopy. Carbon-13 Spectra of Cholic Acids and Hydrocarbons included in Sodium Desoxycholate Solutions. J. Amer. Chem. Soc. **95**, 4996 (1973).

57. APSIMON, J. W., H. BEIERBECK, and J. K. SAUNDERS: Lanthanide Shift Reagents in ^{13}C Nuclear Magnetic Resonance: Quantitative Determination of Pseudocontact Shifts and Assignment of ^{13}C Chemical Shifts of Steroids. Canad. J. Chem. **51**, 3874 (1973).

58. BRIGGS, J., F. A. HART, G. P. MOSS, and E. W. RANDALL: A Ready Method of Assignment for ^{13}C Nuclear Magnetic Resonance Spectra: The Complete Assignment of the ^{13}C Spectrum of Borneol. Chem. Communs. (London) **1971**, 364.

59. PRINZBACH, H., V. FREUDENBERGER, and U. SCHEIDEGGER: Cyclische, gekreuzt-konjugierte Bindungssysteme. XIII. NMR-Untersuchungen am Phenafulven-System. Helv. Chim. Acta **50**, 1087 (1967).

60. BECKER, E. D., J. A. FERRETTI, and T. C. FARRAR: Driven Equilibrium Fourier Transform Spectroscopy. A New Method for Nuclear Magnetic Resonance Signal Enhancement. J. Amer. Chem. Soc. **91**, 7784 (1969).

61. ERNST, R. R.: Sensitivity Enhancement in Magnetic Resonance. Advances in Magn. Reson. **2**, 1 (1966).

62. WAUGH, J. S.: Sensitivity in Fourier Transform NMR Spectroscopy of Slowly Relaxing Systems. J. Molec. Spectr. **35**, 298 (1970).

63. JONES, D. E., and H. STERNLICHT: Fourier Transform Nuclear Magnetic Resonance I. Repetitive Pulses. J. Magn. Res. **6**, 167 (1972).

64. PLATTNER, R. D., G. F. SPENCER, D. WEISLEDER, and R. KLEIMAN: Chromanone Acids in *Calophyllum brasiliense* Seed Oil. Phytochemistry (in press).

65. LINDEMAN, L. P., and J. Q. ADAMS: Carbon-13 Nuclear Magnetic Resonance Spectrometry. Chemical Shifts for the Paraffins through C_9. Anal. Chem. **43**, 1245 (1971).

66. YATES, P., F. N. MacLACHLAN, I. D. RAE, M. ROSENBERGER, A. G. SZABO, C. R. WILLIS, M. P. CAVA, M. BEHFOROUZ, M. V. LAKSHMIKANTHAM, and W. ZEIGER: Haplophytine. A Novel Type of Indole Alkaloid. J. Amer. Chem. Soc. **95**, 7842 (1973).

67. DALLING, D. K., and D. M. GRANT: Carbon-13 Magnetic Resonance IX. The Methyl Cyclohexanes. J. Amer. Chem. Soc. **89**, 6612 (1967).

68. GOUGH, J. L., J. P. GUTHRIE, and J. B. STOTHERS: Stereochemical Assignments in Steroids by ^{13}C Nuclear Magnetic Resonance Spectroscopy: Configuration of the A/B Ring Junction. Chem. Communs. (London) **1972**, 979.

69. WENKERT, E., D. W. COCHRAN, E. W. HAGAMAN, F. M. SCHELL, N. NEUSS, A. S. KATNER, P. POTIER, C. KAN, M. PLAT, M. KOCH, H. MEHRI, J. POISSON, N. KUNESCH, and Y. ROLLAND: Carbon-13 Nuclear Magnetic Resonance Spectroscopy of Naturally Occurrings Substances. XIX. *Aspidosperma* Alkaloids. J. Amer. Chem. Soc. **95**, 4990 (1973).

70. COCHRAN, D. W.: ^{13}C Nuclear Magnetic Resonance Spectroscopy. Indole Alkaloids. Ph. D. Thesis. Indiana Univ. 1971.

71. JENNINGS, H. J., and I. C. P. SMITH: Determination of the Composition and Sequence of a Glucan Containing Mixed Linkages by Carbon-13 Nuclear Magnetic Resonance. J. Amer. Chem. Soc. **95**, 606 (1973).

72. NEUSS, N., C. H. NASH, J. E. BALDWIN, P. A. LEMKE, and J. B. GRUTZNER: Incorporation of (2RS, 3S)-[4—^{13}C] Valine into Cephalosporin C. J. Amer. Chem. Soc. **95**, 3797 (1973).

73. MILAVETZ, B., K. KAKINUMA, K. L. RINEHART, J. P. ROLLS, and W. J. HAAK: Carbon-13 Magnetic Resonance Spectroscopy and the Biosynthesis of Streptovaricin. J. Amer. Chem. Soc. **95**, 5793 (1973).

74. WASSERMAN, H. H., R. J. SYKES, P. PEVERADA, C. K. SHAW, R. J. CUSHLEY, and S. R. LIPSKY: Biosynthesis of Prodigiosin Incorporation Patterns of ^{13}C Labeled Alanine, Proline, Glycine, and Serine Elucidated by Fourier Transform Nuclear Magnetic Resonance. J. Amer. Chem. Soc. **95**, 6874 (1973).

75. BALDWIN, J. E., J. LOLIGER, W. RASTETTER, N. NEUSS, L. L. HUCKSTEP, and N. DE LA HIGUERA: Use of Chiral Isopropyl Groups in Biosynthesis. Synthesis of (2RS)-3S-[4—^{13}C] Valine. J. Amer. Chem. Soc. **95**, 3796 (1973).

76. SETO, H., T. SATO, and H. YONEHARA: Utilization of Carbon-13-Carbon-13 Coupling in Structural and Biosynthetic Studies. An Alternate Double Labeling Method. J. Amer. Chem. Soc. **95**, 8461 (1973).

77. LEVY, G. C.: Carbon-13 Spin-Lattice Relaxation Studies and their Application to Organic Chemical Problems. Accts. Chem. Res. **6**, 161 (1973).

78. LEVY, G. C., J. D. CARGIOLI, and F. A. L. ANET: Carbon-13 Spin-Lattice Relaxation in Benzene and Substituted Aromatic Compounds. J. Amer. Chem. Soc. **95**, 1527 (1973).

79. LEVY, G. C.: On Segmental Motion in Short Aliphatic Chains. J. Amer. Chem. Soc. **95**, 6117 (1973).

80. DODDRELL, D., and A. ALLERHAND: Segmental Motion in Liquid 1-Decanol. Application of Natural-Abundance Carbon-13 Partially-Relaxed Fourier Transform Nuclear Magnetic Resonance. J. Amer. Chem. Soc. **93**, 1558 (1971).

81. GOODMAN, R. A., E. OLDFIELD, and A. ALLERHAND: Assignments in the Natural-Abundance Carbon-13 Nuclear Magnetic Resonance Spectrum of Chlorophyll and a Study of Segmental Motion in Neat Phytol. J. Amer. Chem. Soc. **95**, 7553 (1973).

82. BATES, R. B., and D. J. ECKERT: Nicandrenone, an Insecticidal Plant Steroid Derivative with Ring D Aromatic. J. Amer. Chem. Soc. **94**, 8258 (1972).

(Received March 20, 1974)

Applications of the Chiroptical Techniques to the Study of Natural Products

By P. M. Scopes, Department of Chemistry,
Westfield College (University of London), London, U. K.

With 1 Figure

Contents

I. Introduction

The chiroptical techniques optical rotatory dispersion (o.r.d.) and circular dichroism (c.d.) have now been used by organic chemists for about twenty years. During this time the pace of development has been governed largely by the instrumentation available, which has evolved from manually recording instruments and measurements made over a limited spectral range to sophisticated automatic instruments covering the entire visible and near u.v. regions of the spectrum, 800—185 nm. Most papers which include the words o.r.d. or c.d. in the title deal with fundamental aspects of the subject. As well as the theory of optical activity, these include the study of new chromophores, and of the relationship between the observed Cotton effects (CE) and the underlying electronic transitions, also the formulation and testing of empirical and semi-empirical rules which relate the sign and magnitude of the CE to the geometry of the molecule. Detailed references may be found from the main textbooks and monographs on the subject (*64, 82, 83, 92, 333*) and from the published proceedings of conferences and symposia (*51, 74, 256, 297*).

In addition to this specialist interest, the chiroptical techniques have considerable use as a structural tool for the study of chiral compounds including natural products. About 80% of all applications in structural studies have been concerned with allocation of absolute or relative configuration, about 10% with conformational studies and the remaining 10% with other problems such as the detection of optical activity (to establish the dissymmetry of a molecule), proof of identity of two samples, confirmation of the position of a functional group, and in kinetic studies. These applications of o.r.d. and c.d. are the particular concern of this review. The work is found mainly in papers which do *not* include the words o.r.d. or c.d. in their titles, and which are difficult to trace except by a detailed search of the journals; the scope has been limited to organic compounds, excluding macromolecules and polymers, and the examples considered are mainly from the literature of 1966—1972, i.e. since the publication of the last major book in this field (*83*). The examples are, necessarily, selected but it is hoped they will show how natural product chemists have actually *used* the chiroptical techniques in recent years, *not* how specialists thought they might have been used. Specialist publications for the same period (1966—1972) have been reviewed by the author each year in Annual Reports of the Chemical Society (*286*).

1. Units and Definitions

The chiroptical techniques measure the effects of the interaction of polarised light with a dissymmetric medium. O.r.d. arises from the difference in refraction and c.d. from the difference in absorption by the medium for right and left circularly polarised light. Fig. 1 shows the relationship between the absorption band and the o.r.d. and c.d. of an imaginary compound A. The CE in the c.d. curve of (+)A has a single positive maximum at approximately the same wavelength as the maximum of the absorption band, while in the o.r.d. curve, the CE shows two extrema, one positive (peak) and one negative (trough) (Fig. 1a); in an ideal case, the wavelength at the mid-point of the o.r.d. curve coincides with the wavelength of the c.d. and u.v. maxima. For the enantiomeric compound (−)A, the sign of the CE is necessarily reversed (Fig. 1b).

O.r.d. results are expressed in units of molecular rotation [ϕ] defined by $[\phi] = M \times 10^{-2} [\alpha]$, where M is the molecular weight and [α] the specific rotation at a given wavelength. The amplitude *a*, which is a con-

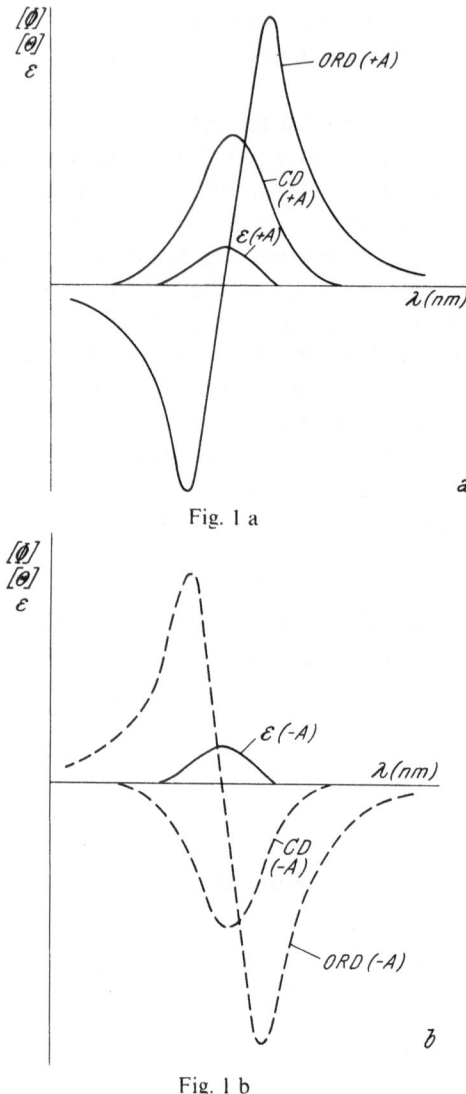

Fig. 1 a

Fig. 1 b

Fig. 1. Optical rotatory dispersion, circular dichroism and ultra-violet absorption curves for an imaginary compound (+)A (Fig. 1 a) and its enantiomer (−)A (Fig. 1 b).

venient measure of the magnitude of the CE, is the difference between the molecular rotation at the extremum of longer wavelength $[\phi]_1$, and that at shorter wavelength, $[\phi]_2$ divided by 100 for convenience. Thus, $a = ([\phi]_1 - [\phi]_2) \times 10^{-2}$. For c.d., two units are in common use, the differential absorption $\Delta\varepsilon = (\varepsilon_L - \varepsilon_R)$ and $[\theta]$ the molecular ellipticity. These two are related by the expression $[\theta] = 3300 \, \Delta\varepsilon$. For perfect

References, pp. 249—265

Gaussian curves, the o.r.d. amplitude a and the c.d. $\Delta\varepsilon$ or $[\theta]$ can be related by the expressions $a = 40.28\ \Delta\varepsilon$ and $a = 0.0122\ [\theta]$.

Since the magnitude and, less frequently, the sign of the CE for a particular compound may vary significantly in different solvents, comparisons of chiroptical data should always be made for the same solvent. The most widely used solvents are methanol (or ethanol), water and hexane (or cyclohexane); others include chloroform (at longer wavelengths), dioxan and acetonitrile. In addition acidic and basic compounds are often studied in solutions of different pH. Until about 1965, only o.r.d. measurements could be made easily, but since then, the availability of good instruments has led to a steady increase in the use of c.d. so that it is now the preferred technique in most cases. The main reason for this is that c.d. curves, being simpler in form, are easier to interpret both empirically and theoretically, particularly for complex molecules with more than one chromophore and with a series of overlapping absorption bands. Furthermore because of their approximately Gaussian form, overlapping c.d. bands can sometimes be analysed by curve resolution, thus providing additional useful information. On the other hand, extensive o.r.d. data are available in the literature, and examples in this review are taken from both o.r.d. and c.d. results.

2. Logic of Applications

Most applications of o.r.d. and c.d. in the study of natural products depend on a comparison, direct or indirect, of the o.r.d./c.d. behaviour of the compound studied (X) with that of a compound of known structure and configuration (A). The simplest correlations are those based on direct identity of two samples, in which (X) is converted into (A) (or its enantiomer) and compared with an authentic sample of (A). More often, (X) is converted to an intermediate (Y), which, though not identical to (A) is sufficiently analogous to be compared with (A). The validity of the comparison depends on the adequacy of the analogy between (Y) and (A), which should have the same chromophore(s) and the same structure in the environment of the chromophore. Furthermore there is an implicit assumption that the molecules concerned are sufficiently similar in structure and in degree and pattern of substitution to adopt the same conformation. If this assumption is not valid, and the molecules are not strictly analogous, then no legitimate correlation can be made.

If arguments of simple analogy are extended to cover a range of compounds, an empirical generalisation may become apparent, relating the sign of CE to the configuration of a particular series of closely related compounds. These empirical rules can then be used directly to study further unknown compounds provided always that they are members of the closely defined series.

Semi-empirical rules arise when empirical results for a series of compounds can be correlated with the geometry of the molecular orbitals and symmetry planes of the chromophore concerned. These rules define regions around the chromophore which are of significance in determining the sign of the CE and may permit assignment of semiquantitative values to the contributions of individual substituents. The best known and most firmly established is the Octant Rule for ketones (236) first described in 1961, but regional rules have been suggested for many other chromophores.

Finally, in a few specialised cases, theoretical treatments of o.r.d./c.d. can be used to calculate, from first principles, the sign of the CE associated with a given enantiomer of the compound studied. In particular, the absolute configurations of compounds of the coupled-oscillator type can be determined from the sign of the CE if the precise geometrical relationship of the chromophores in the molecule and the direction of polarization of the relevant electronic transitions are known. This technique has been applied to a number of natural products including the alkaloids calycanthine (224) and argemonine (225), but in general, these methods are limited in their application and will not be discussed further.

In fact, a survey of published work for the last seven years shows that most applications of the chiroptical techniques (about 80%) are based on empirical comparison between individual compounds or series of compounds, while of the remaining 20%, approximately half are based on the Octant Rule and the other half on one of the other semi-empirical regional rules. The emphasis on empirical correlations is evident in this review for which most material has been selected intentionally from non-specialist papers. Examples are arranged according to the main chromophore and subdivided, where appropriate, into broad classes of natural product.

II. Allocation of Configuration

1. Carbonyl Chromophores in Steroids and Terpenes:
Empirical Correlations
a) Hexahydroindan-1- and 2-ones

One of the earliest applications of o.r.d./c.d. measurements was to determine the relative configuration (and sometimes absolute configuration) at the ring junction of a fused cyclic system by comparison of the sign and magnitude of the carbonyl CE with that of an analogous compound of known configuration. In general, *either* the relative *or* the absolute configuration (but not both) can be determined by a single o.r.d./ c.d. measurement, but in some cases a determination of relative stereochemistry may also be a proof of absolute configuration if the absolute stereochemistry of the rest of the molecule is known.

The steroid ketones $5\alpha,14\alpha$- and $5\alpha,14\beta$-androstan-17-one (1 R = 14αH and 1 R = 14βH) are an important pair of reference compounds for ketones of the hexahydroindanone types. They show positive CEs in the region of the carbonyl n$\rightarrow\pi^*$ transition ($\approx$ 290 nm) with magnitudes $\Delta\varepsilon \approx +3.3$ $(a+140)$ and $\Delta\varepsilon \approx +1$ $(a+40)$ for the $5\alpha,14\alpha$- and $5\beta,14\beta$-compounds respectively; the enantiomers would of course have $\Delta\varepsilon -3.3$ or $\Delta\varepsilon -1$. For example, the degradation product (2) of the fungal metabolite wortmannin has been allotted a *trans* ring junction and the absolute configuration shown, by comparison of the amplitude $a+134$ for its 11,12-dihydro derivative with the standard value for $5\alpha,14\alpha$-androstan-17-one (1 R = 14αH) (*218*). Similarly, hardwickiic acid was assigned the *trans* ring junction and absolute stereochemistry shown (3) from the strong negative maximum ($\Delta\varepsilon -2.5$) of the related ketone (4), which is of the type enantiomeric to (1) (*233*). Zizanoic acid (5) also gives a ketone (6) enantiomeric in type to (1 R = 14αH), and the strong negative CE $(a-104)$ establishes the absolute configuration shown (*184*). In contrast, $\Delta\varepsilon$ values in the range +0.49 to +0.64 have been used to allot relative stereochemistry (14βH *cis* C/D junction) to a series of 17-ketones related to steroid alkaloids and having in common the partial structure (1 R = 14βH) (*119*).

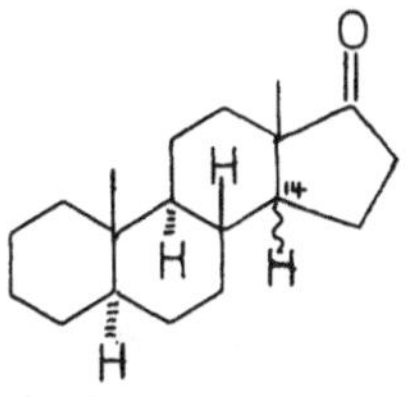

(1)

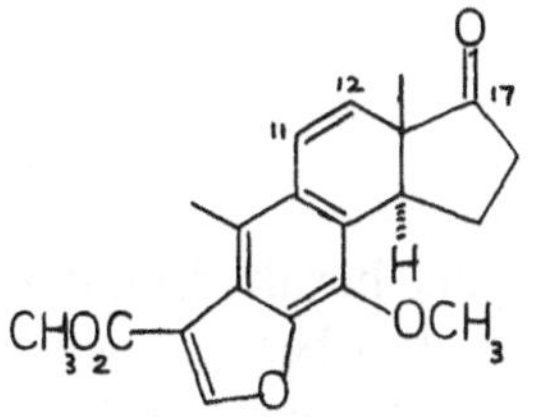

(2)

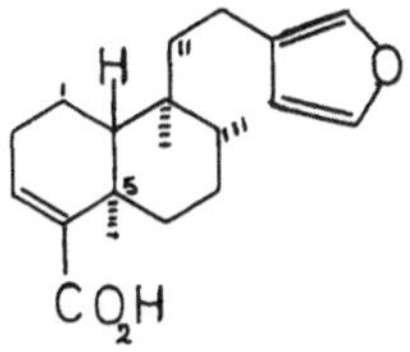

(3)

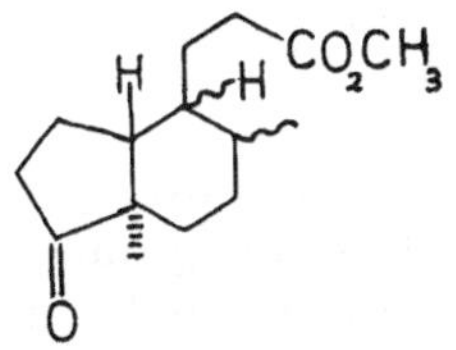

(4)

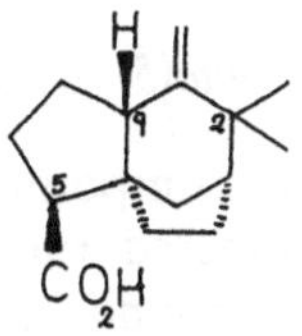

(5)

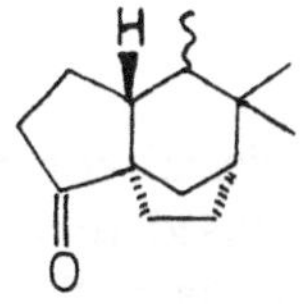

(6)

Analogy with 14α- and 14β-androstan-15-ones has been used to allot relative configurations to 19-oxolupane derivatives. Ketone (**7**) has been assigned the (18α-H) configuration (i.e. *trans* D/E junction) because of the quasi-enantiomeric relationship of its o.r.d. curve (*a* − 115) to that of 14α-androstan-15-one (**8**, 14α-H) for which *a* is + 139. The epimeric compound has been allotted the 18β-H configuration (*cis* D/E junction) because its o.r.d. curve with amplitude *a* + 101 is enantiomeric to that of 14β-androstan-15-one (**8**, 14βH) with *a* − 104 (*17*). A similar analogy confirms the 5β-H configuration of two rearrangement products of withaferin-A having *a* − 101 (**9** R=H) and *a* − 109 (**9** R=OH) respectively, which were originally allotted by chemical evidence and direct comparison with the parent compound (*212*).

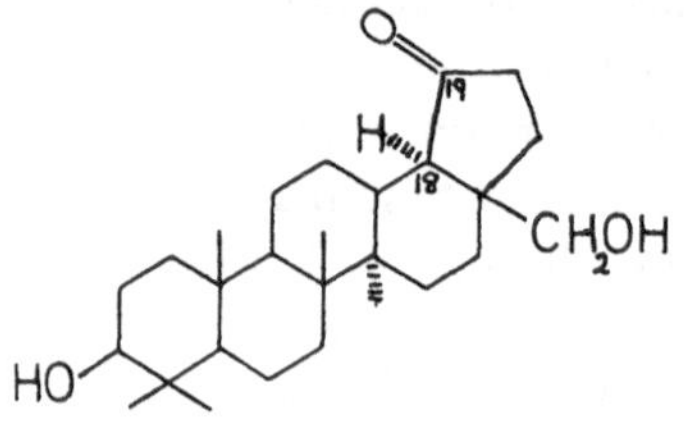

(7)

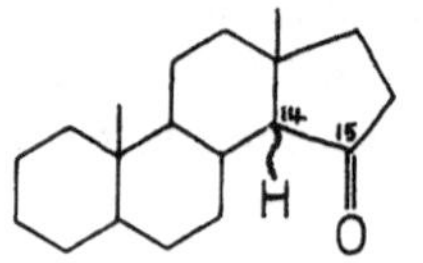

(8)

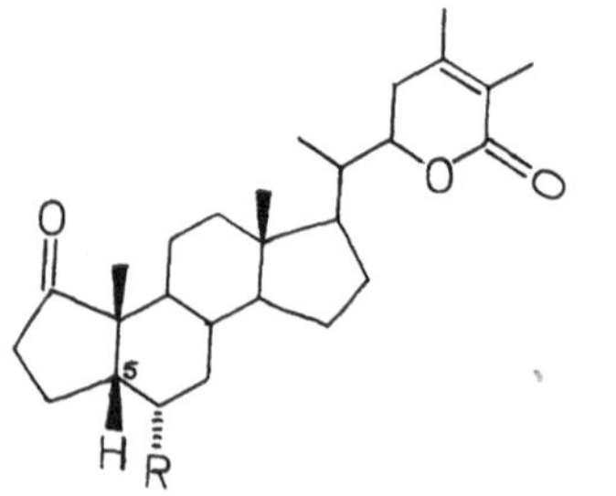

(9)

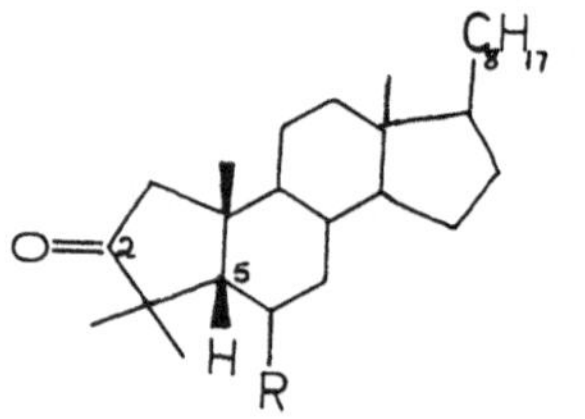

(10)

Among compounds of the hexahydroindan-2-one type, the A-nor steroid 2-ketone (**10** R=OH) has been assigned the 5β-H configuration from empirical comparison of its negative CE Δε − 2.0, with that of a known analogue lacking a hydroxy group at C-6 (**10** R=H), Δε − 2.6 (contrast Δε + 5.2 for the 5α-H compound) (*8*).

b) Bicyclo[5,3,0]-octanones

Empirical correlations for compounds in which a cyclopentanone ring is fused to a seven-membered ring, fall into two groups. Direct correlations with known compounds include, for example, damsinic acid (**11**) which has been allotted the *trans* A/B ring junction because of the similarity of its positive CE to that of damsin (**12**) of known structure (*98*); similarly the *trans* 1α,5β-configuration of tetraneurin-B (**13** R=OAc) has been established by comparison with known coronopilin (**13** R=H) (*349*).

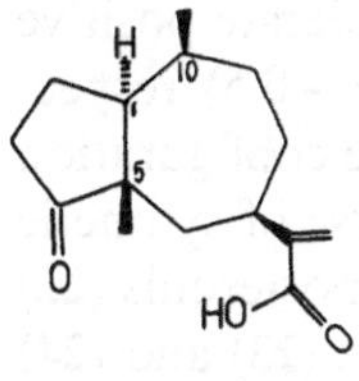

(11) 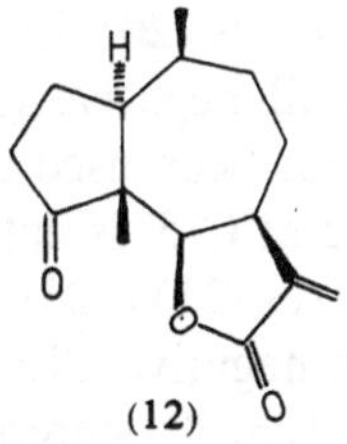   (12) 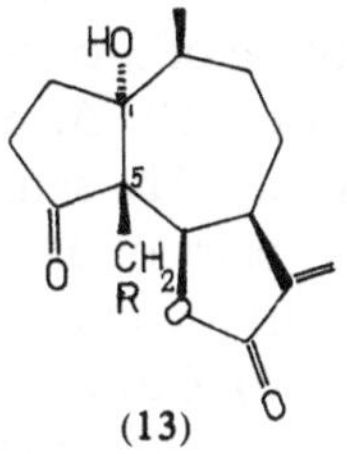   (13)

O.r.d. measurements have also been used to determine the relative configuration at C-1 of α-kessyl ketone (**14**, 1αH) and its C-1 epimer iso-kessyl ketone (**14**, 1βH), which have strong positive and negative CEs respectively (*a* +148 and *a* −191). By analogy with literature data for hexahydroindan-1-ones, the authors infer that these results correspond to the 1α- and 1β-configurations shown (*163*). Comparison with a 16-oxo steroid has been used in a similar way to confirm the *trans*-ring junction of peruvinin (**15**) (*277*). Neither of these two assignments alone could be taken as definitive since the analogy between a 5/6 and 5/7 membered ring system is not wholly adequate (compare early comment *95*); however, other evidence on the absolute stereochemistry is available and the o.r.d. data provide confirmatory evidence.

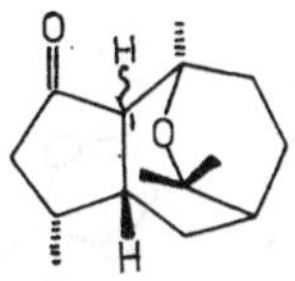

(14) 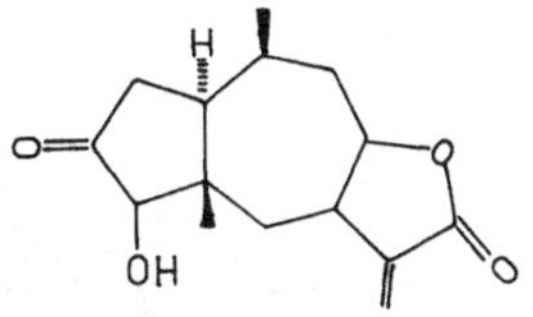   (15)

c) Decalones

The carbonyl chromophore in a fused six-membered ring system has been studied more intensively than any other chromophore in a particular geometric situation. Many steroid and terpene ketones can be regarded as substituted decalones and their stereochemistry may be deduced by analogy with appropriate model compounds.

2-Oxo and 3-oxo-steroids with 5α- and 5β-ring junctions provide simple models for many substituted decalones. For example tetrahydro-nootkatone (16) (217) and dihydrofukinone (17) (254) reduced deriva-tives of known nootkatone (18) and fukinone (19) have been allotted *trans*-fused and *cis*-fused ring systems by comparison of their CEs with those of 3-oxo-5α- and -5β-steroids (20; 5α and 5βH), moderate positive ($a \approx +50$, $\Delta\varepsilon \approx +1.0$) and small negative ($a \approx -20$, $\Delta\varepsilon \approx -0.5$) respec-tively. The same comparison has been used to allot relative configuration to pairs of 3-ketones, epimeric at C-5, isolated in the course of synthetic work on oxidojervane derivatives (21) (223) and on 19-homosteroids (22) (258). Conversely the medium negative CEs of the ketones (23) and (24) are in accordance with the 10α, 5β stereochemistry at the *trans*-ring junctions (158, 191).

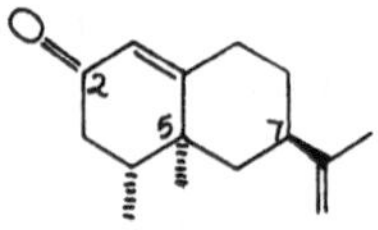

(18) 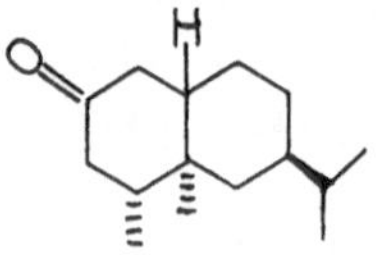   (16)

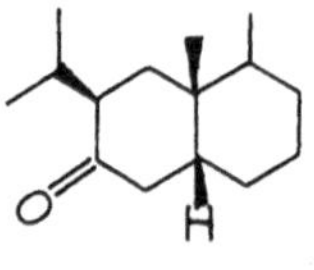

(17) 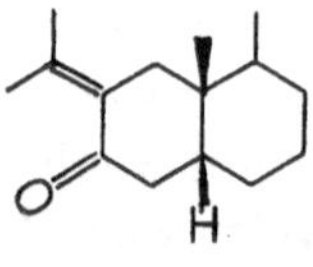   (19)

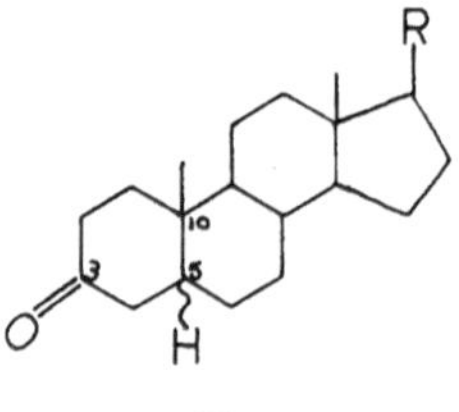

(20) 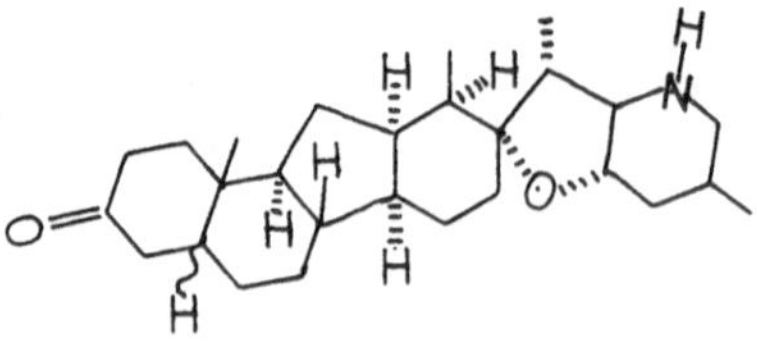   (21) 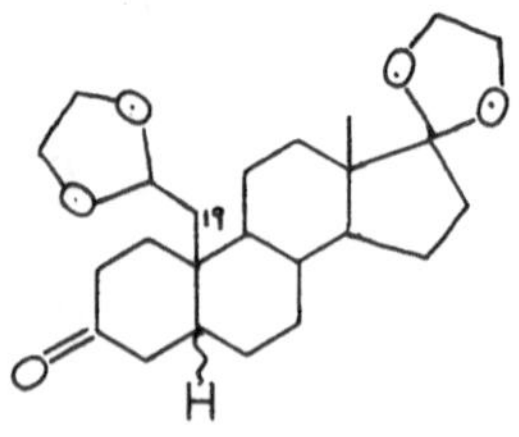   (22)

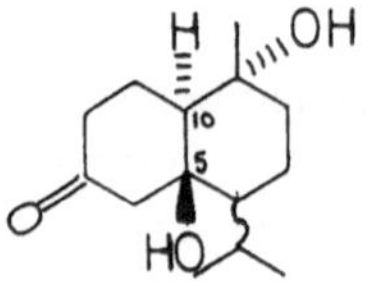

(23) 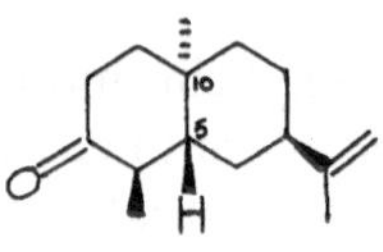   (24)

Direct comparison with a closely analogous compound is particularly necessary when there are significant substituents close to the chromophore and where conformational abnormalities might be expected. The *trans*-ring junction and 4α-configuration of the methyl group in dihydro-α-cyperone (**25**) ($a \approx +78$) have been confirmed by comparison of the large positive carbonyl CE with those of 4β,5αH-tetrahydrosantonin (**26**) and of 4α-methyl-5α-cholestan-3-one (lophanone) (*71*). Conversely the small positive CE of a 4β-methyl-3-oxo steroid $a \approx +21$ has been used to allot the 4β-methyl configuration to the substituent in 4β-methylstigmasta-7,24(28)-dien-3β-ol (**27**), (*307*) and to the 14α-substituent of the ketone (**28**) $a +7$, related to cassaic acid (*75*). Other examples include the assignment of the *trans*-9β,8α-ring junction to des-A-10-oxo-shionane (**29**) for which the o.r.d. and c.d. curves are superimposable on those of des-A-10-oxo-friedelane (**30**); the B-C ring junction in the parent ketone shionone is therefore as shown (**31** R = (CH$_2$)$_2$C:C(CH$_3$)$_2$) (*318*). The virtual identity of the c.d. curves of the saturated analogue shionanone (**31**, R = (CH$_2$)$_3$CH(CH$_3$)$_2$) and the ketone (**32**) establishes the absolute stereochemistry of the latter at C-4, C-5 and C-10 (*37*). The configuration of neoandrographolide (**33**, R^1 = CH$_2$:R^2 = Glc.) has been determined by c.d. comparison of the related ketone (**33**, R^1 = O, R^2 = Glc(Ac)$_4$), Δε + 2.74 (293 nm) with a known compound from the andrographolide series (**34**) Δε + 2.67 (289 nm) (*70*).

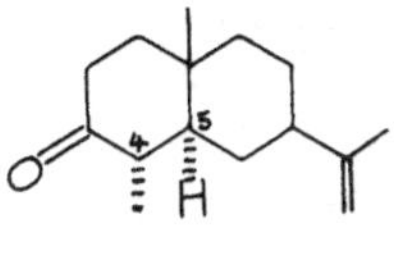

(25)

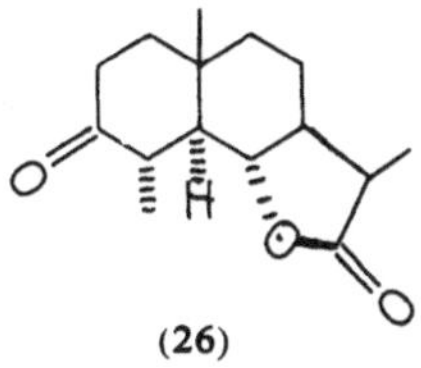

(26)

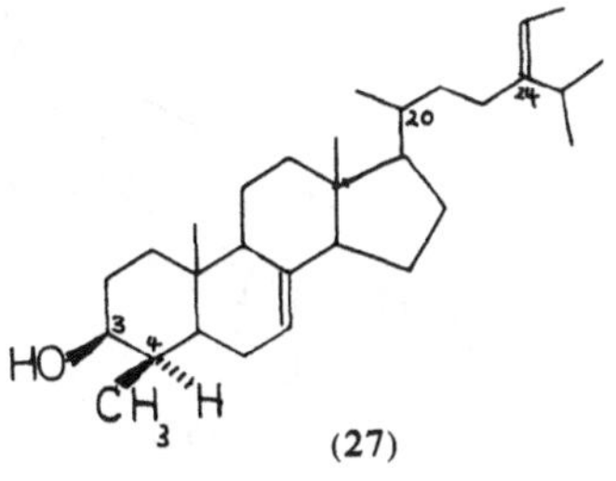

(27)

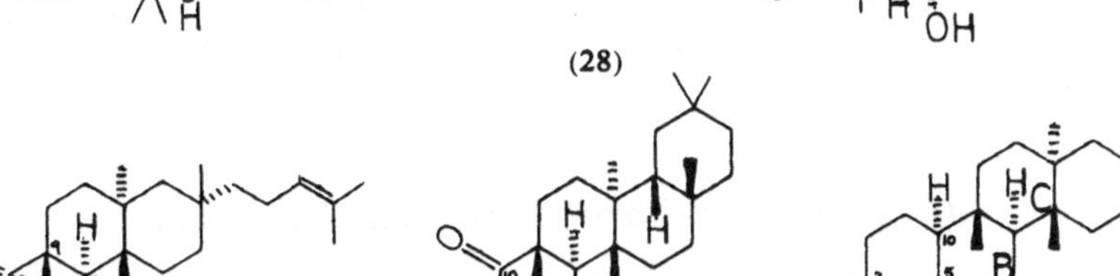

(28)

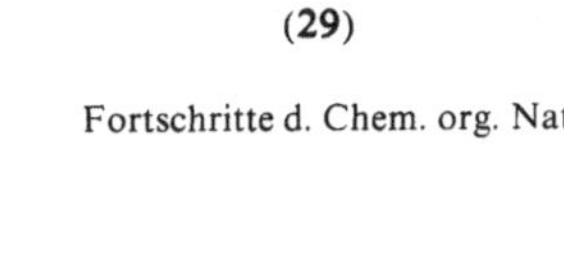

(29)

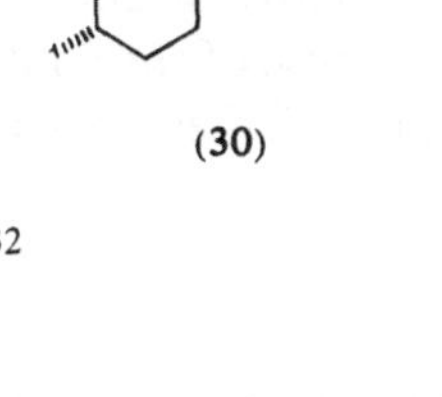

(30)

(31)

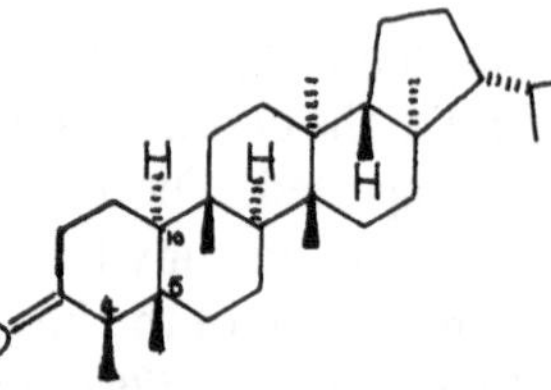

(32)

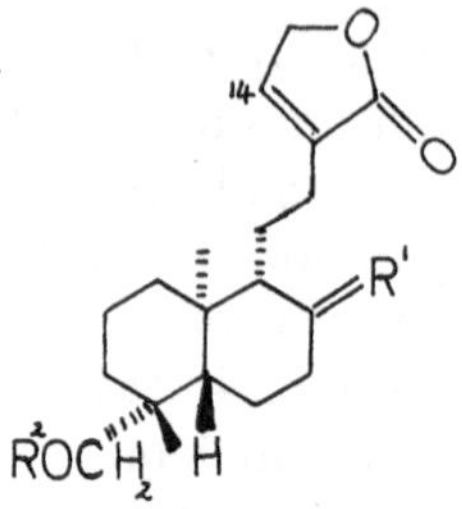

(33)

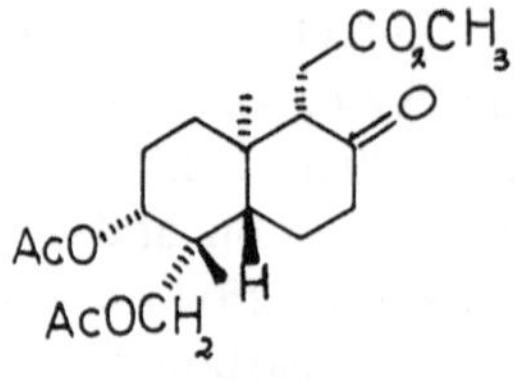

(34)

4-Oxo-steroids with 5α- and 5β-ring junctions provide useful model compounds of the 1-decalone type. Their CEs are different both in sign and magnitude (*a* about −80 to −90 and +5 to +10 respectively) and have been used, for example, to allot the 5β-configuration to 3α-acetoxy-5β-cholestan-4-one (*a* +10) (35) isolated in a study of steroid backbone rearrangements (*49*) and the 5α-configuration to 7β-benzoyloxy-5α-cholestan-4-one (36) (*a* −110) isolated in studies of cyclosteroids (*90*). The *trans*-ring junction in valerianol has been established by comparison of the negative CE of ketone (37) (*a* −27) with that of known 10-methyl-*trans*-1-decalone (*a* −32) (*170*) and the A/B *cis*-junction in (38) has been confirmed by the close similarity of the c.d. curve with that of the known analogue lacking an acetoxy group at C-7 (*189*).

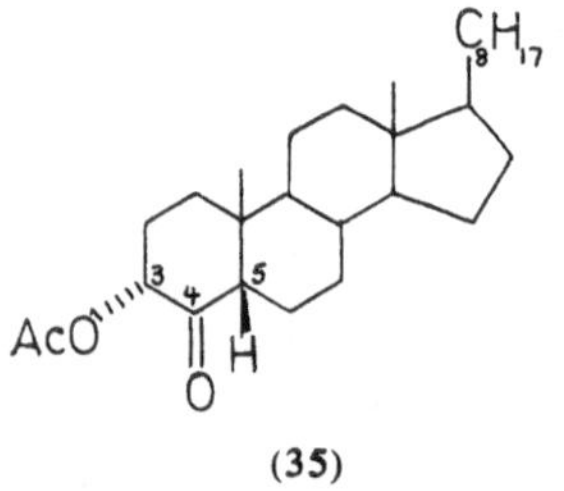

(35)

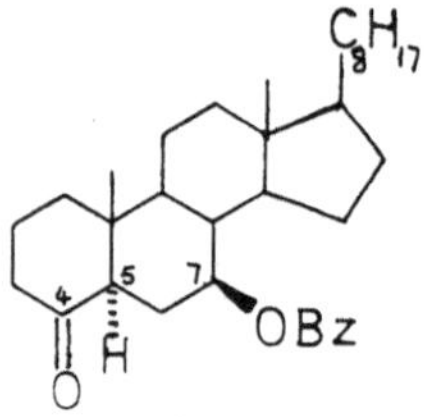

(36)

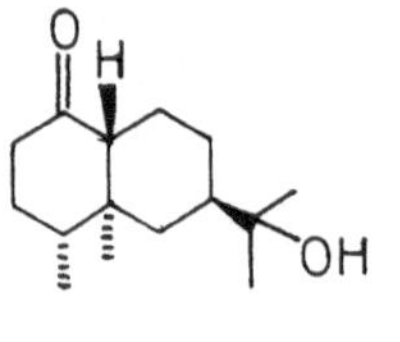

(37)

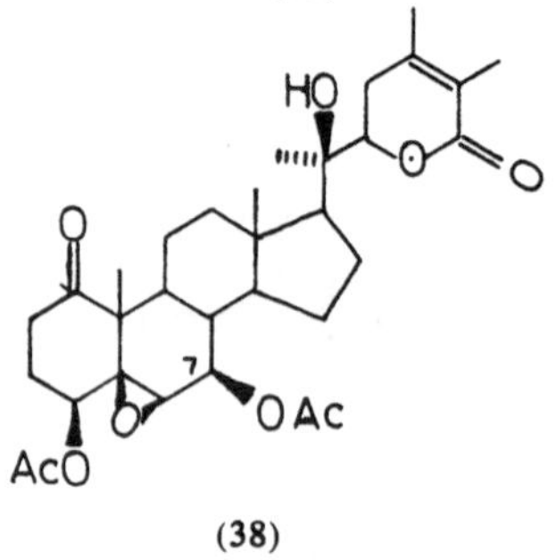

(38)

d) 4,4-Dimethyl-3-oxo-Steroids and Triterpenes

3-Ketones with a *gem*-dimethyl group at C_4 are not directly comparable with other 3-ketones because of the conformational differences which follow from the introduction of these particular substituents (*213*). For saturated compounds with a "normal" *trans* A,B-ring junction (10β,5α) and also for those having a 7,8-olefinic bond the sign of the carbonyl CE is negative, in contrast to the normal 5α-3-ketone without the C-4 dimethyl group which has a positive CE. This characteristic negative CE ($a - 20$ to -40) has been used to confirm the A/B *trans* configuration in many natural products including melianone (**39**) (*211*), ketone (**40**) related to turraeanthin, a product from West African timbers (*39*), and the oxo-acetonide (**41**) derived from odoratol (*77*). The o.r.d. curve of the bis-bicyclic ketone (**42**) has been used to confirm the *trans* substitution pattern and absolute configuration of lansic acid (**43**) (*132*), and a small positive CE ($a + 28$) has been used to allot the pimarane diterpene (**44**) to the enantiomeric 5β,10α,9β series (*66*). For 3-ketones with an additional 8β-methyl substituent (i.e. having a 4,4,8-trimethyl skeleton) and a *trans* A,B-ring junction (10β,5α) the sign of the CE is positive. This has been used, for example, to confirm the partial structure of the A,B-rings in alnincanone (**45**) (*279*).

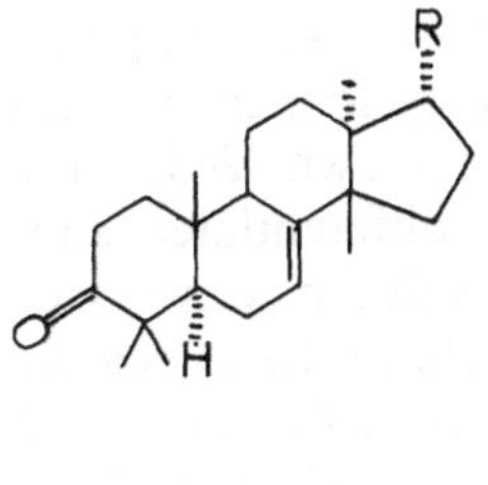

(39) R =

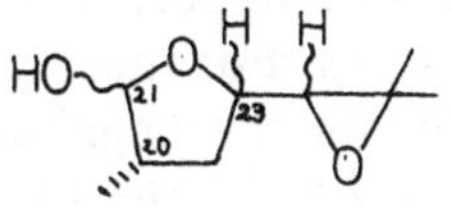

(40) R =

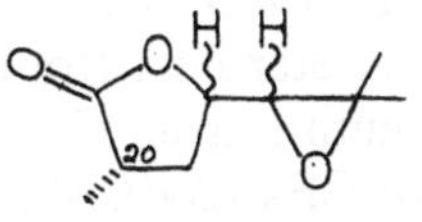

(41) R =

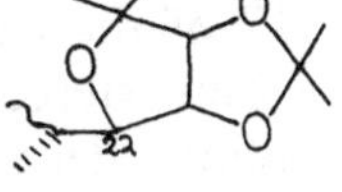

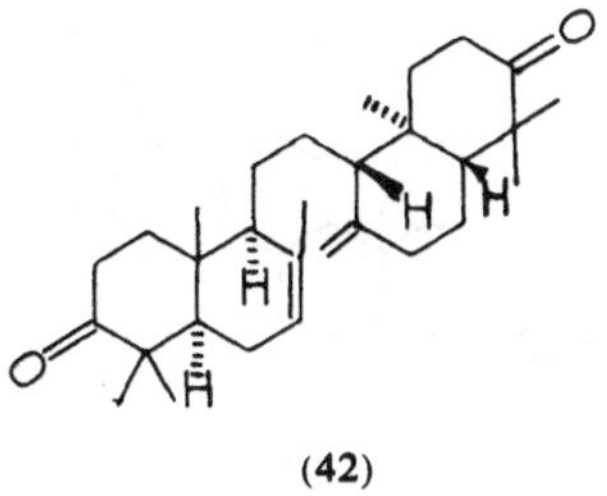

(42)

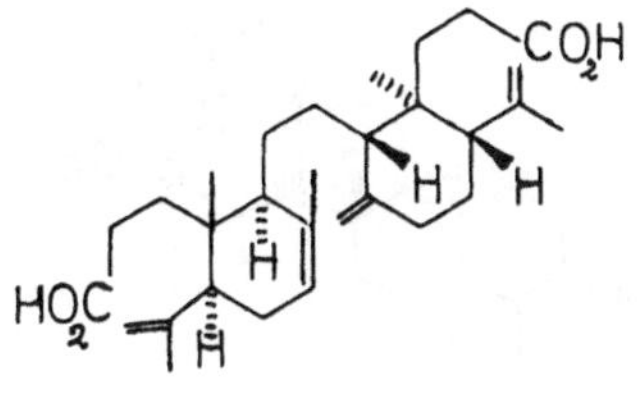

(43)

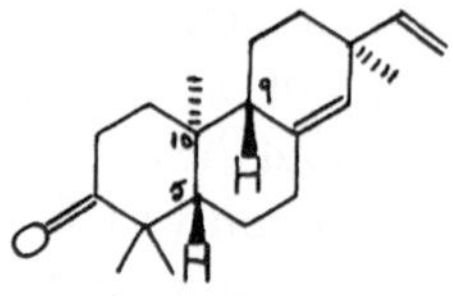

(44)

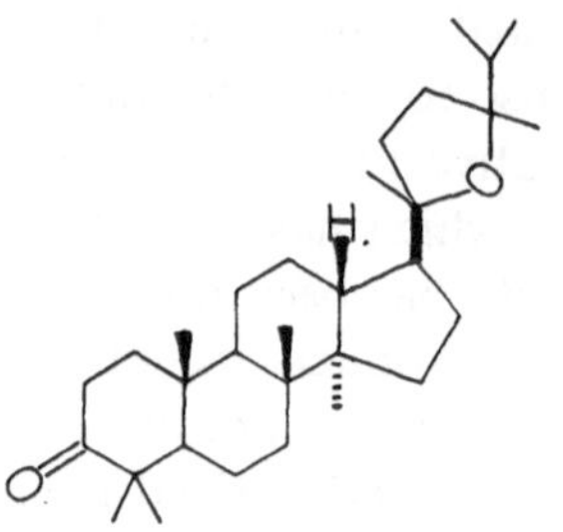

(45)

e) Orientation of the Steroidal Side Chain

The relative configuration at C-17 of the side chain in pregnanes and related steroids can often be determined from an o.r.d. or c.d. curve for the corresponding 20-ketone, since 17β- and 17α-orientations of the side chain lead to positive and negative CEs respectively in both the 14α- and 14β-series. For the 17β-configuration, a further distinction can often be made from the magnitude of the CE, which is strongly positive ($a > +100$; $\Delta\varepsilon > +2.5$) for the 14α-series (rings C/D *trans*-fused); a smaller positive CE ($a < +80$; $\Delta\varepsilon < +2.0$) is indicative of the probable 14β-configuration, i.e. rings C/D *cis*-fused. On this basis the alkaloid cyclomicrobuxine (46) ($a +130$) has been allotted the 17β-configuration shown (248), and the pair of steroid ketones (47) and (48) have been differentiated and the configuration at C-17 assigned from their respective positive and negative CEs (266). Examples in the 14β-series include the allocation of the 14β-,17β-configuration to ketone (49) and thus of the configuration at C-17 of the cardenolide side-chain in alloglaucotoxigenin (53), and of the same stereochemistry to drebyssogenin-F (50) ($a +70$) (42). These correlations depend on the assumption that the conformation of the pregnane side-chain remains similar in a series of closely related compounds and cannot necessarily be applied to 17β,14β bridged compounds.

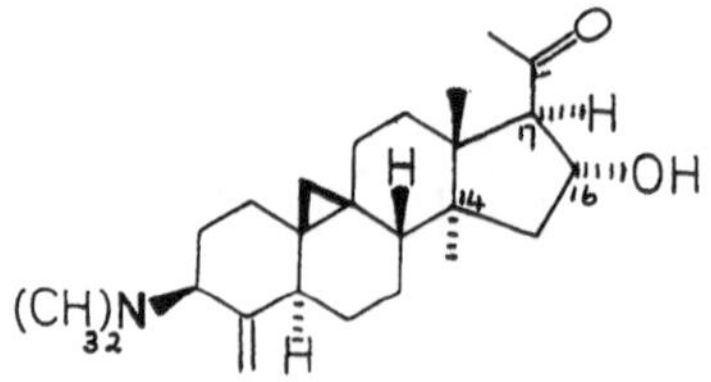

(46)

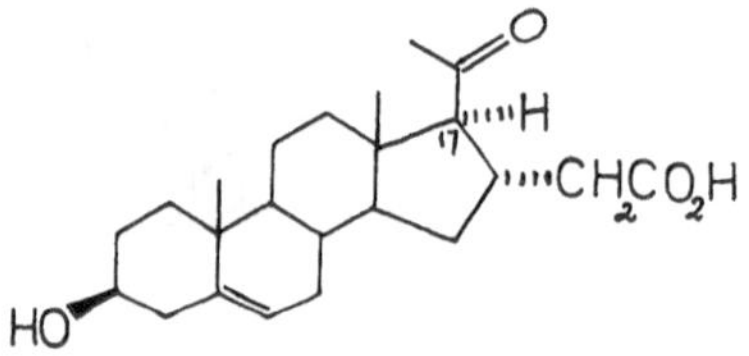

(47)

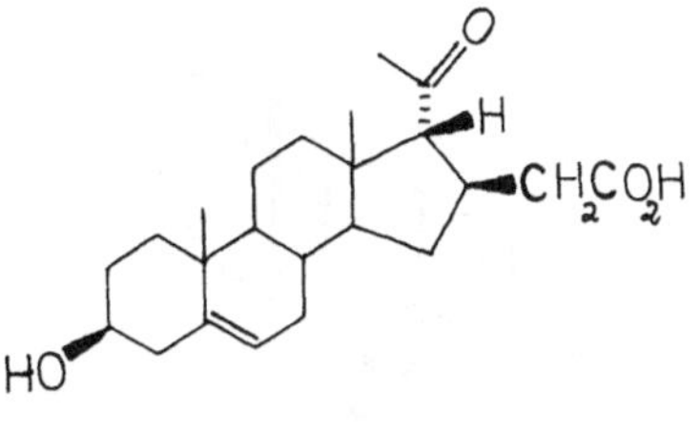

(48)

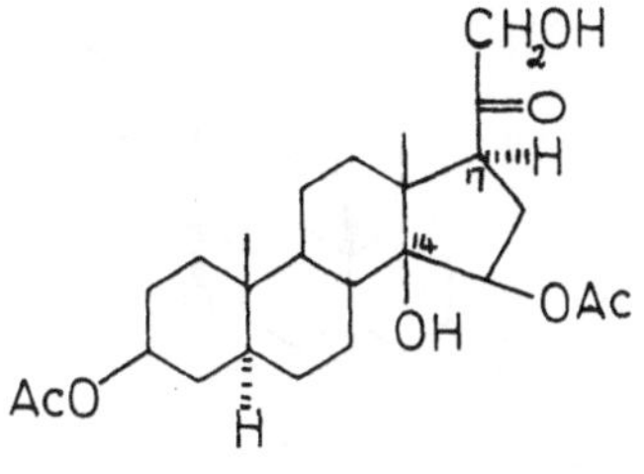

(49)

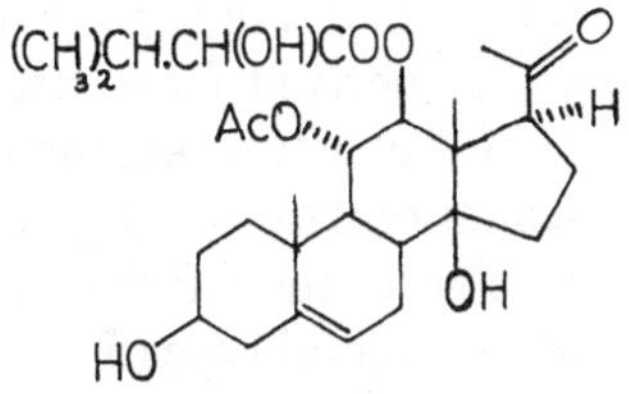

(50)

Although the magnitude of the positive CE can be used (with caution) to establish differences in stereochemistry at C-14 for 17β-acetyl compounds, there is no significant difference in magnitude between the negative CEs of the 14α,17α- and 14β,17α-isomers. For example, the negative ketonic CE of 3β-acetoxy-18-cyano-17α-pregn-5-en-20-one (51) (183) has been used, with other evidence, to establish the configuration at C-17, but *not* at C-14. Other examples include the allocation of the 17α-configuration to the side chain of lineolon (52, $R^1 = R^2 = R^3 = H$) and its derivatives (282, 290), to the 17β-hydroxylated analogue di-*O*-benzoyl-viminolon (52, $R^1 = Bz$, $R^2 = OH$, $R^3 = OBz$) (283), to the polyhydroxylated pregnanes, substance H (53, R = H) and substance J (53, R = OH) (103), and to glycocynanchogenin (54, R = ikemaoyl) (345).

Argument from close analogy is essential in the allotment of configuration by empirical o.r.d. and c.d. comparisons. This can be illustrated by work on the triterpene lactone thurberoginin (55) which was degraded to the ketone (56) and allotted the 19α-configuration shown, by

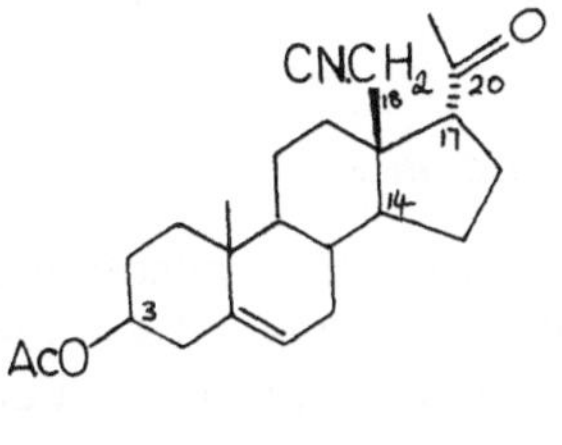

(51)

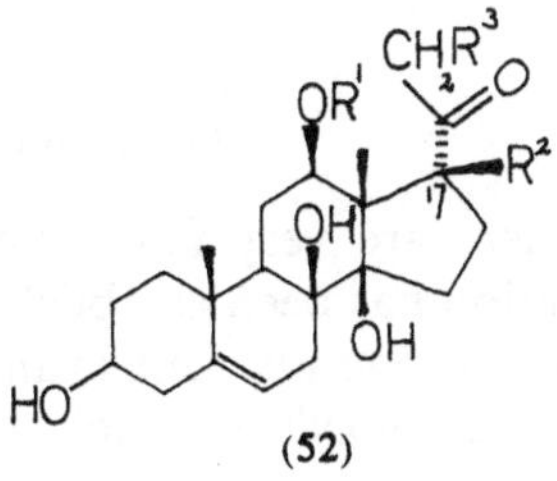

(52)

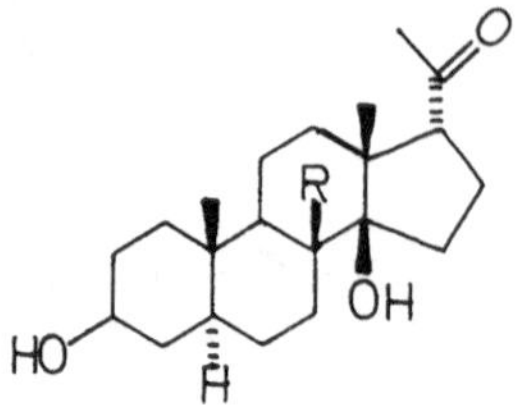

(53)

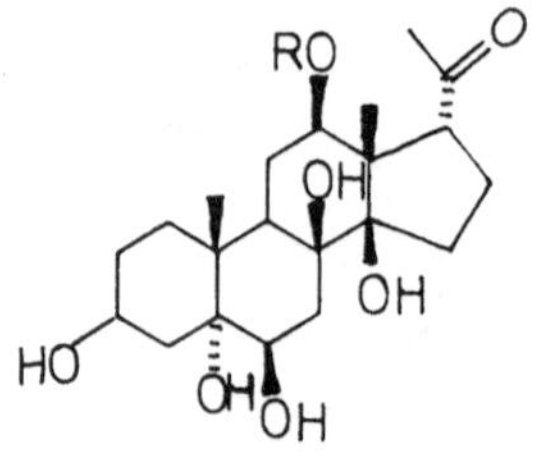

(54)

empirical comparison of its o.r.d. curve with that of a well-chosen analogue, the known ketone (57) derived from betulinic acid (221). This
means that rings D and E of the critical ketone (56) (which may be re-
drawn as (58)) are quasi-enantiomeric to rings C and D of a 14α,17β-
pregnan-20-one and yet the sign of the CE found for (56/58) is the *same*
(positive) as that found for the general type (59). Thus, 14α,17β-pregnan-
20-ones with a 13β-methyl group but lacking a bulky substituent at C-12,
would be inappropriate analogies for other compounds of type (58)
despite the apparent similarities between them.

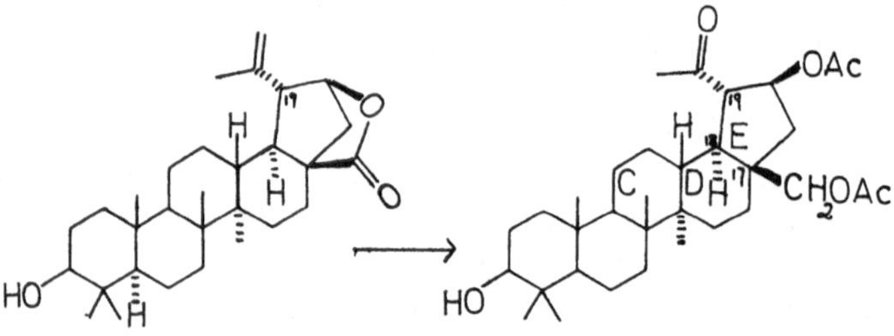

(55) → (56)

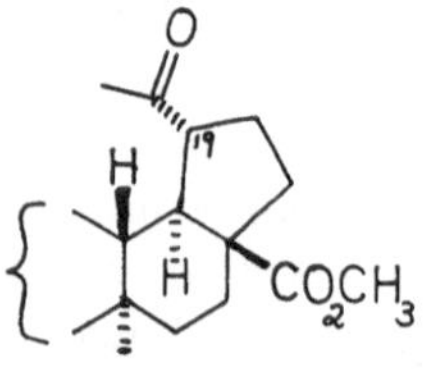

(57) 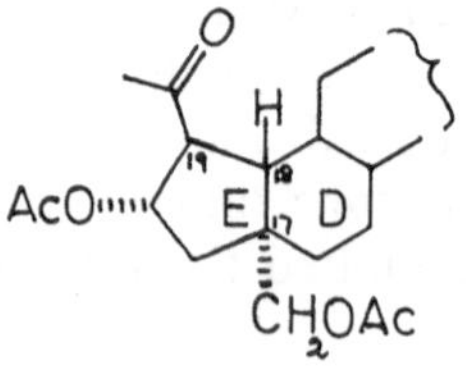 (58) 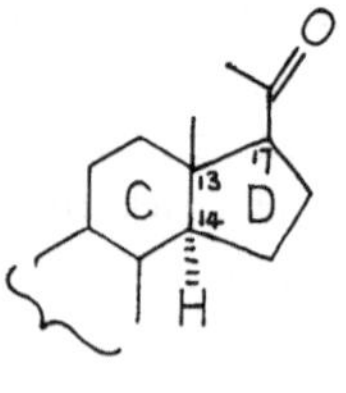 (59)

f) Monocyclic and Acyclic Ketones

There are very few examples in the literature where absolute configuration has been established by comparison of carbonyl CEs in monocyclic compounds. One example concerns (−)-*trans*-2-carboxy-2-methyl-
6-oxocyclohexylacetic acid (60, n = 1) which has a negative ketone CE

closely similar to that of the propionic acid analogue (**60**, n = 2) of known absolute configuration (*93*). This establishes the (1*R*)-configuration for the *trans*-acid (**60**, n = 1), the (1*S*)-configuration for the corresponding (+)-*cis*-2-carboxy-2-methyl-6-oxocyclohexylacetic acid (**61**, n = 1) which shows a positive CE, and the absolute stereochemistry of the A/B rings of rosenonolactone (**62**) from which (**61**) was derived (*93*).

The (*S,S*) and (*R,R*)-configuration have been assigned to (+)-ana-ferine (**63**) and its enantiomer by degradation to (*S*)- and (*R*)-pipecolic acid respectively and by direct comparison of their o.r.d. curves with that of known (*S*)-isopelletierine (**64**) (*106*).

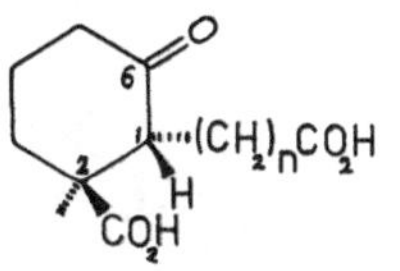

(**60**) 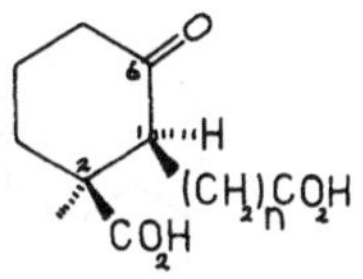   (**61**) 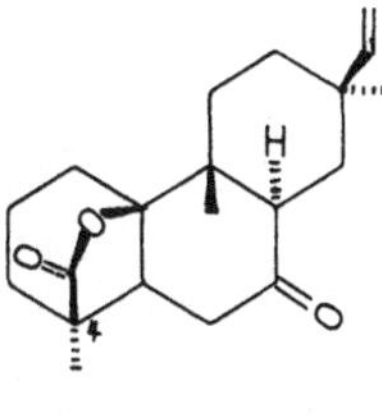   (**62**)

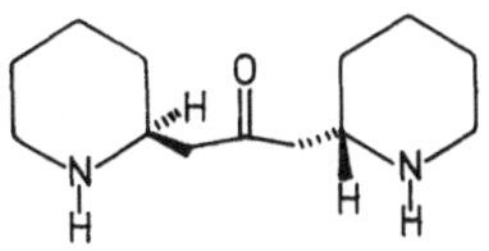

(**63**) 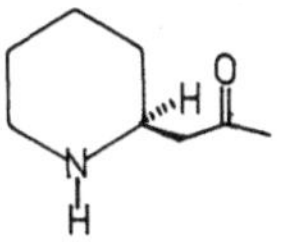   (**64**)

g) Difference Curves in Dicarbonyl Compounds

In general, data on diketones are difficult to interpret, unless the carbonyl groups are so widely separated that there is no possibility of vicinal interaction. Several examples of this latter type involve steroid 20-ketones. The 11α-acetoxy-5β-hydroxy-6β-hydroxymethylene-B-nor-pregn-3,20-dione (**65**) has been allotted the 5β-configuration shown, because the *difference* curve obtained by subtracting the strong positive o.r.d. curve of a 20-ketone from the experimental o.r.d. curve of the dione (**65**) shows a strong negative CE, characteristic of a 5β-B-nor-3-ketone (*338*). A pair of 6,20-ketones epimeric at C-5 (**66**) has been dif-ferentiated by comparing the difference curves obtained by subtracting the CE of a 20-ketone from the experimental curves of each dione. That compound with the larger negative CE in the difference curve was allotted the 5β-configuration (*151*). Other pairs of 6,20-diones have been assigned directly by comparison of the magnitude of their CEs, the compound with the more negative CE being assigned the 5β-configuration, e. g. the 2β-hydroxy-3α-chloropregn-6,20-diones (*332*). A similar approach has been used to differentiate the diketones (**67**), on the basis of the difference between the o.r.d. curves of the two ketones and the two corresponding 3-ketal-20-ketones. That compound which showed a positive CE in the difference curve was allotted the 10βH-configuration by comparison with 19-nor-5α-androstan-3-one, and the epimer the 10α-configuration by analogy with 19-nor-10α,5α-androstan-3-one. However, this analogy may not be valid because of the structural difference between the model compounds (steroids) and the unknown compounds in which ring B is seven membered (*207*).

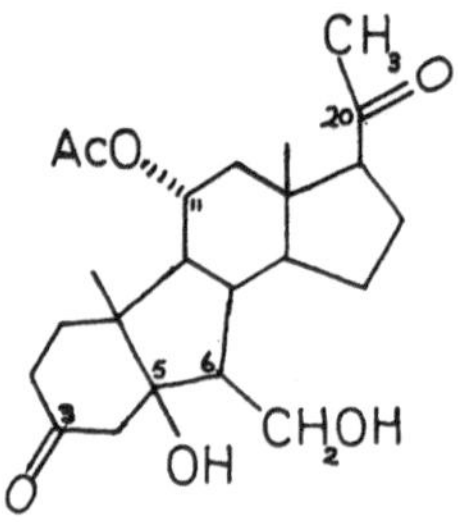

(65) 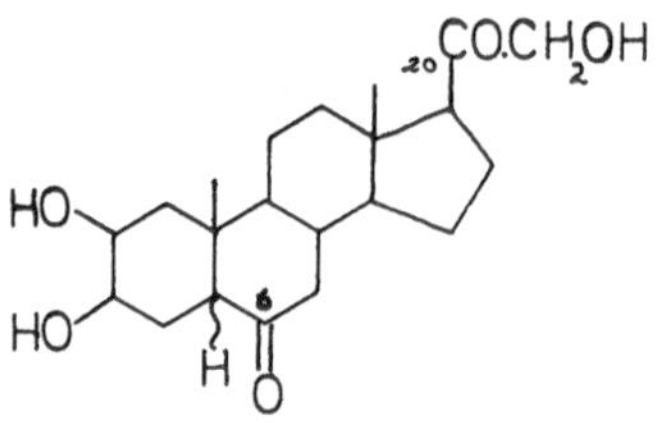 (66) 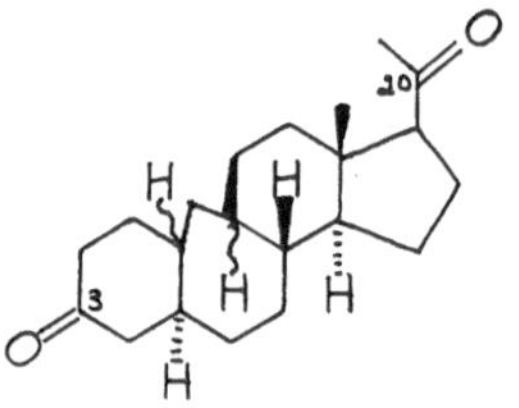 (67)

Difference curves have also been used to establish the *trans* nature of the C/D ring junction in the A-*nor*-D-homo-5β-3,17-dione (**68**) from the similarity of the c.d. difference curve (dione *minus* 3-ketone, Δε −1.8), with that of the model compound (**69** Δε −1.9) (*165*). In another example, the configuration of an isolated asymmetric centre in the side chain has been established by use of difference curves. Labdanolic acid (**70**, $R^1 = CH_3,OH$, $R^2 = OH$) and eperuic acid (**71**, $R^1 = CH_2$: $R^2 = OH$) are enantiomeric with respect to the decalin rings as shown by the oppositely signed CEs of the corresponding keto-acids (**70**, $R^1 = O$, $R^2 = OH$) and (**71**, $R^1 = O$, $R^2 = OH$) (*94*) and by the corresponding diones (**70**, $R^1 = O$, $R^2 = CH_3$) and (**71**, $R^1 = O$, $R^2 = CH_3$) (*263*). However, the dione CEs although opposite in sign, are not equal in magnitude (Δε −2.8 and +2.1 respectively) implying that the ketones are diastereoisomers, with the *same* absolute configuration in the side chain. On the assumption that the contributions of the two carbonyl groups are additive, a small negative CE (Δε −0.3) can be calculated for the contribution of the side chain ketone, corresponding in sign to the CE of (*R*) (+)-4-methylhexan-2-one (**72**) and establishing the (13*R*)-configuration for the eperuic/labdanolic side chain (*263*).

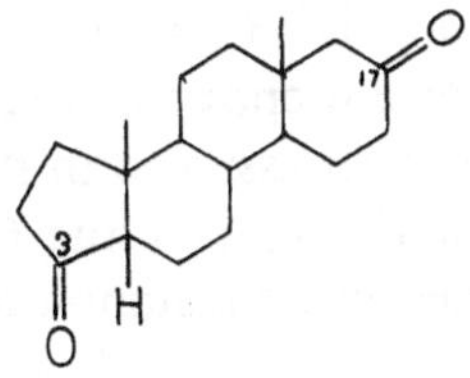

(68)

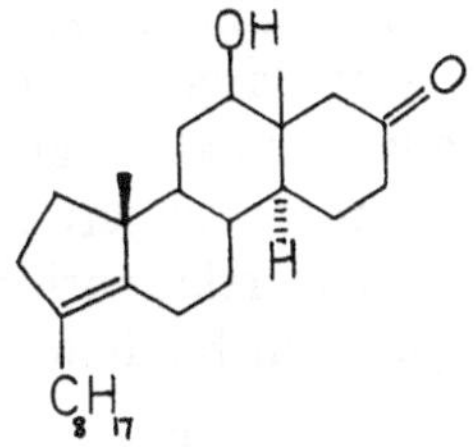

(69)

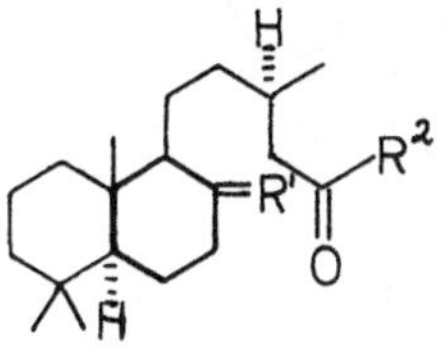

(70)

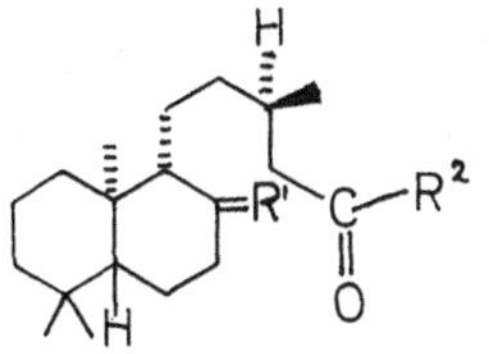

(71)

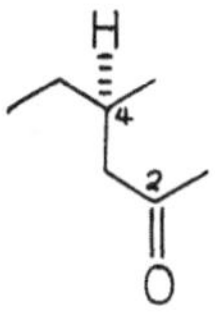

(72)

Although difference curves have been used successfully for configurational assignments in compounds containing two widely-separated carbonyl groups, CEs are not additive in those compounds where the carbonyl groups are sufficiently close for interaction to occur. A recent example concerns steroid 4,7-diketones for which vicinal effects are small in the 5α-series (**73**; 5αH) but appreciable for the 5β-series (**73**; 5βH) (*301*).

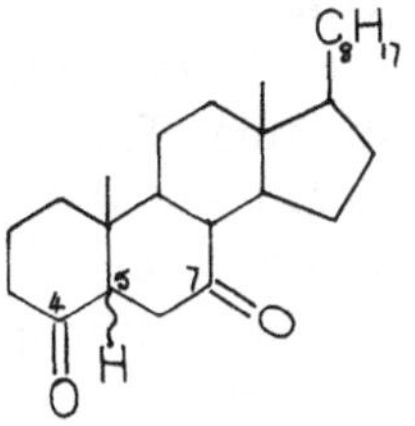

(73)

h) Epoxyketones

The absolute configuration of an epoxy group α,β to a carbonyl group can often be allotted from consideration of the sign of the ketonic CE. The majority of examples depend on empirical comparison of the unknown with a known standard (cf. II 2f for the semi-empirical approach), of which the most common are 4α,5α-epoxy- and 4β,5β-epoxy-3-oxo-steroids (74), with large negative and positive CEs respectively. The relative configurations at C_4 and C_5 of several epoxy ketones produced during photochemical isomerisation, have been assigned in this way (36, 284, 339, 340). The relationship between the sign of the CE and the epoxide configuration holds for substituted cyclopentanones as well as

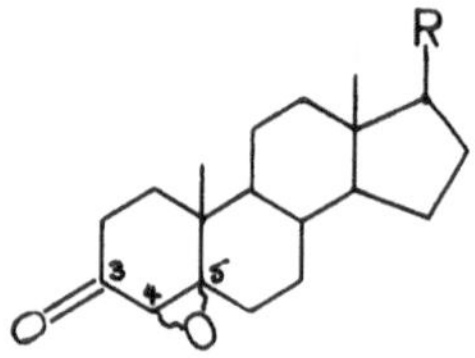

(74)

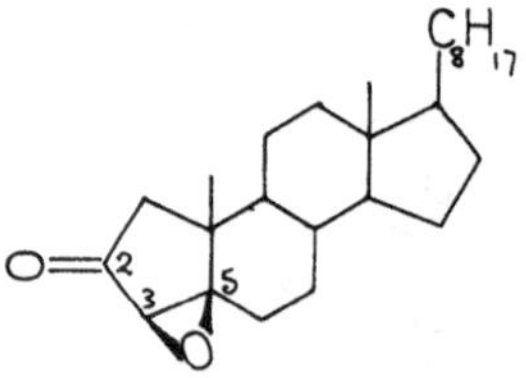

(75)

for cyclohexanones and has been used to establish the relative configuration of the epoxy group in 3β,5-epoxy-A-nor-5β-cholestan-2-one (75) ($\Delta\varepsilon$ +5.0 at 312 nm) (140).

2. Carbonyl Chromophores: Application of the Octant Rule

The relative importance of the octant rule (236) may be attributed first, to the accessibility of the n→π* transition at ca. 290 nm to the earliest spectropolarimeters, second, to the very extensive data available for ketones, and, third, to the comparative simplicity of the carbonyl chromophore, whose two symmetry planes are necessarily two of the regional boundaries in the octant rule. The most suitable compounds for application of the octant rule are those with a rigid cyclic framework and consequently many examples are found among extended decalone systems such as terpenes and steroids. Although it is now possible to measure the CE for the carbonyl π→π* transition at ≈192 nm (187) nearly all applications in the literature refer to the n→π* transition, and all examples given here refer to this longer wavelength band. Recent developments are discussed in detail in Ref. 74 (Chapter 3.1).

a) trans-Decalones

A simple example is the decalone (76) of known relative configuration, obtained by degradation of antibiotic LL-Z1271 (105). This compound shows a negative carbonyl CE ($\Delta\varepsilon$ −0.8) corresponding to the absolute configuration shown, in which the significant angular methyl group lies in the back lower left (negative) octant (76P)*. For corresponding *trans*-1-decalones lacking the angular methyl group a positive CE is found, as in the ketone (77) derived from rupestrol ($\Delta\varepsilon + 2.00$) (52).

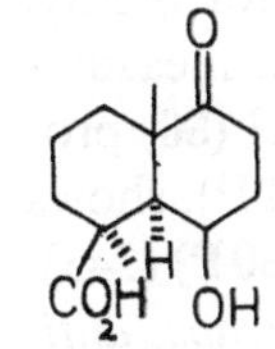

(76) 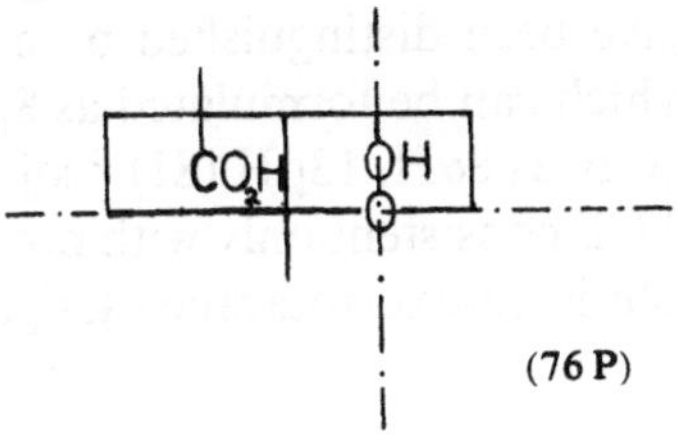   (76 P) 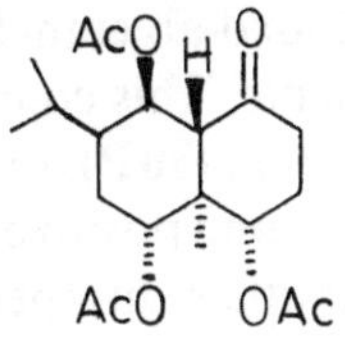   (77)

When there is ambiguity about the relative stereochemistry at a ring junction a single measurement of the carbonyl CE will not necessarily distinguish between the various possibilities. For example, the sesquiterpene ketone (78), of unknown configuration at C-10 shows a positive CE, which according to the octant projections (78P/1) and (78P/2) could imply either the *trans*-configuration of the rings, in which the isopropyl group projects into a front octant, or the *cis*-configuration with the isopropyl group lying in or near a nodal plane. From other evidence the *trans*-configuration is preferred but a distinction cannot be drawn from chiroptical measurements alone (243).

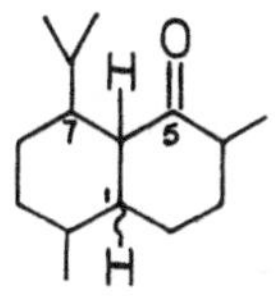

(78) 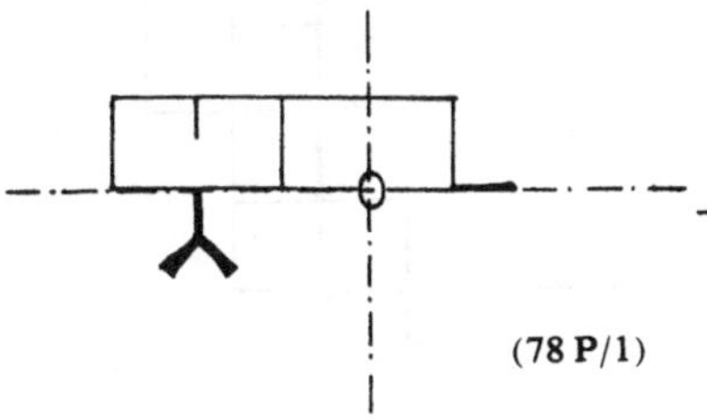   (78 P/1) 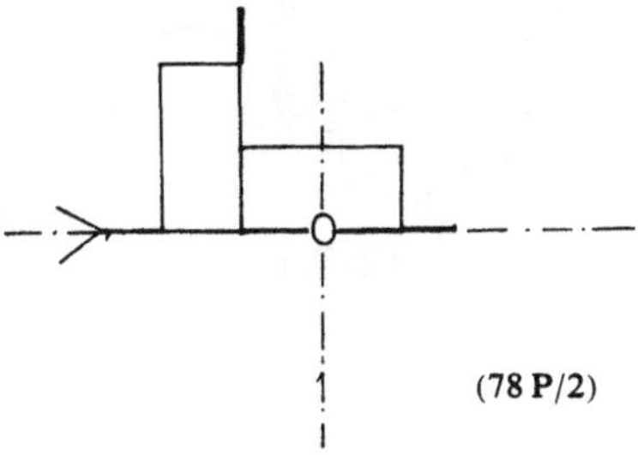   (78 P/2)

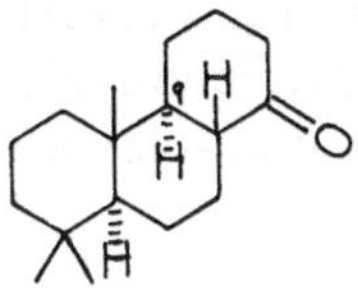

(79) 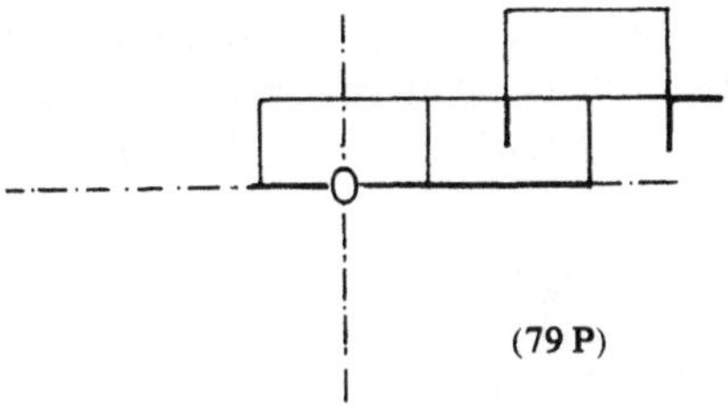   (79 P)

---

* The suffix P in a formula number indicates the octant projection of the molecule concerned. P/1, P/2, etc. are projections for different configurations or conformations of the same molecule.

In more extended systems the laevorotatory ketone (79) isolated during work on total synthesis of phyllocladene was allotted the absolute configuration shown, consistent with the observed negative CE, (79P) (328). [The alternative possibility with a 9βH configuration (rings B/C cis fused) would lead to a positive CE.] Two configurations for methyl 12-oxo-abietan-18-oate have been distinguished by o.r.d./c.d. measurements. This compound, which can be formulated as 8βH, 13αH (80) projection (80P), or alternatively as 8αH, 13βH (81) projection (81P), shows a strong positive carbonyl CE consistent only with projection (80P) (143). Another diterpene example is 13-oxo-totarane, assigned the *trans, anti,*

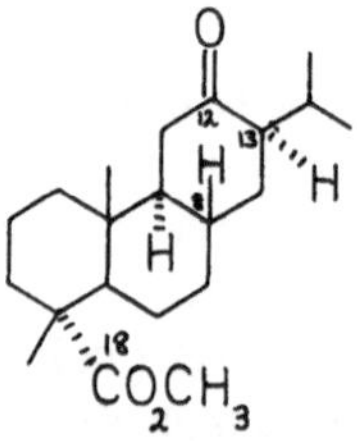

(80)

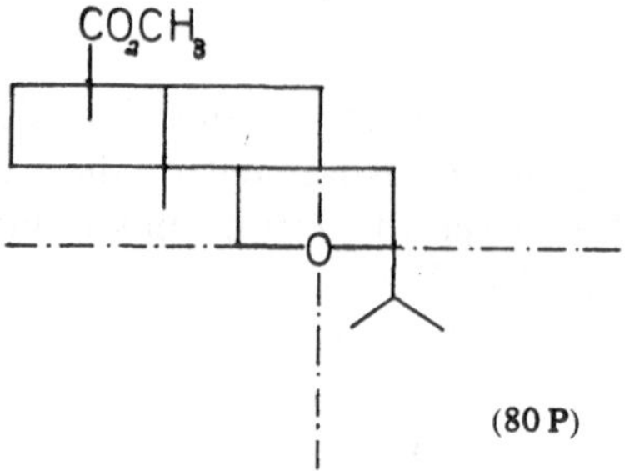

(80 P)

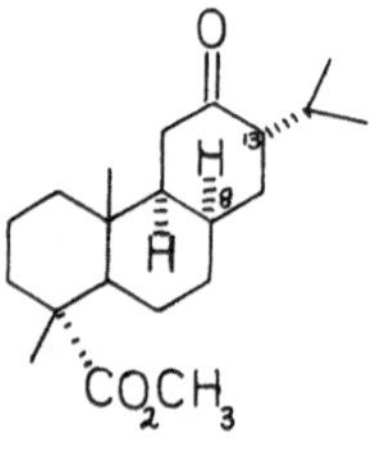

(81)

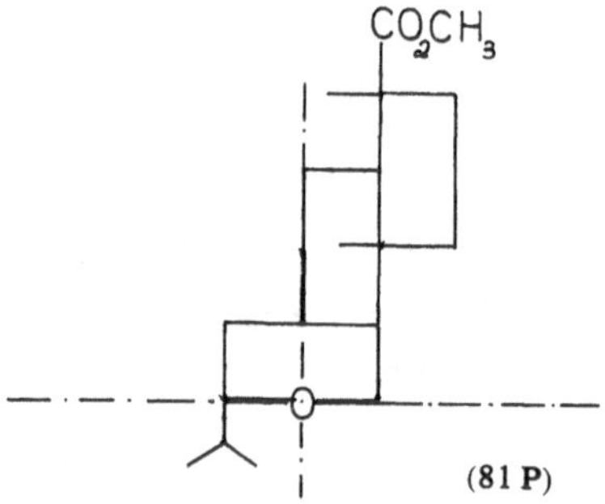

(81 P)

trans stereochemistry with equatorial isopropyl group (14αH) (82) by comparison of its positive CE (a +81) with the prediction of the octant rule (82P) and with data for 13-oxo-podocarpane (a+48) which lacks

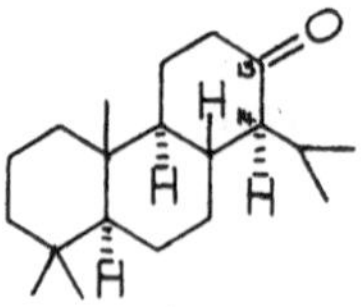

(82)

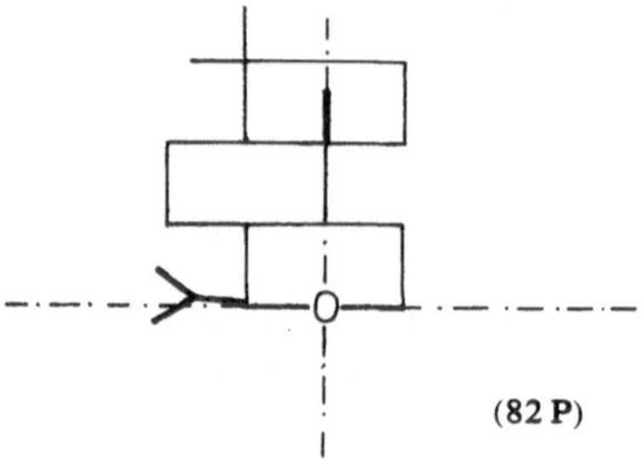

(82 P)

a substituent at C-14, and with 14β-methyl-13-oxo-podocarpane (a +61) (*34*). These data show the increasing positive contribution of the 14β-substituent, which does not lie exactly in the nodal plane but protrudes into the back upper left octant. In the pentacyclic triterpenes, 4-epifriedelin (**83**) has been allotted the 4α-methyl configuration shown, because its CE is much *less* negative (a − 56) than that of friedlin itself, which has the 4β-methyl (equatorial) configuration (a − 130). This reduction in the negative CE is a consequence of the position of the 4α-methyl group which lies in the back lower right (positive) octant (**83P**) (*327*).

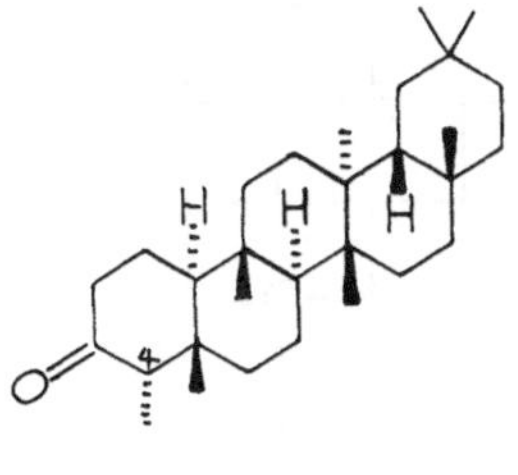

(83)

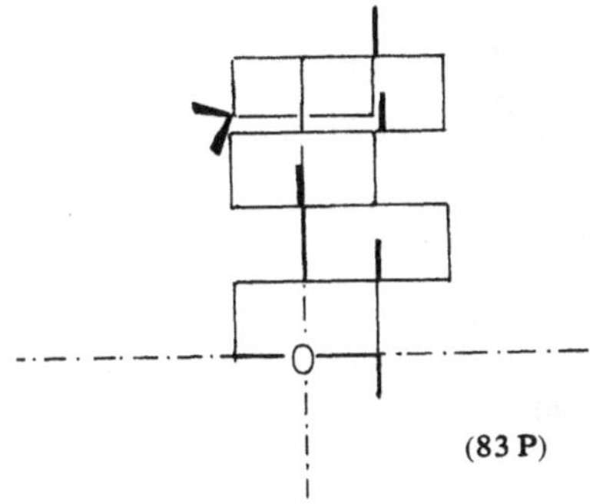

(83 P)

Application of the octant rule may be more difficult in those compounds in which the carbonyl group is in the middle of an extended array of cyclohexane rings. However, the octant diagrams for 6-ketones of 5α (**84P/1**) and 5β-configuration (**84P/2**) related to ecdysone (**84**) rationalise the empirical observation that CEs of 5β-6-ketones, are more negative than those of 5α-6-ketones, and have been used to assign configurations in several cases (*341*). Weak negative CEs similar to those observed for 7-keto-5α-steroids are observed for nomilin (**85**) and related compounds and are in accordance with the prediction of the octant rule (*100*). The

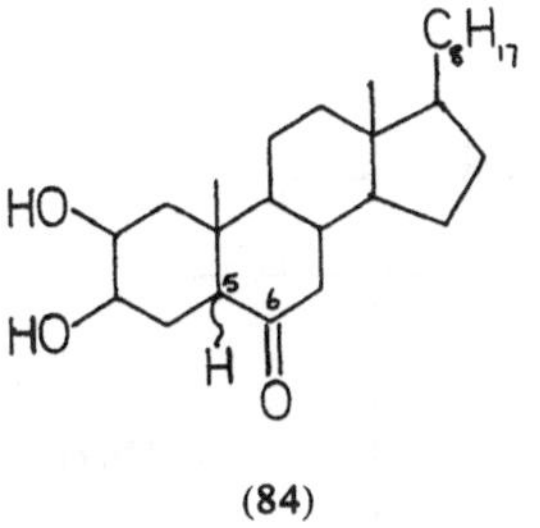

(84)

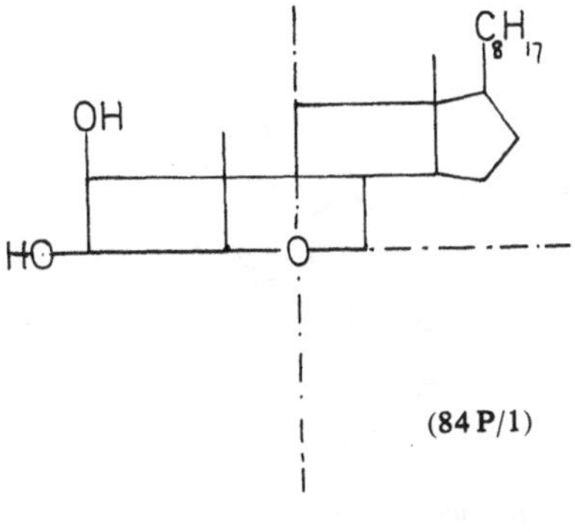

(84 P/1)

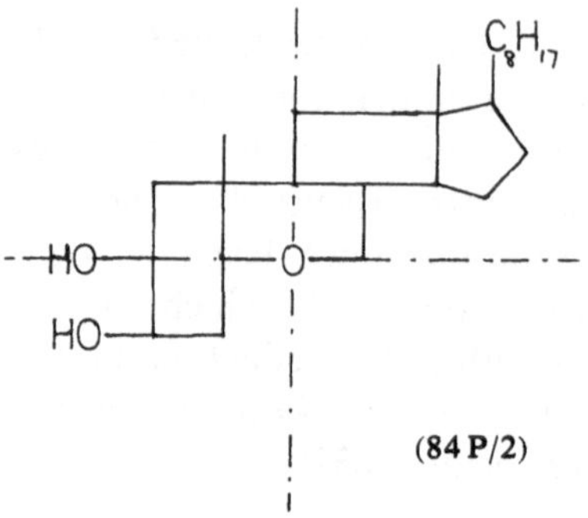

(84 P/2)

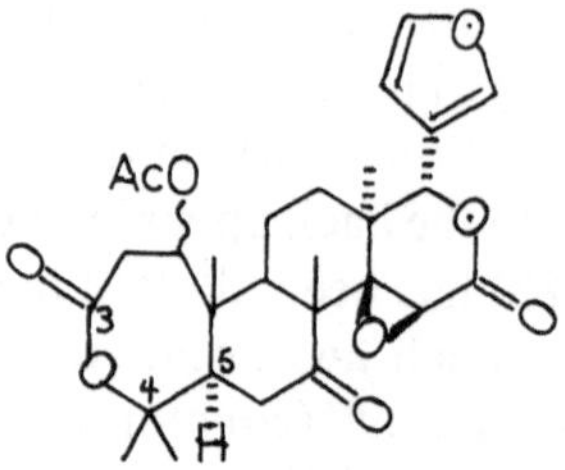

(85)

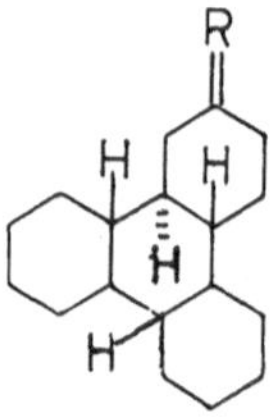

(86)

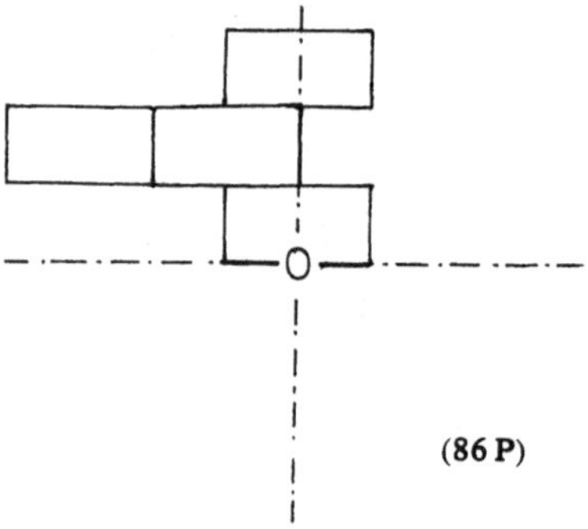

(86 P)

absolute configuration of the *dextro*-rotatory isomer of "all-*trans*" per-hydrotriphenylene (**86**, R = H₂) has been allotted as shown, because the positive CE of the derived ketone (**86**, R = O) corresponds to the octant projection (**86 P**) and not to the enantiomeric arrangement (*109*).

b) cis-Decalones

cis-Decalin itself may adopt two all-chair conformations, but this flexibility is prevented when a *cis*-decalin is *trans*-fused to additional rings. 2-Oxo-5α-cholestane (**87**, 5αH) and 2-oxo-5β-cholestane (**87**, 5βH) for example, can be readily distinguished by their respective positive (**87 P/1**) and negative (**87 P/2**) CEs (*200*). The „abnormal" 10α-confi-

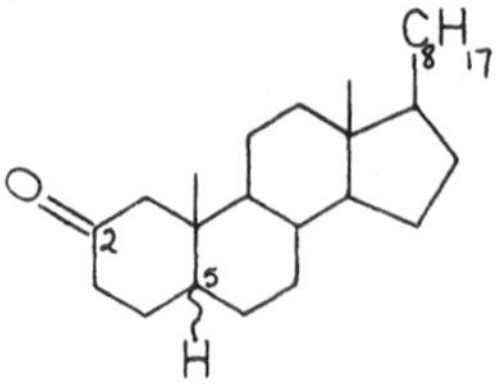

(87)

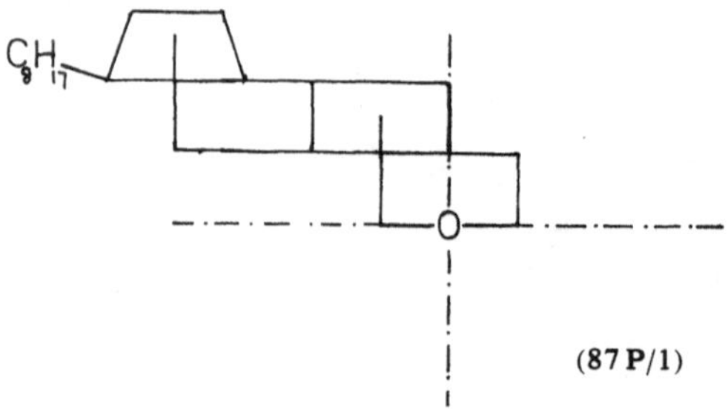

(87 P/1)

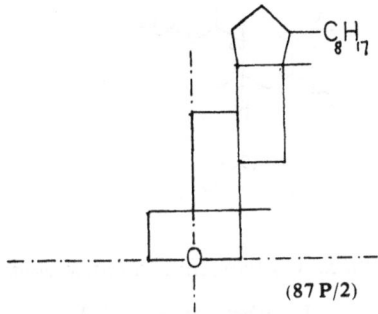

(87 P/2)

guration has been allotted to 10α-cholestan-5α-ol-3-one (**88**) from comparison of the observed small negative CE (*a*−4) with the corresponding octant projection (**88P**) for the 5α,10α *cis* fusion of rings A and B (*101*). The A/B *cis* junction has also been assigned to 12-methoxy-15,16-dinor-5β,10β-podocarpa-8,11,13-trien-4-one (**89**) by application of the octant rule (**89P**) which is in accordance with the observed strong positive

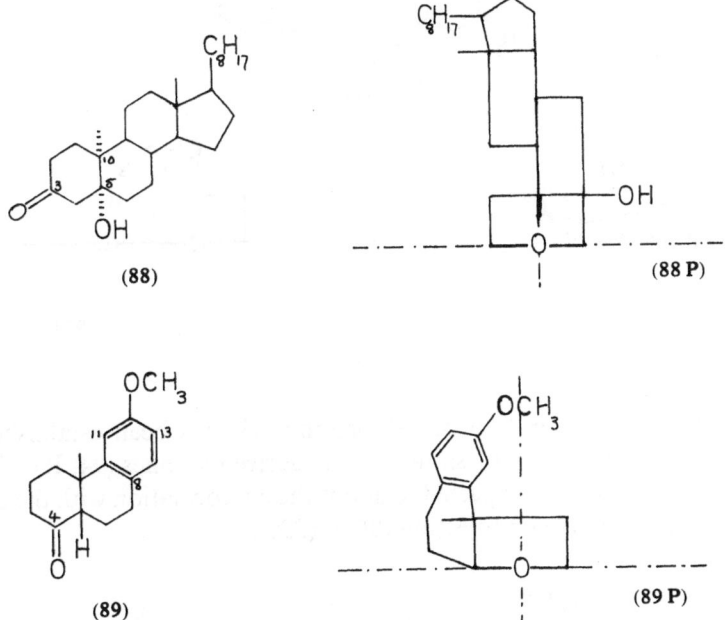

(88)   (88 P)

(89)   (89 P)

CE (*35*). Tetrahydroisopetasol (**90**) may be assumed to adopt that *cis*-decalin conformation in which the bulky 2β-isopropyl group is equatorial; the observed negative CE, characteristic of *cis*-fused 3-oxo-steroids, is in accordance with the 10β,5β-configuration as shown (**90**) (*253*).

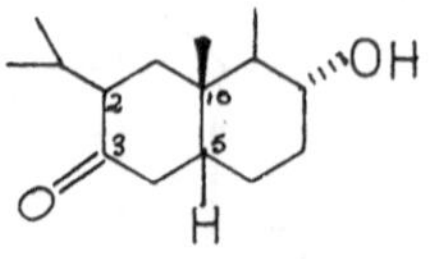

(90)

c) Flexible Monocyclic Cyclohexanones

The CEs shown by simple substituted cyclohexanones may be useful in assigning absolute configuration if the substitution pattern is such that one conformation is clearly preferred. For example, the terpenoid metabolite (91) of known relative configuration has been allotted the absolute stereochemistry shown from consideration of the preferred conformation (92), the observed negative CE and the octant projection (92P) (104).

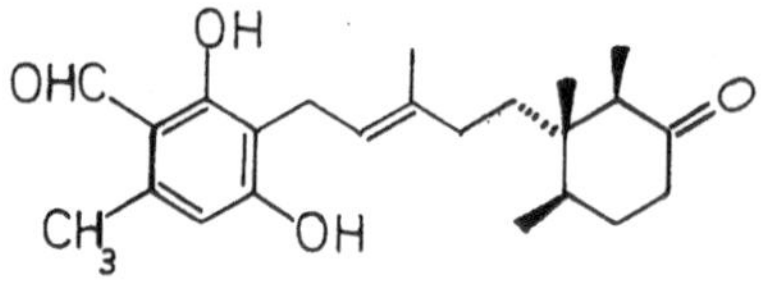

(91)

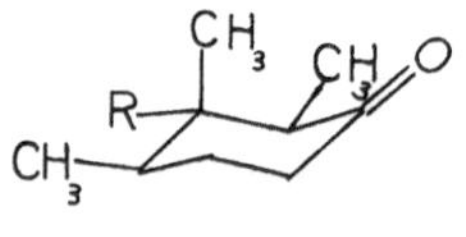

(92)

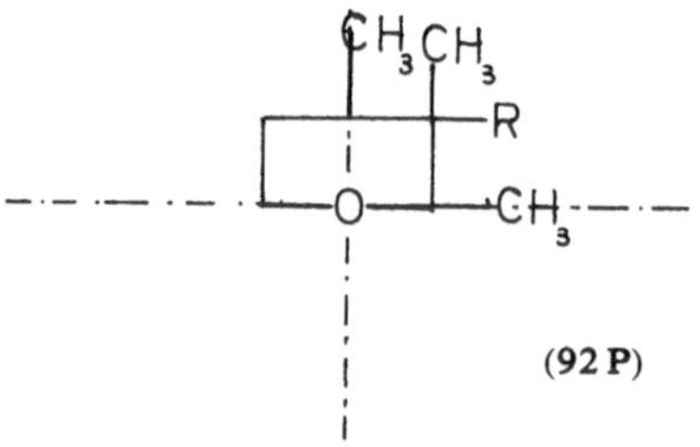

(92 P)

The absolute configuration of delobanone (93) has been established in part from the negative CEs shown by the derived ketones (94, R = H and R = OH) which may be expected to adopt the conformation with the bulky groups equatorial (95), projection (95P) (320).

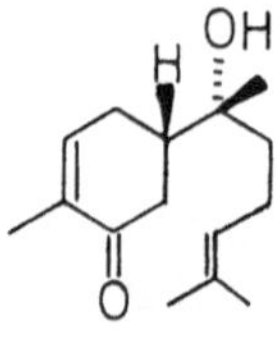

(93)

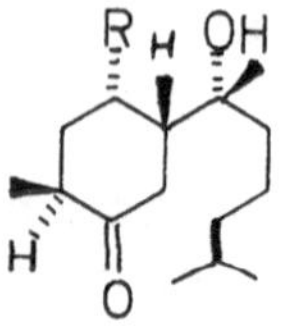

(94)

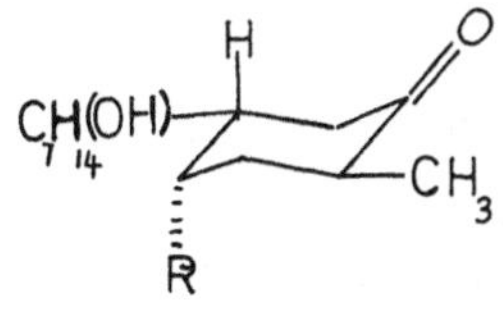

(95)

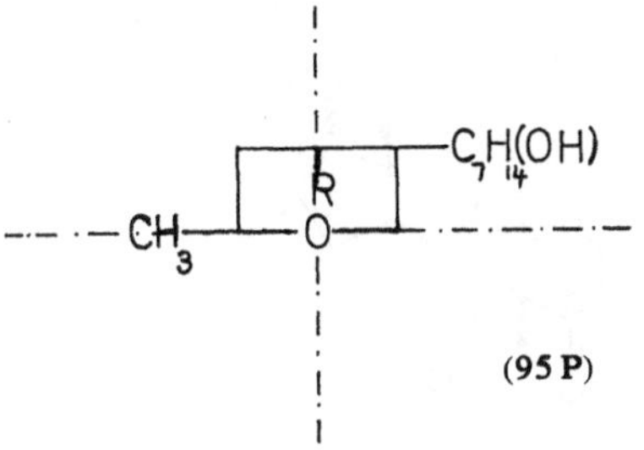

(95 P)

d) Cyclopentanones

The cyclopentanone ring is least strained when it is *cis*-fused to other rings or when it is *not* part of a more extensive ring system. The terminal cyclopentanone ring in ketone (96) for example, is nearly planar and the observed negative CE confirms the 16β,17β-fusion of the ring system which causes the whole of the steroid nucleus to fall in the back upper right (negative) octant (96P) (20). In contrast, the cyclopentanone ring of ketone (97) is twisted because of its fusion to the six-membered C-ring and it is the sense of twist which determines the sign of the observed carbonyl CE (positive) in accordance with the 13α-H configuration; see

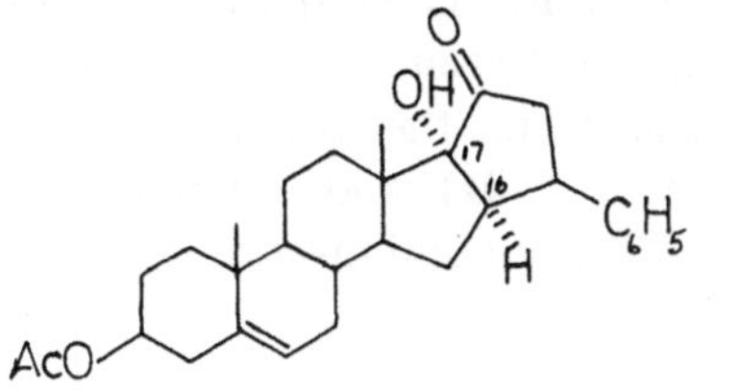

(96)

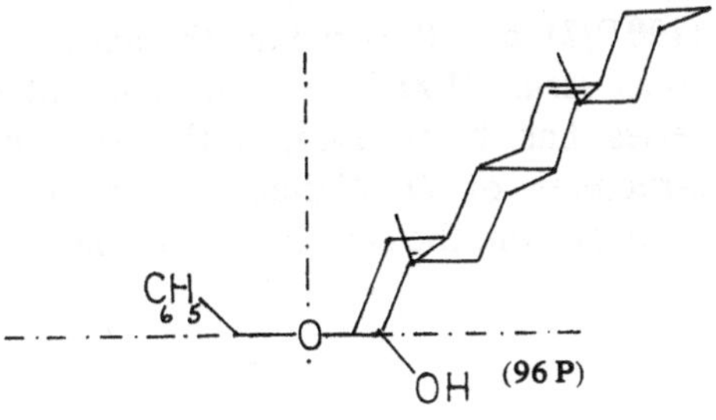

(96 P)

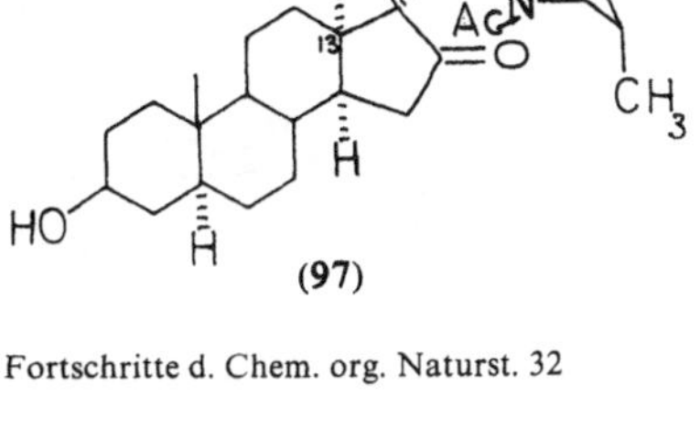

(97)

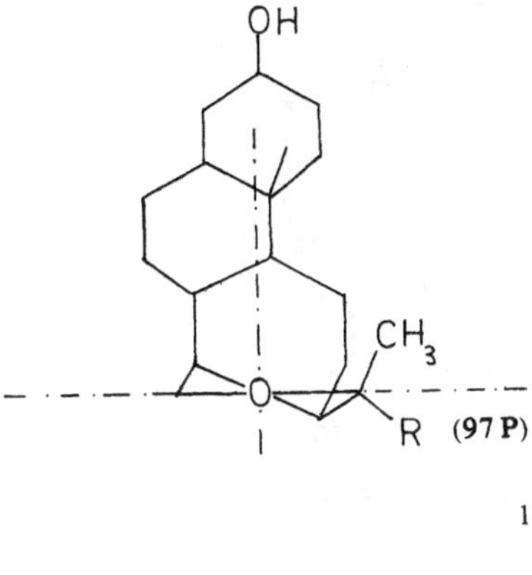

(97 P)

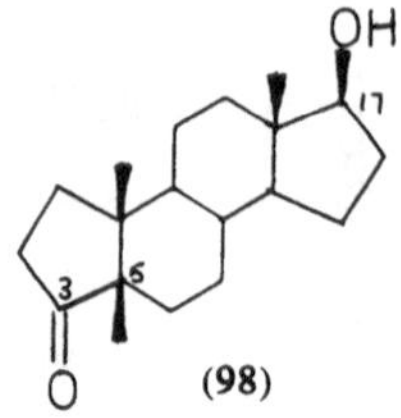

(98)

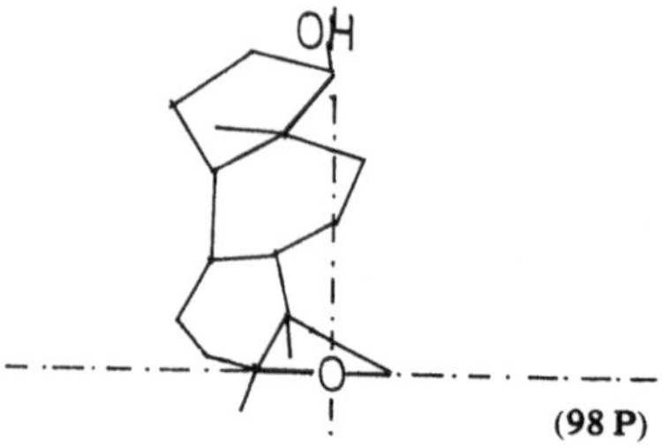

(98 P)

projection (**97P**) (*322*). Similarly for 17β-hydroxy-5-methyl-A-nor-5β-androstan-3-one (**98**), the octant diagram (**98P**) predicts a positive CE, in accordance with experimental results (Δε +1.2 in dioxan) and confirms the *cis* fusion of the A/B rings (*24*).

For *trans*-fused cyclopentanones, the sense of twist of the five-membered ring is also the main feature determining the sign of the CE. For example, A-nor-5α-cholestan-2-one (**99**) shows a very strong positive CE (*a* +228) in accordance with the geometry of the A-ring, as shown by projection (**99P**) (*264*). The C-nor-11-ketones (**100**) have negative CEs both for the C/D *trans* (14αH) and C/D *cis* (14βH) series, but the magnitude for the C/D *cis* series (*a* −87) is less than that for the *trans* (*a* −161) as shown by the octant projections (**100P/1** *trans*) and (**100P/2** *cis*). In both series, the sense of twist of the cyclopentanone is negative but atoms C_{13} and C_{18} lie in a positive octant for the *cis* compound (**100P/2**), and thus reduce the negative magnitude as compared with the *trans* series (**100P/1**) (*206*). This difference between the *cis* and *trans* series has been used, with other evidence, in the revision of the structures of O^3,*N*-diacetyltetrahydrojervine (**101**) and related compounds from the *cis* 12α,14α- to the *trans* 12β,14α-configuration (*222*).

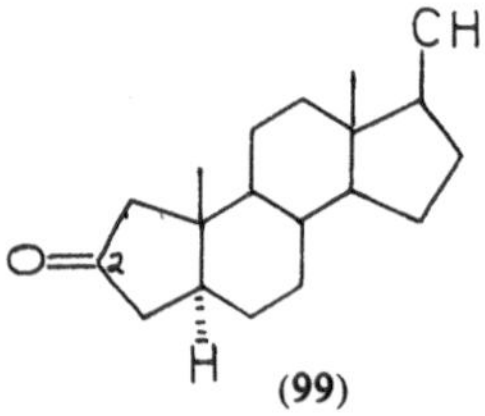

(99)

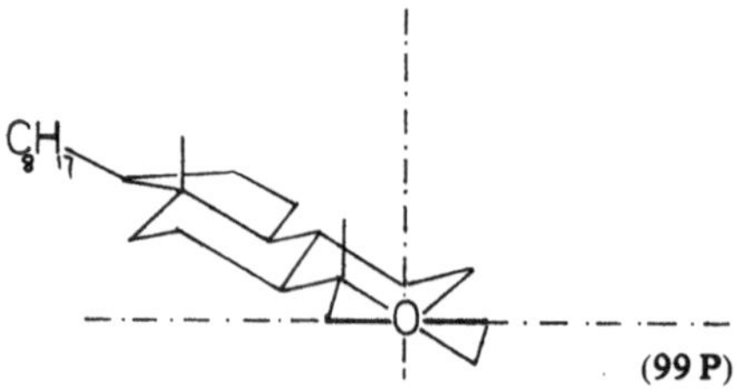

(99 P)

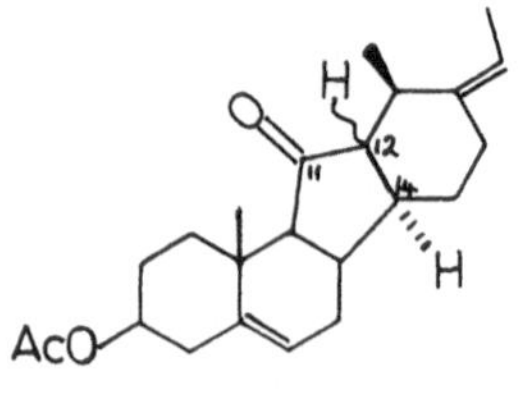

(100)

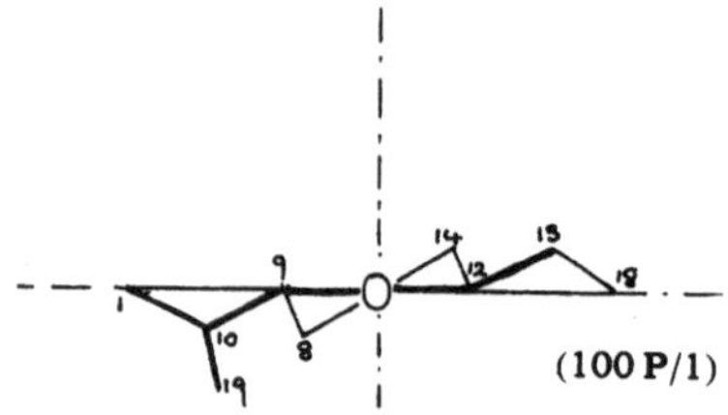

(100 P/1)

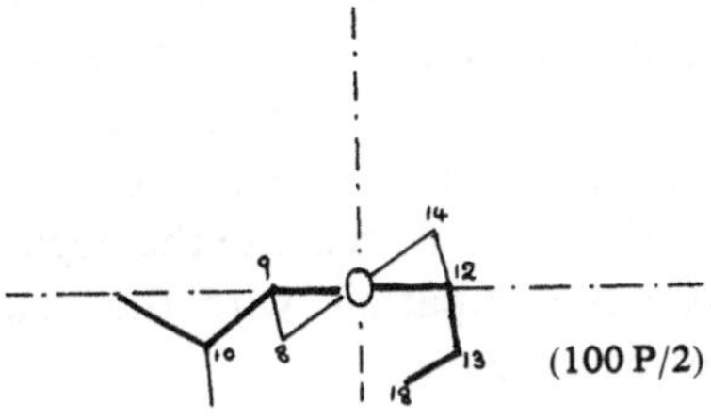

(100 P/2)

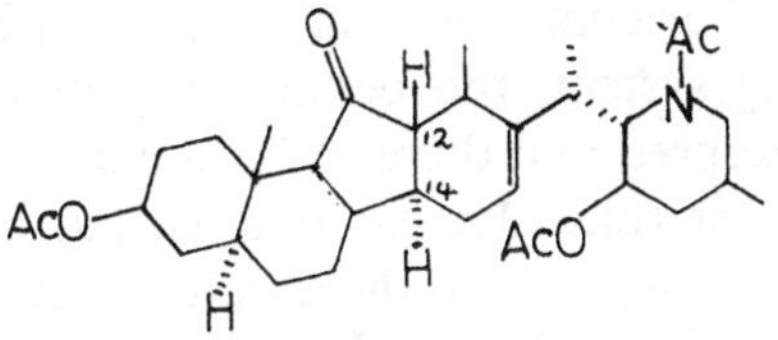

(101)

The sense of twist of the cyclopentanone ring is equally important in compounds which contain fused five and seven membered rings. *trans*-Dehydrodihydroneopulchellin (**102**) shows a strong positive CE (*a* +113) and has been allotted the 1αH,5β-methyl configuration by application of the octant rule (*346*). Similarly allotorilolone (**103**) and its C-1 epimer 1-epiallotorilolone which show strong positive and negative CEs respectively (*a* +210 and −180) have been assigned the 1β- and 1αH-configurations from consideration of their octant projections (*73*), thus confirming the 1βH-configuration in torilin itself.

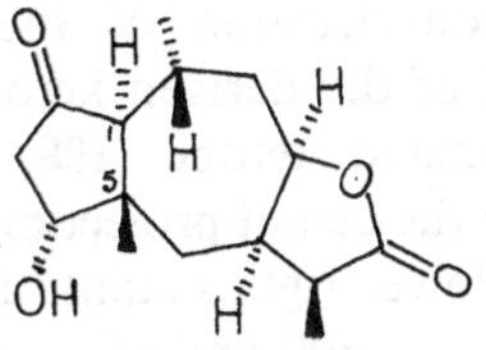

(102)

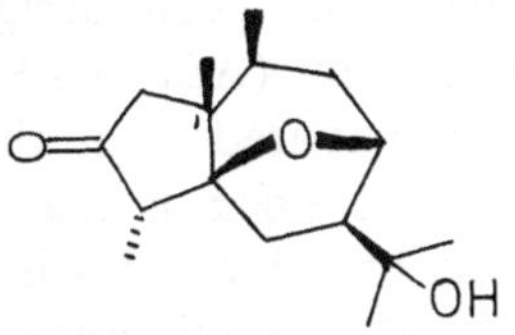

(103)

Problems of uncertain conformation are greatest for flexible monocyclic compounds and o.r.d./c.d. data can only be used to give information about the configuration of a particular compound if the conformation is known. For example the substituted cyclopentanone (**104**) of known relative configuration shows a strong negative CE. On the assumption that the preferred conformation is that in which all three substituents will become most nearly equatorial (cf. **104P**), the absolute configuration shown has been allotted to the ketone (**104**) and thence to the parent compound jasminin (**105**) (*16*).

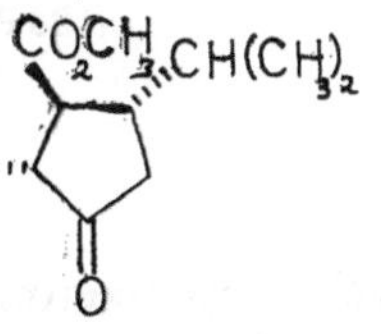

(104)

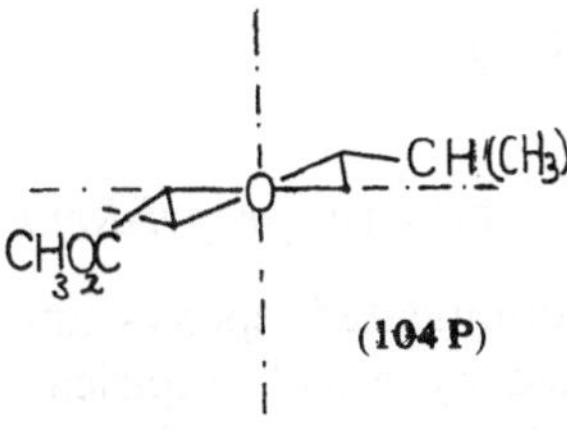

(104 P)

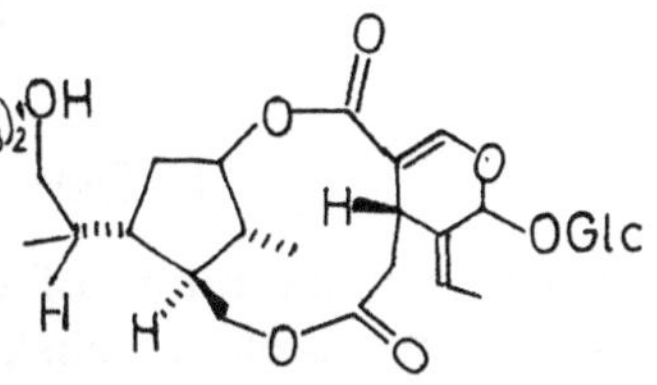

(105)

For compounds in which a carbonyl group acts as a bridge across a ring system, the octant projections may be complex and difficult to interpret, and there are comparatively few successful applications of the octant rule. One example is that of the sesquiterpene seychellene (**106**, R=CH$_2$) allotted the absolute configuration shown, because the negative CE of the derived ketone (**106**, R=O) is compatible with the octant projection (**106P**) rather than the enantiomer (*344*).

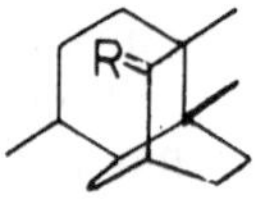

(106) 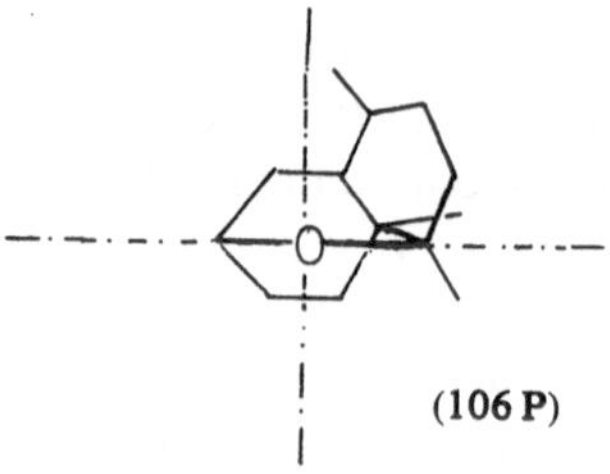   (106 P)

e) Cycloheptanones and Cyclobutanones

The absolute configuration of the angular methyl group in (+)-carotol (**107**) has been assigned from consideration of the strong positive increment ($\Delta\Delta\varepsilon$ +1.9) for the CE of the derived ketone (**108**, R=Br) as compared with the non-brominated ketone (**108**, R=H). This positive increment is in accordance with the octant projection (**108P**) in which the bromine atom is in the back lower right octant, and confirms the absolute configuration of the ketones (**108**) and of (+)-carotol (**107**) (*215*).

Cyclobutanones are rare among natural products but are often formed by addition of dimethylketene to dienes; the signs of the CEs near 300 nm have been used to allot absolute configurations to some ketones formed in this way (*38, 85, 154*).

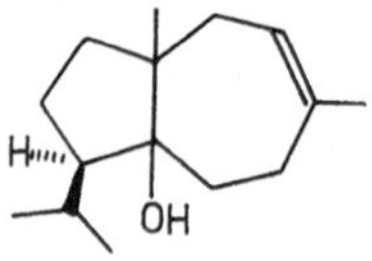

(107) 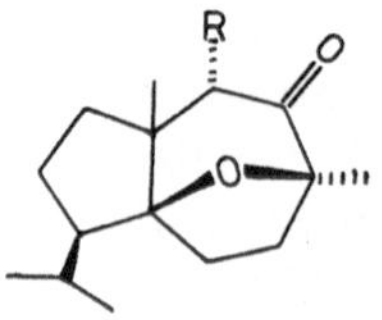   (108) 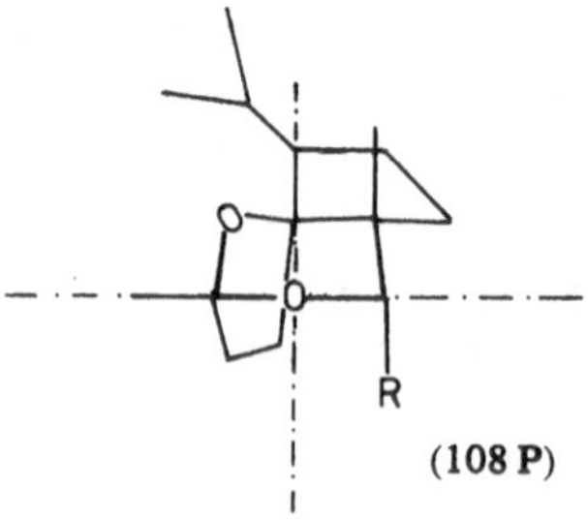   (108 P)

f) Epoxy- and Cyclopropyl-Ketones

The configurations of conjugated epoxy- and cyclopropyl-ketones have mainly been established by direct empirical correlations (see section II 1 h) but a very few correlations have been made by application of the "anti-octant" rule. These include ketone (109) related to picroside-I (*190*) and the ketone (110) related to sirenin and to sesquicarene (*268*). The β-configuration has been assigned to the epoxide ring in ketone (111) derived from fibraurin (*152*) and the 7α,8α-configuration to ketone (112) related to inumakilactone (*164*).

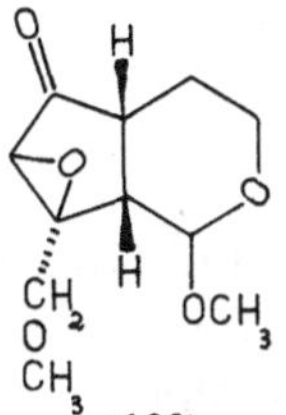

(109)

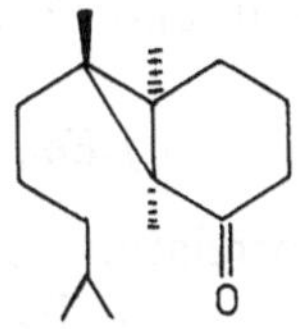

(110)

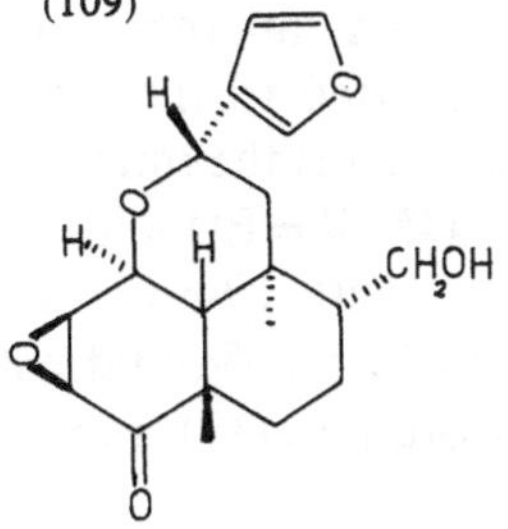

(111)

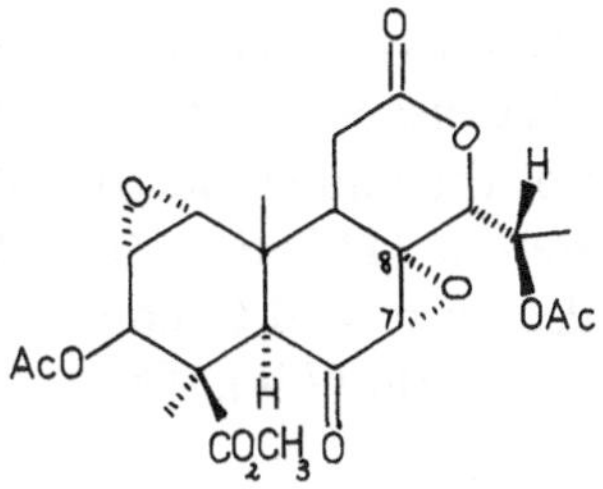

(112)

3. Unsaturated Ketones

Two CEs can be detected for α,β-unsaturated ketones between 400 and 200 nm. The low intensity absorption at longer wavelength (*R* band, 340—350 nm), which can be attributed to the n→π* transition of the carbonyl group, is very sensitive to the polarity of the solvent and to any conformational alterations in the geometry of the chromophore and of the adjacent rings. The band at shorter wavelength (*K*-band, 230—260 nm) is of much greater intensity and may be attributed to a π→π* transition of the carbonyl group. The sign of this latter CE is dependent primarily on the sense of helicity of the chromophore itself, the cisoid (113) and transoid (114) arrangements leading to a positive CE, and the enantiomeric arrangements to a negative CE.

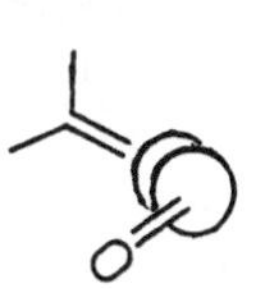

(113)

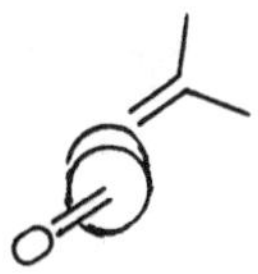

(114)

Chiroptical measurements with unsaturated ketones have been used in one of three ways, (i) to assign configuration at a given centre or centres by empirical comparison between compounds of known and unknown configuration, (ii) to monitor changes in configuration and of conformation in systems with unusual ring-junctions and, (iii) (in a few cases) to assign absolute configuration directly by use of the helicity rule (297) (Chapts. 10 and 13) and (74) (Chapt. 3.2).

a) Empirical Correlations

The simplest correlations are those in which o.r.d. and c.d. curves of the unknown are essentially identical (or opposite) to those of a compound of known structure and configuration. For example (−)-13β-ethyl-17β-hydroxygon-4-en-3-one, isolated from microbiological hydroxylation of a racemic steroid, has been allotted to the "unnatural" series because its o.r.d. curve (positive n→π* CE) is the mirror image of those of the known dextro-rotatory isomer (**115**, R=Et) and of 19-nor-testosterone (**115**, R=Me) (296). The analogy with 19-nor testosterone has also been used to confirm the structure of 19-nor-5α-androst-1-(10)-en-2-one (**116**) (114). The sesquiterpenes β-rotunol (**117**) and α-rotunol

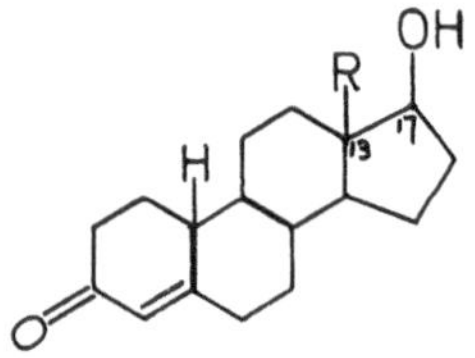

(115)

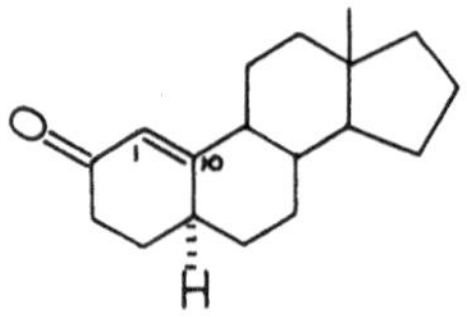

(116)

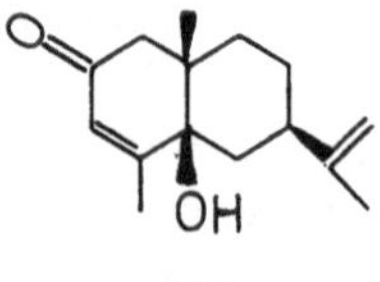

(117)

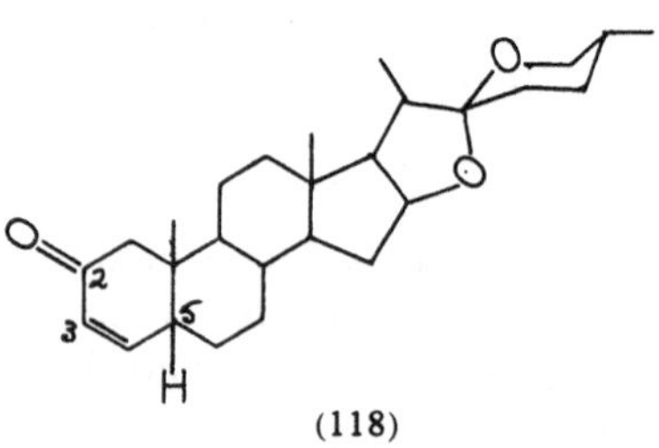

(118)

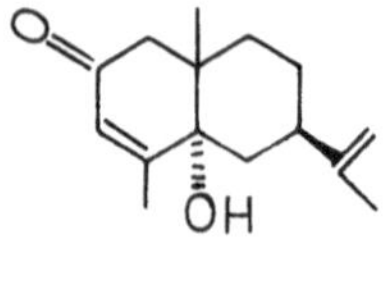

(119)

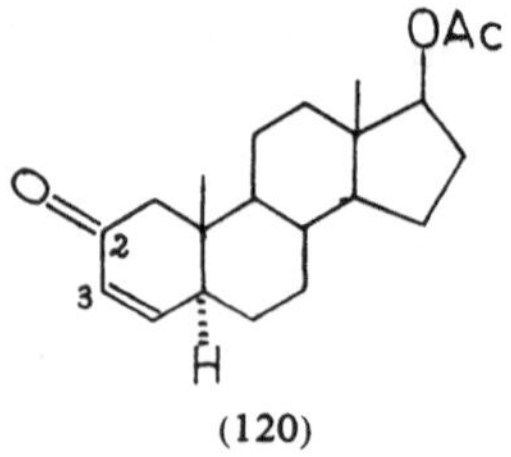

(120)

(119) have been allotted *cis*- and *trans*-fused ring junctions respectively because of the similarity of the n→π* bands in the c.d. curves with those of 5β-spirost-3-en-2-one (118) and 17β-acetoxy-5α-androst-3-en-2-one (120) (*145*). Identity of c.d. curves has also been used to confirm the stereochemistry of the A, B and C rings of grandifoliolenone acetate (121), by comparison with known gedunin (122) (*79*). The insect hormone ecdysone (123) has a positive CE near 340 nm and a strong negative CE near 240 nm, characteristic of a 6-oxo-7-ene with a 5β-H (*cis*-fusion of rings A and B). The configuration of several other closely related hormones of this type including ponasterone B (*247*), rubrosterone (*321*) and muristerone-A (*67*) have been established or confirmed by comparison of their c.d. curves with those of ecdysone. Examples among α,β-un-

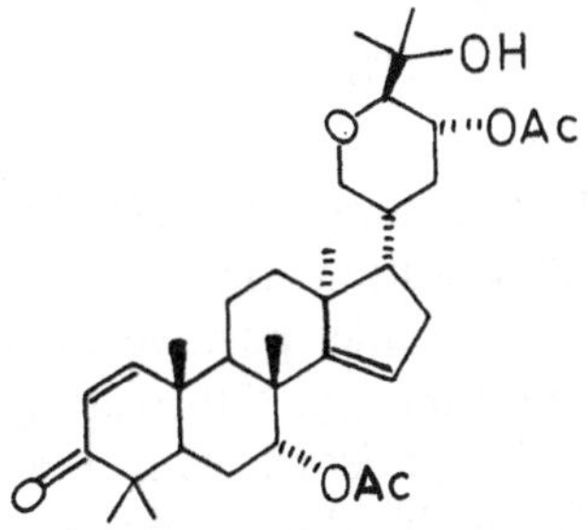

(121)

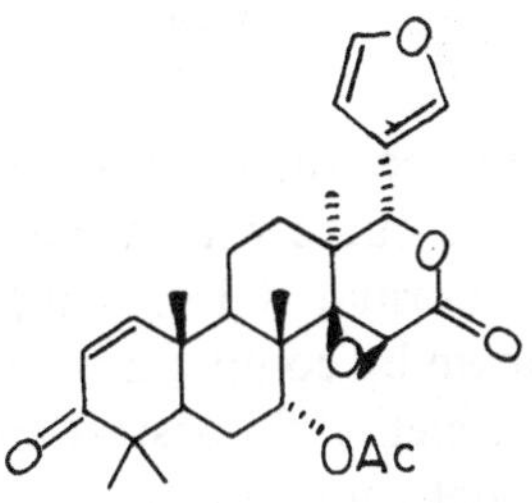

(122)

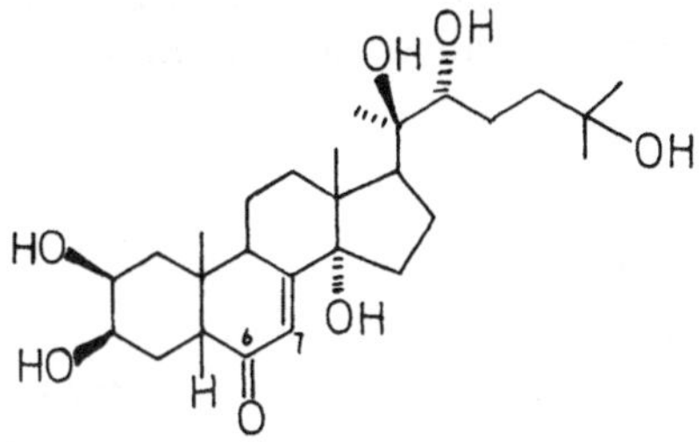

(123)

saturated cyclopentanones include assignment of the 14β-configuration to 14-hydroperoxy-5β,14β-androst-15-en-17-one (124) (*a* +60) by comparison with 3β-acetoxy-14-hydroxy-5α,14β-androst-15-en-17-one (125) (*a* +67) (*291*), and the 10βH- and 14β-methyl configurations allotted to the A-nor 19-nor ketone (126) (*347*) and the "backbone rearrangement" 16-ketone (127) (*50*) respectively from comparison of their negative R-band CEs with that of A-nor-testosterone.

Modification of the normal 10β,9α,8β-configuration of a steroid 4-en-3-one necessarily changes the conformation of rings A and B and this is reflected in the sign and magnitude of the n→π* CE. Retrotest-

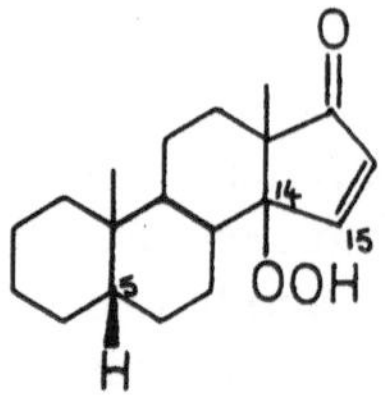

(124)

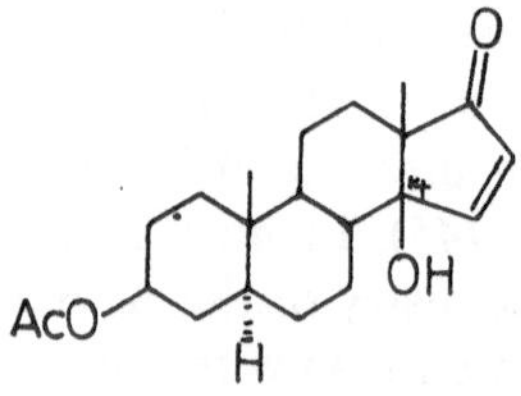

(125)

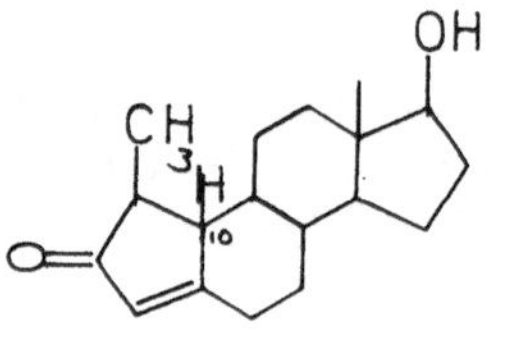

(126)

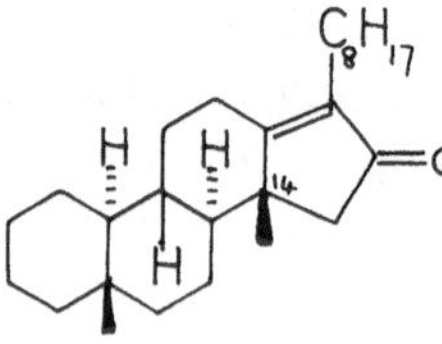

(127)

osterone, for example, with the 10α-CH₃,9βH-configuration (128) gives a c.d. curve with a positive n→π* CE roughly enantiomeric to that of the normal 10β,9α compound (237). A somewhat smaller positive CE is shown by compounds with the 10β,9βH-configuration, e. g. 17β-hydroxy-17α-methyl-9β-oestr-4-en-3-one (129) (110). Comparisons with these model compounds have been used to establish the 9β,10α-configuration of 19(10→9β)abeo-10α-testosterone (130) (60), of the backbone rearranged product (131) (237), and of the 8α,9β,10α,14β-steroid (132) synthesized from oestrone (84), all of which showed positive n→π* CEs near 320 nm.

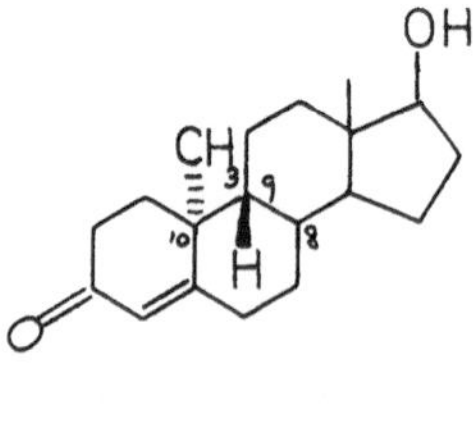

(128)

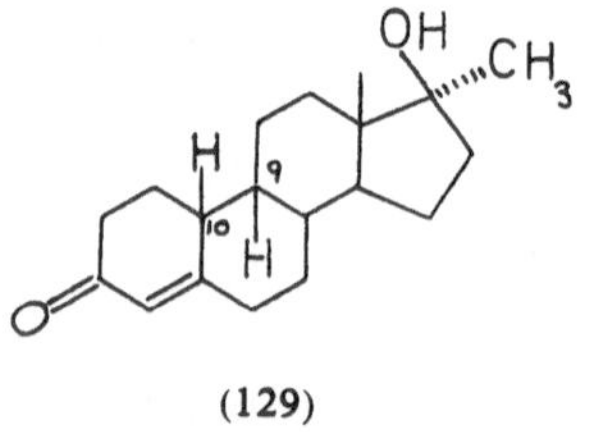

(129)

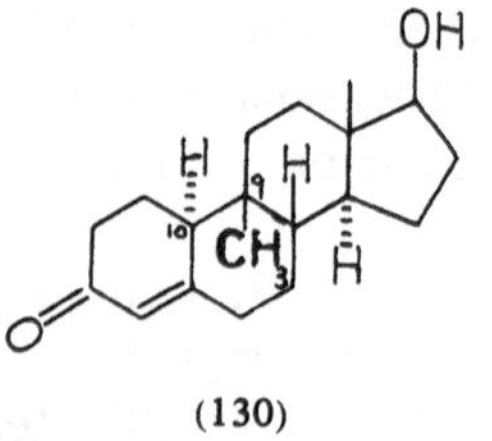

(130)

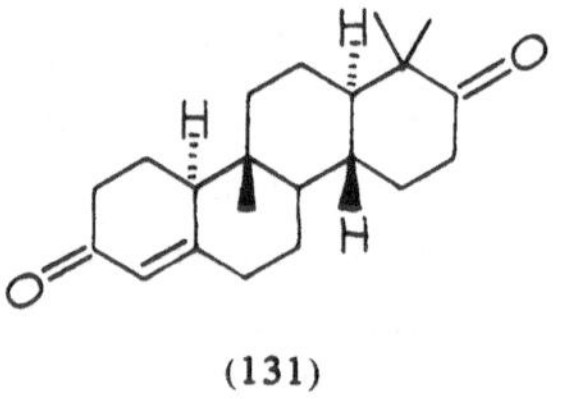

(131)

b) Helicity Rule

As stated above, the helicity of the chromophore in certain conjugated compounds can be related directly to the sign of the corresponding CE and to the absolute configuration of the molecule. Taxicin-I triacetate (**133**) for example, has a strong positive CE corresponding to the $\pi \rightarrow \pi^*$ transition of the unsaturated ketone, and implying a right handed helix for the chromophore. This helicity is consistent only with the absolute configuration shown (**133**) (*108*). In the case of taxicin, the relationship between helicity and absolute stereochemistry is unequivocal because of the rigidity of the molecule as a whole, but this is not always the case. For example, the negative CE observed for the unsaturated ketone in chiloscyphone would be compatible with either the *cis*-fused (**134**, R = βH) or *trans*-fused (**134**, R = αH) ring system, and a decision in favour of (**134**, R = βH) for chiloscyphone was finally made from the positive CE of the reduction product, tetrahydrochiloscyphone (*226*).

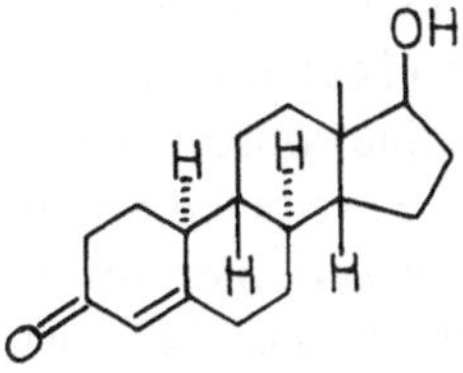

(132) 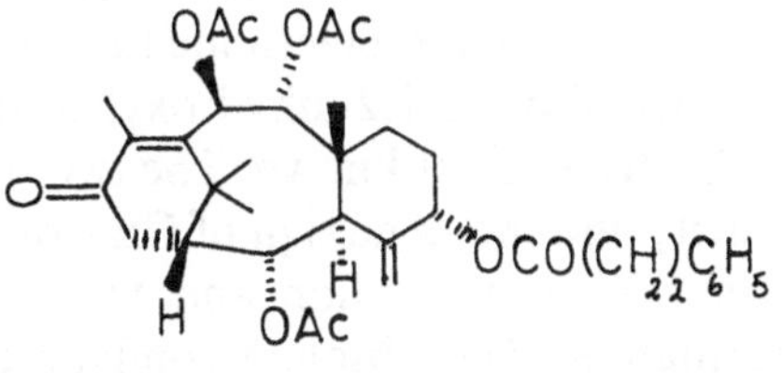   (133) 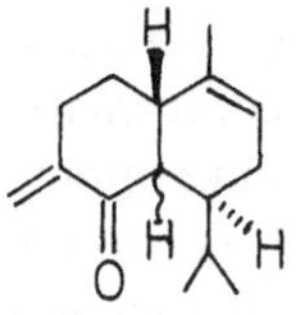   (134)

A helicity rule also governs the signs of the CEs for β,γ-unsaturated ketones, such that right handed and left handed helices give positive and negative CEs respectively (**135**) (*242*). This has been used to establish the configuration at C-4 in the ketopelenolides (**136**), ketopelenolide "a", which gives a strong negative CE at 296 nm, having the 4α-CH₃ and ketopelenolide "b", with a positive CE, the 4β-CH₃-configuration (*311*). Similarly the absolute configuration of ketone (**137**) with a negative CE near 300 nm establishes the absolute configuration of shiromodial diacetate (**138**) (*335*).

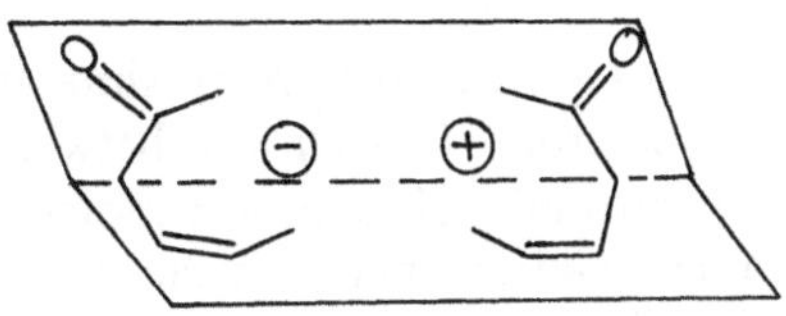

(135) 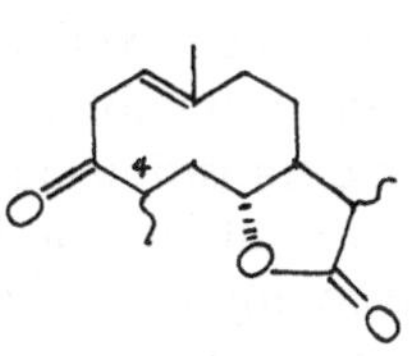   (136)

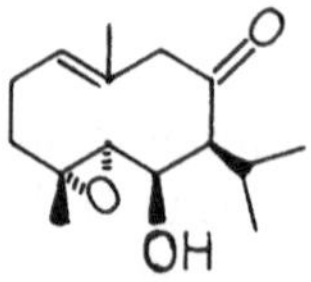

(137)

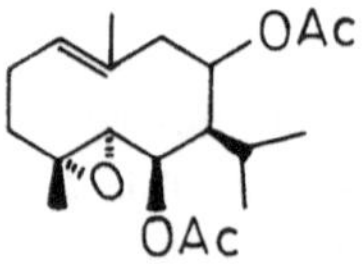

(138)

4. Dienes and Polyenes

a) Helicity Rules for Dienes

A helicity rule also applies to conjugated cisoid dienes (241), right handed (139) and left handed (140) helices corresponding to positive and negative CEs respectively, between 280 and 260 nm. It has been used, for example, to establish the absolute configuration of (−)-aeroplysinin-I (141) (121) which shows a characteristic large negative CE. In contrast, the helicity rule cannot be applied to determine the absolute configuration of (−)-trans-4-chloro-1,2-dihydro-1,2-dihydroxybenzene (142) because the conformation of the diene is not known. The two possible conformations have opposite helicity and the same sign of CE would be predicted for the (−)-compound in one conformation and the (+) enantiomer in the alternative conformation. The absolute configuration shown (142) was therefore established by direct comparison with the dechlorinated analogue of known stereochemistry (168).

(+)

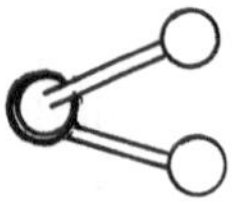

(139)

(−)

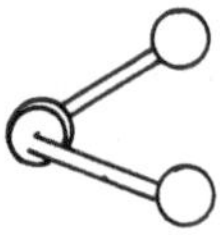

(140)

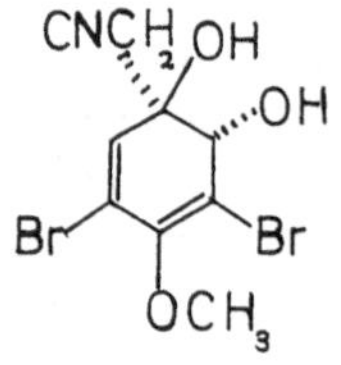

(141)

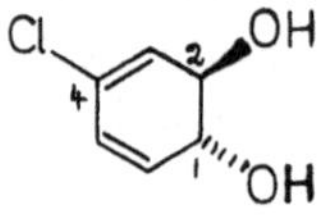

(142)

b) Carotenoids and Related Compounds

Carotenoids exhibit a complex series of absorption maxima in the visible and ultraviolet regions of the spectrum. Correlation between the electronic transitions and observed spectra of such complex molecules is not possible at present, but empirical correlations between the sign of the CE and configuration based on o. r. d. (23) and c. d. measurements (58, 59) have been made successfully. For a review see Ref. 161, Chapt. 5.

For example the c.d. curve of (+)-δ-carotene (**143**) is very similar in sign and shape to that of (+)-α-carotene (**144**, (6′*R*)-β,ε-carotene) of known absolute configuration; consequently the (6*R*) configuration has been allotted to (+)-δ-carotene, i.e. (6*R*)-ε,ψ-carotene (*58*). Furthermore, α-zeacarotene (**145**) (7′,8′-dihydro-ε,ψ-carotene) and (+)-ε-caro-

(143)

(144)

tene (**146**) [(6*R*,6*R*′)-ε,ε-carotene] have c.d. curves of the same pattern (*59*) thus establishing the (6*R*) configuration in the entire series of compounds produced by bio-cyclisation from the straight-chain precursor, neurosporene (**147**) (7,8-dihydro-ψ,ψ-carotene). The absolute configuration of lutein (**148**) [3*R*,3′*S*,6′*R*)-β,ε-carotene-3,3′-diol] has been established by converting it to α-cryptoxanthin [(3*R*,6′*R*)-β,ε-caroten-3-ol] with a c.d. spectrum identical to that of an authentic sample of known absolute configuration (*125*). (For a note on tentative rules for the nomenclature of carotenoids see *161*).

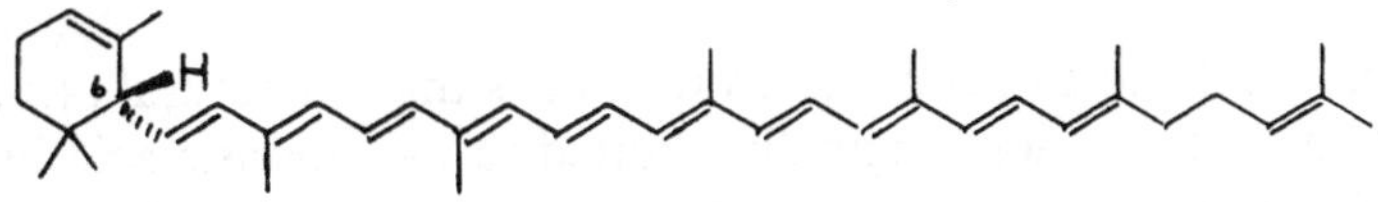

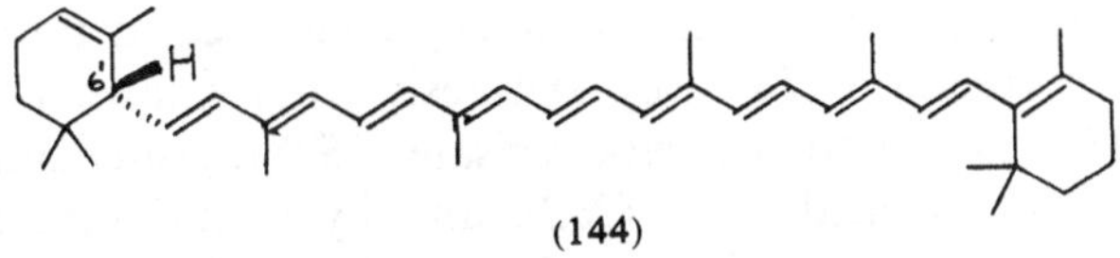

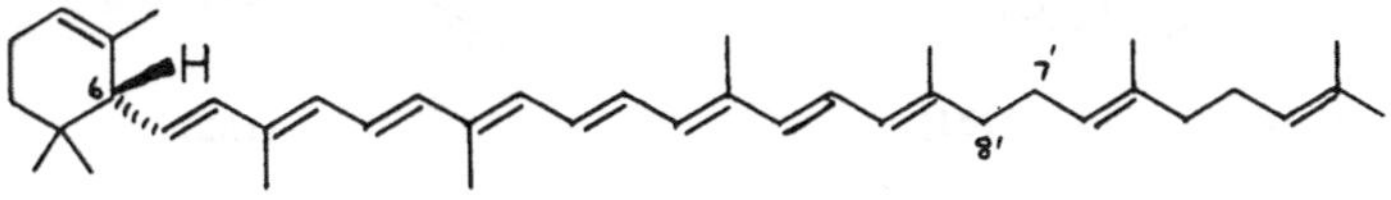

(145)

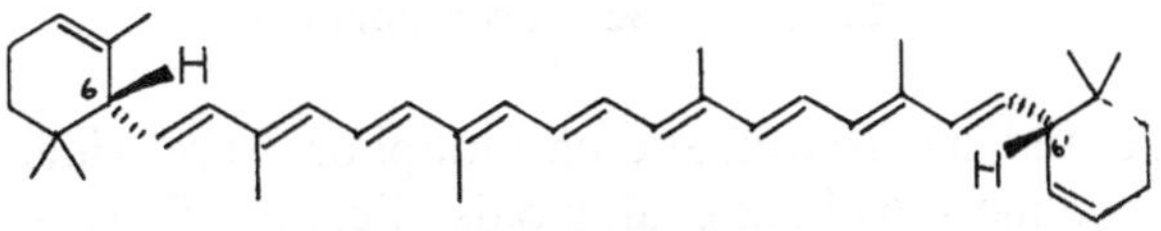

(146)

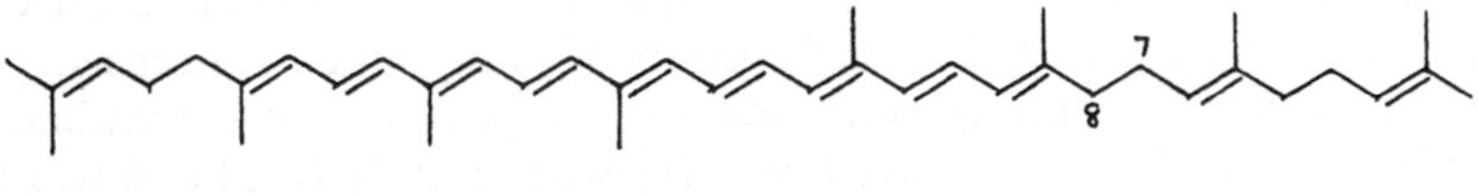

(147)

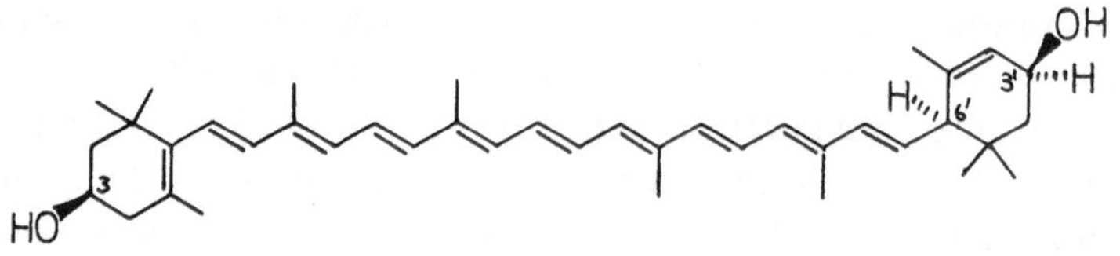

(148)

Closely related to work on carotenoids is that on natural (+)-abscisic acid. The configuration originally allotted to (+)-abscisic acid by use of monochromatic rotations (80) was found to be in conflict with that indicated by o.r.d. work which correlated abscisic acid (62) with violaxanthin (23), and led to the (S)-(+)-configuration shown (149). This latter allottment has been rigorously confirmed in three ways; by a chemical correlation (280), by synthesis from known (−)-α-ionone (150) via (151) to (−) ethyl trans-abscisate (152) quasi-enantiometric to natural (+)-abscisic acid (149) (262), and by quantitative application of the exciton chirality method (135, 201). (For a review of this latter method see 136.)

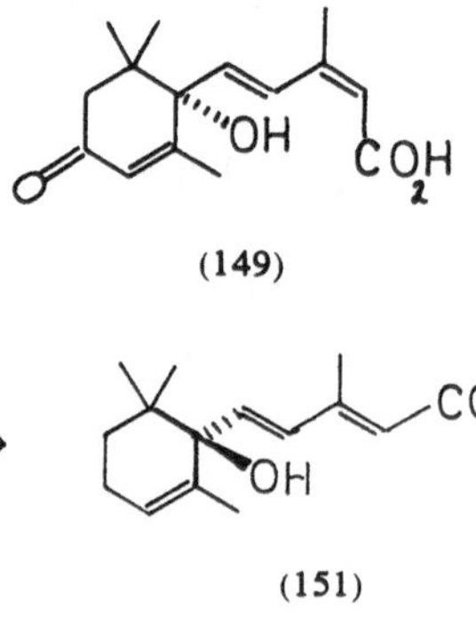

(149)

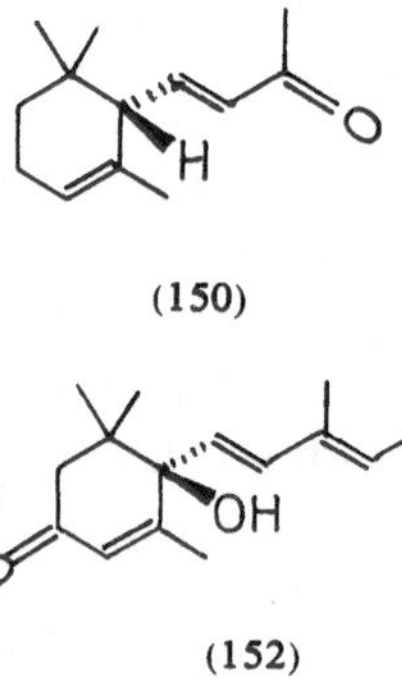

(150)

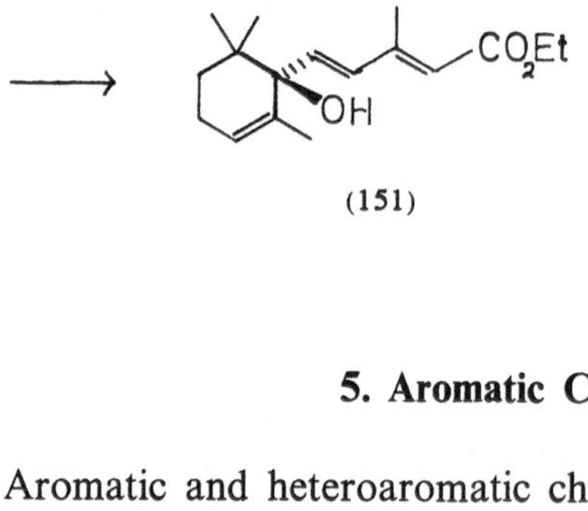

(151)

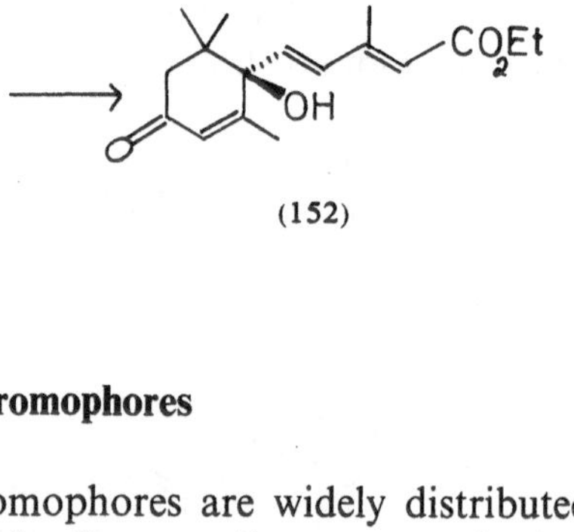

(152)

5. Aromatic Chromophores

Aromatic and heteroaromatic chromophores are widely distributed in natural products including alkaloids, lignans, flavans, porphyrins, nucleosides and other groups. Empirical correlations have been made to relate the sign of CE to absolute configuration, and these are most conveniently considered within each group of compounds concerned.

The aromatic chirality rules and their application to determination of absolute configuration have been reviewed recently (136) and will not be discussed here.

a) Tetrahydroisoquinoline Alkaloids

Alkaloids with the tetrahydroisoquinoline skeleton have only one or two asymmetric carbon atoms and the absolute configuration of a particular compound can often be determined by a single o.r.d./c.d. measurement and by correlation with compounds of known absolute stereochemistry. The first relationships were established with 1-benzyl-tetrahydroisoquinolines of known absolute configuration, (25, 87) which show three positive c.d. maxima near 280, 230 and 205 nm, for compounds of the (1S)-configuration. This correlation has been used to establish the configuration of other benzylisoquinolines, for example, the fullymethylated aglycone of latericine as (1S) (153) (270), and (−)-thalifendlerine (the aglycone of the thalictrum alkaloid (−)-veronamine) as

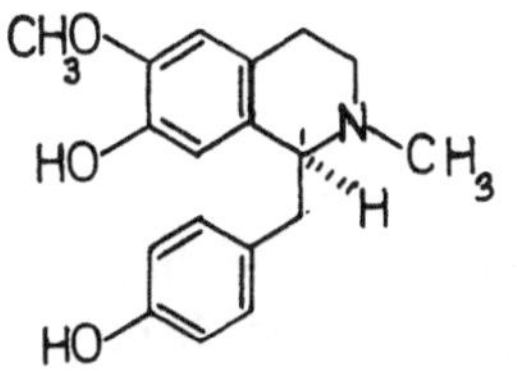

(153) 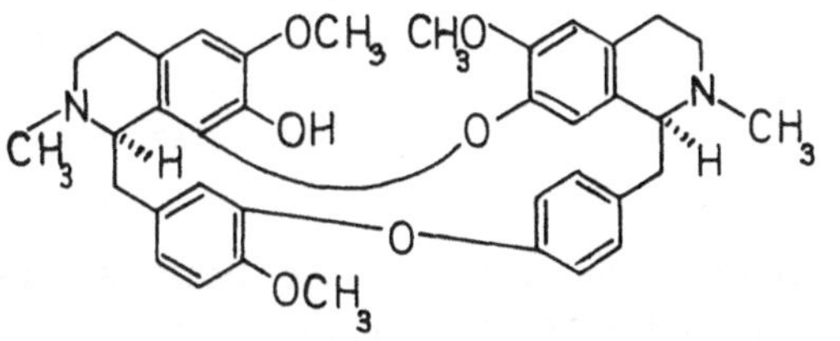   (154)

(1R) (289); also to determine the configuration of bis-benzylisoquinoline alkaloids such as thalrugosine (154) (234), after cleavage to the component monobenzylisoquinoline units.

An empirical correlation has also been established (179) between the sign of the CE and absolute configuration of the morphine alkaloids, by synthesis of the morphine precursor salutaridine (156) and of the enantiomeric sinoacutine from the (1R)- (155) and (1S)-benzylisoquinolines of known absolute stereochemistry. Other alkaloids including pallidine (180), norsinoacutine (308), and flavinine (308), have since been related to (−)-sinoacutine by c.d. correlations.

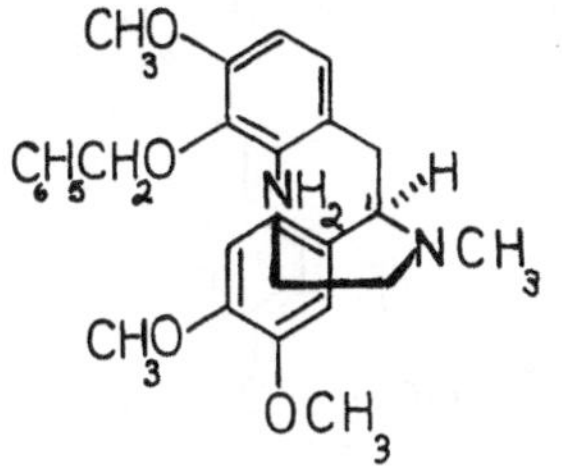

(155) 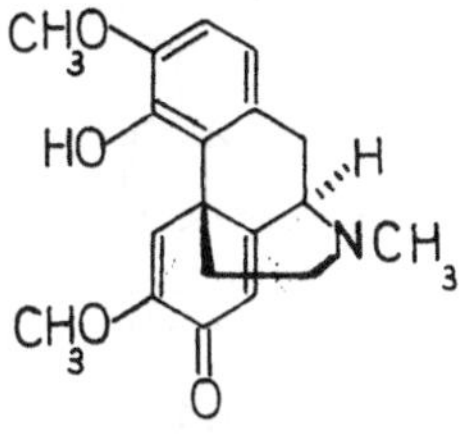   (156)

C.d. curves for alkaloids of the tetrahydroberberine class also show three maxima between 300 and 200 nm. For compounds with the (1S)-configuration the band near 280 nm may be either positive or negative, but the bands near 230 nm and 205 nm are both negative, i.e. opposite in sign to the corresponding benzyltetrahydroisoquinolines of "equivalent" configuration (178, 300); the absolute configuration of kikemanine (157) for example, has been allotted as (1S) on the basis of its negative CE near 230 nm (180).

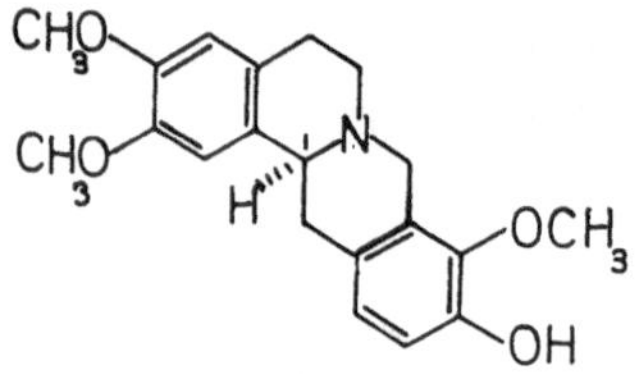

(157)

Similar relationships have been established for the homologous phenethyltetrahydroisoquinolines, by analogy with the benzyl compounds. The phenethylisoquinoline (158) which has a positive CE near 280 nm, was assigned the (1S)-configuration during studies of colchicine biosynthesis (21), and the absolute configuration of (−)-melanthoidine (159) was established by degradation to two phenethylisoquinoline fragments (27). The stereochemistry of androcymbine (161) has been deduced from the positive CE near 270 nm of the reduction product (160) from O-methylandrocymbine (28) and the configuration of some synthetic homotetrahydroberbines, e.g. (162) have been allotted after synthesis from known phenethylisoquinoline precursors (55).

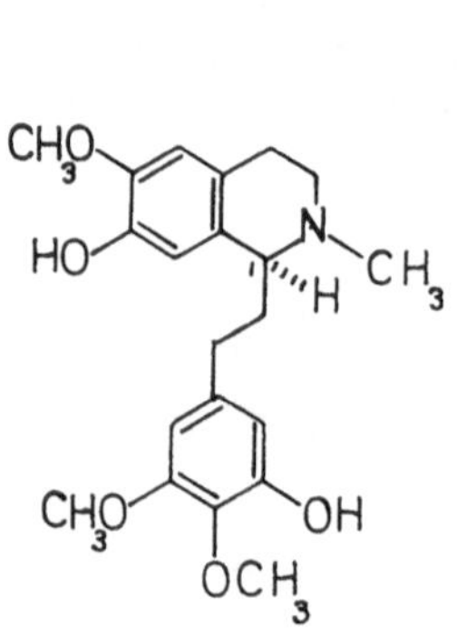

(158)

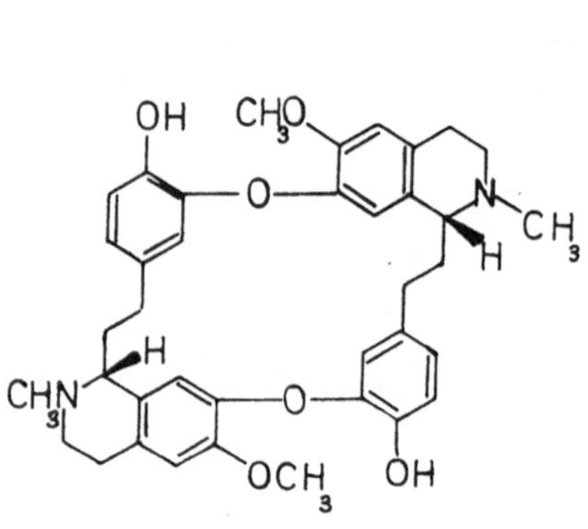

(159)

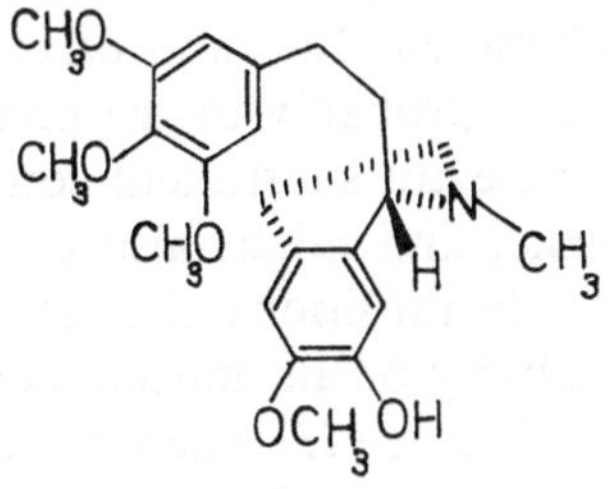

(160)

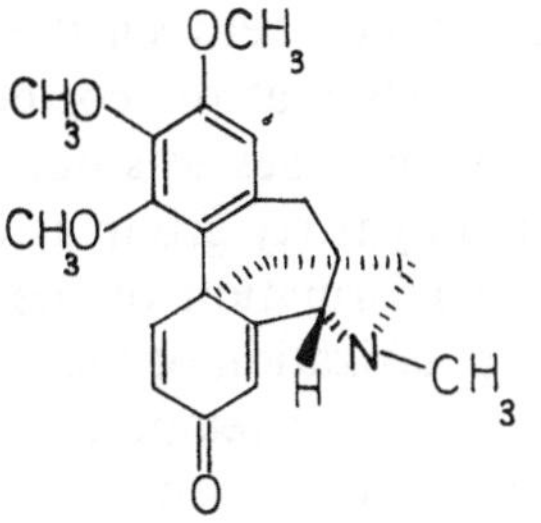

(161)

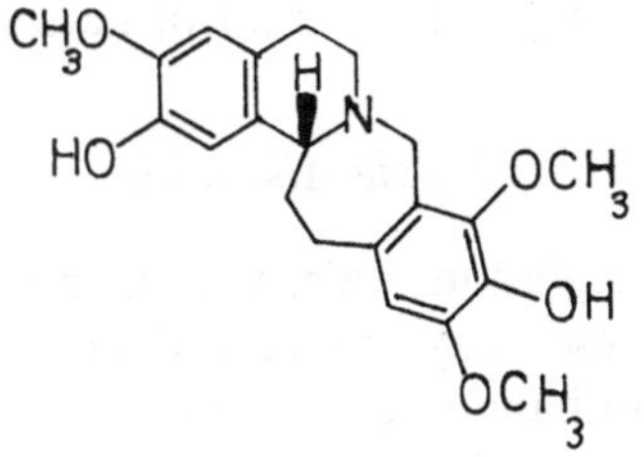

(162)

The difficulties of choosing an adequately analogous compound for the assignment of absolute configuration can be illustrated by reference to the cularine alkaloids. Recently, the absolute configuration of cularine has been unambiguously determined by X-ray analysis of cularine methiodide (Bijvoet technique) as (1S)- (**163**) (*177*). This is in agreement with previous chemical correlations (*205*) but is contrary to the (1R)-assignment made earlier on the basis of n.m.r. and o.r.d. evidence. In this original o.r.d. correlation (*41*), cularine was converted, by hydrogenolysis with sodium and liquid ammonia, to a benzylisoquinoline derivative 1(4,5-dimethoxy-2'-hydroxybenzyl)-7-methoxy-2-methyltetrahydroisoquinoline (**164**), which showed three negative c.d. maxima at 293, 226 and 210 nm. By analogy with previous compounds (*25, 87*) the (1R)-configuration, was assigned to (**164**) and thus to the entire cularine group of alkaloids.

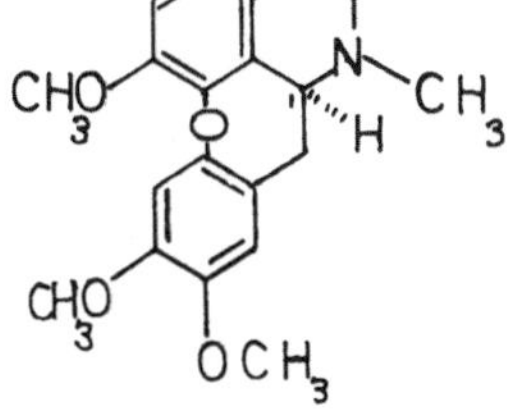

(163)

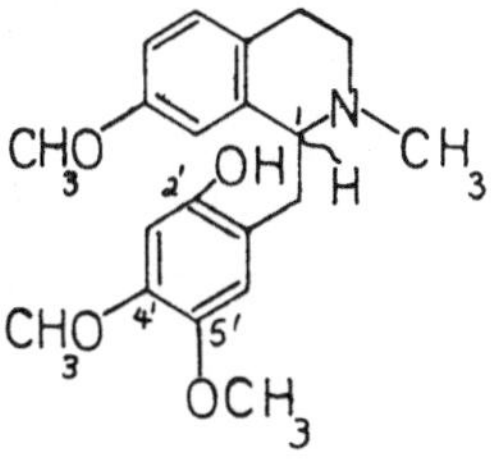

(164)

The discrepancy between the o.r.d. correlation and the later X-ray evidence indicates that the correlation between (164) and other benzyltetrahydroisoquinolines was not valid, due probably to the presence of the additional hydroxy group at C-2′. The extra substituent could easily change the conformation of the molecule, and might also change the direction of polarization of the 'La and 'Lb transitions, thus altering the sign/configuration correlation. The possibility of an unusual conformation was actually discussed in the original paper (41) where it was noted that addition of hydrochloric acid to the solution of (164) caused an inversion of the long wavelength CE, contrary to the behaviour of most benzylisoquinolines, and indicative of conformational change.

b) Indole Alkaloids

Many o.r.d./c.d. correlations exist within the large and varied class of indole alkaloids; these may be considered in two main groups, compounds of the yohimbane and heteroyohimbane type in which the indole nucleus is fully aromatic, and compounds of the aspidospermine type in which the indole nucleus is saturated at C-2 and C-3.

The alkaloids of the yohimbane and corynantheane type which have three or four centres of asymmetry and may have more than one chromophore show a series of CEs between 300 and 200 nm. Several attempts have been made to correlate the sign of the CE with the absolute configuration, but the number of individual absorption bands has led to some confusion, particularly as most early work was based on o.r.d. measurements in which small CEs can easily be obscured by the tail of a neighbouring band. Yohimbane and corynantheane derivatives show one or two o.r.d. CEs above 250 nm, (195). The CE at longest wavelength (295—280 nm) is very small and may not be observed, but there is a correlation between the sign of the second CE (275—250 nm) and the absolute configuration at C-3; for compounds with 15αH, positive and negative CEs correspond to 3α- and 3βH-configurations respectively, independent of the configuration at C-20; for example (−)-reserpic acid (as its methyl ester) (165) has a negative CE near 270 nm corresponding to the 3β-configuration shown.

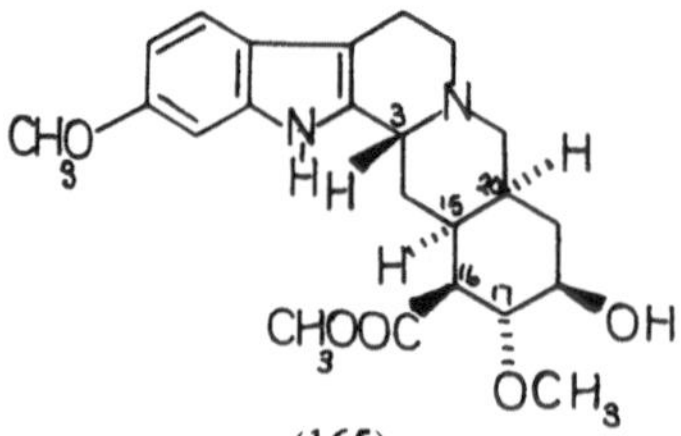

(165)

For the tetracyclic alkaloids related to corynantheidine, the additional α,β-unsaturated acid chromophore absorbs near 250 nm and the c.d. maximum at this wavelength reflects variations in the configuration at C-20. Nevertheless the correlation at longer wavelengths still holds for C-3 and has been used to allot the 3α-configuration to gambirine (**166**) (*228*) and the 3β-configuration to the triol (**167**) related to vincoside (*45*).

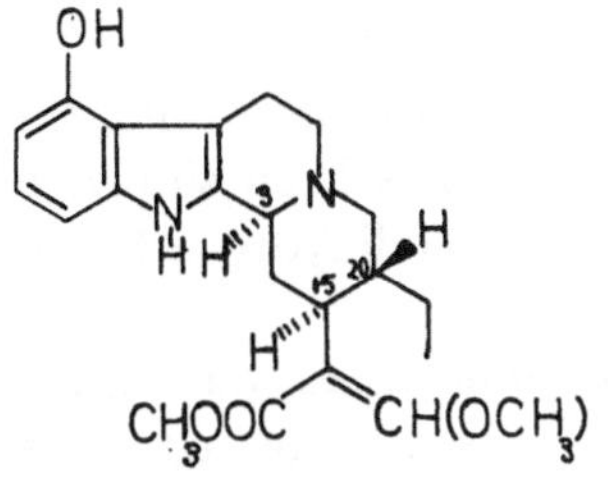

(166) 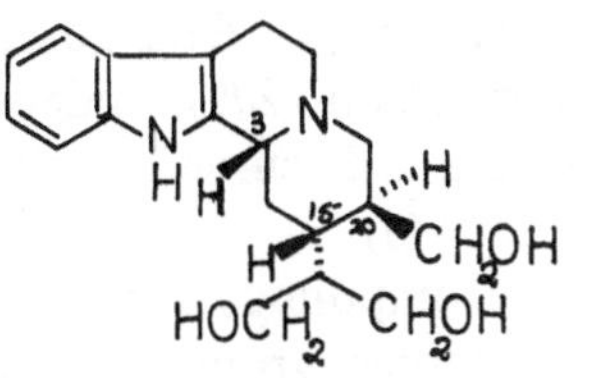   (167)

The absolute configuration at C-21 in (+)-vincamine has been determined by degradation of (+)-vincamine (**168**) to the tetracyclic compound (**169**), which has a positive CE (*a* +69) between 290 and 250 nm (*325*). This links the correlation for C-3 of the yohimbane alkaloids to the equivalent position numbered C-21 in the eburnane series and thus establishes the absolute configuration of several eburnane alkaloids (*46*).

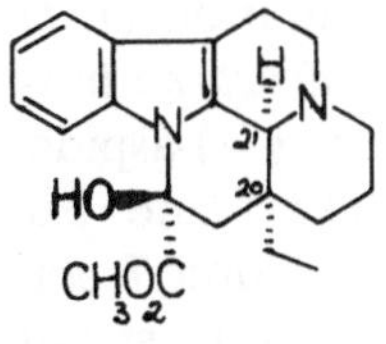

(168) 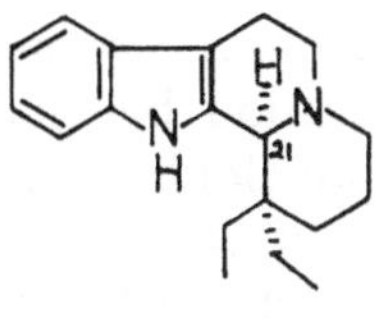   (169) 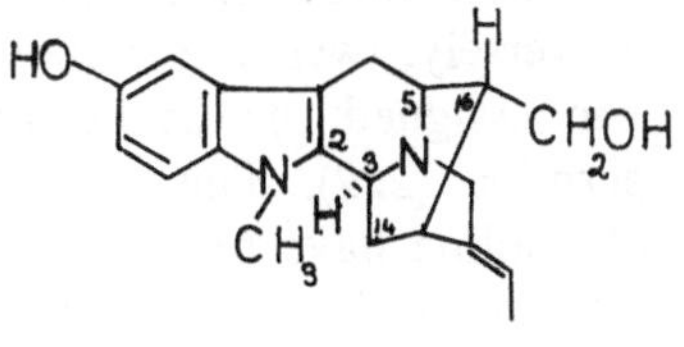   (170)

(+)-Sarpagine and (+)-*N*-methylsarpagine both give positive CEs between 300 and 260 nm, very similar to that of (+)-tombozine of known absolute configuration. This establishes the configuration of (+)-*N*-methylsarpagine (**170**) as 3α- and also extends to the bridged series the correlation established for other indole alkaloids (*336*).

The absolute configurations of the Iboga and Voacanga alkaloids have been allotted by crystallographic analysis, by chemical methods and by chiroptical studies. A recent Bijvoet analysis (*209*) has shown that (+)-coronaridine (**171**, R^1 = Et, R^2 = H) has the same absolute configuration at all ring junctions as (+)-dihydrocatharanthine (**171**, R^1 = H, R^2 = Et) previously known by correlation with cleavamine (*65*); ob-

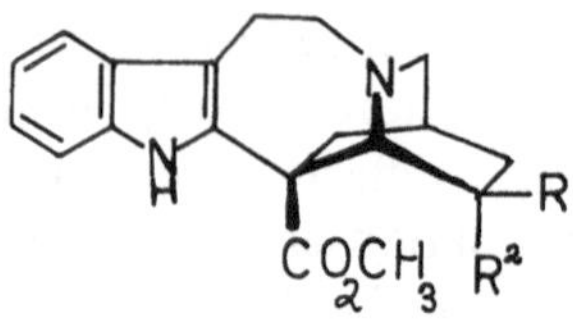

(171)

served o.r.d. data for (+)-coronaridine was in accordance with this correlation (209). Independently, (−)-coronaridine has been shown to give a c.d. curve quasienantiomeric with that of (+)-dihydrocatharanthine (48).

Compounds of the aspidospermine and strychnine series were among the earliest groups of alkaloids studied by o.r.d. (193, 194) and considerable use has been made of the early correlations to establish the stereochemistry at C-2 and C-12 in related compounds. For example (+)-N_a-acetylbeninine (172) (126) has been allotted the configuration with the C-12 "tryptamine bridge" β-oriented, because its strong positive CE at 270—250 nm is enantiomeric to that of (−)-aspidospermine and analogous to that shown by N(a)-acetyl-7-ethyl-5-desethylaspidospermidine (193); the similarity of the o.r.d. curves of samples of (+)-N_a-acetyl-aspidofraktinine (173) derived from (−)-kopsin, (−)-minovincin and (−)-aspidofraktinin has been used to confirm chemical evidence that these three alkaloids have the same absolute configuration (131). Conversely, twelve new alkaloids from Aspidosperma cylindrocarpa have been assigned the 2α,12α-configuration by analogy with (−)-aspidospermine (229). The strychnine alkaloids splendoline (174) and isosplendoline have been allotted the 2β,7β-configuration (position 7 corresponds to 12 in the previous group) on the basis of the very strong positive CE near 250 nm (199) and by analogy with published data for strychnine (193). In the 2,3-unsaturated series (−)-hedrantherine (175) and (−)-17-methoxyhedrantherine have been assigned the absolute configuration shown, by correlation of the pronounced negative CEs between 350 and 300 nm with that of known (−)-vincadifformine (251).

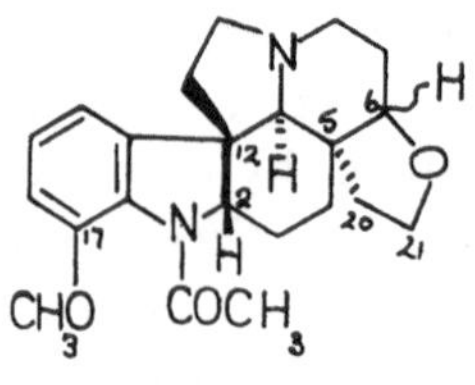

(172)

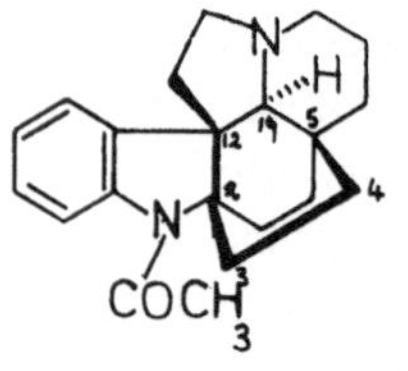

(173)

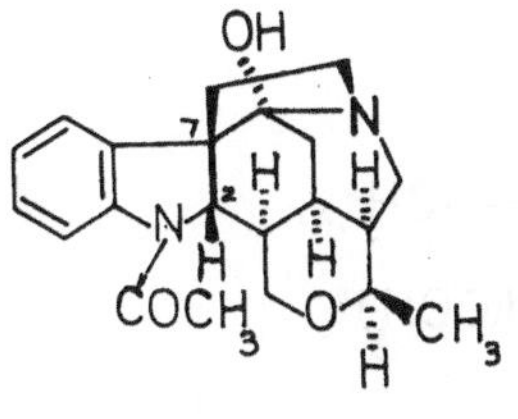

(174)

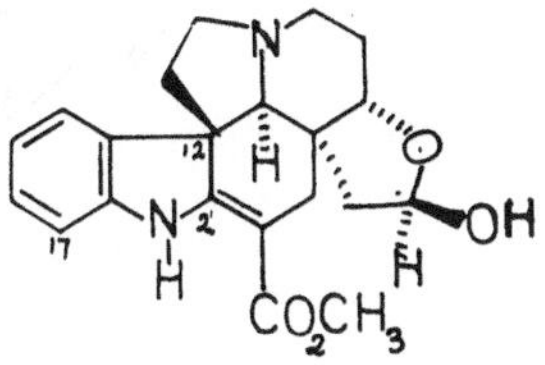

(175)

Within the seredamine group of alkaloids, N_a,O-diacetyl-N_a-de-methylseredamine (**176**) and related compounds show strong negative aromatic CEs near 250 nm; the absolute configuration shown has there-fore been allotted (*133*) by comparison with known compounds (*193, 196*), including comparison of the carbonyl CE of purpeline (**177**) with that of known (+)-21-desoxyajmalone. These o.r.d./c.d. correlations have been confirmed by chemical conversion of (+)-seredamine to natural (+)-ajmaline (**178**).

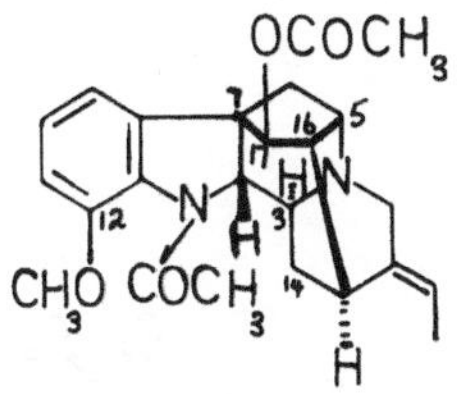

(176)

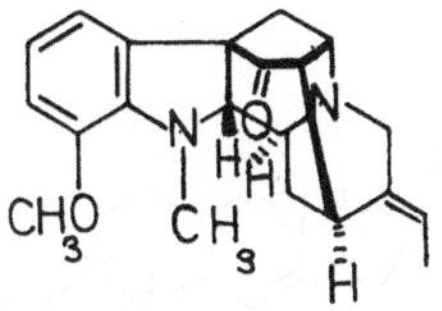

(177)

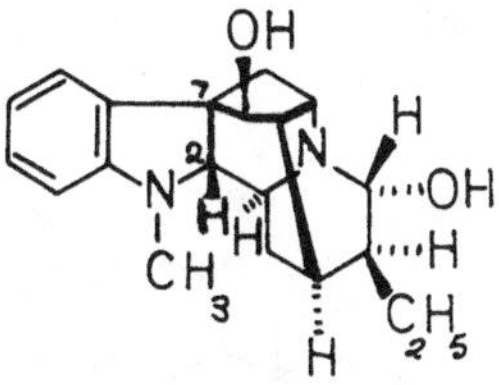

(178)

c) Oxindole and Isoindole Alkaloids

C.d. curves of a series of oxindole alkaloids of known absolute configuration show four maxima between 300 and 200 nm (*269*); the sign of the maximum near 265 nm is controlled by the configuration at C-3, while the signs of the bands at 290 nm and 210 nm are controlled primarily by the configuration of the spiro atom C-7. On this basis the absolute configuration of corynoxine (**179**) has been allotted as shown (*269*). Related work on rhynchophylline-type oxindole alkaloids (*324*) has shown that compounds carrying an −OH or −OCH₃ group at C-9 do not follow the same empirical correlation as other members of the series and must be considered as a separate class when making configurational assignments (*cf.* work on cularine alkaloids Section II 5a where the presence of a hydroxy group alters empirical correlations for the benzyl-tetrahydroisoquinoline alkaloids).

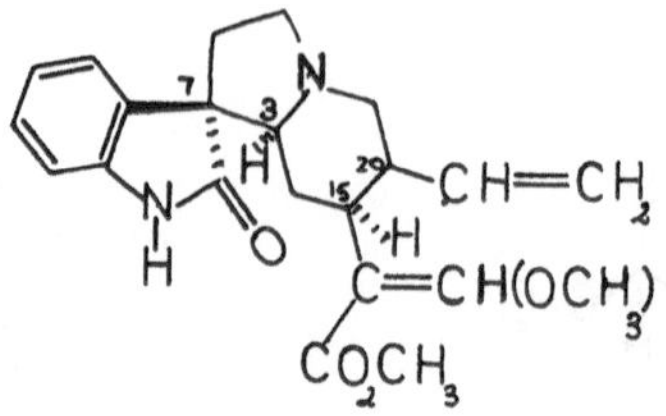

(179)

There is little published work on the c. d. of isoindole alkaloids. A most illogical attempt has been made to "determine" the configuration of C-3 in spiropachysine (180) by analogy with β-hydrastine (181) (185). The argument, which depends on comparison of C-3 in spiropachysine (180) with C-9 in (−)-(1R,9S)-β-hydrastine (181) is wholly invalidated by the dissimilarity of the chromophores (isoindole and aryllactone respectively) and also by the *non-equivalence* of the bonds C−H and $C_9−C_1$ in β-hydrastine with the supposedly less bulky $C_3−C_2$ and more bulky $C_3−C_4$ bonds in spiropachysine. The stereochemistry (182) proposed (185) for spiropachysine cannot be regarded as proven by this argument.

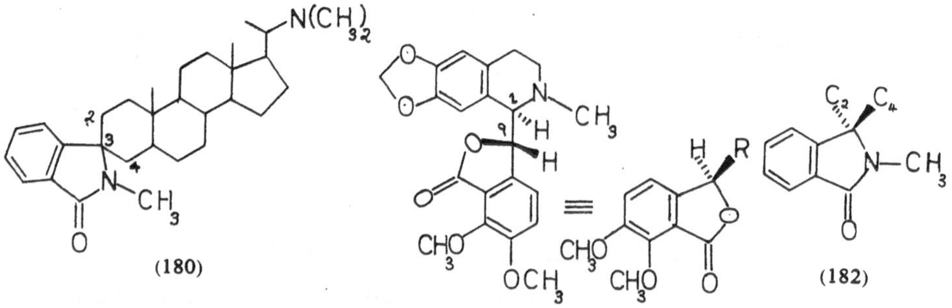

(180)

(181)

(182)

d) Aryltetralins Including Lignans

A correlation has been established (314) for lignans of the 4-aryltetralin class between the sign of the CE at longest wavelength and the chirality at C-4; compounds of general formula (183) show a negative CE at 290—270 nm for the 4β-aryl 4α-H configuration and positive for the compound with 4α-aryl 4β-H. The original correlation was shown to apply to 5-substituted 4-aryltetralins also (lyoniresonol (184) for example has a first negative CE *a* −88), and has been used to allot the absolute configuration shown to the lignan otobain (185) (long-wavelength CE positive $a+89$) (*192*). (−)-Nafenopin[(−)-2-methyl-2-*p*-(1,2,3,4-tetrahydro-1-naphthyl)phenoxypropionic acid] has been allotted the configuration (**186**) from a Bijvoet X-ray analysis on the corresponding *p*-iodobenzoate, and this is confirmed by the positive CE near 280 nm (*33*).

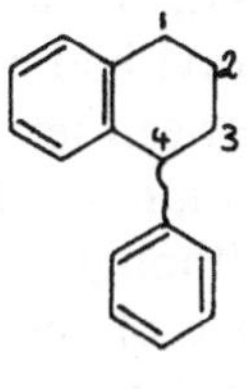

(183)

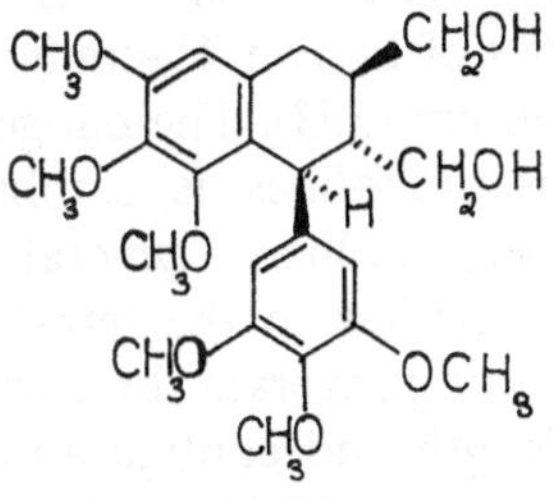

(184)

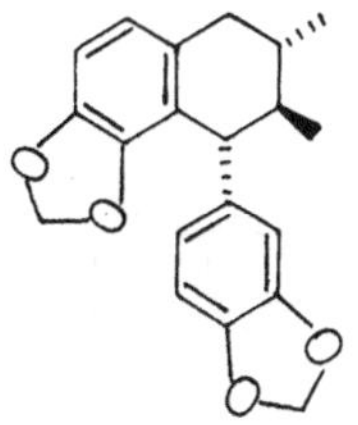

(185)

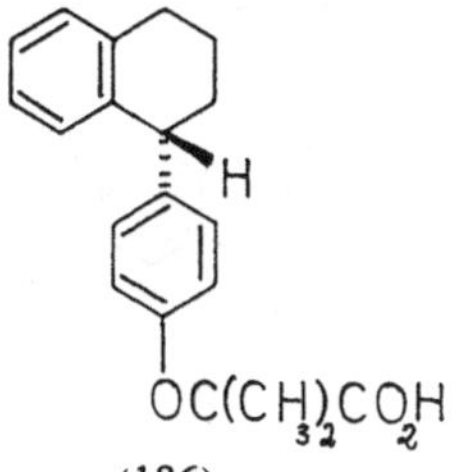

(186)

Plicatic acid of known relative stereochemistry, has been assigned (*315*) the absolute stereochemistry (**187**), both on the grounds of its negative long wavelength CE, and also from the enantiomeric nature of the o.r.d. curves of trimethyl plicatin (**188**) and desoxyisopodophyllotoxin (**189**) of known absolute stereochemistry. (Independent checks suggest that the tertiary hydroxyl groups at C-2 and C-3 in trimethyl plicatin do not significantly affect the correlation.)

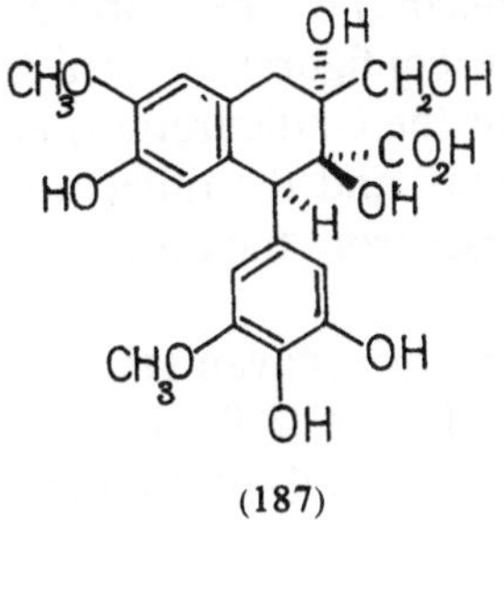

(187)

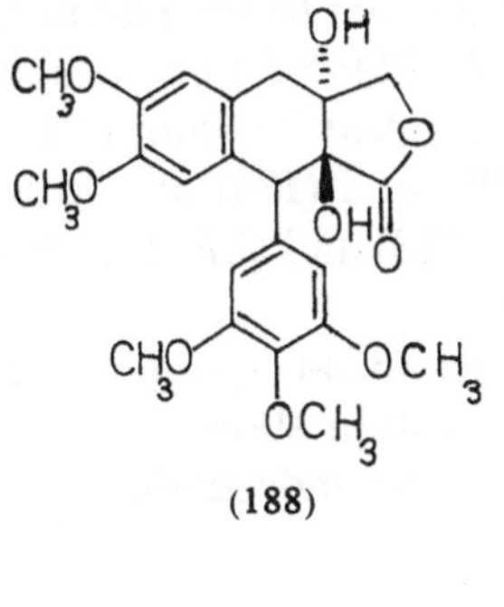

(188)

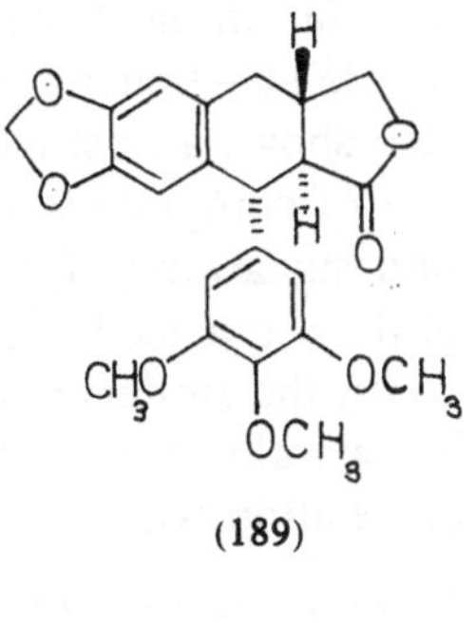

(189)

e) Aryltetrahydroisoquinolines

The alkaloid cherylline (**190**) was originally allotted the (4S)-configuration shown, on the basis of its negative CE at 288 nm (*54*) by analogy with 4-aryltetralins (*314*). In fact the analogy between tetralins and tetrahydroisoquinolines is *not* valid but the correlation was independently substantiated when it was shown that 4-phenyl-1,2,3,4-tetrahydroisoquinolines of known 4S-configuration (Bijvoet X-ray) had a negative CE near 280 nm (*323*). The danger of comparison between tetralins and tetrahydroisoquinolines is emphasised by consideration of 1-phenyltetrahydroisoquinolines, general formula (**191**). Although these are superficially similar to the tetralins (**183**) the presence of the nitrogen atom in the ring reverses the correlation between the sign of the CE at long wavelength and configuration; a negative maximum near 280 nm implies the 1α-aryl 1β-H configuration and a positive maximum the reverse (*182*). This correlation has been used to allot the absolute configuration to (+)-cryptostyline (*181*).

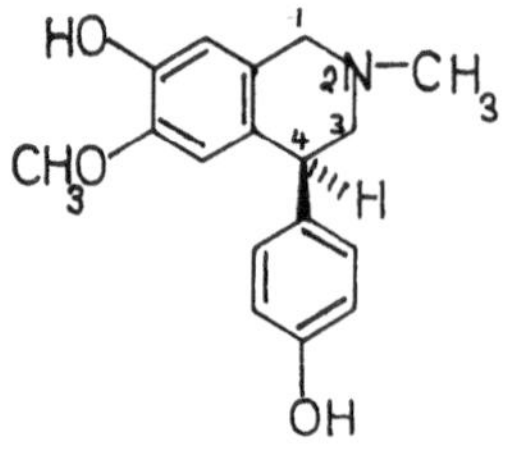

(190) 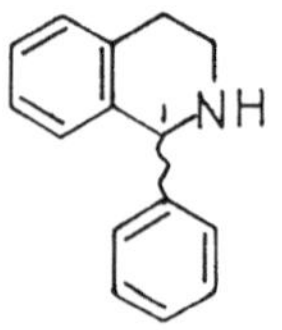   (191)

f) Flavans: Isoflavans and Related Compounds

From a study of common flavanols related to (+)-catechin (**192**) a correlation has been established (*202*) between the absolute configuration at C-2 and the sign of the CE. Compounds of general formula (**193**) with the (2S)-configuration show a positive CE at 280 nm independent of the configuration at C-3, and conversely for the (2R)-configuration. This correlation has been extended to the parent flavans (*68*) by the synthesis of (2R)-(+)-flavan-7-ol (**194**) of known absolute configuration, which also shows a positive CE near 280 nm. (Note that the configurations at C-2 of compounds (**193**) and (**194**) are "equivalent" and the difference in designation (2S) for (**193**) and (2R) for (**194**) arises from the operation of the Sequence Rule.)

In the isoflavan series (3S)-(−)-5,7,3′,4′-tetramethoxyflavan (**195**) of known absolute configuration shows a negative CE near 280 nm. This correlation was used by two independent groups (*208, 334*) to establish

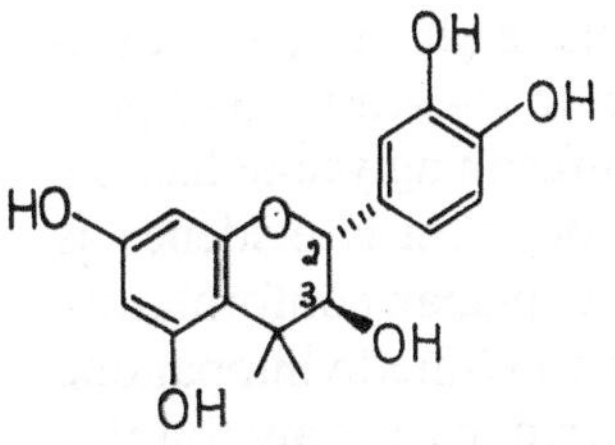

(192) 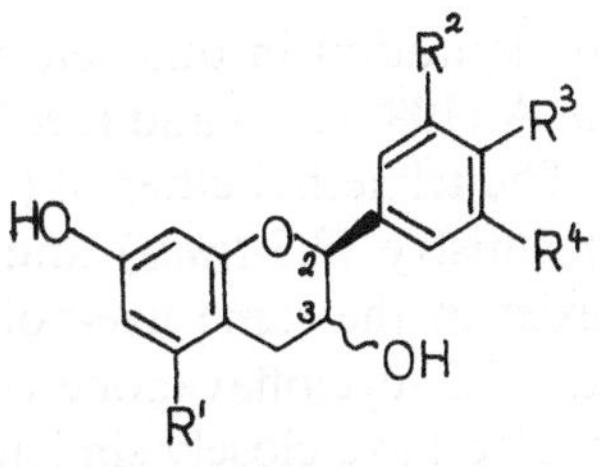   (193) 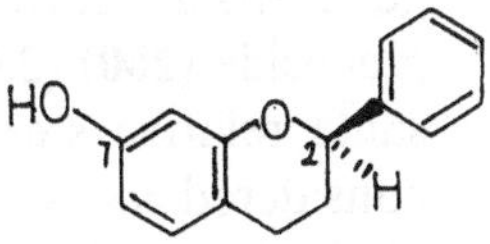   (194)

the (3S)-configuration of (−)-equol [(−)-7,4′-dihydroxyisoflavan] (196) and to correct the (3R)-configuration previously allotted to (−)-equol on the basis of plain o.r.d. curves above 300 nm (76, 313).

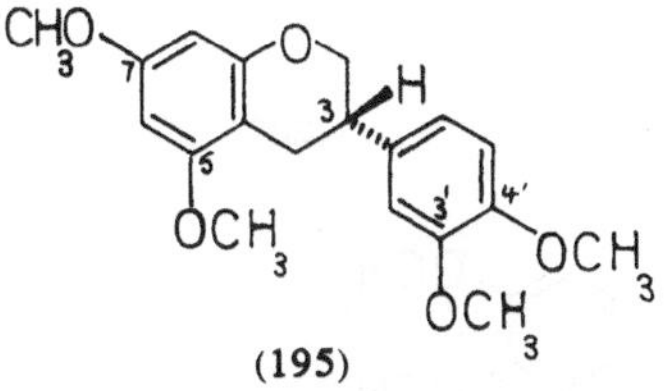

(195) 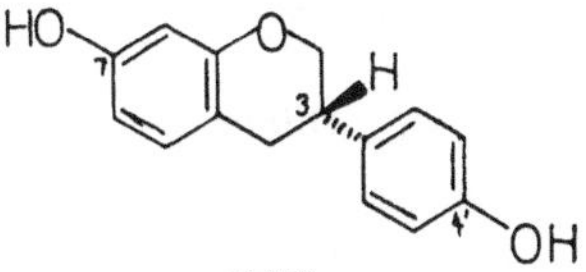   (196)

An extensive collection of c.d. data for flavan-4-ones of known absolute configuration (122) shows that compounds with the (2R)-configuration (general formula 197), and with no substituent at C-3, have a negative CE near 330 nm and a positive CE at 280—290 nm corresponding to the n→π* and π→π* transitions of the arylcarbonyl chromophore. The same relationship holds for 3-hydroxyflavanones of the (2S,3S) configuration (general formula 198). (As for flavans above, these compounds have "equivalent" configurations at C-2, but the description varies due to operation of the Sequence rule.) The presence of a glycosyloxy residue at position 3 does not alter the signs of the CEs, but the π→π* CE is considered more suitable than the longer wavelength n→π* transition for determining aglycone chirality, because of the relative magnitudes of the two bands.

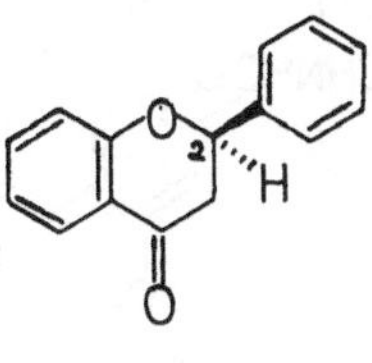

(197) 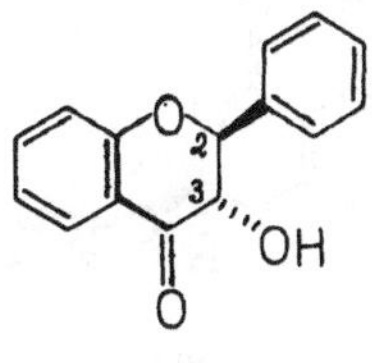   (198)

Assignments of configuration in this field include (2*S*) to (−)-narin-genin from *Helichrysin* A (**199**) (*123*) and (2*R*,3*R*) to (+)-fustin 3-*O*-β-D-glucoside (**200**) (*220*). The trimethyl ether of (+)-fustin aglycone has the same relative stereochemistry (2,3-*trans*) and from n.m.r. evidence is considered likely to exist in the same five-point-coplanar conformation of the heteroring as the 3,2′-cycloflavanone (+)-tri-*O*-methylmopanone (**201**). The compounds also have closely similar o.r.d. curves and on this basis, the (2*R*,3*R*) configuration shown (**201**) has been allotted to mopanone (*99*).

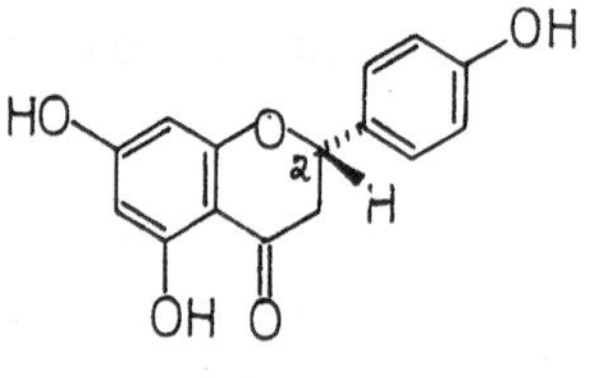

(199)

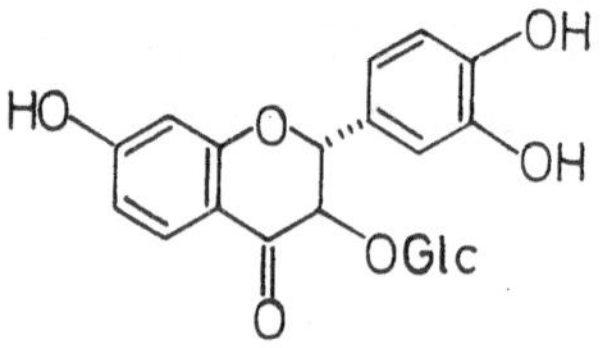

(200)

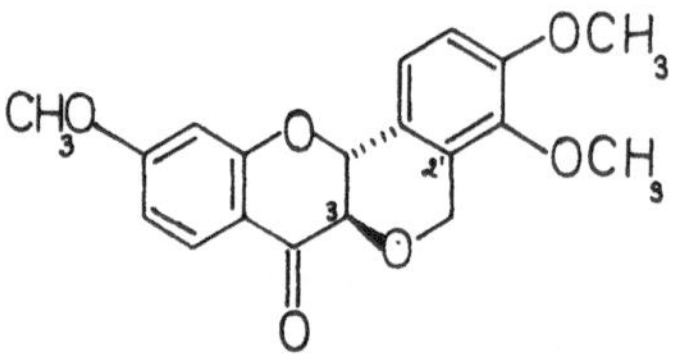

(201)

The dihydroisocoumarins (3*S*)-agrimonolide (**202**) and (3*R*)-ochra-toxin-A (**203**), both of known absolute configuration show CEs of oppo-site sign near 260 nm, positive (*15*) and negative (*157*) respectively. (3*R*)-(−)-Mellein (**205**), also of known absolute configuration, has a negative CE near 260 nm, opposite in sign to that of asperentin, which was therefore assigned the 3*R*-configuration (**204**) (*130*). (Note that (**204**) and (**205**) although of "opposite" configuration at C-3 are both designated (3*R*)- because of the operation of the Sequence Rule.)

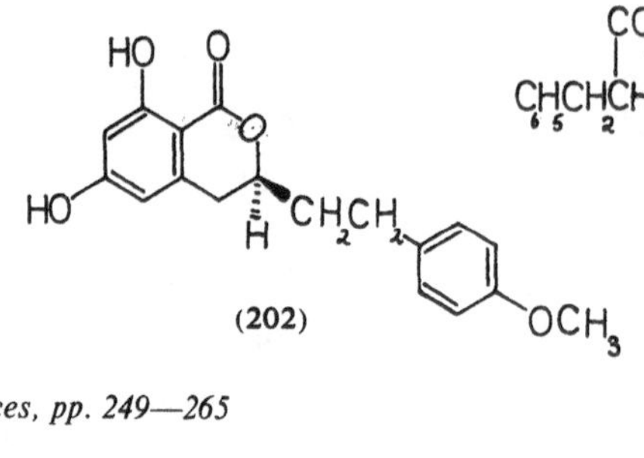

(202)

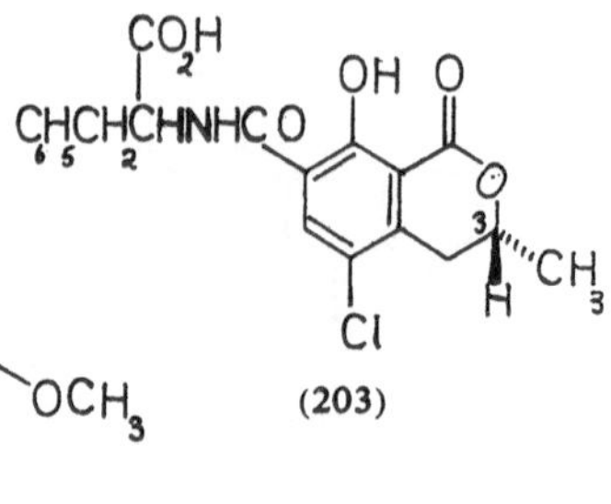

(203)

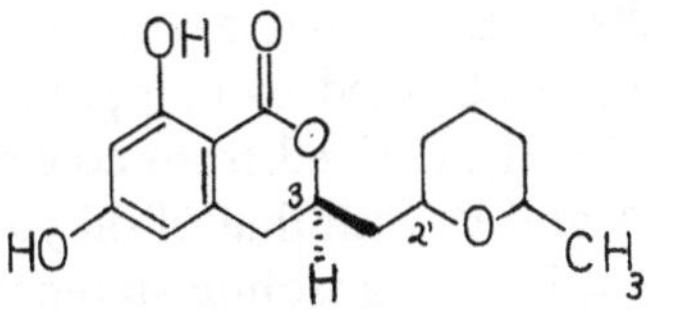

(204) 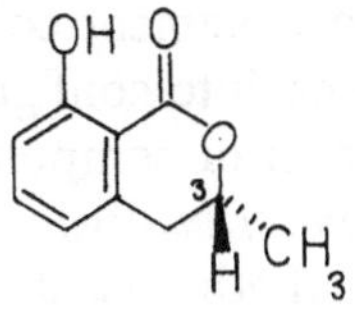   (205)

In the rotenoid series, allocations of configuration have been made by empirical comparison of very similar compounds, for example, dalpanol (206, R=C(CH₃)₂OH) as 6aS,12aS, from the similarity of its CE near 330 nm with that of (−)-rotenone (206, R=C(CH₃)=CH₂) of known 6aS,12aS stereochemistry (5). Other work includes assignment of configuration to (−)-milletone and related compounds (261) and to secalonic acid-D (305).

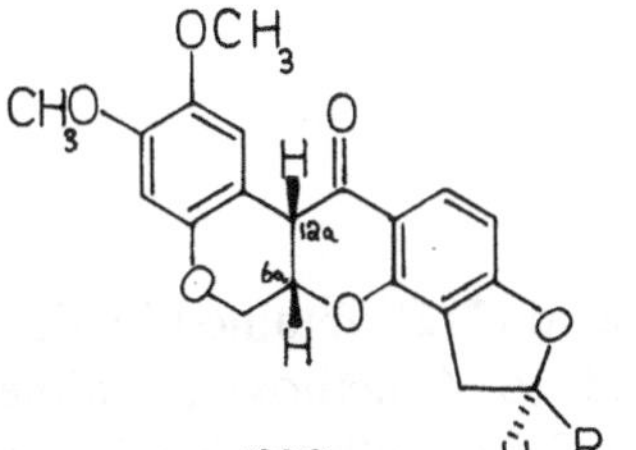

(206)

g) Other Aromatic Correlations

There are comparatively few correlations in which the compounds concerned have only simple aromatic rings as chromophores. An early example is that of latifolin (207) and its derivatives, which were correlated with the quinol diacetate (208) obtained from (R)-4-methoxydalbergione of known absolute configuration. Both compounds show strong positive CEs between 250 and 230 nm and on this evidence the (R)-configuration was allotted to latifolin as shown (97).

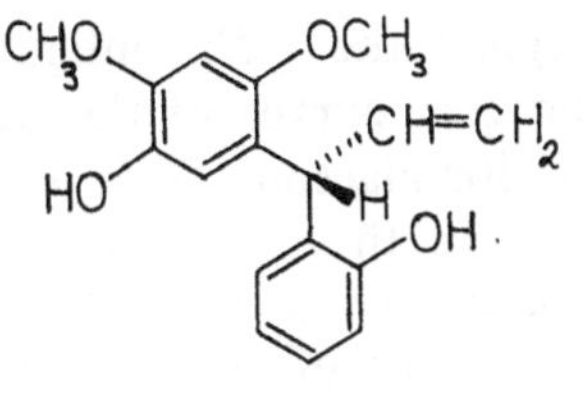

(207) 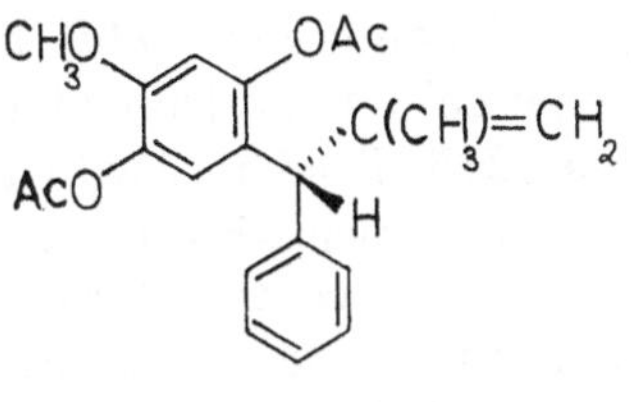   (208)

The yellow pigment coleone-B (**209**) has a single asymmetric carbon atom. The absolute configuration was established by comparison of the o.r.d. spectrum of compound (**210**, R=H), derived from coleone-B and known to have *trans*-fusion of rings A and B, with that of dihydroroyleanone trimethyl ether (**210**, R=CH₃) of known absolute stereochemistry. The close similarity of the o.r.d. curves in sign, shape and magnitude enabled the (10*S*) configuration to be allotted to the derivative (**210**, R=H) and the "equivalent" configuration to coleone-B as shown (**209**) (*272*).

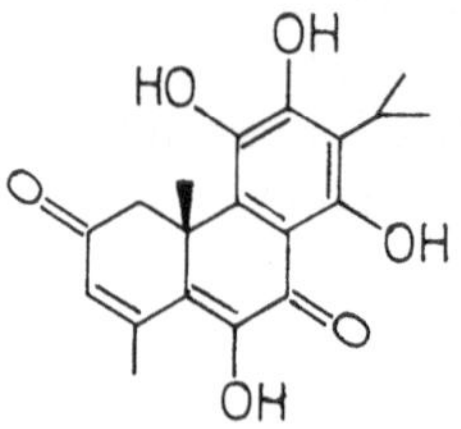

(209) 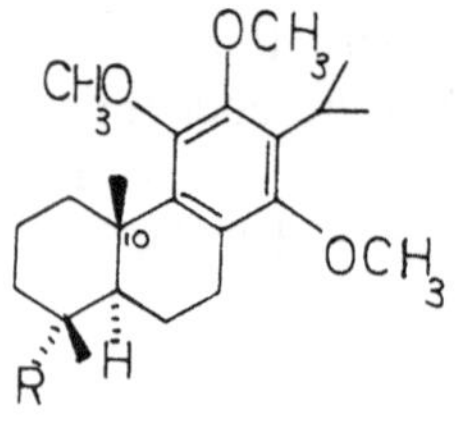   (210)

The dextrorotatory acid (**212**) obtained by degradation of farfugin-A (**211**) can be converted to 4-methoxy-3,7-dimethylindan-1-one (**213**, R=CH₃) and 4-hydroxy-3,7-dimethylindan-1-one (**213**, R=H). Both of these compounds gave o.r.d. curves quasienantiomeric to that of known (3*R*)-(−)-3-methylindan-1-one, thus establishing the (3*S*)-configuration shown for the indan-1-ones (**213**) and the (*S*)-configuration for farfugin-A (*317*).

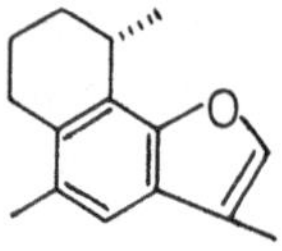

(211) 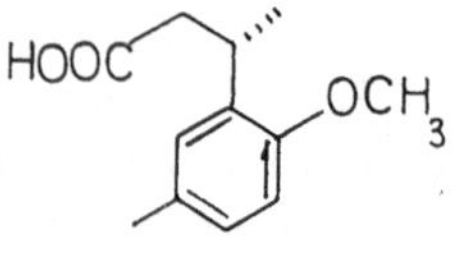   (212) 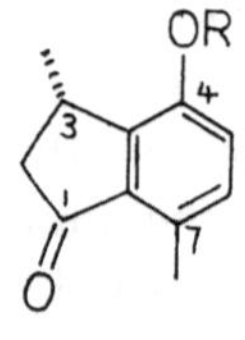   (213)

(−)-Helianthoidin (**214**) has been assigned the (4*S*)-configuration by empirical comparison of the o.r.d. and c.d. curves of the derived lactone (**215**, R=CH₃) with those of (−)-hibalalactone (**215**, R−R= `CH₂′`) of known absolute configuration. Since these two compounds have only one centre of asymmetry and identical oxygen substitution patterns in the aromatic rings, the near identity of their o.r.d./c.d. curves establishes the configuration as shown (*61*).

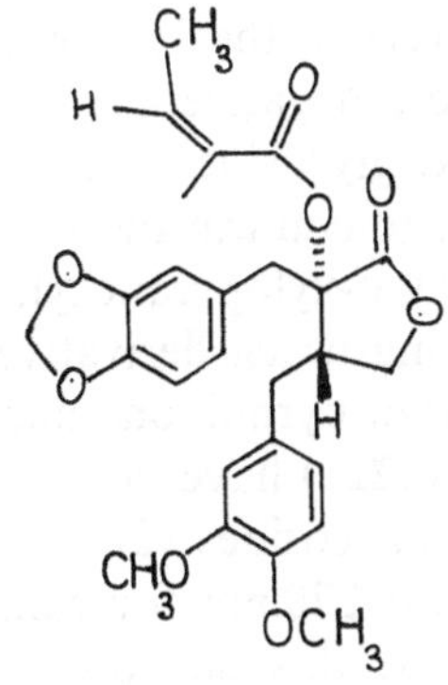

(214) 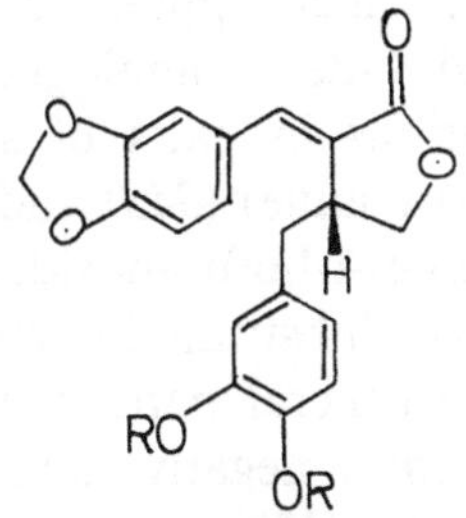   (215)

The sign of the aromatic CEs in the 6-substituted A-ring aromatic steroid oestradiol and related compounds depends on the configuration at C-6. A 6α-hydroxy substituent leads to two negative c.d. maxima corresponding to the $'L_b$ and $'L_a$ transitions of the aromatic A ring and the 6β-hydroxy substituent to two positive maxima. This correlation may be of use in assigning configuration at C-6 (63).

h) Biphenyls

More specialised types of application include those involving bridged biphenyls and allenes (231, 232). The configuration of (−)-9-dimethyl-amino-9,10-dihydro-4,5-dimethylphenanthrene (216) (174) and of methyldecinine (217) (112) have been allotted by comparison with other skewed biphenyls.

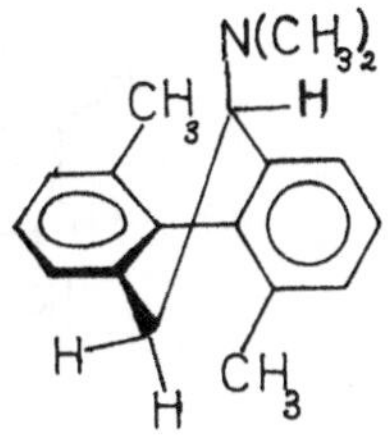

(216)

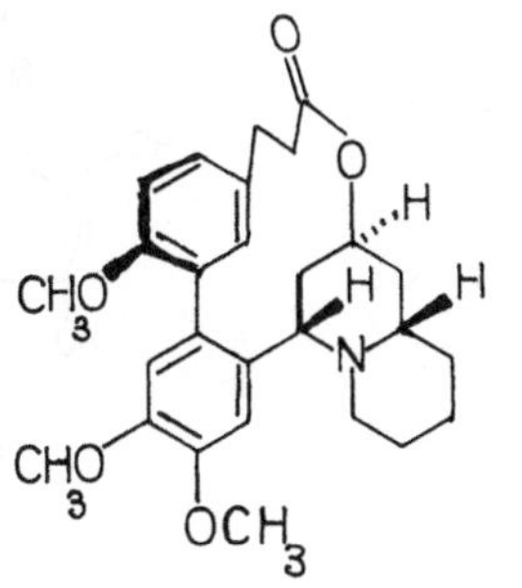

(217)

i) Purine and Pyrimidine Chromophores in Nucleosides

Examination of pairs of nucleosides epimeric at C-1 has established the generalisation that pyrimidine nucleosides with the 1α and 1β configuration give negative and positive CEs respectively at about 270 nm,

while for purine nucleosides the signs are reversed, the 1α-configuration
leading to a positive and 1β-configuration to a negative CE. This generalisation, which applies to fully oxygenated glycosides and to the 2'-
deoxy and 3'-deoxy analogues, has been used to confirm the 1α- and 1β-
configurations in pairs of synthetic 2'-deoxy-D-xylopyranosylthymines
(218, R=thymine) (343) and the corresponding uracil derivatives (342),
in 3'-deoxy-D-ribofuranosides (219) (337) and in pyrimidone nucleosides
(220) (271). Substituted adenine arabinosides (221) have been shown to
follow the general pattern of other purine nucleosides (214). Pyrazole
ribosides show negative and positive CEs near 270 nm for compounds
with the 1β- (222) and 1α-configuration, and this has been used to
establish the configuration at C-1 of the nucleoside antibiotic pyrazomycin (223) (111).

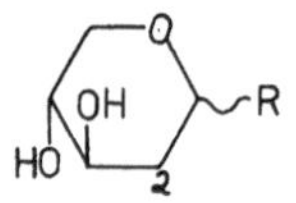

(218)

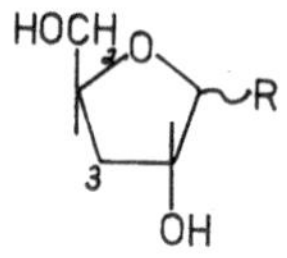

(219)

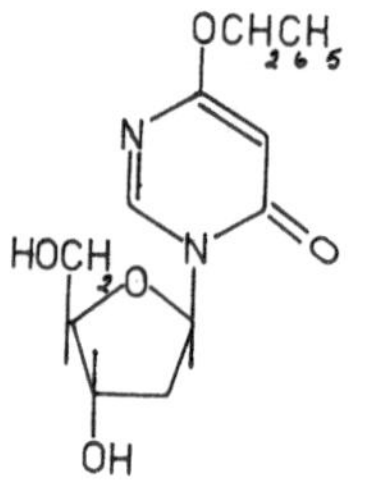

(220)

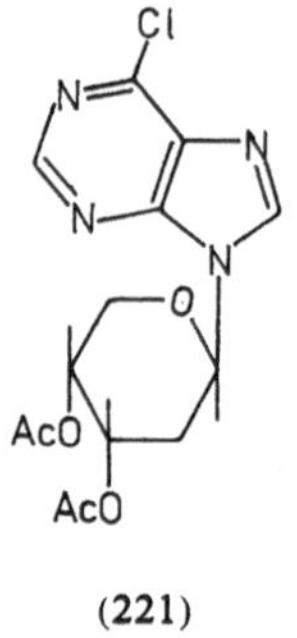

(221)

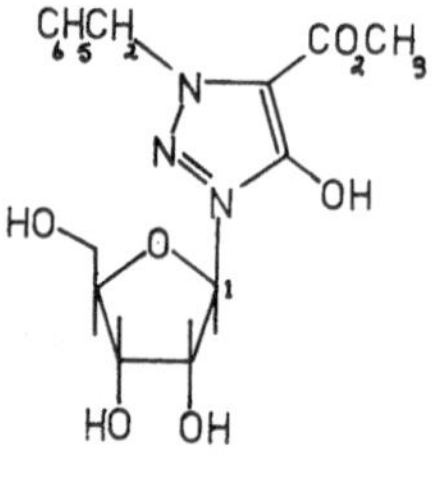

(222)

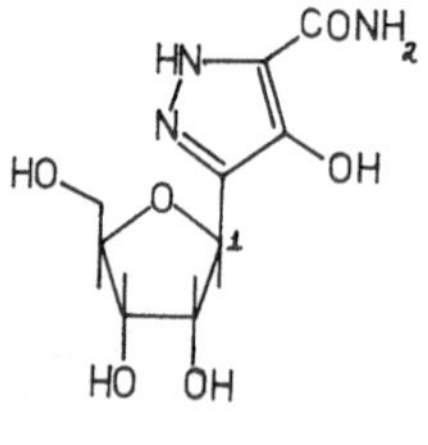

(223)

Differences in the c.d. spectra at shorter wavelength may be
characteristic of different conformations about the C_1-N-bond. For
example, the sequence positive CE (284) negative CE (253) positive CE
(217 nm) is claimed (149) to be characteristic of the anti-conformation
of 1β-anomers of pyrimidine nucleosides.

6. The Carboxyl Chromophore

a) Correlations Involving Acids or Modified Groups as Principal Chromophore

Although the chiroptical behaviour of many series of carboxylic acids has been surveyed, it is the o.r.d./c.d. of α-hydroxy- and α-amino-acids which has found widest application in the study of natural products. In general, α-amino-acids of the L-configuration give positive CEs near 210 nm and those with the D-configuration enantiomeric negative CEs. This generalisation is supported by studies of a wide range of amino acids (*118* and references therein) and has been used, for example, to allot the (2*S*)-configuration to samples of two γ-hydroxynorvalines (2*S*,4*S*)- (**224**, R^1=OH, R^2=H) and (2*S*,4*R*)- (**224**, R^1=H, R^2=OH) isolated from natural sources (*227*). The correlation between the sign of the carboxyl CE and configuration at the α-carbon atom is valid only if the compounds concerned have no other chromophoric groups absorbing in the same spectral region and if the conformation of the compound studied can safely be assumed to be analogous to that of the protein amino acids. The problems can be illustrated by *cis*-3-guanidinoproline (**225**), by alliin (**226**) and by diacetyldopa derivatives (**227**). *cis*-3-Guanidino-

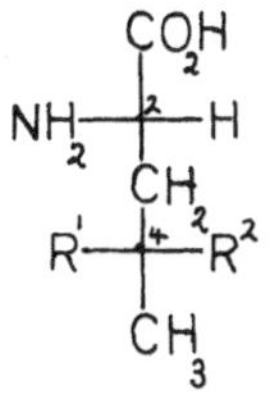

(224)

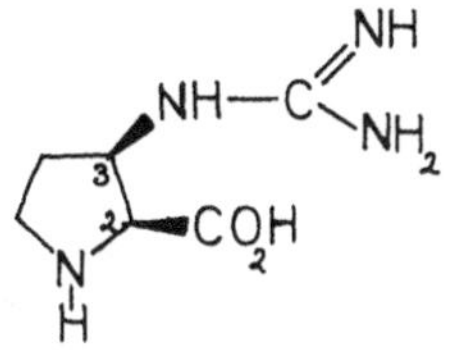

(225)

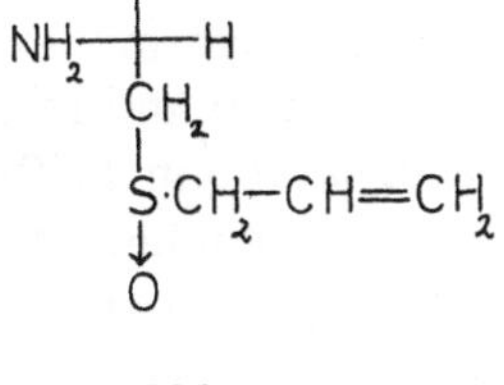

(226)

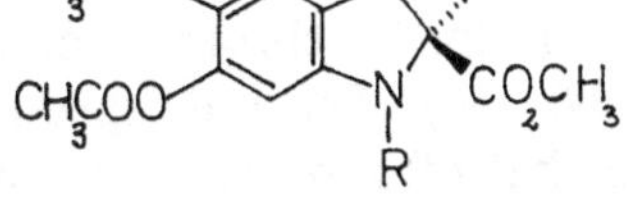

(227)

proline (**225**) has the L-configuration at C-2 but is reported (*124*) to give a negative CE, possibly due to interaction between the carboxyl and guanidine chromophores or to a change of conformation caused by the pair of bulky *cis*-substituents. Alliin (**226**) [(+)-*S*-allyl-L-cysteine sulph-oxide] and its dihydro derivative [(+)-*S*-*n*.propyl-L-cysteine sulphoxide] have virtually enantiomeric o.r.d. curves despite their identical configurations at the α-carbon atom and at sulphur (141) because of the difference between the allyl and propyl substituents*. The diacetyldopa derivatives (227, R = CH₃CO and 227, R = H) have opposite signed CEs near 250 nm whose sign is controlled by the aromatic chromophore; despite the identical configurations at C-2 (116), the N-acyl derivative must be considered to have a different chromophore from the parent compound.

Diketopiperazines show a pronounced CE with a first extremum (o.r.d.) near 230 nm. The CE is negative for the diketopiperazines from L-alanyl-L-alanine and L-alanyl-L-serine (18). By analogy the L,L-configuration has been assigned to echinulin (228, R = (CH₃)₂C = CH.CH₂–) and its synthetic analogue (228, R = H) which both show negative CEs in this region (153). This allotment reverses the L,D-assignment made earlier for echinulin on the basis of plain o.r.d. curves (250) and thus emphasises the inadequacy of comparisons made over a limited wavelength range outside the region of the absorption bands. However, it should also be noted that the presence in echinulin of a highly substituted indole tryptophan residue, makes direct comparison with the simple diketopiperazine of L-seryl-L-alanine unreliable.

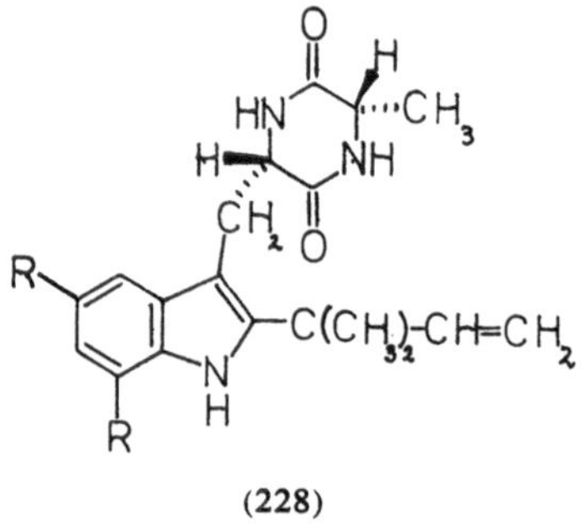

(228)

The piperazinedione system is also present, though modified by the diene and disulphide chromophores in gliotoxin (229) and related compounds (29, 350). The c.d. spectrum of acetylaranotin (230) (246) is similar to that of gliotoxin, and on this basis the configuration at the asymmetric centres of the diketopiperazine ring has been assigned as shown; the allotment has been confirmed by Bijvoet X-ray measurements (81). The dimeric antibiotic verticillin-A has been allotted the absolute configuration shown (231, R¹ = H, R² = OH) (230) because its c.d. curve

---

* More recent c.d. work (118) suggests that there may not be a true anomaly here; the misunderstanding has perhaps arisen from the mismatching of o.r.d. or c.d. bands, but it still underlines the need for caution.

is closely similar to that of chaetocin (**231**, $R^1 = OH$, R^2-H) itself known by Bijvoet analysis (*138*).

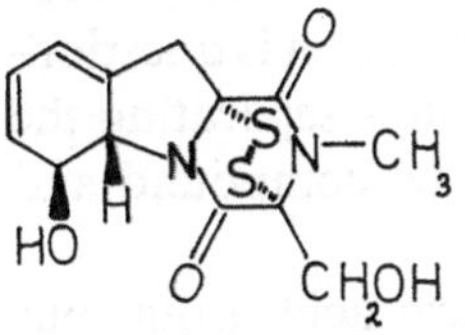

(229) 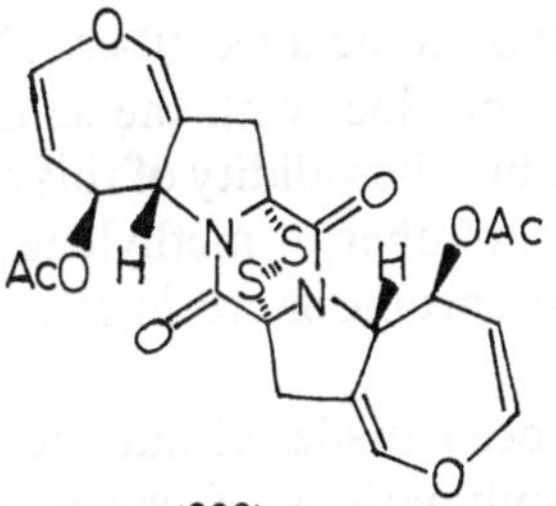   (230) 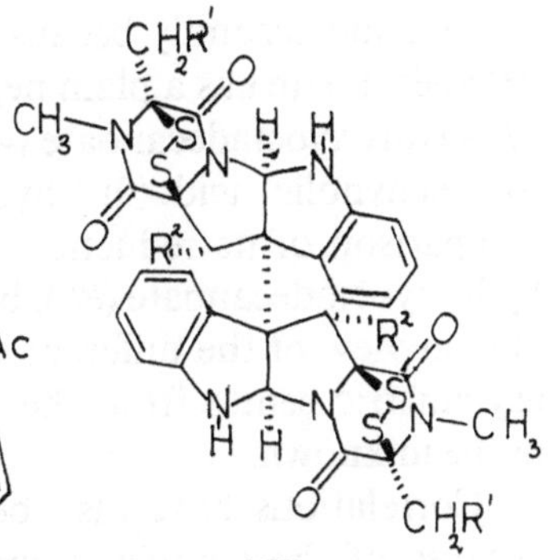   (231)

Other correlations involving single centre compounds include (+)-2-dimethylamino-2-phenylpropionic acid, assigned the (*S*)-configuration (**232**) because its negative CE at 245 nm is quasi-enantiomeric to that of (*R*)-(+)-*O*-methylatrolactic acid methyl ester (**233**) (*89*). (+)-*p*-Hydroxybenzylsuccinic acid (**234**) has been allotted the (*R*)-configuration from the similarity of its positive CE with that of known (*R*)-benzylsuccinic acid (*287*).

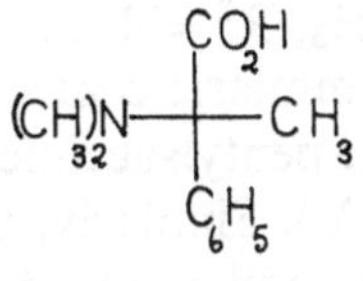

(232) 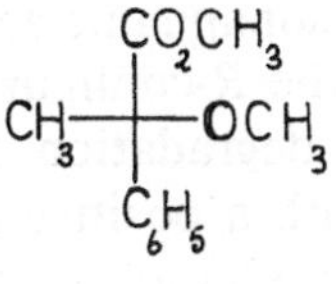   (233)

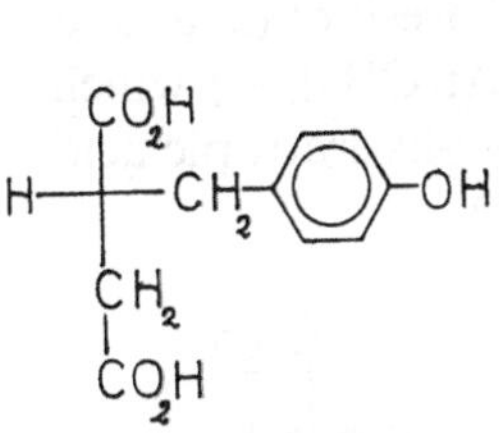

(234) 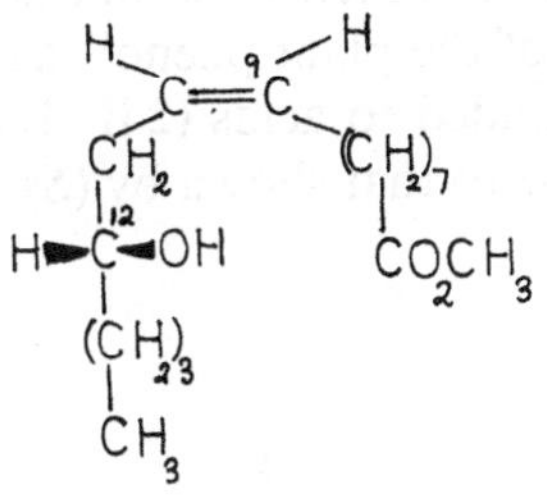   (235) 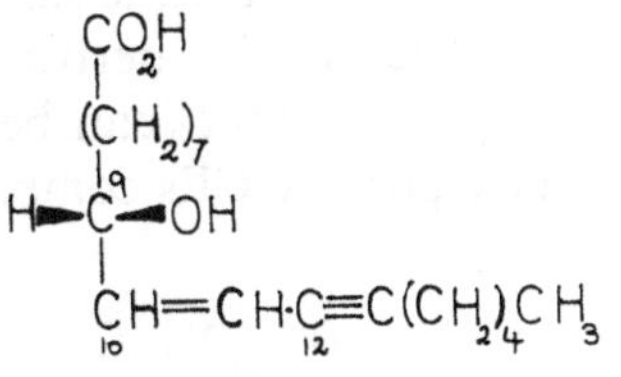   (236)

Allotment of configuration by o.r.d. and c.d. is particularly difficult for long-chain fatty acids in which the asymmetric centre is far removed from the carboxyl chromophore. A good example is that of methyl

12-hydroxy-9-hexadec-*cis*-9-enoate (**235**) which was assigned the D-configuration at C-12, first, on account of its o.r.d. curve which was superimposable on that of known methyl (+)-12-D-hydroxy-9-octadec-*cis*-9-
enoate, and secondly because the reduction product methyl 12-hydroxy-
hexadecanoate has a plain negative curve similar to that of known methyl
12D-hydroxyoctadecanoate (*44*). The (9D)-configuration has been allotted
to helenynolic acid (9D)-hydroxyoctadec-10-en-12-ynoic acid (**236**) by
comparison of its reduction product with the same known methyl 12D-
hydroxyoctadecanoate (*86*), but the validity of this correlation is question-
able in view of the different number of methylene groups separating the
asymmetric centre from the chromophore in the model compound and
in the unknown.

Correlations have also been made within the necic acid group but
because of the complex combinations of chromophores present, these
correlations should be regarded as tentative (*69*).

b) Correlations Involving Degradation to a Carboxylic Acid Fragment

Degradation of a complex natural product often leads to smaller
molecules which preserve the configuration of the parent compound at
one or more centres and which can be used to determine its absolute
stereochemistry. Carboxylic acids obtained by oxidative degradation
are frequently used in this way. For example, papuanic acid (**237**, $R^1 = H$,
$R^2 = CH_3$) and isopapuanic acid (**237**, $R^1 = CH_3$, $R^2 = H$) have both been
shown to have the *R*-configuration at the asymmetric centre in the side-
chain (*306*) by degradation to known (*R*)-*n*-pentylsuccinic acid (**238**,
$R = C_5H_{11}-$) with a positive carboxyl CE (*120*). Similarly, the absolute
configuration at C-3 of (+)-2-(3′,4′-dimethoxyphenyl)-4,6-dimethoxy-
3-methylcoumaran (**239**) has been established (*13*) by the identity of the
o.r.d. curve of the methylsuccinic acid obtained by degradation with
that of an authentic sample of (*R*)-methylsuccinic acid (**238**, $R = CH_3$).

The dihydro-derivatives of the plant phenols agatharesinol (**240**) and
sequirin-C (**242**) can be degraded to acids (**241**, $R = Ar(CH_2)_2-$) which
give positive CEs comparable to that shown by (*S*)-(+)-hydratropic acid

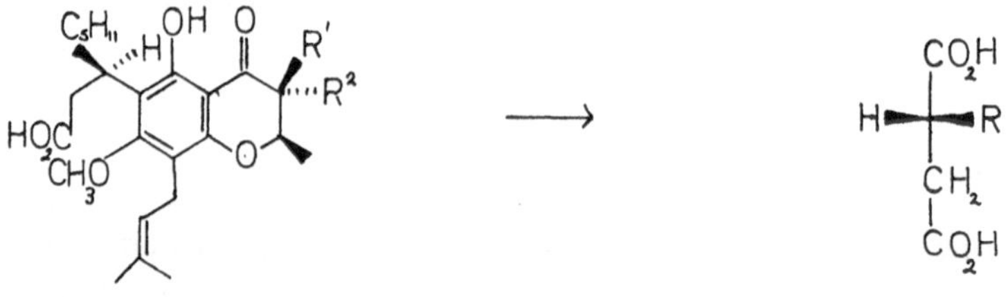

(237)   (238)

(238) ⟵

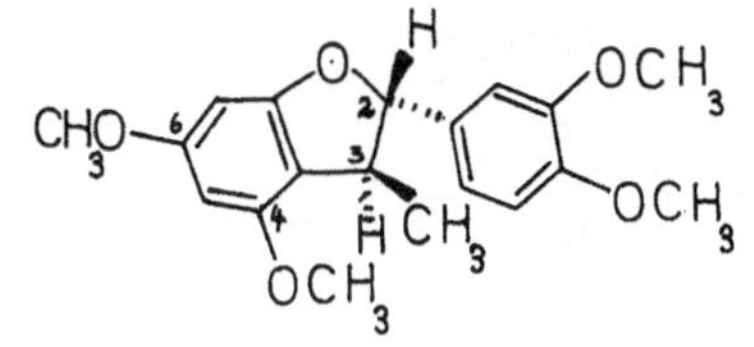

(239)

(241, R = Me), thus establishing the configuration at C-3 in the original compounds (107, 137). The absolute configurations shown have been allotted to (+)-hexoestrol (243) (160) and to the metabolite of phenan-

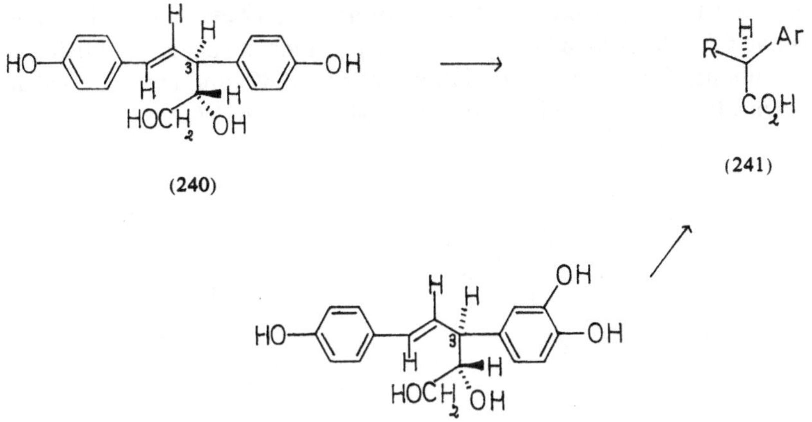

(240) ⟶

(241)

(242)

threne (245) (235) by degradation to (2R,3R)-diethylsuccinic acid (244, R = Et) and (2R,3R)-diacetoxysuccinic acid (244, R = OAc) respectively, and by comparison of the o.r.d. and c.d. curves of the latter (as their bromphenacyl esters) with those of authentic samples.

The single asymmetric centre of the alkaloid brevicolline (246) has been allotted the (S)-configuration (47) by degradation to N-methyl-

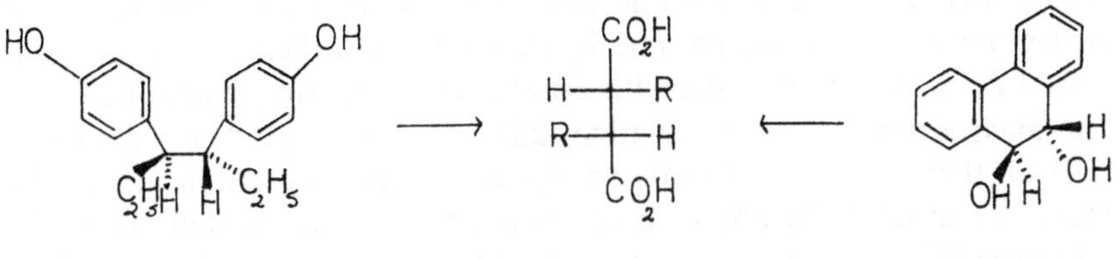

(243) ⟶ (244) ⟵ (245)

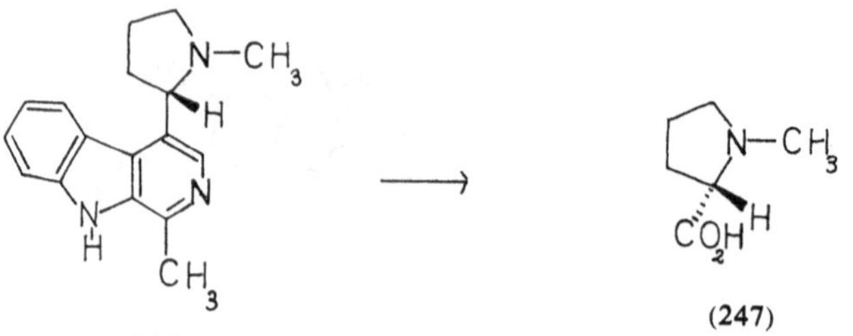

(246)   (247)

proline (247) with an o.r.d. curve identical to that of the material obtained by methylating natural (S)-proline. By a longer series of transformations, (−)-humulone (248) has been converted via ozonolysis of tetrahydro-humulone to 2-hydroxy-5-methylhexanoic acid (249). This acid gave a positive CE which by analogy with other α-hydroxy acids establishes the (S)-configuration, in the degradation product (249) and the "equivalent" configuration at the single centre in (−)-humulone (91).

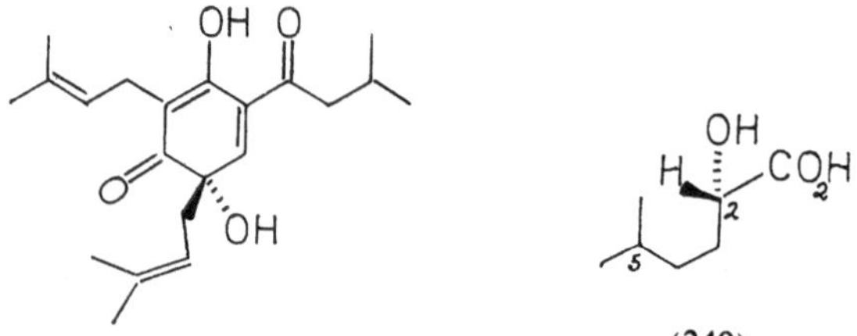

(248)   (249)

c) Empirical Correlations in Complex Lactones and Lactams

Many correlations are concerned with complicated molecules for which useful assignments of configuration can be made only by direct comparison with a closely similar molecule of known stereochemistry. For example phyllanthine (250, R = OCH₃) can be allotted the same absolute stereochemistry as securinine (250, R = H) because their o.r.d. curves in three solvents are virtually superimposable (265). Reduction products (251) of virosecurinine have been allotted the *trans* fusion of the lactone and cyclohexane rings by comparison of their c.d. curves with those of related compounds with established *trans* ring junctions (249). The macrocyclic dilactone azimine (252, n = m = 5) has been shown to have all three ring substituents on the piperidine ring *cis* (n.m.r.). The negative carboxyl CE observed for azimine is similar to that shown by carpaine (252, n = m = 7) of known absolute configuration, thus establish-

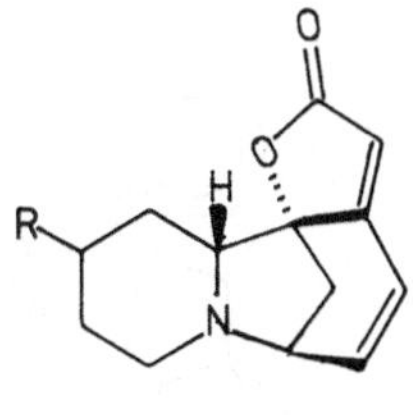

(250)

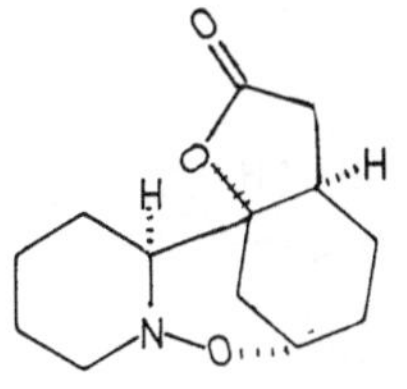

(251)

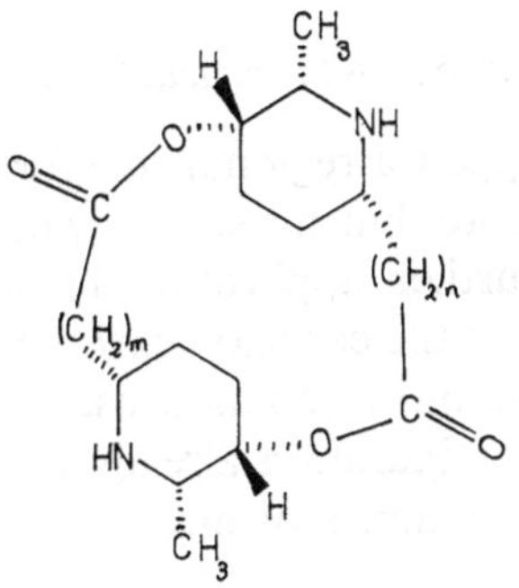

(252)

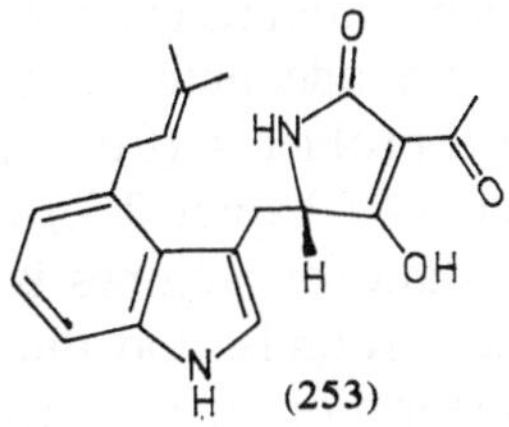

(253)

ing the absolute stereochemistry of azimine (*294*). The single asymmetric centre in bissecodehydrocyclopiazonic acid (**253**) has been allotted the (*S*)-configuration because the CEs observed at 280 and 325 nm are of the same sign and order of magnitude as those observed for tenuazonic acid (**254**), itself correlated with (2*S*)-isoleucine (**255**) (*150*).

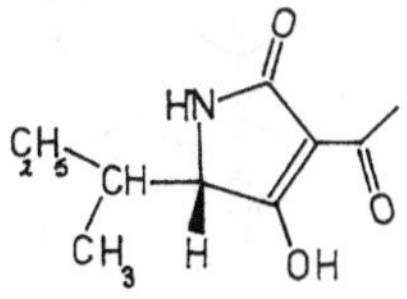

(254)

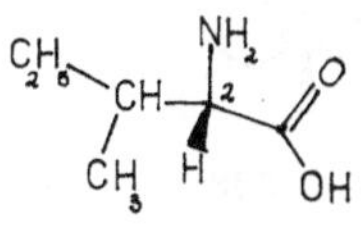

(255)

In some cases c.d. curves give information about the skeleton of a complex molecule without necessarily covering every point of stereochemistry. For example conocarpin, of gross structure (**256**) gives o.r.d. and c.d. curves which are quasienantiomeric to those of leucodrin of known absolute configuration (**257**). This information suggests that the two compounds are enantiomeric with respect to the main ring system, but does not establish the stereochemistry of the three substituents (*204*).

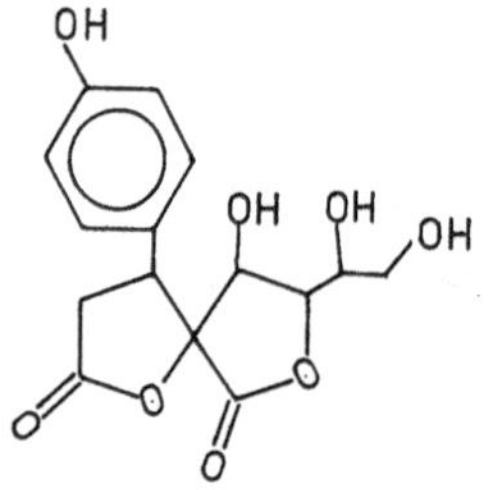

(256)

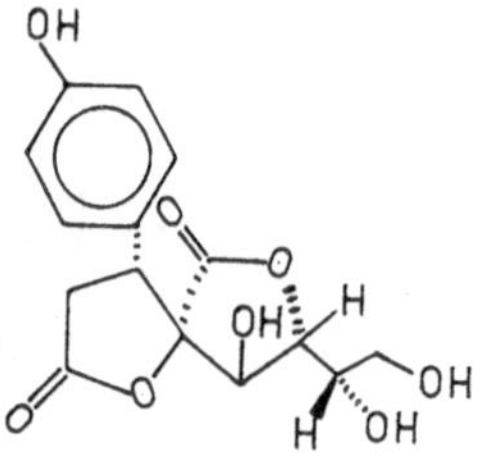

(257)

d) Saturated Lactones: Correlations by Semi-Empirical Rules

Several attempts have been made to suggest a regional rule for compounds containing the carboxyl chromophore, but no single "rule" has proved an adequate generalisation for further application to stereochemical problems. (For a review of work on the carboxyl chromophore see Ref. *74* Chapter 3.3.) Consequently the most common use of chiroptical data for lactones is to confirm a configuration already assigned on other grounds, and there are very few examples of direct allocation of configuration by a carboxyl regional rule.

One example is the allotment (*139*) of the 5β-methyl configuration to 5-methyl-3-oxa-A-nor-5β-cholestan-2-one (**258**) which shows a negative CE in accordance with the prediction of the sector rule (*167*) for *cis* fused 1-oxahexahydroindan-2-ones of this type. The signs of the lactone CEs and the prediction of the sector rule have also been used to confirm

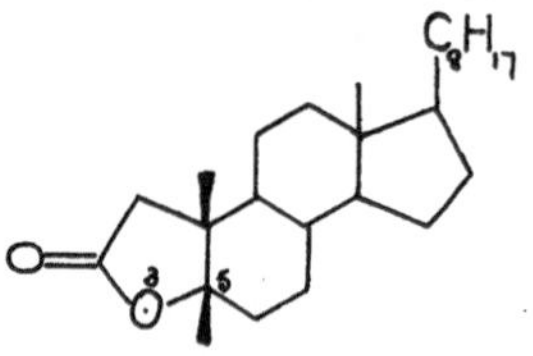

(258)

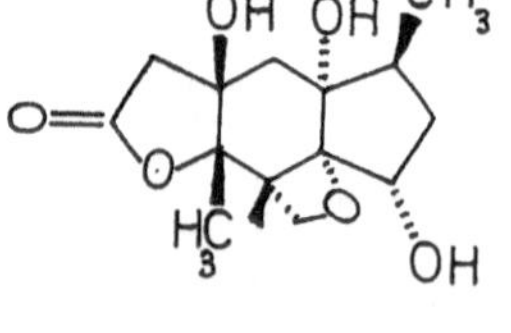

(259)

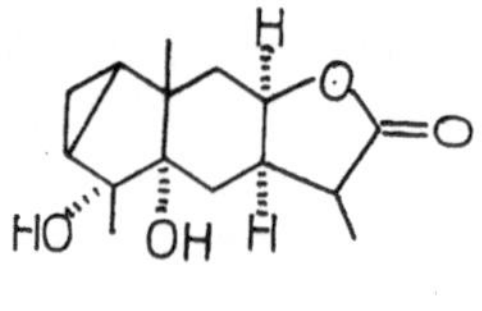

(260)

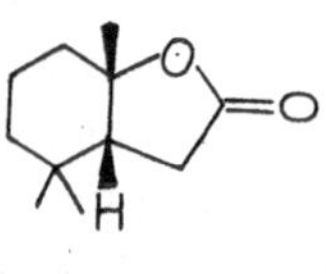

(261)

the *cis*-fusion of the lactone ring allotted from other evidence, in compounds (**259**) (*259*), (**260**) (*319*) and (**261**) (*273*) related to pseudoanisatin, isolinderoxide and tetrahydroactinidiolide respectively. Other examples occur in work on 16-oxasteroids (*19*) and on fusidic acid (*203*). Hydrolysis of the peptide detoxin-D_1 (**262**) yields L-alanine, L-phenylalanine, (*S*)-2-methylbutyric acid and a new amino acid detoxinine (**263**), which can be converted to lactone (**264**) shown by n.m.r. to have *cis*-fusion of the two rings. The lactone CE is negative and in accordance with the sector rule, leads to the absolute configuration shown (*176*).

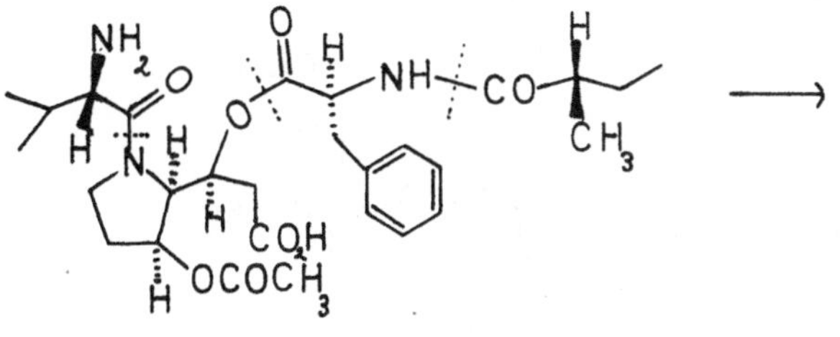

(262)

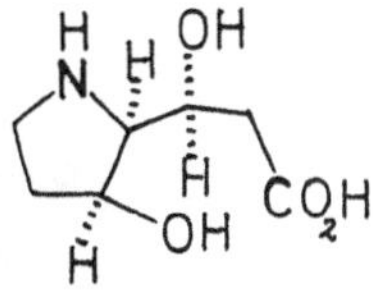

(263)

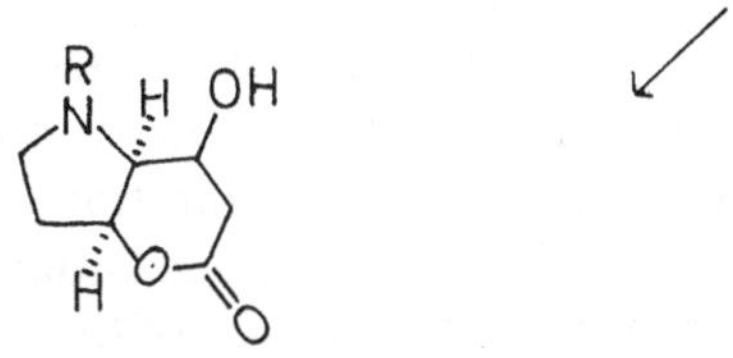

(264)

The alternative sector or "comet" rule (*303*) which seeks to emphasise the difference between the two oxygen functions in the carboxyl group, has also been used to establish configuration in some cases. For example the 25S-configuration has been allotted to the lactone (**265**) of known relative stereochemistry obtained by degradation of cyasterone, (**266**) (*147*) and the absolute configuration of portentol (**267**) has been established from the sign of the carboxyl CE for the corresponding reduction product (*1*).

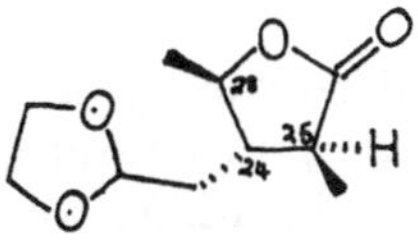

(265)

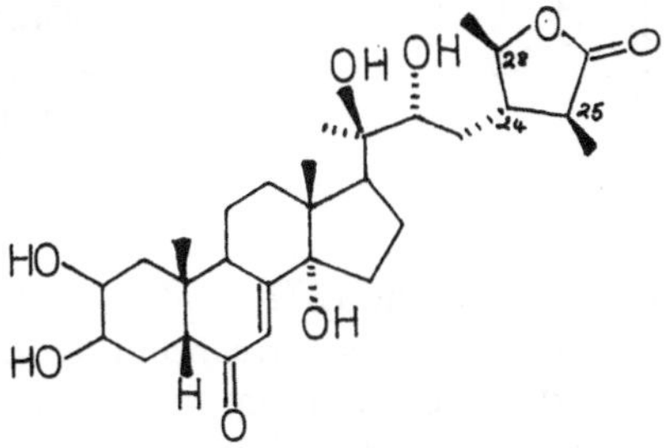

(266)

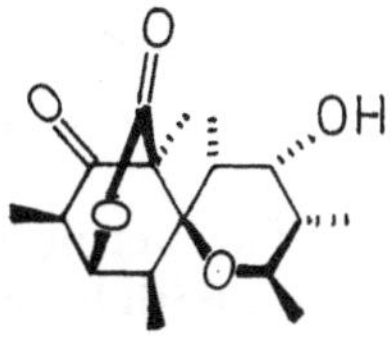

(267)

e) α,β-Unsaturated Lactones

(+)-Parasorbic acid [(+)-hex-2-en-5-olide] (**268**, $R^1 = CH_3$, $R^2 = H$) is a key compound for the assignment of absolute configuration in six-membered α,β-unsaturated lactones; it has been allotted the (5S)-configuration by chemical correlation (*216*) and exhibits a positive CE near 250 nm. The configuration of (R)-(−)-massoia lactone (**268**, $R^1 = H$, $R^2 = C_5H_{11}$), which exhibits a negative CE has been allotted by direct comparison (*88*) with parasorbic acid, on the assumption that the conformation of the pentenolide ring is the same for both compounds, i.e. that in which the alkyl group at C-5 is pseudoequatorial.

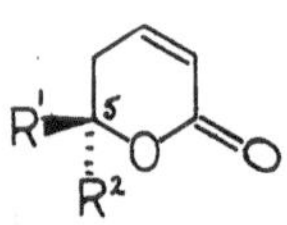

(268)

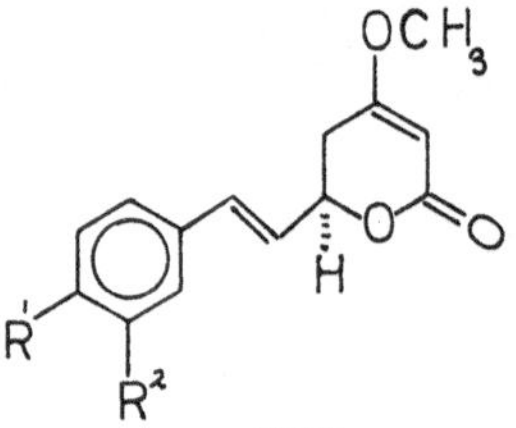

(269)

The dihydro derivatives of kawain (**269**, $R^1 = R^2 = H$) and methysticin (**269**, $R^1 = R^2 = OCH_2O$) both show strong positive CEs at 246 nm corresponding to the same configuration as (+)-parasorbic acid (**268**, $R^1 = CH_3$, $R^2 = H$) (*298*) and in accordance with the semi-empirical helicity rule for the n → π* transition of the carbonyl group in unsaturated lactones (the R-band) (*304*). Dihydrokawain-5-ol which has the two substituents at C-5 and C-6 *cis* (n.m.r.) and the C-6 group equatorial, also has a positive CE at 247 nm and has been allotted the (6S)-configuration (**270**) (*2*). Conversely, the fungal lactone (**271**) has a negative CE near 250 nm, and a *trans*-configuration in which the C-5-CH$_3$ is pseudo-equatorial (n.m.r.); it has therefore been assigned the (5R,6R)-configuration (*278*).

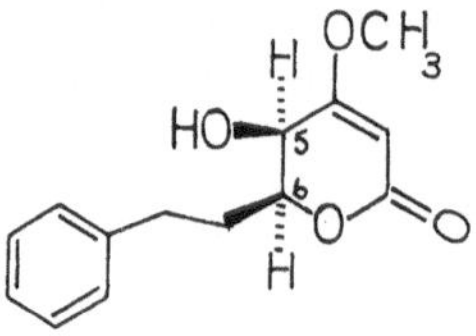

(270)

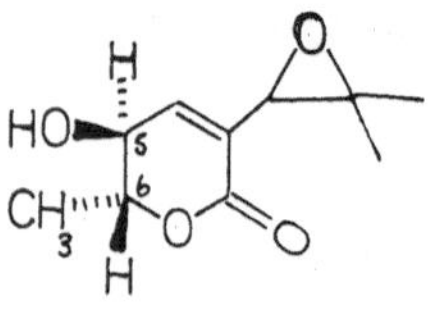

(271)

The steroid side-chain lactone (**272**) and the related 1-hydroxy derivative both have small positive CEs corresponding to the α,β-unsaturated lactone absorption, and have therefore been allotted the (22R)-configuration by analogy with parasorbic acid (*3*). (The difference in description arises from the operation of the R,S-system.) 23-Deoxy-antheridiol (**273**) and its C-22 epimer have been distinguished on the basis of the difference curves obtained by subtracting from their c.d. curves, the c.d. of 7-ketocholesterol (3β-hydroxycholest-5-en-7-one). The difference curve for (deoxyantheridiol *minus* 7-ketocholesterol) gave a positive CE near 250 nm and that for the C-22 epimer a negative CE corresponding to the 22R and 22S configurations respectively (*128*).

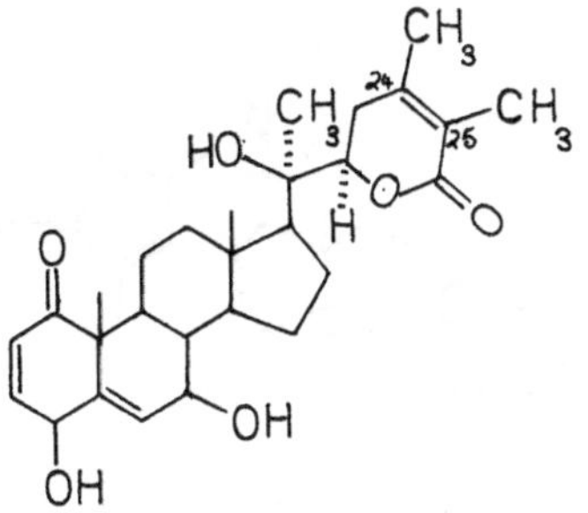

(272) 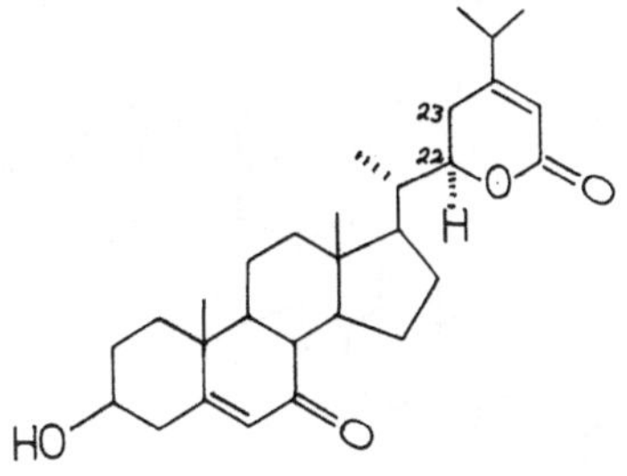   (273)

The correlations discussed above all depend on the evidence (or assumption) that the conformation of the lactone ring is that in which the bulkier substituent at C_α' is equatorial and the hydrogen is axial. This problem has been discussed in detail (*30*) for the complex spiro-lactone dioscorine (**274**). The positive dichroism observed for dioscorine is in accordance with the absolute configuration shown (**274**) if C-13 is pseudo-equatorial, but following the helicity rule (*304*), implies the enantiomeric configuration if C-13 is pseudo-axial. X-ray evidence shows that C-13 is pseudo-equatorial in the crystalline state, and on the *assumption* that this conformation will probably persist in solution, dioscorine has been allotted the configuration (**274**) as shown.

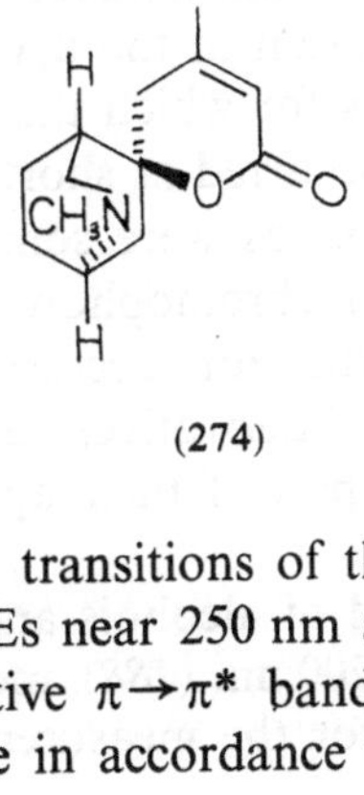

(274)

The n→π* and π→π* transitions of the carbonyl chromophore in butenolides give rise to CEs near 250 nm and 220 nm respectively; for example, the strong negative π→π* bands of grandisolide (**275**) and cyclograndisolide (**276**) are in accordance with the (23R)-configuration established by Bijvoet X-ray (*10*). This correlation has been used to allot the absolute configuration at C-23 to antheridiol (**277**) and its C-22 epimer. Both compounds give strong positive CEs near 200 nm, *opposite* in sign to the negative π→π* CE of cyclograndisolide (**276**) and have therefore been allotted the "opposite" configuration at C-23 (*102*). (Note that this work assumes that the C-22-hydroxy group does not affect the correlation between cyclograndisolide and antheridiol, and this may not be justified.)

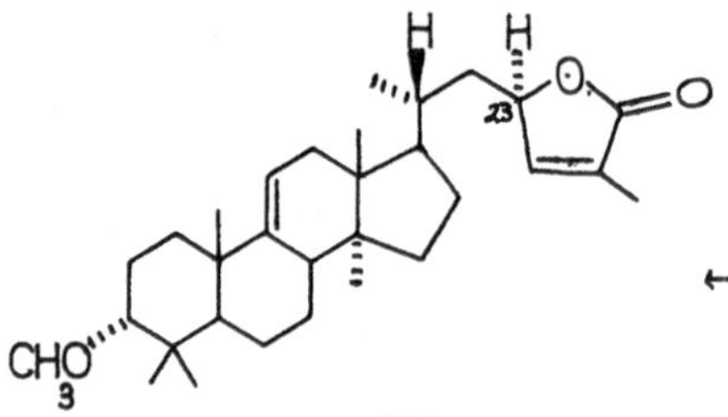

(275) 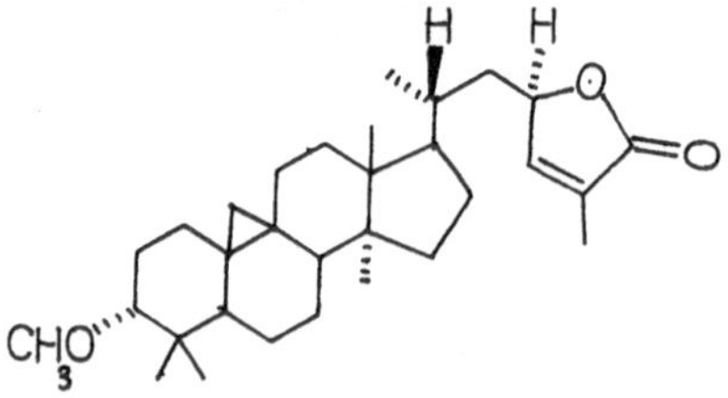   (276)

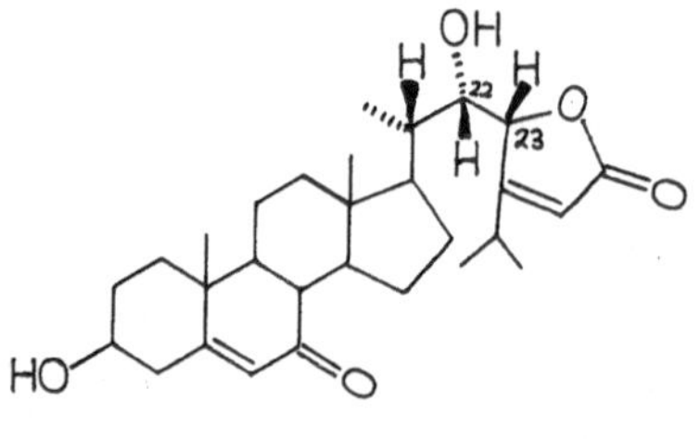

(277)

7. Chromophoric Derivatives

Earlier instruments for measurement of chiroptical data were hampered by lack of penetration in the u.v., making it difficult or impossible to study compounds for which the first main absorption band and the corresponding CE occurred at short wavelengths. To overcome this difficulty, parent compounds were sometimes studied in the form of a derivative containing a chromophore which absorbed at longer wavelengths and which was therefore accessible for chiroptical measurements. Many "chromophoric" derivatives have been suggested but there are comparatively few examples of their application to stereochemical problems in the literature.

The first absorption band of alcohols and the corresponding CE in chiral alcohols occur below 200 nm (*188*) and chromophoric derivatives which have been suggested for the measurement of chiroptical data at longer wavelengths include xanthate esters. These esters give significant CEs near 360 nm (293), which can be correlated with the configuration of an adjacent asymmetric centre, for example, in establishing the absolute configuration at C-2 of putranjic acid by conversion of the methyl ester (278, R=H) to the corresponding xanthate (278, R=CSSCH₃) of positive CE (288). The absolute stereochemistry of the single asymmetric centre in pulvilloric acid (279) has also been assigned from the negative CE of the derived xanthate (280), opposite in sign to the CE of the xanthate (281) derived from citrinin (282) of known absolute configuration (22). The correlation between the two xanthates (280) and (281) is acceptable because the sign of the xanthate CE is determined primarily by the adjacent asymmetric centre; direct correlation between pulvilloric acid (279) and citrinin (282) would not have been valid because of the differing effects of mono and disubstitution on the chromophore.

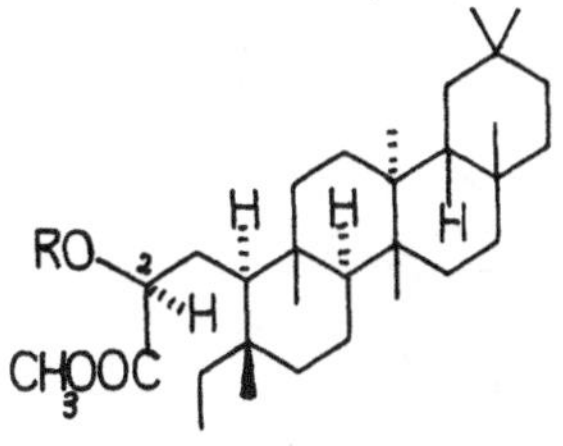

(278) 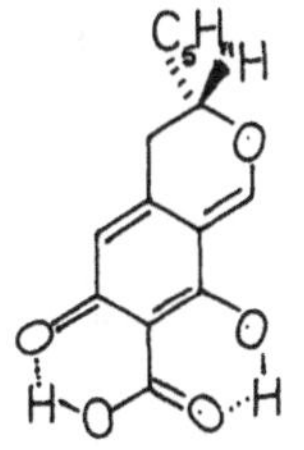 (279) 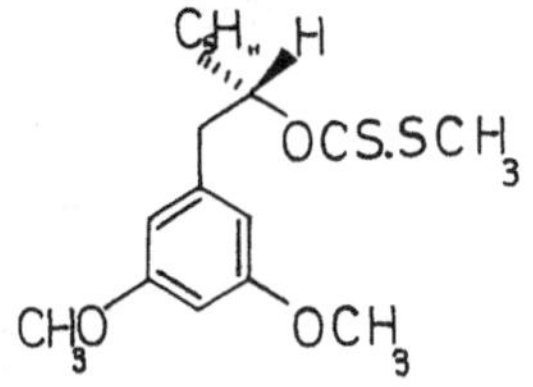 (280)

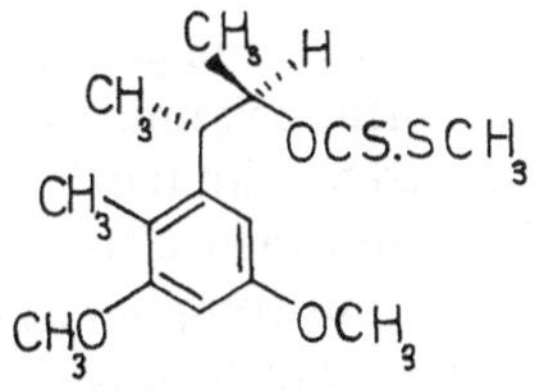

(281) 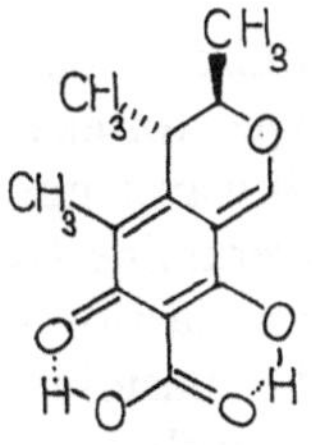 (282)

Another example is the conversion of an olefinic double bond to a pyrazoline by the action of diazomethane (299, 310). The pyrazoline has a strong absorption band near 320 nm and the sign of the corresponding CE can be used to assign the stereochemistry of the molecule on the assumption that the olefine is attacked by diazomethane on the less hindered face. For example, the positive CE found for the pyrazoline (283) of the unsaturated lactone floribundin corresponds to the relative configuration shown, in which the lactone ring is cis-fused to the seven-membered ring (7αH, 8αH) and diazomethane attacks on the α-face. Conversely the negative CE of the pyrazoline (284) of the C-8 epimer vermeerin, corresponds to the *trans*-fused ring system and β-attack of the diazomethane (*142*). Similarly the strong negative CE for the pyrazoline of axivalin (**285**) permits the allotment of the *cis*-fusion for the lactone ring in a situation enantiomeric to that of floribundin (**283**) (*11*). The absolute configuration of the aglycone (**286**) has been established by conversion to the pyrazoline derivative (**287**) which has a negative CE corresponding to the stereochemistry shown, and which shows no evidence of H-bonding in the i.r. spectrum, in accordance with the *trans* geometry of the hydroxy and nitrogen substituents (*326*).

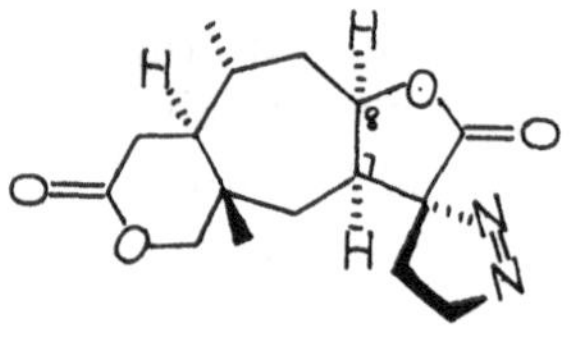

(283) 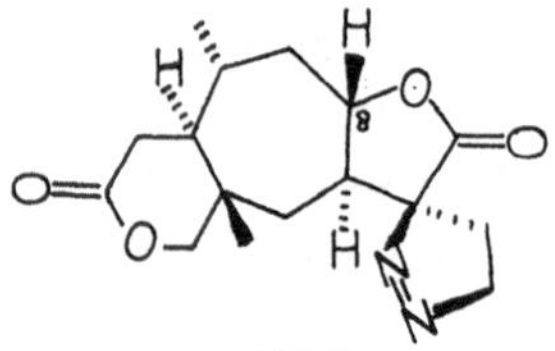   (284) 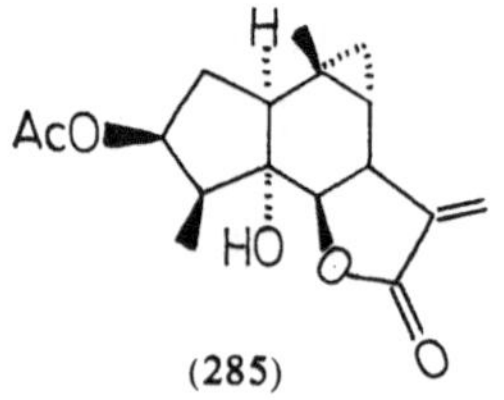   (285)

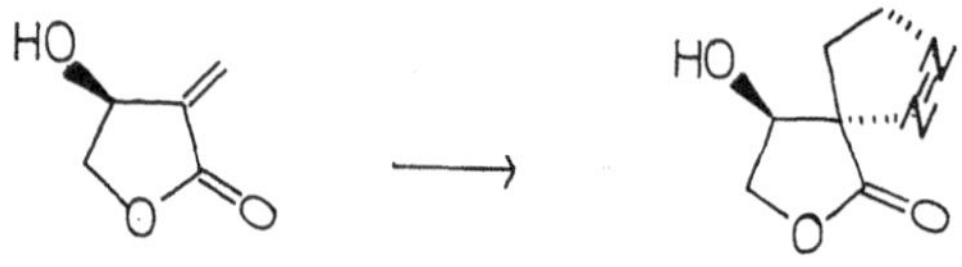

(286)   (287)

Many chromophoric derivatives have been suggested for amines. *N*-nitroso derivatives, for example, absorb near 350 nm, and the negative CE observed at this wavelength has been used to allot the (22S)-configuration to *N*-nitrosoveramine (**288**, R = NO) and thence to veramine itself (**288**, R = H), by comparison with the analogous compounds in the tomatidine (22S) and solasidine (22R) series, which give negative and positive CEs respectively (*4*). The configuration at C-22 in other solanum alkaloids has been assigned from measurements on the corresponding *N*-chloroamines (*275*). Other amino derivatives include those with *N*-salicylidene and *N*-dimedonyl groups. An example of application of the former is the allocation of configuration to certain phenylnorbornane derivatives (**289**) and (**290**) (*295*) and of the latter, the allocation of the 3α-configuration to the amino group of the alkaloid gitingensine (**291**) (*6*).

Dithiocarbamates have also been used, particularly as chromophoric derivatives of amino acids (*96*). An example of their application is the allocation of configuration at C-5' in the nucleoside antibiotic polyoxein C (**292**) from the positive CE of the corresponding *N*-dithiocarbethoxy-2',3'-*O*-isopropylidene deoxypolyoxein-C derivative (**293**) (*162*).

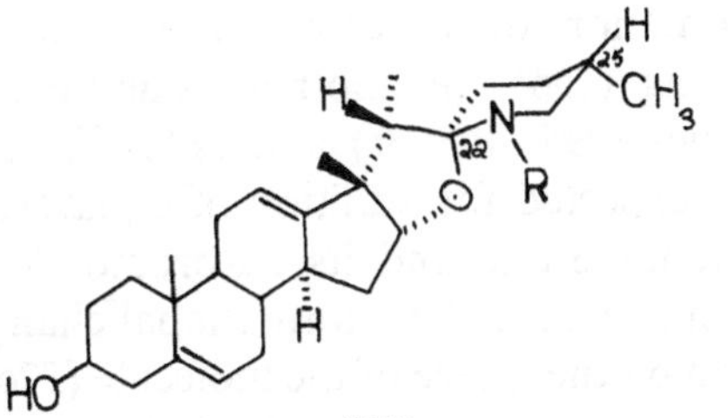

(288)

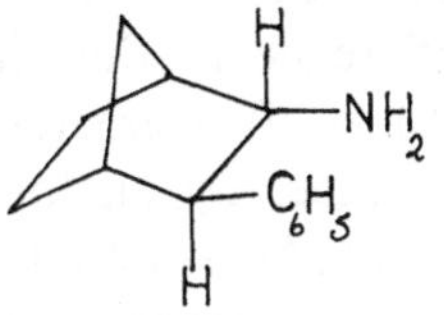

(289)

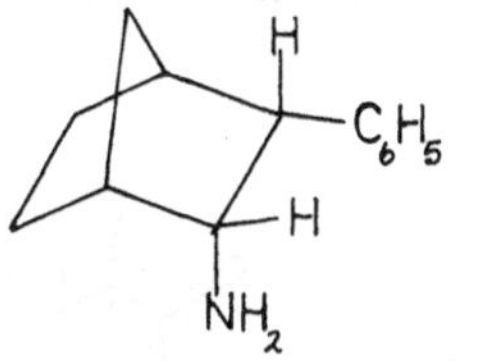

(290)

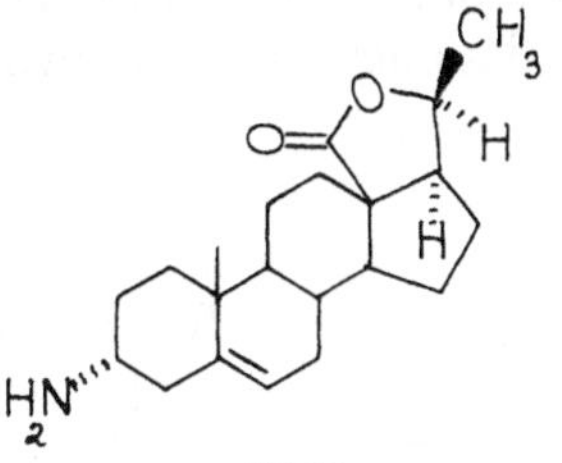

(291)

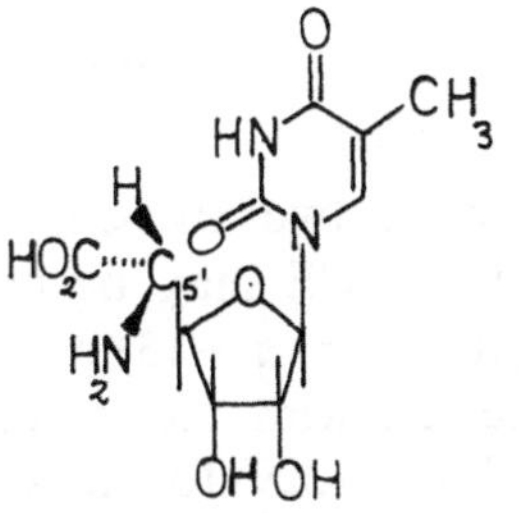

(292)

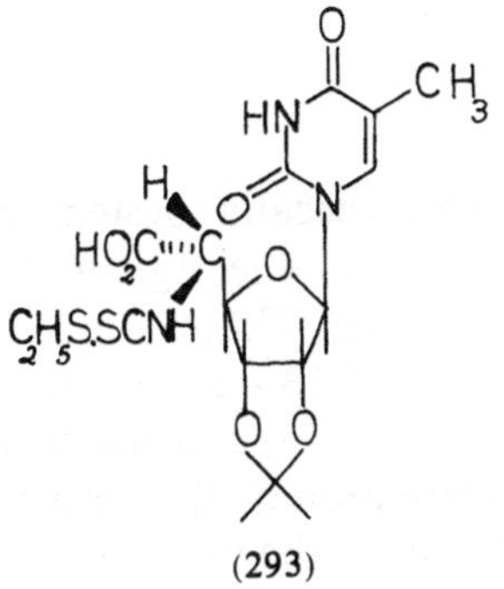

(293)

III. Studies of Conformation

The use of chiroptical measurements for study of conformation has been limited; most applications are concerned with saturated ketones, though a few refer to α,β-unsaturated ketones and dienes. Some examples including 4,4-dimethyl-3-oxo-steroids have already been discussed in previous sections. The simplest examples are those in which an o.r.d. or c.d. measurement for a particular compound is either in agreement with, or in contradiction to the result expected for a given configuration and conformation. For example, the negative CE of ketone (294) is of the same magnitude as that for analogous compounds lacking the 4α-methyl group. Since the axial methyl substituent would be expected to make an appreciable positive contribution if the molecule were in an all-chair conformation, the observed CE is evidence for flattening or twisting of the ring (*156*). In another example the difference in the CEs of the ketones (**295, R = Br**) and (**295, R = H**) ($\Delta a + 20$) is greater than would be expected for addition of equatorial bromine, by analogy with 5α-cholestan-3-one and its 2α-bromo derivative ($\Delta a - 5.2$). This is evidence for some small conformational change possibly caused by steric compression on the α-face of the molecule (*276*). Similarly, the unexpectedly small positive CE of 1α,3α-dichloro-5α-cholestan-2-one (**296**) is evidence for distortion of the steroid ring A, caused by interaction of the two axial chlorine substituents (*292*).

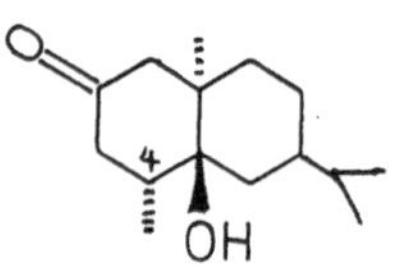

(294) 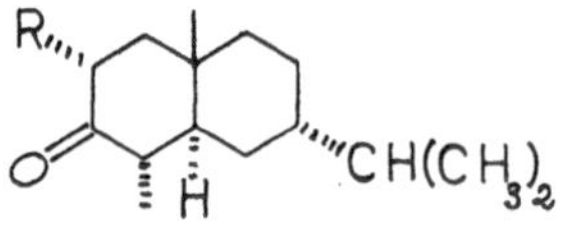   (295) 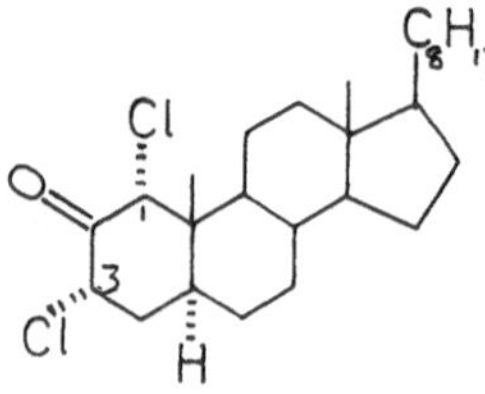   (296)

cis-Fused decalin system can adopt one of two two-chair conformations, and the sign and magnitude of the CE may be used to distinguish them. For example, the small negative CE ($a - 13$) of 18-nor-5β-abietan-8,11,13-trien-3-one (**297**) is in keeping with the steroid-like conformation (*155*) and the strong negative CE ($a - 229$) for the steroid-like conformation of 4β-acetoxy-B-nor-5β-cholestan-3-one (**298, 4β-OAc**) (*175*). However, the positive CE ($a + 55$) for the 4α-acetoxy-B-nor-steroid (**298, 4α-OAc**) is thought to imply that ring A takes up a boat conformation with the 4α-substituent equatorial. The negative CE for the 3-ketone (**299, R = H**) is in accordance with the prediction of the octant rule for a 5β-3-oxosteroid but the positive increment $\Delta a + 10$ for the addition of a 2β-methyl substituent (**299, R = CH₃**) is *not* as predicted for an equatorial methyl group and suggests a conformational change in the molecule (*238*).

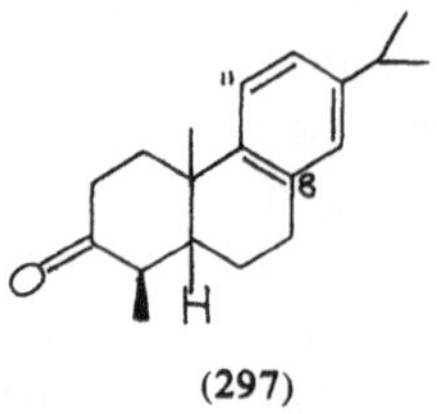

(297) 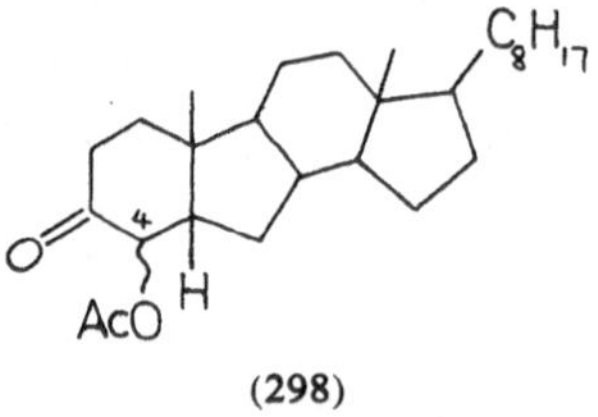   (298) 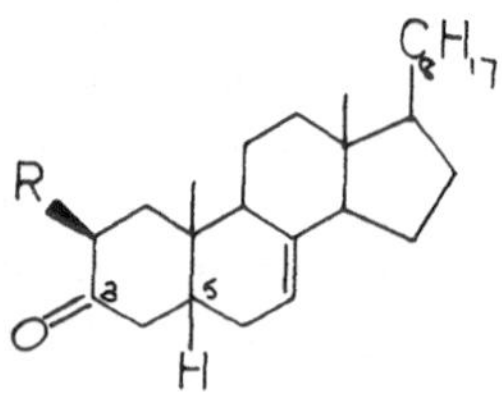   (299)

In many cases a ketone of known absolute configuration can adopt one of two (or more) conformations. If, according to the octant rule, CEs of opposite signs are predicted for these conformations, then measurement of the experimental CE may enable the preferred conformation to be established. The ketone (**300**) for example, can exist in two conformations (**301**) and (**302**) for which the corresponding octant projections (**301 P**) and (**302 P**) indicate negative and positive CEs respectively. The observed small negative CE suggests that (**301**) is the preferred (though not necessarily the exclusive) conformation (9). For the monocyclic (2*S*)-methyl-(1*R*)-acetylcyclohexane (**303**) there are three probable conformations, in each of which the carbonyl group eclipses a C–H or C–C bond. From the three corresponding octant projections, (**303 P/1, P/2** and **P/3**), the observed negative CE $\Delta\varepsilon - 1.27$ identifies the preferred conformation as that in which the carbonyl group eclipses the $C_1 - C_6$ bond (**303 P/2**) (*144*).

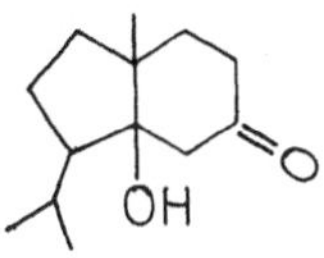

(300)

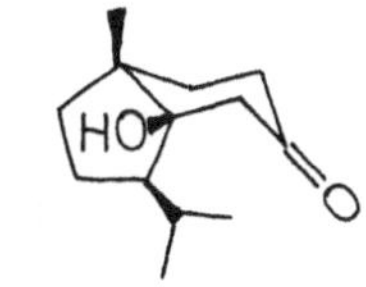

(301)

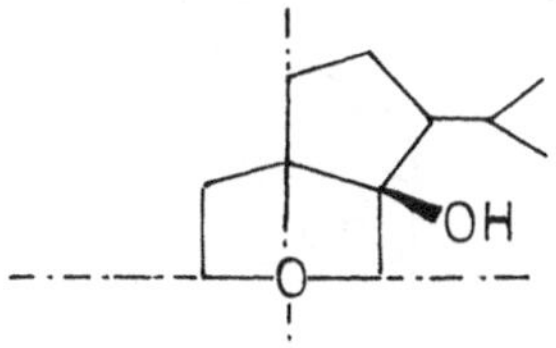

(301 P)

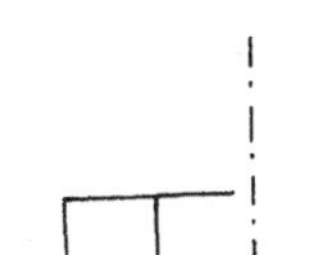

(302)

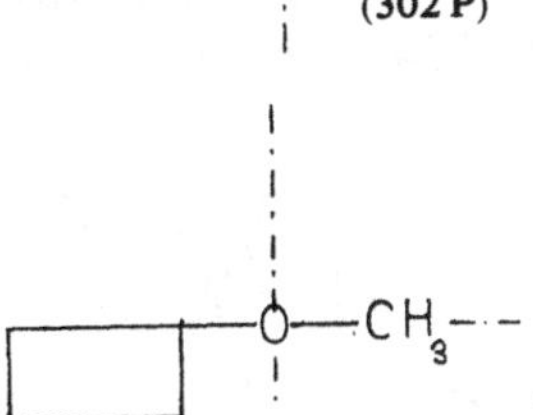

(302 P)

(303)

(303 P/1)

(303 P/2)

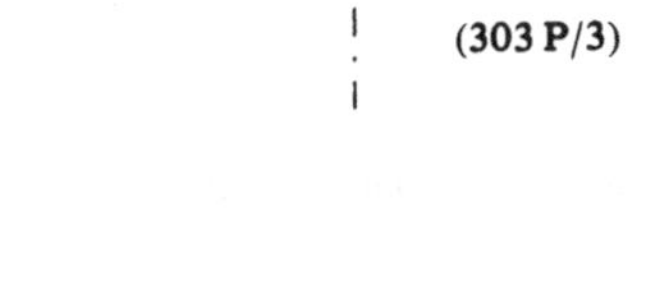

(303 P/3)

Unfortunately there are many cases in which uncertainties exist about both configuration and conformation in a particular compound, and a single measurement of a chiroptical property can rarely be used to establish both unknown facts. For example in a study of methyl angolensate, (**304**) both the configuration at C-1 and the conformation of ring A were in doubt and could *not* be established from a study of the ketone CE (*40*). The 16-oxosteroid (**305**) derived from the veratrum alkaloids shows a very small positive CE (*a* +4) which is different from either the strong negative CE of a 16-oxo 13β,14α- C/D *trans* compound (*a* ≈ −280) or the strong positive CE of the 13β,14β-*cis* isomer (*a* ≈ +140). In fact, an X-ray crystallographic analysis shows 13α,14α C/D *cis*-fusion, and simultaneous existence of both chair and boat conformers in the crystalline state is claimed (*148*).

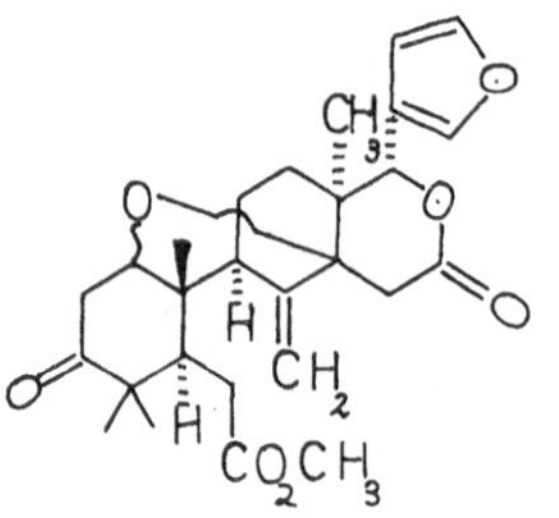

(304)

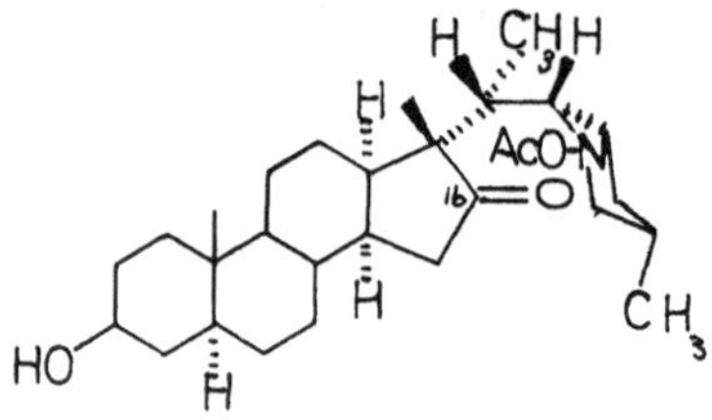

(305)

Additional information about conformational changes, particularly in flexible molecules, can sometimes be obtained by studies of c. d. curves at varying temperatures. Chiograsterone acetate (**306**) which shows two c.d. maxima, positive at 318 nm and negative at 288 nm has been studied at room temperature and at − 185°. No significant temperature dependence was observed for the negative maximum at 288 nm which was therefore attributed to that carbonyl group (at C-6) which is part of a rigid cyclic structure. Conversely the observed temperature dependence of the positive maximum at 318 nm was attributed to decreasing free rotation for the side chain C-24 carbonyl group at lower temperatures (*281*).

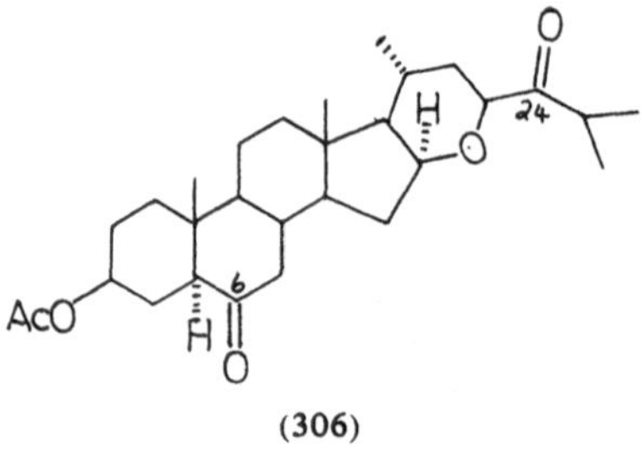

(306)

Occasionally, the sign of a CE is changed on passing from a polar to a non-polar solvent and for flexible molecules this may be due to a conformational change dependent on the change in solvent polarity. For example, (1R,4R)-(−)-4-hydroxymenthone which has a strong negative CE in isooctane, shows two CEs of opposite sign moderately strong negative and weak positive in methanol, and a small negative CE with a strong positive CE in dimethylsulphoxide. It has been suggested that this progression corresponds to the gradual change from intra-molecular hydrogen bonding in non-polar solvents [i.e. equatorial OH (**307**)], to intermolecular H-bonds in polar solvents [axial OH, (**308**)] (*312*). In contrast the diastereoisomeric (1R,4S)-(+)-4-hydroxyisomenthone which does *not* show inversion of the sign of the CE with changing solvent polarity is considered to exist preferentially in conformation (**309**) in all solvents.

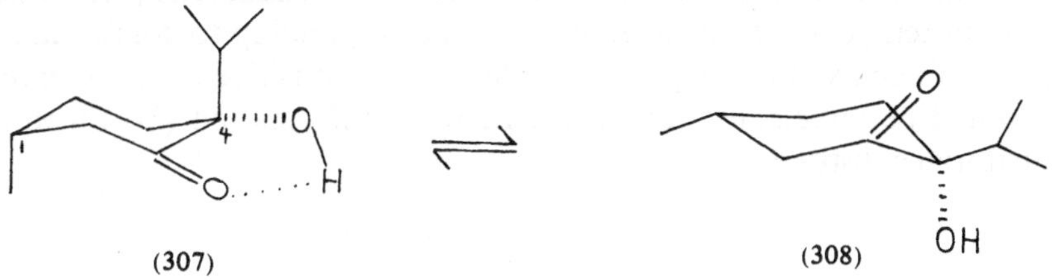

(**307**)   (**308**)

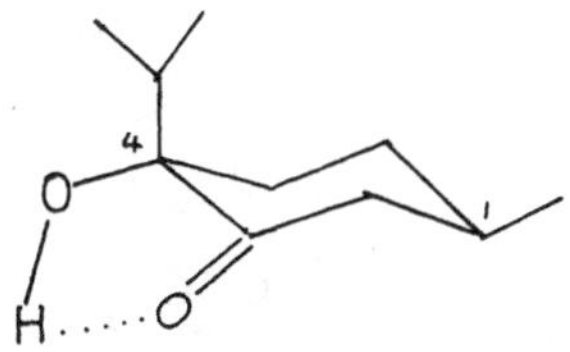

(**309**)

There are very few examples of the application of chiroptical measurements to study the conformation of αβ-unsaturated ketones or dienes. The unsaturated carbonyl chromophore in nimbin (**310**) gives rise to two CEs both negative, near 330 and 230 nm. This is consistent with ring A adopting a twisted conformation such that the 4β-methyl group is more nearly equatorial than axial (*252*). (+)-Mesembrenone (**311**) can adopt one of two conformations, for which CEs of opposite sign would be predicted. The observed negative CE supports conformation (**312**) and the magnitude of the CE, which is greater than the range of values found for 3-keto-Δ⁴-steroids, suggests that the eneone system has slightly greater deformation from coplanarity in mesembrenone than in typical steroid enones (*166*). The diene (**313**) is known from n.m.r. to adopt a non-steroid conformation and this is corroborated by the observed negative CE corresponding to skewing of the cisoid butadiene

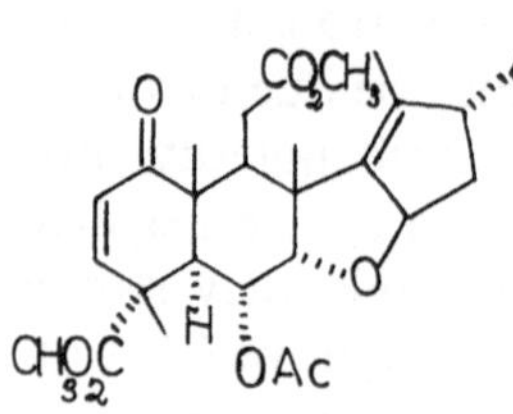

(310) 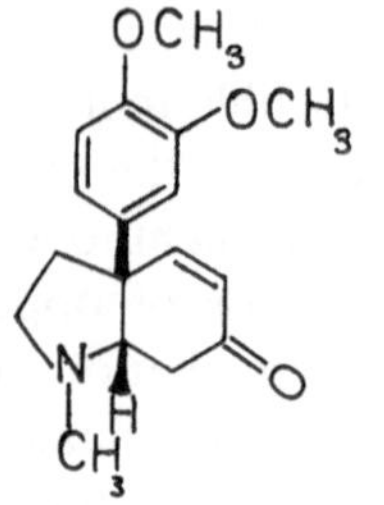   (311) 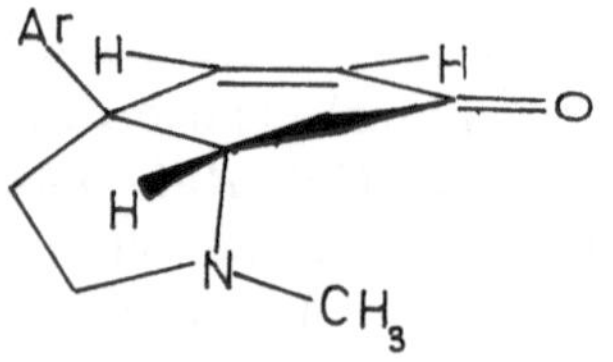   (312)

in a left-handed helix (*12*). Diene (**314**) shows a very intense negative CE, comparable to that of laevopimaric acid; this suggests that the two compounds have the same folded conformation at the B/C ring junction, as a result of interaction between the 11βH- and 10β-CH₃-groups, (*32*). Low temperature c.d. has been used to study α-phellandrene (**315**) which at room temperature has a positive CE corresponding to conformation (**316**) with an axial isopropyl group. At −185° a positive CE is observed, as would be anticipated for the alternative conformation (**317**) with an equatorial substituent (*302*).

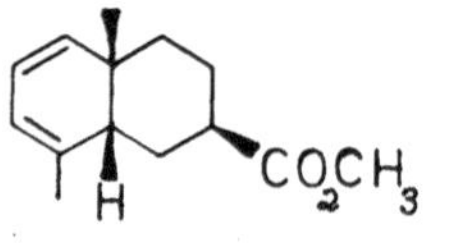

(313) 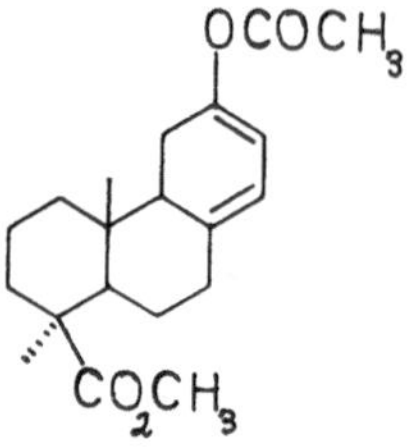   (314) 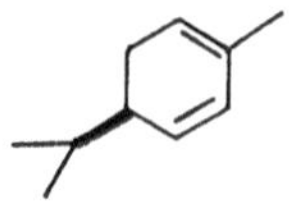   (315)

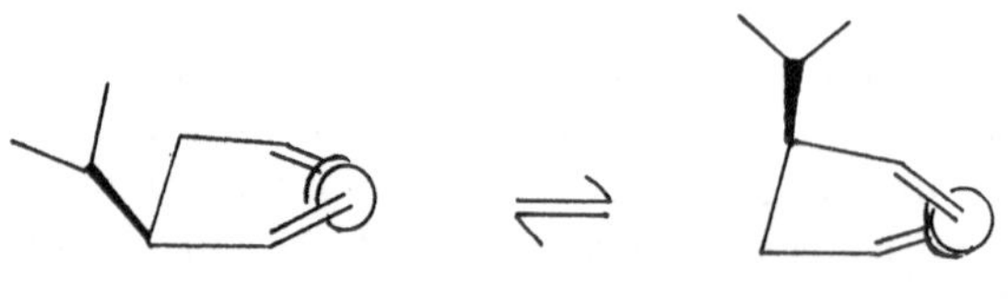

(316)   (317)

IV. Position of Functional Groups

O.r.d. or c.d. measurements can sometimes be used to establish the position of a functional group in a polycyclic molecule. In particular, carbonyl groups (or hydroxy groups which can be oxidised to the corresponding ketone) can often be located from the sign and magnitude of the carbonyl CE. The simplest examples are those in which a polycyclic skeleton (commonly steroid or terpene) has a single oxygen substituent in the ring and the sign and magnitude of the corresponding CE either establishes or confirms the position of the group.

The carbonyl group in ketone (**318**) of known 10α,5β configuration was placed at C-1 from consideration of its positive CE (*78*) and that in canophyllol (**319**) (also established 10α,5β) was placed at C-3 from its negative CE, superimposable on that of friedelin (*127*). The position of an OH-group in a new hydroxysandaracopimaric acid was established as C-6 from the CE of the corresponding ketone (**320**) (*316*), and carbonyl CEs have also been used to determine the position of the ketone group at C-6 in new spirostanes (**321**) (*274*) and in the *cis*-fused-A-nor-cholestanone (**322**) (*267*). The negative carbonyl CE, comparable to that observed for 21-oxoserratene derivatives, confirms the C-21 ketone in the triterpene (**323**) (*210*), in which the hindered nature of the carbonyl group is also demonstrated by negligible hemiketal formation in the presence of acid.

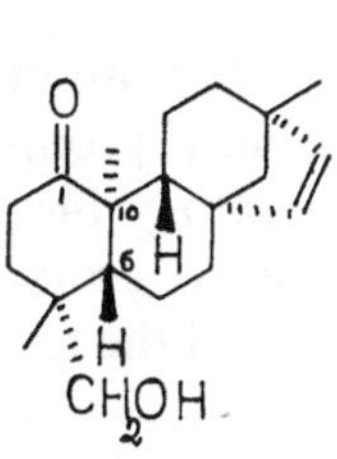

(318) 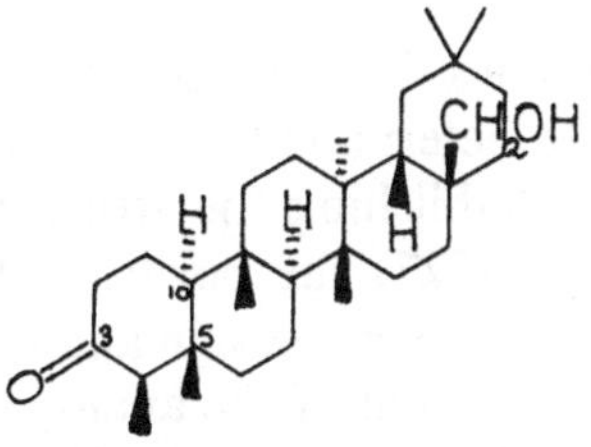   (319) 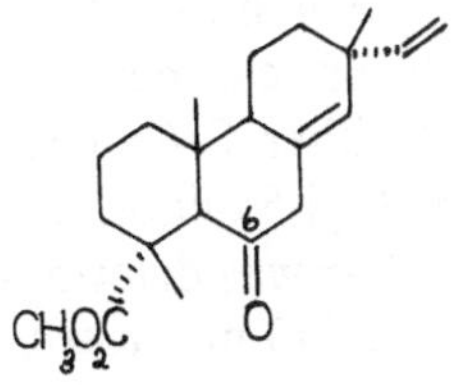   (320)

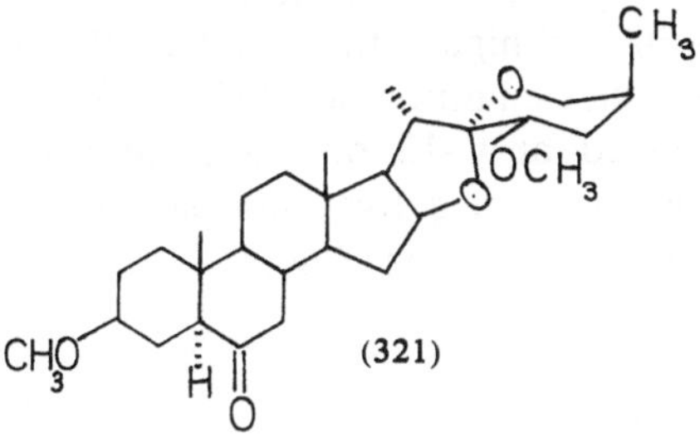

(321)

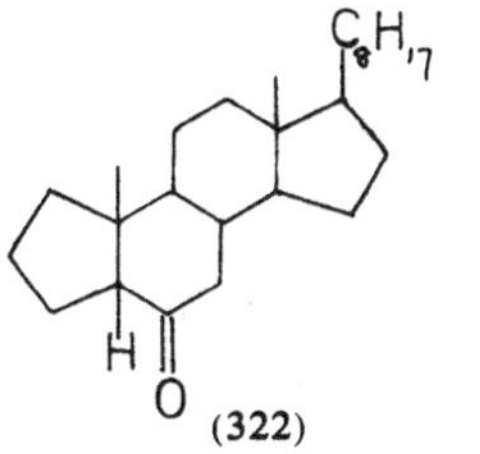

(322) 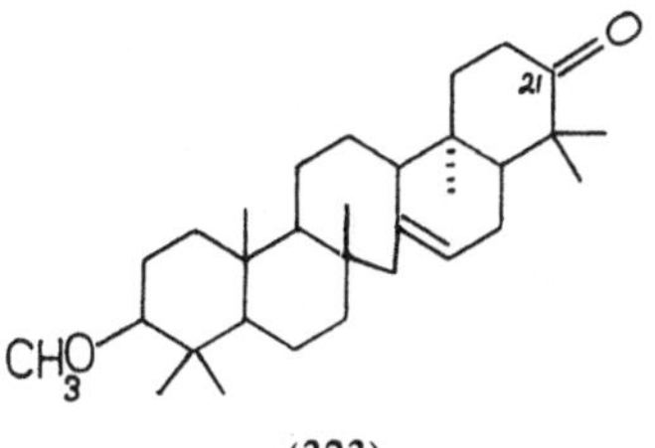   (323)

More complicated examples include compounds with two carbonyl (or potential carbonyl) groups. The dioxolactone (**324**) was partially reduced at C-3 and the negative CE of the resulting monoketone confirmed the position of the second carbonyl group at C-6 (*239*). The monoketone (**325**) also gave a negative CE similar to that of the known ketone (**326**) derived from panaxatriol, and different from that of the alternative C-12 ketone, thus establishing that the carbonyl group is at C-6 in (**325**) (*245*).

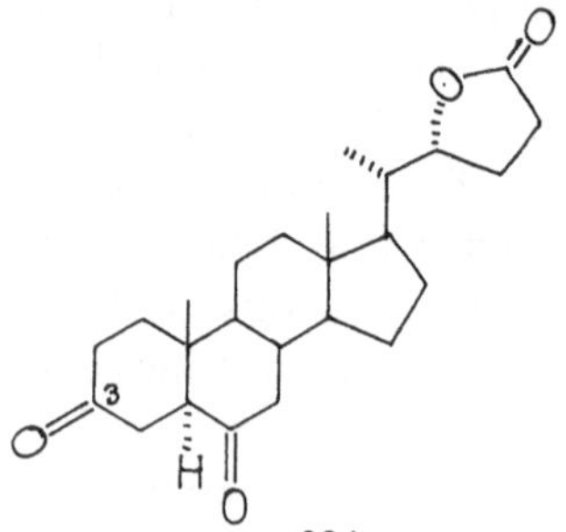

(324) 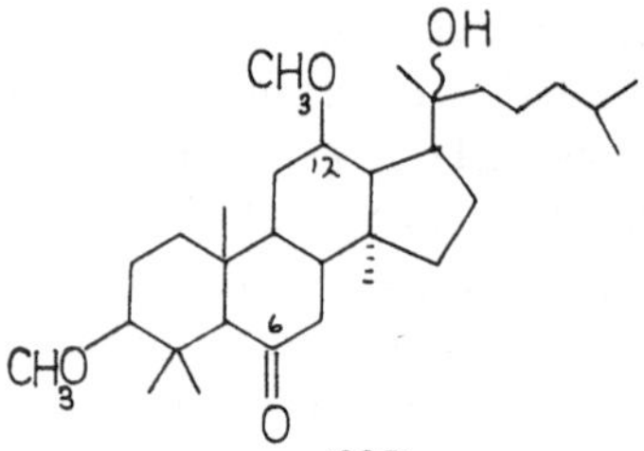   (325) 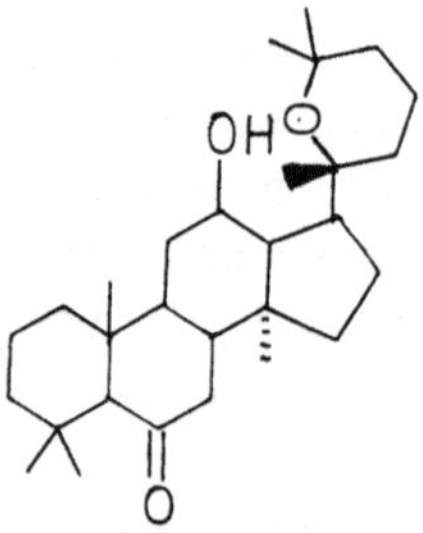   (326)

Among unsaturated ketones there are several examples of the use of chiroptical measurements to determine the position of a carbonyl group at either C-6 or C-12. In establishing the structure of 3β-acetoxy-14α-methyl-5α-cholest-7-en-6-one (**327**) the position of the carbonyl group at C-6 was determined from the positive sign and fine structure of the CE near 340 nm nearly identical to that of the analogous compound lacking a 14α-methyl substituent (*197*). Carpesterol (**328**) (*31*) and sengosterone (**329**) (*146*) also contain 7-en-6-one α,β-unsaturated ketones characterised by positive CEs, though the amplitude for the compound in the *trans* series (**328**) ($a \approx +360$) is higher than that of the *cis* compound (**329**) ($a \approx +107$). In the quasi-enantiomeric 9(11)-ene-12-oxo type, the keto group has been placed at C-12 of 3α-methoxylanost-9(11)-en-12-one (**330**) because of the strong negative multiple CE similar to that of 3β-acetoxy-18α-olean-9(11)-en-12-one (*329*).

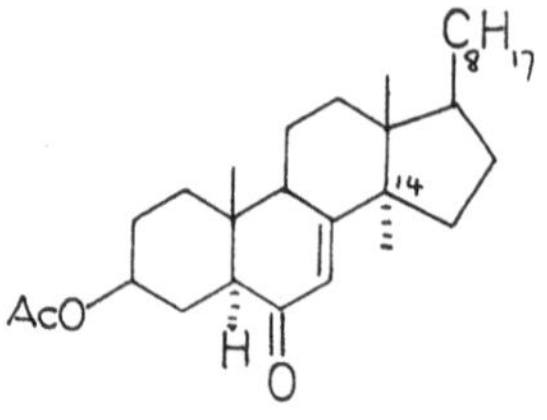

(327) 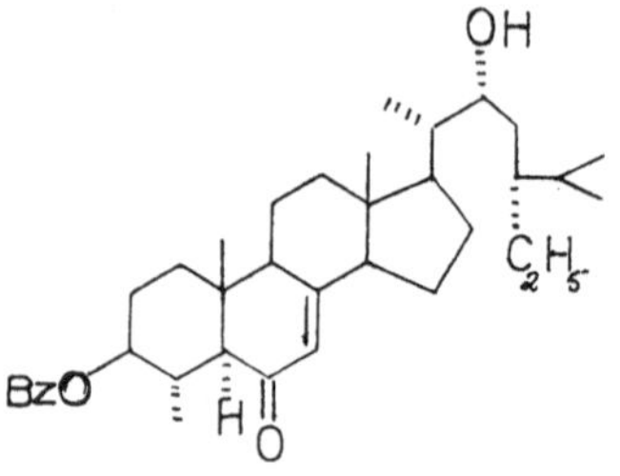   (328)

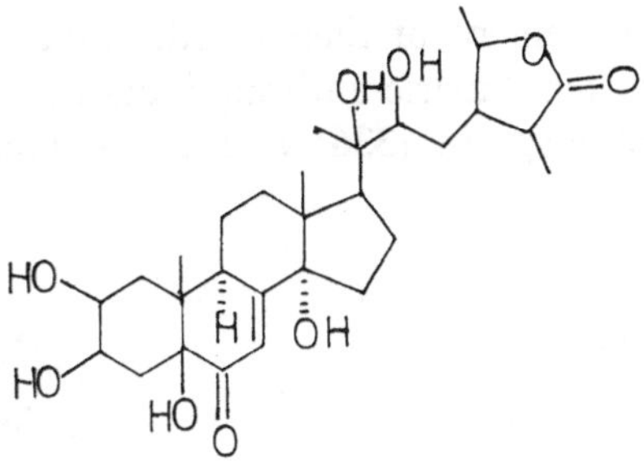

(329) 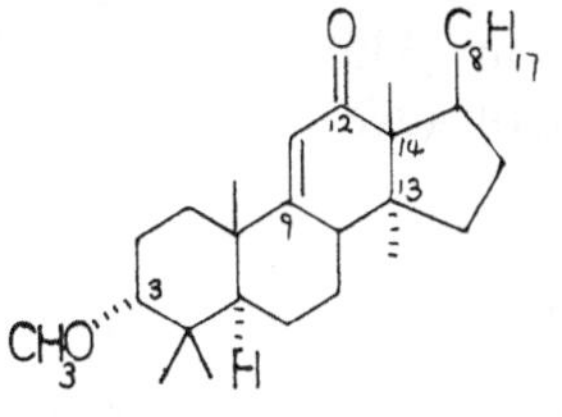　　　　　　　　　　　　　　(330)

The position of the double bond in ketone (331) has been established from measurement of the very small positive carbonyl CE characteristic of the 3-oxo-4,4-dimethyl-8(9)-ene structure (14); the C-5 epimer (5βH) gives a very small negative CE also in keeping with the structure shown. For macdougallin (332) the position of the 14α-methyl group was established by evidence which included the elimination of C-4 as a possible site. The CE for the 3-ketone from macdougallin was reduced to zero on addition of acid, thus proving that the 3-ketone was unsubstituted at neighbouring carbon atoms, and requiring the additional methyl group to be placed elsewhere (198).

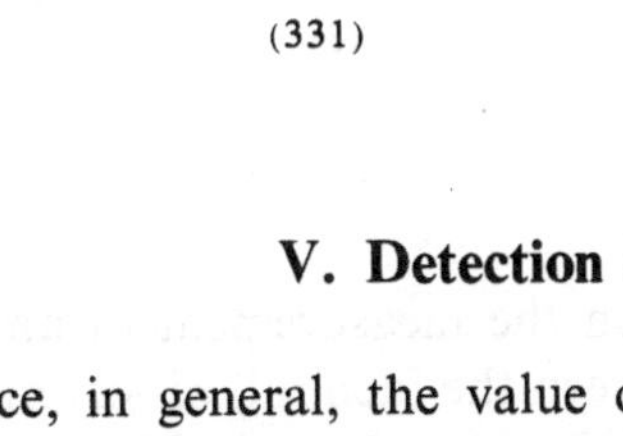

(331) 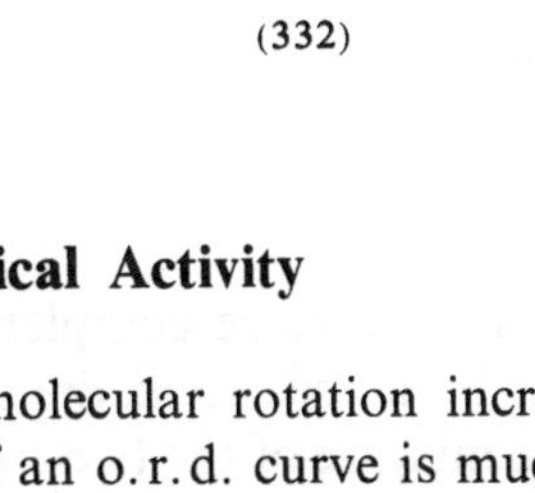　　　　　　　　　　　　　　(332)

V. Detection of Optical Activity

Since, in general, the value of the molecular rotation increases at shorter wavelengths, the measurement of an o.r.d. curve is much more sensitive for the detection of optical activity than a single measurement of monochromatic rotation at 589 nm. Consequently, o.r.d. curves have been used occasionally to prove that a molecule is chiral, particularly in cases where chirality arises from restricted rotation. For example, gossypol (333) which has a pronounced o.r.d. curve, was the first example of a naturally occurring ββ'-dinaphthyl derivative to be shown to be chiral (186). Similarly, the biflavanone (334) and related compounds

16*

were shown to be chiral by measurement of their o.r.d. curves (*159*); other examples include a series of ten dimeric anthraquinones including roseoskyrin (**335**) (*257*), and isodiospyrin (**336**) and the related bis-isodiospyrin (*348*).

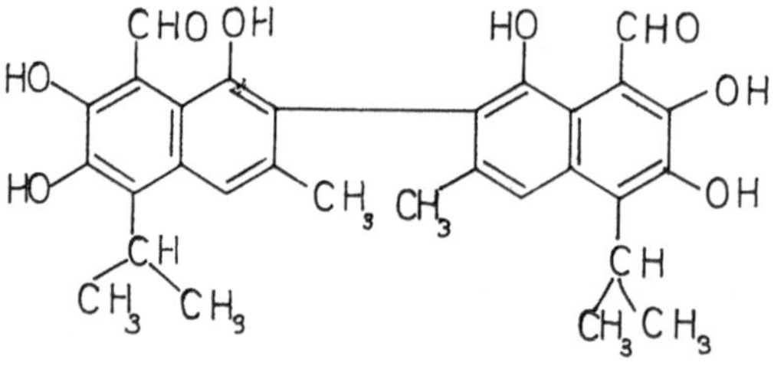

(333)

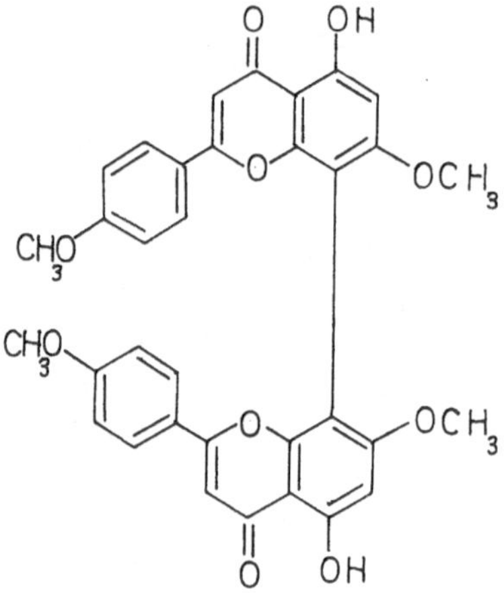

(334)

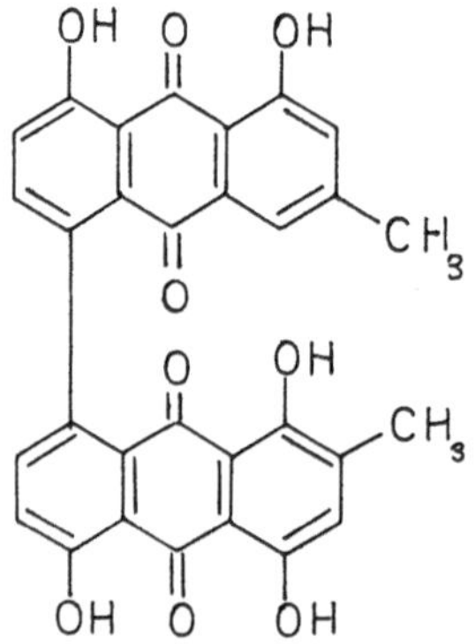

(335)

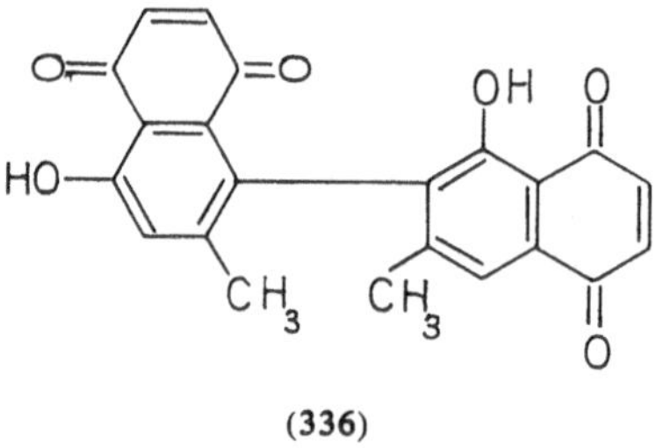

(336)

In a slightly more complex situation the measurement of an o.r.d. curve has been used to distinguish between the (achiral) *cis*-ozonide and a racemic mixture of the two *trans* ozonides of symmetrical 2,5-dimethyl-hex-3-ene. Each solution was treated separately with less than one molecular proportion of brucine. Since the brucine reacted preferentially with one or other of the two enantiomeric *trans* forms to give a precipitate, the remaining solution became enriched in the other enantiomer of the ozonide. After some time both solutions were filtered; that solution which gave an o.r.d. curve contained the *trans* isomer and that which was optically inactive the *cis* (*244*).

VI. Identity of Two Samples

Like other physical properties, o.r.d. or c.d. curves can be used as a characteristic of a particular compound for its identification and for comparison of synthetic and naturally occurring samples, or of naturally occurring samples from different plant or animal sources. This application of chiroptical measurements parallels the older use of specific rotation as a physical characteristic but the comparison is much more extensive because of the spectral range covered.

O.r.d. and c.d. curves have been used, together with other physical data to identify kopsanone (337) (115) from *Aspidosperma* species, dalbergione (338) (260) from kuhlmanniquinol, and ergosta-4,6,8(14),22-tetraen-3-one (339) (285) from "Fungus Lancis", by comparison with authentic samples. In work on the correlation of loganin and verbenalin, two samples of the degradation product (340) one from loganin and one from verbenalin, were shown to be identical by evidence which included o.r.d. curves (26). (+)-Eburnamenine (341) was shown to be one component of a new dimeric indole alkaloid umbellamin, by comparison of its mass spectrum, Rf values and o.r.d. curves with those of an authentic sample (240), and the alkaloid "goziline" was shown to have the same absolute configuration as the previously known 2′,3′-dihydroanhydro-vobtusin (342) by evidence which included the great similarity of the o.r.d. curves (7). Similarly, two carotenoids, named neoxanthin and foliaxanthin by independent groups of workers, have been shown to be identical (343) from a study of physical properties including o.r.d. curves (219). O.r.d. has also been used in characterisation of compounds with the eremophilane skeleton, e.g. (344) (72).

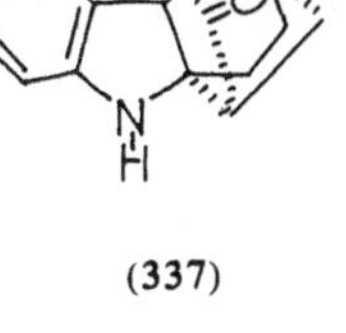

(337)

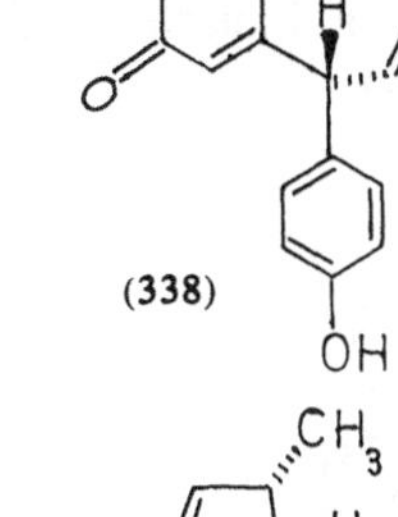

(338)

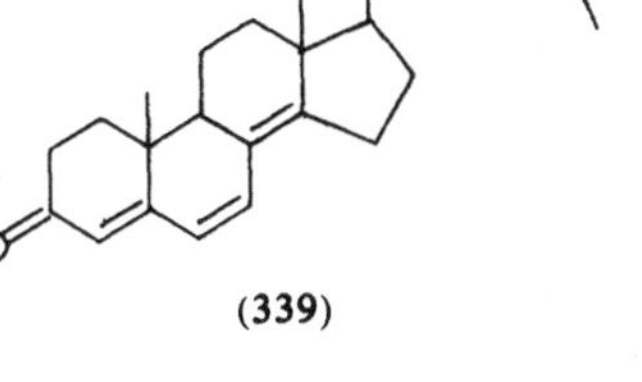

(339)

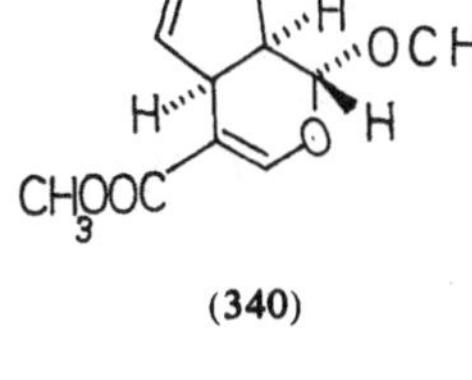

(340)

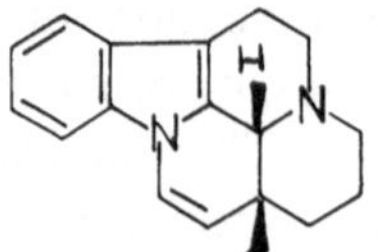

(341)

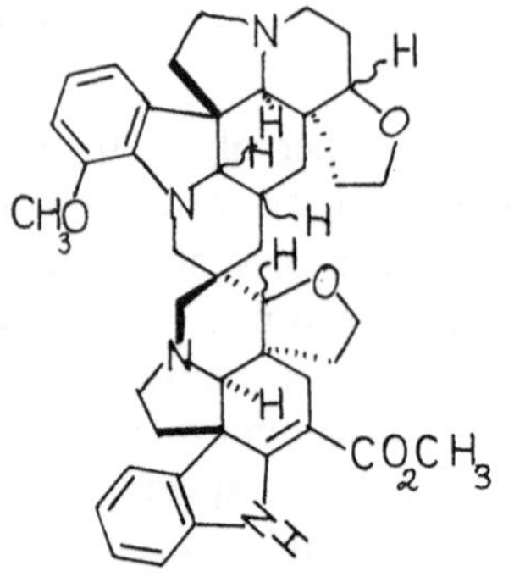

(342)

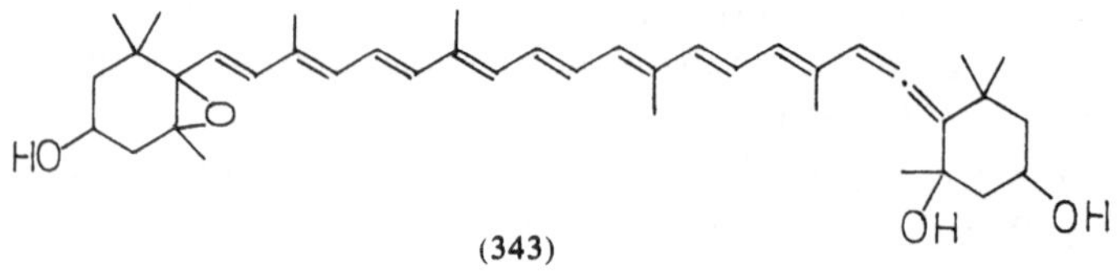

(343)

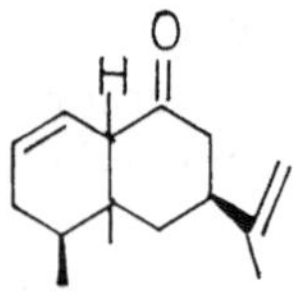

(344)

Comparison of the o.r.d. or c.d. curves of a naturally occurring compound with those of a synthetic sample has often been used, in conjunction with other physical evidence, to prove the identity of a synthetic end product. Examples include the alkaloids (−)-cherylline (**190**) (*56*) and (−)-petaline (**345**) (*129*), and the diterpene sandaracopimaradiene (**346**) (*113*). The stereochemistry of desacetylconfertifolin at C-1,5,6 and 10 has been established by its conversion to isodamsin (**347**) identical with an authentic sample (*117*) and synthetic circulin A has been shown to be identical to the natural peptide by evidence which included the similarity of the o.r.d. curves of the nickel complexes of synthetic and natural material (*309*).

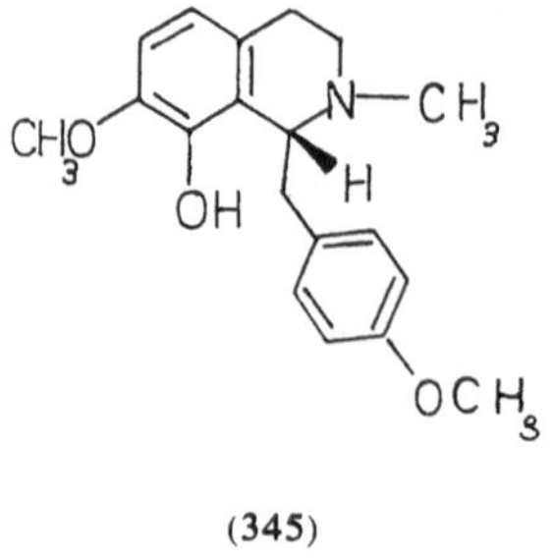

(345)

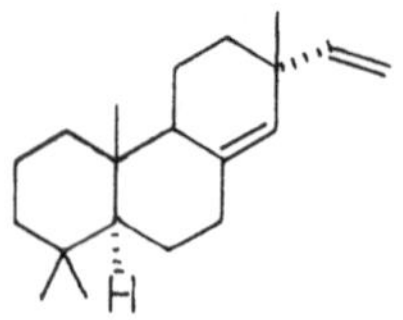

(346)

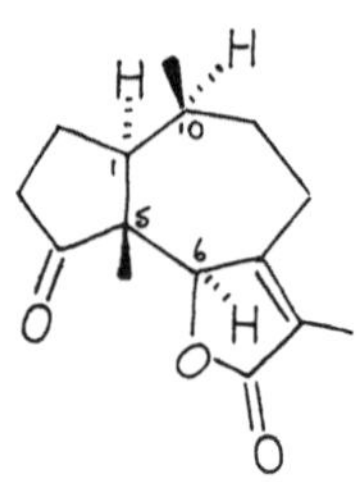

(347)

VII. Reaction Kinetics and Equilibria

Chiroptical measurements have had only a limited role in studies of reaction mechanisms, and applications in this field are comparatively rare. One example is the quantitative measurement of the magnitude of the CE to estimate the position of equilibrium between two compounds. This has been used particularly for steroid-6-ketones to study the equilibrium between the 5α- (348) and 5β- (349) isomers. Both epimers show negative CEs but the magnitude for the 5β-isomer is approximately three times that of the 5α-compound, and the magnitude of the CE of the

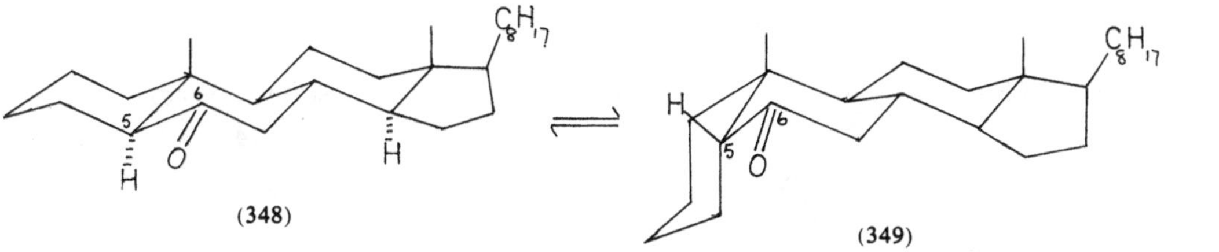

(348)   (349)

equilibrium mixture can be used to determine the proportion of each isomer present at equilibrium (43, 173). The effect on the equilibrium of substituents in ring A, has been studied for 3α-methyl- and 3α-chloro-cholestan-6-ones (172) for a series of androstan-6-ones with varying substituents at 17β (171) and for 2β,3β-dihydroxy-6-keto-steroids with various substituents at C-17 (331). O.r.d. curves have been used in exactly the same way to study the base catalysed equilibration of 14α- and 14β,15-keto steroids (330) and it has been suggested that chiroptical measurements could be applied to study the isomerisation of 2′-benzoyl and 3′-benzoyl nucleosides, by measurement of $\Delta\varepsilon$ at 230 nm (255).

O.r.d./c.d. measurements can also be used in principle to show whether a reaction proceeds with inversion or with retention of configuration. For example, in studying the reactions of (R)-benzyl-p-tolyl-sulphoxide (350), it was converted to the corresponding (R)-O-ethyl fluoroborate salt (351) by the action of triethyloxonium fluoroborate. This reaction was reversed by the action of base, but the product (352) was the enantiomer of the starting material as shown by the mirror image o.r.d. curves. Repeating the cycle of reactions (352)→(353)→(350) regenerated the starting material via a second inversion at the sulphur atom (169).

In some cases the criterion of racemisation can be used to test the validity of a set of mechanistic proposals. For example, the optically-active dicarboxylic acid (354, R=CO_2H) was decarboxylated by two different processes to the monocarboxylic acid (354, R=H). The product of a series of chemical transformations retained its chirality as shown by

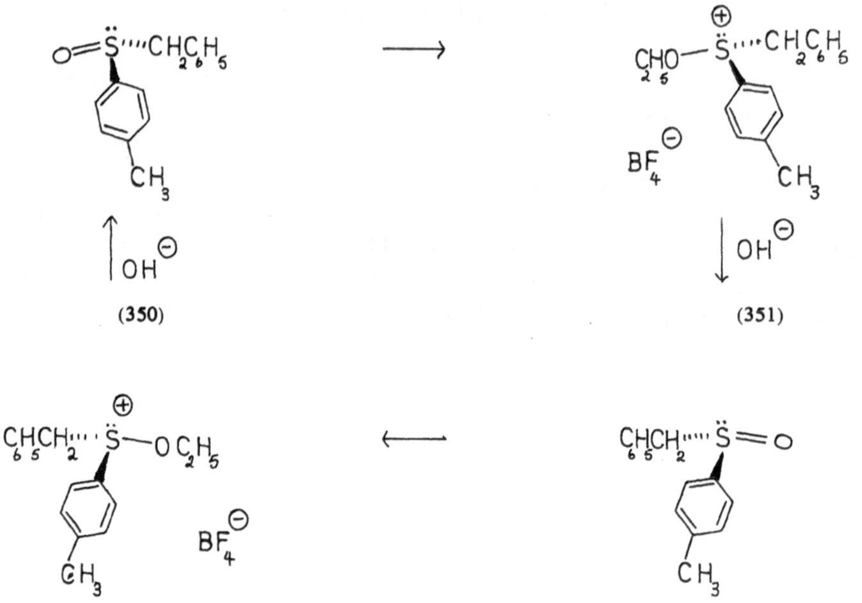

(350)

(351)

(353)

(352)

the positive carbonyl CE, but the corresponding product obtained by thermal decarboxylation was totally racemic, showing that the mechanism of the thermal reaction *cannot* involve a bridgehead double bond and therefore Bredt's rule is not infringed (*57*).

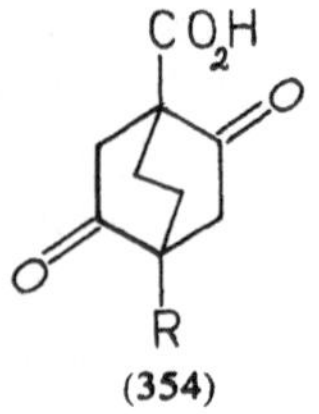

(354)

Finally, chiroptical measurements can be used very occasionally to monitor the progress of a reaction; an example is the reduction of nitro steroids to the corresponding oximino derivatives. The first extrema occur at 280 nm and 236 nm respectively for the nitro group and for the oxime and the course of the reaction can therefore be followed by periodic recording of an o.r.d. curve (*134*).

Acknowledgement

I am indebted to Professor W. Klyne, who read this chapter in preparation, for many helpful suggestions, also to Mrs. Beryl Tracey for her help in preparation of formulae. We are grateful to the Science Research Council for its generous support of our o.r.d./ c.d. work since 1962.

References

1. ABERHART, D. J., and K. H. OVERTON: An unusual polypropionate from the lichen *Roccella portentosa*. J. Chem. Soc. (C) (London) **1970**, 1612.

2. ACHENBACH, H., and G. WITTMANN: Dihydrokawain-5-ol. Ein neuer Alkohol aus Rauschpfeffer (*Piper methysticum* Forst.). Tetrahedron Letters **1970**, 3259.

3. ADAM, G., and M. HESSE: Strukturaufklärung eines C-28 Steroidlactons vom Witha-ferin-typ aus *Dunalia Australis* (Griseb) Sleum. Tetrahedron **28**, 3527 (1972).

4. ADAM, G., K. SCHREIBER, J. TOMKO, Z. VOTICKY, and A. VASSOVA: Veramine, a novel type of veratrum alkaloid with a 17β-methyl-18-nor-17-isospirosolane skeleton. Tetrahedron Letters **1968**, 2815.

5. ADINARAYANA, D., M. RADHAKRISHNIAH, J. RAJASEKHARA RAO, R. CAMPBELL, and L. CROMBIE: Dalpanol, a new 6'-hydroxyrotenoid from a *Dalbergia* species. J. Chem. Soc. (C) (London) **1971**, 29.

6. AGUILAR-SANTOS, G., E. SANTOS, and P. CRABBÉ: Stereochemistry of the alkaloid gitingensine. J. Organ. Chem. (USA) **32**, 2642 (1967).

7. AGWADA, V., M. B. PATEL, M. HESSE, and H. SCHMID: Die Alkaloide aus *Hedranthera barteri (Hook. f.) Pichon*. Helv. Chim. Acta **53**, 1567 (1970).

8. ALAIS, J., J. LEVISALLES, and I. TKATCHENKO: Stéréochimie XXX. Réarrangements benzilique et semi-benzilique. VI. Bull. soc. chim. France **1969**, 3189.

9. ALLARD, M., J. LEVISALLES, and H. RUDLER: Terpénoides IV. Carotol III. Corrélation entre la (−)dihydrocarvone et le (+)carotol; remarques sur l'acétolyse de tosylates cétoniques. Bull. soc. chim. France **1968**, 303.

10. ALLEN, F. H., J. P. KUTNEY, J. TROTTER, and N. D. WESTCOTT: The structures and absolute stereochemistry of cyclograndisolide and epicyclograndisolide, novel triter-pene lactones from *Abies grandis*. Tetrahedron Letters **1971**, 283.

11. ANDERSON, G. D., R. S. McEWEN, and W. HERZ: Relative and absolute configuration of axivalin and its congeners. Tetrahedron Letters **1972**, 4423.

12. ANDO, M., K. NANAUMI, T. NAKAGAWA, T. ASAO, and K. TAKASE: Syntheses of (−)-occidentalol and its C-7 epimer. Tetrahedron Letters **1970**, 3891.

13. ANIRUDHAN, C. A., D. W. MATHIESON, and W. B. WHALLEY: A novel rearrangement in the catechin series. J. Chem. Soc. (C) (London) **1966**, 634.

14. APLIN, R. T., H. R. ARTHUR, and W. H. HUI: The structure of the triterpene simi-arenol (an E:B-*friedo*hop-5-ene) from the Hong Kong species of *Rhododendron Simiarum*. J. Chem. Soc. (C) (London) **1966**, 1251.

15. ARAKAWA, H., N. TORIMOTO, and Y. MASUI: Die absolute Konfiguration des (−)β-Tetralols und des Agrimonolides. Tetrahedron Letters **1968**, 4115.

16. ASAKA, Y., T. KAMIKAWA, and T. KUBOTA: Total synthesis of (−)4-hydroxy-2-hydroxymethyl-3α-dimethylcyclopentane ethanol, the iridane part of jasminin. Tetrahedron Letters **1972**, 1597.

17. BADDELEY, G. V., R. A. EADE, J. ELLIS, P. HARPER, and J. J. H. SIMES: Oxidation by mercuric acetate in the lup-20(29)-ene and related series. Tetrahedron **25**, 1643 (1969).

18. BALASUBRAMANIAN, D., and D. B. WETLAUFER: Optical rotatory properties of diketo-piperazines. J. Amer. Chem. Soc. **88**, 3449 (1966).

19. BANERJEE, A. K., and M. GUT: 16-Oxasteroid. Synthesis and structural assignment. J. Organ. Chem. (USA) **34**, 1614 (1969).

20. BARBIERI, W., A. CONSONNI, and R. SCIAKY: Acid catalysed cyclisation of 21-benzy-lidene-16-dehydropregnenolone acetate. J. Organ. Chem. (USA) **33**, 3544 (1968).

21. BARKER, A. C., A. R. BATTERSBY, E. McDONALD, R. RAMAGE, and J. H. CLEMENTS: Biosynthesis of colchicine; ring expansion and later stages. Structure of speciosine. Chem. Commun. **1967**, 390.

22. BARRETT, G. C., J. F. W. McOMIE, S. NAKAJIMA, and S. W. TANENBAUM: Absolute configuration of pulvilloric acid. J. Chem. Soc. (C) (London) **1969**, 1068.

23. BARTLETT, L., W. KLYNE, W. P. MOSE, P. M. SCOPES, G. GALASKO, A. K. MALLAMS, B. C. L. WEEDON, J. SZABOLCS, and G. TÓTH: Optical rotatory dispersion of carotenoids. J. Chem. Soc. (C) (London) **1969**, 2527.

24. BASCOUL, J., and A. CRASTES DE PAULET: Synthèse partielle d'énantiomères de stéroides naturels. V. Bull. soc. chim. France **1969**, 189.

25. BATTERSBY, A. R., I. R. C. BICK, W. KLYNE, J. P. JENNINGS, P. M. SCOPES, and M. J. VERNENGO: Optical rotatory dispersion. Part XIV. Bisbenzyltetrahydroisoquinoline alkaloids. J. Chem. Soc. (London) **1965**, 2239.

26. BATTERSBY, A. R., E. S. HALL, and R. SOUTHGATE: Alkaloid biosynthesis Part XIII. The structure, stereochenistry and biosynthesis of loganin. J. Chem. Soc. (C) (London) **1969**, 721.

27. BATTERSBY, A. R., R. B. HERBERT, L. MO, and F. ŠANTAVÝ: 1-Phenethylisoquinoline alkaloids Part 1. Structure and synthesis of (−)melanthioidine a bisphenethylisoquinoline alkaloid. J. Chem. Soc. (C) (London) **1967**, 1739.

28. BATTERSBY, A. R., R. B. HERBERT, L. PIJEWSKA, F. ŠANTAVÝ, and P. SEDMERA: Alkaloid biosynthesis. Part XVII. The structure and chemistry of androcymbine. J. Chem. Soc. Perkin I (London) **1972**, 1736.

29. BEECHAM, A. F., and A. McL. MATHIESON: The circular dichroism of gliotoxin. Tetrahedron Letters **1966**, 3139.

30. BEECHAM, A. F., H. H. MILLS, F. B. WILSON, C. B. PAGE, and A. R. PINDER: The absolute configuration of dioscorine. Tetrahedron Letters **1969**, 3745.

31. BEISLER, J. A., and Y. SATO: The chemistry of carpesterol, a novel sterol from *Solanum xanthocarpum*. J. Organ. Chem. (USA) **36**, 3946 (1971).

32. BELL, R. A., and M. B. GRAVESTOCK: Enol acetylation of methyl 12-oxopodocarp-13-en-19-oate and methyl 12-oxopodocarp-8(14)-en-19-oate. J. Organ. Chem. (USA) **37**, 1065 (1972).

33. BENCZE, W. L., B. KISIS, R. T. PUCKETT, and N. FINCH: The absolute configuration of a hypolipidemic 1-aryltetralin, nafenopin. Tetrahedron **26**, 5407 (1970).

34. BENNETT, C. R., and R. C. CAMBIE: Chemistry of the Podocarpaceae XI. Reductions of totarol. Tetrahedron **22**, 2845 (1966).

35. BENNETT, C. R., and R. C. CAMBIE: Chemistry of the Podocarpaceae XII. Oxidation of *O*-methylpodocarpic acid with lead tetra-acetate. Tetrahedron **23**, 927 (1967).

36. BERNER, H., L. BERNER-FENZ, R. BINDER, W. GRAF, T. GRÜTTER, C. PASCUAL, and H. WEHRLI: Die Synthese von 5β*O*,19*N*-Ep(oxyethanoimino)-Steroiden. Helv. Chim. Acta **53**, 2252 (1970).

37. BERTI, G., F. BOTTARI, and A. MARSILI: Structure and stereochemistry of a triterpenoid epoxide from *Adiantum Capillus-Veneris*. Tetrahedron **25**, 2939 (1969).

38. BERTRAND, M., J. L. GRAS, and J. GORÉ: Stereochimie de l'addition du dimethylcetene et du cyclononadiene-1,2. Tetrahedron Letters **1972**, 1189.

39. BEVAN, C. W. L., D. E. U. EKONG, T. G. HALSALL, and P. TOFT: West African timbers. Part XX. The structure of turraeanthin an oxygenated tetracyclic triterpene monoacetate. J. Chem. Soc. (C) (London) **1967**, 820.

40. BEVAN, C. W. L., J. W. POWELL, D. A. H. TAYLOR, T. G. HALSALL, P. TOFT, and M. WELFORD: West African Timbers Part XIX. The structure of methyl angolensate, a ring-B-seco, tetranortetracyclic triterpene of the Meliacin family. J. Chem. Soc. (C) (London) **1967**, 163.

41. BHACCA, N. S., J. C. CRAIG, R. H. F. MANSKE, S. K. ROY, M. SHAMMA, and W. A. SLUSARCHYK: The configuration and conformation of cularine. Tetrahedron **22**, 1467 (1966).

42. BHATNAGAR, A. S., W. STÖCKLIN, and T. REICHSTEIN: Die Pregnanderivate der Samen von *Dregea abyssinica* (Hochst.) K. Schum *(Asclepiadaceae)* II. Strukturbestimmungen. Helv. Chim. Acta **51**, 133 (1968).

43. BIELLMANN, J. F., and W. S. JOHNSON: Détermination expérimentale de l'énergie-A: le group méthyle. Bull. soc. chim. France **1965**, 3500.

44. BINDER, R. G., and A. LEE: Hydroxymonoenoic acids of *Lesquerella densipila* seed oil. J. Organ. Chem. (USA) **31**, 1477 (1966).

45. BLACKSTOCK, W. P., R. T. BROWN, and G. K. LEE: Configuration at C-3 in vincoside. Chem. Commun. **1971**, 910.

46. BLÁHA, K., K. KAVKOVÁ, Z. KOBLICOVÁ, and J. TROJÁNEK: On alkaloids XX. Absolute configuration of alkaloides of the eburnane series; optical rotatory dispersion. Collect. Czech. Chem. Comm. **33**, 3833 (1968).

47. BLÁHA, K., Z. KOBLICOVÁ, J. POPÍŠEK, and J. TROJÁNEK: Absolute configuration of the alkaloid brevicolline. Collect. Czech. Chem. Comm. **36**, 3448 (1971).

48. BLÁHA, K., Z. KOBLICOVÁ, and J. TROJÁNEK: Absolute configuration of the iboga and voacanga alkaloids: chiroptical approach. Tetrahedron Letters **1972**, 2763.

49. BLUNT, J. W., J. M. COXON, M. P. HARTSHORN, and D. N. KIRK: Reactions of Epoxides-XIII. The backbone rearrangement of 3α-acetoxy-4α,5-epoxy-5α-cholestanes with Lewis acid. Tetrahedron **23**, 1811 (1967).

50. BLUNT, J. W., M. P. HARTSHORN, and D. N. KIRK: Reactions of epoxides XVII. Backbone rearrangements of cholest-5-ene and 5,6α-epoxy-5α-cholestane. Tetrahedron **25**, 149 (1969).

51. BONNETT, R., and J. G. DAVIS (Eds.): Some newer physical methods in Structural chemistry. London United Trade Press Ltd. 1967.

52. BOX, V. G. S., W. R. CHAN, and D. R. TAYLOR: The structure of rupestrol — a new sesquiterpenoid from *Verbesina rupestris (Urb.)* Blake. Tetrahedron Letters **1971**, 4371.

53. BRANDT, R., W. STÖCKLIN, and T. REICHSTEIN: Alloglaucotoxigenin, Strukturbestimmung. Helv. Chim. Acta **49**, 1662 (1966).

54. BROSSI, A., G. GRETHE, S. TEITEL, W. C. WILDMAN, and D. T. BAILEY: Cherylline, a 4-phenyl-1,2,3,4-tetrahydroisoquinoline alkaloid. J. Organ. Chem. (USA) **35**, 1100 (1970).

55. BROSSI, A., and S. TEITEL: Homoberbine, eine neue Klasse heterocyclischer Verbindungen. Helv. Chim. Acta **52**, 1228 (1969).

56. BROSSI, A., and S. TEITEL: The synthesis of cherylline. J. Organ. Chem. (USA) **35**, 3559 (1970).

57. BUCHANAN, G. L., N. B. KEAN, and R. TAYLOR: Bredt's Rule: an anomalous decarboxylation. J. Chem. Soc. (Chem. Commun.) (London) **1972**, 201.

58. BUCHECKER, R., and C. H. EUGSTER: Absolute Konfiguration von δ-carotin aus der Tomatenmutante Del/Del 65-3-54-5. Helv. Chim. Acta **54**, 327 (1971).

59. BUCHECKER, R., and C. H. EUGSTER: Absolute Konfiguration von α-Zeacarotin, α-Apo-8-carotinal und α-Apo-8-carotinol. Helv. Chim. Acta **56**, 1124 (1973).

60. BULL, J. R., and A. TUINMAN: Stereoselective synthesis of 19(10→9β) *abeo*-10α-testosterone. J. Chem. Soc. (Chem. Commun.) (London) **1972**, 921.

61. BURDEN, R. S., L. CROMBIE, and D. A. WHITING: The extractives of *Heliopsis scabra:* constitution of two new lignans. J. Chem. Soc. (C) (London) **1969**, 693.

62. BURDEN, R. S., and H. F. TAYLOR: The structure and chemical transformations of xanthoxin. Tetrahedron Letters **1970**, 4071.

63. BURROWS, E. P., D. L. DI PIETRO, and H. E. SMITH: 6α- and 6β-Hydroxyestriol. Synthesis, configurational assignments and spectral properties. J. Organ. Chem. (USA) **37**, 4000 (1972).

64. CALDWELL, D. J., and H. EYRING: The theory of optical activity. New York: Wiley-Interscience. 1971.

65. CAMERMAN, N., and J. TROTTER: The structure of cleavamine: X-ray analysis of cleavamine methiodide. Acta Crystallogr. **17**, 384 (1964).

66. CANDY, H. A., J. M. PAKSHONG, and K. H. PEGEL: Pimarane diterpenes from *Cleistanthus schlechteri.* J. Chem. Soc. (C) (London) **1970**, 2536.

67. CANONICA, L., B. DANIELI, I. WEICZ-VINCZE, and G. FERRARI: Structure of muristerone-A, a new photoecdysone. J. Chem. Soc. (Chem. Commun.) (London) **1972**, 1060.

68. CARDILLO, G., L. MERLINI, G. NASINI, and P. SALVADORI: Constituents of dragon's blood. Part 1. Structure and absolute configuration of new optically active flavans. J. Chem. Soc. (C) (London) 1971, 3967.

69. ČERVINKA, O., and L. HUB: Asymmetric Reactions XXVIII. Absolute configurations of some necic acids. Collect. Czech. Chem. Comm. 33, 2933 (1968).

70. CHAN, W. R., D. R. TAYLOR, C. R. WILLIS, and H. W. FEHLHABER: The structure of neoandrographolide — a diterpene glucoside from Andrographis Paniculata Nees. Tetrahedron Letters 1968, 4803.

71. CHETTY, G. L., G. S. KRISHNA RAO, SUKH DEV, and D. K. BANERJEE: Studies in Sesquiterpenes XXV. A synthesis of α-selinine. Tetrahedron 22, 2311 (1966).

72. CHETTY, G. L., L. H. ZALKOW, and R. A. MASSY-WESTROPP: 7a(H)-Eremophila-1,11-dien-9-one. A new sesquiterpene of the eremophilane type. Tetrahedron Letters 1969, 307.

73. CHIKAMATSU, H., M. MAEDA, and M. NAKAZAKI: Structure of torilin. Tetrahedron 25, 4751 (1969).

74. CIARDELLI, F., and P. SALVADORI (Eds.): Fundamental aspects and recent developments in optical rotatory dispersion and circular dichroism. London: Heyden. 1973.

75. CLARKE, R. L., S. J. DAUM, P. E. SHAW, and R. K. KULLNIG: The structure of cassaic acid. J. Amer. Chem. Soc. 88, 5865 (1966).

76. CLARK-LEWIS, J. W., I. DAINIS, and G. C. RAMSAY: Flavan derivatives XIV. The absolute configurations of some 1,2-diarylpropane derivatives and of some isoflavans. Austral. J. Chem. 18, 1035 (1965).

77. CONNOLLY, J. D., K. L. HANDA, R. MCCRINDLE, and K. H. OVERTON: Tetranortriterpenoids and related substances. Part XI. Odoratol and its congeners from Cedrela glaziovii. J. Chem. Soc. (C) (London) 1968, 2230.

78. CONNOLLY, J. D., and A. E. HARDING: Constituents of erythroxylon species Part VII. Diterpenoids from Erythroxylon australe. J. Chem. Soc. Perkin I (London) 1972, 1996.

79. CONNOLLY, J. D., and R. MCCRINDLE: Tetranortriterpenoids and related substances. Part XIII. The constitution of grandifoliolenone, an apo-tirucallol derivative from Khaya grandifoliola (Meliaceae). J. Chem. Soc. (C) (London) 1971, 1715.

80. CORNFORTH, J. W., W. DRABER, B. V. MILBORROW, and G. RYBACK: Absolute stereochemistry of (+)-Abscisin-II. Chem. Commun. 1967, 114.

81. COSULICH, D. B., N. R. NELSON, and J. H. VAN DEN HENDE: Crystal and molecular structure of LL-S88α, an antiviral epidithiapiperazinedione derivative from Aspergillus terreus. J. Amer. Chem. Soc. 90, 6519 (1968).

82. CRABBÉ, P.: Optical rotatory dispersion and circular dichroism in organic chemistry. San Francisco: Holden-Day. 1965.

83. CRABBÉ, P.: Applications de la dispersion rotatoire optique et du dichroisme circulaire optique en chimie organique. Paris: Gauthier-Villars. 1968.

84. CRABBÉ, P., A. CRUZ, and J. IRIARTE: Synthesis of enantiomeric steroids from oestrone. Chem. and Ind. 1967, 1522.

85. CRAGG, G. M. L.: Cycloaddition of dichloroketen to some steroid olefins. J. Chem. Soc. (C) (London) 1970, 1829.

86. CRAIG, J. C., S. K. ROY, R. G. POWELL, and C. R. SMITH: Optical rotatory dispersion and circular dichroism VI Structure and absolute configuration of helenynolic acid. J. Organ. Chem. (USA) 30, 4342 (1965).

87. CRAIG, J. C., M. M. SMITH, S. K. ROY, and J. B. STENLAKE: Optical rotatory dispersion and absolute configuration VIII. Petaline and other benzyltetrahydroisoquinoline alkaloids. Tetrahedron 22, 1335 (1966).

88. CROMBIE, L., and P. A. FIRTH: Biosynthesis of parasorbic acid (hex-2-en-5-olide) by the rowan berry (Sorbus aucuparia L.). J. Chem. Soc. (C) (London) 1968, 2852.

89. DAHN, H., J. A. GARBARINO, and C. O'MURCHU: Über die Stereochemie der Hydrogenolyse von N-Benzylbindungen 1. Die Hydrogenolyse von Derivaten der 2-Amino-2-phenylpropionsäure. Helv. Chim. Acta 53, 1370 (1970).

90. DAVIES, A. R., and G. H. R. SUMMERS: 5α,7α-Cyclosteroids. J. Chem. Soc. (C) (London) 1967, 909.

91. DE KEUKELEIRE, D., and M. VERZELE: The structure and the absolute configuration of (−)humulone. Tetrahedron 26, 385 (1970).

92. DJERASSI, C.: Optical rotatory dispersion. New York: McGraw-Hill. 1960.

93. DJERASSI, C., B. GREEN, W. B. WHALLEY, and C. G. DE GRAZIA: The chemistry of fungi Part LIII. Optical rotatory dispersion studies. Part CV. The absolute configuration of rosenono- and rosololactones. J. Chem. Soc. (C) (London) 1966, 624.

94. DJERASSI, C., and D. MARSHALL: Optical rotatory dispersion studies — XII. Absolute configurations of eperuic and labdanolic acids. Tetrahedron 1, 238 (1957).

95. DJERASSI, C., J. OSIECKI, and W. HERZ: Optical rotatory dispersion studies XIII. Assignment of absolute configuration to certain members of the guaianolide series of sesquiterpenes. J. Organ. Chem. (USA) 22, 1361 (1957).

96. DJERASSI, C., K. UNDHEIM, R. C. SHEPPARD, W. G. TERRY, and B. SJÖBERG: Optical Rotatory Dispersion studies XLVI. Stereochemical assignments through anomalous rotatory dispersion curves of α-amino acid derivatives. 3-Phenyl-2-thiohydantoins and N-thionocarbethoxy amino acids. Acta Chem. Scand. 15, 903 (1961).

97. DONNELLY, D. M. X., B. J. NANGLE, P. B. HULBERT, W. KLYNE, and R. J. SWAN: Optical rotatory dispersion Part XLVIII. Some 3,3-diarylprop-1-enes and related quinones: the absolute configuration of latifolin. J. Chem. Soc. (C) (London) 1967, 2450.

98. DOSKOTCH, R. W., and C. D. HUFFORD: The structure of damsinic acid, a new sesquiterpene from Ambrosia ambrosioides (Cav.) Payne. J. Organ. Chem. (USA) 35, 486 (1970).

99. DREWES, S. E., and D. G. ROUX: Stereochemistry and biogenesis of mopanols and peltogynols and associated flavanoids from Colophospermum mopane. J. Chem. Soc. (C) (London) 1966, 1644.

100. DREYER, D. L.: Citrus bitter principles VIII. Application of o.r.d. and c.d. to stereochemical problems. Tetrahedron 24, 3273 (1968).

101. EDWARD, J. T., and N. E. LAWSON: The conversion of cholesterol into 10α-cholesterol. J. Organ. Chem. (USA) 35, 1426 (1970).

102. EDWARDS, J. A., J. SUNDEEN, W. SALMOND, T. IWADARE, and J. H. FRIED: A new synthetic route to the fungal sex hormone antheridiol and the determination of its absolute stereochemistry. Tetrahedron Letters 1972, 791.

103. ELBER, R., EK. WEISS, and T. REICHSTEIN: Cardenolide und Pregnanderivate aus den Wurzeln von Trachycalymma fimbriatum (Weimarck) (Bullock). Helv. Chim. Acta 52, 2583 (1969).

104. ELLESTAD, G. A., R. H. EVANS, and M. P. KUNSTMANN: Some new terpenoid metabolites from an unidentified Fusarium species. Tetrahedron 25, 1323 (1969).

105. ELLESTAD, G. A., R. H. EVANS, M. P. KUNSTMANN, J. E. LANCASTER, and G. O. MORTON: Structure and chemistry of antibiotic LL-Z1271α an antifungal carbon-17 terpene. J. Amer. Chem. Soc. 92, 5483 (1970).

106. EL-OLEMY, M. M., and A. E. SCHWARTING: The resolution and absolute configuration of the racemic isomer of anaferine. J. Organ. Chem. (USA) 34, 1352 (1969).

107. ENZELL, C. R., Y. HIROSE, and B. R. THOMAS: The chemistry of the order Araucariales-6. Absolute configurations of agatharesinol, hinokiresinol and sugiresinol. Tetrahedron Letters 1967, 793.

108. EYRE, D. H., J. W. HARRISON, and B. LYTHGOE: Taxine Part VI. The stereochemistry of taxicin-I and taxicin-II. J. Chem. Soc. (C) (London) 1967, 452.

109. FARINA, M., and G. AUDISIO: Stereochemistry of perhydrotriphenylene-II. Absolute rotation and configuration of optically active anti-trans-anti-trans-anti-trans-perhydrotriphenylene. Tetrahedron 26, 1839 (1970).

110. FARKAŠ, E., J. M. OWEN, and D. J. O'TOOLE: The preparation and chemistry of 9β-estra-4-en-3-ones. J. Organ. Chem. (USA) 34, 3022 (1969).

111. Farkaš, J., Z. Flegelová, and F. Šorm: Synthesis of pyrazomycin. Tetrahedron Letters **1972**, 2279.

112. Ferris, J. P., C. B. Boyce, R. C. Briner, U. Weiss, I. H. Qureshi, and N. E. Sharpless: Lythraceae alkaloids. X. Assignments of absolute stereochemistries on the basis of chiroptical effects. J. Amer. Chem. Soc. **93**, 2963 (1971).

113. Fétizon, M., and M. Golfier: Corrélation entre stéroides et diterpènes du groupe pimarique III. Synthèse du sandaracopimaradiène. Bull. soc. chim. France **1966**, 870.

114. Fétizon, M., and J.-C. Gramain: Synthèses de céto-2-stéroides. Bull. soc. chim. France **1968**, 3301.

115. Filho, J. M. F., B. Gilbert, M. Kitagawa, L. A. Paes Leme, and L. J. Durham: Four heptacyclic alkaloids from *Aspidosperma* species. J. Chem. Soc. (C) (London) **1966**, 1260.

116. Fischer, N., and A. S. Dreiding: Preparation of a 2-methylcyclodopa derivative. Helv. Chim. Acta **53**, 1937 (1970).

117. Fischer, N. H., and T. J. Mabry: New pseudoguaianolides from *Ambrosia Confertiflora* DC. (Compositae). Tetrahedron **23**, 2529 (1967).

118. Fowden, L., P. M. Scopes, and R. N. Thomas: Optical rotatory dispersion and circular dichroism. Part LXX. The circular dichroism of some less common amino-acids. J. Chem. Soc. (C) (London) **1971**, 833.

119. Frappier, F., M. Païs, and F.-X. Jarreau: Alcaloides stéroidiques. CXXVII (18) Isomérisation spinale des aminostéroides IX (19) Bull. soc. chim. France **1972**, 610.

120. Fredga, A., J. P. Jennings, W. Klyne, P. M. Scopes, B. Sjöberg, and S. Sjöberg: Optical rotatory dispersion Part XV. Monosubstituted succinic acids. J. Chem. Soc. (London) **1965**, 3928.

121. Fulmor, W., G. E. van Lear, G. O. Morton, and R. D. Mills: Isolation and absolute configuration of the aeroplysinin I. Enantiomorphic pair from *Ianthella Ardis*. Tetrahedron Letters **1970**, 4551.

122. Gaffield, W.: Circular dichroism, optical rotatory dispersion and absolute configuration of flavanones, 3-hydroxyflavanones and their glycosides. Determination of aglycone chirality in flavanone glycosides. Tetrahedron **26**, 4093 (1970).

123. Gaffield, W., and A. C. Waiss: Optical rotatory dispersion and absolute configuration of flavanones, 3-hydroxyflavanones and their glycosides. Chem. Commun. **1968**, 29.

124. Gallina, C., C. Marta, C. Colombo, and A. Romeo: Capreomycidine and 3-guanidinoproline from viomycidine. Synthesis of *cis*- and *trans*-3-aminoprolines. Tetrahedron **27**, 4681 (1971).

125. Goodfellow, D., G. P. Moss, and B. C. L. Weedon: The absolute configuration of lutein. Chem. Commun. **1970**, 1578.

126. Gorman, A. A., V. Agwada, M. Hesse, U. Renner, and H. Schmid: Zur Chemie des Beninins und des Vobtusins. Helv. Chim. Acta **49**, 2072 (1966).

127. Govindachari, T. R., N. Viswanathan, B. R. Pai, U. R. Rao, and M. Srinivasan: Triterpenes of *Calophyllum Inophyllum* Linn. Tetrahedron **23**, 1901 (1967).

128. Green, D. M., J. A. Edwards, A. W. Barksdale, and T. C. McMorris: The isolation and structure of 23-deoxyantheridiol and the synthesis of its C-22-epimer. Tetrahedron **27**, 1199 (1971).

129. Grethe, G., M. Uskoković, and A. Brossi: Total synthesis of petaline. J. Organ. Chem. (USA) **33**, 2500 (1968).

130. Grove, J. F.: New metabolic products of *Aspergillus flavus* Part 1. Asperentin, its methyl ethers and 5'-hydroxyasperentin. J. Chem. Soc. Perkin I (London) 1972, 2400.

131. Guggisberg, A., A. A. Gorman, B. W. Bycroft, and H. Schmid: Ein neuer Abbau des Indolalkaloids Kopsin; chemische Korrelierung der Alkaloide von Kopsin und Pleiocarpin-Typ mit Minovincin. Helv. Chim. Acta **52**, 76 (1969).

132. Habaguchi, K., M. Watanabe, Y. Nakadaira, K. Nakanishi, A. K. Kiang, and F. Y. Lim: The full structures of lansic acid and its minor congener, an unsymmetric onoceradienedione. Tetrahedron Letters **1968**, 3731.

133. HANAOKA, M., M. HESSE, and H. SCHMID: Na-Demethylseredamin, ein neues Alkaloid aus *Rauwolfia sumatrana;* absolute Konfiguration von Seredamin. Helv. Chim. Acta 53, 1723 (1970).

134. HANSON, J. R., and E. PREMUZIC: Applications of chromous chloride II. The reduction of some steroidal nitro compounds. Tetrahedron 23, 4105 (1967).

135. HARADA, N.: Absolute configuration of (+)-*trans*-abscisic acid as determined by a quantitative application of the exciton chirality method. J. Amer. Chem. Soc. 95, 240 (1973).

136. HARADA, N., and K. NAKANISHI: The exciton chirality method and its application to configurational and conformational studies of natural products. Accts. Chem. Res. 5, 257 (1972).

137. HATAM, N. A. R., and D. A. WHITING: The constituents of Californian redwood: the constitution, absolute stereochemistry and chemistry of sequirin-B and sequirin-C. J. Chem. Soc. (C) (London) 1969, 1921.

138. HAUSER, D., H. P. WEBER, and H. P. SIGG: Isolierung und Strukturaufklärung von Chaetocin. Helv. Chim. Acta 53, 1061 (1970).

139. HECKENDORN, R., and CH. TAMM: Synthese von 5-Methyl-3-oxa-A-nor-5β-cholestan und 3-Oxa-Δ⁴-Cholesten-2-on. Helv. Chim. Acta 50, 1499 (1967).

140. HECKENDORN, R., and CH. TAMM: Synthese der 2,5-Diole des A-nor-5α-Cholestans und A-nor-5β-Cholestans. Helv. Chim. Acta 50, 1964 (1967).

141. HENSON, P. D., and K. MISLOW: Optical rotatory dispersion and circular dichroism of diastereoisomeric S-allyl-L-cysteine S-oxides. Chem. Commun. 1969, 413.

142. HERZ, W., K. AOTA, M. HOLUB, and Z. SAMEK: Sesquiterpene lactones and lactone glycosides from *Hymenoxys* species. J. Organ. Chem. (USA) 35, 2611 (1970).

143. HERZ, W., and J. J. SCHMID: Resin Acids XVI. Some transformations of methyl 12α-hydroxy-13β-abiet-8(9)en-18-oate. J. Organ. Chem. (USA) 34, 2775 (1969).

144. HEYMES, A., M. DVOLAITZKY, and J. JAQUES: Conformation et réactivité des sites exocycliques. Sur la stéréochimie de la réduction par les hydrures des méthyl-2 acétylcyclohexanes *cis* et *trans*. Bull. soc. chim. France 1968, 2898.

145. HIKINO, H., K. AOTA, D. KUWANO, and T. TAKEMOTO: Structure of α-rotunol and β-rotunol. Tetrahedron Letters 1969, 2741.

146. HIKINO, H., K. NOMOTO, and T. TAKEMOTO: Sengosterone, an insect metamorphosing substance from *Cyathula capitata:* Structure. Tetrahedron 26, 887 (1970).

147. HIKINO, H., K. NOMOTO, and T. TAKEMOTO: Cyasterone, an insect metamorphosing substance from *Cyathula capitata:* absolute configuration. Tetrahedron 27, 315 (1971).

148. HÖHNE, E., I. SEIDEL, G. ADAM, K. SCHREIBER, and J. TOMKO: Stereochemie von Veratrum Alkaloiden. Ord und Röntgenstrukturanalytische Beweise für eine C-BBotkonformation in Tetrahydroveralkalminderivaten. Tetrahedron 28, 409 (1972).

149. HOLÝ, A.: Nucleic acid components and their analogues CXLVII. Preparation of 5-ethoxycarbonyluridine, 5-carboxyuridine and their nucleotidic derivatives. Collect. Czech. Chem. Comm. 37, 1555 (1972).

150. HOLZAPFEL, C. W., R. D. HUTCHINSON, and D. C. WILKINS: The isolation and structure of two new indole derivatives from *Penicillium Cyclopium* Westling. Tetrahedron 26, 5239 (1970).

151. HORA, J.: On Steroids CXX. Insect chemosterilants derived from 21-hydroxypregnane-6,20-dione. Collect. Czech. Chem. Comm. 34, 344 (1969).

152. HORI, T., A. K. KIANG, K. NAKANISHI, S. SASAKI, and M. C. WOODS: The structures of fibraurin and a minor product from *Fibraurea chloroleuca*. Tetrahedron 23, 2649 (1967).

153. HOUGHTON, E., and J. E. SAXTON: Studies in the echinulin series. Preliminary synthetical studies and the absolute configuration of echinulin. Tetrahedron Letters 1968, 5475.

154. HUBER, U. A., and A. S. DREIDING: Synthese von terpenartigen bicyclischen Systemen über die Cycloaddition von Dimethylketen an Methylcyclopentadien. Helv. Chim. Acta 53, 495 (1970).

155. Huffman, J. W.: Studies on resin acids VII. Isomerisation of 19-norabietatetraenes. J. Organ. Chem. (USA) **37**, 17 (1972).

156. Huffman, J. W.: Attempted duplication of the methyl shift in eremophilane biosynthesis. J. Organ. Chem. (USA) **37**, 2736 (1972).

157. Hutchison, R. D., P. S. Steyn, and D. L. Thompson: The isolation and structure of 4-hydroxyochratoxin-A and 7-carboxy-3,4-dihydro-8-hydroxy-3-methylisocoumarin from *Penicillium viridicatum*. Tetrahedron Letters **1971**, 4033.

158. Iguchi, M., A. Nishiyama, H. Koyama, S. Yamamura, and Y. Hirata: Isolation and structure of isocalamendiol. Tetrahedron Letters **1969**, 3729.

159. Ilyas, M., J. N. Usmani, S. P. Bhatnagar, M. Ilyas, W. Rahman, and A. Pelter: WB1 and W11, the first optically active biflavones. Tetrahedron Letters **1968**, 5515.

160. Inhoffen, H. H., D. Kopp, S. Marić, J. Bekurdts, and R. Selimoglu: Untersuchungen an hochsubstituierten Athylenen und Glykolen X. Zur Stereochemie der Hexoestrole. Tetrahedron Letters **1970**, 999.

161. Isler, O. (Ed.): Carotenoids. Basel: Birkhäuser Verlag. 1971.

162. Isono, K., and S. Suzuki: The structure of polyoxin-C. Tetrahedron Letters **1968**, 203.

163. Itô, S., M. Kodama, T. Nozoe, H. Hikino, Y. Hikino, Y. Takeshita, and T. Takemoto: Structure and absolute configuration of α-kessyl alcohol and kessyl glycol. Tetrahedron **23**, 553 (1967).

164. Itô, S., M. Kodama, M. Sunagawa, T. Takahashi, H. Imamura, and O. Honda: Structure of inumakilactone-A; a bisnorditerpenoid. Tetrahedron Letters **1968**, 2065.

165. Jacquesy, J. C., J. Levisalles, and J. Wagnon: Stéréochimie XXXV. Stéroides fluorés VI. Réarrangement spinal de l'androstène-5 diol-3β,17β. Bull. soc. chim. France **1970**, 670.

166. Jeffs, P. W., G. Ahmann, H. F. Campbell, D. S. Farrier, G. Ganguli, and R. L. Hawks: Alkaloids of *Sceletium* species III. The structures of four new alkaloids from *S. Strictum*. J. Organ. Chem. (USA) **35**, 3512 (1970).

167. Jennings, J. P., W. Klyne, and P. M. Scopes: Optical Rotatory Dispersion Part XXIV. Lactones. J. Chem. Soc. (London) **1965**, 7211.

168. Jerina, D. M., H. Ziffer, and J. W. Daly: The role of the arene oxide-oxepin system in the metabolism of aromatic substrates IV. Stereochemical considerations of dihydrodiol formation and dehydrogenation. J. Amer. Chem. Soc. **92**, 1056 (1970).

169. Johnson, C. R., and D. McCants: Nucleophilic displacement on sulphur. The inversion of sulphoxide configurations. J. Amer. Chem. Soc. **87**, 5404 (1965).

170. Jommi, G., J. Křepinský, V. Herout, and F. Šorm: The structure of valerianol, a sesquiterpenic alcohol of eremophilane type from valeriana oil. Tetrahedron Letters **1967**, 677.

171. Jones, D. N., and R. Grayshan: Conformational free-energy differences in steroids Part V. Equilibration of 5α- and 5β-androstan-6-one derivatives structurally modified in ring D. J. Chem. Soc. (C) (London) **1970**, 2421.

172. Jones, D. N., R. Grayshan, and D. E. Kime: Conformational free-energy differences in steroids Part II. Equilibration of 3α-methyl-5α- and 5β-cholestan-6-ones and of 3α-chloro-5α- and 5β-cholestan-6-ones. J. Chem. Soc. (C) (London) **1969**, 48.

173. Jones, D. N., and D. E. Kime: Conformational free energy differences in steroids Part 1. Equilibration of 3β-substituted 5α- and 5β-cholestan-6-ones. J. Chem. Soc. (C) (London) **1966**, 846.

174. Joshua, H., R. Gans, and K. Mislow: Stereochemistry of 9-dimethylamino-9,10-dihydro-4,5-dimethylphenanthrene. J. Amer. Chem. Soc. **90**, 4884 (1968).

175. Joska, J., J. Fajkoš, and F. Šorm: On steroids C. The B-nor-steroid hydroxy ketones and their o.r.d. curves. Collect. Czech. Chem. Comm. **31**, 2745 (1966).

176. Kakinuma, K., N. Otake, and H. Yonehara: The structure of detoxin-D_1. A selective antagonist of blasticidin-S. Tetrahedron Letters **1972**, 2509.

177. Kametani, T., T. Honda, H. Shimanouchi, and Y. Sasada: Absolute configuration of the alkaloid cularine: an X-ray structure determination. Chem. Commun. **1972**, 1072.

178. KAMETANI, T., and M. IHARA: The circular dichroism and optical rotatory dispersion of protoberberines. J. Chem. Soc. (C) (London) **1968**, 1305.

179. KAMETANI, T., M. IHARA, K. FUKUMOTO, and H. YAGI: Studies on the syntheses of heterocyclic compounds. Part CCC. Syntheses of salutaridine, sinoacutine and thebaine. Formal total syntheses of morphine and sinomenine. J. Chem. Soc. (C) (London) **1969**, 2030.

180. KAMETANI, T., M. IHARA, and T. HONDA: The alkaloids of *Corydalis pallida* var. tenuis (Yatabe) and the structures of pallidine and kikemanine. J. Chem. Soc. (C) (London) **1970**, 1060.

181. KAMETANI, T., H. SUGI, and S. SHIBUYA: The absolute configuration of cryptostyline-III. Studies on the syntheses of heterocyclic compounds. CCCXCVII. Tetrahedron **27**, 2409 (1971).

182. KAMETANI, T., H. SUGI, H. YAGI, K. FUKUMOTO, and S. SHIBUYA: Synthesis and stereochemistry of 1,2,3,4-tetrahydro-6-methoxy-2-methyl-1-phenyl-isoquinoline and related compounds. J. Chem. Soc. (C) (London) **1970**, 2213.

183. KASAL, A., and V. ČERNÝ: On Steroids. CVIII. Preparation and properties of some androstan-18-carboxylic acid derivatives. Collect. Czech. Chem. Comm. **32**, 3733 (1967).

184. KIDO, F., H. UDA, and A. YOSHIKOSHI: The stereochemistry of zizanoic acid. Tetrahedron Letters **1968**, 1247.

185. KIKUCHI, T., T. NISHINAGA, and K. KURIYAMA: The stereochemistry of spiropachysine. Tetrahedron Letters **1969**, 2519.

186. KING, T. J., and L. B. DE SILVA: Optically active gossypol from *Thespesia Populnea*. Tetrahedron Letters **1968**, 261.

187. KIRK, D. N., W. KLYNE, W. P. MOSE, and E. OTTO: Circular dichroism of ketones at 185–195 nm. J. Chem. Soc. Chem. Commun. (London) **1972**, 35.

188. KIRK, D. N., W. P. MOSE, and P. M. SCOPES: Circular dichroism of saturated chiral alcohols. J. Chem. Soc. Chem. Commun. (London) **1972**, 81.

189. KIRSON, I., D. LAVIE, S. M. ALBONICO, and H. R. JULIANI: The withanolides of *Acnistus Australis* (Griseb.). Tetrahedron **26**, 5063 (1970).

190. KITAGAWA, I., K. HINO, T. NISHIMURA, E. MUKAI, I. YOSIOKA, H. INOUYE, and T. YOSHIDA: Picroside I. A bitter principle of Picrorhiza Kurrooa. Tetrahedron Letters **1969**, 3837.

191. KLEIN, E., and W. ROJAHN: (−)7β,10α-Selina-4,11-dien und (+)-5β,7β,10α-Selina-3,11-dien. Zwei neue Sesquiterpene der Eudesmanreihe. Tetrahedron Letters **1970**, 279.

192. KLYNE, W., R. STEVENSON, and R. J. SWAN: Optical rotatory dispersion Part XXVIII. The absolute configuration of otobain and derivatives. J. Chem. Soc. (C) (London) **1966**, 893.

193. KLYNE, W., R. J. SWAN, B. W. BYCROFT, and H. SCHMID: Ermittlung der absoluten Konfiguration von Indolinalkaloiden durch Vergleiche der Optischen Rotationsdispersionen ihrer *N*(a)-Acylderivate. Helv. Chim. Acta **49**, 833 (1966).

194. KLYNE, W., R. J. SWAN, B. W. BYCROFT, D. SCHUMANN, and H. SCHMID: Absolute Konfiguration von Alkaloiden der Aspidospermin-Gruppe. Helv. Chim. Acta **48**, 443 (1965).

195. KLYNE, W., R. J. SWAN, N. J. DASTOOR, A. A. GORMAN, and H. SCHMID: Optische Rotationsdispersion von Indolalkaloiden der Yohimban-Corynanthean- und Quebrachamin-Gruppe. Helv. Chim. Acta **50**, 115 (1967).

196. KLYNE, W., R. J. SWAN, A. A. GORMAN, A. GUGGISBERG, and H. SCHMID: Optische Rotationsdispersion von Indolinalkaloiden mit Ketogruppen. Helv. Chim. Acta **51**, 1168 (1968).

197. KNIGHT, J. C., and G. R. PETTIT: Oxidation of 3β-acetoxy-14α-methyl-5α-cholest-7-ene. J. Organ. Chem. (USA) **33**, 1684 (1968).

198. KNIGHT, J. C., D. J. WILKINSON, and C. DJERASSI: The structure of the catcus sterol, macdougallin (14α-methyl-Δ⁸-cholestene-3β,6α-diol). A novel link in sterol biogenesis. J. Amer. Chem. Soc. **88**, 790 (1966).

199. Koch, M., M. Plat, and J. Lemen: Loganiacées de la Côte d'Ivoire-VIII. Stereochemie des alcaloides du *Strychnos splendens* Gilg. Tetrahedron **25**, 3377 (1969).

200. Koga, T., and M. Tomoeda: Studies on conformation and reactivity-VII. The synthesis and transannular cyclisation of 2α-hydroxy-5β-cholestane, leading to 2α,9α-epoxy-5β-cholestane; a new functionalisation reaction at 9 C in the steroid nucleus. Tetrahedron **26**, 1043 (1970).

201. Koreeda, M., G. Weiss, and K. Nakanishi: Absolute configuration of natural (+)-abscisic acid. J. Amer. Chem. Soc. **95**, 239 (1973).

202. Korver, O., and C. K. Wilkins: Circular dichroism spectra of flavanols. Tetrahedron **27**, 5459 (1971).

203. Krakower, G. W., H. A. van Dine, P. A. Diassi, and I. Bacso: Transformations of fusidic acid Part III. 17-Oxa-4α,8,14-trimethyl-D-homo-18-norandrostanes. J. Organ. Chem. (USA) **32**, 184 (1967).

204. Kruger, P. E. J., and G. W. Perold: Conocarpin, a leucodrin-type metabolite of *Leucospermum conocarpodendron* (L) Buek. J. Chem. Soc. (C) (London) **1970**, 2127.

205. Kunitomo, J., K. Morimoto, K. Yamamoto, Y. Yoshikawa, K. Azuma, and K. Fujitani: The absolute configuration of cularine: a chemical correlation to L(S)-laudanosine. Chem. Pharm. Bull. (Japan) **19**, 2197 (1971).

206. Kupchan, S. M., and M. J. Abu El-Haj: Steroid hormone analogs III. The synthesis and stereochemistry of C-nor-D-homoprogesterone analogs. J. Organ. Chem. (USA) **33**, 647 (1968).

207. Kupchan, S. M., E. Abushanab, K. T. Shamasundar, and A. W. By: Buxus alkaloids XIII. A synthetic approach to the 9(10→19) *abeo*-pregnane system. J. Amer. Chem. Soc. **89**, 6327 (1967).

208. Kurosawa, K., W. D. Ollis, B. T. Redman, I. O. Sutherland, O. R. Gottlieb, and H. M. Alves: The absolute configurations of the animal metabolite equol, three naturally occurring isoflavans and one natural isoflavanquinone. Chem. Commun. **1968**, 1265.

209. Kutney, J. P., K. Fuji, A. M. Treasurywala, J. Fayos, J. Clardy, A. I. Scott, and C. C. Wei: Structure and absolute configuration of (+)coronaridine hydrochloride. A comment on the absolute configuration of the Iboga alkaloids. J. Amer. Chem. Soc. **95**, 5407 (1973).

210. Kutney, J. P., and I. H. Rogers: Novel triterpenes from Sitka Spruce *(Picea Sitchensis* [*Bong.*] *Carr)*. Tetrahedron Letters **1968**, 761.

211. Lavie, D., M. K. Jain, and I. Kirson: Terpenoids-V. Melianone from *Melia Azedarach* L. Tetrahedron Letters **1966**, 2049.

212. Lavie, D., Y. Kashman, E. Glotter, and N. Danieli: Constituents of *Withania sommifera* Dun. Part VII. Rearrangements in Withaferin A. J. Chem. Soc. (C) (London) **1966**, 1757.

213. Lehn, J.-M., J. Levisalles, and G. Ourisson: Étude de cétones cycliques (XII). La conformation des céto-3-triterpènes. Bull. soc. chim. France **1963**, 1096.

214. Leutzinger, E. E., W. A. Bowles, R. K. Robins, and L. B. Townsend: Purine Nucleosides XVIII. The direct utilisation of unsaturated sugars in nucleoside syntheses. The conformation and structure of certain 9-(2'-deoxyribopyranosyl) purines prepared from D-arabinal. J. Amer. Chem. Soc. **90**, 127 (1968).

215. Levisalles, J., and H. Rudler: Terpénoides II — Carotol I. Stéréochimie du carotol. Bull. soc. chim. France **1967**, 2059.

216. Lukeš, R., J. Jarý, and J. Němec: Über Lactone VII. δ-Lacton der 4,6-Didesoxy-L-ribohexonsäure und die absolute Konfiguration der Parasorbinsäure. Collect. Czech. Chem. Comm. **27**, 735 (1962).

217. Macleod, W. D.: The constitution of nootkatone, nootkatene and valencene. Tetrahedron Letters **1965**, 4779.

218. Macmillan, J., T. J. Simpson, A. E. Vanstone, and S. K. Yeboah: Fungal products

Part II. Structure and stereochemistry of the acid $C_{18}H_{16}O_5$, a degradation product of wortmannin. J. Chem. Soc. Perkin I (London) **1972**, 2892.

219. MALLAMS, A. K., E. S. WAIGHT, B. C. L. WEEDON, L. CHOLNOKY, K. GYÖRGYFY, J. SZABOLCS, N. I. KRINSKY, B. P. SCHIMMER, C. O. CHICHESTER, T. KATAYAMA, L. LOWRY, and H. YOKOYAMA: The identity of neoxanthin and foliaxanthin. Chem. Commun. **1967**, 484.

220. MARKHAM, M. R., and T. J. MABRY: The structure and stereochemistry of two new dihydroflavonol glycosides. Tetrahedron **24**, 823 (1968).

221. MARX, M., J. LECLERCQ, B. TURSCH, and C. DJERASSI: Terpenoids LX. Revised structures of the cactus triterpene lactones, stellatogenin and thurberogenin. J. Organ. Chem. (USA) **32**, 3150 (1967).

222. MASAMUNE, T., K. ORITO, and A. MURAI: Revision of the configuration of an acetolysis product of 3-O,N-diacetyltetrahydrojervine. Tetrahedron Letters **1969**, 251.

223. MASAMUNE, T., N. SATO, K. KOBAYASHI, I. YAMAZAKI, and Y. MORI: Syntheses and n.m.r. spectra of 22,27-imino-17,23-oxidojervane derivatives. Tetrahedron **23**, 1591 (1967).

224. MASON, S. F., and G. W. VANE: The circular dichroism and absorption spectra of alkaloids containing the aniline chromophore. The absolute configuration of calycanthine. J. Chem. Soc. (B) (London) **1966**, 370.

225. MASON, S. F., G. W. VANE, K. SCHOFIELD, R. J. WELLS, and J. S. WHITEHURST: The circular dichroism and absolute configuration of Tröger's base. J. Chem. Soc. (B) (London) **1967**, 553.

226. MATSUO, A., and S. HAYASHI: The stereostructure of chiloscyphone. Tetrahedron Letters **1970**, 1289.

227. MATZINGER, P., PH. CATALFOMO, and C. H. EUGSTER: Isolierung von (2S,4S)-(+)-γ-Hydroxynorvalin and (2S,4R)-(−)-γ-Hydroxynorvalin aus *Boletus satanas Lenz.* Helv. Chim. Acta **55**, 1478 (1972).

228. MERLINI, L., R. MENDELLI, G. NASINI, and M. HESSE: Gambirine, a new indole alkaloid from *Uncaria gambier* Roxb. Tetrahedron Letters **1967**, 1571.

229. MILBORROW, B. V., and C. DJERASSI: Alkaloid Studies Part LXI. The structure of twelve new alkaloids from *Aspidosperma Cylindrocarpon.* J. Chem. Soc. (C) (London) **1969**, 417.

230. MINATO, H., M. MATSUMOTO, and T. KATAYAMA: Verticillin-A, a new antibiotic from *Verticillum sp.* Chem. Commun. **1971**, 44.

231. MISLOW, K., E. BUNNENBERG, R. RECORDS, K. WELLMAN, and C. DJERASSI: Inherently dissymmetric chromophores and circular dichroism II. J. Amer. Chem. Soc. **85**, 1342 (1963).

232. MISLOW, K., M. A. W. GLASS, R. E. O'BRIEN, P. RUTKIN, D. H. STEINBERG, J. WEISS, and C. DJERASSI: Configuration, conformation and rotatory dispersion of optically active biaryls. J. Amer. Chem. Soc. **84**, 1455 (1962).

233. MISRA, R., R. C. PANDEY, and SUKH DEV: The absolute stereochemistry of hardwickiic acid and its congeners. Tetrahedron Letters **1968**, 2681.

234. MITSCHER, L. A., WU-NAN WU, R. W. DOSKOTCH, and J. L. BEAL: Antibiotics from higher plants. *Thalictrum rugosum.* A new bisbenzylisoquinoline alkaloid active vs. *Mycobacterium smegmatis.* Chem. Commun. **1971**, 589.

235. MIURA, R., S. HONMARU, and M. NAKAZAKI: The absolute configurations of the metabolites of naphthalene and phenanthrene in mammalian systems. Tetrahedron Letters **1968**, 5271.

236. MOFFITT, W., R. B. WOODWARD, A. MOSCOWITZ, W. KLYNE, and C. DJERASSI: Structure and the optical rotatory dispersion of saturated ketones. J. Amer. Chem. Soc. **83**, 4013 (1961).

237. MONNERET, C., Q. KHUONG-HUU, and R. GOUTAREL: Transpositions spinales de D-homoandrostène-13-dione-3,17. Bull. soc. chim. France **1972**, 291.

238. Morand, P., J. M. Lyall, and H. Stollar: Reactions of Δ^7-steroids Part III. Conformational transmission effects in the methylation of 5β-cholest-7-en-3-one. J. Chem. Soc. (C) (London) **1970**, 2117.

239. Mori, H., K. Shibata, K. Tsuneda, and M. Sawai: Synthesis of ecdysone VI. Synthesis of ecdysone from stigmasterol. Tetrahedron **27**, 1157 (1971).

240. Morita, Y., M. Hesse, and H. Schmid: Umbellamin, ein neues „dimeres" Indolalkaloid. Helv. Chim. Acta **52**, 89 (1969).

241. Moscowitz, A., E. Charney, U. Weiss, and H. Ziffer: Optical activity in skewed dienes. J. Amer. Chem. Soc. **83**, 4661 (1961).

242. Moscowitz, A., K. Mislow, M. A. W. Glass, and C. Djerassi: Optical rotatory dispersion associated with dissymmetric non-conjugated chromophores. An extension of the octant rule. J. Amer. Chem. Soc. **84**, 1945 (1962).

243. Motl, O., M. Romanuk, and V. Herout: On terpenes CLXXVIII. Composition of the oil from *Amorpha fruticosa* L. fruits: structure of (−)γ-amorphene. Collect. Czech. Chem. Comm. **31**, 2025 (1966).

244. Murray, R. W., R. D. Youssefyeh, and P. R. Story: An unequivocal ozonide stereoisomer assignment. J. Amer. Chem. Soc. **88**, 3655 (1966).

245. Nagai, Y., O. Tanaka, and S. Shibata: Chemical studies on the oriental plant drugs. — XXIV. Structure of ginsenoside-Rg₁. A neutral saponin of ginseng root. Tetrahedron **27**, 881 (1971).

246. Nagarajan, R., N. Neuss, and M. M. Marsh: Aranotin and related metabolites III. Configuration and conformation of acetylaranotin. J. Amer. Chem. Soc. **90**, 6518 (1968).

247. Nakanishi, K., M. Koreeda, M. L. Chang, and H. Y. Hsu: Insect hormones V. The structures of ponasterones B and C. Tetrahedron Letters **1968**, 1105.

248. Nakano, T., and M. Hasegawa: Buxus Alkaloids Part VI. The constitutions of cyclomicrobuxine, cyclomicrobuxinine and alkaloid-L. J. Chem. Soc. (London) **1965**, 6688.

249. Nakano, T., S. Terao, K. H. Lee, Y. Saeki, and L. J. Durham: Some observations on the oxidation of virosecurinine with monoperphthalic acid. J. Organ. Chem. (USA) **31**, 2274 (1966).

250. Nakashima, R., and G. P. Slater: The configuration of echinulin. Tetrahedron Letters **1967**, 4433.

251. Naranjo, J., M. Hesse, and H. Schmid: Indolalkaloide aus den Blättern von *Hedranthera barteri. (Hook.f.)* Pichon. Helv. Chim. Acta **55**, 1849 (1972).

252. Narayanan, C. R., and R. V. Pachapurkar: The structure of nimbinic acid. Tetrahedron Letters **1966**, 553.

253. Naya, K., and I. Takagi: The structure of petasitin. A new sesquiterpene from *Petasites Japonicus* Maxim. Tetrahedron Letters **1968**, 629.

254. Naya, K., I. Takagi, Y. Kawaguchi, Y. Asada, Y. Hirose, and N. Shinoda: The structure of fukinone; a consituent of *Petasites Japonicus* Maxim. Tetrahedron **24**, 5871 (1968).

255. Nikolenko, L. N., W. N. Nesawibatko, A. F. Usatiy, and M. N. Semjenowa: Acylierung von Nucleosiden in wässrigen Lösungen. Tetrahedron Letters **1970**, 5193.

256. Nyholm, R. S. (editor): A discussion on circular dichroism: electronic and structural principles. Proc. Roy. Soc. (London) **297 A**, 1 (1967).

257. Ogihara, Y., N. Kobayashi, and S. Shibata: Further studies on the bianthraquinones of *Penicillium Islandicum* Sopp. Tetrahedron Letters **1968**, 1881.

258. Oka, K., Y. Ike, and S. Hara: The skeletal synthesis of the early proposed cycloneosamandione-I. The synthesis of 19-homosteroids. Tetrahedron Letters **1969**, 4543.

259. Okigawa, M., and N. Kawano: The structure of pseudoanisatin. Tetrahedron Letters **1971**, 75.

260. Ollis, W. D., B. T. Redman, R. J. Roberts, I. O. Sutherland, and O. R. Gottlieb: New neoflavanoids from *Machaerium kuhlmannii* and *Machaerium nictitans* and the recognition of a new neoflavanoid type the neoflavenes. Chem. Commun. **1968**, 1392.

261. OLLIS, W. D., C. A. RHODES, and I. O. SUTHERLAND: The extractives of *Millettia dura* (Dunn.). The constitutions of durlettone, durmillone, milldurone, millettone and millettosin. Tetrahedron 23, 4741 (1967).

262. ORITANI, T., and K. YAMASHITA: Synthesis of optically active abscisic acid and its analogs. Tetrahedron Letters 1972, 2521.

263. OVERTON, K. H., and A. J. RENFREW: The configuration at C-13 in labdanolic and eperuic acid. J. Chem. Soc. (C) (London) 1967, 931.

264. PARANJAPE, B. V., and J. L. PYLE: The Dieckmann cyclisation as a route to A-nor steroids. Evidence concerning stereochemistry. J. Organ. Chem. (USA) 36, 1009 (1971).

265. PARELLO, J.: Structure de la phyllanthine, alcaloide mineur du *Phyllanthus discoides* Muell. Arg. (Euphorbiacées). Bull. soc. chim. France 1968, 1117.

266. PÉREZ, G., O. HALPERN, J. IRIARTE, and P. CRABBÉ: Préparation de quelques dérivés du pregnane substitués en C-16 et de leurs produits de cyclisation. Bull. soc. chim. France 1968, 4063.

267. PETE, J. P., and M. L. VIRIOUT-VILLAUME: Réarrangement photochimique d'α-époxy cétones I. Transformation thermique et photochimique des époxy-4,5β-cholestanone-6 et époxy-5,6β-cholestanone-4. Bull. soc. chim. France 1971, 3699.

268. PLATTNER, J. J., and H. RAPOPORT: The synthesis of d- and l-sirenin and their absolute configurations. J. Amer. Chem. Soc. 93, 1758 (1971).

269. POUSSET, J. L., J. POISSON, and M. LEGRAND: Application des spectres de dichroisme circulaire a l'étude des alcaloides oxindoliques: structure de la corynoxine. Tetrahedron Letters 1966, 6283.

270. PREININGER, V., A. D. CROSS, and F. ŠANTAVÝ: Isolation and chemistry of the alkaloids from some plants of the genus *Papaver* XXXIII. On the constitution of latericine. Collect. Czech. Chem. Comm. 31, 3345 (1966).

271. PRYSTAŠ, M., and F. ŠORM: Nucleic acid components and their analogues CVI. Synthesis of substituted 1-glycosyl-4-benzyloxy-6(1*H*)-pyrimidinones by the mercuri process. Collect. Czech. Chem. Comm. 33, 210 (1968).

272. RIBI, M., A. CHANG SIN-REN, H. P. KÜNG, and C. H. EUGSTER: Struktur und Reaktionen von Coleon B. Helv. Chim. Acta 52, 1685 (1969).

273. RIBI, M., and C. H. EUGSTER: Über die Chiralität der enantiomeren *cis*-Tetrahydroactinidiolide. Helv. Chim. Acta 52, 1732 (1969).

274. RIPPERGER, H., and K. SCHREIBER: Struktur von Paniculonin A und B: zwei neuen Spirostanglykosiden aus *Solanum paniculatum L.* Chem. Ber. 1968, 2450.

275. RIPPERGER, H., F. J. SYCH, and K. SCHREIBER: Solanum Alkaloide XCVI. Synthese von Solafloridin und 25-iso-Solafloridin. Tetrahedron 28, 1619 (1972).

276. ROBINSON, D. L., and D. W. THEOBALD: Sesquiterpenoids-VII. The synthesis of some 2-ketoeudesmanes. Tetrahedron 24, 5227 (1968).

277. ROMO, J., P. J. NATHAN, A. R. DE VIVAR, and C. ALVAREZ: The structure of peruvinin. A pseudoguaianolide isolated from *Ambrosia Peruviana* Willd. Tetrahedron 23, 529 (1967).

278. ROSENBROOK, W., and R. E. CARNEY: A new metabolite from an unidentified *Aspergillus* species. Tetrahedron Letters 1970, 1867.

279. RYABININ, A. A., L. H. MATYUKHINA, I. A. SALTIKOVA, F. PATIL, and G. OURISSON: Structure de l'alnincanone. Un triterpène demmaranique en C_{31}. Bull. soc. chim. France 1968, 1089.

280. RYBACK, G.: Revision of the absolute stereochemistry of (+)-abscisic acid. J. Chem. Soc. (Chem. Commun.) (London) 1972, 1190.

281. SAUER, G., A. SHIMAOKA, and K. TAKEDA: Components of *Chionographis japonica* Maxim Part IV. Structures of chiograsterone and isochiograsterone, two new sterols from *Chionographis japonica* Maxim. J. Chem. Soc. (C) (London) 1970, 910.

282. SCHAUB, F., H. KAUFMANN, W. STÖCKLIN, and T. REICHSTEIN: Die Pregnanglykoside der oberirdischen Teile von *Sarcostemma viminale* (L. R. Br.). Helv. Chim. Acta 51, 738 (1968).

283. Schaub, F., W. Stöcklin, and T. Reichstein: Die Struktur der Genine G. (12-O-Benzoyldesacetylmetaplexigenin) und H (Di-O-benzoylviminolon) aus den Stengeln von *Sarcostemma viminale (L. R. Br.)*. Helv. Chim. Acta **51**, 767 (1968).

284. Saboz, J. A., T. Iizuka, H. Wehrli, K. Schaffner, and O. Jeger: Zur photochemischen Umwandlung eines α,β-ungesättigten γ,δ-Epoxyketons; 3-Oxo-6α,7α-oxido-17β-acetoxy-Δ⁴-androsten. Helv. Chim. Acta **51**, 1362 (1968).

285. Schulte, K. E., G. Rücker, and H. Fachmann: Ergosta-4,6,8(14)22-tetraenon-3 als Inhaltsstoff des Lärchenschwammes. Tetrahedron Letters **1968**, 4763.

286. Scopes, P. M.: Annual Reports of the Chemical Society for 1966, 1967, 1968, 1969, 1970, 1971 and 1972.

287. Sélégny, E., and M. Vert: Dédoublement de deux phénoldiacides succiniques racémiques, caractérisation des isomères optiques et détermination de leur configuration par dispersion rotatoire. Bull. soc. chim. France **1968**, 2549.

288. Sengupta, P., and A. K. Dey: Terpenoids and related compounds XXI. The structures and stereochemistry of putranjivic acid and putranjic (putric) acid. Tetrahedron **28**, 1307 (1972).

289. Shamma, M., M. G. Kelly, and M. A. Podczasy: Thalictrum alkaloids VI. (−)-Veronamine, a glycosidic benzylisoquinoline. Tetrahedron Letters **1969**, 4951.

290. Shimizu, Y., and H. Mitsuhashi: Studies on the components of Asclepiadaceae Plants — XXII. Structures of cynanchogenin and sarcostin. Tetrahedron **24**, 4143 (1968).

291. Shoppee, C. W., and B. C. Newman: Steroids and Walden inversion. Part LXVI. 5β,14α-Androst-15-en-17-one. J. Chem. Soc. (C) (London) **1969**, 2767.

292. Shoppee, C. W., and S. C. Sharma: Steroids and Walden inversion Part LXI. Chlorination of 5α-cholestan-2-one. J. Chem. Soc. (C) (London) **1967**, 2385.

293. Sjoberg, B., D. J. Cram, L. Wolf, and C. Djerassi: Optical rotatory dispersion studies LXX. Anomalous rotatory dispersion of xanthates; application in stereochemical studies of alcohols. Acta Chem. Scand. **16**, 1079 (1962).

294. Smalberger, T. M., G. J. H. Rall, H. L. de Waal, and R. R. Arndt: The structures and configuration of azimine and azcarpine. Tetrahedron **24**, 6417 (1968).

295. Smith, H. E., and T. C. Willis: The absolute configurations of two 3-phenylnorborn-2-ylamine hydrochlorides and some related phenylnorbornanes. Chem. Commun. **1969**, 873.

296. Smith, L. L., G., Greenspan, R. Rees, and T. Foell: Microbiological hydroxylation of steroids of unnatural configuration. J. Amer. Chem. Soc. **88**, 3120 (1966).

297. Snatzke, G. (Ed.): Optical rotatory dispersion and circular dichroism in organic chemistry. London: Heyden. 1967.

298. Snatzke, G., and R. Hansel: Die Absolutkonfiguration der Kawa-lactone. Tetrahedron Letters **1968**, 1797.

299. Snatzke, G., and J. Himmelreich: Circulardichroismus-XXIII. Cotton-Effekt einiger heterocyclisch substituierter Steroidketone, deren Vorstufen und des Azo-chromophors. Tetrahedron **23**, 4337 (1967).

300. Snatzke, G., J. Hrbek, L. Hruban, A. Horeau, and F. Šantavý: Absolute configuration and chiroptical properties of chelidonine and tetrahydroberberine alkaloids. Tetrahedron **26**, 5013 (1970).

301. Snatzke, G., and K. Kinsky: Circular dichroismus-XLIX Vicinaleffekt bei 4,7-Diketosteroiden. Tetrahedron **28**, 295 (1972).

302. Snatzke, G., E. Sz. Kováts, and G. Ohloff: Circulardichroismus von α-Phellandren. Tetrahedron Letters **1966**, 4551.

303. Snatzke, G., H. Ripperger, C. Horstmann, and K. Schreiber: Circulardichroismus XIX. Nitrosamine und Nitrosaminoketale Sektorregel für Nitrosamine, Nitroverbindungen und Lactone. Tetrahedron **22**, 3103 (1966).

304. Snatzke, G., H. Schwang, and P. Welzel in Ref. *51*, p. 159.

305. Steyn, P. S.: The isolation, structure and absolute configuration of secalonic acid-D, the toxic metabolite of *Penicillium oxalicum*. Tetrahedron **26**, 51 (1970).

306. STOUT, G. H., G. K. HICKERNELL, and K. D. SEARS: *Calophyllum* Products IV. Papuanic and isopapuanic acids. J. Organ. Chem. (USA) **33**, 4191 (1968).

307. ST. PYREK, J.: A new 4β-methyl sterol from marigold flowers. Chem. Commun. **1969**, 107.

308. STUART, K. L., and C. CHAMBERS: Flavinine a new morphinandienone alkaloid from *Croton flavens* L. Tetrahedron Letters **1967**, 2879.

309. STUDER, R. O., W. LERGIER, and K. VOGLER: Synthesen in der Polymyxinreihe. Die Synthese von Circulin A. Helv. Chim. Acta **49**, 974 (1966).

310. SUCHÝ, M., L. DOLEJŠ, V. HEROUT, F. ŠORM, G. SNATZKE, and J. HIMMELREICH: On terpenes CXCVII. The constitution of jurineolide, a new germacranolide from *Jurinea Cyanoides* (L) RCHB. Collect Czech. Chem. Comm. **34**, 229 (1969).

311. SUCHÝ, M., Z. SAMEK, V. HEROUT, R. B. BATES, G. SNATZKE, and F. ŠORM: On terpenes CLXXXVIII. Constitution and configuration of pelenolides. A new group of sesquiterpene lactone germacranolides. Collect. Czech. Chem. Comm. **32**, 3917 (1967).

312. SUGA, T., T. SHISHIBORI, and T. MATSURA: Stereochemical studies of monoterpene compounds II. The conformation of 4-hydroxymenthones. J. Organ. Chem. (USA) **32**, 965 (1967).

313. SUGINOME, H.: The absolute and relative stereochemical correlation of isoflavanone, sophorol and naturally occurring isoflavan derivatives. Bull. Chem. Soc. Japan **39**, 1544 (1966).

314. SWAN, R. J., and W. KLYNE: Optical rotatory dispersion curves of lignans: the 4-aryltetralin group. Chem. and Ind. **1965**, 1218.

315. SWAN, R. J., W. KLYNE, and H. MACLEAN: Optical rotatory dispersion studies, XLI. The absolute configuration of plicatic acid. Canad. J. Chem. **45**, 319 (1967).

316. TABACIK-WLOTZKA, C., and Y. LAPORTHE: Diterpenes de *Juniperus Phoenicea* LI. Acide hydroxy-6α-sandaracopimarique. Tetrahedron Letters **1968**, 2531.

317. TADA, M., Y. MORIYAMA, Y. TANAHASHI, and T. TAKAHASHI: Absolute configuration of Farfugin A. Tetrahedron Letters **1972**, 5251.

318. TAKAHASHI, T., Y. MORIYAMA, Y. TANAHASHI, and G. OURISSON: The structure of shionone. Tetrahedron Letters **1967**, 2991.

319. TAKEDA, K., H. MINATO, I. HORIBE, and M. MIYAWAKI: Components of the root of *Lindera strychnifolia Vill.* Part XII. The structure of isolinderoxide. J. Chem. Soc. (C) (London) **1967**, 631.

320. TAKEDA, K., K. SAKURAWI, and H. ISHII: Sesquiterpenes of *Lauraceae* plants — III. Structure and absolute configuration of delobanone and acetoxydelobanone from *Lindera Triloba*. Tetrahedron **27**, 6049 (1971).

321. TAKEMOTO, T., Y. HIKINO, H. HIKINO, S. OGAWA, and N. NISHIMOTO: Structure of rubrosterone, a novel C-19 metabolite of insect moulting substances from *Achyranthes rubrafusca*. Tetrahedron Letters **1968**, 3053.

322. TOMKO, J., A. VASSOVÁ, G. ADAM, K. SCHREIBER, and E. HÖHNE: Veralkamine. A novel type of steroidal alkaloid with a 17β-methyl-18-nor-17-isocholestane carbon skeleton. Tetrahedron Letters **1967**, 3907.

323. TOOME, V., J. F. BLOUNT, G. GRETHE, and M. USKOKOVIĆ: Determination of the absolute configuration of 4-phenyl-1,2,3,4-tetrahydroisoquinolines utilizing o.r.d. and c.d. spectroscopy. Tetrahedron Letters **1970**, 49.

324. TRAGER, W. F., C. M. LEE, J. D. PHILLIPSON, R. E. HADDOCK, D. DWUMA-BADU, and A. H. BECKETT: Configurational analysis of rhynchophylline-type oxindole alkaloids. The absolute configuration of ciliaphylline, rhynchociline, specionoxeine, isospecionoxeine, rotundifoline and isorotundifoline. Tetrahedron **24**, 523 (1968).

325. TROJÁNEK, J., Z. KOBLICOVÁ, and K. BLÁHA: On alkaloids XIX. Absolute configuration of vincamine and some related alkaloids; optical rotatory dispersion. Collect. Czech. Chem. Comm. **33**, 2950 (1968).

326. TSCHESCHE, R., F.-J. KÄMMERER, and G. WULFF: Über die Struktur der antibiotisch aktiven Substanzen der Tulpe *(Tulipa gesneriana L.)*. Chem. Ber. **1969**, 2057.

327. Tsuyuki, T., R. Aoyagi, S. Yamada, and T. Takahashi: 4-epifriedelin and 4-epishionone. Tetrahedron Letters 1968, 5263.
328. Turner, R. B., K. H. Gänshirt, P. E. Shaw, and J. D. Tauber: The total synthesis of phyllocladene. J. Amer. Chem. Soc. 88, 1776 (1966).
329. Uyeo, S., J. Okada, S. Matsunaga, and J. W. Rowe: The structure and the stereochemistry of abieslactone. Tetrahedron 24, 2859 (1968).
330. Van-Horn, A. R., and C. Djerassi: Optical rotatory dispersion studies. CVII. Factors governing the relative stability of hydrindanones. Syntheses of 17-alkyl-15-keto steroids. J. Amer. Chem. Soc. 89, 651 (1967).
331. Velgová, H., V. Černý, and F. Šorm: On steroids CXXXVI. Effect of 17-substituent in base-catalysed equilibration of steroidal $2\beta,3\beta$-disubstituted 6-ketones. Collect. Czech. Chem. Commun. 37, 1015 (1972).
332. Velgová, H., V. Černý, F. Šorm, and K. Sláma: On steroids CXXVI. Further compounds with antisclerotization effect on Pyrrhocoris apterus L. Larvae; structure and activity correlations. Collect. Czech. Chem. Comm. 34, 3354 (1969).
333. Velluz, L., M. Legrand, and M. Grosjean: Optical circular dichroism. Weinheim: Verlag-Chemie. 1965.
334. Verbit, L., and J. W. Clark-Lewis: Optically active aromatic chromophores VIII. Studies in the isoflavanoid and rotenoid series. Tetrahedron 24, 5519 (1968).
335. Wada, K., and K. Munakata: Insect antifeedants from Parabenzoin Trilobum II. The absolute configuration of shiromodiol diacetate. Tetrahedron Letters 1968, 4677.
336. Waldner, E. E., M. Hesse, W. I. Taylor, and H. Schmid: Über die Konstitution des Macralstonidins. Helv. Chim. Acta 50, 1926 (1967).
337. Walton, E., F. W. Holly, G. E. Boxer, and R. F. Nutt: 3'-Deoxynucleosides IV. Pyrimidine 3'-deoxynucleosides. J. Organ. Chem. (USA) 31, 1163 (1966).
338. Wechter, W. J.: Reactions of a β-hydroxy nitrene. The synthesis of B-nor steroids. J. Organ. Chem. (USA) 31, 2136 (1966).
339. Wehrli, H., C. Lehmann, T. Iizuka, K. Schaffner, and O. Jeger: Photoisomerisierung von α,β-Epoxyketonen II. Der sterische Verlauf der Umlagerung von 3-Oxo-4,5-oxido-Steroiden. Helv. Chim. Acta 50, 2403 (1967).
340. Wehrli, H., C. Lehmann, P. Keller, J. J. Bonet, K. Schaffner, and O. Jeger: Photoisomerisierung von α,β-Epoxyketonen I. Die Umlagerung von 3-Oxo-4,5-oxido-Steroiden zu 3,5-Dioxo-10-(5→4)-abeo-Derivaten. Helv. Chim. Acta 49, 2218 (1966).
341. Wiechert, R., U. Kerb, P. Hocks, A. Furlenmeier, A. Fürst, A. Langemann, and G. Waldvogel: Zur Synthese des Ecdysons. Synthesen von $2\beta,3\beta$-Dihydroxy-6-keto-A/B-cis-Steroiden. Helv. Chim. Acta 49, 1581 (1966).
342. Wittenburg, E.: Synthese der anomeren 2'-Desoxy-D-ribopyranosyl-uracile. Chem. Ber. 1968, 2132.
343. Wittenburg, E., G. Etzold, and P. Langen: Thymin-nucleoside der 2-Desoxy-D-xylopyranose. Chem. Ber. 1968, 494.
344. Wolff, G., and G. Ourisson: Le seychellene. Isolement et structure. Tetrahedron 25, 4903 (1969).
345. Yamagishi, T., K. Hayashi, R. Kiyama, and H. Mitsuhashi: Structure of glycocynanchogenin, a novel polyoxypregnane from Cyanchum caudatum Max. Tetrahedron Letters 1972, 4005.
346. Yanagita, M., S. Inayama, and T. Kawamata: Neopulchellidine and neopulchellin. Tetrahedron Letters 1970, 3007.
347. Yoshida, K.: Acid catalysed rearrangements of 17β-acetoxy-A-nor-5α-androstan-$1\alpha,2\alpha$-epoxide. Tetrahedron 25, 1367 (1969).
348. Yoshihira, K., M. Tezuka, and S. Natori: Naphthoquinone derivatives from Diospyros. Spp. Bisisodiospyrin, a tetrameric naphthoquinone. Tetrahedron Letters 1970, 7.

349. YOSHIOKA, H., H. RÜESCH, E. RODRIGUEZ, A. HIGO, J. A. MEARS, T. J. MABRY, J. G. CALZADA A., and X. A. DOMINGUEZ: Tetraneurin-B, -C and -D, New C_{14}-oxygenated pseudoguaianolides from *Parthenium* (Compositae). Tetrahedron **26**, 2167 (1970).
350. ZIFFER, H., U. WEISS, and E. CHARNEY: Optical activity of non-planar conjugated dienes IV. Interacting chromophores in gliotoxin. Tetrahedron **23**, 3881 (1967).

(Received February 28, 1974)

Chemistry and Biosynthesis of Plant Galactolipids

By H. C. VAN HUMMEL, Botanisch Laboratorium,
Faculteit der Wiskunde en Natuurwetenschappen, Katholieke Universiteit,
Nijmegen, Netherlands

Contents

I. Introduction

Together with sulphoquinovosyl diglyceride (sulpholipid) the galactosyl diglycerides (galactolipids) are the glycolipids occurring in the photosynthetic tissue of algae and higher plants. In higher algae and plants they are mainly confined to the chloroplasts (1, 76, 108) where they form a structural part of the thylakoid membranes. Apart from this structural function in chloroplasts several other functions have been attributed to the galactolipids. These vary from a function as a carbohydrate reservoir to a function in the electron transport chain in photosynthesis. This review is intended to summarize the present knowledge of the synthetic and degradative chemistry of the galactolipids and of their functions in the photosynthesis of algae and higher plants.

II. Occurrence and Localization

In certain parts of higher plants such as flowers, storage organs (e. g. bulbs, roots and potatoes) and fruits, galactolipids only occur in minor quantities (*31, 32, 49, 55, 56, 97*). In leaves of higher plants, however, they are among the predominant classes of lipids (*2, 107*) (Table 1). In tobacco leaves 83% of the total cellular monogalactosyl diglyceride and 88% of the digalactosyl diglyceride is present in the chloroplasts (*76*). In leaves of *Zea mays* and *Antirrhinum majus* 62% ($2 \cdot 10^9$ molecules) of the total leaf galactolipids is found in the chloroplasts (*53*). Within chloroplasts the majority of galactolipids is located in the lamellae and osmiophilic grana (*7, 91*). Ultrasonicated chloroplast lamellae agglutinated a monospecific antiserum to monogalactosyl diglyceride to a greater extent than did untreated stroma-free chloroplasts. The antigen determinants of the monogalactosyl diglyceride are therefore located on the outer surface of the thylakoid membrane (*81*). After the chloroplast envelope membranes were stripped by osmotic shock, spinach chloroplasts lost 17% of their total lipid. However, the lipids of the envelope membranes were not qualitatively different from those remaining in the lamellae of the stripped chloroplasts (*80*). Quantitative differences in lipid composition between envelope membranes and lamellae were observed by Mackender and Leech (*60*). The galactolipid : phospholipid ratio was 18 : 1 in the lamellae and 2.6 : 1 in the envelopes. This was interpreted to reflect the functional differences between lamellae and envelope. Wintermans could not find evidence to support models in which digalactosyl diglyceride is supposed to be associated specifically with either photosystem I or II (*108*).

Table 1. *Galactolipid Composition of Chloroplasts*

Plant	Lipid (% of total lipids)		References
	Monogalactosyl diglyceride	Digalactosyl diglyceride	
Sugar beet	44	24	(*107*)
Tobacco	42	31	(*76*)
Spinach (lamellae)	38	25	(*1*)

Polygalactolipids, i.e. tri- and tetragalactosyl diglycerides were also identified in spinach chloroplasts. Their concentrations are too small for them to play an important role in the thylakoid structure (*100*).

In chloroplast containing algae such as *Chlorella vulgaris* and *Euglena gracilis* the lipid composition is comparable to that found in leaves of higher plants. The galactolipid content of these algae depends on the way the cells are grown (*68*). On illumination of dark grown cells of *Euglena gracilis* the monogalactosyl diglyceride content rose from 2 μmoles in 100 mg of dark grown cells to 27 μmoles in fully green cells; the digalactosyl diglyceride content rose from 1 μmole to 11 μmoles (*84, 85*).

The blue-green algae have a relatively simple lipid composition, because they contain very little subcellular structures other than those of the photosynthetic lamellae (*29*). They contain the four classes of lipids also observed in the photosynthetic systems of higher algae and plants, namely mono- and digalactosyl diglyceride (**1, 2**), sulpholipid (**3**), and phosphatidyl glycerol (**4**) (*70*).

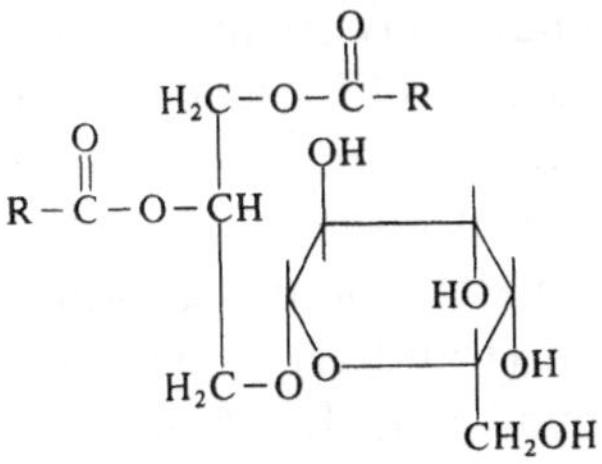

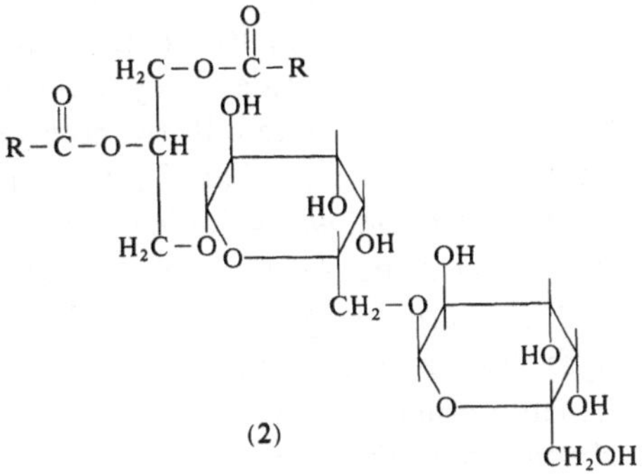

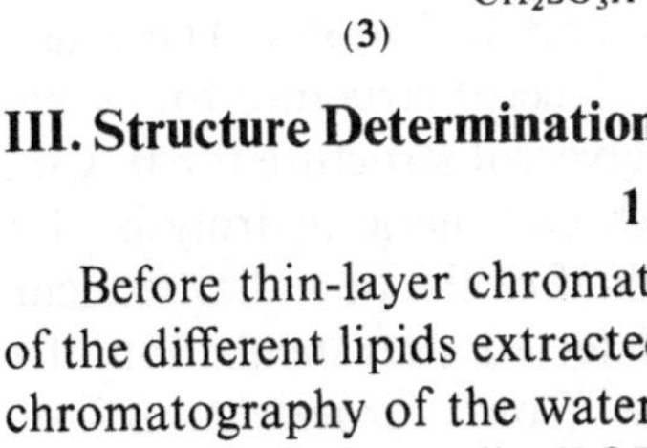

III. Structure Determination and Synthesis of Galactosyl Diglycerides

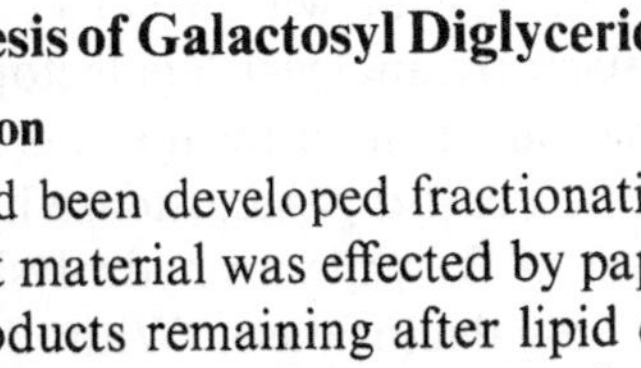

1. Fractionation

Before thin-layer chromatography had been developed fractionation of the different lipids extracted from plant material was effected by paper chromatography of the water soluble products remaining after lipid deacylation in methanolic KOH (*10, 11, 12*). Also counter current distribution of a benzene extract from wheat flour followed by chromatography of the fractions on a carbon column was used (*20*). Only partial fractionation of glycolipids extracted from plants can be effected by chromatography on columns of silicic acid (*104*). The separation of polar lipids on columns of DEAE cellulose followed either by thin-layer chromatography (*71*) or column chromatography on silica gel (*1*) is very much

better. At present the fastest and most accurate separations of glycolipids
are achieved by one- or two-dimensional thin-layer chromatography
with a variety of solvent systems (54, 67, 70, 79, 93). Preparative isolation
of mono- and digalactosyl diglycerides has been achieved using column
chromatography on Sephadex LH-20 (45) and preparative thin-layer
chromatography (37).

2. Identification and Structure Determination of Galactosyl Diglycerides

Over 60 years ago Winterstein and coworkers isolated lipid fractions
from various plant material. After acid hydrolysis of these fractions,
sugars were obtained which included galactose. It seems very probable
that galactolipids were involved. The highest percentage of sugar was
found in lipids from wheat flour. From this same source Carter and coworkers (19, 20, 21) isolated, identified and characterized mono- and
digalactosyl diglyceride. They made a benzene extract of bleached wheat
flour and fractionated it by counter current distribution between n-
heptane and 95% methanol into 4 fractions. One of these, a lipocarbo-
hydrate fraction, was soluble in warm acetone and after cooling separated
into an acetone-insoluble and a soluble fraction (I and II). Fractions I
and II were hydrolyzed in boiling sodium hydroxide for 5 hr. The
hydrolysates were acidified and the free fatty acids extracted with petroleum ether. The aqeous solutions containing the carbohydrates were
chromatographed on a carbon-column giving 2 fractions (A and B),
which were chromatographically homogeneous in 3 solvent systems.
Fractions A and B were crystallized from absolute methanol and ethanol-
water respectively, and on acid hydrolysis in hydrochloric acid at 100°
for 1 hr both fractions gave galactose and glycerol.

On oxidation with periodate, periodate reacting with vicinal hydroxyl
groups, A consumed 3 moles of periodate and B 5 moles. These data
were consistent with a galactopyranosyl-1-glycerol structure for A and
a galactopyranosyl-1,6-galactopyranosyl-1-glycerol structure for B. Confirmation of these results was obtained with enzymatic hydrolysis of A
and B. Yeast α-galactosidase liberated one half of the potential reducing
power of B, while no hydrolysis could be detected after incubation of A
with α-galactosidase. With β-galactosidase from *Escherichia coli* no
hydrolysis could be detected after incubation with B; however, with A
hydrolysis took place. From these results it was concluded that galactose
was linked β to glycerol and that the second galactose was linked α to
the first one. Supplementary evidence for the nature of the glycosidic
linkages was obtained from optical rotation data (Table 2), since β-
galactosides usually have very low specific rotations, whereas α-galac-
tosides generally have specific optical rotations between 125 and 200°.

Table 2. *Physico-Chemical Constants of Mono- and Digalactosyl Glycerol*

| | Monogalactosyl glycerol | | Digalactosyl glycerol | | |
	melting point °C	optical rotation $[\alpha]_D$	melting point °C	optical rotation $[\alpha]_D$	References
WICKBERG	140.5—141.5	−7	195—197	—	(106)
CARTER et al.	139—140	+3.77	182—184	+86.4	(19)
SASTRY and KATES	138—138.5	—	188—189	—	(88)
SMITH and WOLFF	132—138	—	—	—	(94)
STEIM	138.1—139.2	—	—	—	(95)
HEINZ	141.5	−7.7	—	—	(42)
BRUNDISH and BADDILEY	138—140	0	—	—	(17)
MYRHE	—	−6	—	—	(65)
WEHRLI and POMERANZ	135—138	0	—	—	(102)
HEINZ	137	−8.12	—	—	(43)
SHVETS et al.	136—137	−7.9	—	—	(92)

Based on these results A was assigned the structure β-D-galactopyranosyl-1-glycerol und B the structure α-D-galactopyranosyl-1,6-β-D-galacto-pyranosyl-1-glycerol.

Methylation of A and B with silver oxide and methyliodide, followed by alkaline deacylation in methanolic-aqeous sodium hydroxide resulted in products which consumed 1 mole of periodate. Acid hydrolysis in aqeous sulphuric acid at 100° for 16 hr released 2,3,4-tri-O-methyl-D-galactose and glycerol from A and 2,3,4-tri-O-methyl-D-galactose, 2,3,4,6-tetra-O-methyl-D-galactose and glycerol from B. On the basis of these results it was concluded that in both mono- and digalactosyl-glycerol lipids 2 vicinal esterified hydroxyls were present in the glycerol moiety.

Finally, the configuration of the glycerol moiety was determined by means of the infra red spectrum of monogalactosyl glycerol, the "finger-print region" (730—960 cm^{-1}) of which was identical with the infra red spectrum of synthetic O-β-D-galactopyranosyl-(1→1)-D-glyceritol obtained by WICKBERG (106). Thus, the structures assigned to the 2 lipocarbohydrates extracted from wheat flour were determined as 2,3-diacyl-1-β-D-galactopyranosyl-D-glycerol   [= 1,2-O-diacyl-3-O-β-D-galacto-pyranosyl-sn-glycerol (1)] and 2,3-diacyl-1-(α-D-galactopyranosyl-1,6-β-D-galactopyranosyl)-D-glycerol   [= 1,2-diacyl-3-O-(α-D-galactopyrano-syl-1,6-β-D-galactopyranosyl)-sn-glycerol (2)].

3. Synthesis of Galactosyl Glycerol and Galactosyl Diglycerides

The chemical synthesis of galactosyl glycerols using the Koenig-Knorr reaction was carried out by WICKBERG (105, 106) in order to compare deacylated natural glycolipids from the red algae *Polysiphonia fastigiata* and *Corallina officinalis* with synthetic compounds. Therefore, special emphasis was laid on the preparation of the desired configuration of the glycerol moiety. The compounds synthesized included 1-O-α-D- (**9**), 1-O-β-D- (**10**), 3-O-α-D- (**11**) and 3-O-β-D-galactopyranosyl-*sn*-glycerol (**12**). Some ten years later Boos *et al.* (*16*) showed that β-galactosidase could transfer the galactose moiety from a β-galactoside like o-nitrophenyl-β-D-galactopyranoside (**13**) to 1,2-O-isopropylidene glycerol (**14**) to give 1,2-isopropylidene-3-O-β-D-galactopyranosyl glycerol (**15**), which, after mild acid hydrolysis yielded 42% of (**12**) (*Scheme 1*).

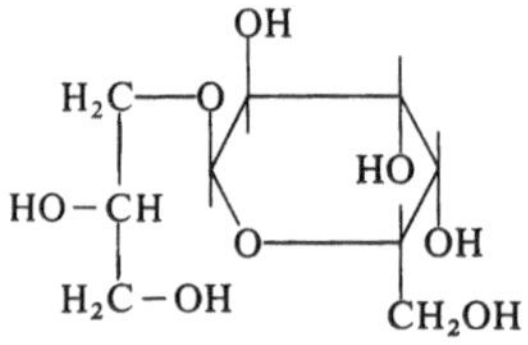

(9)

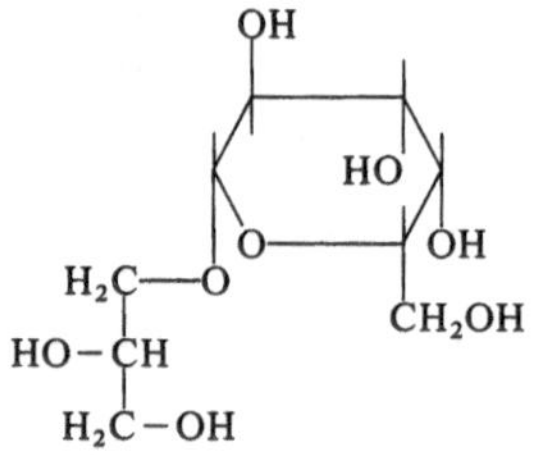

(10)

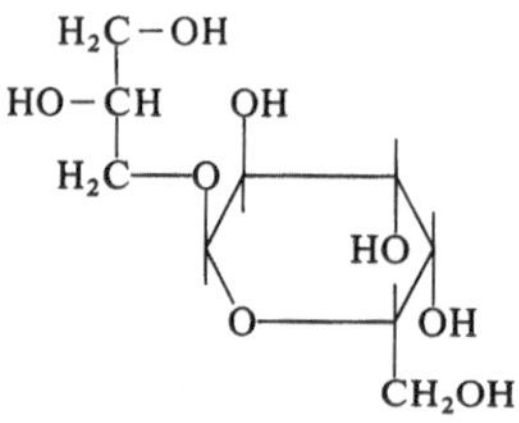

(11)

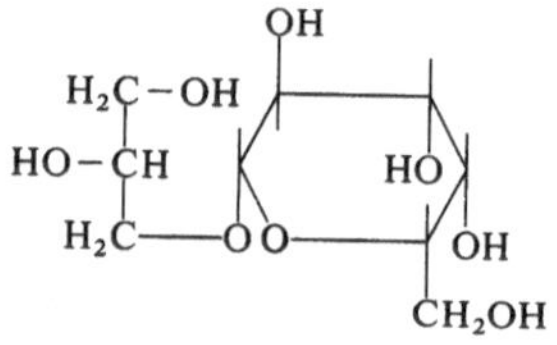

(12)

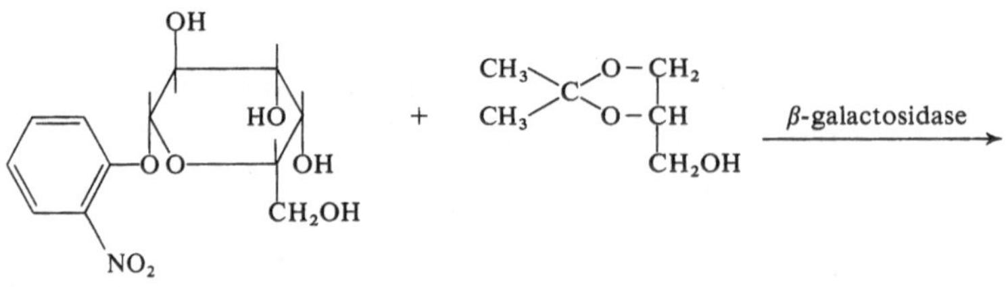

(13) (14)

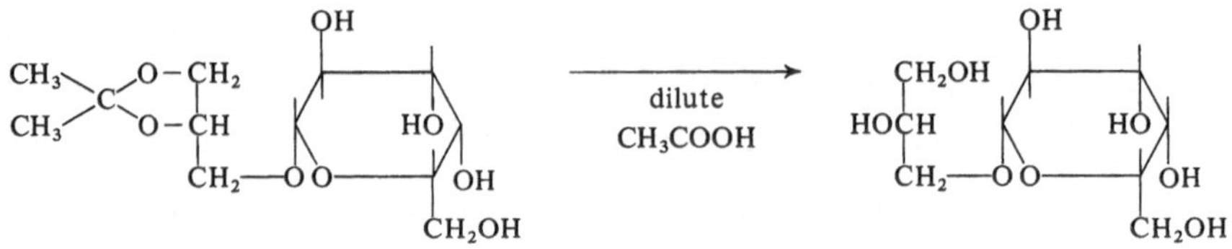

(15) (12)

Scheme 1. Synthesis of galactosyl glycerol

A method for the synthesis of galactosyl diglycerides also starting with (14) was investigated by WEHRLI and POMERANZ (*102*) (*Scheme 2*). Reaction of (14) with α-acetobromogalactose (16) yields racemic 1,2-isopropylidene-3-O-β-D-tetraacetogalactopyranosyl glycerol (17). Under

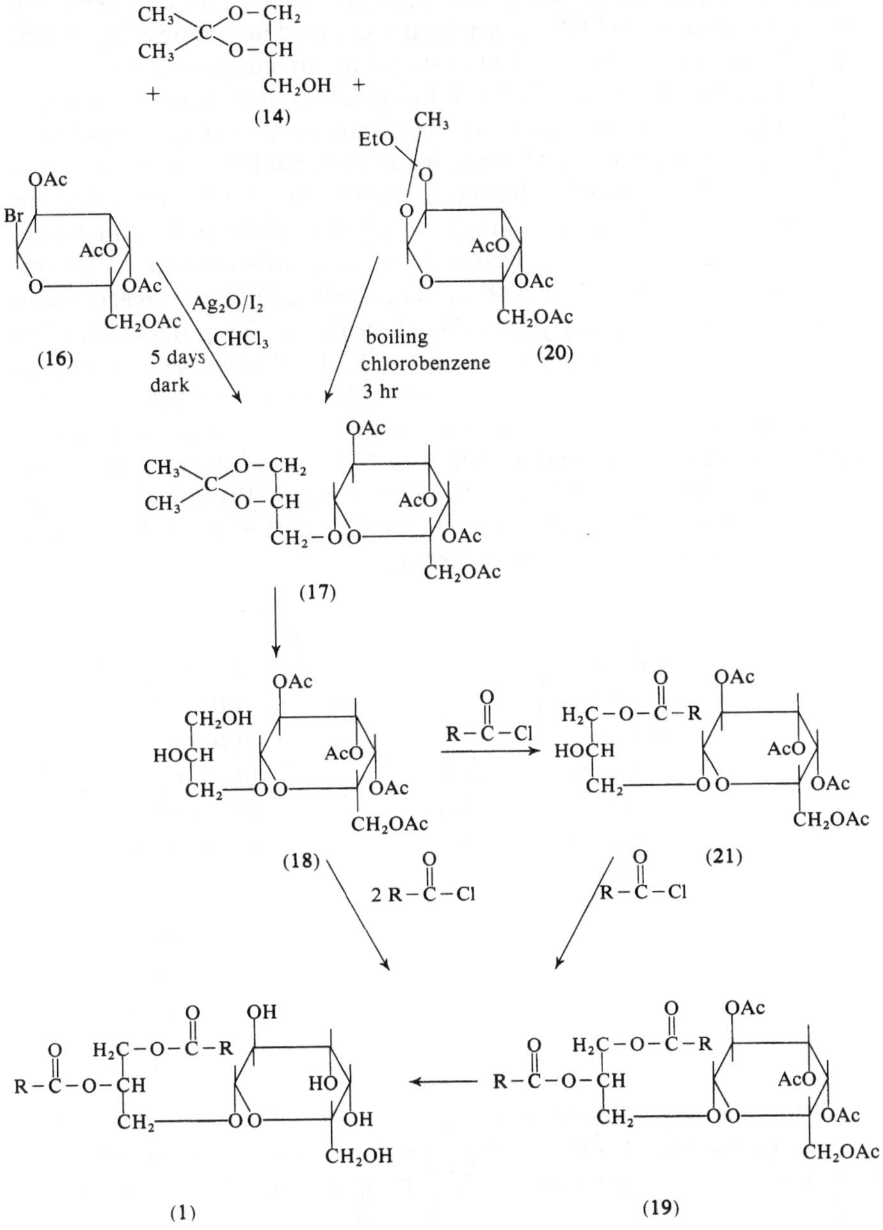

Scheme 2. Synthesis of monogalactosyl diglyceride (*92, 102*)

the conditions of this reaction the optical activity at C-2 of (14) is not preserved. After acid hydrolysis (pH 2) in hot 95% ethanol to rac-3-O-β-D-2,3,4,6-tetra-O-acetylgalactopyranosyl glycerol (18), fatty acid chlorides are attached to C-1 and C-2 of the glycerol moiety giving rac-1,2-di-O-acyl-3-O-(2,3,4,6-tetra-O-acetyl-β-D-galactopyranosyl)glycerol (19). The acetylated galactolipid (19) is then hydrazinolyzed in boiling 85% ethanol (15 min) to rac-1,2-di-O-acyl-3-O-β-D-galactopyranosylglycerol (1).

    If the preparation of galactosyl diglyceride is undertaken to compare the synthetic material with naturally occurring galactosyl diglycerides, this method is not very suitable because of racemization in the conversion of (14) to (17). Recently, however, SHVETS et al. (92) published an alternative method for the synthesis of glycosyl diglycerides which uses the orthoester method of glycosylation. Ethylorthoacetates of several glycoses [D-mannose, D-glucose and D-galactose (20)] were reacted with 1,2-O-isopropylidene-sn-glycerol (14) (Scheme 2). The product (17) was hydrolyzed in 50% aqeous acetic acid to (18). If (18) was acylated in benzene at 55—60° a monoacylic derivative (21) was produced. However, acylation of (18) in boiling toluene led to quantitative yields of the diacyl derivative (19). The second acylation of (21) also took place in boiling toluene yielding the diacyl derivative (19). Finally (19) was deacetylated by hydrazinolysis in boiling 85% ethanol (25 min) giving 1,2-di-O-acyl-3-O-β-D-galactopyranosyl-sn-glycerol (1).

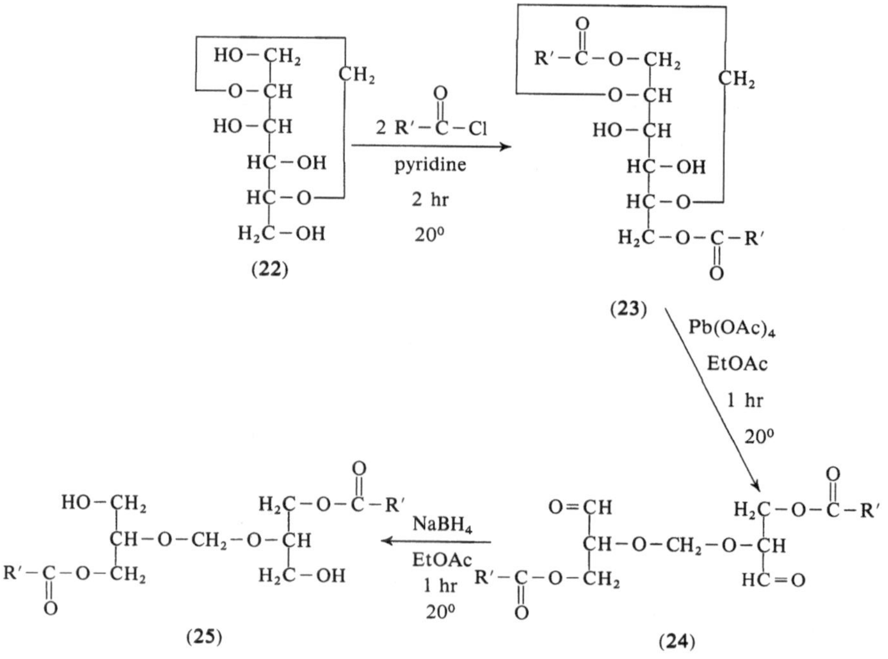

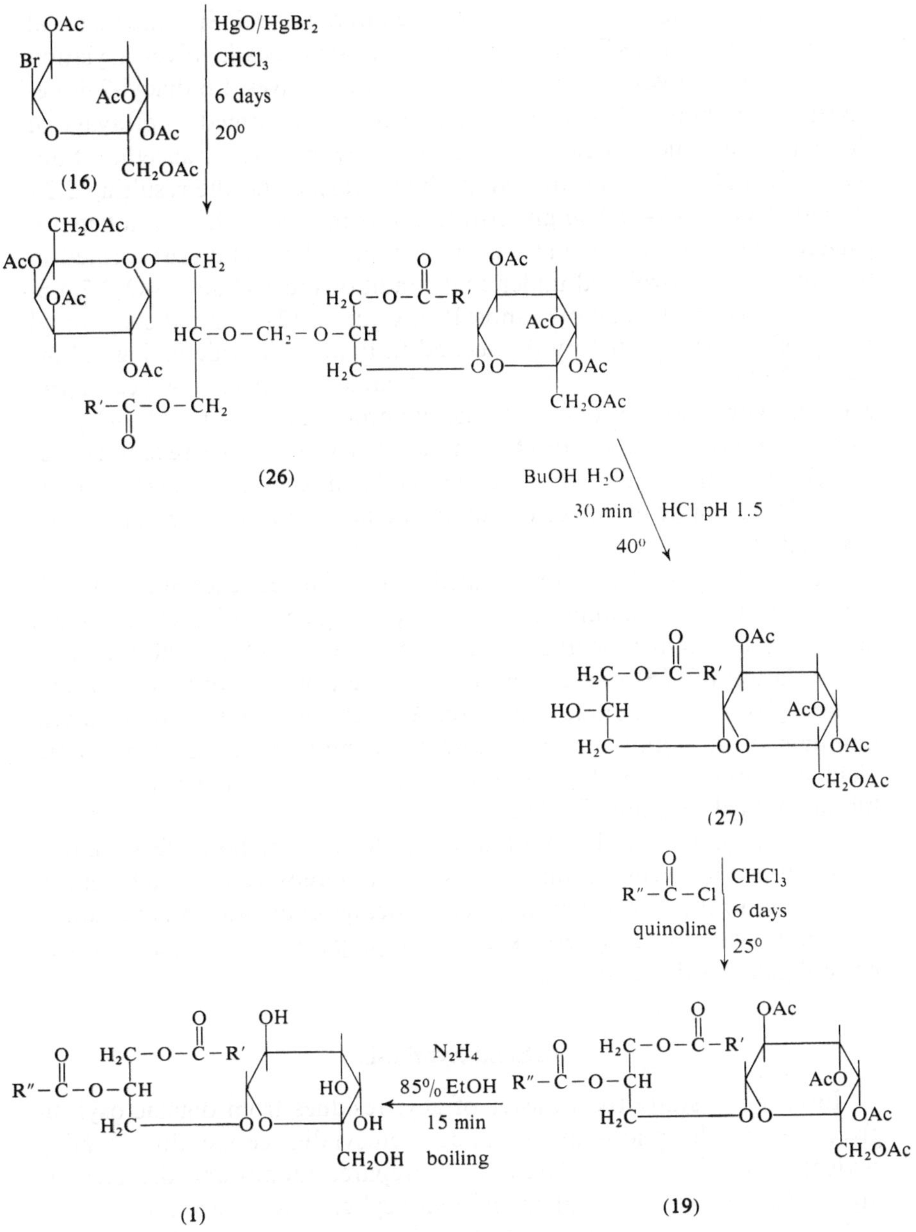

Scheme 3. Synthesis of monogalactolipid (*102*)

In order to preserve the optical activity at C-2 of the glycerol moiety and to be able to attach two different fatty acids specifically at the C-1 and C-2 positions of the *sn*-glycerol WEHRLI and POMERANZ (*102*) investigated the reaction sequence presented in *Scheme 3*. Starting material here is 2,5-methylene-D-mannitol (**22**). The synthesis involves acylation of the primary hydroxyl groups of (**22**) which gives 1,6-diacyl-2,5-me-thylene-D-mannitol (**23**). Then an aldehyde (**24**) is formed by cleavage of the mannitol moiety between C-3 and C-4 with lead tetraacetate. Subsequently, (**24**) is reduced by sodium borohydride and the resulting 2,2'-O-methylene-bis-(1-acyl-*sn*-glyceritol) (**25**) is reacted with α-acetobromo-galactose (**16**) by the Koenig-Knorr reaction. The acetal (**26**) formed in this manner is hydrolyzed under acid conditions to 1-O-acyl-3-O(2,3,4,6-tetra-O-acetyl-β-D-galactopyranosyl)-*sn*-glycerol (**27**), and the hydroxyl group of (**27**) is acylated by the second fatty acid chloride giving 1,2-di-O-acyl-3-O-(2,3,4,6-tetra-O-acetyl-β-D-galactopyranosyl)-*sn*-glycerol (**19**), which is hydrazinolyzed (*39, 99*) to the end product 1,2-di-O-acyl-3-O-β-D-galactopyranosyl-*sn*-glycerol (**1**). The overall yield of the reactions (**22**) to (**27**) was only 2.3% and the deacetylated synthetic galactosyl di-glyceride was optically inactive due to the non-stereospecific reaction of (**25**) and (**16**).

SHVETS *et al.* (*92*) too investigated a reaction sequence starting with 2,5-O-methylene-D-mannitol. The reaction sequence is largely the same as the one shown in *Scheme 3* starting from (**25**) except that the 1,2-*trans*-glycosylation of (**25**) was carried out with tert-butyl orthoesters of ace-tylated glycoses in boiling chlorobenzene in the presence of catalytic amounts of 2,6-lutidinium perchlorate. Starting from (**25**) the overall yield of monogalactosyl diglyceride was 34%, while the products ex-hibited optical rotation (Table 2).

In conclusion it can be said that the orthoester method offers the best possibilities for glycosylation because of the stereospecific attachment of the glycose to the glycerol moiety. The Koenig-Knorr attachment cannot be used if synthetic diglycerides are to be prepared for comparison with natural galactosyl diglycerides.

4. Semi-Synthesis

In order to study the transfer of acyl residues from digalactosyl di-glyceride and phospholipids to monogalactosyl diglyceride, thus forming acylgalactosyl diglyceride, HEINZ (*42*) prepared mono- and digalactosyl diglycerides having identical or different acyl groups attached to C-1 and C-2 of the glycerol moiety by starting from natural lipids. Basically the method achieves an acyl exchange. The advantage of this semi-synthetic method is that the stereochemistry of the anomeric bond between the galactose and glycerol moieties is preserved. The starting material con-sisted of naturally-occurring galactolipids which were extracted from *Sinapsis alba* and *Spinacia oleracea* leaves. The acyl residues of the galactolipid are hydrogenated (*43*). Next the hydroxyl hydrogens of the galactose moiety are substituted by O-(1-methoxyethyl)groups giving 1,2-di-O-acyl-3-O-[2,3,4,6-tetra-O-(1-methoxyethyl)-β-D-galactopyranosyl]-*sn*-glycerol (**28**) (*Scheme 4*). Subsequently the acyl residues in (**28**) were split off with sodium methoxide yielding 3-O-[2,3,4,6-tetra-O-(1-methoxyethyl)-β-D-galactopyranosyl]-*sn*-glycerol (**29**). At this point in the reaction sequence the two paths of preparing galactosyl diglycerides with the same or two different acyl residues at C-2 and C-3 of the glycerol moiety diverge. In order to prepare galactosyl diglycerides with the same fatty acyl residues (**29**) is reacted with fatty acid chloride in pyridine at room temperature. The protecting groups of the product (**30**) are then hydrolyzed by heating with boric acid under anhydrous conditions, a procedure which gives rise to 1,2-di-O-acyl-3-O-β-D-galactopyranosyl-*sn*-glycerol (**1**). In this way monogalactosyl diglycerides were prepared having 2 acetyl (**31**), palmitoyl (**32**), or oleoyl (**33**) groups. Yields varied between 20—33%. With digalactosyl diglycerides the same procedure can be followed; compounds synthesized by HEINZ (*42*) include digalactosyl diglycerides with 2 acetyl (**34**) or oleoyl (**35**) groups, the yields being 18% and 24% respectively.

In preparing galactosyl diglycerides with two different acyl groups use has been made of a reaction described by DE HAAS and VAN DEENEN (*40*) for the synthesis of lysolecithins (*Scheme 5*). They reacted racemic glycerol-1-benzylether (**36**) with stearoyl chloride at low temperature and found

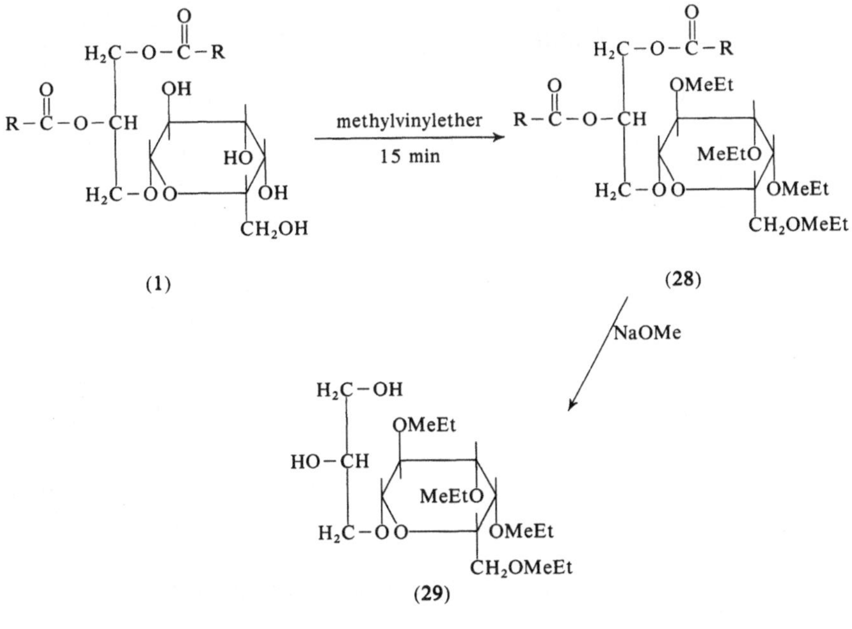

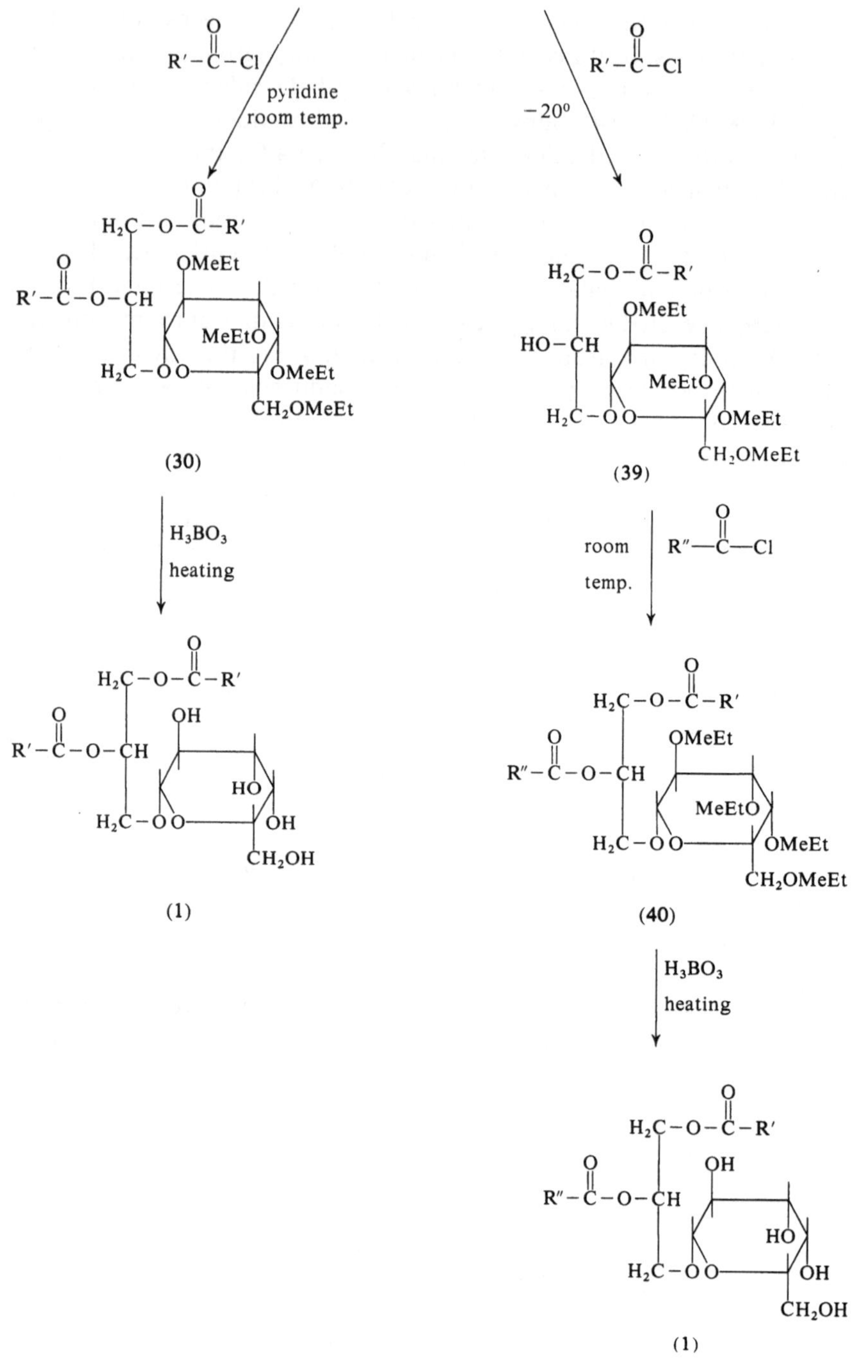

Scheme 4. Synthesis of different monogalactolipids (42)

the products to be rac-3-stearoyl- (**37**) and rac-2-stearoyl-glycero-1-benzylether (**38**) in a molar ratio of 15:1. In the reaction sequence of HEINZ (*Scheme 4*) (**29**) is reacted with fatty acid chloride at $-20°$ giving the lyso-compound (**39**), which after purification by column chromatography is then acylated with a second fatty acid chloride at room temperature. The product (**40**) is subsequently hydrolyzed with boric acid giving a monogalactosyl diglyceride with two different fatty acyl residues (**1**). By this method a mono- and digalactosyl diglyceride with oleate at C-1 and a palmitoleate at C-2 (**41**, **42**) were made.

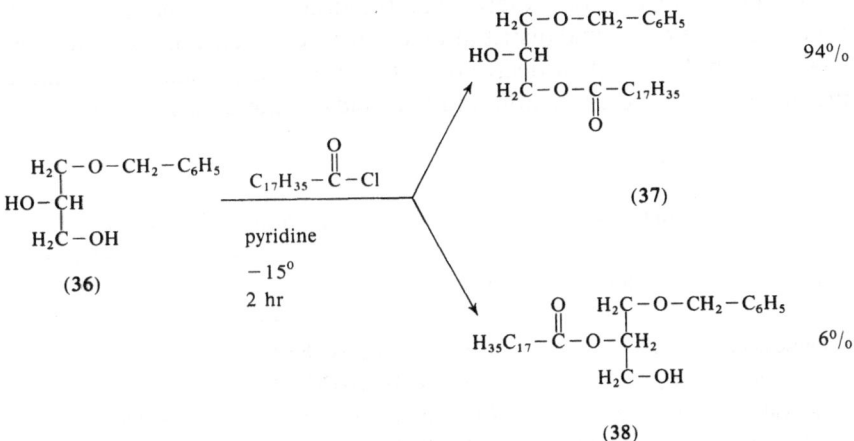

$$H_2C-O-R_1$$
$$OH$$
$$R_2-O-CH$$
$$HO$$
$$H_2C-O O OH$$
$$CH_2OR_3$$

(31) $R_1 = R_2 =$ acetyl, $R_3 =$ H

(32) $R_1 = R_2 =$ palmitoyl, $R_3 =$ H

(33) $R_1 = R_2 =$ oleoyl, $R_3 =$ H

(34) $R_1 = R_2 =$ acetyl, $R_3 = a$-D-galactopyranosyl

(35) $R_1 = R_2 =$ oleoyl, $R_3 = a$-D-galactopyranosyl

(41) $R_1 =$ oleoyl, $R_2 =$ palmitoleoyl, $R_3 =$ H

(42) $R_1 =$ oleoyl, $R_2 =$ palmitoleoyl, $R_3 = a$-D-galactopyranosyl

$$H_2C-O-CH_2-C_6H_5$$
$$HO-CH \qquad 94\%$$
$$H_2C-O-C-C_{17}H_{35}$$
$$O$$

(37)

$$H_2C-O-CH_2-C_6H_5 \qquad O$$
$$HO-CH \qquad C_{17}H_{35}-C-Cl$$
$$H_2C-OH$$

(36)

pyridine

$-15°$

2 hr

$$O \quad H_2C-O-CH_2-C_6H_5$$
$$H_{35}C_{17}-C-O-CH_2 \qquad 6\%$$
$$H_2C-OH$$

(38)

Scheme 5. Synthesis of stearoyl-glycero-1-benzylethers (*40*)

Later, Heinz (44) used these routes to synthesize radioactive mono-galactosyl diglyceride by using lipids that were labelled with ^{14}C from leaves of *Canna indica*. The end product was 1,2-di-O-oleoyl-3-O-β-D-(U-^{14}C)galactopyranosyl-sn-(U-^{14}C)glycerol. For synthesis of digalacto-syl diglyceride he used unlabelled leaf lipids as starting material. In *Scheme 4* (**29**) was reacted with (9,10-$^3H_2$)oleoyl chloride yielding in the end 1,2-di-O-(9,10-$^3H_2$)oleoyl-3-O-(6-O-α-D-galactopyranosyl-β-D-galac-topyranosyl)-sn-glycerol. By doing so he was able to discriminate between acyl-transfer from mono- to monogalactosyl diglyceride and from di- to monogalactosyl diglyceride.

Another approach was investigated by Shvets *et al.* (*92*). Glycosyl diglycerides were prepared by direct glycosylation of 1,2-di-O-palmitoyl-sn-glycerol with orthoesters of several glycoses under the conditions used for the reaction of (**20**) to (**17**) in *Scheme 2*. Deacetylation of the product (**19**) (with R = palmitoyl) was effected in the usual way by boiling in a solution of hydrazine in 85% ethanol for a short time.

5. Fatty Acid Content

The galactolipids of photosynthetic tissues contain large amounts of polyenoic fatty acids (Table 3). In leaves of higher plants more than 80% of the fatty acids of mono- and digalactosyl diglycerides contain 3 double bonds. α-Linolenic acid ($C_{18:3}$) is the fatty acid most abundant in the galactolipids of chloroplasts of higher plants (Table 4). Linolenic acid occurs only in minor amounts in lower photosynthetic organisms like the "red tide" organism *Gonyaulax polyedra* (*77*), marine and freshwater diatoms (*51*) and red and brown algae (*52*). In these organisms long chain fatty acids with 4, 5 or 6 double bonds are found ($C_{18:4}$, $C_{20:5}$, and $C_{22:6}$).

In the monogalactosyl diglyceride fraction of *Chlorella vulgaris* and other algae the positional distribution of fatty acids depends more on their chain length than on their degree of unsaturation, C_{18} acids accumulating preferentially in the 1-position and C_{16} acids in the 2-position (*87*).

Table 3. *Structures of Fatty Acids Occurring in Galactolipids of Higher Plants*

Name	Symbol	Structure
Palmitic acid	$C_{16:0}$	$CH_3-(CH_2)_{14}-COOH$
Stearic acid	$C_{18:0}$	$CH_3-(CH_2)_{16}-COOH$
Oleic acid	$C_{18:1}(9c)$	$CH_3-(CH_2)_7-CH \overset{C}{.} CH-(CH_2)_7-COOH$
Linoleic acid	$C_{18:2}(9c,12c)$	$CH_3-(CH_2)_3-(CH_2-CH \overset{C}{.} CH)_2-(CH_2)_7-COOH$
α-Linolenic acid	$C_{18:3}(9c,12c,15c)$	$CH_3-(CH_2-CH \overset{C}{.} CH)_3-(CH_2)_7-COOH$

Table 4. *Fatty Acid Composition of Galactolipids*

Plant	Fatty acids (% of total)							References
	16:0	16:1	16:3	18:0	18:1	18:2	18:3	
Runner bean leaves								(88)
MGDG	2	t	—	t	t	2	96	
DGDG	5	t	—	1	t	1	93	
Spinach leaves								(2)
MGDG	t	t	30	—	1	1	67	
DGDG	6	—	3	1	4	3	84	
Castor leaves								(73)
etiolated								
MGDG	7	3	—	—	7	14	67	
DGDG	14	2	—	—	8	21	51	
after 20 hrs illumination								
MGDG	4	—	—	—	3	4	88	
DGDG	15	2	—	—	7	7	65	

Abbreviations: MGDG = monogalactosyl diglyceride, DGDG = digalactosyl diglyceride, t = trace

IV. Biosynthesis

The first studies of galactolipid biosynthesis were carried out by BENSON's group in 1958 (*12, 30*). They found rapid incorporation of $^{14}CO_2$ into monogalactosyl diglyceride and slower entry of label into digalactosyl diglyceride of photoautotrophically growing *Chlorella pyrenoidosa*. Within 5 min more than half of the label incorporated into lipids was found in the galactosyl moiety of the galactolipids. Based upon these findings they proposed the following scheme for the biosynthesis of mono- and digalactosyl diglycerides in plants:

$$CO_2 \;\rightarrow\; PGA \rightarrow\rightarrow\rightarrow\; \text{UDP-glucose} \rightarrow \text{UDP-galactose} \xrightarrow{\text{diglyceride}}$$

$$\text{monogalactosyl diglyceride} \xrightarrow{\text{UDP-galactose (5)}} \text{digalactosyl diglyceride}$$

This scheme was confirmed by NEUFELD and HALL (*66*) who found that galactose from UDP-galactose was transferred to an endogenous acceptor in spinach chloroplasts with the formation of mono-, di-, tri-, and tetra-galactosyl diglycerides. The transfer was inhibited by bi- and tri-

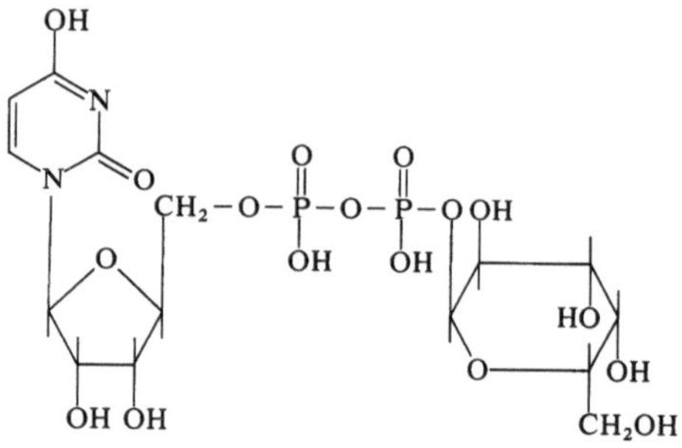

OH
$$\text{(structure of CDP-galactose)}$$

(5)

valent cations and by UTP. Only UDP-glucose of several other nucleotide glycoses studied was also able to function as a glycosyl donor, presumably after epimerization to UDP-galactose by UDP-galactose-4'-epimerase. The enzyme and the endogenous acceptor were separated from each other by preparation of an acetone powder from spinach chloroplasts. This acetone preparation was also used by ONGUN and MUDD (75) who observed incorporation of (^{14}C)-galactose from UDP-(^{14}C)-galactose into monogalactosyl diglyceride, but not into digalactosyl diglyceride when diolein was used as acceptor. Dipalmitine could not serve as acceptor. For digalactosyl diglyceride formation the preferred acceptor was monogalactosyl diglyceride. Because of experiments with fragmented chloroplasts and stroma in which the monogalactosyl/digalactosyl diglyceride ratio after incorporation of labelled galactose from UDP-galactose into galactolipid was different they concluded that the enzyme responsible for the synthesis of monogalactosyl diglyceride is more tightly bound to the chloroplast membranes than that responsible for the formation of digalactosyl diglyceride. Later, MUDD et al. (64) showed that in their preparation the pH optimum for galactolipid synthesis from UDP-galactose was 7.2. At higher pH values the proportion of monogalactosyl diglyceride increased and that of di- and trigalactosyl diglyceride decreased. Maximal incorporation of galactose from UDP-galactose into galactolipids was attained at 45°, the proportion of monogalactosyl diglyceride increasing with temperature. They also showed that the biosynthesis of monogalactosyl diglyceride in the presence of spinach acetone powder proceeds most efficiently with more highly unsaturated diglyceride acceptors. Addition of diglyceride acceptors to chloroplast suspensions did not have any effect.

Similar data were obtained by BAJWA and SASTRY (8). These workers found that dialysed spinach leaf homogenates and chloroplast preparations incorporated 1-^{14}C-labelled palmitate, stearate and linolenate into monogalactosyl diglyceride in a reaction that was enhanced by CoA and ATP. A higher rate of incorporation was observed with unsaturated fatty acids than with saturated fatty acids. Addition of galactosyl glycerol significantly stimulated only the incorporation of oleate and linoleate.

Evidence that the formation of monogalactosyl diglyceride does not depend on the degree of unsaturation of the diglyceride acceptor of the galactose moiety was presented by ECCLESHALL and HAWKE (28). Unlike MUDD et al. (64), they observed the same incorporation of ^{14}C-galactose from UDP-^{14}C-galactose into galactolipids when synthetic diglycerides, which differed in the degree of unsaturation of their fatty acids, were added to acetone powders of spinach chloroplasts. With all the diglycerides tested, monogalactosyl diglyceride was the main product with little accompanying synthesis of digalactosyl diglyceride. It was concluded that the galactosylation of diglycerides is not specifically directed towards polyunsaturated diglycerides. The polyunsaturated monogalactosyl diglyceride could arise either by desaturation of the fatty acyl residues subsequent on monogalactosyl diglyceride formation or by transacylation.

A galactolipid synthesizing system in a soluble subchloroplast fraction of spinach was also described by CHANG and KULKARNI (23). With UDP-galactose as the sole substrate the K_m for the monogalactosyl diglyceride activity was 4.0×10^{-4} mM and that for the digalactosyl diglyceride activity was 2.2×10^{-4} mM. The enzyme was completely inhibited by Hg^{2+}, but this inhibition could be abolished by addition of mercaptoethanol (22). It is therefore plausible that the active centre of this enzyme contains one or more sulphydryl groups.

Inhibition of galactolipid biosynthesis by sulphydryl reagents (N-ethylmaleimide, $CdCl_2$ and p-hydroxymercuribenzoate) was also observed by MUDD et al. (63). The inhibition of the galactolipid biosynthesis was in the order tri- > di- > monogalactosyl diglyceride. Galactolipid biosynthesis was also inhibited by ozone as a consequence of the peroxidation of the double bonds in the unsaturated fatty acids.

APPELQVIST et al. (5) observed an increase of ^{14}C-incorporation in the linolenic acid residues of the monogalactosyl diglyceride fraction of greening barley leaves, while there was no increase of label in the linolenic acid residues of the digalactosyl diglyceride fraction. Only after two cell generations was label also found in the linolenic acid residues of the digalactosyl diglyceride fraction.

Noting the clear difference in fatty acid composition between mono- and digalactosyl diglyceride, with monogalactosyl diglyceride containing more highly unsaturated fatty acyl residues than digalactosyl diglyceride, BLOCH et al. (14) concluded that direct galactosylation of mono- to digalactosyl diglyceride was highly improbable, unless it is assumed that the highly unsaturated fatty acyl residues in monogalactosyl diglyceride are also chemically reduced to a considerable extent in the course of this process.

On the other hand, Bolling and El Bayã (*15*), studying the fatty acid content of wheat galactolipids during various stages of growth, observed appreciable changes in the content of digalactosyl diglyceride during maturation. This led them to conclude that for the biosynthesis of digalactosyl diglyceride UDP-galactose prefers monogalactosyl diglycerides that contain few polyunsaturated acyl moieties.

Consistent with the observations of both Bloch *et al.* (*14*) and Bolling and El Bayã (*15*) were the results of Gurr (*38*). On labelling studies with ^{14}C-acetate he showed that newly synthesized galactosyl diglycerides contain mainly saturated fatty acyl residues. Subsequent to *de novo* synthesis a series of alterations of fatty acid structure can take place within the same molecule.

Extracts of the green alga *Euglena gracilis* specifically catalyse the incorporation of acyl groups from acyl carrier protein (ACP) thiolesters into monogalactosyl diglycerides (*82*). This reaction is stimulated by α-glycerophosphate. CoA thiolesters are also transferred into monogalactosyl diglycerides, but these thiolesters are incorporated into phospholipids too. In 1970 Douce *et al.* (*27*) observed incorporation of ^{14}C-glycerophosphate into phosphatidate in a reaction catalyzed by chloroplasts from spinach and *Zea mays* and from etioplasts from *Zea mays*. The reaction required the addition of fatty acids, CoASH and ATP. If UDP-galactose was also included in the reaction mixture, label was found in the monogalactosyl diglyceride fraction, but not in the digalactosyl diglycerides. They concluded therefore that the galactosylation of diglyceride is spatially separated from that of monogalactosyl diglyceride in accordance with earlier observations of Ongun and Mudd (*75*).

The galactolipid synthesizing system in photoautotrophic *Euglena gracilis* chloroplasts differs from that in spinach chloroplasts (*58, 61*). In an 1 hour incubation of *Euglena gracilis* chloroplasts with UDP-^{14}C-galactose 70% of the total amount of incorporated galactose was transferred within 2 min, while in spinach chloroplasts only 15% was transferred within the same time. The molar ratio of mono- to digalactosyl diglyceride was 1 : 2 and 3 : 1 in *Euglena* and spinach chloroplasts respectively. With the *Euglena* enzyme the same effect of Hg^{2+} and mercaptoethanol was observed. After centrifugation at $35,000 \times g$ of an ultrasonicated suspension of *Euglena* cells and incubation of the pellet and supernatant fractions with UDP-galactose, the ratio of mono- to digalactosyl diglyceride was 0.79 and 1.82 respectively. This indicated that the digalactosyl diglyceride synthesizing enzyme was more tightly membrane bound than that involved in monogalactosyl diglyceride synthesis in contrast with the results obtained with spinach chloroplasts (*75*).

An acetone powder prepared from *Euglena* chloroplasts did not exhibit galactosyltransferase activity when incubated with UDP-galactose and monogalactosyl diglyceride (*58*). A direct galactosylation of mono- to digalactosyl diglyceride seemed therefore impossible. An alternative way of synthesizing digalactosyl diglyceride would involve the transfer of two galactosyl moieties from UDP-galactose to an endogenous lipophilic acceptor, which, in turn, would transfer the digalactosyl unit to diglyceride:

2 UDP-galactose + acceptor → gal-gal-acceptor + 2 UDP

gal-gal-acceptor + diglyceride → digalactosyl diglyceride + acceptor

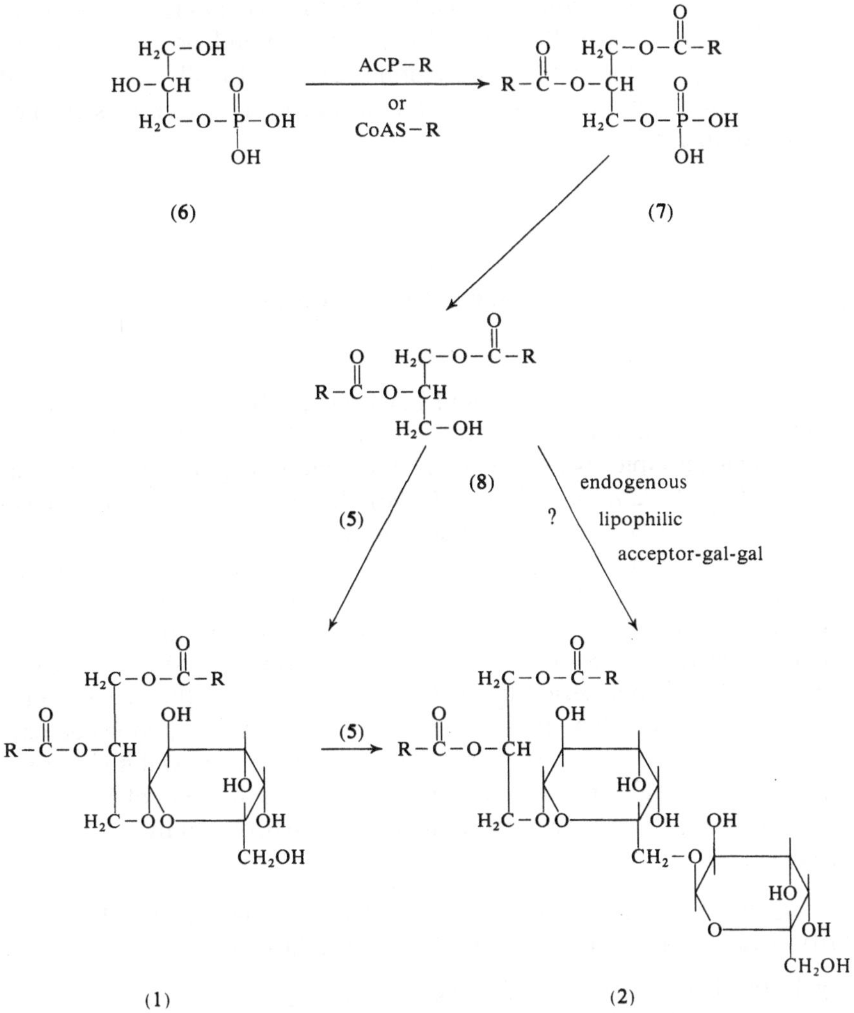

Scheme 6. Biosynthesis of mono- and digalactosyl diglyceride (*58, 82*)

It was suggested by Lin and Chang that phosphatidic acid, which was found by Renkonen and Bloch (*82*) to be an intermediate in monogalactosyl diglyceride synthesis, could play the role of endogenous acceptor of the two galactosyl units in analogy with similar lipophilic intermediates found in the biosynthesis of bacterial cell wall polysaccharides (*3, 109*).

At present all data which have been obtained so far on the biosynthesis of galactolipids are compatible with a pathway for galactolipid formation that involves transfer of saturated (*38*) fatty acyl groups from ACP or CoA thiolesters to α-glycerophosphate (**6**) (*8, 27, 82*) with formation of phosphatidic acids (**7**) (*Scheme 6*). After dephosphorylation of the products to 1,2-diglycerides (**8**), these compounds can either react with a galactosyl moiety from UDP-galactose to give monogalactosyl diglyceride (**1**) or with a digalactosyl unit from an endogenous lipophilic acceptor to give digalactosyl diglyceride (**2**) (*58*). Finally the saturated fatty acyl residues in the mono- and digalactosyl diglycerides formed by this route can be desaturated.

V. Enzymatic Hydrolysis

It was found by Kates (*49*) and Ferrari and Benson (*30*) that galactolipids have a high turnover rate. The enzymes necessary for the breakdown of galactolipids were shown to be present in plant tissues by various authors (*32, 34, 36, 46, 47, 62, 89, 90*). In leaves of the runner bean (*Phaseolus multiflorus*) galactolipase activity was observed by Sastry and Kates (*89, 90*). Galactolipase activity was associated with the chloroplasts, but was also shown to be present in a soluble form in the cytoplasm. The pH optima for the mono- and digalactosyl hydrolyzing enzyme were found to be 7.5 and 5.9 respectively. The chloroplast associated activity for digalactosyl diglyceride decreased after storage of the enzyme preparation at 4° for a few days. However, the activity for monogalactosyl diglyceride did not decrease. The soluble galactolipase exhibited hydrolyzing activity only for digalactosyl diglyceride. On the basis of these data they concluded that the two galactolipids were hydrolyzed by two distinct enzymes:

monogalactosyl diglyceride → (monogalactosyl monoglyceride) + free fatty acid → monogalactosyl glycerol + free fatty acid

digalactosyl diglyceride → (digalactosyl monoglyceride) + free fatty acid → digalactosyl glycerol + free fatty acid

Further hydrolysis of the galactosyl glycerols is catalyzed by α- and β-galactosidases as follows:

digalactosylglycerol $\xrightarrow{\text{α-galactosidase}}$ monogalactosyl glycerol

$\xrightarrow{\text{β-galactosidase}}$ glycerol + galactose

In the above scheme mono- and digalactosyl monoglyceride are formed as intermediate products. However, the authors could not detect these intermediates in their preparations. SASTRY and KATES also found low galactolipase activities in spinach leaves, namely 1% for the monogalactosyl diglyceride hydrolyzing enzyme and 3% for the digalactosyl diglyceride hydrolyzing enzyme after 30 min incubation at 30°. Galactolipase activity for monogalactosyl diglyceride in preparations from young spinachleaves was found by HELMSING (46) to be of the same order of magnitude. However, the activity for digalactosyl diglyceride was much higher. He also found a decrease in activity for digalactosyl diglyceride after a few days in storage at 4° but the activity for monogalactosyl diglyceride reached a maximum after 10—11 days. Later, HELMSING (47) purified the mono- and digalactolipase activity from the cytoplasmic fraction of runner bean leaves remaining after centrifugation at 105,000 × g. At each step of the purification procedure the relative specific activities towards both substrates increased by the same factor. Electrophoresis of the purified preparation at two different pH's and on a 5 M urea gel resulted in a single band. His conclusion was, therefore, that the two hydrolase activities were combined within one molecule. The pH optima for mono- and digalactolipase activity were found to be 7.0 and 5.6 respectively. He assumed, therefore, that the enzyme would undergo a reversible allosteric transformation under the influence of the pH. Both enzyme activities were completely inhibited by cysteine but activated by strong reductants like sodium dithionite ($Na_2S_2O_4$) and sodium metabisulfite ($Na_2S_2O_5$). This activation could be due to reduction of quinones that are formed during oxidation of phenolic compounds which can be responsible for enzyme inhibition (4, 59). The molecular weight of the galactolipase was calculated to be 110,000.

In the particle-free supernatant fraction from potato tubers, GALLIARD (34) observed phospholipid- and galactolipid-acyl hydrolase activity, which was associated with acyl transferase activity. Monogalactosyl diglyceride was particularly susceptible to hydrolysis and monogalactosyl monoglyceride was detected as an intermediate product. After partial purification (35) he observed that the acylhydrolase activities for phospholipids, galactolipids and mono- and diglycerides showed many similarities with respect to subcellular localization, molecular size and charge and their behaviour with substrates, inhibitors and detergents.

VI. Function

Since galactolipids are the major lipids in chloroplasts, it is evident that they play an important role in the structure of thylakoids (*85, 96*). Apart from this they have been assigned several other functions including that of being an intermediate in fatty acid synthesis and functioning in the electron transport system of chloroplasts. In this chapter these and other functions will be discussed in some detail.

1. Galactolipids in Chloroplast Membrane Structure

Chloroplast membranes are composed of 45% protein and 55% lipophilic material, 80% of which is galactolipid (*74, 101*). In the model of Weier and Benson thylakoid membranes consist of lipoprotein subunits, consisting of a protein core surrounded by compounds determined by the nature and environment of the membrane (*101*). A similar model was proposed by Branton and Park (*18*). Since the stroma and the loculi of chloroplasts contain aqeous materials, it was proposed by Weier and Benson (*101*) that the membranes bordering these spaces bind the surface active glycolipids. Their polar moieties protrude from the surface into the aqeous phase and the highly unsaturated fatty acyl residues occurring in the galactolipids are associated with the hydrophobic regions of the lipoprotein units. The firmness of binding of digalactosyl diglyceride into the thylakoid membranes seems to be greater than that of monogalactosyl diglyceride. After centrifugation of suspensions of thylakoids and grana stacks in buffers of different pH values the monogalactosyl diglyceride/digalactosyl diglyceride ratio in the supernatant was found to increase with decreasing proton concentration (*41*). This result was confirmed by a study of Costes et al. (*26*) on differential extraction of lipids from lyophilized chloroplasts. Rosenberg (*83*) suggested that the hydrophobic regions of the polyunsaturated fatty acyl residues with 3 cis double bonds form a key which fits into the lock provided by the 3 methyl side-groups of the phytyl group of chlorophyll. This hydrophobic association would keep the chlorophyll molecules in the proper orientation required for an efficient photoreceptive surface (*48*).

When green cells of *Euglena gracilis* are kept in the dark for as long as 100 hrs, the amount of galactosyl diglyceride diminishes until the chlorophyll/galactosyl diglyceride ratio is 1 : 2. Thereafter, the organisms begin to disintegrate because chlorophyll and galactosyl diglyceride disappear simultaneously (*83*).

2. Galactolipids and the Synthesis of Fatty Acids

Much attention has focused on the fatty acid composition of galactosyl diglycerides during greening of chloroplasts. The fatty acid composition is strongly dependent on light and on non-lipidic exogenous metabolites

(84). When *Chlorella vulgaris* cells are grown in the dark, fatty acid synthesis from (2-^{14}C)-acetate is almost entirely limited to the production of saturated and monoenoic acids. In light-incubated cells both saturated and unsaturated fatty acids are synthesized. In *Chlorella vulgaris* NICHOLS et al. (72) found that monogalactosyl diglyceride belongs to a group of lipids with very high fatty acid turnover rate, while digalactosyl diglyceride belongs to a group with slow turnover rate for fatty acids. They suggested, therefore, that the group of lipids with rapid fatty acid turnover rate could be involved in sequences of saturated and unsaturated fatty acid synthesis, and thus could have an intermediate function in fatty acid biosynthesis (69).

After incorporation of $^{14}CO_2$ and (1-^{14}C)-acetate into the phospho- and glycolipids of pumpkin leaves the highest specific activity is found in phosphatidyl choline. This activity rapidly decreased which indicates a high turnover of fatty acids. The turnover rate of mono- and digalactosyl diglycerides is much slower. It is therefore suggested that α-linolenic acid is transferred from phosphatidyl choline to the galactosyl diglycerides. This transfer can be catalyzed by acyl transferases (86).

In the blue-green alga *Anabaena variabilis*, however, monogalactosyl diglyceride seem to act as an intermediate in linoleate biosynthesis (6). Incubation of dark grown *Euglena gracilis* cells with sodium-(1-^{14}C)-octanoate in the dark for 6 hr resulted in a specific incorporation of radioactivity into neutral lipids. After illumination radioactivity in phospholipids increased during 24 hr and thereafter decreased, while radioactivity in sulpholipid, phosphatidyl glycerol and the galactolipids steadily increased (78). Here too, there seems to be a transfer of fatty acids from neutral lipids via phospholipids to the galactolipids. In contrast with the evidence for transfer of fatty acids to galactolipids is the finding of GURR (38). He showed that newly synthesized galactosyl diglycerides have mainly saturated fatty acids, which can be desaturated subsequently without transacylation being involved.

3. Galactolipids and Electron Transport

As a consequence of the unusual high amount of polyunsaturated fatty acyl residues in mono- and digalactosyl diglycerides and their location in the chloroplast, it has been thought that they have a function in the photosynthetic process which leads to the evolution of oxygen. It is proposed that galactolipids form relatively non-specific micellar components which constitute an organized medium of low dielectric constant. In this medium the transport chains that are inhibited by water can function properly (14, 25, 48).

In 1965 Chang and Lundin (24) found that in chloroplasts the photoreduction of cytochrome-c was increased by the addition of mono- and digalactosyl diglyceride, as well as by a number of redox compounds such as flavine mononucleotide, coenzyme Q_6 and α-tocopherol. Their conclusion was that galactolipids can take part in the light reaction in which the short wavelength system is involved. Udel'nova and Boichenko (98) isolated a complex containing 12% Mn, digalactosyl diglyglyceride and flavine from plants differing in their mode of metabolism. The complex had a high oxidation-reduction potential, so it was proposed that the complex could participate as an oxidizer in photosynthetic oxygen evolution. However, Wessels (103) has been able to isolate a particulate fraction from digitonin-treated chloroplasts which contained 75% protein and 25% lipid. The lipid portion was found to contain 8.4% nonpigment lipids. The fraction was able to reduce $NADP^+$ in the light, suggesting that this fraction represented a purified photosystem I preparation. The percentage of non-pigment lipids is significantly higher in whole chloroplasts and quantosome preparations (57). Because detergents, like digitonin, can apparently substitute in part for lipids in providing the photochemical apparatus with the appropriate medium and structure, such results do not support an active role for the galactolipids in photosynthesis.

4. Other Functions

Two other functions have been proposed for galactolipids. First a function as carbohydrate reservoir was proposed by the group of Benson (11) because of labelling studies with $^{14}CO_2$ which showed a rapid turnover of the sugar moiety. This energy storage function has also been proposed by the group of Nichols (72). Later, Benson's group proposed that galactolipids have a function in sugar transport through membranes (9, 30). This proposal was also based on an observed rapid turnover of the galactose residues. However, Roughan (86) was unable to detect significant turnover of galactose residues in galactolipids of spinach and pumpkin leaves.

In pine needles galactosyl diglycerides seem to be able to protect the thylakoid membranes against frost injury (13). On transfer of the trees from $-5°$ to $32°$ conversion of a fraction of digalactosyl diglyceride to monogalactosyl diglyceride was observed. This fraction contained very long-chain fatty acids ($C_{26:0}$). It was proposed that the very long-chain fatty acids may link different sub-units of the chloroplast and may contribute to an additional layer of H-bonded water on the membrane, thus forming a protective layer against frost injury.

References

1. ALLEN, C. F., P. GOOD, H. F. DAVIS, P. CHISUM, and S. D. FOWLER: Methodology for the Separation of Plant Lipids and Application to Spinach Leaf and Chloroplast Lamellae. J. Am. Oil Chem. Soc. **43**, 223 (1966).

2. ALLEN, C. F., P. GOOD, H. F. DAVIS, and S. D. FOWLER: Plant and Chloroplast Lipids. I. Separation and Composition of Major Spinach Lipids. Biochem. Biophys. Res. Comm. **15**, 424 (1964).

3. ANDERSON, J. S., M. MATSUHASKI, M. A. HASKIN, and J. L. STROMINGER: Lipid-phosphoacetylmuranyl-pentapeptide and Lipid-phosphodisaccharide-pentapeptide: Presumed Membrane Transport Intermediates in Cell Wall Synthesis. Proc. Nat. Acad. Sci. (USA) **53**, 881 (1965).

4. ANDERSON, J. W., and K. S. ROWAN: Extraction of Soluble Leaf Enzymes with Thiols and Other Reducing Agents. Phytochem. **6**, 1047 (1967).

5. APPELQVIST, L.-A., J. E. BOYNTON, P. K. STUMPF, and D. VON WETTSTEIN: Lipid Biosynthesis in Relation to Chloroplast Development in Barley. J. Lipid Res. **9**, 425 (1968).

6. APPLEBY, R. S., R. SAFFORD, and B. W. NICHOLS: The Involvement of Lecithin and Monogalactosyl Diglyceride in Linoleate Synthesis by Green and Blue-green Algae. Biochim. Biophys. Acta **248**, 205 (1971).

7. BAILEY, J. L., and A. G. WHYBORN: The Osmiophylic Globules of Chloroplasts. II. Globules of the Spinach-Beet Chloroplast. Biochim. Biophys. Acta **78**, 163 (1963).

8. BAJWA, S. S., and P. S. SASTRY: Incorporation of Fatty Acids into Monogalactosyl Diglycerides by Spinach Leaf Cell-free Preparations. Biochem. J. **128**, 44 P (1972).

9. BENSON, A. A.: The Plant Sulfolipid. Adv. Lipid Res. **1**, 387 (1963).

10. BENSON, A. A., H. DANIEL, and R. WISER: A Sulfolipid in Plants. Proc. Nat. Acad. Sci. (USA) **45**, 1582 (1959).

11. BENSON, A. A., J. F. G. M. WINTERMANS, and R. WISER: Chloroplast Lipids as Carbohydrate Reservoirs. Plant Physiol. **34**, 315 (1959).

12. BENSON, A. A., R. WISER, R. A. FERRARI, and J. A. MILLER: Photosynthesis of Galactolipids. J. Am. Chem. Soc. **80**, 4740 (1958).

13. BERVAES, J. C. A. M., P. J. C. KUIPER, and A. KYLIN: Conversion of Digalactosyl Diglyceride (Extra Long Carbon Chain Conjugates) into Monogalactosyl Diglyceride of Pine Needle Chloroplasts upon Dehardening. Physiol. Plant. **27**, 231 (1972).

14. BLOCH, K., G. CONSTANTOPOULOS, C. KENYON, and J. NAGAI: Lipid Metabolism of Algae in the Light and in the Dark. In: T. W. GOODWIN, Biochemistry of Chloroplasts, Vol. II, p. 197. London: Academic Press. 1966.

15. BOLLING, H., and A. W. EL BAYÃ: Veränderungen der Fettsäurezusammensetzung in den Galaktolipiden des Weizens während der Reife. Chem. Phys. Lipids **8**, 102 (1972).

16. BOOS, W., J. LEHMANN, and K. WALLENFELS: Asymmetrischer Galaktosyltransfer auf Glycerin mit β-Galactosidase aus *E. coli*. Carbohydrate Res. **1**, 419 (1966).

17. BRUNDISH, D. E., and J. BADDILEY: Synthesis of Glucosylglycerols and Diglucosylglycerols and Their Identification in Small Amounts. Carbohydrate Res. **8**, 308 (1968).

18. BRANTON, D., and R. B. PARK: Subunits in Chloroplast Lamellae. J. Ultrastructure Res. **19**, 283 (1967).

19. CARTER, H. E., R. A. HENDRY, and N. Z. STANACEV: Wheat Flour Lipids. III. Structure of the Mono- and Digalactosyl Glycerol Lipids. J. Lipid Res. **2**, 223 (1961).

20. CARTER, H. E., R. H. McCLURER, and E. D. SLIFER: Lipids of Wheat Flour. I. Characterization of Galactosyl Glycerol Components. J. Am. Chem. Soc. **78**, 3735 (1956).

21. CARTER, H. E., K. OHNO, S. NOJIMA, C. L. TIPTON, and N. Z. STANACEV: Wheat Flour Lipids. II. Isolation and Characterization of Glycolipids of Wheat Flour and Other Plant Sources. J. Lipid Res. **2**, 215 (1961).

22. Chang, S. B.: Sulfhydril Nature of Galactosyl Transfer Enzymes of Spinach Chloroplasts. Phytochem. **9**, 1947 (1970).

23. Chang, S. B., and N. D. Kulkarni: Enzymatic Reactions for Galactolipid Synthesis with a Soluble, Sub-chloroplast Fraction from *Spinacia oleracea*. Phytochem. **9**, 927 (1970).

24. Chang, S. B., and K. Lundin: Specificity of Galactolipids in Photochemical Reactions Coupled with Cytochrome-c. Biochem. Biophys. Res. Comm. **21**, 424 (1965).

25. Constantopoulos, G., and K. Bloch: Effect of Light Intensity on the Lipid Composition of *Euglena gracilis*. J. Biol. Chem. **242**, 3538 (1967).

26. Costes, C., R. Bazier, and D. Lechevallier: Rôle structural des lipides dans les membranes de chloroplastes de Blé. Physiol. Vég. **10**, 291 (1972).

27. Douce, R., and T. Guillot-Salomon: Sur l'incorporation de la radioactivité du sn-glycerol-3-phosphate-^{14}C dans le monogalactosyl-diglyceride des plastes isolés. FEBS **11**, 121 (1970).

28. Eccleshall, T. R., and J. C. Hawke: Biosynthesis of Monogalactosyl Diglyceride by Chloroplasts from *Spinacia oleracea* and from Some Gramineae. Phytochem. **10**, 3035 (1971).

29. Echlin, P., and I. Morris: The Relationship between Blue-Green Algae and Bacteria. Biol. Rev. **40**, 143 (1965).

30. Ferrari, R. A., and A. A. Benson: The Path of Carbon in Photosynthesis of the Lipids. Arch. Biochem. Biophys. **93**, 185 (1961).

31. Galliard, T.: Aspects of Lipid Metabolism in Higher Plants. I. Identification and Quantitative Determination of the Lipids in Potato Tubers. Phytochem. **7**, 1907 (1968).

32. — Aspects of Lipid Metabolism in Higher Plants. II. The Identification and Quantitative Analysis of Lipids from the Pulp of Pre- and Post-Climacteric Apples. Phytochem. **7**, 1915 (1968).

33. — The Isolation and Characterization of Tetragalactosyl Diglyceride from Potato Tubers. Biochem. J. **115**, 335 (1969).

34. — The Enzymic Breakdown of Lipids in Potato Tuber by Phospholipid- and Galactolipid-Acyl Hydrolase Activities and by Lipoxygenase. Phytochem. **9**, 1725 (1970).

35. — The Enzymic Deacylation of Phospholipids and Galactolipids in Plants. Purification and Properties of a Lipolytic Acyl-hydrolase from Potato Tubers. Biochem. J. **121**, 379 (1971).

36. — Hydrolytic and Oxidative Breakdown of Acyllipids in Plants. 15th Int. Conf. Biochem. Lipids: Enzymes in Lipid Biochemistry. p. 62. 1972.

37. Gardner, H. W.: Preparative Isolation of Mono- and Digalactosyl Diglyceride by Thin-layer Chromatography. J. Lipid Res. **9**, 139 (1968).

38. Gurr, M. I.: The Biosynthesis of Polyunsaturated Fatty Acids in Plants. Lipids **6**, 266 (1971).

39. Gurr, M. I., P. P. M. Bonsen, J. A. F. Op den Kamp, and L. L. M. van Deenen: The Chemical Synthesis of Glucosaminyl-phosphatidylglycerol. Comparison with a New Phospholipid Isolated from *Bacillus megaterium*. Biochem. J. **108**, 211 (1968).

40. de Haas, G. H., and L. L. M. van Deenen: Structural Identification of Isomeric Lysolecithins. Biochim. Biophys. Acta **106**, 315 (1965).

41. Heise, K.-P.: Die Freisetzung von Lipiden aus Thylakoiden isolierter Chloroplasts von *Spinacia oleracea*. Ein Beitrag zur Kenntnis der Lokalisation und der funktionellen Beteiligung von Lipiden in Thylakoidmembranen. Thesis. Göttingen. 1972.

42. Heinz, E.: Acylgalaktosyldiglycerid aus Blatthomogenaten. Biochim. Biophys. Acta **144**, 321 (1967).

43. — Semisynthetic Galactolipids of Plant Origin. Biochim. Biophys. Acta **231**, 537 (1971).

44. — Some Properties of the Acyl Galactosyl Diglyceride-forming Enzyme from Leaves. Z. Pflanzenphysiol. **69**, 359 (1973).

45. HELMSING, P. J.: Isolation and Separation of Mono- and Digalactosyl Diglycerides from Spinach Leaves with Sephadex LH-20. J. Chromatogr. **28**, 131 (1967).

46. — Hydrolysis of Galactolipids by Enzymes in Spinach Leaves. Biochim. Biophys. Acta **144**, 470 (1967).

47. — Purification and Properties of Galactolipids. Biochim. Biophys. Acta **178**, 519 (1969).

48. JAMES, A. T., and B. W. NICHOLS: Lipids of Photosynthetic Systems. Nature **210**, 372 (1966).

49. KALRA, S. K., and J. L. BROOKS: Lipids of Ripening Tomato Fruit and its Mitochondrial Fraction. Phytochem. **12**, 487 (1973).

50. KATES, M.: Chromatographic and Radioisotopic Investigations of the Lipid Components of Runner Bean Leaves. Biochim. Biophys. Acta **41**, 315 (1960).

51. KATES, M., and B. E. VOLCANI: Lipid Components of Diatoms. Biochim. Biophys. Acta **116**, 264 (1965).

52. KLENK, E., W. KNIPPRATH, D. EBERHAGEN, and H. P. KOOF: Über die ungesättigten Fettsäuren der Fettstoffe von Süßwasser- und Meeresalgen. Hoppe-Seyler's Z. Physiol. Chem. **334**, 44 (1963).

53. KOENIG, F.: Konzentration einiger Lipide in den Chloroplasten von *Zea mays* und *Antirrhinum majus*. Z. Naturforsch. **26**, 1180 (1971).

54. LEPAGE, M.: Isolation and Characterization of an Esterified Form of Steryl Glucoside. J. Lipid Res. **5**, 587 (1964).

55. — Identification and Composition of Turnip Root Lipids. Lipids **2**, 244 (1967).

56. — The Lipid Components of White Potato Tubers *(Solanum tuberosum)*. Lipids **3**, 477 (1968).

57. LICHTENTHALER, H. K., and R. B. PARK: Chemical Composition of Chloroplast Lamellae from Spinach. Nature **198**, 1070 (1963).

58. LIN, M. F., and S. B. CHANG: Biosynthesis of Galactolipids in Photoautotrophic *Euglena gracilis* Chloroplasts. Phytochem. **10**, 1543 (1971).

59. LOOMIS, W. D., and J. BATTAILE: Plant Phenolic Compounds and the Isolation of Plant Enzymes. Phytochem. **5**, 423 (1966).

60. MACKENDER, R. O., and R. M. LEECH: The Isolation and Characterization of Plastid Envelope Membranes. Proc. II. Intern. Congr. Photosynth. Res. Vol. **2**, p. 1431. Stresa. 1971.

61. MATSON, R. S., M. FEI, and S. B. CHANG: Comparative Studies of Biosynthesis of Galactolipids in *Euglena gracilis* strain Z. Plant Physiol. **45**, 531 (1970).

62. MCCARTY, R. E., and A. T. JAGENDORF: Chloroplast Damage due to Enzymatic Hydrolysis of Endogenous Lipids. Plant Physiol. **40**, 725 (1965).

63. MUDD, J. B., T. T. MCMANUS, A. ONGUN, and T. E. MCCULLOGH: Inhibition of Glycolipid Biosynthesis in Chloroplasts by Ozone and Sulfhydryl Reagens. Plant Physiol. **48**, 335 (1971).

64. MUDD, J. B., H. H. D. M. VAN VLIET, and L. L. M. VAN DEENEN: Biosynthesis of Galactolipids by Enzyme Preparations from Spinach Leaves. J. Lipid Res. **10**, 623 (1969).

65. MYRHE, D. V.: Glycolipids of Soft Wheat Flour. I. Isolation and Characterization of 1-O-(6-O-acyl-β-D-galactopyranosyl)-2,3-di-O-acyl-D-glyceritols and Phytosteryl 6-O-acyl-β-D-glucopyranosides. Can. J. Chem. **46**, 3071 (1968).

66. NEUFELD, E. F., and C. W. HALL: Formation of Galactolipids by Chloroplasts. Biochem. Biophys. Res. Comm. **14**, 503 (1964).

67. NICHOLS, B. W.: Separation of Plant Phospholipids and Glycolipids. In: A. T. JAMES and L. J. MORRIS, New Biochemical Separations, p. 321. London: Van Nostrand. 1964.

68. — Light Induced Changes in the Lipids of *Chlorella vulgaris*. Biochim. Biophys. Acta **106**, 274 (1965).

69. — Fatty Acid Metabolism in the Chloroplast Lipids of Green and Blue-Green Algae. Lipids **3**, 354 (1968).

70. Nichols, B. W., R. V. Harris, and A. T. James: The Lipid Metabolism of Blue-Green Algae. Biochem. Biophys. Res. Comm. **20**, 256 (1965).
71. Nichols, B. W., and A. T. James: The Lipids of Plant-Storage Tissues. Fette, Seifen, Anstrichmittel **66**, 1003 (1964).
72. Nichols, B. W., A. T. James, and J. Breuer: Interrelationships between Fatty Acid Biosynthesis and Acyl-Lipid Synthesis in *Chlorella vulgaris*. Biochem. J. **104**, 486 (1967).
73. Nichols, B. W., J. M. Stubbs, and A. T. James: The Lipid Composition and Ultra-structure of Normal Developing and Degenerating Chloroplasts. In: T. W. Goodwin, Biochemistry of Chloroplasts, Vol. **II**, p. 677. London: Academic Press. 1967.
74. O'Brien, J. S.: Cell Membranes: Composition-Structure-Function. J. Theoret. Biol. **15**, 307 (1967).
75. Ongun, A., and J. B. Mudd: Biosynthesis of Galactolipids in Plants. J. Biol. Chem. **243**, 1558 (1968).
76. Ongun, A., W. W. Thomson, and J. B. Mudd: Lipid Composition of Chloroplasts Isolated by Aqeous and Nonaqeous Techniques. J. Lipid Res. **9**, 409 (1968).
77. Patton, S., G. Fuller, A. R. Loeblich, and A. A. Benson: Fatty Acids of the "Red Tide" Organism, *Gonyaulax polyedra*. Biochim. Biophys. Acta **116**, 577 (1966).
78. Pohl, P.: Some Evidence for Light Induced Transfers of Fatty Acids in *Euglena gracilis*. 15th Int. Conf. Biochem. Lipids: Enzymes in Lipid Biochemistry, p. 96. 1972.
79. Pohl, P., H. Glasl, and H. Wagner: Zur Analytik pflanzlicher Glyko- und Phospho-lipoide und ihrer Fettsäuren. I. Eine neue Dünnschicht-chromatographische Methode zur Trennung pflanzlicher Lipoide und quantitativen Bestimmung ihrer Fettsäure-Zusammensetzung. J. Chromatogr. **49**, 488 (1970).
80. Poincelot, P. R.: Differences in Lipid Composition between Intact and Membrane-stripped Spinach Chloroplasts. Biochim. Biophys. Acta **239**, 57 (1971).
81. Radunz, A.: Localisation with Specific Antisera of the Thylakoid Membrane Lipids MG, SL and PG. 15th Int. Conf. Biochem. Lipids: Enzymes in Lipid Biochemistry, p. 98. 1972.
82. Renkonen, O., and K. Bloch: Biosynthesis of Monogalactosyl Diglycerides in Photoautotrophic *Euglena gracilis*. J. Biol. Chem. **244**, 4899 (1969).
83. Rosenberg, A.: Galactosyl Diglycerides: Their Possible Function in *Euglena* Chloroplasts. Science **157**, 1191 (1967).
84. Rosenberg, A., and J. Gouax: Monogalactosyl and Digalactosyl Diglycerides from Heterotrophic, Hetero-autotrophic, and Photobiotic *Euglena gracilis*. J. Lipid Res. **7**, 733 (1966).
85. — Quantitative and Compositional Changes in Mono- and Digalactosyl Diglyceride during Light-induced Formation of Chloroplasts in *Euglena gracilis*. J. Lipid Res. **8**, 80 (1967).
86. Roughan, P. G.: Turnover of the Glycerolipids of Pumpkin Leaves. The Importance of Phosphatidyl choline. Biochem. J. **117**, 1 (1970).
87. Safford, R., and B. W. Nichols: Positional Distribution of Fatty Acids in Mono-galactosyl Diglyceride Fractions from Leaves and Algae. Biochim. Biophys. Acta **210**, 57 (1970).
88. Sastry, P. S., and M. Kates: Lipid Components of Leaves. V. Galactolipids, Cerebrosides and Lecithin of Runner-bean Leaves. Biochem. **3**, 1271 (1964).
89. — — Hydrolysis of Monogalactosyl and Digalactosyl Diglycerides by Specific Enzymes in Runner-bean Leaves. Biochem. **3**, 1280 (1964).
90. — — Monogalactosyl and Digalactosyl Diglyceride Acyl Hydrolase. In: S. P. Colo-wick and N. O. Kaplan, Methods in Enzymology, Vol. **XIV**, p. 204. London: Academic Press. 1969.
91. Shibuya, I., and B. Maruo: Surfactant Lipids of Plant Quantasomes. Nature **207**, 1096 (1965).

92. SHVETS, V. I., A. I. BASHKATOVA, and R. P. EVSTIGNEEVA: Synthesis of Glycosyl Diglycerides. Chem. Phys. Lipids **10**, 267 (1973).

93. SINGH, H., and O. S. PRIVETT: Incorporation of ^{33}P in Soybean Phosphatides. Biochim. Biophys. Acta **202**, 200 (1970).

94. SMITH, C. R., and J. A. WOLFF: Glycolipids of *Briza spicata* Seed. Lipids **1**, 123 (1966).

95. STEIM, J. M.: Monogalactosyl Diglyceride: a New Neurolipid. Biochim. Biophys. Acta **144**, 118 (1967).

96. TEVINI, M.: Die Phospho- und Glykolipid-Änderungen während des Ergrünens etiolierter *Hordeum*-Keimlinge. Z. Pflanzenphysiol. **65**, 266 (1971).

97. THOMPSON, A. C., R. D. HENSON, J. P. MINYARD, and P. A. HEDRIN: Fatty Acid Composition of Polar Lipids of Cotton Buds. Lipids **3**, 373 (1968).

98. UDEL'NOVA, T. M., and E. A. BOICHENKO: Manganese in Combination with Galactolipids of Leaves. Biokhimiya **32**, 644 (1967).

99. VERHEY, H. M., P. F. SMITH, P. P. M. BONSEN, and L. L. M. VAN DEENEN: The Chemical Synthesis of a Phosphatidylglucose. Biochim. Biophys. Acta **218**, 97 (1970).

100. WEBSTER, D. E., and S. B. CHANG: Polygalactolipids in Spinach Chloroplasts. Plant Physiol. **44**, 1523 (1969).

101. WEIER, T. S., and A. A. BENSON: The Molecular Organization of Chloroplast Membranes. Amer. J. Bot. **54**, 389 (1967).

102. WEHRLI, H. P., and Y. POMERANZ: Synthesis of Galactosyl Glycerides and Related Lipids. Chem. Phys. Lipids **3**, 357 (1969).

103. WESSELS, J. S. C.: Isolation and Properties of two Digitonin-soluble Pigment-Protein Complexes from Spinach. Biochim. Biophys. Acta **153**, 497 (1968).

104. WHEELDON, L. W.: Composition of Cabbage Leaf Phospholipids. J. Lipid Res. **1**, 439 (1960).

105. WICKBERG, B.: Structure of a Glyceritol Glycoside from *Polysiphonia fastigiata* and *Corallina officinalis*. Acta Chem. Scand. **12**, 1183 (1958).

106. — Synthesis of 1-Glycerital-D-Galactopyranosides. Acta Chem. Scand. **12**, 1187 (1958).

107. WINTERMANS, J. F. G. M.: Concentrations of Phosphatides and Glycolipids in Leaves and Chloroplasts. Biochim. Biophys. Acta **44**, 49 (1960).

108. — On the Galactolipid Composition of Subchloroplast Fragments. Biochim. Biophys. Acta **248**, 530 (1971).

109. WRIGHT, A., M. DANKERT, and P. W. ROBBINS: Evidence for an Intermediate Stage in the Biosynthesis of the *Salmonella* O-Antigen. Proc. Nat. Acad. Sci. (USA) **54**, 235 (1965).

(Received October 8, 1973)

Recent Advances in Polynucleotide Synthesis

By H. Kössel, Institut für Biologie III der Universität
Freiburg i. Br., Federal Republic of Germany, and
H. Seliger, Institut für Makromolekulare Chemie der Universität
Freiburg i. Br., Federal Republic of Germany

Contents

Acknowledgement. We are delighted to express our gratitude to many colleagues who supported our work by sending us reprints, preprints and/or unpublished results for incorporation into this review. We wish also to acknowledge several grants from the Deutsche Forschungsgemeinschaft for the support of our own work to which reference is made here.

Introduction

During the past decade solutions of an increasing variety of problems in the field of molecular genetics have rested on the availability of synthetic polynucleotides. Thus, to cite only a few examples, the elucidation of the genetic code was based on synthesis of the 64 possible trinucleoside diphosphates and on the preparation of polynucleotides containing repeating sequences (*186, 249*). More recently the development of synthetic procedures has culminated in the total synthesis of two tRNA-genes (*2, 188, 190*). A further useful application has been demonstrated in the use of synthetic oligomers of specific base sequence as specific primers for DNA sequencing (*247, 366, 379, 467*). Because of the many problems which remain with respect to our understanding in gene function or to future gene manipulation, it seems not surprising that the effort for finding new methods or for improving earlier methods in polynucleotide synthesis still continues or even increases in many laboratories all over the world.

Accordingly a large number of contributions has appeared during the past three or five years which we will try to summarize in the present review. However, although we have attempted to give a broad survey of the entire field, comprehensiveness – deemed a hopeless task not only in the field of synthetic polynucleotides – could not be our major aim. We therefore offer our apologies to authors, whose contributions we could not recognize properly for reasons of space limitation or because their contributions are positioned more towards the periphery of the field. We have also limited ourselves to a compilation of those contributions made during the last 3—5 years, as the earlier literature is accessible through several excellent monographs (*45, 184, 197, 258*) and review articles (*61, 65, 69, 185, 186, 319*).

Abbreviations and Symbols

The system of abbreviations used in this review is principally that which has been suggested by the IUPAC-IUB commission in J. Mol. Biol. **55**, 299 (1971). Thus, a monosubstituted terminal phosphoric acid residue is represented by a small p. Internal phosphoric diester 3'—5'-linkages are represented by a small p between the respective nucleoside symbols or by hyphens.

Nucleosides or nucleoside residues are represented by the following symbols: A adenosine, C cytidine, G guanosine, U uridine, T thymidine, I inosine, X xanthosine, Pu unspecified purine nucleoside, Py unspecified pyrimidine nucleoside, N or M unspecified nucleoside. The common 2'-deoxyribonucleosides are designated by the same symbols, modified in one of the following ways: small d is used as prefix preceding each residue or preceding whole chains, or small d is used as subscript at individual nucleoside symbols.

The diesterified phosphate residue, represented by a hyphen or by small p is considered to be attached to the oxygen atom of the 3'-carbon on its left and to that of the 5'-carbon on its right. For other types of linkage, the numerical form, as in 2'—5' or 5'—5' is used.

Examples of oligonucleotides: A–G–Up or ApGpUp represents a trinucleoside of the ribo series with internal 3'—5'-linkages and with a 3'-terminal phosphate. A–G–U > p represents the same trinucleotide but with terminal 2':3'-cyclic phosphate. pA–G–U is the same, but commencing with a 5'-phosphate and terminating in a uridine with unsubstituted 2'- and 3'-hydroxyls. d (pG–A–C–T) or dpG–A–C–T is a tetranucleotide (all deoxy), with 5'-terminal phosphate on G. $(rA)_6-T_d-T_d$ represents an octanucleotide with unsubstituted 5'- and 3'-terminal hydroxyls; the six 5'-terminal A-residues belong to the ribo series, whereas the two 3'-terminal T-residues belong to the deoxy series.

In polymerized nucleotides the prefix "poly" is usually substituted by the subscript n as in $(dU)_n$ which stands for poly dU. Non-covalent association between two polynucleotide chains, such as that ascribed to hydrogen-bonding, is indicated by a centre dot as in $(rI)_n \cdot (r2thioC)_n$.

Symbols for N-protecting groups are: bz for benzoyl; an for anisoyl; ac for acetyl; ibu for isobutyryl. They are placed immediately before the single capital letters representing the nucleoside or nucleoside residue. In other cases they appear beginning with capital letters above the nucleoside symbols as in A^{bz}, dpG^{ibu}, or $dpG^{ac}-C^{an}$.

Symbols for O-protecting groups at the ribose or deoxyribose residues are: (MeOTr) or MMTr for monomethoxytrityl, [(MeO)$_2$Tr] or DMTr for di-methoxytrityl, (Thp) or THP for tetrahydropyranyl, (Ac) or O-Ac for acetyl.

The condensing agents are commonly abbreviated by:

DCC for N,N'-dicyclohexyl-carbodiimide,
MS   for mesitylene-sulfonyl-chloride,
TPS for 2,4,6-tri-isopropyl-benzene-sulfonyl-chloride.

Additional symbols for blocking groups etc. are indicated in the respective sections.

1. Protecting Groups

1.1. General Considerations

In natural polynucleotides the nucleotide monomers (Fig. 1.1) are exclusively linked by 3'—5'-phosphodiester linkages (Fig. 1.2). The formation of this linkage is normally the goal of the work done in chemical and enzymic oligo- and polynucleotide synthesis. In enzymic synthesis the specificity of the enzymes will only allow the "right" connection of the units. In a sequence-specific chemical synthesis several problems have to be solved in order to achieve a natural internucleotidic bond:

The intermediates have to be suitably protected.

Phosphorylation methods, suitable for the formation of internucleotidic bonds, have to be developed.

Techniques for the separation of reactants, products and by-products have to be elaborated.

The first three sections of this review describe recent advances toward the solution of these three problems.

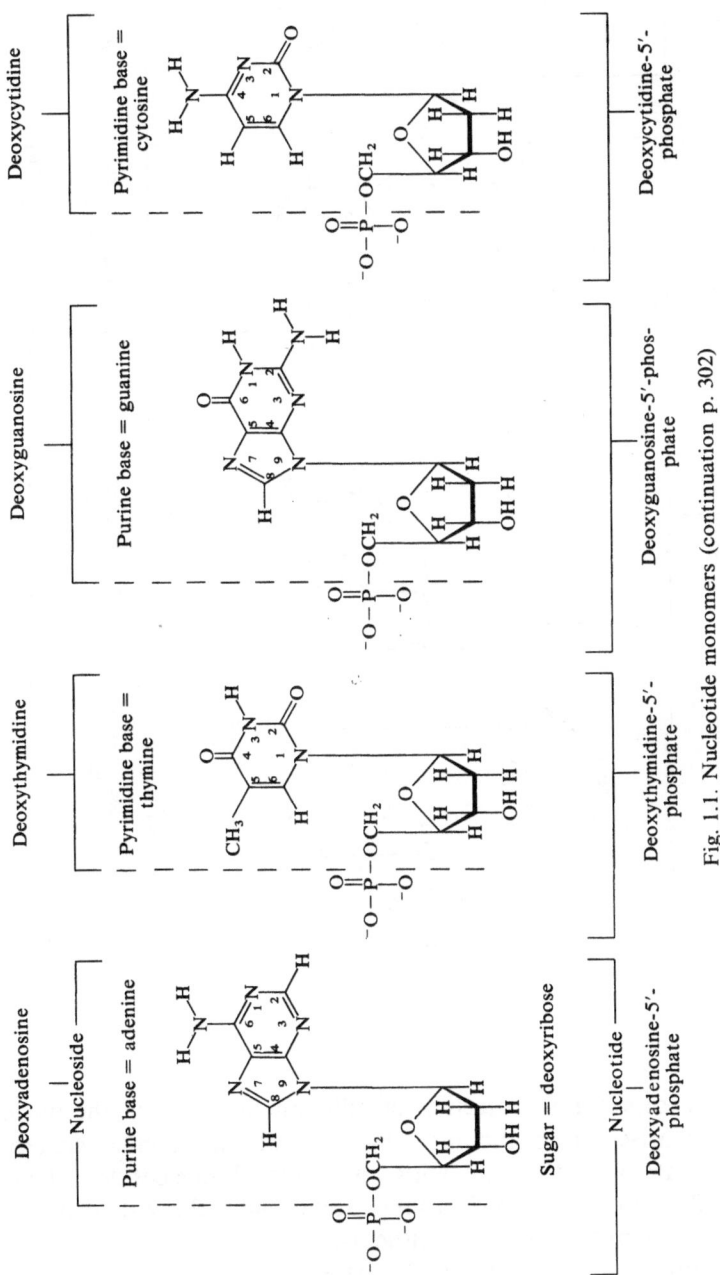

Fig. 1.1. Nucleotide monomers (continuation p. 302)

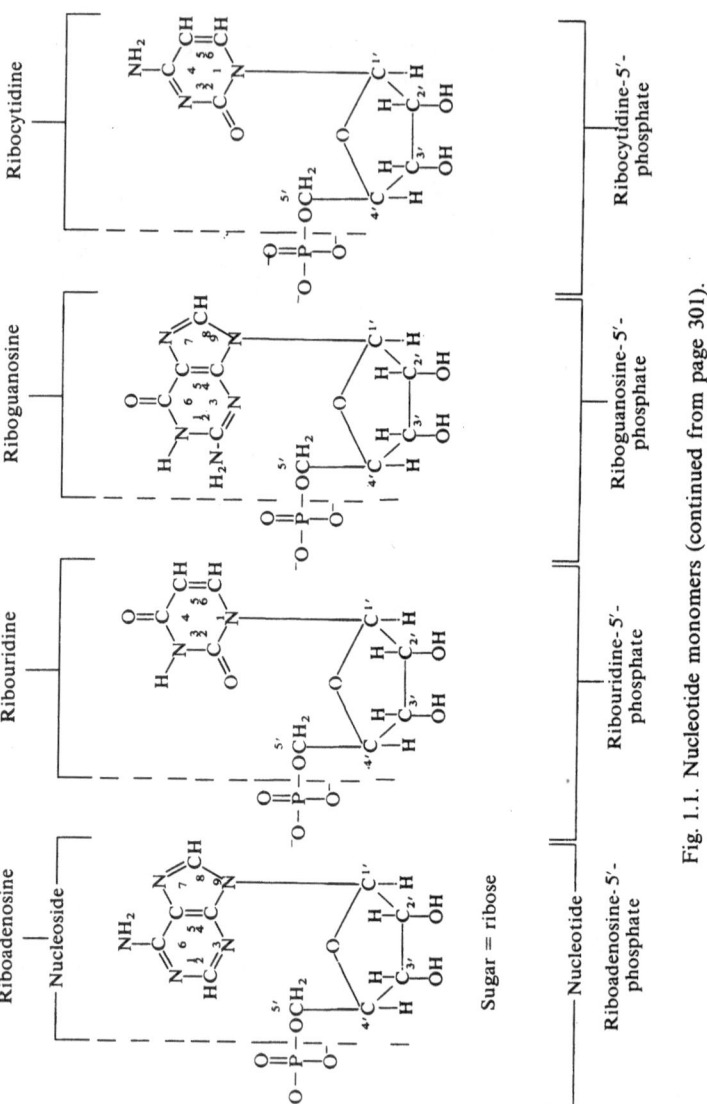

Fig. 1.1. Nucleotide monomers (continued from page 301).

The necessity for protection of different functions of the nucleotide molecule arises from the fact that several nucleophilic centers (see Fig. 1.1) are able to react with an activated nucleotide. These are:

the 3'- and 5'-hydroxyl groups, additionally the 2'-hydroxyl group in ribonucleic acid constituents,

the amino groups of the nucleobases,

the phosphate residue (with formation of pyrophosphates or branched-chain oligonucleotides).

Other nucleophilic centers, e.g. N^3 of pyrimidine bases, are generally not protected.

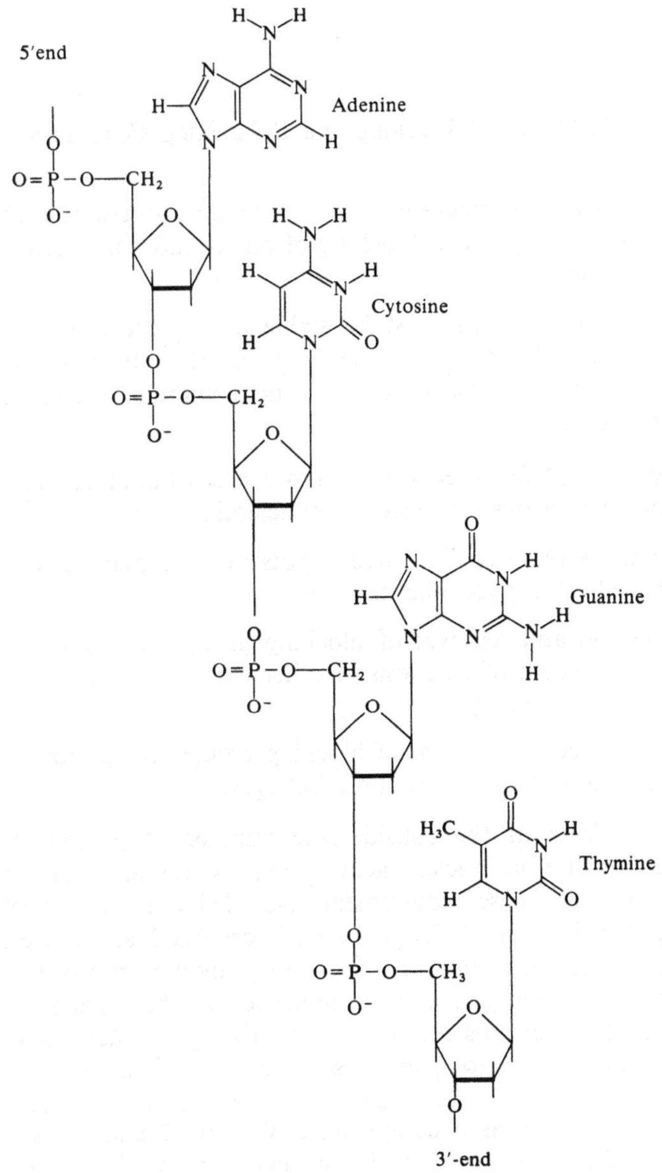

Fig. 1.2. Linkage of nucleotide monomers

The blocking groups used for the different functions of the nucleoside or nucleotide molecule and the conditions for their introduction and removal will be discussed. Earlier work in this field (*322*) has been reviewed in several articles and monographs cited in the introduction. Recently, an excellent treatment of the protecting groups for phosphoric acids, incl. nucleotides, has been given by F. Eckstein (*82*).

1.2. Choice of Blocking and Deblocking Conditions

The following considerations, in principle, govern the choice of reagents for blocking and deblocking of oligo- and polynucleotides and their constituents:

1. The protecting groups must be stable during the formation of an internucleotide linkage. At a later stage of the oligonucleotide synthesis it must be possible to remove them without alteration of the original function.

2. Protection and deprotection must proceed without rupture or isomerization of previously formed internucleotide bonds.

3. The same is required for "weak spots" of the nucleotide molecule itself, especially the glycosidic linkage.

4. Introduction and removal of blocking groups should be, as far as possible independent of reactions of other blocking groups of the same molecule, and *vice versa*.

5. Steric and electronic effects of blocking groups should not be adverse to the formation of internucleotidic linkages.

A closer look at the considerable number of protecting groups described for use in nucleic acid chemistry reveals that only very few will meet all these requirements (see Table 1.1 and Section 1.3 for a detailed discussion). Even the basic criteria 1 and 2 are not met in all cases. Thus, for example, the strongly alkaline removal of N-acyl groups may be accompanied by deamination of the cytosine base, and the use of benzyl groups for protection of the internucleotidic linkage was abandoned due to partial scission of the latter on anionic debenzylation. The use of such groups or conditions may, nevertheless, be necessary as a compromise. Criteria 2 and 3 also exclude a number of groups from use in the deoxy resp. ribo series due to the strongly acid resp. alkaline conditions necessary for their removal (see Sections 1.3 and 1.6). This is because the glycosidic linkage of N-substituted purine-deoxyribonucleotides is very sensitive to media of pH <4, and the ribo-internucleotide bond is easily cleaved at pH >10 due to the neighbouring effect of the 2'-OH. Criterion 4 is of importance in the design of specifically blocked intermediates for stepwise oligonucleotide synthesis. As discussed in Sections 1.5 and 1.6, the number of groups, which fulfil all requirements of selectivity is very small, and various strategies for the selective blocking or deblocking of certain functions have been developed to overcome this difficulty. Criterion 5 is hard to take into account, since little is known for instance about conformations of blocked nucleotides in non-aqueous media and steric effects of blocking groups on intermediate states of the phosphorylation reaction. Such effects have been described, for example for the "shielding" of 3'-hydroxyl groups of ribonucleosides by blocking groups at the 2'-function and *vice versa* (see Section 1.6).

Very recently attention has been drawn to the use of blocking groups in stepwise enzymatic oligonucleotide synthesis. This new approach, realized with the enzyme polynucleotide phosphorylase, is described in Section 5. The blocking groups used in this case have been selected by testing a great number of candidates. Selection of the most suitable protecting agent for this purpose would be greatly simplified by a more detailed knowledge of the steric and electronic environment of the active site of the enzymes in question.

1.3. Survey of Blocking Groups

A list of blocking groups for use in nucleotide chemistry is given in Table 1.1. Of course, it is impossible to take into account every group tested for blocking purposes in laboratories all over the world or to go into detail on all the different conditions elaborated for introduction and removal of certain groups. However, the information given in Table 1.1 is not restricted to the most commonly used groups, but attempts to draw attention also to other groups which have been described, but perhaps not fully exploited as to their potentiality. This is certainly justified by the fact that any discrimination between "useful" and "unuseful" blocking groups would be arbitrary. Moreover, protecting groups rejected for the synthesis of one intermediate may well serve for the preparation of another. The discussion in Sections 1.5 and 1.6 will demonstrate how such intermediates for different approaches to oligonucleotide synthesis are built up by combination of different groups.

Table 1.1 is subdivided in the following way:

In column 1 the protecting groups are numbered according to structural similarity and degree of substitution. These numbers will be referred to in the further discussion. In column 2 the blocking groups are named and classified according to structural similarity. Column 3 lists standard abbreviations. These are a) recommendments of the IUPAC-IUB Commission on Biochemical Nomenclature, for amino blocking groups, resp. b) for terminal radicals; c) abbreviations as used in publications of H. G. Khorana and coworkers, if different from a) and b); d) abbreviations as used by other authors cited as references, if different from a) and b). Column 4 gives representations of the structural formulae. In column 5 the functions are listed, for which the resp. blocking groups can be used. It is convenient to distinguish between the following functions to be blocked: phosphomonoester resp. phosphodiester residues (i.e. terminal phosphate or internucleotidic bonds), hydroxyl- and amino groups in general (including the 2'- and/or 3'-hydroxyl groups of ribosides, if separately blocked), the vicinal diol group of ribosides in cases, in which it reacts as one unity. Cases, in which the blocking group is introduced selectively into one out of several similar functions, are indicated by e. g. "selectively 5'-OH" etc. Column 6 gives the appropriate reagents for blocking. More detailed reaction condi‹ tions are listed only if necessary for reasons of selectivity. Similarly in columns 7 and 8 the reagents for deblocking and the groups which are deblocked under these conditions are listed. The latter are identical with the blocked moieties except for cases of selective deblocking, which, thus, can be easily discerned. Column 9 points to special applications of blocking groups. The following cases are listed: a) blocking groups which allow solvent extraction of oligonucleotides, b) blocking groups rendering possible the separation of oligonucleotides by affinity chromatography, c) activable blocking groups, d) blocking groups for enzymatic monoaddition substrates. A detailed discussion follows in Section 1.4.

Column 10 of Table 1.1 contains all literature references pertinent to the different blocking groups. Since this column provides ample literature information, we will, in the following sections, limit ourselves to citing only those publications, which may serve to illustrate those points which are especially stressed and discussed.
It should be said in conclusion, that this survey does not include all those groups and reagents, which are used for other than blocking purposes, e.g. groups used for selective base modification in tRNA or other polynucleotides, even if they would, in principle, meet some of the requirements for protecting groups listed in Section 1.2.

Table 1.1. *Blocking Groups in Nucleotide and Polynucleotide Chemistry**

No.	Blocking group	Abbreviation	Structural formula	Blocked moiety	Conditions for blocking	Deblocked moiety	Deblocking conditions	Special applications	References
1	*β-substituted ethyl esters:* β-cyanoethyl-	CNEt[b] CE[c]	$-O-CH_2-CH_2-CN$	*phosphate* phospho-mono-ester, phospho-diester	nucleoside, β-cyano-ethyl-phos-phate, DCC nucleotide, hydracrylo-nitrile, DCC	phospho-mono-ester, phospho-diester	mild alkali		(6, 42, 74, 113, 189, 197, 235, 417, 418, 419)
2	2-cyano-1-methyl-ethyl		$-O-CH-CH_2-CN$ $\quad\ \ \ CH_3$	phospho-mono-ester	nucleotide, 1-cyano-propanol-2, DCC	phospho-mono-ester	mild alkali		(65)
3	2-acetyl-2-methyl-ethyl		$-O-CH_2-CH-C-CH_3$ (=O) $\quad\quad CH_3$	phospho-mono-ester	nucleotide, 1-hydroxy-2-methyl-3-butanone, DCC	phospho-mono-ester	alkali		(65)
4	2-acetyl-1-methyl-ethyl		$-O-CH-CH_2-C-CH_3$ (=O) $\quad\ \ \ CH_3$	phospho-mono-ester	nucleotide, 2-hydroxy-4-pentanone, DCC	phospho-mono-ester	alkali		(65)

* For explanations see text of Section 1.3 (p. 305).

Table 1.1 (continued)

No.	Blocking group	Abbreviation	Structural formula	Blocked moiety	Conditions for blocking	Deblocked moiety	Deblocking conditions	Special applications	References
5	2-sulfolen-4-yl		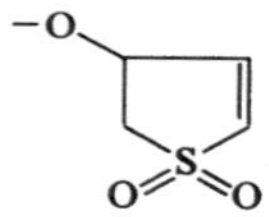 [ring structure with $-O-$, $S(=O)(=O)$]	phospho-mono-ester	nucleotide, 4-hydroxy-2-sulfolene, DCC	phospho-mono-ester	alkali		(65)
6	2(α-pyridyl-)ethyl		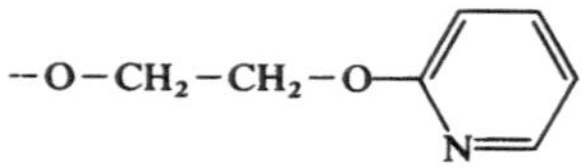 $-O-CH_2-CH_2-O-$ (pyridyl)	phospho-mono-ester	nucleotide, α-pyridyl-ethanol, DCC	phospho-mono-ester	$NaOCH_3$ in methanol/ pyridine		(97, 98)
7	2-(phenyl-carbomyl-)ethyl		$-O-CH_2-CH_2-\overset{O}{\overset{\|}{C}}-NH-$ (phenyl)	phospho-mono-ester	nucleotide, phenyl-hydra-crylamide, DCC	phospho-mono-ester	alkali	affinity	(7, 289, 290)
8	2-(p-methoxy-phenylcarba-moyl-)ethyl	MPH[d]	$-O-CH_2-CH_2-\overset{O}{\overset{\|}{C}}-NH-$ (phenyl)$-O-CH_3$	phospho-mono-ester	nucleotide, p-methoxy-phenyl-hydracryl-amide, DCC	phospho-mono-ester	alkali	affinity	(289, 290)
9	2-(benzyl-carbamoyl-)ethyl		$-O-CH_2-CH_2-\overset{O}{\overset{\|}{C}}-NH-CH_2-$ (phenyl)	phospho-mono-ester	nucleotide, benzylhydra-crylamide, DCC	phospho-mono-ester	alkali	affinity	(289, 290)

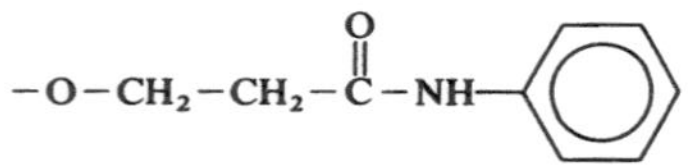

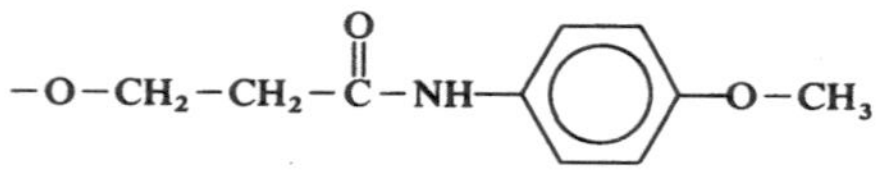

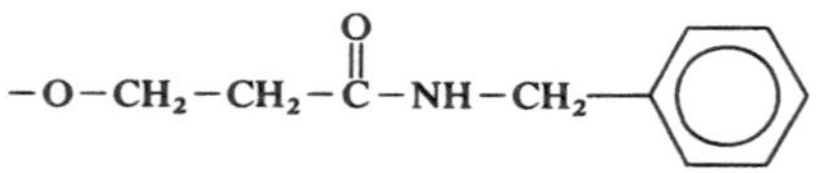

No.	Name	Abbr.	Structure	Type	Introduction	Protected group	Removal		References
10	2-(phenyl-mercapto-)ethyl	PME[d]	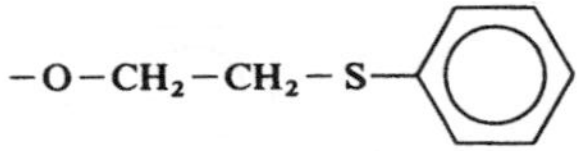	phospho-mono-ester	nucleotide, 2-phenyl-mercapto-ethanol, DCC	phospho-mono-ester	1) periodate 2) alkali	affinity extraction	(7, 290a, 422, 465)
11	9-fluorenyl-methyl		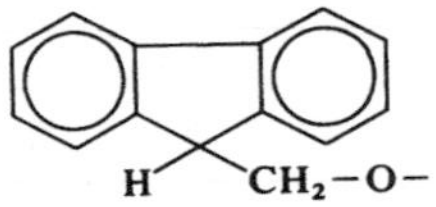	phospho-mono-ester	nucleotide, 9-fluorenyl-methanol, TPS	phospho-mono-ester	alkali	extraction	(176)
12	2′,3′-dimethoxy-benzylidene-)uridinyl		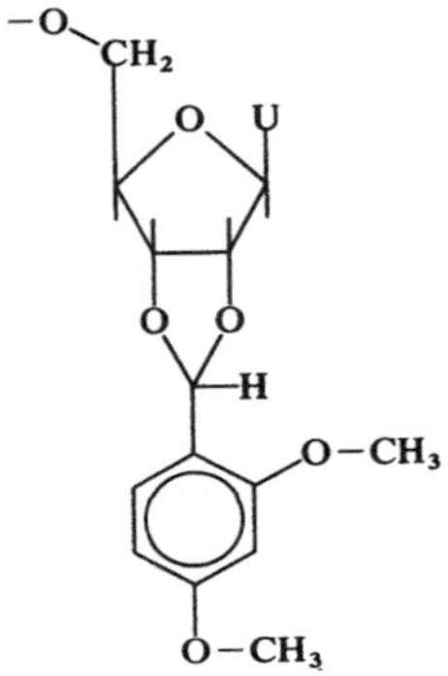	phospho-mono-ester	nucleotide, 2′,3′-(2,4-dimethoxy-benzylidene-)uridine, DCC	phospho-mono-ester	1. mild acid 2. NaJO$_4$ 3. alkali		(177, 178, 398a)
13	β,β,β-tri-chloroethyl	Cl$_3$Et[d]	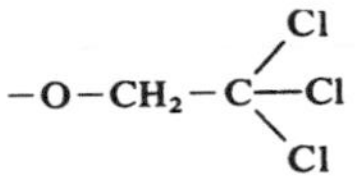	phospho-mono-ester, phospho-diester	nucleotide, β,β,β-tri-chloro-ethanol, DCC	phospho-mono-ester, phospho-diester	Zu/Cu in DMF		(42, 75, 76, 78, 94, 197, 296, 298)

Table 1.1 (continued)

No.	Blocking group	Abbrev-iation	Structural formula	Blocked moiety	Conditions for blocking	Deblocked moiety	Deblocking conditions	Special applications	References
					nucleoside, β,β,β-tri-chloroethyl-phosphate TPS				
					nucleoside, β,β,β-tri-chloroethyl-phosphoro-dichloridate				
					nucleoside, β,β,β-tri-chloroethyl-β-cyano-ethyl-phospho-chloridate				
misc. ester groups:									
14	phenyl-		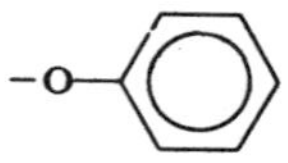	phospho-diester	nucleoside, phenyl-phosphodi-chloridate nucleoside, phenyl-phosphate, TPS	phospho-diester	strong alkali		(31, 73 350, 352)

15	o-chloro-phenyl-	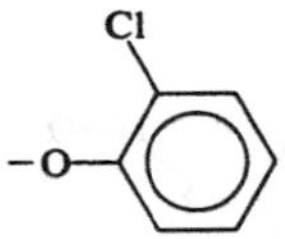	phospho-diester	nucleoside, o-chloro-phenyl-phosphate, TPS	phospho-diester	alkali	(352)
16	m-chloro-phenyl-	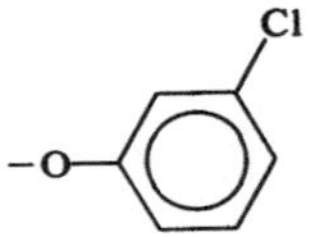	phospho-diester	nucleoside, m-chloro-phenyl-phosphate, TPS	phospho-diester	alkali	(352)
17	o-fluoro-phenyl-	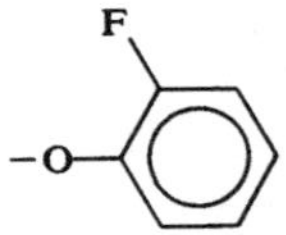	phospho-diester	nucleoside, o-fluoro-phenyl-phosphate, TPS	phospho-diester	alkali	(352)
18	4-chloro-2-nitro-phenyl-	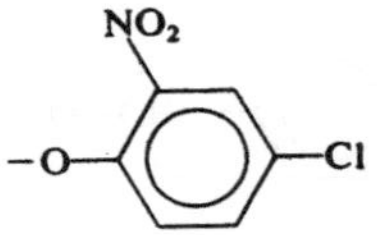	phospho-mono-ester	nucleotide, 4-chloro-2-nitro-phenol, DCC	phospho-mono-ester	strong alkali	(289)
19	4-nitro-2-chloro-methyl-phenyl-	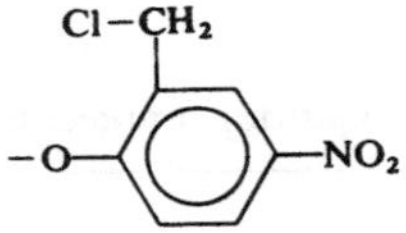	phospho-mono-ester	nucleoside, 4-nitro-2-chloro-methyl-phenyl-phosphate, DCC	phospho-mono-ester	aqueous pyridine	activation (281)

Table 1.1 (continued)

No.	Blocking group	Abbreviation	Structural formula	Blocked moiety	Conditions for blocking	Deblocked moiety	Deblocking conditions	Special applications	References
20	benzyl-	Bzl[b]	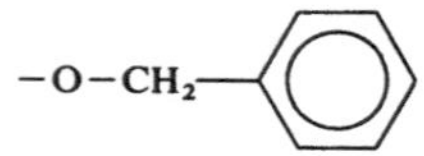 $-O-CH_2-$ (phenyl)	phospho-mono-ester	nucleoside, benzyl-phospho-dichloridate	phospho-mono-ester	Pd/H$_2$		(197)
				phospho-diester	nucleotide, phenyl-diazo-methane	phospho-diester	NaJ in acetonitrile		(369)
21	benzhydryl-		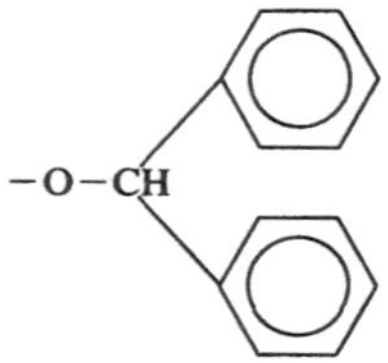 $-O-CH$ (diphenyl)	phospho-mono-ester	nucleotide, diphenyl-diazo-methane	phospho-mono-ester	acid		(63, 64)
22	benzaldoxime ester		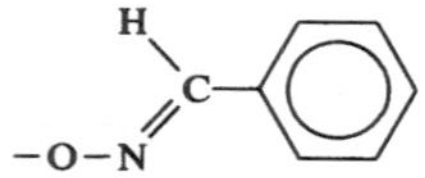 $H-C, -O-N$ (phenyl)	phospho-mono-ester	benzal-doxime + nucleoside-5'-phosphor-morpho-lidate	phospho-mono-ester	alkali	affinity	(290)

No.	Name	Structure		Reagents		Deprotection		References
23	ethylthio-	EtS[d]	$-S-CH_2-CH_3$	phospho-mono-ester	S-ethyl-phosphoro-thioate, nucleoside, DCC	phospho-mono-ester	J₂/pyridine — activation		(55, 56, 58, 129, 146, 488)
24	t-butyl-		structure: $-O-C(CH_3)(CH_3)-CH_3$	phospho-mono-ester	t-butanol, DCC, nucleotide	phospho-mono-ester	acid		(63, 489)
25	1-oxido-2-picolyl-		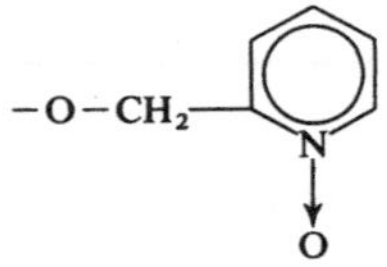 structure: $-O-CH_2-$ (1-oxido-pyridin-2-yl)	phospho-mono-ester	nucleotide, 1-oxido-pyridine-2-yl-diazo-methane	phospho-mono-ester	1) acetic anhydride 2) methanol. ammonia		(83, 271)

phosphoramidate groups:

No.	Name	Structure		Reagents		Deprotection		References
26	anilidate	PhNH	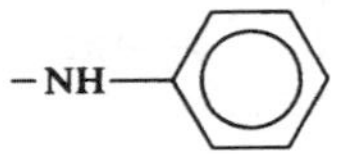 structure: phenyl $-NH-$	phospho-mono-ester	nucleotide, aniline, DCC	phospho-mono-ester	isoamyl-nitrite, pyridine/ acetic acid		(309, 310, 312)
27	p-hydroxy-anilidate		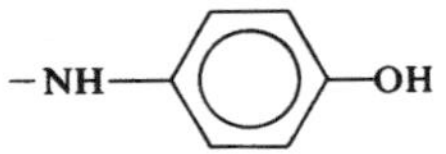 structure: $-NH-$ (p-hydroxyphenyl) OH	phospho-mono-ester	nucleotide, p-hydroxy-aniline, DCC	phospho-mono-ester	isoamyl-nitrite, pyridine/ acetic acid — activation		(316)
28	p-methoxy-anilidate		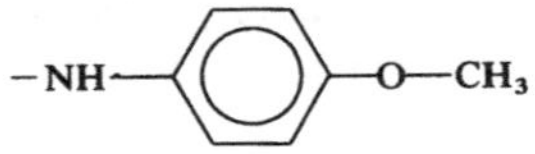 structure: $-NH-$ (p-methoxyphenyl) $O-CH_3$	phospho-mono-ester	nucleotide, p-methoxy-aniline, DCC	phospho-mono-ester	isoamyl-nitrite, pyridine/ acetic acid		(309)

Table 1.1 (continued)

No.	Blocking group	Abbreviation	Structural formula	Blocked moiety	Conditions for blocking	Deblocked moiety	Deblocking conditions	Special applications	References
29	p-(trityl-)anilidate	TPM[c]	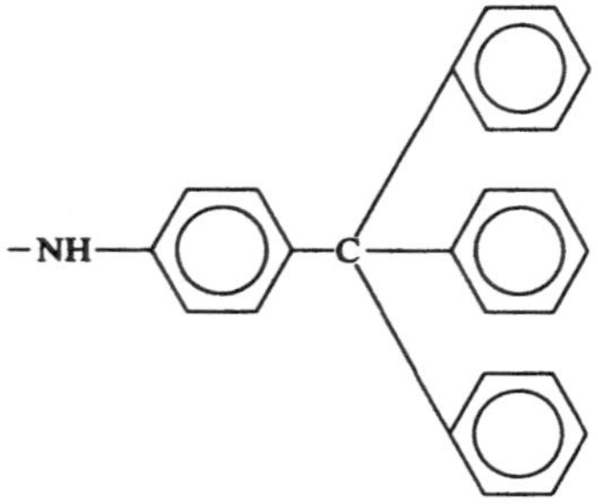	phospho-mono-ester	nucleotide, p-amino-phenyl-triphenyl-methane, DCC	phospho-mono-ester	isoamyl-nitrite, pyridine, acetic acid	affinity extraction	(3)
30	p-(N,N-dimethylamino-anilidate		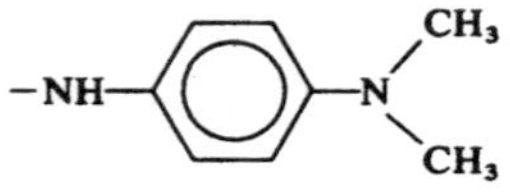	phospho-mono-ester	nucleotide, N,N-di-methyl-p-phenylene diamine, DCC	phospho-mono-ester	isoamyl-nitrite, pyridine, acetic acid	affinity	(136, 138, 435)

ester groups:

		reagent	sugar, base	conditions		ref
31 formyl-	$-\overset{H}{\underset{}{C}}=O$	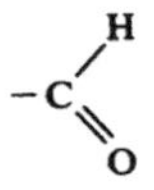 formic acetic anhydride formic acid N-formyl-imidazole	OH	mild alkali, aqueous pyridine	OH	(63, 99, 332, 398, 418, 419, 477)
		trimethyl-orthoformate p-toluene-sulfonic acid (via methoxy-methylidene)	sel. 2'/3'-OH			(117)
				sel. 5'-OH methanol		(477)
32 benzoyl-formyl	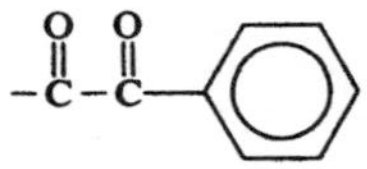 (benzene ring)–$\overset{O}{\underset{}{C}}$–$\overset{O}{\underset{}{C}}$	benzoyl-formyl-chloride	-OH	aqueous pyridine	-OH	(234)
33 acetyl- ac[a] Ac[b]	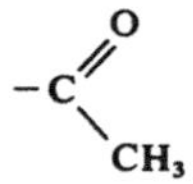 $-\overset{O}{\underset{}{C}}-CH_3$	acetic anhydride	-OH, -NH$_2$	alkali, ammonia	-OH -NH$_2$	(184, 197, 258)
		acetic anhydride, H$_2$O	sel. -OH			(43, 429a)
		acetic anhydride, BF$_3$-ether	sel. -OH			(346)

Table 1.1 (continued)

No.	Blocking group	Abbreviation	Structural formula	Blocked moiety	Conditions for blocking	Deblocked moiety	Deblocking conditions	Special applications	References
33	acetyl (continued)			-OH	dioxane, acetonitrile, HCl				(482)
				sel. NH₂	acetic anhydride, DMF, tri-n-butylamine				(329, 330)
				sel. 2'/3'-OH	trimethyl-orthoacetate (via methoxy-ethylidene-)				(99)
				sel. 3'-OH	8-hydroxy-quinoline N-acetate				(268)
				sel. -NH₂	5-(acetyl-oxymino-)2,6-dioxo-4-(methyl-imino-)1,3-dimethyl-hexahydro-pyrimidine				(28)

	sel. 5'-OH			acetic anhydride, diethylazo-dicarboxylate, triphenylphosphine	(264)
	sel. α-NH₂ of amino-acyl-nucleoside			5-chloro-8-hydroxy-quinoline-O-acetate	(47)
	sel.-OH	strong alkali			(184, 197, 258)
34 methoxy-acetyl	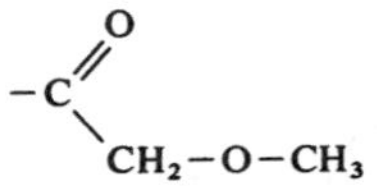 -OH	alkali		methoxy-acetic anhydride, trimethyl-methoxy-orthoacetate	(351, 352, 353)
35 triphenyl-methoxy-acetyl	-OH	mild alkali	trac[d] affinity		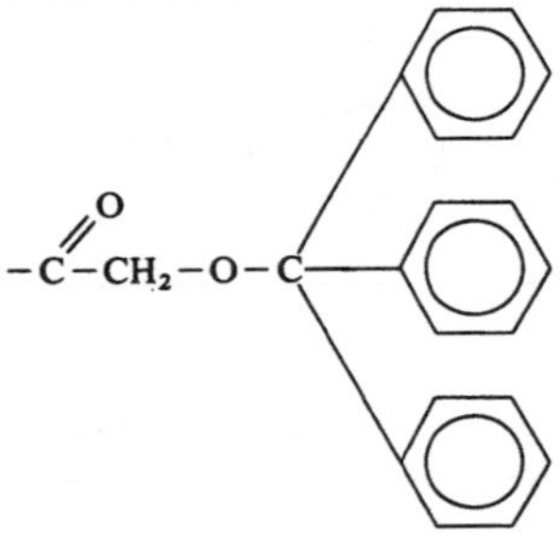 triphenyl-methoxy-acetic acid, triisopropyl-benzene-sulfonyl-chloride	(463)

Structure 34: $-\!\overset{O}{\overset{\|}{C}}-CH_2-O-CH_3$

Structure 35: $-\!\overset{O}{\overset{\|}{C}}-CH_2-O-C(C_6H_5)_3$

Table 1.1 (continued)

No.	Blocking group	Abbreviation	Structural formula	Blocked moiety	Conditions for blocking	Deblocked moiety	Deblocking conditions	Special applications	References
36	phenoxyacetyl		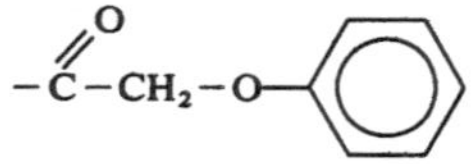	-OH	phenoxy acetic anhydride	-OH	alkali		(351)
37	p-chloro-phenoxy-acetyl		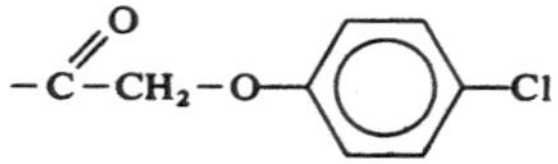	-OH	p-chloro-phenoxy-acetic anhydride	-OH	alkali		(353)
38	chloroacetyl		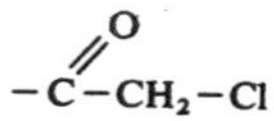	-OH	chloroacetic anhydride	-OH	alkali; 2-mercapto-ethylamine, neutral		(57)
39	trichloro-acetyl		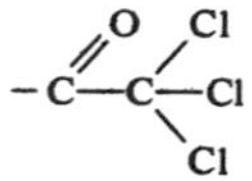	-OH	trichloro-acetic anhydride	-OH	alkali		(197)
40	diphenyl-chloroacetyl		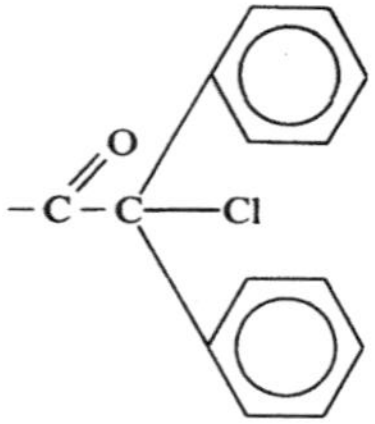	-OH	diphenyl-chloro-acetyl-chloride	-OH	alkali; thiourea, neutral		(57)

No.	Name	Structure	Protected group	Reagent	Protected group	Removal	Ref.
41	trifluoro-acetyl	F_3CCO^b —C(=O)—CF₃	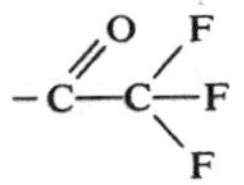 —OH	trifluoro-acetic anhydride	—OH	alkali	(197)
42	propionyl	—C(=O)—CH₂—CH₃	—OH, —NH₂	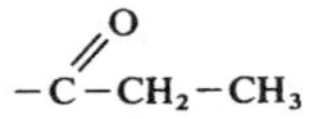 propionic anhydride	—OH, —NH₂	alkali, ammonia	(159, 160)
43	dihydrocinnamoyl	—C(=O)—CH₂—CH₂— (phenyl)	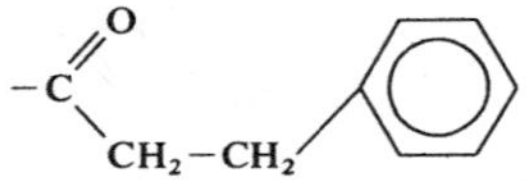 —OH	dihydro-cinnamoyl chloride (anhydride)	—OH	chymotrypsin in acetonitrile/phosphate buffer pH 7	(363, 437)
44	β-benzoyl-propionyl-	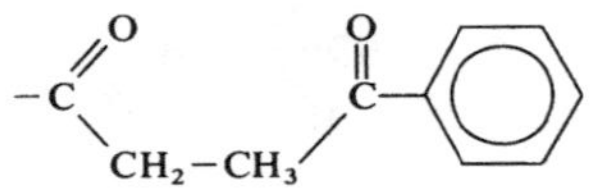 βB^d —C(=O)—CH₂—CH₃ (benzoyl)	—OH	β-benzoyl-propionic acid DCC	—OH	hydrazine in pyridinium acetate	(112, 230, 470)
45	n-butyryl	—C(=O)—CH₂—CH₂—CH₃	—OH, —NH₂	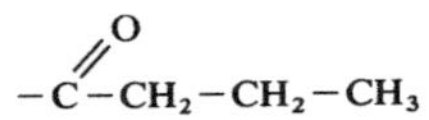 butyric anhydride	—OH, —NH₂	alkali, ammonia	(159, 160)
46	isobutyryl-	iB^a iBu^c —C(=O)—C(H)(CH₃)CH₃	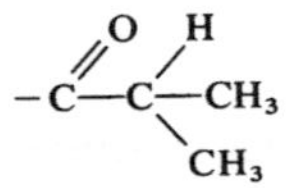 —OH, —NH₂	isobutyric anhydride	—OH, —NH₂ sel. —OH	alkali, ammonia strong alkali	(37, 457)

Table 1.1 (continued)

No.	Blocking group	Abbreviation	Structural formula	Blocked moiety	Conditions for blocking	Deblocked moiety	Deblocking conditions	Special applications	References
47	2-methyl-butyryl	mB[a]	$-C(=O)-CH(CH_3)-CH_2-CH_3$	$-OH$, $-NH_2$	2-methyl-butyric anhydride	$-OH$, $-NH_2$ sel. $-OH$	alkali, ammonia strong alkali		(41)
48	isovaleryl		$-C(=O)-CH_2-CH(CH_3)_2$	$-OH$	isovaleryl chloride	$-OH$	alkali	mono-addition substrate	(179, 180)
49	pivaloyl-(trimethyl-acetyl)		$-C(=O)-C(CH_3)_3$	$-OH$, $-NH_2$	pivaloyl-chloride	$-OH$, $-NH_2$	alkali, ammonia		(99)
50	octanoyl-		$-C(=O)-(CH_2)_6-CH_3$	$-OH$, $-NH_2$	octanoic anhydride	$-OH$, $-NH_2$	alkali, ammonia		(159, 160)
51	linoleyl-		$-C(=O)-C_{17}H_{29}$	$-OH$, $-NH_2$	linoleic anhydride	$-OH$, $-NH_2$	alkali, ammonia		(159, 160)

52	benzoyl-	bz[a] Bz[b]	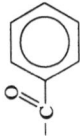				
				—OH, —NH₂	benzoyl-chloride	alkali, ammonia n-butylamine for G^bz	(184, 197, 258, 315, 457)
				sel.—OH		strong alkali	
				sel.—NH₂	benzoic acid-N-hydroxy-succinimide ester		(303)
				sel. 5'-OH	benzoic acid, diethylazodi-carboxylate, triphenyl-phosphine		(264)
						sel.—NH₂ hydrazine in pyridine-acetate	(233)
				—OH, —NH₂	benzoyl cyanide		(157)
				sel.—OH	benzoic acid anhydride/H₂O		(43)

Table 1.1 (continued)

No.	Blocking group	Abbreviation	Structural formula	Blocked moiety	Conditions for blocking	Deblocked moiety	Deblocking conditions	Special applications	References
52	benzoyl (continued)			sel. 2'(3')-OH	trimethyl-orthobenzoate, p-toluenesulfonic acid (via methoxy benzylidene)				(99)
53	anisoyl	an[a] An[b]	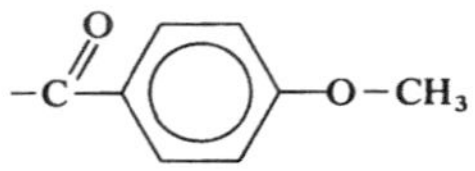	-OH, -NH$_2$	anisoyl chloride	-OH, -NH$_2$ sel. -OH	alkali, ammonia strong alkali		(197)
54	dinitro-benzoyl-		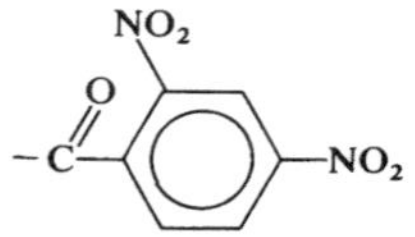	-OH, -NH$_2$	dinitro-benzoyl-chloride	-OH, -NH$_2$	alkali, ammonia		(197)
55	adamantoyl-		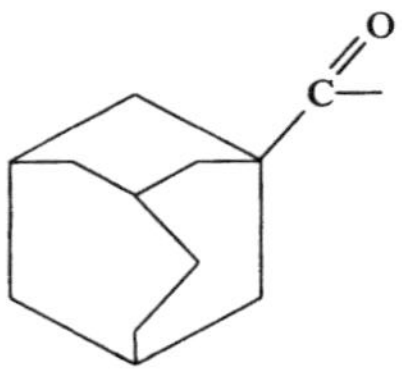	-OH	adamantane-carbonyl-chloride	adamantane- -OH	alkali		(253, 405)

56 dinitro-benzene-sulfenyl-	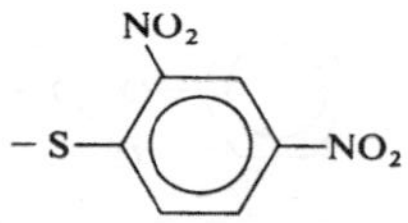	-OH	2,4-dinitro-benzene-sulfenyl-chloride	-OH	thiophenol, neutral	(112, 197)
57 mesyl-	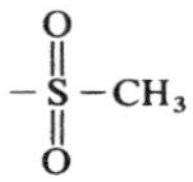	-OH,	methane-sulfonyl-chloride	-OH, -NH$_2$	alkali	(197)
58 tosyl Tos[b]	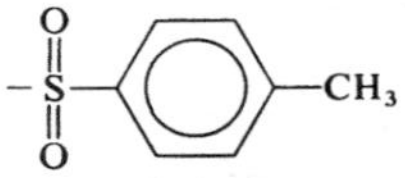	-OH	p-toluene-sulfonyl-chloride	-OH	alkali	(10, 197, 162a, 162b, 162c)
59 trimethyl-silyl TMS[b]	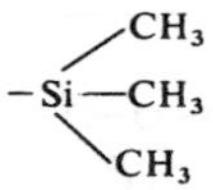	-OH, -NH$_2$	trimethyl-chlorosilane, bis-trimethyl-trifluoro-acetamide	-OH, -NH$_2$	weak acid or weak alkali	(197)
60 t-butyl-dimethyl-silyl	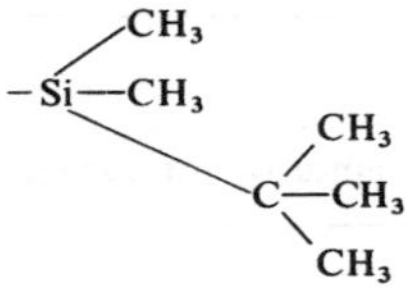	-OH	t-butyl-dimethyl-silyl-chloride	-OH	NR$_4^+$ F$^-$ neutral	(307)

21*

Table 1.1 (continued)

No.	Blocking group	Abbreviation	Structural formula	Blocked moiety	Conditions for blocking	Deblocked moiety	Deblocking conditions	Special applications	References
61	ethyloxycarbonyl-		$-\overset{\text{O}}{\overset{\|}{\text{C}}}-\text{O}-\text{CH}_2-\text{CH}_3$	$-\text{OH}$	nucleoside chloroformate + ethanol	$-\text{OH}$	alkali		(391)
62	trichloroethyloxycarbonyl		$-\overset{\text{O}}{\overset{\|}{\text{C}}}-\text{O}-\text{CH}_2-\overset{\text{Cl}}{\underset{\text{Cl}}{\overset{\|}{\underset{\|}{\text{C}}}}}-\text{Cl}$	$-\text{OH},$ $-\text{NH}_2$	trichloroethylchloroformate	$-\text{OH},$ $-\text{NH}_2$	Zn/acetic acid (methanol)		(183, 466)
63	tribromoethyloxycarbonyl-		$-\overset{\text{O}}{\overset{\|}{\text{C}}}-\text{O}-\text{CH}_2-\overset{\text{Br}}{\underset{\text{Br}}{\overset{\|}{\underset{\|}{\text{C}}}}}-\text{Br}$	$-\text{OH}$	tribromoethylchloroformate	$-\text{OH}$	Zn/Cu in acetic acid		(54)
64	isobutyloxycarbonyl	BOC[b]	$-\overset{\text{O}}{\overset{\|}{\text{C}}}-\text{O}-\text{CH}_2-\text{CH}(\text{CH}_3)_2$	$-\text{OH},$ $-\text{NH}_2$ sel. 5'OH	isobutylchloroformate	$-\text{OH},$ $-\text{NH}_2$ / sel. $-\text{OH}$	alkali ammonia / strong alkali		(236, 304)
65	phenyloxycarbonyl-		$-\overset{\text{O}}{\overset{\|}{\text{C}}}-\text{O}-\!\!\bigcirc$	$-\text{OH}$	phenylchloroformate	$-\text{OH}$	alkali		(14)

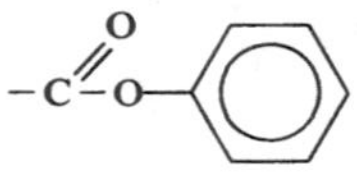

No.	Name	Structure		Method		Removal		Ref.
66	p-nitrophenyl-oxycarbonyl-	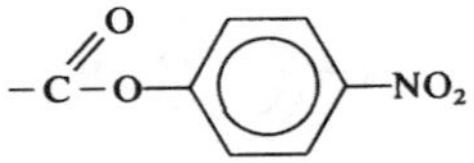	–OH	p-nitrophenyl-chloroformate; nucleoside chloroformate + p-nitrophenol	–OH	alkali, ammonia	affinity	(229, 391)
67	p-phenylazo-phenyloxy-carbonyl-	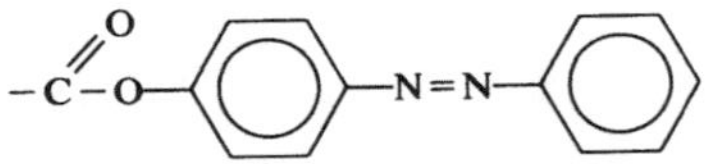	–OH, sel. 5′-OH	nucleoside chloroformate + p-phenylazophenol	–OH	alkali	affinity	(391)
68	piperidine-carbamoyl-	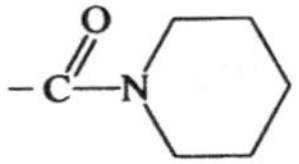	–OH	nucleoside chloroformate + piperidine	–OH	alkali		(391)
69	naphthyl-carbamoyl-	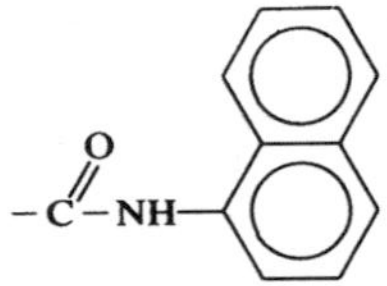	–OH	naphthyliso-cyanate	–OH	alkali	affinity	(4)

Table 1.1 (continued)

No.	Blocking group	Abbreviation	Structural formula	Blocked moiety	Conditions for blocking	Deblocked moiety	Deblocking conditions	Special applications	References
	Schiff bases, orthoesters and deriv.								
71	bis-(2-chloroethyl)-orthoformate		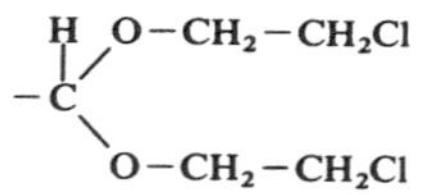 H–C with $O–CH_2–CH_2Cl$ and $O–CH_2–CH_2Cl$	–OH	2-chloroethylorthoformate	–OH	acid		(133)
72	N,N-dimethyl-amino-methylidene-	DMM^d	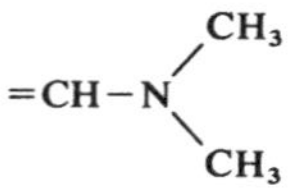 $=CH–N$ with CH_3, CH_3	–NH$_2$ selective	dimethyl-formamide-acetals	–NH$_2$	alkali, mild acid		(153, 154, 155, 197, 397, 398, 420)
73	p-nitro-benzylidene		 $=CH$– (C$_6$H$_4$)–NO_2	–NH$_2$ (concurrent with No. 123)	1. nitro-benzal-dehyde, HC(OEt)$_3$, F$_3$C–COOH, DMF 2. benzoyl-chloride	–NH$_2$	mild acid		(484)

ether groups:

74	benzyl-	bzl[a] Bzl[b]	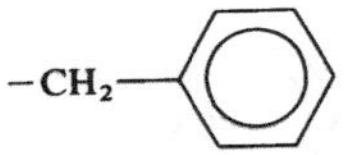 −CH₂⟨phenyl⟩	−OH, −NH₂	benzyl-chloride/alkali	H₂/Pd	(197)
				−OH, −NH₂, sel. 2′-OH	benzyl-chloride/NaH		(25, 26, 27, 163, 192)
				−OH, −NH₂	phenyldiazo-methane, SnCl₂		(52)
				sel. −NH₂	nucleoside-Na⁺-salt + benzyl-chloride		(390)
75	triphenyl-methyl-	tr[a] Tr[b] 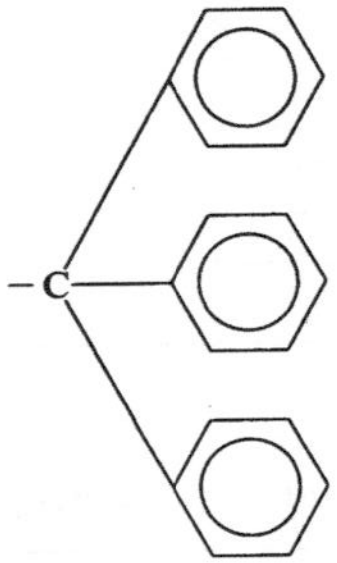	⟨triphenyl-C⟩	−OH, −NH₂, selective 5′-OH	triphenyl-methyl-chloride	acid	(184, 197, 354)
				−OH, −NH₂		silicagel	(225)
				sel. 2′-OH		acid	(197)
				sel. 3′-OH		acid	(197)
				sel. 5′-OH		acid	(88, 218a, 354)
						affinity	(41)

Table 1.1 (continued)

No.	Blocking group	Abbreviation	Structural formula	Blocked moiety	Conditions for blocking	Deblocked moiety	Deblocking conditions	Special applications	References
76	p-methoxy-triphenyl-methyl	mmt[a] MeOTr[b] MMTr[c]	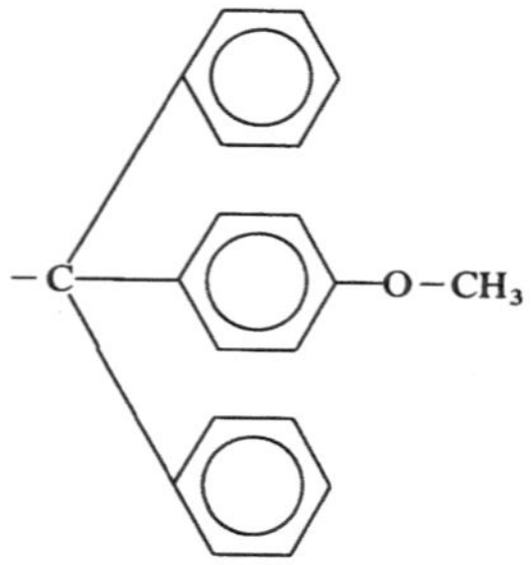	-OH, -NH$_2$ selective 5'-OH	p-methoxy-triphenyl-methyl-chloride	-OH, -NH$_2$	mild acid		(184, 197)
								affinity	(41, 398b)
77	p,p'-di-methoxy-triphenyl-methyl-	dmt[a] (MeO)$_2$Tr[b] DMTr[c]	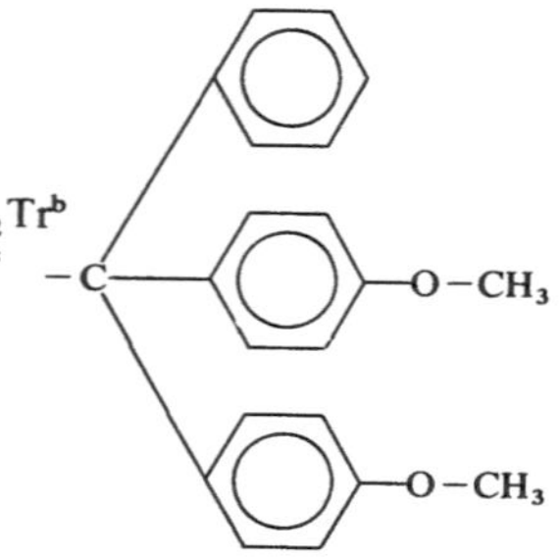	-OH, -NH$_2$ selective 5'-OH	p,p'-di-methoxy-triphenyl-methyl-chloride	-OH, -NH$_2$	very mild acid	affinity	(41, 197)

No.	Name		Structure	Protects	Reagent	Removed from	Conditions	Ref.
78	p,p',p''-tri-methoxy-triphenyl-methyl		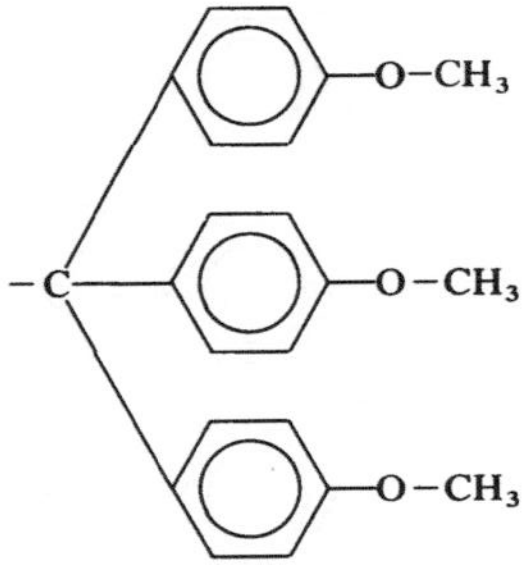	-OH, -NH₂, selective 5'OH	p,p',p''-tri-methoxy-triphenyl-methyl-chloride	-OH, -NH₂	extremely mild acid	(197)
79	p-hydroxy-trityl	pHOTr[d]	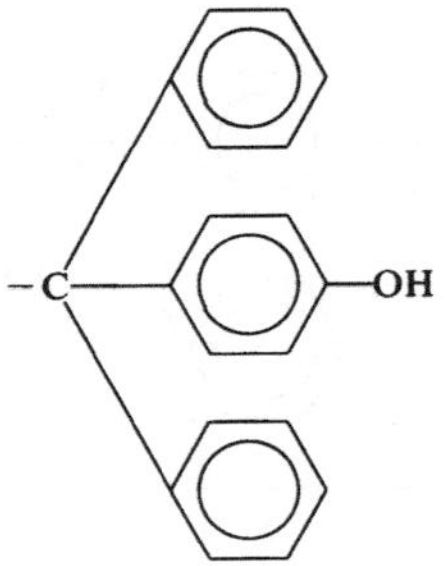	OH- sel. 5'-OH	p-hydroxy-phenyl-diphenyl-methyl-chloride	OH-	mild acid	(436)

Table 1.1 (continued)

No. Blocking group	Abbrev- iation	Structural formula	Blocked moiety	Conditions for blocking	Deblocked moiety	Deblocking conditions	Special appli- cations	References
80 p-acetoxy- trityl	pAcOTr[d]	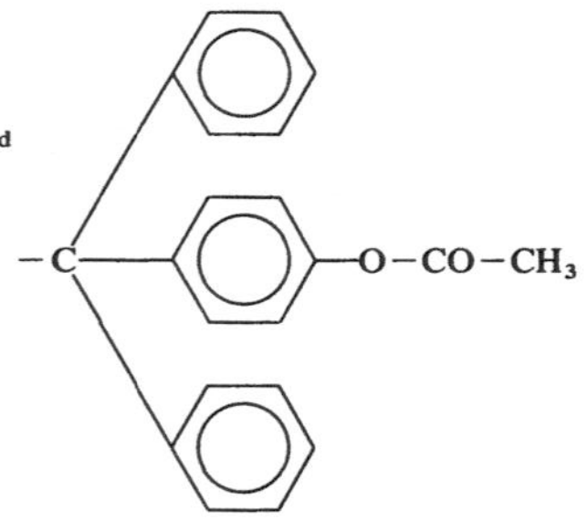	OH– sel. 5′-OH	p-acetoxy- phenyl- diphenyl- methyl- chloride	OH–	mild acid		(436)
81 m-hydroxy- trityl	mHOTr[d]	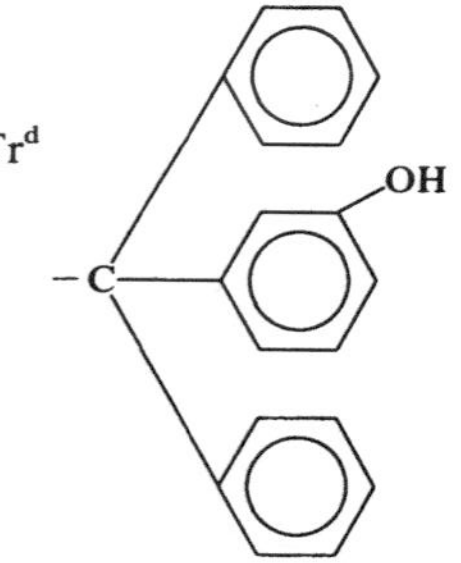	OH– sel. 5′-OH	m-hydroxy- phenyl- diphenyl- methyl- chloride	OH–	mild acid		(436)

82 m-acetoxy-trityl | mAcOTr[d] | OH– sel. 5'-OH | m-acetoxy-phenyl-diphenyl-methyl-chloride | OH– | mild acid | (436)

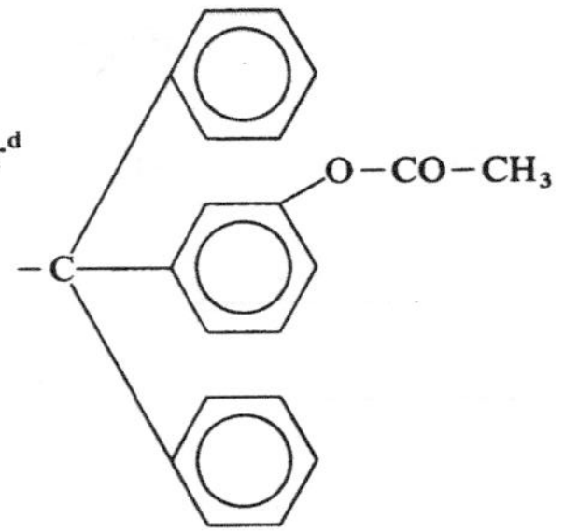

83 di-(p-benzyloxy-)trityl | DPTr[d] | OH– sel. 5'-OH | di(benzyloxyphenyl-)phenyl-methyl-chloride | OH– | very mild acid | (436)

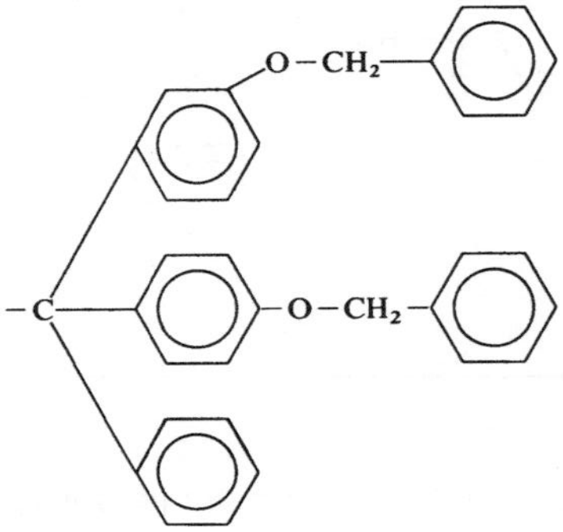

Table 1.1 (continued)

No.	Blocking group	Abbreviation	Structural formula	Blocked moiety	Conditions for blocking	Deblocked moiety	Deblocking conditions	Special applications	References
84	bromo-phenacyl-trityl-	BPTr[d]	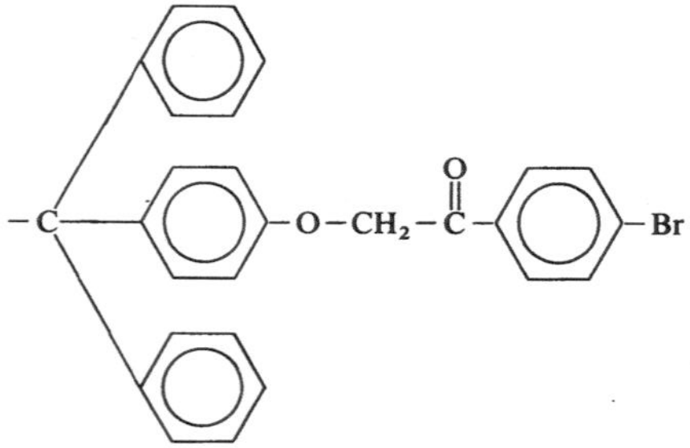	–OH selective 5′-OH	p-bromo-phenacyloxy-phenyl-phenyl-diphenyl-methyl-chlorid	–OH	very mild acid		(436)
acetal, ketal groups									
85	α-(methoxy-ethyl)		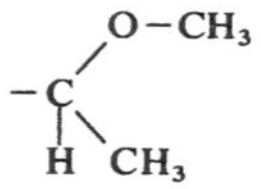	–OH	methylvinyl-ether p-toluene-sulfonic acid	–OH	mild acid	mono addition substrate	(252)
					acetaldehyde + methanol in DMF				(386a, 387)

No.	Name	Abbrev.	Structure		Reagent		Removal	Ref.
86	α-ethoxy-ethyl	EtOEt[b] EE[d]	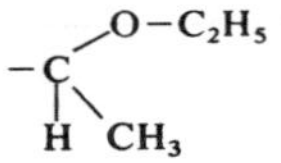 $-C$ with $O-C_2H_5$, H, CH_3	$-OH$, $-NH_2$	ethylvinyl ether	$-OH$, $-NH_2$	mild acid	(197)
87	n-butoxy-ethyl		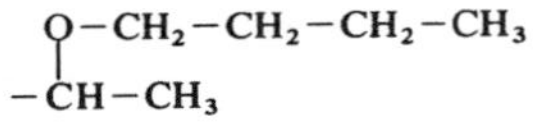 $O-CH_2-CH_2-CH_2-CH_3$ $-CH-CH_3$	$-OH$,	n-butylvinyl-ether, trifluoro-acetic acid; acetaldehyde + n-butanol in DMF	$-OH$	mild acid	(386a, 387)
88	sec-butoxy-ethyl		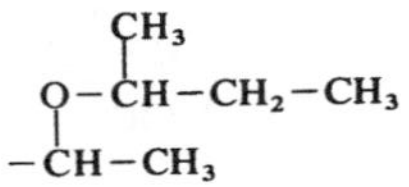 CH_3 $O-CH-CH_2-CH_3$ $-CH-CH_3$	$-OH$	sec-butyl-vinyl ether, trifluoro-acetic acid; acetaldehyde + sec-butanol in DMF	$-OH$	mild acid	(386a, 387)
89	t-butoxy-ethyl		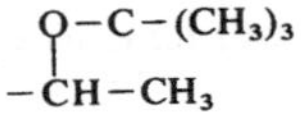 $O-C-(CH_3)_3$ $-CH-CH_3$	$-OH$	tert-butyl vinyl ether, trifluoro-acetic acid	$-OH$	mild acid	(386a, 387)

Table 1.1 (continued)

No.	Blocking group	Abbreviation	Structural formula	Blocked moiety	Conditions for blocking	Deblocked moiety	Deblocking conditions	Special applications	References
90	α-methoxyiso-propyl-		$-C\begin{smallmatrix}OCH_3\\-CH_3\\CH_3\end{smallmatrix}$	-OH sel. 5'-OH	2-methoxy-propylene acid 2,2-di-methoxy-propane, diphenyl-phosphate, dimethyl-acetamide	-OH	mild acid		(197)
91	isopropoxy-isobutyl-		$-CH-CH(CH_3)_2$ $O-CH(CH_3)_2$	-OH	isopropoxy-isobutylene trifluoro-acetic acid; isobutyric aldehyde + iso-propanol in DMF	-OH	mild acid		(386a, 387)

92	n-butoxy-isobutyl	$O-CH_2-CH_2-CH_2-CH_3$ / $-CH-CH(CH_3)_2$	-OH	isobutyric aldehyde + n-butanol in DMF / n-butoxy-isobutylene-trifluoro-acetic acid	-OH	mild acid	(386a, 387)
93	isobutoxy-isobutyl-	$O-CH_2-CH(CH_3)_2$ / $-CH-CH(CH_3)_2$	-OH	isobutoxy-isobutylene, trifluoro-acetic acid / isobutyric aldehyde + isobutanol in DMF	-OH	mild acid	(386a, 387)
94	2-methoxy-ethoxy-isobutyl	$O-CH_2-CH_2-O-CH_3$ / $-CH-CH(CH_3)_2$	-OH	isobutyr-aldehyde + 2-methoxy-ethanol, trifluoro-acetic acid	-OH	mild acid	(387)
95	1-methoxy-cyclohexyl-	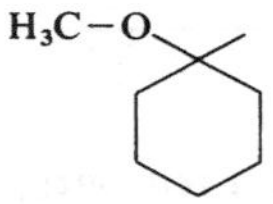 H_3C-O- (cyclohexyl)	-OH	1-methoxy-cyclohexene, acid	-OH	mild acid	(197)

Table 1.1 (continued)

No.	Blocking group	Abbreviation	Structural formula	Blocked moiety	Conditions for blocking	Deblocked moiety	Deblocking conditions	Special applications	References
96	tetrahydropyranyl-	thp[a] Thp[b] THP[c]	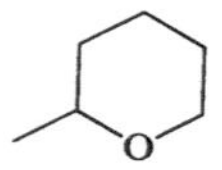	-OH, -NH$_2$	dihydropyran p-toluenesulfonic acid		acid		(184, 197, 258)
						sel. -OH	mild acid		(297, 299)
97	methoxytetrahydropyranyl-		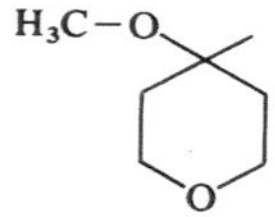 H$_3$C-O	-OH	4-methoxy-dihydropyran, p-toluenesulfonic acid	-OH	mild acid		(116, 352)
98	methoxytetrahydrothiopyranyl		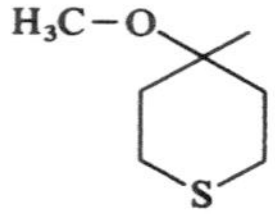 H$_3$C-O	-OH	4-methoxy-5,6-dihydro-4H-thiopyran, mesitylene sulfonic acid	-OH	mild acid		(32)
99	corresp. sulfone of No. 98		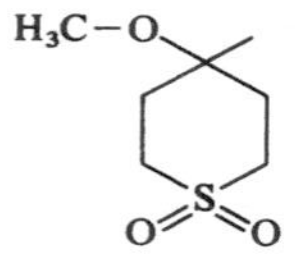 H$_3$C-O	-OH	methoxy-tetrahydrothio-pyranyl-nucleoside + m-chloroperbenzoic acid	-OH	acid		(32)

acetal, ketal groups		structure	Sugar vicinal diol			
100 isopropyl-idene	$>CMe_2^a$ 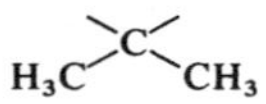	$H_3C-C-CH_3$	$2',3'-(-OH)_2$	acetone, HCl $(2',3'-OH)_2$ / 2,2-di-methoxy-propane, p-toluene-sulfonic acid	strong acid	(99, 100, 101, 184, 197, 258)
101 diethylmethyl-idene		$H_5C_2-C-C_2H_5$	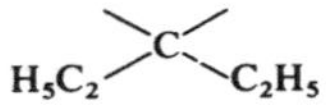 $2',3'-(-OH)_2$	diethyl-ketone, HCl $2',3'-(-OH)_2$	strong acid	(197)
102 methyl-t-butyl-methylidene-		$H_3C-C-C(CH_3)_3$	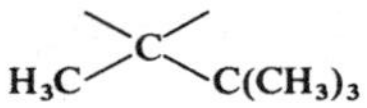 $2',3'-(-OH)_2$	methyl-t-butylketone, HCl $2',3'-(-OH)_2$	strong acid	(197)
103 diphenyl-methylidene-		(diphenyl)	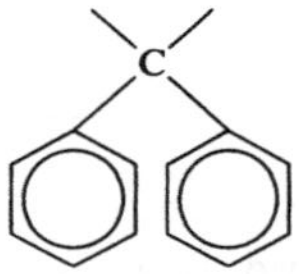 $2',3'-(-OH)_2$	diphenyl-ketone, HCl $2',3'-(-OH)_2$	strong acid	(197)

Table 1.1 (continued)

No.	Blocking group	Abbrev-iation	Structural formula	Blocked moiety	Conditions for blocking	Deblocked moiety	Deblocking conditions	Special appli-cations	References
104	2-phenyl-ethylidene-		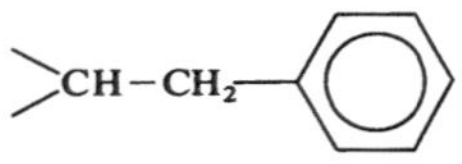	2',3'-(–OH)$_2$	2-phenyl-acetal-dehyde (di)ethyl-phosphoro-thioate, 2,2-di-methoxy-propane, dimethyl-formamide	2',3'-(–OH)$_2$	strong acid		(100, 101)
105	2-chloro-1-methylethyl-idene		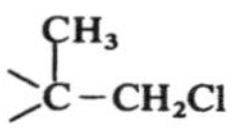	2',3'-(–OH)$_2$	methyl chloro-me-thylketone, (di)ethyl phosphoro-thioate, 2,2-di-dimethoxy-propane, dimethyl-formamide	2',3'-(–OH)$_2$	acid		(100, 101)

			2',3'-(-OH)$_2$		2',3'-(-OH)$_2$		
106	*n*-propylidene-	>CH−CH$_2$−CH$_3$	2',3'-(-OH)$_2$	propion-aldehyde, (di)ethyl-phosphoro-thioate, 2,2-di-methoxy-propane, dimethyl-formamide	2',3'-(-OH)$_2$	strong acid	(100, 101)
107	*sec*-butyl-idene-	CH$_3$ / >C−CH$_2$−CH$_3$	2',3'-(-OH)$_2$	methyl-ethyl-ketone, (di)ethyl-phosphoro-thioate, 2,2-di-methoxy-propane, dimethyl-formamide	methyl-ethyl- 2',3'-(-OH)$_2$	strong acid	(100, 101)
108	1,3-dimethyl-n-butylidene-	CH$_3$ CH$_3$ / >C−CH$_2$−CH−CH$_3$	2',3'-(-OH)$_2$	methyliso-butyl-ketone, (di)ethyl-phosphoro-thioate, 2,2-di-methoxy-propane, diemthyl-formamide	methyliso- 2',3'-(-OH)$_2$	strong acid	(100, 101)

Table 1.1 (continued)

No. Blocking group	Abbreviation	Structural formula	Blocked moiety	Conditions for blocking	Deblocked moiety	Deblocking conditions	Special applications	References
109 1-ethyl-n-propylidene		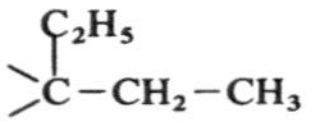 $\underset{\overset{\displaystyle C_2H_5}{\big\vert}}{\text{C}}-CH_2-CH_3$	2',3'-(-OH)$_2$	diethyl-ketone, (di)ethyl-phosphoro-thioate, 2,2-di-methoxy-propane, dimethyl-formamide	2',3'-(-OH)$_2$	strong acid		(100, 101)
110 1-methyl-n-nonylidene-		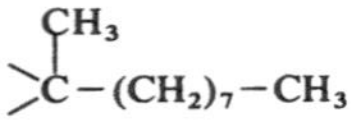 $\underset{\overset{\displaystyle CH_3}{\big\vert}}{\text{C}}-(CH_2)_7-CH_3$	2',3'-(-OH)$_2$	methyl-n-octyl-ketone, (di)ethyl-phosphoro-thioate, 2,2-di-methoxy-propane, dimethyl-formamide	2',3'-(-OH)$_2$	strong acid		(100, 101)
111 Cyclopentyl-idene-		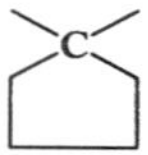	2',3'-(-OH)$_2$	cyclopenta-none, HCl	2',3'-(-OH)$_2$	strong acid		(100, 101, 197)

112 cyclo-heptylidene-	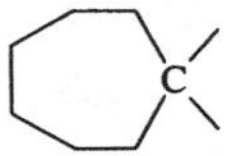	$2',3'$-$(-OH)_2$	cyclohepta-none, HCl	$2',3'$-$(-OH)_2$	strong acid	(197)
113 cyclo-octylidene	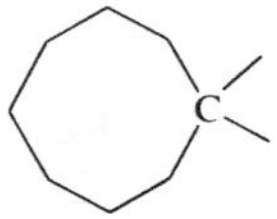	$2',3'$-$(-OH)_2$	cyclo-octanone, HCl	$2',3'$-$(-OH)_2$	strong acid	(197)
114 benzylidene	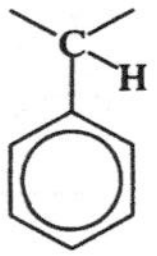	$2',3'$-$(-OH)_2$	benzal-dehyde, p-toluene-sulfo-acid	$2',3'$-$(-OH)_2$	acid	(99, 197)
115 *p*-methyl-benzylidene		$2',3'$-$(-OH)_2$	p-methyl-benzal-dehyde, (di)ethyl-phosphoro-thioate, 2,2-di-methoxy-propane, dimethyl-formamide	$2',3'$-$(-OH)_2$	acid	(100, 101)

Table 1.1 (continued)

No. Blocking group	Abbreviation	Structural formula	Blocked moiety	Conditions for blocking	Deblocked moiety	Deblocking conditions	Special applications	References
116	4-methoxy-benzylidene	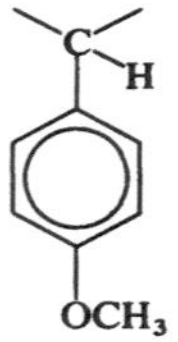	2',3'-(-OH)₂	4-methoxy-benzal-dehyde, acid	2',3'-(-OH)₂	acid		(197)
117	4-dimethyl-aminobenzyl-idene-	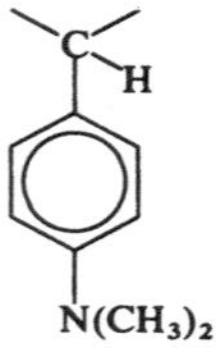	2',3'-(-OH)₂	4-dimethyl-amino-benzal-dehyde, trifluoro-acetic acid	2',3'-(-OH)₂	mild acid		(197)
118	2,4-di-methoxy-benzylidene-	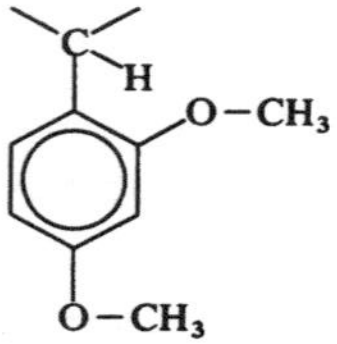	2',3'-(-OH)₂	2,4-di-methoxy-benzal-dehyde, tri-fluoroacetic acid	2',3'-(-OH)₂	mild acid		(197)

119 4-chloro-benzylidene		2',3'-(-OH)$_2$	4-chloro-benzal-dehyde, acid	2',3'-(-OH)$_2$	acid	(197)
120 *p*-nitro-benzylidene		2',3'-(-OH)$_2$	*p*-nitro-benzal-dehyde, tri-fluoroacetic acid	2',3'-(-OH)$_2$	acid	(484)
121 *p*-(N-methyl-N-β-chloro-ethyl)-amino-benzylidene	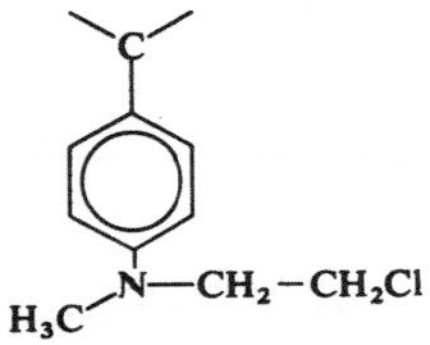	2',3'-(-OH)$_2$	*p*-(N-methyl-N-(β-chloro-ethyl-)amino-benzal-dehyde, p-toluene-sulfonic acid	2',3'-(-OH)$_2$	mild acid	(118)

Table 1.1 (continued)

No.	Blocking group	Abbreviation	Structural formula	Blocked moiety	Conditions blocking	Deblocked moiety	Deblocking conditions	Special applications	References
	orthoesters and deriv.								
122	methoxy-methylidene		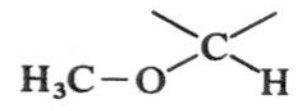	2',3'-(-OH)$_2$	trimethyl-ortho-formate, p-toluene-sulfonic acid	1. 2',3'-(-OH)$_2$ 2. 2'(3')-OH	1. mild acid→ formate 2. alkali		(117, 352)
123	ethoxy-methylidene concurrent with p-nitro-benzylidene		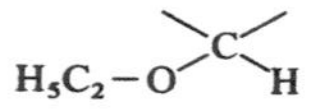	2',3'-(-OH)$_2$ + -NH$_2$	p-nitro-benzaldehyde + ethylortho-formate, trifluoro-acetic acid	1. 2',3'-(-OH)$_2$ 2. 2'(3')-OH -NH$_2$	1. mild acid→ formate 2. alkali, acid		(479, 484)
124	dimethoxy-methylidene-		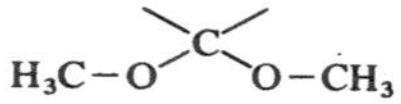	2',3'-(-OH)$_2$	tetramethyl-ortho-carbonate, p-toluene-sulfonic acid	1. 2',3'-(-OH)$_2$ 2. 2'(3')-OH	1. mild acid→ carbonate 2. alkali		(197)

125 methoxy-ethylidene	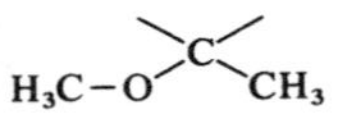	2',3'-(-OH)$_2$	dimethyl-ortho-acetate, p-toluene-sulfonic acid	1. 2',3'-(-OH)$_2$ 2. 2'(3')-OH	1. mild acid→ acetate 2. alkali	(99, 352)
126 methoxy-benzylidene	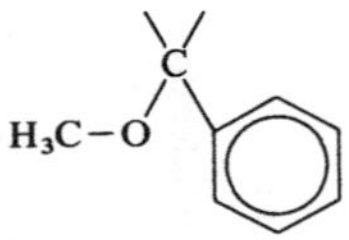	2',3'-(-OH)$_2$	trimethyl-ortho-benzoate, p-toluene-sulfonic acid	1. 2',3'-(-OH)$_2$ 2. 2'(3')-OH	1. mild acid→ benzoate 2. alkali	(99)
127 phenyl-boronate	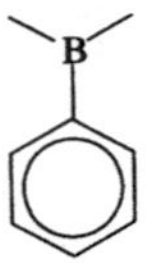	2',3'-(-OH)$_2$	phenyl-boronic acid	2',3'-(-OH)$_2$	propane-diol-1,3 in DMF/water	(84, 202, 203, 474, 475)

1.3.1. Protecting Groups for the Phosphate Moiety

Phosphate protecting groups can be introduced in three ways:

1. Reaction of an appropriate alcohol or amine with a nucleotide, using a condensing agent.

2. Reaction of a nucleoside with an activated phosphoric acid ester or amidate of the blocking agent, e. g. the respective phosphorodichloridate.

3. Addition of nucleotides to blocking reagents, which are unsaturated systems, e. g. diphenyldiazomethane.

Depending on the deblocking conditions phosphate protecting groups as well as alcohol and amino protecting groups can be classified into alkali-labile groups, acid-labile groups and others which are removable at near neutral pH by specific reagents. The mechanism of deblocking of phosphate residues can follow two pathways which are outlined in Figs. 1.3 and 1.4. In pathway 1 the blocking group is removed by

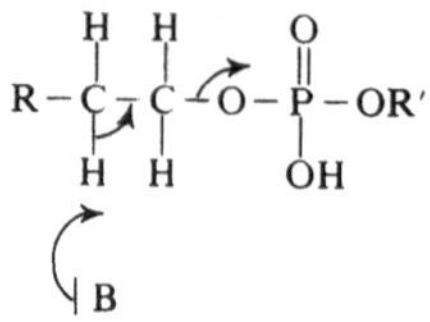

R = blocking substituent
R′ = nucleoside or oligonucleotide

Fig. 1.3

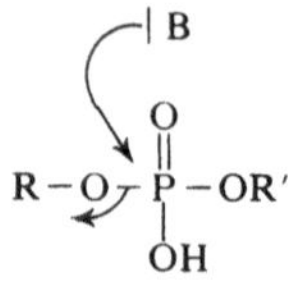

R = blocking substituent

R′ = nucleoside or oligonucleotide

Fig. 1.4

rupture of the O–C_α-bond with liberation of a phosphate anion. Pathway 2 involves the attack of a nucleophile, in most cases water, on the phosphorus atom with rupture of the P–O-bond. This transfer of a phosphoryl moiety to another nucleophile is, in fact, the same as is used in the phosphorylation of alcohols (see Section 2.1). Alcoholysis generally necessitates a higher degree of activation, i.e. electron withdrawal from the phosphoryl moiety, but a sharp distinction between groups used for protection resp. activation is not possible (compare for example phosphoroanilidates and phosphoromorpholidates). Thus, in a few cases protecting groups have, indeed, been transformed into activating groups (see Section 1.4).

The most numerous and widely applied class of phosphate protecting groups is that of β-substituted ethyl esters (Table 1.1, no. 1—13). β-Substituents are introduced, which are electron withdrawing and allow alkaline cleavage with C–O-bond rupture according to Fig. 1.3 (for example R′ = –CN for the β-cyanoethyl group). In the special case of groups no. 10 and 12 (phenylmercaptoethyl resp. uridinyl) the β-substituent has to be rendered electron-withdrawing by oxidation before the group becomes alkali labile (178, 290a). The trichloroethyl group (no. 13) is deblocked by reduction with a zinc/copper couple in a neutral medium, the zinc atom acting as electron donor instead of alkali (Fig. 1.5) (76, 78).

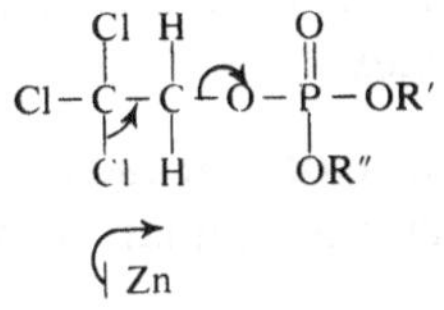

R′ = nucleoside or oligonucleotide

R″ = H, nucleoside or oligonucleotide

Fig. 1.5

Groups no. 14—25 are also ester groups, but their cleavage can involve liberation of either phosphate or phosphoryl groups. Of these the phenyl groups (no. 14—17) have become interesting for protection of the internucleotidic bond (see "triester method", Section 4) (352). Three other groups of this series (no. 19, 20 and 23) offer a possibility of cleavage in neutral medium. Especially useful in polynucleotide synthesis is the ethylthio group (no. 23) which has been investigated by A. L. NUSSBAUM and his colleagues (55). Scission of the P–S-bond is effected by iodine oxidation with possible formation of a phosphoroiodinate intermediate, which allows not only hydrolysis, but also alcoholysis (Scheme 1.6).

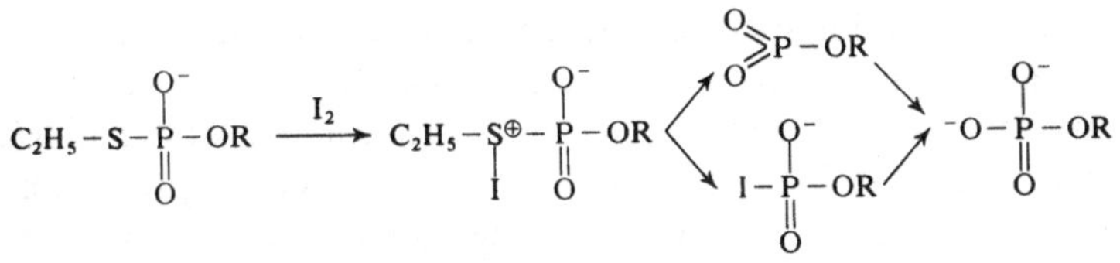

R = 3′-O-acetylthymidine-5′-

Scheme 1.6. Removal of the ethylthio group

A route to a third class of phosphate protecting groups, namely the phosphoramidate type (no. 26—30) was opened up by the finding of E. Ohtsuka and coworkers, that acidic hydrolysis of phosphoro-anilidates, normally accompanied by glycoside cleavage, can be carried out under very mild conditions, if isoamylnitrite is added to the medium (310). Similar to the ethylthiophosphate case, electrophilic attack on the nitrogen atom with intermediate nitrosation is believed to be responsible for the easy removal of this group. All phosphoranilidate groups thus belong to the type of residue cleaved according to Fig. 1.4. By appropriate substitution of the anilidate residue several groups with special properties have been developed, useful, for example, for activation (no. 27) (316) solvent extraction (no. 29) (3) or affinity separation (no. 30) (136) of nucleotides or oligonucleotides.

1.3.2. Protecting Groups for the Hydroxyl and Amino Functions of the Sugar and Base Moieties

The introduction of blocking groups into the hydroxyl and amino functions of sugar and bases proceeds by reactions similar to the ones discussed for phosphate protection, namely

1. Reaction of activated derivatives of carboxylic acids, carbonic or carbamic acid and of highly electrophilic alkyl halides with sugar and/or bases,

2. Reaction of "activated alcohol derivatives" of nucleosides with protecting agents,

3. Acid-catalyzed acetalization, ketalization or transacetalization of nucleosides,

4. Addition of nucleosides and nucleotides to compounds with polarized double bonds.

By far the greatest number of protecting groups is attached by reaction type 1. Esterification of nucleotide hydroxyl and amino groups with acid chlorides or preferably anhydrides generally proceeds with excellent, often quantitative yields. The alternative route, reaction with a carboxylic acid and a condensing agent, is less attractive because yields are lower and is used only when activated acid derivatives are not available (e.g. benzoylpropionic acid, no. 44). Trityl and benzyl groups are similarly introduced by the action of trityl and benzyl halides.

Reaction 2 is still an exceptional case for alcohol protection. The activation of hydroxyl groups of the sugar moiety by triphenyl-phosphine and azodicarboxylate has been described by O. Mitsunobu and coworkers (264). Activated alcohol groups can be reacted with carboxylic acids as well as with phosphoric acid (see Section 2.2); for steric reasons the 5'-OH group is specifically substituted. The blocking of OH-groups by reaction of nucleoside chloroformates with alcohols and amines may, to a certain extent, also be counted among reactions of this type. On the whole, activation of the alcohol groups of nucleosides is an interesting addition to the possibilities of protection, which merits further investigation.

Examples of reactions with aldehydes, acetals and orthoesters are equally scarce in the protection of "isolated" OH- and NH_2-groups (for the blocking of vicinal diol groups see Section 1.3.3). The only case of importance is the formation of Schiff bases specifically with exocyclic amino groups of the bases by treatment with dimethylformamide acetals.

Reaction with enol ethers (type 4) leads to introduction of acetal and ketal groups, which are of great importance for the protection of ribo nucleosides and nucleotides. As acid catalysis is necessary for this reaction, their use is restricted to the ribo series.

A more detailed discussion of ester type protecting groups has to start with the acetyl and benzoyl groups (no. 33 and 52), both of which are standard protecting substituents in sugar chemistry. Acetylation is normally carried out with acetic anhydride, whereas the more reactive acyl halides are preferred for benzoylation. Treatment with these reagents normally leads to substitution of all OH- and NH_2-functions; partial deblocking of the hydroxyl groups can subsequently be effected by strong alkali under controlled conditions due to the relative stability of carboxamide relative to ester moieties at high pH (197). The conditions of these acylation reactions are subject to a great deal of variation, and some of these variations are noteworthy for reasons of selectivity. Thus, the amino group of cytosine nucleoside and nucleotide could be selectively acetylated by acetic anhydride in dimethylformamide (329, 330) and benzoylated by benzoic acid –N-hydroxysuccinimide ester (303). The 5′-hydroxyl-selective acylation by for example, benzoic acid, triphenylphosphine and azodicarboxylate (264) was discussed above. Selective acylation of either 2′- or 3′-OH in ribosides or ribotides can be achieved by acid treatment of orthoester derivatives (no. 125 and 126) (99). Suitable substitution of the acetyl or benzoyl substituent gives blocking groups which are more readily (e. g. no. 34, 35, 36, 37 and 54) or less readily (e. g. no. 46–49, 53) cleaved by alkali than the parent groups. The first alternative is preferred in oligoribonucleotide synthesis; the second is sought for in the design of amino protecting groups for deoxyoligonucleotides. The methoxyacetyl, isobutyryl and anisoyl groups are among the most widely used in oligonucleotide synthesis.

Ester groups, which offer a possibility of cleavage in neutral medium, are the chloroacetyl (57) resp. benzoylpropionyl and benzoylformyl groups (230, 234) (no. 32, 38, 44), all displaced by a cyclisation mechanism with thiourea or 2-mercaptoethylamine in the case of no. 38 and hydrazine resp. o-phenylenediamine in the two latter cases (Scheme 1.7). Hydrazine treatment in a pyridinium acetate buffer

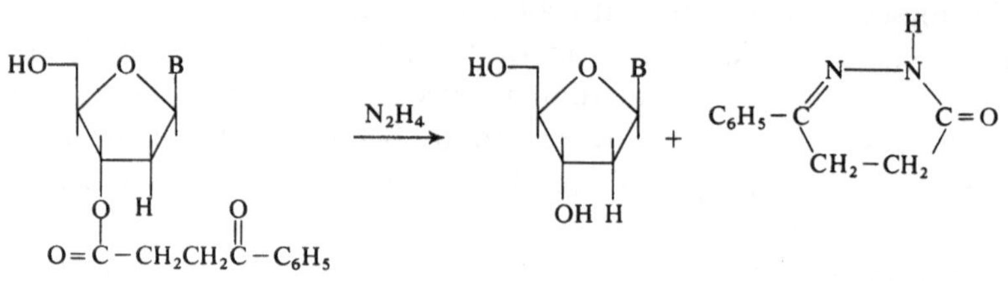

Scheme 1.7. Removal of the benzoylpropionyl group

will also selectively remove benzoyl or anisoyl groups from the nucleobase (233). The 2,4-dinitrobenzenesulfenyl group is readily cleaved in a neutral medium by thiophenol (197), however, its use is restricted by alkali-sensitivity. Similarly, the formyl group, although very useful in polypeptide synthesis, has not been widely used in the polynucleotide field due to its extreme lability in weakly alkaline media. Recent investigations, however, show, that 3'-O-formyl esters of deoxynucleotides and -nucleoside polyphosphates can be readily synthesized in quantitative yield by action of formic acid or formic acetic anhydride on base-protected nucleotides and directly used for oligonucleotide synthesis without purification. The formyl group is stable in pyridine solution and is lost during aqueous workup of the condensation mixture (332, 398).

A completely different approach to selective cleavage in neutral medium, namely the enzymatic hydrolysis by esterases, has been investigated by A. TAUNTON-RIGBY and N. A. STARKOVSKY for the case of the dihydrocinnamoyl substituent (no. 43) (363).

Blocking groups of the carbonate or carbaminate type are among the most widely used in the peptide field, and a number of such groups has also been of use in oligonucleotide synthesis (no. 61—69). Of these the isobutyloxycarbonyl group (no. 64) can be selectively attached to the 5'-hydroxyl group due to steric hindrance (236, 304); the trichloro- or tribromophenyloxycarbonyl group can be cleaved in near neutral medium by Zn/Cu in methanol/acetic acid by a mechanism similar to the one described in Fig. 1.5, section 1.3.1 (54, 183, 466). It appears that the full scope of possibilities for selective introduction and cleavage is not yet fully exploited for this class of protecting groups in the nucleotide field. The possibility of introducing a variety of alcohol or amine substituents by addition to nucleoside chloroformates

(*391*) may simplify the search for new solutions. An interesting new development in this area is the naphthylcarbamoyl group (no. 69), described by K. L. AGARWAL *et al.* (*4*) for the blocking of unreacted 3'-OH ends by a group, which permits separation of truncated sequences by affinity chromatography (see Section 1.4).

Orthoester groups, widely used for the blocking of the vicinal diol function in ribonucleosides (see 1.3.3) are occasionally also employed for the protection of other hydroxyl and amino functions. Of the examples given in Table 1.1 (no. 71—73) the dimethylaminomethylidene group (no. 72) is of special interest, because it is the only group which can be selectively introduced into the base (*197*). Its application in oligonucleotide synthesis is, however, limited by the fact that it is labile to mild acid as well as alkaline conditions. Dimethylformamide dimethyl- and dineopentyl acetal are generally used to introduce this group. The second reagent is preferred in many cases, since the dimethyl acetal can also act as an alkylating (*481*) or dephosphorylating agent (*480*).

A second major category of hydroxyl and amino blocking groups (no. 74—84) includes those which have ether, acetal and ketal substituents. Benzyl ethers have been used for hydroxyl protection since the beginnings of oligonucleotide chemistry; however, their use is restricted by the fact that deprotection through catalytic hydrogenation may affect the pyrimidine bases (*258*). Apart from this group all other ether substituents are of the trityl type. As cleavage of unsubstituted trityl ethers requires relatively strong acidic conditions, e. g. 80% acetic acid for several hours at reflux, *p*-methoxy substituents have been introduced to facilitate their removal. Each *p*-methoxy substitution produces a ten-fold acceleration of the rate of acidic hydrolysis, as was found by chromatographic (*6*) and NMR (*354*) investigations. Mono- and di-*p*-methoxytrityl groups are at optimum as to stability and deblocking conditions and are widely used in polynucleotide synthesis, especially as, under appropriate conditions, they are selectively introduced into the 5'-hydroxyl group of N-protected nucleosides and nucleotides (*184, 197*).

Acetal and ketal groups (no. 85—99) need acid catalysis for introduction and removal and, therefore, are more suitable for work in the ribo series. Several acyclic and cyclic alkyl vinyl ethers have been employed as blocking agents. The ethoxyethyl- (*197*), tetrahydropyranyl- (*197*) and methoxytetrahydropyranyl groups (*352*) (no. 86, 96 and 97) have been successfully employed in oligoribonucleotide synthesis for several years. Recently the methoxytetrahydrothiopyranyl group (no. 98) has been described; its acid stability can be regulated by oxidation of the sulfide moiety (*32*). Methoxyethyl groups (no. 85) have been introduced for the purpose of enzymatic monoaddition (252) (see Section 4). Since the cytosine base is protonated under the conditions of reaction with dihydropyran, this could be used to selectively introduce the tetrahydropyranyl group into the 3'-position of deoxycytidine monophosphate as an intermediate step in the preparation of N-benzoyl-deoxycytidylic acid (197). This is one of the few examples, in which such blocking groups habe been applied in the deoxy series.

1.3.3. Protecting Groups for the Vicinal Diol Group of Ribonucleic Acid Constituents

Acetal, ketal and orthoester substituents (no. 100—127) are used for protection of the 2',3'-diol function in ribotides. They are generally introduced by acid-catalyzed reaction of the diol group with appropriate aldehydes, ketones or orthoesters. Isopropylidene groups (no. 100), standard protecting agents in sugar chemistry, are often used (197). However, they are very stable to acidic hydrolysis, and in order to avoid an isomerization of the internucleotidic bond more acid-labile groups, such as 2,4-dimethoxybenzylidene and 4-dimethylaminobenzylidene (no. 117 and 118) (197) are preferred for stepwise oligoribonucleotide synthesis.

The orthoester substituents (no. 122—126) possess the additional feature of undergoing isomerization to 2'- or 3'-acylates on mild

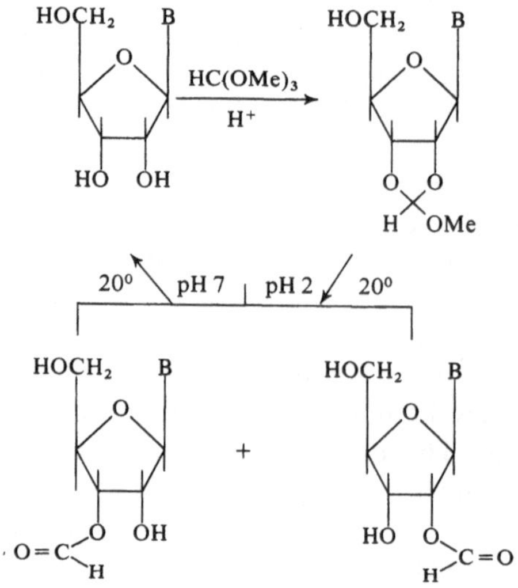

Scheme 1.8. Removal of orthoester substituents

treatment with aqueous acid (*117, 352*). This offers a possibility for differentiation between the two hydroxyl groups which are of equal reactivity in internucleotide bond formation (for further discussion see Section 1.6). Final removal of these protecting groups then necessitates mild alkali or ammonia treatment (Scheme 1.8).

One example of a diol protecting group which can be removed in a neutral medium, has been reported: the phenylboronate substituent (no. 127), described by A. M. YURKEVITCH and coworkers (*475*). It is easily introduced by treatment with phenylboronic acid. Cleavage is effected by propanediol-1, in dimethylformamide. Although this group is stable during the formation of internucleotidic bonds, its lability in aqueous acidic to neutral media is disadvantageous, since the diol protection must be stable throughout all steps of a ribooligonucleotide synthesis (see Section 1.6).

In concluding this section it should, again, be emphasized that alternatively, the diol moiety can be blocked by two substituents of the type discussed in Section 1.3.2, e.g. by two acyl groups. Several investigators (see Section 1.6) have employed this as the method of choice for oligoribonucleotide synthesis.

1.4. Protecting Groups with Special Applications

In several more recent publications blocking groups have been described which perform an additional job, such as opening up new ways of separation or activating the nucleotides or oligonucleotides to which they are attached. Some of the most significant developments in this area will be highlighted in this section.

1.4.1. Protecting Groups for Solvent Extraction

The introduction of a blocking group will in most cases change the solubility properties of polynucleotides and their constituents, a change which most often goes in the direction of enhanced hydrophobicity. In extreme cases nucleotides or oligonucleotides can become water-insoluble and extractable into water-immiscible organic media. Examples have been described of oligonucleotide synthesis using *p*-aminophenyltriphenylmethane (*3*) or 9-fluoroenylmethanol (*176*) for protection of the phosphate component (groups no. 11 and 29). The use of these groups in the preparation of short-chain oligonucleotides will be discussed in more detail in Sections 3.2 and 4.1.1.

1.4.2. Protecting Groups for Separations by Affinity Chromatography

Mono-, oligo- and polynucleotides with hydrophobic substituents were shown to be retained on appropriate affinity columns. Chromatographic materials used for this purpose are, for example, tritylated or naphthoylated cellulose (*41*) or benzoylated DEAE cellulose (*259, 289*). The demands placed on the hydrophobicity of groups for this technique are not as restricting as for the solvent extraction procedures; this permits, in principle, the use of a wide variety of substituents for such separations. H. G. Khorana and coworkers, for instance, have separated tritylated oligonucleotides from untritylated oligomers and monomers on trityl cellulose., Similarly, homologs of shorter chain length are removed during stepwise oligonucleotide synthesis after protection of their free 3'-OH ends by naphthylisocyanate (no. 69) (*4, 6*). Hydrophobic protecting groups, mostly of the β-substituted ethyl phosphate type (no. 7—10), have been studied by S. A. Narang and coworkers for stepwise oligonucleotide syntheses with purification on benzoylated DEAE (*289, 290, 290a, 465*) or benzoylated DEAE-Sephadex (*259*). Affinity separations, based on the interaction of oligonucleotides blocked with the *p*-dimethylammoniumanilidate residue (no. 30) with cation exchangers, have been employed for stepwise oligonucleotide synthesis by T. Hata *et al.* (*136, 138, 435*).

A different approach has been studied by H. Seliger and coworkers (*396, 398a, 398b*). When nucleotides or nucleosides, blocked by affinity groups, are copolymerized with unprotected (resp. only N-protected)

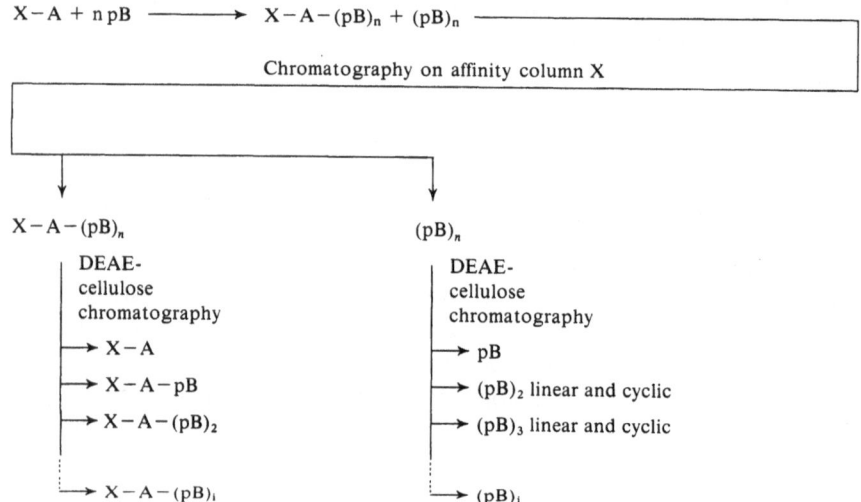

Scheme 1.9a. Binary cooligocondensation and separation of nucleotides blocked by affinity groups

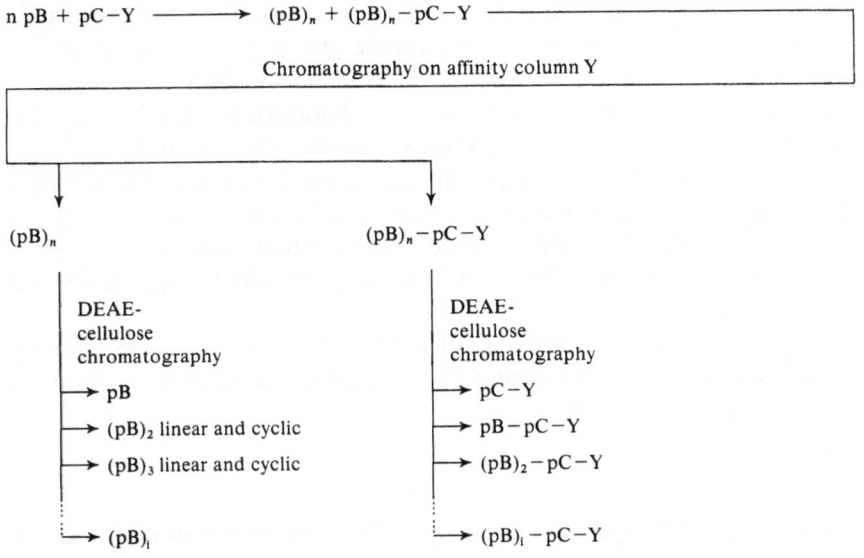

$$n\,pB + pC-Y \longrightarrow (pB)_n + (pB)_n-pC-Y$$

Chromatography on affinity column Y

$(pB)_n$ $(pB)_n-pC-Y$

DEAE-
cellulose
chromatography

→ pB

→ $(pB)_2$ linear and cyclic

→ $(pB)_3$ linear and cyclic

⋯→ $(pB)_i$

DEAE-
cellulose
chromatography

→ pC-Y

→ pB-pC-Y

→ $(pB)_2-pC-Y$

⋯→ $(pB)_i-pC-Y$

Scheme 1.9b. Binary cooligocondensation and separation of nucleotides blocked by affinity groups

nucleotides, two types of homologous sequences are produced, namely those that bear affinity end groups and others that do not. Two types of affinity groups have been used, namely hydrophobic groups, such as methoxy trityl or phenylazophenyloxycarbonyl (no. 67 and 76) for the 5'-terminus and the uridinyl group (no. 12) for the 3'-terminus. Separations are carried out on trityl cellulose resp. boronate celluloses. A general reaction and separation scheme for binary copolymerizations of the components X–A and pB resp. pB and pC–Y (where X and Y are affinity groups and A, B and C are nucleosides) is given in Scheme 1.9. The two types of homologs, X–A–B_n and B_n resp. B_n and B_n–C–Y are subsequently separated into individual sequences according to charge, thus yielding a variety of building fragments for polynucleotide synthesis through one reaction.

1.4.3. Activable Protecting Groups

In the discussion of phosphate protecting groups (Section 1.3.1) we have pointed out that there may be no fundamental difference between protecting and activating residues. Hence, several phosphate protecting groups can be modified to become strongly activating groups, thus allowing alcoholytic and phosphorolytic attack on the phosphorus atom. Early work by F. CRAMER and H. SELIGER (*389*) showed that functions, such as enol esters on phosphoric acids become strongly activating on bromination. Similar results with nucleotides have been obtained only recently. Thus, the ethylphosphorothioate group (no. 23) on oxidation with iodine yields an activated intermediate, which can undergo alcoholysis (55) (compare Scheme 1.6). Oxidative activation was also demonstrated for the p-hydroxyanilidate residue (no. 27) by E. Ohtsuka et al. (316) (see Scheme 4.21). In this case internucleotide linkages could be formed on activation in presence of a nucleoside, although in moderate yield. The 4-nitro-2-chloromethylphenyl residue (no. 19) is activated in the presence of tertiary bases (e. g. pyridine) by quaternization (281).

A fourth area of special application of blocking groups, namely in the formation of enzymatic monoaddition substrates, is reviewed in more detail in Section 5.1.

1.5. Strategy of Consecutive Blocking or Deblocking of Several Functions

The last two sections of this chapter deal with different approaches to the preparation of intermediates for nucleotide polycondensation and stepwise oligonucleotide synthesis. Initially some general considerations governing the choice of blocking groups for certain functions will be discussed.

We have seen in earlier sections, that differences in nucleophilicity between the hydroxyl groups of the sugar and the exocyclic amino groups of the bases are relatively small. This means that, except for the reaction of nucleotide phosphate groups with alcohols and amines as blocking reagents, all other reactions used for the introduction of blocking groups tend to be unspecific. Nevertheless, it is possible to block selectively one of several alcohol or amino functions by almost any group by adopting one of the following routes:

1. If a selective reagent is available, this can be directly used to block the desired function.

2. If a selective reagent is available for the other functions, these can be blocked first. The function in question is left free to react with an unspecific reagent. Afterwards the other blocking groups are removed, if necessary.

3. An unspecific reagent is used to introduce a blocking group into all functions. All but the desired function are then selectively deblocked.

4. If this is not possible, total blocking can be followed by selective deblocking of the desired function, which is then reblocked with the reagent of choice.

These four routes can be illustrated for the example of the acetyl protecting group:

Route 1: Selective acetylation of the 5'-OH group is affected by acetic acid / azodicarboxylate / triphenylphosphine (264). For the exocyclic amino group of the cytosine nucleotides acetic anhydride in dimethylformamide / tri-n-butylamine has been described as selective reagent (329, 330). Finally, 2' or 3'-OH groups of ribonucleosides can be acetylated with trimethylorthoacetate via a methoxyethylidene intermediate (99).

Route 2: The selective 3'-O-acetylation of nucleosides has been described by introducing first a dimethylaminomethylene group into the base, then a methoxytrityl group into the 5'-hydroxyl. Acetylation followed by acid treatment leaves acetyl as only protecting group in 3'-position (197).

Route 3: The most widely used approach to N-acetylation of nucleosides and nucleotides involves unspecific reaction with acetic anhydride in pyridine, followed by selective removal of O-acetyl groups on short treatment with strongly alkaline media (197).

Route 4: N-benzoyl-3'-O-acetyl nucleotides are usually synthesized by perbenzoylation of nucleotides, followed by selective O-debenzoylation in strong alkali and treatment with acetic anhydride (184, 197).

A detailed description of all possibilities for selective introduction or removal of the different blocking groups listed in Table 1.1 would by far exceed the scope of this review. We must limit ourselves to some general observations. Thus, for example, sterically hindered reagents react more readily with amino and 5'-hydroxyl groups than with 2'- and 3'-hydroxyl. The rate of hydrolytic cleavage can also be higher for 5'- than 3'-substituents as was demonstrated for nucleosides containing several trityl groups. In the case of acylations the reaction conditions may be chosen so as to give specific substitution of the nucleobase in cytidylic acid. The 3'- resp. 2'-OH groups as sterically most hindered functions can generally not be blocked in neutral or weakly basic media without prior protection of the amino and (if unphosphorylated) 5'-hydroxyl groups. More possibilities for selective reactions are contained in Table 1.1, and the information given there may be of help in designing new approaches.

1.6. General Blocking Schemes for Intermediates of Polynucleotide Synthesis

The basis for chemical synthesis of oligo- and polynucleotides, as of other complex organic molecules, is the adoption of a certain strategy for protecting the intermediates. Since the preparation of sufficient quantities of blocked intermediates constitutes a major part of the work involved in the synthesis of longer oligo- and polynucleotides, the groups working in this field tend to pursue one strategy, once they have developed it, for the duration of one or several synthetic projects, like for example, the synthesis of a biologically important polynucleotide. Evidently most groups which tackle the preparation of a polynucleotide have started out by developing a new technique, often a new protecting group or a system of protecting groups. Although this means, that most groups have a different approach and that the authorship of a sequence can often be predicted from the type of intermediates used in the synthesis, some generalizations can be made and will be discussed in the following.

In planning the synthesis of a deoxyoligonucleotide we first have to decide whether the diester or triester method shall be used (see Section 4.1). In the diester method 5′-nucleotides are mostly taken for chain extension, nucleosides only as terminal units. The triester method allows chain elongation reaction with a blocked phosphomonoester, followed by condensation with a nucleoside. The advantage of the latter approach in the sector of blocking groups is that only one type of intermediate is needed for terminal and intrachain units and nucleosides are generally cheaper starting compounds. However, the large-scale preparation of $N,O^{3′}$-protected nucleosides is more time-consuming than the synthesis of analogous nucleotide derivatives, as is clear from the preceding section.

Next the blocking groups for nucleotides and nucleosides have to be selected. Both kinds of monomer units contain three functions, two of which have to be blocked for stepwise oligonucleotide synthesis, namely either

5′-OH resp. -phosphate and bases

or

3′-OH resp. -phosphate and bases.

In an ideal situation these three functions would have to be blocked by three independently-removable blocking groups. This has not yet been realized. Although there are, in principle, three types of deblocking conditions, namely alkaline, acid and neutral with selective reagents, the choice is limited by the circumstance that acid labile groups are preferred as end groups only (deglycosidation hazard, see Section 1.2) and selectively cleavable groups are mostly also acid or alkali-labile. One feature is common to all approaches: The amino groups of the nucleobases are protected most strongly, usually by a group which is removed only by prolonged alkaline treatment. Benzoyl and anisoyl have been widely used for deoxyadenosine resp. deoxycytidine and their nucleotides; acetyl and more recently isobutyryl, isobutyloxycarbonyl, methylbutyryl and also benzoyl have been advocated for deoxyguanosine and dGMP (6, 184, 197, 205, 315, 457).

For the protection of phosphate and hydroxyl functions three different strategies have been generally applied.

1. The strategy developed mainly by H. G. KHORANA and coworkers (6, 184, 197, 319, 457) uses preferably compounds A of Table 1.2 as intermediates. The methoxy- or dimethoxytrityl group serves as acid-labile 5'-end group. The 5'-phosphate group of nucleotides is protected by the β-cyanoethyl moiety, labile to brief treatment in alkali. The acetyl group, removable by alkali or ammoniacal treatment, is taken for 3'-end protection. Thus, 5'-terminal units or building fragments for polynucleotide synthesis contain an acid-labile and an alkali-labile end.

Chain extension is effected by monomers of building blocks containing a 5'-phosphate and a 3'-hydroxyl end. All these building fragments such as a dinucleotide (Fig. 1.12), on construction from units

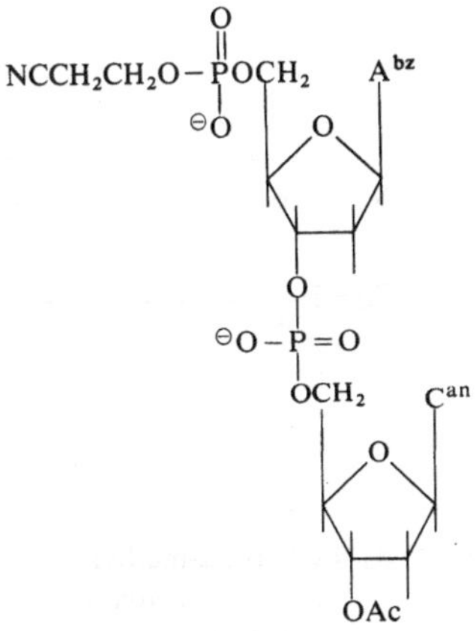

Fig. 1.12

A 1+2 of Table 1.2, will have two alkali-labile blocking groups at the ends. Their removal leaves both ends free. Reblocking by acetylation gives a new building block for lengthening towards the 3'-end, β-cyanoethylation affords a fragment for lenghthening towards the 5'-end. This strategy has been used successfully throughout the work on the synthesis of two genes and is described explicitly by H. BÜCHI, H. WEBER and H. G. KHORANA (37, 457).

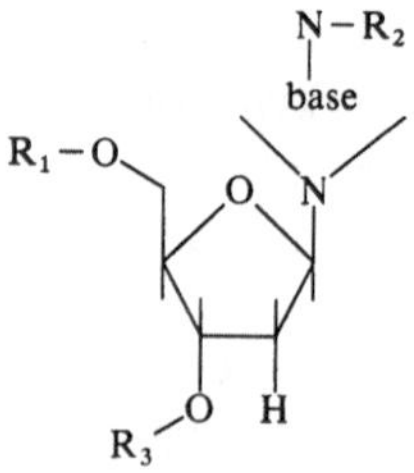

Table 1.2. *Some Examples of Blocked*

Me-thod	5'-terminal units:	
	1	1'
A	Fig. 1.10 R_1 = methoxy-trityl- = dimethoxy-trityl- R_2 = acyl- R_3 = H-	
B	Fig. 1.10 R_1 = methoxy-trityl- = dimethoxy-trityl- R_2 = acyl- R_3 = H-	
C	Fig. 1.10 R_1 = methoxytrityl- R_2 = acyl- R_3 = H- ⟶	Fig. 1.10 R_1 = methoxytrityl- R_2 = acyl- R_3 = OH | $-P=O$ | $O-CH_2-CH_2-CN$
D*	Fig. 1.10 R_1 = trityl- R_3 = H- ⟶	Fig. 1.10 R_1 = trityl- R_3 = O-H | $-P=O$ | $O-CH_2-CCl_3$
E*	Fig. 1.10 R_1 = 2,4-(bis-2-methyl-butyl-2-)phenyl-oxyacetyl- R_3 = H- ⟶	Fig. 1.10 R_1 = 2,4-(bis-2-methyl-butyl-2-)phenyl-oxyacetyl- R_3 = OH | $-P=O$ | $O-C_6H_5$

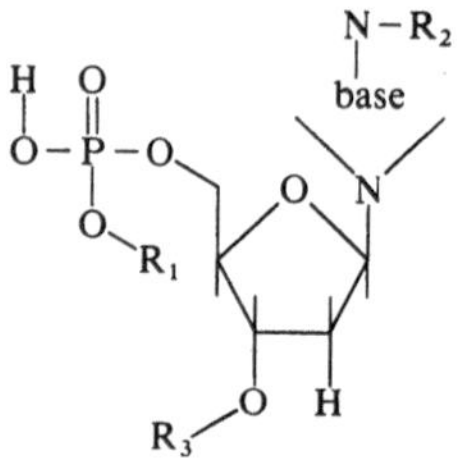

* R_2 is not given, where thymidine derivatives were the only ones used.

Fig. 1.10

Fig. 1.11

Intermediates for Oligodeoxyribonucleotide Synthesis

Intrachain units:		3'-terminal units:
2	2'	3
Fig. 1.11 $R_1 = \beta$-cyanoethyl-		Fig. 1.11 $R_1 = $H-
$R_2 = $acyl-		$R_2 = $acyl-
$R_3 = $H-		$R_3 = $acetyl-
Fig. 1.11 $R_1 = $trichloroethyl-, $= $anilidate-, $= $ethylthio-, $= $phenylmercaptoethyl-		Fig. 1.11 $R_1 = $H-
$R_2 = $acyl		$R_2 = $acyl-
$R_3 = $H-		$R_3 = $acetyl-
Fig. 1.10 $R_1 = $H-	Fig. 1.10 $R_1 = $H-	Fig. 1.10 $R_1 = $H-
$R_2 = $acyl-	$R_2 = $acyl-	$R_2 = $acyl-
$R_3 = \beta$-benzoylpropionyl- $\longrightarrow$	$R_3 = $ OH $\mid$ $-$P$=$O $\mid$ O$-$CH$_2$$-CH_2$$-$CN	$R_3 = $methoxytrityl-, $= \beta$-benzoylpropionyl-
Fig. 1.10 $R_1 = $H-	Fig. 1.10 $R_1 = $H-	same as intrachain
$R_2 = $acyl-	$R_2 = $acyl-	unit D$_2$
$R_3 = $acetyl- $\longrightarrow$	$R_3 = $ OH $\mid$ $-$P$=$O $\mid$ O$-$CH$_2$$-CCl_3$	
Fig. 1.10 $R_1 = $H-	Fig. 1.10 $R_1 = $H-	same as interchain
		unit E$_2$
$R_3 = $methoxytetrahydropyranyl- $\longrightarrow$	$R_3 = $ OH $\mid$ $-$P$=$O $\mid$ O$-$C$_6$H$_5$	

2. The necessity for reblocking of oligonucleotide fragments is over-come by using instead of unit A 2 an intermediate B 2 (Table 1.2), in which the phosphate residue is blocked by a group which is stable to alkali and removable in neutral medium by selective reagents. Groups that have been especially useful for this purpose are the trichloroethyl group (no. 13), introduced by F. Eckstein and coworkers (94, 197), the anilidate residue (no. 26), described as selective labile group by E. Ohtsuka, M. Ikehara and coworkers (310), and the ethylthio group (no. 23), investigated by A. L. Nussbaum and colleagues (55). The phenylmercaptoethyl group (no. 10), described by S. A. Narang and coworkers (465), is also a development along this line, although its complete removal, after oxidation, affords alkali and leads to the loss of all alkali-labile protecting groups on 3′-OH and bases. All of these protecting groups have been successfully used in the sequence specific preparation of oligonucleotides, the latter two in projects pertaining to gene synthesis.

3. In the phosphotriester method blocked nucleosides are the inter-mediates for chain extension, as was discussed above. The 5′-terminus is again masked by trityl groups. The blocking of the 3′-end depends on the type of triester substituent. If the blocking group for the inter-nucleotidic linkage is alkali labile, as is the β-cyanoethyl group in the approach of R. L. Letsinger and coworkers (232, 234, 235), the 3′-protection must be removed in neutral medium. The β-benzoylpropionyl group is well suited for this purpose. The protected intermediates one needs are represented in Table 1.2, line C. Alternatively, groups like trichloroethyl and benzyl (cleaved in neutral medium by reduction resp. anionic debenzylation) have been used to protect the internucleotidic linkage, thus allowing the well-developed scheme of 3′-O-acetylation to be retained (intermediates D in Table 1.2) (76, 78, 369). Recently, phenyl groups have been advocated as triester blocking groups (350, 352). As these are, again, alkali-labile, an acid-labile protecting group has been chosen for the (growing) 3′-terminus (intermediates E in Table 1.2). Building fragments (Fig. 1.13) for the preparation of longer oligonucleotide chains have been synthesized on this basis (73).

In the synthesis of oligoribonucleotides an additional complication is introduced by the presence of the 2′-OH group. Since the hydroxyl groups at the 2′- and 3′-position are about equally reactive towards electrophilic attack some kind of differentiation is necessary to obtain intermediates which allow an internucleotidic bond to be formed specifically at O–C³′. This is most simply achieved by using suitably blocked nucleoside-3′-phosphates, which are available either by sub-stitution of 3′-ribonucleotides or by phosphorylation of ribonucleosides

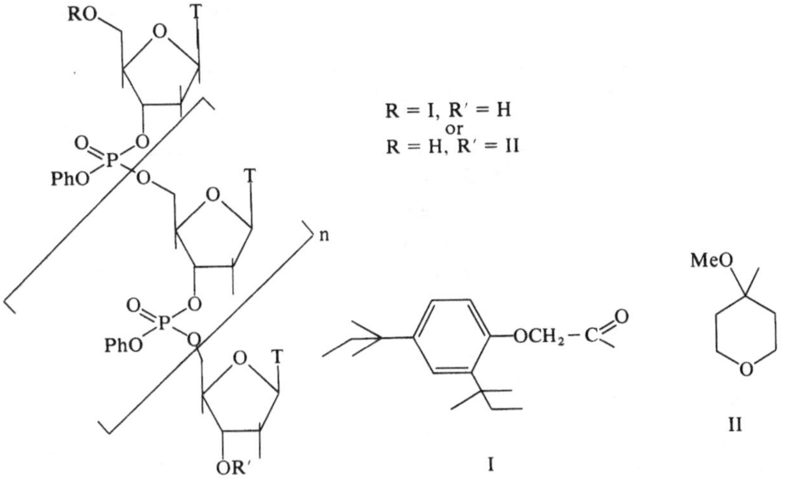

R = I, R' = H
or
R = H, R' = II

Fig. 1.13

with selectively unblocked 3'-OH. Several routes to the latter compounds have been described.

The choice of blocking groups for the different functions again differs among various laboratories. For the preparation of ribonucleotide triplets, H. G. KHORANA and coworkers have used the intermediates F of Table 1.3 (249). They all contain alkali-labile groups for those functions that remain protected until completion of the sequence. Acid-labile trityl groups are used for the 5'-termini. Because the 3'-phosphate remains unprotected, this approach necessitates chain extension from the 3'- to the 5'-end. In a variation described by E. OHTSUKA et al. (309, 313, 319), protection is provided for the 3'-phosphate through the anilidate residue, which can be selectively removed by isoamylnitrite as discussed in Section 1.3.1, and monomers can now be added to both ends of the intermediate G 2 in Table 1.3.

In a different method elaborated by the group at Prague (152) the permanently blocked positions (2'-OH and terminal diol) contain acid-labile groups. Since acid-labile protection of the bases is unusual, the amino groups and the 5'-hydroxyl function in intermediate H 1 (Table 1.3) are protected in an alkali-labile fashion. Chain lenghthening is again done from the 3'- to the 5'-terminus. On deblocking of the 5'-end both alkali-labile groups are lost and the NH2-functions can be selectively reprotected by treatment with dimethylformamide acetals. The intermediates H 2 and H 3 are also N-dimethylaminomethylene

Table 1.3. *Some Examples of Blocked*

R_1-O

N$-R_2$

base

R_4 R_3

Fig. 1.14

R_1-O

N$-R_2$

base

$R_4-O-P=O$ R_3

OH

Fig. 1.15

Me-thod		5′-terminal units:	
		1	1′
F	Fig. 1.15	R_1 = trityl- R_2 = acyl- R_3 = acyl- R_4 = H-	
G	Fig. 1.14	R_1 = methoxytrityl- R_2 = acyl- R_3 = acyl- R_4 = H-	
H	Fig. 1.15	R_1 = acetyl- R_2 = acetyl- R_3 = ethoxyethyl- R_4 = H-	Fig. 1.15 R_1 = methoxytrityl-, = acetyl- R_2 = acetyl- R_3 = tetrahydropyranyl- R_4 = β-cyanoethyl- = β,β,β-trichloro- ethyl-
I	Fig. 1.14	R_1 = trityl-, = acyl-, = methoxytetra- hydropyranyl R_2 = acyl- R_3 = methoxytetra- hydropyranyl- R_4 = H- ⟶	Fig. 1.14 R_1 = methoxytetra- hydropyranyl- R_2 = acyl- R_3 = methoxytetra- hydropyranyl- R_4 = OH | −P=O | O−C$_6$H$_5$
K	Fig. 1.14	R_1 = trityloxyacetyl- R_2 = acyl- R_3 = tetrahydropyranyl- R_4 = H- ⟶	Fig. 1.14 R_1 = trityloxyacetyl- R_2 = acyl- R_3 = tetrahydropyranyl- R_4 = OH | −P=O | O−CH$_2$−CCl$_3$

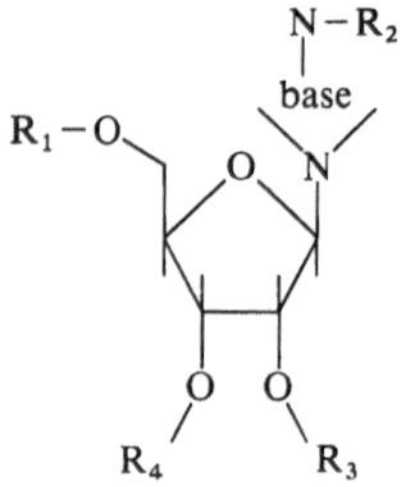

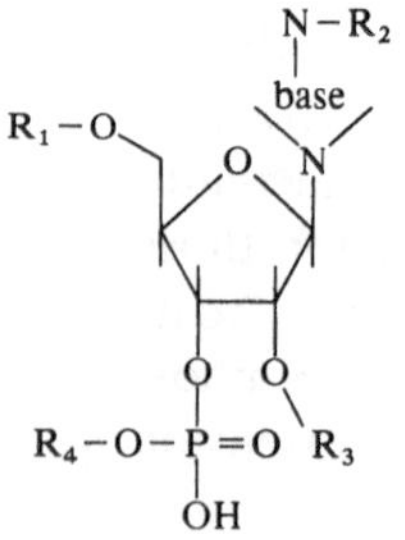

Intermediates for Oligoribonucleotide Synthesis

Intrachain units:		3'-terminal units:
2	2'	3
Fig. 1.15 R_1 = trityl-		Fig. 1.14 R_1 = H-
R_2 = acyl-		R_2 = benzoyl-
R_3 = acyl-		R_3 = benzoyl-
R_4 = H-		R_4 = benzoyl-
Fig. 1.15 R_1 = H-		Fig. 1.15 R_1 = H-
R_2 = acyl-		R_2 = acyl-
R_3 = acyl-		R_3 = acyl-
R_4 = anilidate-		R_4 = anilidate-
Fig. 1.15 R_1 = H-		Fig. 1.14 R_1 = H-
R_2 = dimethylamino- methylene-	same as 5'-terminal unit H_1,	R_2 = dimethylamino- methylene-
R_3 = ethoxyethyl-		R_3 = ethoxymethylene-
R_4 = H-		R_4 = ethoxymethylene-
Fig. 1.14 R_1 = H- = -PO_3H_2		Fig. 1.14 R_1 = H-
R_2 = acyl-		R_2 = acyl-
R_3 = methoxytetra- hydropyranyl-		R_3 = methoxymethyl- idene-
R_4 = H-		R_4 = methoxymethyl- idene-
Fig. 1.14 R_1 = H-	Fig. 1.14 R_1 = H-	same as intrachain
R_2 = acyl-	R_2 = acyl-	unit K_2
R_3 = tetrahydropyranyl-	R_3 = tetrahydropyranyl-	

R_4 = H- $\longrightarrow$	R_4 = OH
	           |
	        - P = O
	           |
	       O–CH_2–CCl_3	

derivatives. The phosphate-blocked intermediate H 1' was recently used by J. Smrt for triester syntheses (418, 419). Condensation of H 1' with H 3 gave a fully protected dinucleoside phosphate, from which the 5'-blocking group was selectively cleaved prior to chain lenghthening, which was done again with H 1'.

C. B. Reese and coworkers (352) have extensively studied the synthesis of intermediates for oligoribonucleotide synthesis in the diester and more recently also in the triester fashion. The units I of Table 1.3 are proposed as intermediates for stepwise oligoribonucleotide synthesis. They all contain acyl or methoxytetrahydropyranyl groups for the bases resp. 2'-OH. The methoxymethylidene group is generally employed to block the diol end. In the diester approach the chain is lenghthened by adding units I 2 to the 3'-end of the 5'-terminal unit or fragment. For syntheses by the triester variation the 3'-end is first reacted with phenyl- or substituted phenyl phosphate, then the oligonucleotide is extended with intermediates I 2' or I 3.

Further improvements were introduced by T. Neilson and coworkers (296, 298). They use exclusively the triester method, starting the synthesis from the 5'-end, which is blocked by the trityloxyacetyl moiety. The chain is lenghthened by first adding trichloroethylphosphate, then the intermediates K 2 of Table 1.3. Most remarkably these intermediates are unblocked at the 3'-position, since it was found that the bulky nucleoside-3'-trichloroethyl phosphate would not react with the 3'-OH of another nucleoside, due to "shielding" by the neighbouring tetrahydropyranyl group. This triester approach has been the first one leading to the synthesis of a longer oligoribonucleotide chain (300), as will be discussed in Section 4.1.2.

In concluding this section it should be made clear that the intermediates discussed here are only examples of some more widely used blocking schemes. Nearly all of the blocking groups listed in Table 1.1 have been tested during the "evolution" of one or the other system of selective blocking, and it is impossible, in this review, to retrace all the different lines of development. In polynucleotide, as in polypeptide synthesis, the search for better and more straightforward solutions continues. New selectively labile protecting groups are still much in demand, and the possibility of enzymatic cleavage is a further valuable addition in this sector. New blocking groups for nucleobases stable to all conditions of polynucleotide synthesis and selectively removable in neutral medium would be helpful. New developments in the direction of "multipurpose" blocking groups are to be foreseen, and it can be hoped, that these developments will lead to simpler solutions not only for the field of protection, but for all questions involved in polynucleotide synthesis.

2. Phosphorylation Methods in the Synthesis of Mono- and Oligonucleotides

Polynucleotides are – from the standpoint of polymer chemistry – polyphosphodiesters. All internucleotide linkages are thus phospho-diester groups, whereas the chain termini can be either nucleosides or phosphomonoesters. The formation of a phosphoric acid ester linkage, i.e. the phosphorylation, can be effected chemically

1. by transfer of a phosphoryl group onto an alcohol with formation of a P–O-bond,

2. by transfer of a phosphate group onto an alcohol with formation of a C–O-bond,

3. by oxidation of a phosphite.

Phosphorylation by phosphoryl transfer is still the most widely used phosphorylation method in the synthesis of nucleotides and oligonucleotides. Reagents and mechanisms for phosphoryl transfer have been very extensively studied in the early 1960s. Since the development of the common phosphorylation techniques by H. G. KHORANA, F. CRAMER and others nearly ten years ago, no major advance in this field has been reported, and as several excellent reviews of this earlier work have appeared (35, 61, 65, 70, 184, 258, 451) we can limit ourselves to giving a few guidelines on the selection of phosphorylating agents and a brief mechanistic discussion.

In recent years an increasing number of instances have been described, where not a phosphoryl, but a phosphate group is transferred, as, for example, during studies of prebiotic or "thermal" phosphorylations and of phosphate transfer to activated nucleoside hydroxyl groups. Although phosphoryl transfer is still the route generally employed in internucleotide bond formation, this mechanistically different approach certainly merits continued interest. The same can be said of phosphorylations involving the oxidation of a nucleoside phosphite. Although this is one of the oldest phosphorylation methods, it has not furnished a breakthrough for internucleotide bond formation. Nevertheless, this pathway is often reinvestigated and results in the development of new reagents and techniques.

An elegant route, if applicable, is enzymatic attachment of a phosphoryl moiety. Since enzymatic methods of internucleotide bond formation will be described in Section 5 the discussion in Section 2.4 can be limited to the description of several kinases.

2.1. Transfer of a Phosphoryl Group

The transfer of a phosphoryl group to water, alcohols or amines proceeds through nucleophilic attack of these compounds on the phosphorus atom of a phosphorylating agent. Before we discuss the question of what should be defined as a phosphorylating agent, we should first clarify some basic steric and electronic aspects of phosphate chemistry.

In orthophosphoric acid and its derivatives the phosphorus atom occupies the centre, the four ligands the corners of a tetrahedron (161). Nucleophilic substitution reactions on phosphorus can be described in the same way as nucleophilic substitutions on a saturated carbon atom. Thus, there are two principal mechanistic pathways. The first, similar to the S_{N2} reaction on carbon, involves direct attack by the nucleophile on the phosphorus atom with displacement of one of the four ligands. This mechanism must involve an inversion of the ligands with intermediacy of a pentacoordinate complex. Alternatively, one of the ligands can dissociate prior to nucleophilic attack, a pathway parallelling the S_{N1} mechanism of substitution at carbon. In carbon chemistry this results in the intermediate formation of a carbonium ion; analogously a phosphoryl cation would be the primary dissociation product. Both cations demand stabilization; and a special way of stabilization exists in phosphate esters, which have at least one residual free acid function. In this case a proton can be expelled from the acid function with formation of a derivative of metaphosphoric acid – HPO_2. Monomeric metaphosphoric acid derivatives have been trapped and characterized by F. Westheimer and coworkers (464), but they are highly unstable and tend to form oligomeric or polymeric derivatives, of which the trimetaphosphates are best characterized. Such oligomeric metaphosphates are hypothesized to be intermediates in the transfer of phosphoryl groups derived from phosphoric acid and its monoesters (458). The two alternative mechanisms are shown in schemes 2.2 and 2.3 (see below).

The electron density around the phosphorus atom of phosphoric acid and its esters will be examined next. Three of the four ligands can be either –OH, i.e. acid functions, or ester groups. The fourth is an oxygen atom, and the resulting $P=O$-bond is polarized by electron withdrawal of the oxygen similar to the $C=O$-bond in esters of carboxylic or carbonic acids. However, the electron density around the phosphorus atom is not lowered to the same degree as that of carbon due to two facts (53):

1. The phosphorus atom contains empty d-orbitals, which can overlap with p-orbitals of neighbouring oxygen or nitrogen atoms

containing a lone electron pair. The partial $p\pi$-$d\pi$-bonds formed in this manner serve to distribute the positive charge from phosphorus to the neighbouring atoms.

2. In phosphomono- and diesters acid functions remain. The p_{KA} values for the two acid functions in phosphomonoesters are around 1 and 6, the p_{KA} of the residual acid function in phosphodiesters is between 1 and 2 (258). Both types of compounds are, therefore, relatively strong acids which fully deprotonate in alkaline media. The resulting negative charge is, of course, "smeared" over the $O\dot{=}P\dot{=}O$-system. The increase in electron density on the P atom and the charge repulsion are responsible for the remarkable stability of phosphate esters to alkaline hydrolysis, which increases in the series phosphotriester < phosphodiester < phosphomonoester (35, 258).

If simple phosphoric acid esters, like ethyl or phenyl, are not very susceptible to nucleophilic attack, what can be done to facilitate the transfer of a phosphoryl group, i.e. to generate a phosphorylating agent? We have to attach an activating group. The structural requirements for such groups have been brilliantly generalized by V. M. CLARK et al. (53). According to their basic scheme, all potential phosphorylating agents possess a function described by the general structural formula P–X–Y–Z (X, Y and Z being any element, preferentially C, H, N, O, S, halogen). Z must be (or must be convertible into) a strong electron acceptor, and the X–Y system must be capable of mediating an electron shift from the P–X-bond to Z. Since, in the cases we are looking at, X is mostly oxygen or nitrogen, i.e. atoms containing lone electron pairs, we have to arrange for these electron pairs to be incorporated into a $p\pi$-$p\pi$-bond of the X–Y–Z system in order to reduce the stabilization of the P–X-bond by $p\pi$-$d\pi$-overlap. This is illustrated in Fig. 2.1. The effect of the X–Y–Z system will be then to produce

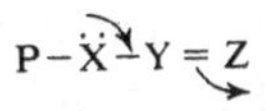

$$P\text{--}\ddot{X}\text{--}Y = Z$$

Fig. 2.1

an "energy-rich" bond between P and X, which favors nucleophilic substitution of this group by the result of a negative reaction enthalpy and eventually a positive entropy change due to fragmentation of the X–Y–Z system. Both pathways of decomposition of an activated phosphate, by direct nucleophilic attack or by monomolecular dissociation, are shown in Schemes 2.2 and 2.3.

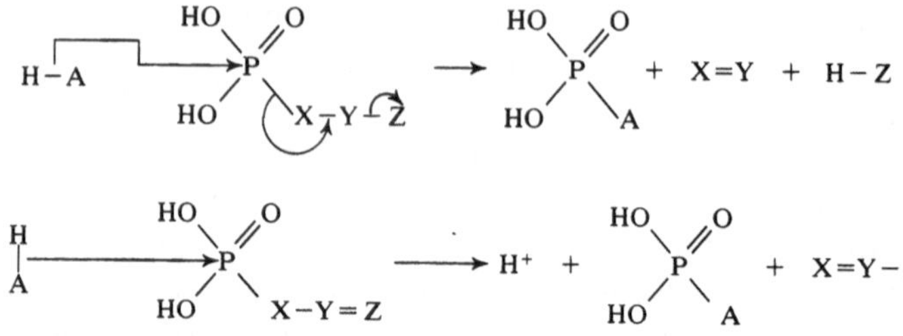

Scheme 2.2. Decomposition of an activated phosphate by nucleophilic attack on phosphorus

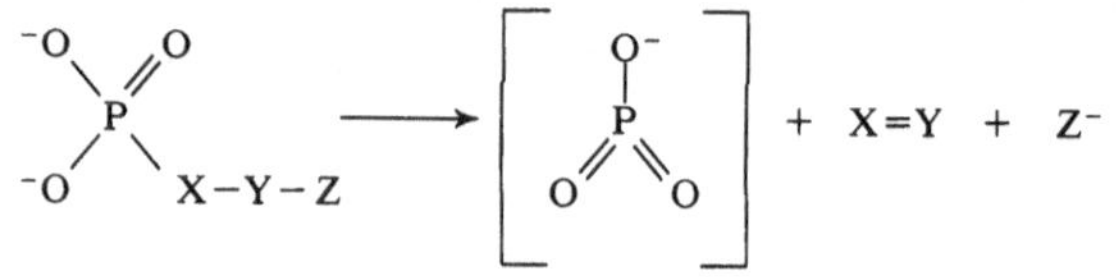

Scheme 2.3. Decomposition of an activated phosphate with formation of metaphosphate

Although this scheme may serve well for designing new potential phosphorylating agents, no prediction is possible as to whether these will be useful in nucleotide chemistry. Activated phosphates, which may easily react with water and ethanol, may be sluggish in phosphorylating the sterically much more hindered nucleosides or oligonucleotides, and only a few of the most powerful activating agents allow the formation of an internucleotide bond. But not only the nature of the attacking nucleophile· influences the phosphorylation reaction. The other substituents on the phosphorus atom, and solvents, catalysts and salts, have an effect as well. Generally, as A. M. MICHELSON has pointed out, it is difficult to make a clear distinction between "high-energy" and "low-energy" phosphates in chemical reactions (258). We have already mentioned this difficulty, when discussing protecting groups for the phosphate moiety of nucleotides in Section 1.3.1.

Some phosphorylating agents that have been of aid in the synthesis of nucleotides, oligonucleotides or related biologically active derivatives, are compiled in Table 2.1. The activated phosphates or activating agents are shown in columns 2, 3 and 4 together with the primary activated intermediates they are postulated to produce. Of course, we have to differentiate between two cases: In the first case the phosphorylating agent is a stable activated intermediate, e.g. a nucleoside phosphorochloridate. Then columns 3 and 4 must be identical. In the second case an unstable activated intermediate has to be formed first by reaction of a "low-energy" phosphate, such as a nucleoside phosphate, with a "condensing agent", e.g. dicyclohexylcarbodiimide. Then the structure of the activating agent, which is listed in column 3 of Table 2.1, is different from the structure of the activated intermediate. We will see later in this section, that in the case of dicyclohexylcarbodiimide an imidoylphosphate is assumed to be the primary activated intermediate. The structure of this latter compound is, therefore, shown in column 4. However knowledge of the structure of the primary activated intermediate does not necessarily imply knowledge of the actual pathway of phosphorylation, i. e. whether it follows the mechanism of Scheme 2.2 or the one of Scheme 2.3. This is in most cases unknown, it depends not only on the nature of the activated intermediate, but also on reaction conditions, solvents etc. A thorough discussion of possible reaction paths will be given later for the case of the two best studied condensing agents, dicyclo-hexylcarbodiimide and sulfonylchlorides.

A problem arises from the fact that, depending on what derivative of phosphoric acid we have, we can activate up to 3 functions of the phosphate molecule. Since, in most cases, we wish to form only one phosphate ester linkage at a time, we can solve this problem either by introducing blocking groups into all functions which are not supposed to react or by adopting a route which prevents the participation of other unblocked functions. We will find examples of both strategies in the following discussion of phosphoryl halide reagents.

The phosphorylating agents in Table 2.1 are grouped into different structural types. The first of these groups includes the phosphoryl halide reagents, i. e. derivatives of $POCl_3$. Since $POCl_3$ itself is a trifunctional acid chloride, care had to be taken to form only one ester linkage. This could be done by using derivatives, which had only one residual acid chloride function and two blocking groups (no. 7—15 in Table 2.1), such as diphenyl- (*132, 258*) or bis-(βββ,trichloroethyl-) phosphorochloridate (*95*). In another line of development multifunctional phosphorochloridates, even $POCl_3$ itself, could be used avoiding side reactions by a careful selection of reaction conditions and basic catalysts, e. g. 2,6-lutidine (no. 1—6) (*216, 350*). Nucleotide phosphoro-fluoridates have been used in the synthesis of oligonucleotides. The chain extension necessitated the use of nucleoside alcoholates (see Section 4.3) as nucleophilic partners (*440*).

Mixed anhydrides have been of great interest in peptide chemistry, and they are so also in oligonucleotide chemistry. In analogy to bio-logical phosphorylations, where derivatives of di- and triphosphoric acid play an important role, similar compounds have been investi-gated for their use in chemical phosphorylations. These include trimetaphosphate (*258, 364, 385*), "polyphosphoric acid ester" (prepared from P_2O_5 and stoichiometric amounts of alcohol) (*19, 258*) and triester-

Table 2.1. *Reagents for the Chemical Transfer of Phosphoryl Groups*

No.	Phosphoryl-ating agent	Structural formula	Postulated primary activated intermediate	Application to the synthesis of				References
				Nucleotides	Nucleotide-poly-phosphates coenzymes etc.	Inter-nucleotide bonds	nucleo-phosphates	
Phosphoryl halides								
1	phosphoryl chloride	$POCl_3$		+	+	+	+	*(35, 216, 258, 124 sel. 5'-phos-phoryl.: 262, 325, 326, 411, 424, 425, 471, 472)*
2	pyro-phos-phoryl-chloride	$Cl_2P(O)-O-P(O)Cl_2$		+	+	+		*(35, 258) sel. 5'-phos-phoryl.: (262, 411, 424, 471)*
3	methyl-phosphoro-dichlori-date	$H_3CO-P(O)Cl_2$		+		+		*(415)*

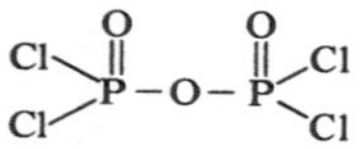

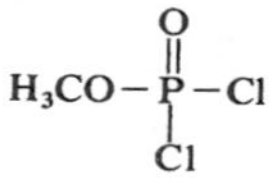

No.		Structure				Ref.
4	phenyl-phosphoro-dichloridate	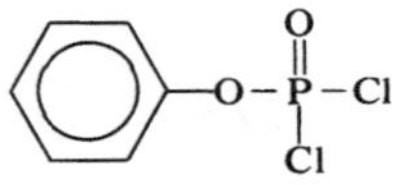 $\underset{\text{Cl}}{\overset{\text{O}}{\underset{\|}{\text{P}}}}\text{—Cl}$ (phenyl–O–)	+	+	+	(109, 258, 350)
5	2-chloro-methyl-4-nitro-phenyl-phosphoro-dichloridate	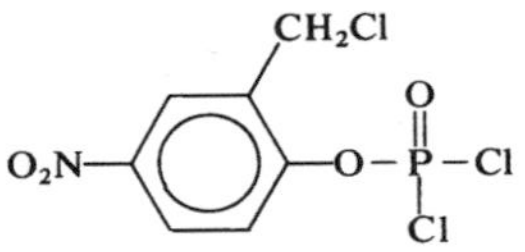 O_2N-substituted phenyl with CH_2Cl, $\text{O—}\overset{\text{O}}{\underset{\text{Cl}}{\overset{\|}{\text{P}}}}\text{—Cl}$	+			(134, 135, 282)
6	P^1-Phenyl-P^2-morpholino-pyro-phosphoro-dichloridate	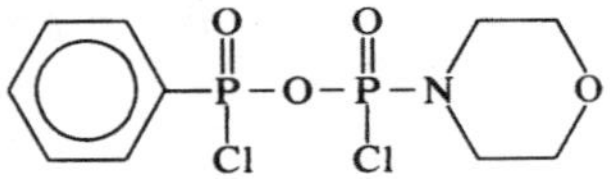 phenyl–$\overset{\text{O}}{\underset{\text{Cl}}{\overset{\|}{\text{P}}}}$–O–$\overset{\text{O}}{\underset{\text{Cl}}{\overset{\|}{\text{P}}}}$–N(morpholino)	+	+		(162)
7	diethyl-phosphoro-chloridate	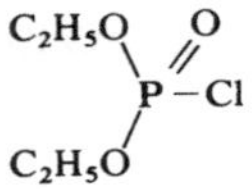 C_2H_5O–$\overset{\text{O}}{\underset{C_2H_5O}{\overset{\|}{\text{P}}}}$–Cl	+			(258)
8	di-β-cyano-ethyl-phosphoro-chloridate	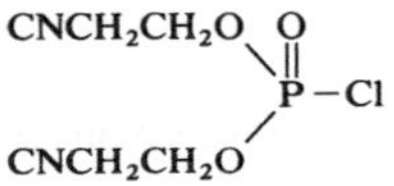 $CNCH_2CH_2O$–$\overset{\text{O}}{\underset{CNCH_2CH_2O}{\overset{\|}{\text{P}}}}$–Cl	+			(258)

Table 2.1 (continued)

No.	Phosphorylating agent	Structural formula	Postulated primary activated intermediate	Application to the synthesis of			References
				nucleotides	nucleotide-polyphosphates coenzymes etc.	inter-nucleonucleo bonds	
9	bis-(β,β,β-trichloroethyl-)phosphorochloridate	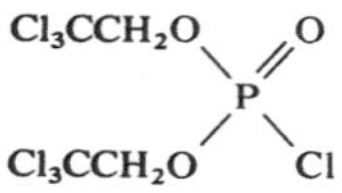 Cl_3CCH_2O, Cl_3CCH_2O — $P(=O)$ — Cl		+ sel. 5'-OH			(95)
10	dibutylphosphorochloridate	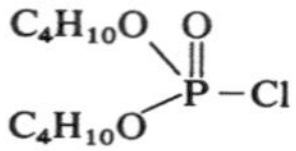 $C_4H_{10}O$, $C_4H_{10}O$ — $P(=O)$ — Cl		+			(258)
11	diphenylphosphorochloridate	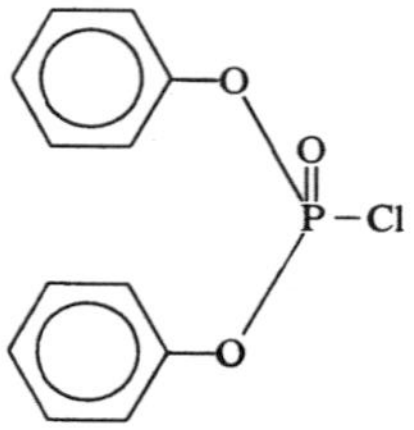		+		+	(258)

12	dibenzyl-phosphoro-chloridate	+	+	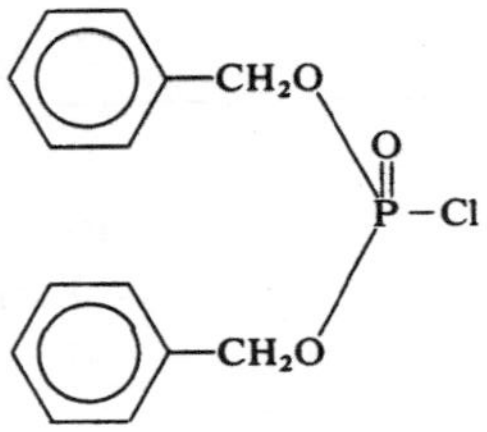 (258)
13	bis-(p-nitro-phenyl-)phosphoro-chloridate		+	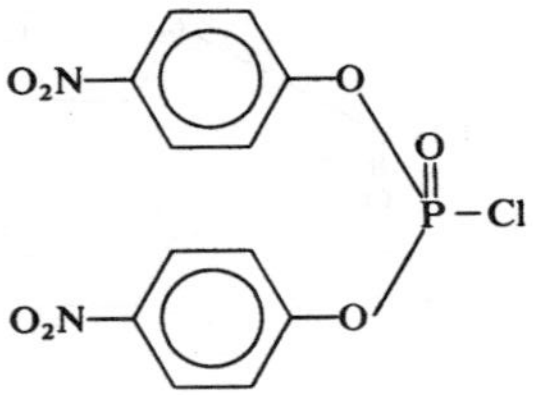 (132)
14	O-phenylene-phosphoro-chloridate	+		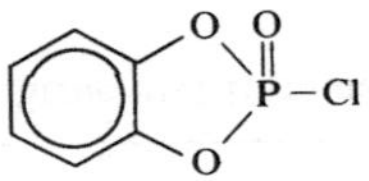 (191)

Table 2.1 (continued)

No.	Phosphorylating agent	Structural formula	Postulated primary activated intermediate	Application to the synthesis of			References
				nucleotides	nucleotide-polyphosphates coenzymes etc.	internucleotide bonds	
15	di-morpho-lidic-phosphoro-chloridate (bromidate)	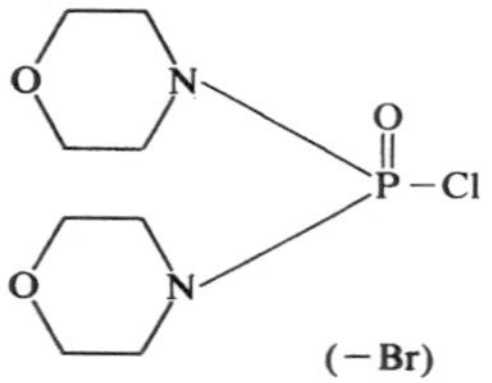 (−Br)				+	(258)
15a	nucleoside-phosphoro-fluoridate	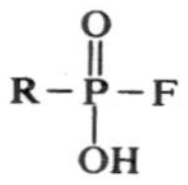			+	+	(426, 440)

mixed anhydrides

No.	Compound	Structure			References
16	trimeta-phosphoric acid and esters	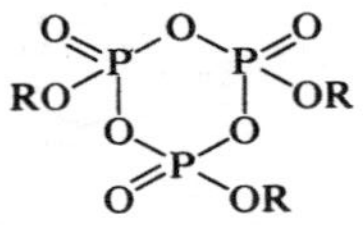 R = H, Na, Me, Ph	+	+	(166, 258, 458) sel. 2',3'-OH: (364, 385, 386)
17	poly-phosphoric acid and esters	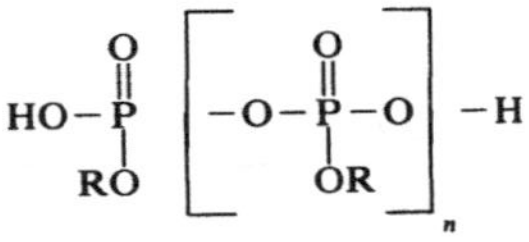 $\mathrm{HO-P(=O)(RO)-[-O-P(=O)(OR)-O-]_n-H}$ R = H, Me, Ph	+	+	(19, 258, 344)
18	tetra-p-nitro-phenyl-pyrophosphate	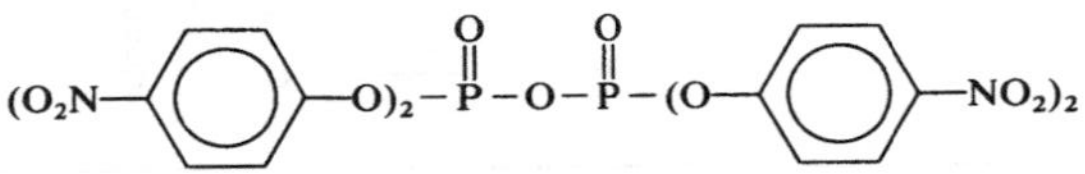 $\mathrm{(O_2N-C_6H_4-O)_2-P(=O)-O-P(=O)-(O-C_6H_4-NO_2)_2}$	+	+	(258)
19	P'-nucleo-sidyl-P²-diethyl-pyrophosphate	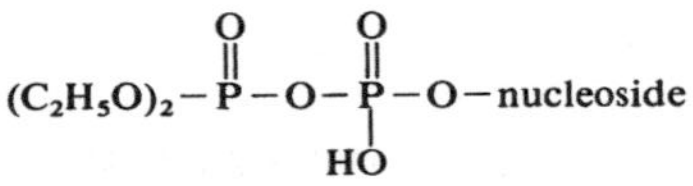 $\mathrm{(C_2H_5O)_2-P(=O)-O-P(=O)(OH)-O-nucleoside}$	+	+	(61)

Table 2.1 (continued)

No.	Phosphorylating agent	Structural formula	Application to the synthesis of			References
			nucleotides	nucleotide-polyphosphates coenzymes etc.	inter-nucleotide bonds	
19a	benzyl-, or dibenzyl-, or nucleoside-, phosphoric acid benzoic acid anhydride	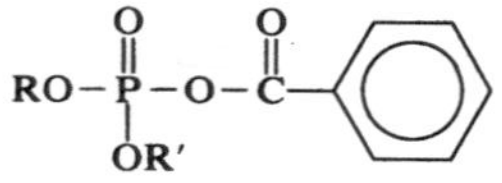 RO—P—O—C structure, OR'; R = benzyl, nucleoside; R' = H, benzyl	+		+	(269, 270)
20	mesitoyl-chloride	structure (mesitoyl chloride)	+		+	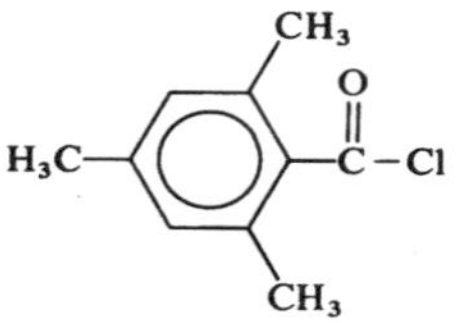 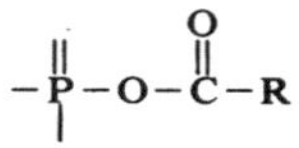 (274)
21	p-toluene-sulfonyl-chloride	—SO$_2$Cl structure	+	+	+	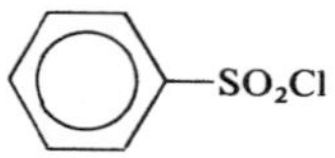 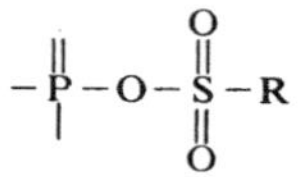 (258)

22	1,3,5-tri-methyl-benzene-sulfonyl-chloride (MS)		+	+	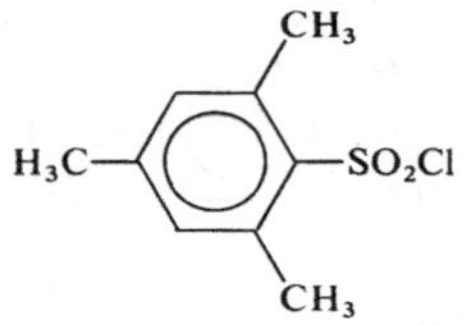 (166)
23	1,3,5-tri-isopropyl-benzene-sulfonyl-chloride (TPS)		+	+	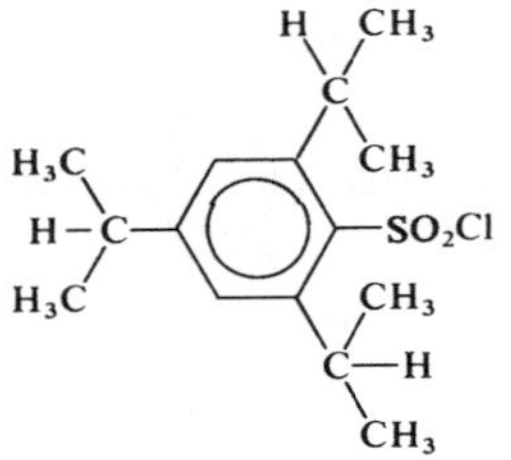 (248)
24	poly-3,5-diethyl-styrene-sulfonyl-chloride		+	+	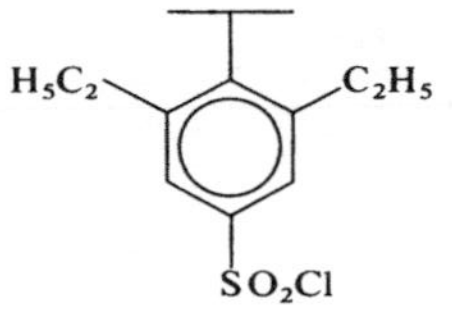 (362)

Table 2.1 (continued)

No.	Phosphorylating agent	Structural formula	Postulated primary activated intermediate	Application to the synthesis of					References
				nucleotides	nucleotide-poly-phosphates coenzymes etc.	nucleotide-nucleotide bonds	inter-nucleotide bonds		
25	trimethyl-benzene-sulfonyl-imidazolide			+		+			(20)
25a	trimethyl-benzene-sulfonyl-1,2,4-triazolide			+		+			(176a)
25b	triisopropyl-benzene-sulfonyl-1,2,4-triazolide			+		+			(176a)

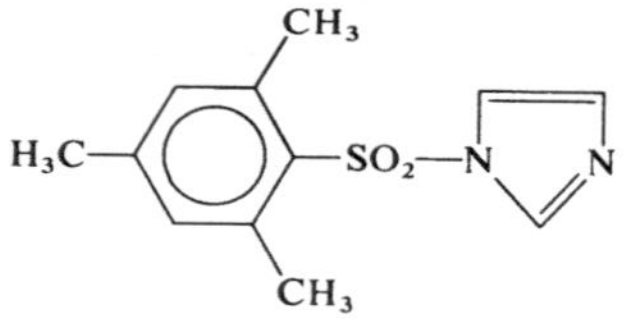

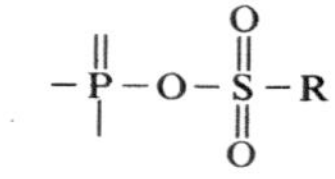

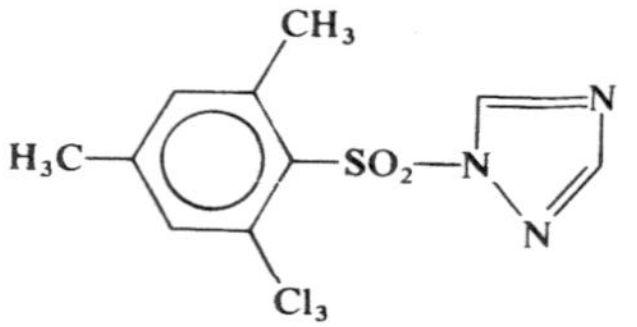

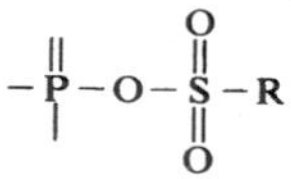

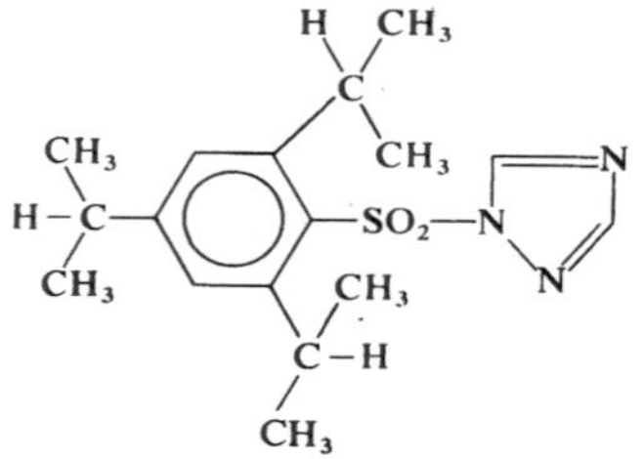

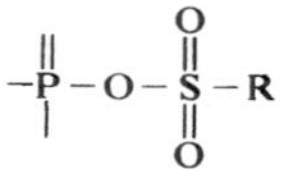

activated esters

26	*p*-nitrophenyl-phosphate		+	+	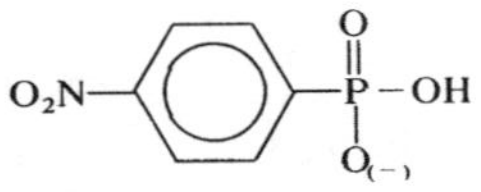 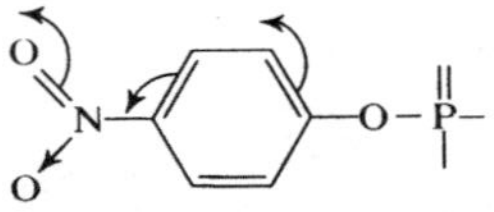 (48, 108, 137)
27	picryl chloride		+	+	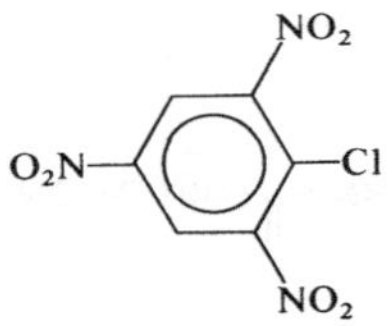 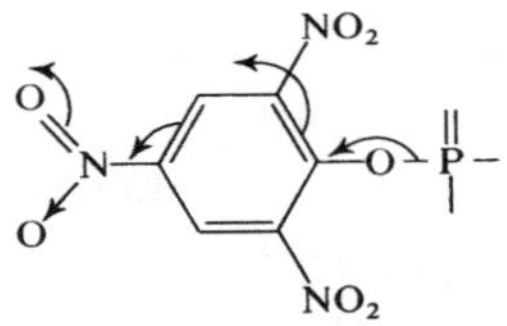 (65)
28	Catechol cyclic phosphate		+	+	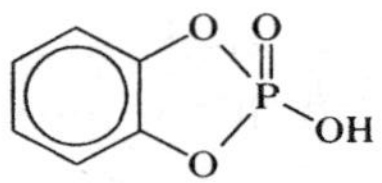 (258)
29	α-hydroxy-pyridine-phosphates	+	+		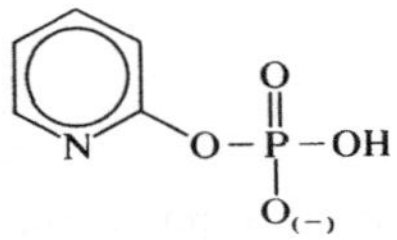 (65)

Table 2.1 (continued)

No.	Phosphoryl-ating agent	Structural formula	Postulated primary activated intermediate	Application to the synthesis of — nucleotides	nucleotide-poly-phosphates coenzymes etc.	inter-nucleotide bonds	References
30	diethyl-(1-ethoxy-2-carbethoxy-vinyl-)phosphate	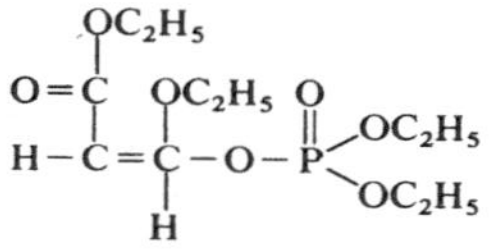 $O=C\!\begin{smallmatrix}OC_2H_5\\\\\end{smallmatrix}$ OC_2H_5 $H-C=C-O-P=O$ with OC_2H_5, OC_2H_5, H	pyrophosphoric acid triester see 19	+		+	(61)
31	α-bromo-α-cyano-acetamide + triphenyl-phosphine	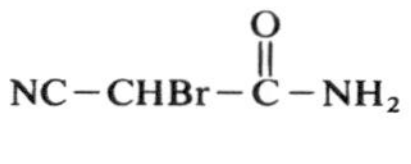 $NC-CHBr-\overset{O}{\overset{\|}{C}}-NH_2$	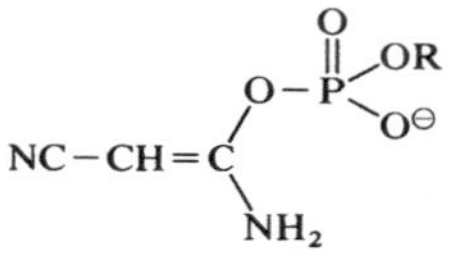 $NC-CH=\overset{OR}{\underset{O^\ominus}{\overset{\|}{C}}}$ ... $O-\overset{O}{\overset{\|}{P}}-O^\ominus$; $O-C\begin{smallmatrix}\\NH_2\end{smallmatrix}$	+		+	(65, 66)
32	2-methylthio-4H-1,3,2-benzodioxa-phosphorin-2-oxide	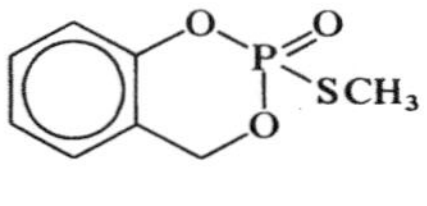	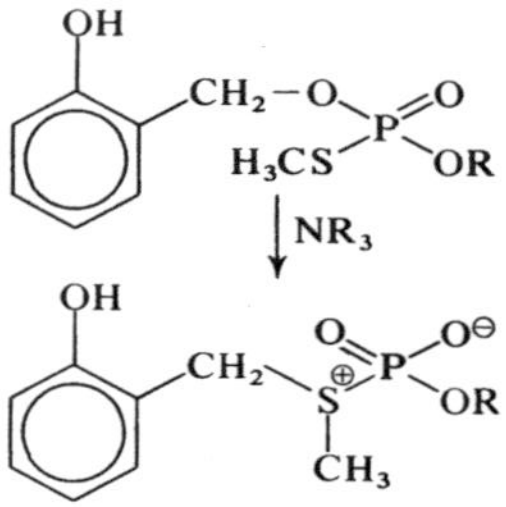	+			(85)

imidoylphosphates

No.	Compound	Structure				Ref.
33	dicyclohexyl-carbodiimide (DCC)		+	+	+	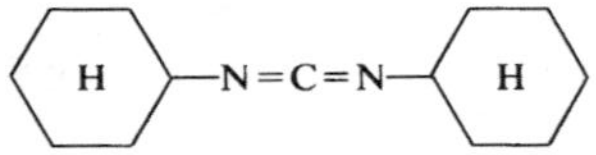 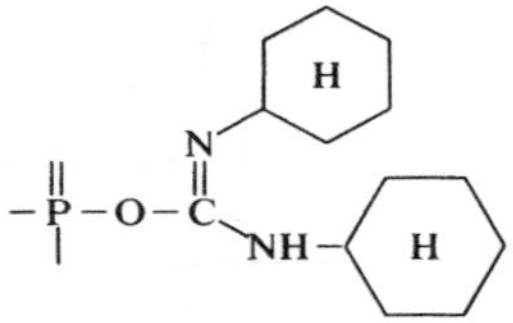 (166, 184, 258, 458)
34	di-*p*-tolyl-carbodiimide		+	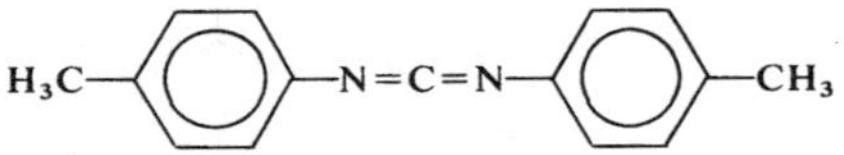 compare no. 33		(258)
35	1-cyclohexyl-3-(2N-methyl-morpholino-ethyl-)carbodiimide methosulfate		+	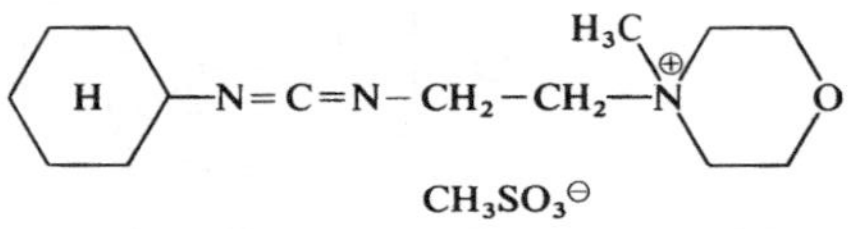 compare no. 33	+	(12, 460)
36	1-ethyl-3-(3-dimethyl-aminopropyl-)carbodiimide hydrochloride		+	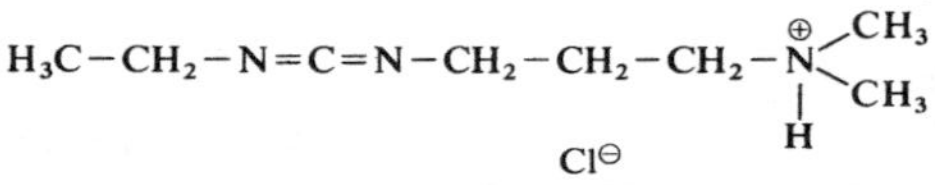 compare no. 33	+	(46, 72)

Table 2.1 (continued)

No. Phosphorylating agent	Structural formula	Postulated primary activated intermediate	Application to the synthesis of			References
			nucleotides	nucleotide-polyphosphates coenzymes etc.	internucleotide bonds	
37 phosgene + dimethylformamide	$\left[\begin{array}{c} H_3C \\ H_3C \end{array} \overset{\oplus}{N} = C \begin{array}{c} H \\ Cl \end{array} \right]$ Cl⁻	$=\!\!P\!-\!O\!-\!\overset{H}{\underset{R}{C}}\!=\!\overset{\oplus}{N}\!\!\begin{array}{c}R\\R\end{array}$ Cl⊖	+		+	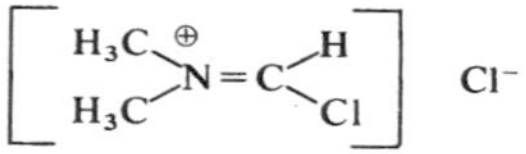 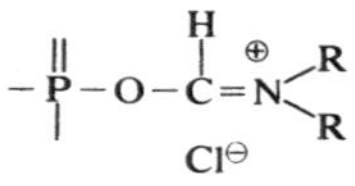 (61, 166)
38 N-ethyl-(methyl-)5-phenyl-isoxazolium-fluoroborate	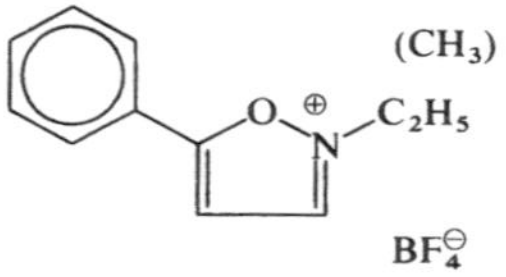 (structure: 2-ethyl(methyl)-5-phenyl-isoxazolium, BF₄⊖)	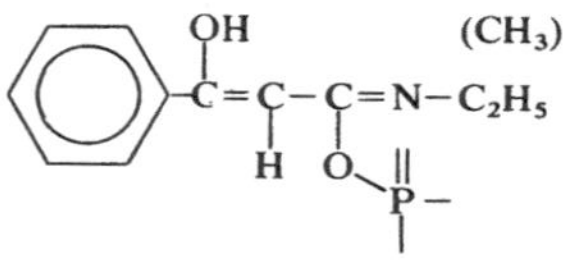 OH ... $C\!=\!C\!-\!C\!=\!N\!-\!C_2H_5$ (CH₃) ... $\overset{\parallel}{\underset{H}{O}}\!-\!P=$	+		+	(62, 166)
39 trichloroacetonitrile	$\underset{Cl}{\overset{Cl}{\underset{|}{\overset{|}{C}}}}\!-\!C\!\equiv\!N$	$Cl_3C\!-\!C\!=\!NH$ $\overset{\parallel}{\underset{O}{}}\!-\!P=$		+		(65, 258)

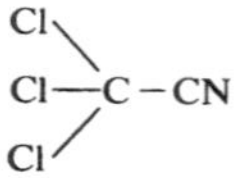

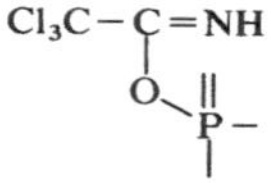

activated phosphoramidates

No.	Name	Structure				Ref.
40	benzyl-hydrogen phosphor-amidate	H₂N–P–OCH₂ (O=, O⁻) phenyl	+	+		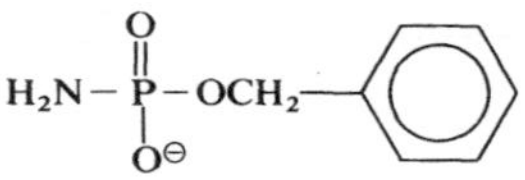 (35)
41	phosphoro-morpholi-date	N–P–OR (O=, OH) morpholine	+	+		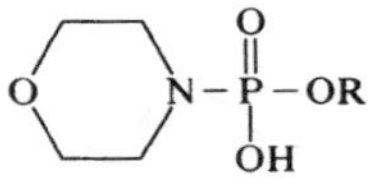 (184, 252)
42	phosphoryl-imidazole-phosphate + carbonyl-diimidazole	imidazolyl–C=O; =P– imidazolyl	+	+	+	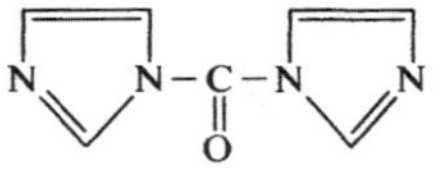 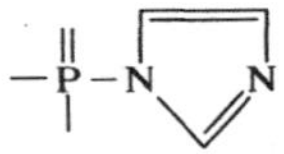 (59, 246, 258, 324)
43	diimidazolyl-phosphinic acid and derivatives	imidazolyl–P–imidazolyl (O=, O⁻)	+			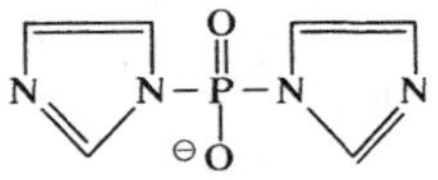 (79, 258)

oxidative phosphorylation

Table 2.1 (continued)

No.	Phosphorylating agent	Structural formula	Postulated primary activated intermediate	Application to the synthesis of				References
				nucleotides	nucleotide-phosphates	poly-phosphates coenzymes etc.	inter-nucleotide bonds	
44	naphtohydro-quinone-phosphate + Br_2				+			(35)
45	phosphoro-p-hydroxy-anilidate + Br_2				+		+	(316)
46	nucleoside-S-ethyl-thio-phosphates + iodine						+	(55, 56)
47	triphenyl-phosphine + dipyridyl-disulfide		$Ph_3P^{\oplus} - O - \overset{O}{\underset{OH}{P}} - OR$		+		+	(280)

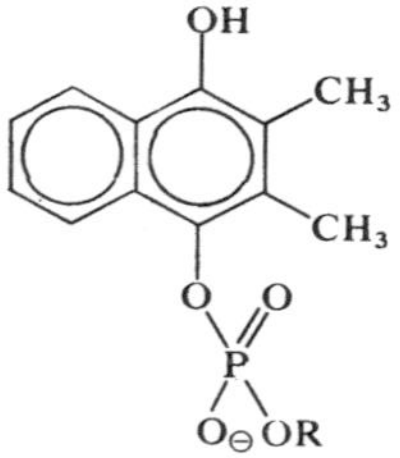

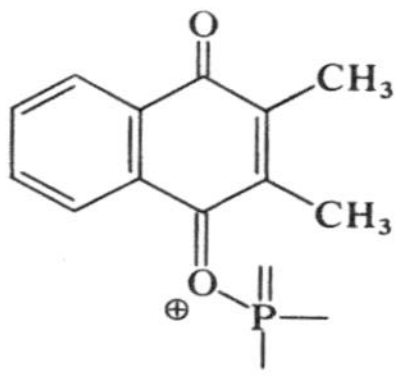

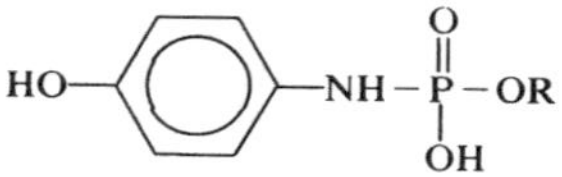

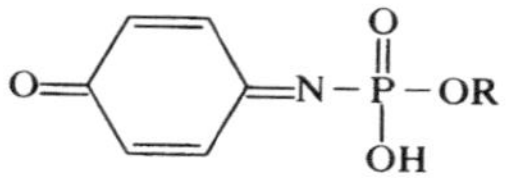

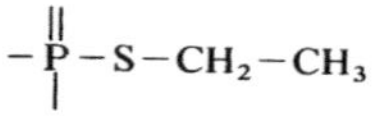

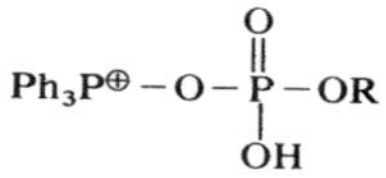

pyrophosphate (*61*), accessible *via* a reaction of an enol phosphate with a nucleoside (no. 16—19). F. CRAMER and coworkers have shown in several studies, that in contrast to symmetrical diesters of pyrophosphoric acid, which show a maximum of resonance stabilization, the triesters of pyrophosphoric acid are relatively unstable substances which easily react to transfer the monoester part to a nucleophile (Scheme 2.4) (*61*).

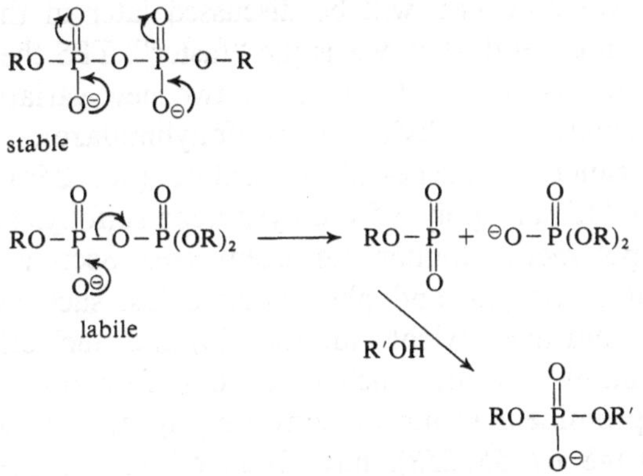

Scheme 2.4. Comparative reactivity of di- and triesters of pyrophosphoric acid

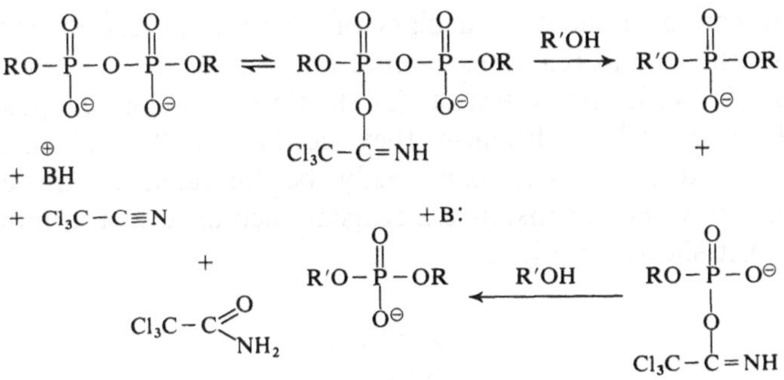

Scheme 2.5. Activation of a pyrophosphate by trichloroacetonitrile

This mechanism also explains the activation of symmetrically substituted pyrophosphates by condensing agents, such as trichloroacetonitrile (Scheme 2.5). In spite of these interesting mechanistic aspects pyro- and polyphosphate reagents play only a minor role in nucleoside phosphorylations and, especially, in the stepwise synthesis of oligo-nucleotides.

Among the most widely used phosphorylating agents, however, are mixed anhydrides of phosphoric and sulfonic acids. *p*-Toluenesulfonyl chloride (no. 21) has long been known as a condensing agent, but it has the disadvantage that it can concurrently tosylate and thus block the alcohol reactant (258). In order to suppress this side reaction, sterically hindered sulfonylchloride reagents, such as mesitylene-sulfonyl-chloride (abbreviation: MS) (no. 22) (166) and triisopropylbenzene-sulfonylchloride (abbreviation: TPS) (no. 23) (248) have been developed by H. G. Khorana and coworkers, and, are, at the moment, the most popular activating agents in polynucleotide chemistry. The mechanism of action of MS and TPS will be discussed later in this section, but it can be said in general, that MS is the "faster", TPS the more selective of the two reagents (248, 274). Recently two new variations have been described, namely trimethylbenzenesulfonylimidazole (no. 25) (20) respectively trimethylbenzene-1,2,4-triazolide (no. 25a) (176a) as an alternative to MS and poly-3,5-diethylstyrene-sulfonylchloride (no. 24) (362) as a polymeric condensing agent (see Section 4.2.4). Mixed anhydrides of carboxylic and phosphoric acids, such as the activated intermediate from mesitoylchloride (no. 20) and nucleotides (274) were found less suitable for internucleotide bond formation.

Another prominent group of activating agents, the activated esters (no. 26—31) (48, 61, 65, 258), have been of limited use in nucleotide and internucleotide bond synthesis. This is in contrast to peptide bond formation, where activated esters are of great use as stable and readily available reagents. On the basis of their structure all activated esters belong or can be related to a class of compounds called enol phosphates. Syntheses and reactions of enol phosphates, have been reviewed earlier by F. W. Lichtenthaler (245). Enol phosphates possess an activated ester residue because they contain a "quasi" preformed aldehyde or ketone which can easily be liberated as an excellent leaving group, when for instance a reagent such as a proton induces the electron shift shown in Fig. 2.6.

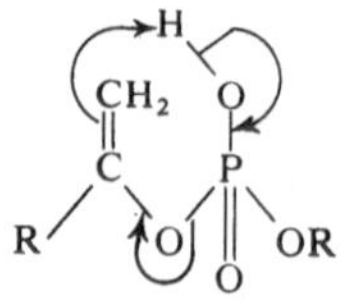

Fig. 2.6

A slight modification of the enol phosphate concept leads us to an extremely useful class of activating agents, the imidoylphosphate inter-mediates. Imidoylphosphates are analogous to enol phosphates, but the Y–Z system is a $C=N$ double bond instead of a $C=C$ double bond. The $C=N$ bond is much more easily protonated, e.g. by pyridinium ions in a water-free pyridine medium. Protonation produces a very labile species which easily fragments with liberation of metaphosphate.

The most important of these reagents are the carbodiimides (no. 33—36), among which the dicyclohexylcarbodiimide (no. 33) (*144, 258, 274*) in particular has been a standard condensing agent in nucleotide as well as peptide chemistry for many years. Compared to the sulfonyl-chloride reagents discussed above, dicyclohexylcarbodiimide (abbreviation: DCC) has two disadvantages: in order to obtain good yields one needs significantly longer reaction times and the by-product, dicyclo-hexylurea, is difficult to remove, since it is only slightly soluble in a few solvents. However, the carbodiimides are not very sensitive to water around neutral pH, so it was possible to construct and successfully use water-soluble carbodiimides, e.g. no. 35 and 36 of Table 2.1, for condensations in aqueous media (*46, 72, 460*). Other condensing agents, designed to give imidoyl phosphate intermediates (no. 37—39) (*61, 62, 65, 166*) have not been as successful in the stepwise synthesis of oligonucleotides, however, reagents such as trichloroacetonitrile (no. 39) (*65, 258*) and also picryl chloride (no. 27) (*65*) have shown good results in the polycondensation of nucleotides.

Whereas all activations we have discussed so far proceed with P–O-cleavage, a P–N bond is broken in the nucleophilic displacement of activated phosphoramidates (no. 40—43) (*6, 35, 184, 258*). Although the activation of these compounds proceeds by protonation similar to the imidoylphosphates, they are not as "energy-rich" as these latter derivatives, because the basicity of the nitrogen atom is reduced by $p\pi$–$d\pi$-overlap to the phosphorus atom as discussed earlier. The main range of application of the activated phosphoramidates, especially the phosphoromorpholidates, lies in the synthesis of nucleotide coenzymes.

A final group of compounds for phosphoryl transfer consists of reagents useful for oxidative phosphorylation. Oxidative phosphorylation is one of the basic phosphorylation processes of biological systems, and thus quinol phosphates have also been tested for their ability to mediate chemical phosphorylations. Although interesting from a mechanistic standpoint, these studies have not yielded any phosphorylating agents of major importance. In the case of the activating residues no. 44 and 45, the quinol phosphates were activated by bromine oxidation (*35, 316*). Through oxidation of the *p*-hydroxyanilidate residue (no. 45) a moderate yield of dTpdT could be obtained; thus this residue could, be used as an activatable blocking group (see Section 1.4) (*316*). However, the need for bromine oxidation precludes a more general use of this method. Another example of an activation during blocking group cleavage is the oxidative removal of the S-ethyl-phosphorothioate residue with iodine (no. 46; see also Section 1.4 and Scheme 1.6) (*55*).

H. Kössel and H. Seliger:

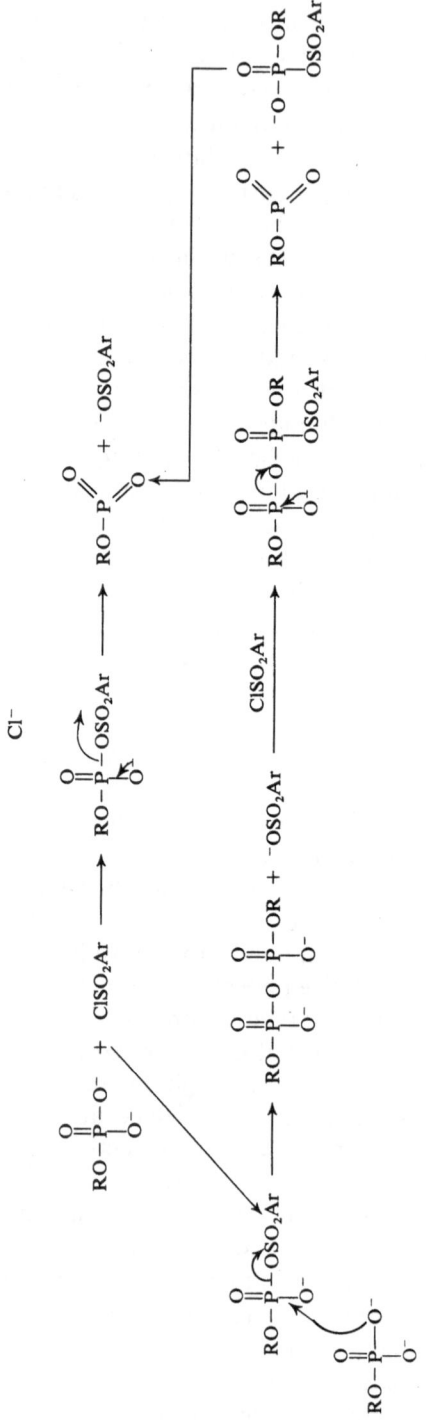

Scheme 2.7. Possible mechanisms for formation of a metaphosphate with sulfonyl chlorides

A similar "activated" phosphorothioate intermediate seems to be present in phosphorylations with 2-methylthio-4H-1,3,2-benzodioxa-phosphorin-oxide (no. 32) (85), although in this case an ester rearrangement rather than a redox reaction produces the activated species. Another recent addition to the spectrum of phosphorylating agents is triphenyl-phosphine, which, in the presence of dipyridyldisulfide as oxidising agent, can form an intermediate activated phosphonium phosphate (no. 47) (280). Triphenylphosphine as activating agent for alcohols will be discussed in the following section.

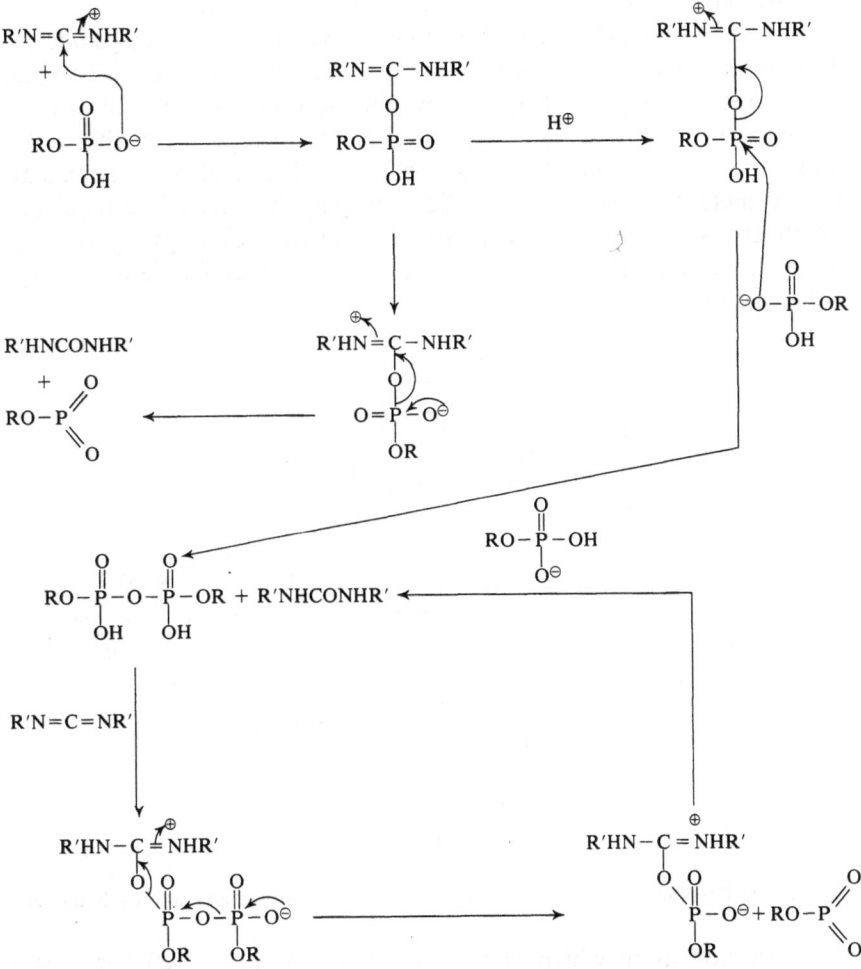

Scheme 2.8. Possible pathways for formation of a metaphosphate with dicyclohexylcarbodi-imide

In concluding this section we will take a closer look at the mechanism of action of the two most frequently used phosphorylating agents, namely sulfonyl chlorides and dicyclohexylcarbodiimide. Reaction schemes have been advanced by different authors (35, 166, 248, 258), the picture which emerges is, however, still relatively complex. Most certainly, more than one mechanistic pathway is possible. For the sulfonyl chlorides some mechanistic alternatives are shown in Scheme 2.7 (166, 258). One could be termed a "direct" route: A phosphomonoester is activated by sulfonylation and subsequent decomposition of the mixed anhydride to give metaphosphate. Alternatively metaphosphate could be formed by an "indirect" route via a symmetrical pyrophosphate, which can again be activated by sulfonylchlorides, as discussed earlier.

The same two alternatives are shown in Scheme 2.8 for dicyclohexylcarbodiimide activation. Once the monomeric metaphosphate is formed, it will most likely be trapped by the solvent pyridine to give a phosphoryl-pyridinium ion. This pyridinium complex (or the metaphosphate itself) could then react with all other nucleophiles present, namely metaphosphate to give oligo- or polyphosphate intermediates, the phosphoric acid starting compound to revert back to pyrophosphate or, finally an alcohol resp. nucleoside to give a phosphodiester linkage (Scheme 2.9).

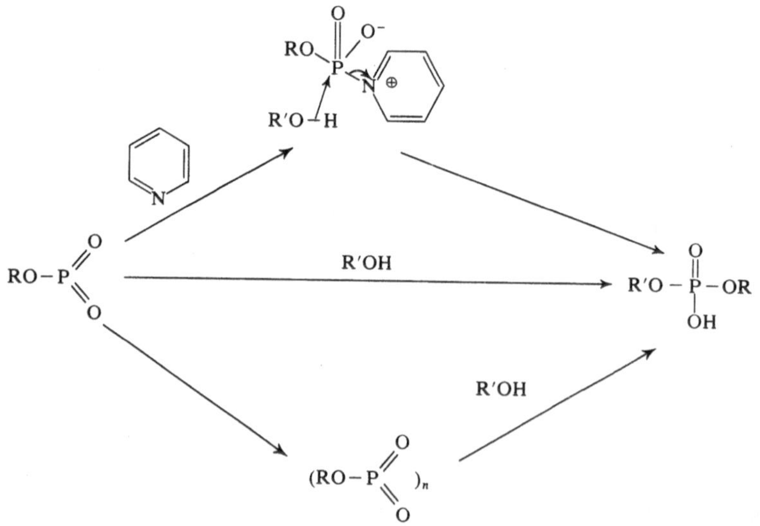

Scheme 2.9. Possible pathways for conversion of a metaphosphate to a phosphodiester

A decision as to which of these different routes is preferred is still difficult. Recent studies by G. M. Blackburn and coworkers (22, 23, 24) showed, that no phosphodiester formation would occur, when the phosphomonoester component was linked to a polymer support, thus preventing the formation of di- or polyphosphate intermediates. On the other hand, excellent yields of phosphotriester were obtained by R. L. LETSINGER and coworkers, when support-bound phosphodiesters (blocked nucleoside-monophosphates) were activated, in the presence of a nucleoside, with TPS (not, however, with DCC) (*230, 238*). Obviously, different mechanistic routes prevail with different phosphorylating agents as well as with different reactants.

2.2. Transfer of a Phosphate Group

Phosphate transfer to nucleosides or nucleotides has been observed to occur on activation of nucleoside hydroxyl groups. The phosphate transfer to activated hydroxyl groups proceeds by a nucleophilic attack of phosphate or nucleotide on the carbon atom next to the hydroxyl group which is displaced, i.e. in most cases on $C^{5'}$. The mechanism can be formulated as an S_{N2}-type, as shown in Fig. 2.10. Activation of the

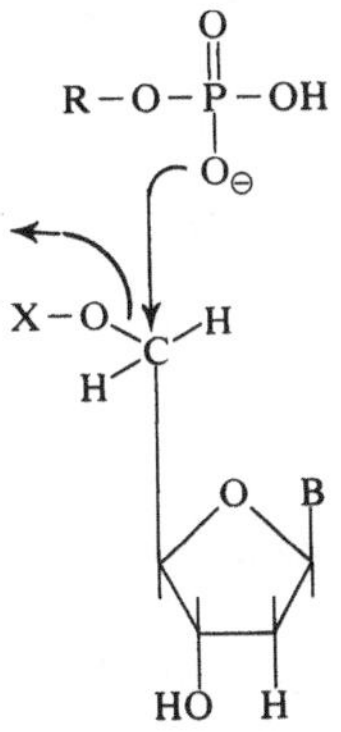

(X = activating residue)

Fig. 2.10

hydroxyl group must significantly lower the electron density on the carbon atom in question, thus facilitating nucleophilic attack, and must also transform the alcohol group into a good leaving group. This has been achieved by O. MITSUNOBU and coworkers (*263*) by activation with triphenylphosphine and azodicarboxylate, a triphenylphosphonium derivative of the nucleoside being the presumed intermediate. E. W. HAEFFNER (*122*) has reported, alcohol activation as the result of mesylation which results in nucleoside phosphorylation and even, in small yield, in the formation of the dinucleoside phosphate dTpdT; however, this could be done only at elevated temperature. A similar example from the older literature is the phosphorylation of 3'-O-acetyl-5'-iodo-5'-deoxythymidine with silver dibenzyl phosphate (258). Along similar lines, the synthesis of dinucleoside phosphates from 5'-chloro-5'-deoxynucleoside and tri-n-butylamonium-3'-uridylate was reported more recently (429).

A series of investigations has dealt with anionic attack of phosphates and nucleotides on cyclic anhydro nucleosides. The formation of an ether or sulfide bridge between $C^{5'}$ or $C^{3'}$ and C^2 or C^8 of a pyrimidine resp. purine base makes these carbon atoms of the sugar ring susceptible to nucleophilic attack (Scheme 2.11). This has been

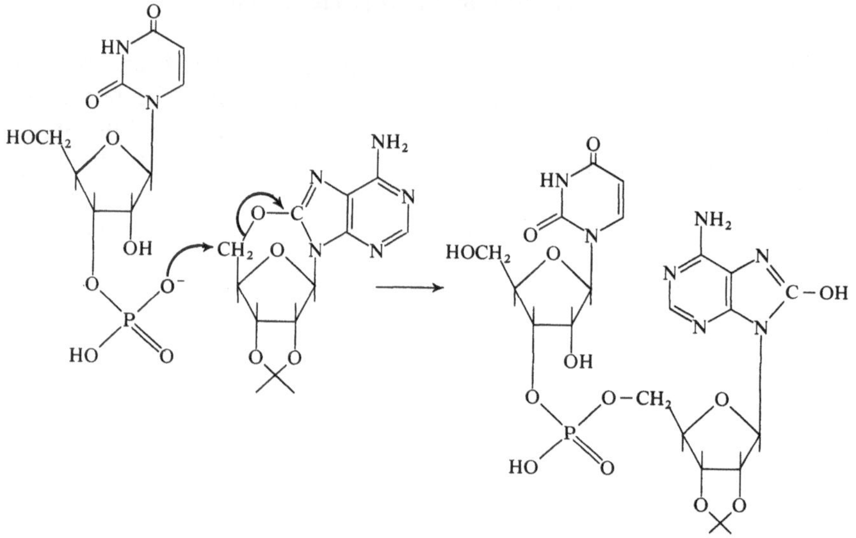

Scheme 2.11. Internucleotide bond formation with anhydro nucleosides

used for the formation of several dinucleoside phosphates (1, 269, 270, 283, 427, 428, 478). Generally the yields are moderate even at higher temperatures. To obtain a good yield of rCprA from $O^2,C^{3'}$-anhydrocyclocytidine and 5'-AMP, the nucleotide was additionally activated as a phosphoric acid – benzoic acid anhydride (269, 270). Of course, the formation of natural 3'-5'-internucleotide linkages implies that the reaction proceeds with complete inversion at $C^{3'}$, and this was not completely the case.

2.3. Miscellaneous Chemical Phosphorylation Reactions

This section describes several approaches, which are either mechanistically completely different from the ones reviewed in Sections 2.1 and 2.2 or not yet well enough understood to allow an unequivocal classification.

The phosphorylation methods mainly to be dealt with are

1. phosphorylations involving oxidation of an intermediate phosphorous acid ester of a nucleoside,

2. so-called "thermal" phosphorylations, and

3. phosphorylations under possible prebiotic conditions, involving "prebiotic phosphorylating agents", mineral surfaces etc.

Different reasons have led to these investigations. The oxidation of nucleoside phosphites has since long been known as a good method for the introduction of the phosphate moiety into nucleosides which are not sensitive to oxidising agents; however, there seems to be no significant advantage as compared with more straightforward methods for introducing the phosphoryl moiety such as the use of DCC or TPS. The direct "thermal" conversion of nucleosides into nucleotides by inorganic phosphate with no need for protection or activation could be an approach unsurpassed in its simplicity; however, as long as a mixture of isomers results and the yields of a single product cannot be significantly improved, the main value lies in a one-step synthesis of labelled nucleotides. Phosphorylations on mineral surfaces, finally, have been studied mainly with regard to solving questions about the prebiotic formation of nucleotides. None of these three methods is as yet of any importance in oligonucleotide synthesis.

2.3.1. Phosphorylation by Oxidation of Nucleoside Phosphites

The phosphorylation of alcohols and amines by a mixture of tetrachloromethane and phosphorous acid dialkyl esters was described as early as 1945 by Lord TODD and coworkers. It was demonstrated that a diesterphosphorochloridate is formed as intermediate (35, 258). Nucleoside phosphorylations were similarly done with O-benzylphosphorous-O-diphenylphosphoric anhydride, a reagent prepared from

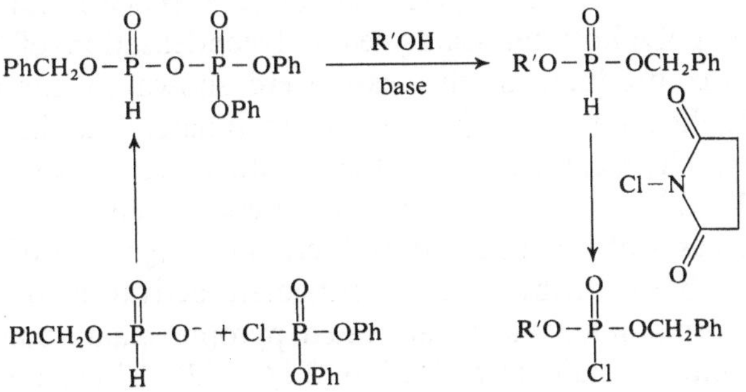

Scheme 2.12. Preparation of an alkyl-benzyl-phosphorochloridate

diphenylphosphorochloridate and monobenzyl phosphite (*35*). Alcohols react with this reagent by nucleophilic attack at the less acidic component of the anhydride to give the alkyl benzyl phosphite, which is then chlorinated with N-chlorosuccinimide to the alkyl benzyl phosphorochloridate. This can either be hydrolyzed to the corresponding phosphate or used directly as an activated nucleotide derivative (Scheme 2.12). In other cases the phosphite residue was introduced by reaction with phosphorus trichloride (*469, 473*), phosphorous acid + DCC (*35*) or trichloromethane phosphonic acid dialkylester (*158*).

2.3.2. *"Thermal" Phosphorylation*

Since 1965 several investigations have been published, in which inorganic phosphate was transferred to nucleosides without introduction of activating groups into any of the reaction partners. Since this could be done in most cases only at relatively high temperature, this approach has been named "thermal" phosphorylation (*164, 275, 276, 277, 278, 345*). Typically, a nucleoside and phosphoric acid (often applied as the tri-n-butylammonium salt) are heated at reflux in dry dimethylformamide for several hours, the resulting generally complex mixture of products being separated according to charge. Other sources of phosphate can be used, such as pyrophosphoric acid or the nucleotides themselves. In an example of the latter case, studied by T. UEDA and I. KAWAI (*452*), 5'-AMP was refluxed in DMF to give predominantly adenosine and adenosine-2',3'-cyclic phosphate.

A thermal phosphorylation resulting in the conversion of nucleotides into a homologous mixture of oligo- and polynucleotides has been studied by O. PONGS and P. O. P. Ts'o (*339, 340*). The reaction is done in refluxing dimethylformamide with catalysis by β-imidazolyl-4(5-)propionic acid, triethylamine hydrochloride or other proton donors. The polycondensations were achieved with unblocked nucleotides. In contrast to earlier experiments of H. SCHRAMM and coworkers (*258*) on the polyphosphate-catalyzed polycondensation of unblocked nucleotides the products in this case were shown to contain nearly 95% of 3'-5'-phosphodiester linkages. Unfortunately, neither the yields nor the stereochemical purity of the product are as yet sufficiently high to make this very simple approach preparatively workable.

All these cases of "thermal" introduction or migration of phosphate residues have been attributed to intermediate activation of phosphate by dimethylformamide, but an activated pyrophosphate could also be a plausible intermediate (see Section 2.1) (*452*). Thus, although the overall reaction would suggest phosphate attack, the moiety which really is transferred seems to be a phosphoryl group. However, as the mechanistic aspects await further clarification, we have preferred to describe this approach in the present section. It should be added, that this reaction need not necessarily be a thermal one, since it has been shown that the reaction of inorganic phosphate with nucleosides proceeds very well at room temperature with predominant formation of 5′-nucleotides, if formamide is used as reaction medium (333).

2.3.3. Prebiotic Phosphorylations

For several years an increasing number of studies has been devoted to shedding light on the manner in which nucleic acid components could have been formed under primitive earth conditions. Of course, the considerations governing work of this type are completely different from those valid for preparative organic chemistry. The main question is: "Is it possible that all the assumed reaction partners might have been present in reasonable quantity and close contact in a primitive earth environment." Yields of products, on the other hand, are not as important as in preparative reactions, since the time at the disposal of nature to accumulate a certain product is incomparably longer than the observation time of a laboratory. Since this review is concerned primarily with preparative aspects of polynucleotide chemistry, it may suffice to retrace two lines of development in the field of prebiotic phosphorylation.

Most probably inorganic phosphates must have been the source of phosphoric acid and nucleosides or sugars the phosphate acceptors in such reactions. The working hypotheses differ in answers to the question how phosphorylation could have been mediated. One line of work is based on the assumption, that primitive oceans could have contained "prebiotic" phosphorylating agents such as dicyanogen, malonitrile or acrylonitrile, all of which have been demonstrated to allow the formation of nucleotides or sugar phosphates from nucleosides and D-ribose (126, 367). A second line of development has attempted to demonstrate that nucleotides could have been formed in contact with mineral surfaces. The mineral surface may act merely as a catalyst, but especially high yields of nucleotides have been obtained when the mineral present was a phosphate donor, such as hydroxylapatite, and when urea and ammonium chloride were added as adjuvants (250, 302, 323).

2.4. Enzymic Phosphorylation

Kinases and phosphorylases are biological catalysts for the introduction and transfer of a phosphate moiety. Kinases catalyze the phosphate addition to a biological intermediate, phosphorylases the phosphorolytic cleavage of such compounds.

The conversion of nucleosides to nucleotides catalyzed by nucleoside kinases was found to be an intermediate step in the metabolic pathway leading from nucleosides to nucleoside triphosphates according to Scheme 2.13 (*331*). The overall reaction, for the example of thymidine

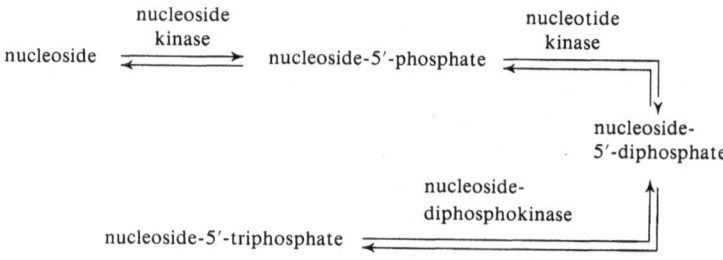

Scheme 2.13. Enzymatic conversion of nucleosides to nucleoside triphosphates

$$\text{deoxythymidine + nucleoside triphosphate} \xrightleftharpoons{\text{deoxythymidine kinase}}$$
$$\text{deoxythymidine-5'-phosphate + nucleoside diphosphate}$$

Scheme 2.14

kinase, is shown in Scheme 2.14. Kinases are known for several of the common nucleosides and three species, namely adenosine- (*204*), thymidine- (*320, 321*) and deoxycytidine (*273*) kinase have been purified from *E. coli*, calf thymus and other sources. Substrat specificity was relatively high in the case of adenosine- and thymidine kinase, whereas deoxycytidine kinase could phosphorylate also ara-cytidine, deoxyadenosine and deoxyguanosine. Nucleoside triphosphates act as phosphate donors, whereby mostly the end product of the metabolic chain, i.e. the respective triphosphate of the preferred substrate, is a strong inhibitor. The products are specifically 5'-nucleotides.

A "low-energy" phosphate transfer has been demonstrated for plant and animal tissues by E. Chargaff (*36*), who recently succeeded in purification of an enzyme from *E. coli*, which transfers phosphate from low-energy organic donors to nucleosides, nucleoside-5'-phosphates and deoxynucleoside-5'-triphosphates. With the exception of adenosine, the nucleosides are converted almost exclusively into 2'- and 3'-nucleotides. Thymidine and its derivatives are the best acceptors of phosphate groups. The fact that nucleoside-2'(3')-phosphates are the main products suggests a regulatory role for this enzyme rather than an involvement in the biosynthesis of nucleoside polyphosphates. A phosphotransferase from carrot was similarly used to prepare a variety of 5′-nucleotides and analogs for DNA polymerase binding studies (*129*). A low-energy phosphate transfer catalyzed by cells of *Pseudomonas mesophilus* has been described as a preparative approach to guanylic and inosinic acid in a Japanese patent (*267*).

Enzyme preparations from cells of *B. ammoniagenes* have been used in several studies, especially in the synthesis of labelled nucleotides (*93, 284, 285, 301*). The enzyme catalyzes the transfer of a phosphoribosyl group to nucleobases. Uracil is converted nearly quantitatively into 5′-uridylic acid. Phosphoribosylation of adenine takes place with lower efficiency, whereas guanine and cytosine are practically not phosphoribosylated at all. Labelled orotidine-5′-phosphate was also prepared.

A different approach to enzymatic phosphorylation of nucleosides has been described by A. HOLY and G. KOWOLLIK (*156*). The authors reacted nucleosides with guanosine-2′,3′-cyclic phosphate under catalysis by T$_1$ RNase (compare Section 5.3). Subsequent cleavage of the internucleotide linkage by snake venom phosphodiesterase transfers the phosphate residue to the 5′-position of the starting nucleoside. Although the yields did not exceed 30%, the method may be interesting as an extremely mild procedure for the phosphorylation of very labile nucleoside analogs. Also this method, like other enzymatic phosphorylations of nucleosides, has the general advantage, that specific substitutions can be obtained with unblocked starting compounds.

The enzyme polynucleotide kinase, which transfers orthophosphate from ATP to polynucleotides, oligonucleotides and even nucleoside-3′-phosphates, has been isolated from T4-bacteriophage infected *E. coli* by C. C. RICHARDSON (*357*). The overall reaction, as shown in Scheme 2.15, is a specific 5′-phosphorylation of the oligo- or poly-

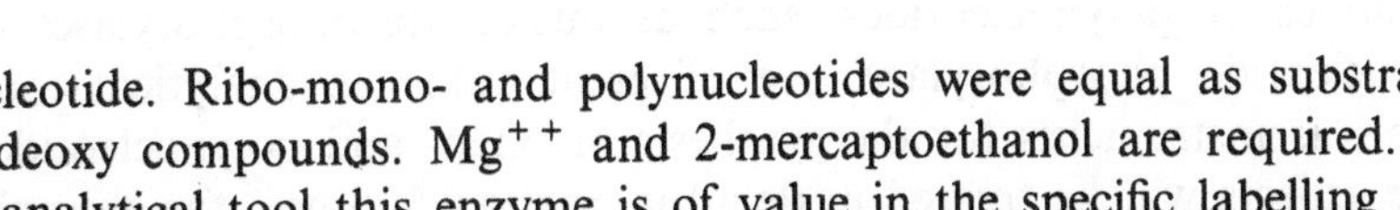

Scheme 2.15

nucleotide. Ribo-mono- and polynucleotides were equal as substrates to deoxy compounds. Mg^{++} and 2-mercaptoethanol are required. As an analytical tool this enzyme is of value in the specific labelling and analysis of the 5′-terminus of polynucleotides and in testing the specificity of exonucleases. It has been of even greater interest in the chemical synthesis of polynucleotides, since it allows the phosphorylation of oligonucleotide fragments, which can then be joined by polynucleotide ligase (see Section 4.1.1) (6, 188). The *in vivo* role of this enzyme is not quite clear. Since it does not accept nucleosides its action is different from that of the above-mentioned nucleoside kinases. The possibility of an *in vivo* production and condensation of activated oligonucleotide fragments has been suggested.

Complementary to the use of polynucleotide kinase is a method described by H. Kössel and R. Roychoudhury (208) for the specific addition of phosphate to the 3'-end of an oligodeoxynucleotide chain. This is done by the sequence of enzymatic and chemical reactions shown in Scheme 2.16. This method is especially useful for attaching a radio-

deoxyoligonucleotide (I) $\xrightarrow{\text{terminal deoxynucleo-tidyl transferase}}$ deoxyoligonucleotidyl-$[^{32}P]p\ A_r$ (II)

$\quad\quad\quad +$ +

$[a\text{-}^{32}P]$ ATP   deoxyoligonucleotidyl-$[^{32}P]p\ A_r[^{32}P]p\ A_{r\text{-}}$ (III)

                                        +

                             pyrophosphate

II $\xrightarrow{\text{OH}^-}$ II $\xrightarrow{\text{phosphatase}}$ II + $[^{32}P]p$ + riboadenosine

and   and

III   deoxyoligonucleotidyl-$[^{32}P]p\ A_r[^{32}P]p$ (IV)

                           +

        riboadenosine

II $\xrightarrow{\text{IO}_4^-\ \text{and}\ \text{cyclohexylamine}}$ deoxyoligonucleotidyl-$[^{32}P]p$ (V)

Scheme 2.16. Phosphorylation method of Kössel and Roychoudhury

actively-labelled phosphate to the 3'-terminus. By spleen phosphodiester-ase digestion the label is transferred to the 3'-terminal unit of the sequence, thus allowing an end group determination of polydeoxynucleo-tides.

Enzymes which catalyze phosphorolytic cleavage of nucleosides, nucleotides or polynucleotides, such as nucleoside phosphorylases or polynucleotide phosphorylase, are of great value as analytical tools and/or, as instruments for the synthesis of oligo- and polynucleotides. These aspects will be treated in detail in Section 5. Since, for example, the displacement of an internucleotidic linkage by inorganic phosphate is not a preparative approach to nucleoside polyphosphates, we can abstain from discussing these enzymes in this section.

In concluding this section it should be emphasized that we have looked at chemical and enzymic phosphorylation reactions mainly from the point of view of their utility in the synthesis of oligo- or polynucleotides. Different considerations are valid, if one wishes to produce just mononucleotides. In this case – and this is just to briefly outline the possibilities – the workup of material from natural sources is still by far the predominant route, and nearly all of the commercially available nucleotides are made in this way. Chemical phosphorylation of nucleosides is an alternative, of interest mainly for obtaining blocked nucleotides or nucleotides with rare or modified bases and sugars. A third approach, namely the synthesis of nucleotides by fusion of glycosyl halides with (silyl-) nucleobases has been described (*11, 406*), but is not a method of general use.

3. Separation Techniques

Purification and characterization of products are major time consuming steps in most, if not all, of the conventional synthetic procedures. The possible reduction of these steps to the products of the final reaction has been one main motivation for the efforts already invested in polymer support synthesis. At the same time, however, several new separation techniques applicable to conventional synthetic procedures could be developed for large scale preparations as well as for work on an analytical scale. In addition to the introduction of well-established absorbents as, for instance, Sephadex or Biogel for polnucleotidic mixtures, improvements could also be achieved by the development of new adsorbent types specific for certain functional groups of protected or unprotected oligonucleotides.

Furthermore, progress could be made by introduction of new solvent systems for elution or for chromatography, or by new combinations of already known solvent systems. Finally entirely new lipophilic protecting groups have been devised which allow the specific extraction of intermediates whereby time consuming column steps can be avoided or simplified.

3.1. Column Procedures

3.1.1. Column Chromatography on Conventional Adsorbent Types

Application of Sephadex column chromatography for the preparative separation of oligonucleotide mixtures has been reported in several

26

studies (5, 39, 129, 286, 287, 288, 315, 318, 347, 400, 433, 454). This method appears attractive as the desired synthetic products – usually those of the highest chain lengths within a reaction mixture – will be eluted within the exclusion volume, whereby rapid isolation within a comparatively small elution volume is guaranteed. A further advantage consists in the low buffer concentration (usually triethyl-ammonium bicarbonate) of the eluent which simplifies further work-up of the isolated compounds. Although successful separations of products resulting from single nucleotide additions have been reported (287, 288, 315), a prerequisite to satisfactory separation seems to be the condition that the compounds to be separated differ maximally from one to another in size (288, 347). Even if this is achieved by adjusting the synthetic plan so that approximate doubling of the chain lengths occurs during any of the reaction steps, additional complications may arise from other factors. Thus, with protected nucleotides retardation has been observed to increase in the following order: pT < dpbzA < dpanC < dpibuG (347), and this order is reflected in the elution patterns of derived oligo-mers. Conformational influences on the elution behaviour of oligo-nucleotides during Sephadex column chromatography have been documented by the successful separation of 2′—5′ dinucleoside mono-phosphates from their 3′—5′ isomers (433). The observation that certain nucleotide derivatives – notably d-panC and its relatives are eluted in two peaks may also reflect conformational influence (347). In spite of all these possible complications, the gel permeation technique at least in selected cases seems to compare favourably with the more time-consuming DEAE-cellulose column chromatography of protected oligonucleotides and several cases have been reported where the products isolated by Sephadex column chromatography were sufficiently pure for further condensation steps (287, 288, 315). Gel permeation tech-niques on Sephadex or Biogel have found especially widespread application for the separation of unprotected oligo- (144, 221, 433) and polynucleotides (2, 39, 129, 130, 195, 399, 400, 402, 446, 454). Particularly suitable for this technique seem to be product mixtures resulting from polynucleotide ligase catalized reactions as on the one hand conventional DEAE cellulose column chromatography does not provide the necessary resolution power for chain lengths in the range of 20 and more nucleotides and as on the other hand the prerequisite of a relatively large difference in the chain lenghts between the fragments to be coupled and the products is always fulfilled.

DEAE-cellulose column chromatography in spite of its drawbacks is still the technique most widely used for the separation of protected oligonucleotides (5, 37, 38, 40, 218a, 219, 315, 341, 457). As the absorption is largely governed by ionic forces, the elution order of a polymeric mixture primarily reflects the number of negative charges of the various components. This therefore results essentially in separation according to chain length. Complications however, arise from an additional retardation order (T < dpbzA < dpanC < dpibuG) which is obviously due to nonionic interaction and which is reflected also in the elution patterns of derived oligomers. A similar interaction seems to be even stronger with compounds containing the highly lipophilic monomethoxytrityl group at the 5'-end. While ethanol gradients super-imposed on the salt gradients can be applied in order to selectively retard monomethoxytrityl containing products (37, 40, 205, 457), use of salt gradients containing the more hydrophilic methanol effect complete absorption of monomethoxytrityl protected components even in the presence of high salt. Based on this observation a simple technique using two successive salt gradients with methanol and ethanol on the same column was devised for the separation of 5'-monomethoxytrityl containing oligomers (in the deoxy series usually the starting block and the desired product) from the remaining products of a block condensation mixture (379, 380).

One major drawback of DEAE cellulose consists in its reduced resolution power for protected oligomers of higher chain length. In this range (ten nucleotides and longer) it is therefore necessary to increase the chain length by several nucleotide units at a time (block conden-sation) in order to obtain satisfactory resolution. On the other hand, analytical DEAE-cellulose columns in the presence of 7 M urea, have proven to be useful tools for the characterization or final purification of unprotected oligomers in the chain length range of 8 to 20, especially as the standard paper chromatographic procedures also are severely limited in this size range. As evidenced by its routine use in recent work on the synthesis of the tRNA-Ala-Gene (37, 38, 40, 218a, 219, 315, 457), the disadvantage of this time-consuming column tech-nique which requires in addition the complete deprotection of the oligomers to be characterized, seems to be fully counterbalanced by its high resolution power. While in most cases neutral salt gradients were applied, use of acidic ammonium formate (pH 3.5) seems also to be possible even in the case of purine-containing deoxyoligonucleotides (37).

A promising technique for the rapid analytical separation of unpro-tected or fully protected oligonucleotide mixtures appears to be high pressure liquid chromatography on a pellicular weak anion exchanger consisting of a polymeric aliphatic amine (131). The time necessary for separation of one optical density unit of a condensation mixture was reported to be less than 30 minutes.

3.1.2. Column Chromatography on Newly Developed Adsorbent Types

The nonionic interaction between the highly lipophilic 5'-O-pro-
tecting monomethoxytrityl group and cellulose is strongly increased
if the cellulose matrix itself is modified by naphthoylation (4, 5) or by
tritylation (6, 41). *Tritylated cellulose* seems to be an especially powerful
tool for the selective adsorbance of monomethoxytrityl-containing
compounds (in the deoxy series usually the starting block and the desired
product). After elution of the nontritylated components of a given
condensation mixture in the presence of low alcohol concentration,
elution of the trityl-containing components is effected simply by a
switch to higher alcohol concentration in the eluent. More recently this
separation principle has been extended to other lipophilic 5'-O-protect-
ing groups such as the 2-S-naphthylmercaptoethyl group (7). The fact
that only low salt concentrations are necessary in the eluents simplifies
further workup of the compounds isolated by this procedure. The
elegance of this technique appears, however, to be counterbalanced
to some extent by the requirement for further fractionation of the monomethoxytrityl-containing compound mixtures which are usually composed of the unreacted starting block and of the desired product. A
general scheme for selective blocking of the 3'-hydroxyl group of
unreacted starting material by naphthylisocarbamoyl has been proposed
which would allow further condensation steps at the 3'-hydroxyl group
of the desired products exclusively (6); as a consequence further
fractionation of the mixtures containing the tritylated oligonucleotidic
products could be avoided.

Increased affinity of oligonucleotidic compounds containing aromatic
protecting groups is also observed with *benzoylated DEAE-Sephadex*
(259, 289, 290, 290a). Specific aromatic 5'-phosphate protecting groups
such as benzhydracrylamidyl or 2-phenylmercaptoethyl cause adsorption
of the respective nucleotide derivatives in the absence of alcohol
even at high salt concentrations. After elution of the components free
of aromatic 5'-phosphate protecting groups in the presence of aqueous
salt gradients, isolation of the 5'-phosphate protected derivatives is
achieved by addition of 50% ethanol to the eluent. Maintenance of high
salt concentrations is, however, also necessary for effective elution.
As in the case of trityl cellulose, this ethanolic fraction (containing
the desired product and one of the unutilized condensation components)
has to be further fractionated, and separation of the two main components (and other minor side products) could be achieved by Sephadex
column chromatography (290a). In view of the two column steps
necessary for effective purification of the final product, the usefulness
of this technique seems limited.

Cellulose and polymethacrylic acid gels, to which dihydroxyboryl groups had been attached covalently, have been described as column chromatographic adsorbents, specific for the cis-diol group of a variety of ribonucleotidic derivatives (377, 381, 460). Thus, due to complex formation of the dihydroxyboryl groups of the respective matrices with the cis-diol group of ribonucleoside residues at slightly alkaline pH, specific adsorption of ribonucleosides, of 5'-ribonucleotides, of ribooligonucleotides lacking a 3'-phosphate group, of 3'-ribonucleoside terminated oligodeoxynucleotides and of free tRNA is observed, whereas deoxynucleosides, deoxynucleotides, deoxyoligonucleotides, ribooligonucleotides with a 3'-terminal phosphate and aminoacylated tRNA are eluted within the void volume. The adsorbed compounds can subsequently be recovered from the column matrix by lowering the pH of the eluent buffer to neutrality. This technique has proved especially useful for the preparative separation of mixtures containing the components $(pT)_n$ and $(pT)_{n-1}pU_r$ where the lack of differences in size or in net charge does not allow separation by any of the other conventional techniques (381).

3.2. Extraction Procedures

The use of extraction procedures in the chemical synthesis of protected dinucleoside monophosphates from a 5'-trityl protected nucleoside and a protected nucleoside 5'-phosphate has been reported earlier (205). After extraction of the unreacted nucleoside derivative by ethyl acetate, the protected dinucleoside monophosphate can be separated from the unreacted mononucleotide derivative by extraction with chloroform. The extractability of the protected dinucleoside monophosphate seems to depend on the presence of the lipophilic 5'-O-trityl protecting group and on the presence of lipophilic counterions of the product such as pyridinium or triethylammonium cations. More recently this extraction principle could successfully be extended to the separation of a protected trinucleoside diphosphate and even of a protected pentanucleoside tetraphosphate (38, 40, 315). In order to eploit extraction procedures also for the isolation of protected mono- and dinucleotides containing 5'-terminal phosphate groups lipophilic protecting groups containing one or more aromatic rings have been developed for the protection of 5'-phosphomonoester groups (3, 7, 176). Thus, synthesis and organic solvent extraction of all 16 possible deoxydinucleotides in the protected form has been reported after p-aminophenyltriphenylmethane was used for the amidation of the respective 5'-terminal phosphate groups (3). Introduction of the 2-S-phenylmercaptoethyl or of the 2-S-naphthylmercaptoethyl group for the protection of 5′-terminal phosphates seems also to allow selective solvent extraction of protected mononucleotide derivatives (7).

There seems to be no doubt that extraction procedures do simplify the total workup of condensation reactions in many cases and that therefore, whenever possible, extraction procedures should be used in order to avoid the more time-consuming column separation steps. Care is, however, necessary to make sure that products isolated by extraction procedures are pure enough for subsequent condensation steps. In addition, due to the detergent effect of protected oligo-nucleotides, it is sometimes difficult to separate the two solvent layers after thorough mixing and the resulting emulsions are occasionally stable enough to resist even prolonged centrifugation at maximum speed.

3.3. Miscellaneous Techniques

The use of thin layer chromatography for the separation of oligo-nucleotides has been reviewed recently (348). In the meantime a few additional systems for the anylytical separation of oligomeric mixtures have been published (3, 5, 37, 123, 219, 223, 260, 291, 293, 318, 457). In the triester approach (see below) extensive use has been made of pre-parative as well as analytical thin layer chromatography (42, 76, 78, 236, 300, 350, 418, 419) and of short column chromatography (73). The latter technique seems especially useful for the rapid separation and purifi-cation of large quantities of protected oligonucleotides.

Synthetic homooligomers from the series $d(pT)_n$, $d(pA)_n$ and $d(pC)_n$ and a number of oligomers of varying base composition have been characterized by analytical ultracentrifugation (251). Though this method would in principle allow also separation of oligomers from each other, its application seems to be useful mainly for the chain length character-ization of oligomers.

4. Formation of Internucleotide Linkages by Chemical Synthesis

4.1. Conventional Methods

4.1.1. Synthesis in the Deoxy Series

4.1.1.1. Synthesis via Phosphodiester Intermediates

During the past ten years chemical synthesis of deoxyribopolynucleo-tides up to a chain length of twenty nucleotide units has become feasible owing to the pioneering work of KHORANA and co-workers (*185, 186, 187, 188, 319*). The crucial steps, the formation of the phosphodiester linkages, are achieved by successive condensations of the 5'-phosphomonoester group of nucleotidic components (**1**; Scheme 4.1) with the 3'-hydroxyl of the respective nucleosidic components (**2**; Scheme 4.1). While dicyclohexylcarbodiimide (DCC) has been used as condensing agent in most of the earlier work, its application has now become limited more or less to the preparation of protected dinucleoside monophosphates, as in contrast to the aromatic sulfonyl chlorides no anionic products are introduced by DCC during the course of the

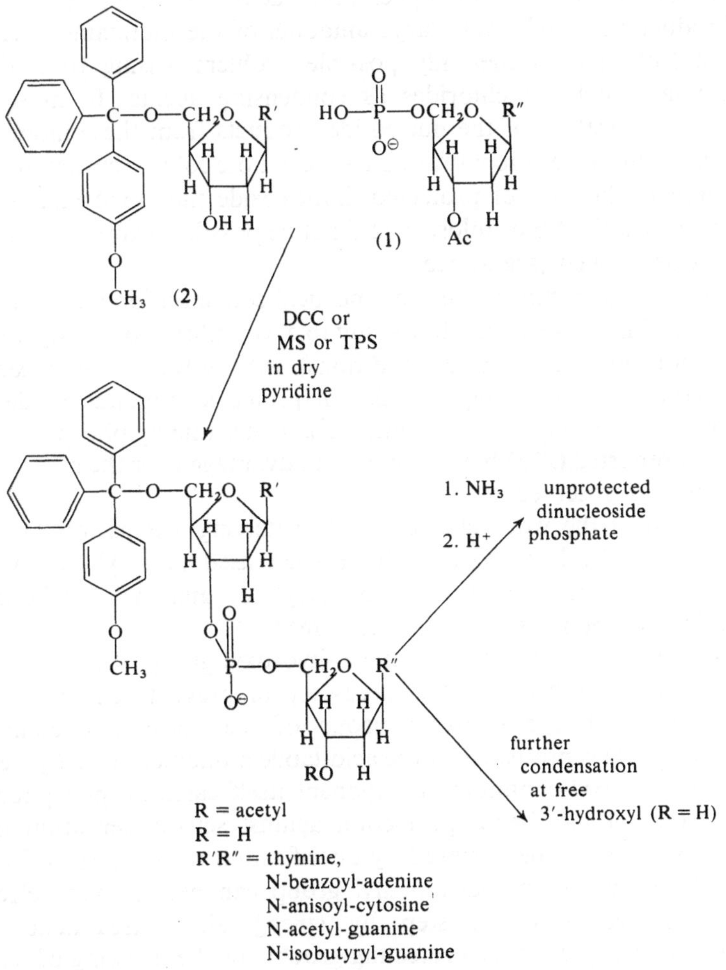

Scheme 4.1. Formation of internucleotide linkages in the deoxy series *via* phosphodiester intermediates

reaction (monovalent anions such as chloride and arylsulfonates produced by hydrolysis of aromatic sulfonylchlorides would not be separable from protected dinucleoside monophosphates by conventional DEAE cellulose column chromatography as the latter also possess one negative net charge). Although DCC in principle seems an effective condensing agent for the preparation of longer chains (207) also, there are major drawbacks because of the required longer reaction times and the necessity for complete absence of even traces of strongly basic amines in the reaction mixtures. Therefore preferential use has been made of aromatic sulfonyl chlorides such as mesitylene sulfonyl chloride (MS) and triisopropyl sulfonyl chloride (TPS) as condensing agents for the synthesis of higher oligonucleotides (166, 248).

Production of relatively large amounts of the monoanions chloride and arylsulfonate is the only possible problem arising from the use of aromatic sulfonyl chlorides as condensing agents. In most cases, however, separation of the nucleotide products from these monoanions is readily achieved by preparative DEAE cellulose column chromatography and even in the case of protected dinucleoside monophosphates where separation on DEAE cellulose would not be possible, extraction procedures can be applied (see above).

The solvent routinely used for chemical condensation reactions is dry pyridine. This solvent combines suitable volatility (for ready removal by evaporation under mild conditions) with solvent power generally satisfactory for even longer chains of protected oligonucleotides. The use of other solvents such as lutidine and hexamethylphosphotriamide has been reported (207) but no apparent advantage over the more volatile pyridine was observed.

In most of the work classical acyl protecting groups for the amino functions of the base residues have been used (anisoyl for cytosine, benzoyl for adenine, acetyl or isobutyryl for guanine) for both condensation components (185, 186, 187, 188).

Protection by acetylation of the 3′-hydroxyl group of the nucleotidic component (1, Scheme 4.1) is necessary to prevent self-condensation of the latter. The 5′-hydroxyl group of the nucleosidic component (2, Scheme 4.1) is protected by the acid-labile monomethoxytrityl residue. If the 3′-hydroxyl-containing component itself carries a phosphomonoester group at its 5′-end, protection against self-condensation is also necessary. This can be achieved by esterification with excess of β-cyanoethanol; the resulting β-cyanoethyl group can be removed selectively after each condensation step by strong alkali treatment. More recently a variety of other protecting groups for the masking of terminal phosphate residues has been introduced to permit selective removal without cleavage of the N-acyl groups which protect the bases. From this, two general routes have been developed for the synthesis of longer oligonucleotide chains in the deoxy-series: the first approach, outlined in Scheme 4.2, for the synthesis of an undecamer (*379*) consists in repetitive condensation steps at the 3'-end of a growing chain containing a protected 5'-hydroxyl group. This approach, already developed earlier for poly-nucleotide synthesis in relation to the genetic code (*186, 187*), has been used extensively in more recent work dealing with the synthesis of various gene segments including two complete sets corresponding to tRNA genes (Table 4.1). Virtually throughout all this work the acid labile mono-methoxytrityl group has been used for the 5'-hydroxyl protection. While the 5'-end of the growing chain remains blocked by this group throughout the entire reaction sequence (except for the very last deprotection step), the 3'-terminal O-acetyl groups, still present at the extended chains immediately after each condensation step, are selectively removed by alkali treatment before every subsequent condensation step is carried

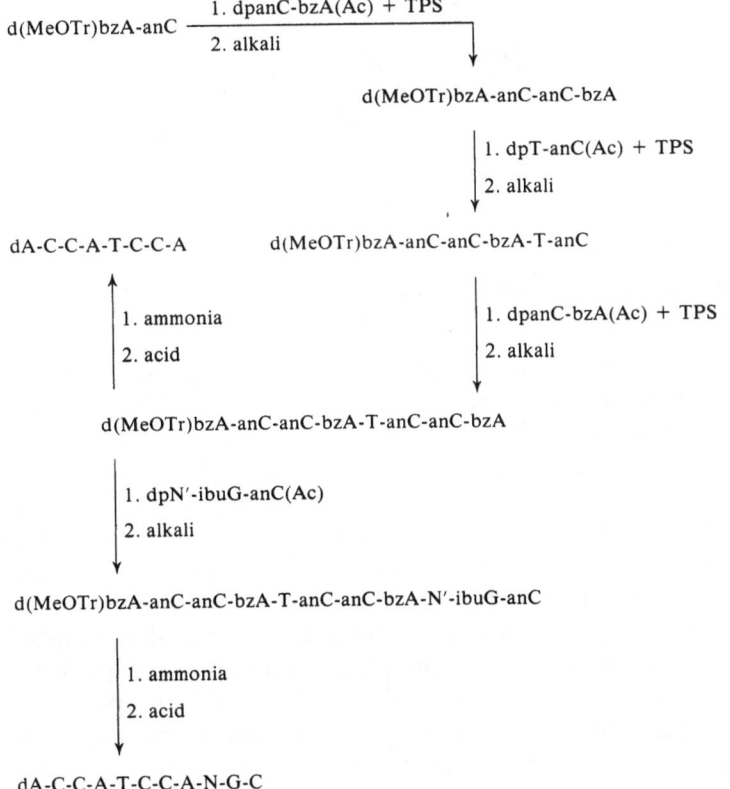

Scheme 4.2. Synthesis of the octanucleotide dA-C-C-A-T-C-C-A and of the undeca-nucleotides dA-C-C-A-T-C-C-A-N-G-C

out. Thus, an overall growth direction from the left (5'-end) to the right (3'-end) is the result. After final deprotection by ammonia and acid treatment products without 5'-terminal phosphate residues are isolated.

In contrast to this, the second general route leads to the synthesis of 5'-terminal phosphate containing nucleotide chains. As outlined in Scheme 4.3 protection of the 5'-terminal phosphate group present in the 3'-hydroxyl bearing components is then necessary. In many cases

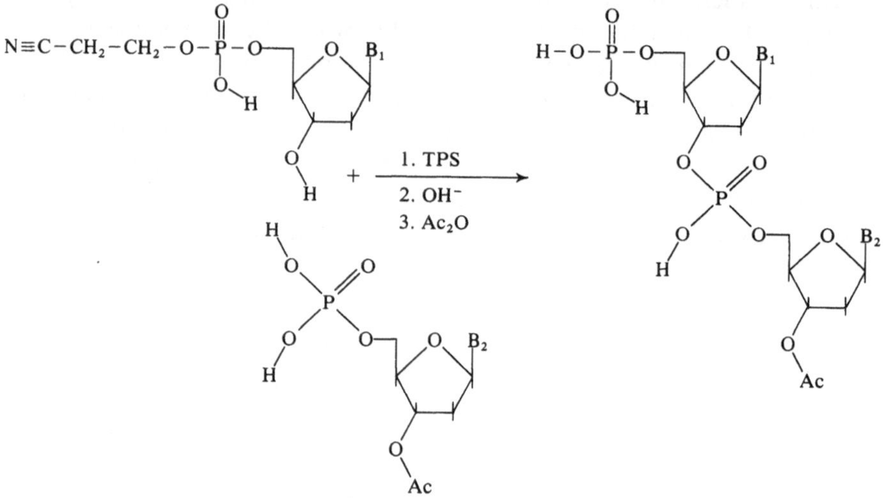

B₁, B₂ = T, A^{bz}, C^{an}, G^{ac}, G^{ibu}

Scheme 4.3. Synthesis of 5'-terminal phosphate containing nucleotide chains in the deoxy series

the cyanoethyl group has served as phosphate protecting group in such reactions and the preparation of "blocks" by reactions analogous to the one outlined in Scheme 4.3 has become almost routine (see references in Table 4.1). A major disadvantage of the cyanoethyl (and of other alkali-labile) phosphate protecting group is that it is removed by the alkali treatment necessary for the hydrolysis of the 3'-terminal O-acetyl groups after each condensation step. Thus, cyanoethylation of the 5'-terminal phosphate residues has to be carried out before every subsequent reaction step. In order to eliminate this complication, a number of alkali stable phosphate protecting groups has been developed more recently, which (like the 5'-O-trityl group) can remain at the 5'-phosphate group throughout the entire reaction sequences necessary to build up the longer nucleotide chains (see foregoing chapter on protecting groups). This approach seems to be useful not only for the preparation of oligonucleotide blocks (3, 7, 55, 56, 176, 290a, 312), as the synthesis of comparatively long chains has also been reported. Thus, by using the phosphothioethyl group three dodeca-nucleotides and one tridecanucleotide of specific sequences correspond-ing to fragments of a DNA coding for a derivative of S-peptide of ribonuclease A could be synthesized (*58, 129, 146, 341*) in addition to a dodecanucleotide sequence constituting the 5′-terminus of the r-strand of λ-phage DNA (*145*) (see Table 4.1). In analogous reaction sequences but using the phenylmercaptoethyl group for the protection of the 5′-terminal phosphate, the synthesis of two deoxyribopolynucleotide fragments of chain lenghts nine and twelve has been reported which contain a natural sequence of the phage T4 lysozyme gene (*291*) (see Table 4.1).

Table 4.1. *Oligodeoxynucleotides of Specific Sequence Synthesized Chemically by the Phosphodiester Method**

No.	Sequence	Resp. Gene	References
1	T–G–G–T–G–G–A–C–G–A–G–T	tRNA$_\text{yeast}^\text{Ala}$	*(315)*
2	p C–C–A–C–C–A	tRNA$_\text{yeast}^\text{Ala}$	*(315)*
3	C–C–G–G–A–C–T–C–G–T	tRNA$_\text{yeast}^\text{Ala}$	*(219a)*
4	C–C–G–G–A–A–T–C	tRNA$_\text{yeast}^\text{Ala}$	*(219a)*
5	C–C–G–G–T–T–C–G–A–T–T	tRNA$_\text{yeast}^\text{Ala}$	*(219a)*
6	G–A–A–C–C–G–G–A–G–A–C–T–C–T–C–C–C–A–T–G	tRNA$_\text{yeast}^\text{Ala}$	*(457)*
7	p A–G–A–G–T–C–T	tRNA$_\text{yeast}^\text{Ala}$	*(219)*
8	G–C–T–C–C–C–T–T–A–G–C–A–T–G–G–G–A–G–A–G	tRNA$_\text{yeast}^\text{Ala}$	*(37)*
9	C–T–A–A–G	tRNA$_\text{yeast}^\text{Ala}$	*(219)*
10	G–G–A–G–C–G–C–G–C–T	tRNA$_\text{yeast}^\text{Ala}$	*(5)*
11	T–C–G–G–T–A–G–C–G–C	tRNA$_\text{yeast}^\text{Ala}$	*(40)*
12	A–C–C–G–A–C–T–A–C–G	tRNA$_\text{yeast}^\text{Ala}$	*(40)*
13	T–G–G–C–G–C–G–T–A–G	tRNA$_\text{yeast}^\text{Ala}$	*(40)*
14	C–G–C–C–A–C–A–C–G–C–C–C	tRNA$_\text{yeast}^\text{Ala}$	*(38)*
15	G–G–G–C–G–T–G	tRNA$_\text{yeast}^\text{Ala}$	*(38)*
16	C–T–A–C–C–G–A–C–T–A–C–G	tRNA$_\text{yeast}^\text{Ala}$	*(5)*
17	C–T–A–A–G–G–G–A–G	tRNA$_\text{yeast}^\text{Ala}$	*(219)*
18	T–C–T–C–C–G–G–T–T	tRNA$_\text{yeast}^\text{Ale}$	*(219)*
19	C–G–A–G	tRNA$_\text{E. coli}^\text{tyr, su}$	*(190)*

* The blocks used are indicated by underlining of the respective partial sequences. As chain growth occurs always from left to right, the sequences of the intermediates can also be deduced. Thus, for the synthesis of 1, the oligonucleotides TG, TGG, TGGT, TGGTG, TGGTGG, TGGTGGA, and TGGTGGACG in the protected form are intermediates whereas for the synthesis of 2 the intermediates pCC and pCCA in the protected form are derived.

Table 4.1 (continued)

No.	Sequence	Resp. Gene	References
20	C–C–C–C–A–C–C–A–C–C–A	$tRNA_{E.\ coli}^{tyr,\ su}$	(188, 190)
21	T–C–G–A–A–T–C–C–T–T–C	$tRNA_{E.\ coli}^{tyr,\ su}$	(188, 190)
22	T–G–G–G–G–G–A–A–G–G–A	$tRNA_{E.\ coli}^{tyr,\ su}$	(188, 190)
23	T–T–C–G–A–A–C–C–T	$tRNA_{E.\ coli}^{tyr,\ su}$	(188, 190)
24	T–T–C–G–A–A–G–G–T	$tRNA_{E.\ coli}^{tyr,\ su}$	(188, 190)
25	C–G–T–C–A–T–C–G–A–C	$tRNA_{E.\ coli}^{tyr,\ su}$	(188, 190)
26	C–T–A–A–A–T–C–T–G–C	$tRNA_{E.\ coli}^{tyr,\ su}$	(188, 190)
27	T–C–G–A–A–G–T–C–G–A	$tRNA_{E.\ coli}^{tyr,\ su}$	(188, 190)
28	T–G–A–C–G–G–C–A–G–A	$tRNA_{E.\ coli}^{tyr,\ su}$	(188, 190)
29	T–T–T–A–G–A–G–T–C–T	$tRNA_{E.\ coli}^{tyr,\ su}$	(188, 190)
30	G–C–T–C–C–C–T–T–T–G	$tRNA_{E.\ coli}^{tyr,\ su}$	(188, 190)
31	G–C–C–G–C–T–C–G–G–G–A–A	$tRNA_{E.\ coli}^{tyr,\ su}$	(188, 190)
32	C–C–C–C–A–C–C–A–C–G–G	$tRNA_{E.\ coli}^{tyr,\ su}$	(190)
33	G–A–G–C–A–G–A–C–T	$tRNA_{E.\ coli}^{tyr,\ su}$	(188, 190)
34	C–G–G–C–C–A–A–A–G–G	$tRNA_{E.\ coli}^{tyr,\ su}$	(188, 190)
35	G–G–G–T–T–C–C–C–G–A–G	$tRNA_{E.\ coli}^{tyr,\ su}$	(188)
36	G–G–T–G–G–G–G–T–T–C–C	$tRNA_{E.\ coli}^{tyr,\ su}$	(190)
37	A–T–T–A–C–C–C–G–T	$tRNA_{E.\ coli}^{tyr,\ su}$	(190)
38	A–G–T–A–A–A–A–G–C	$tRNA_{E.\ coli}^{tyr,\ su}$	(190)
39	G–G–A–G–C–A–G–G–C–C	$tRNA_{E.\ coli}^{tyr,\ su}$	(190)
40	G–C–T–T–C–C–C–G–A–T–A–A–G	$tRNA_{E.\ coli}^{tyr,\ su}$	(190)
41	G–T–A–A–T–G–C–T–T–T	$tRNA_{E.\ coli}^{tyr,\ su}$	(190)
42	T–A–C–T–G–G–C–C–T	$tRNA_{E.\ coli}^{tyr,\ su}$	(190)
43	G–C–T–C–C–C–T–T–A–T–C–G	$tRNA_{E.\ coli}^{tyr,\ su}$	(190)
44	G–G–A–A–G–C	$tRNA_{E.\ coli}^{tyr,\ su}$	(190)
45	p T–T–A–A–T–T–A–C–A–A–T–A	Bovine insuline chain A	(288)
46	p A–T–T–T–T–C–C–A–A–T–T–G	Bovine insuline chain A	(288)
47	p A–T–A–C–A–A–A–C–T–A–C–A	Bovine insuline chain A	(288)
48	p A–T–T–A–A–G–T–G–A–T–G–G	T$_4$-Lysozyme	(291)
49	p A–C–T–T–T–T–T–G–T	T$_4$-Lysozyme	(291)
50	p A–A–G–A–C–A–G–C–A–T–A–T	Pancreatic RNase A (S-peptide)	(341)

51	p T–T–A–A–T–C–C–A–T–A–T–G–C	Pancreatic RNase A (S-peptide)	(58)
52	p T–G–C–T–A–A–A–T–T–T–G–A	Pancreatic RNase A (S-peptide)	(146)
53	p A–A–A–T–T–T–G–A–A–A	Pancreatic RNase A (S-peptide)	(146)
54	p T–G–T–C–T–T–T–C–A–A–A–T	Pancreatic RNase A (S-peptide)	(129)
55	p T–T–A–G–C–A–G–C–C–G–C–A–G	Pancreatic RNase A (S-peptide)	(131)
56	p A–G–G–T–C–G–C–C–G–C–C–C	Sticky end of phage λ	(145)
57	A–C–C–A–T–C–C–A–A–G–C	Coat protein, phage fd.	(379)
58	A–C–C–A–T–C–C–A–C–G–C	Coat protein, phage fd.	(379)
59	A–C–C–A–T–C–C–A–G–G–C	Coat protein, phage fd.	(379)
60	A–C–C–A–T–C–C–A–T–G–C	Coat protein, phage fd.	(379)
61	A–C–C–A–T–T–C–A–A–G–C	Coat protein, phage fd.	(380)
62	A–C–C–A–T–T–C–A–C–G–C	Coat protein, phage fd.	(380)
63	A–C–C–A–T–T–C–A–G–G–C	Coat protein, phage fd.	(380)
64	A–C–C–A–T–T–C–A–T–G–C	Coat protein, phage fd.	(380)
65	A–G–A–A–A–T–A–A–A–A	Ribosomal binding site of phage ΦX174	(383)
66	C–A–G–T–T–T–G–A–G–C–A–T	Endolysine of phage λ	(467)
67	A–G–T–C–C–A–T–C–A–C–T–T	T₄-Lysozyme	(467a)
68	A–G–T–C–C–A–T–C–A–C–T–T–A–A	T₄-Lysozyme	(467a)
69	p C–C–A–A–A–C–C–A–A–A	T₄-Lysozyme	(467a)
70	G–T–T–C–T–G	—	(20)
71	p G–G–T–T–T–C–G–T–G–G	—	(20)

Polycondensation reactions. If mononucleotides protected only on the base residues are reacted with a condensing agent, polycondensation to a homologous series of polynucleotides occurs (Scheme 4.4). During

$$dpN \longrightarrow dpN(pN)_npN$$

Scheme 4.4

synthetic work related to the genetic code (*186, 187*) this technique has been extended to the polymerization of preformed di-, tri-, and tetranucleotide blocks (Scheme 4.5; *286, 308*). By this technique

$$dpN_1pN_2 \longrightarrow d(pN_1pN_2)_n$$
$$dpN_1pN_2pN_3 \longrightarrow d(pN_1pN_2pN_3)_n$$
$$dpN_1pN_2pN_3pN_4 \longrightarrow d(pN_1pN_2pN_3pN_4)_n$$

Scheme 4.5

complete oligonucleotide series including members of comparatively large size are readily accessible within short time. The method, however, is generally limited to the synthesis of polynucleotides containing *repeating* nucleotide sequences (in most cases homopolymers).

More recently polycondensation has been carried out with a mixture of 5'-thymidylic acid and 2',3'-O-dibenzoyl uridine-5'-phosphate which after ammonia treatment leads to a mixture of the series $(pT)_n$ and $(pT)_npU_r$ (*381*). As the components can be separated by a combination of conventional DEAE cellulose column chromatography and chromatography on a borate-containing matrix (*359, 381*), this approach seems generally applicable for the synthesis of 3'-ribonucleoside terminated homooligodeoxynucleotides. Recently polycondensation techniques have also been utilized for the synthesis of copolymers of specific sequence (*396, 398a, 398b;* see also section 1.4.2).

A characteristic feature of the condensation methods described in this chapter is that the phosphodiester functions, already present in the reaction components or formed during the course of the reaction, are left unprotected. This approach, generally called *diester approach*, introduces two main disadvantages: first, dialkyl phosphate anions are not chemically inert to the conditions necessary for synthesis and therefore side reactions such as pyrophosphate formation or cleavage of internucleotide bonds by pyridine can occur (*166*). These – and perhaps other – side reactions seem to be increasingly severe in the synthesis of longer chains as a substantial decrease in the general yields has quite regularly been encountered in the preparations of higher oligonucleotides. Secondly, the partially protected intermediates are insoluble in organic solvents (with the exception of protected dinucleoside monophosphates and other shorter oligonucleotides carrying highly lipophilic protecting groups, see the section on protecting groups and on extraction methods) and it is therefore necessary to use laborious fractionation procedures such as DEAE cellulose chromatography to purify them after each condensation step. It seemed likely that both these disadvantages would be overcome if the internucleotidic linkages of the intermediates

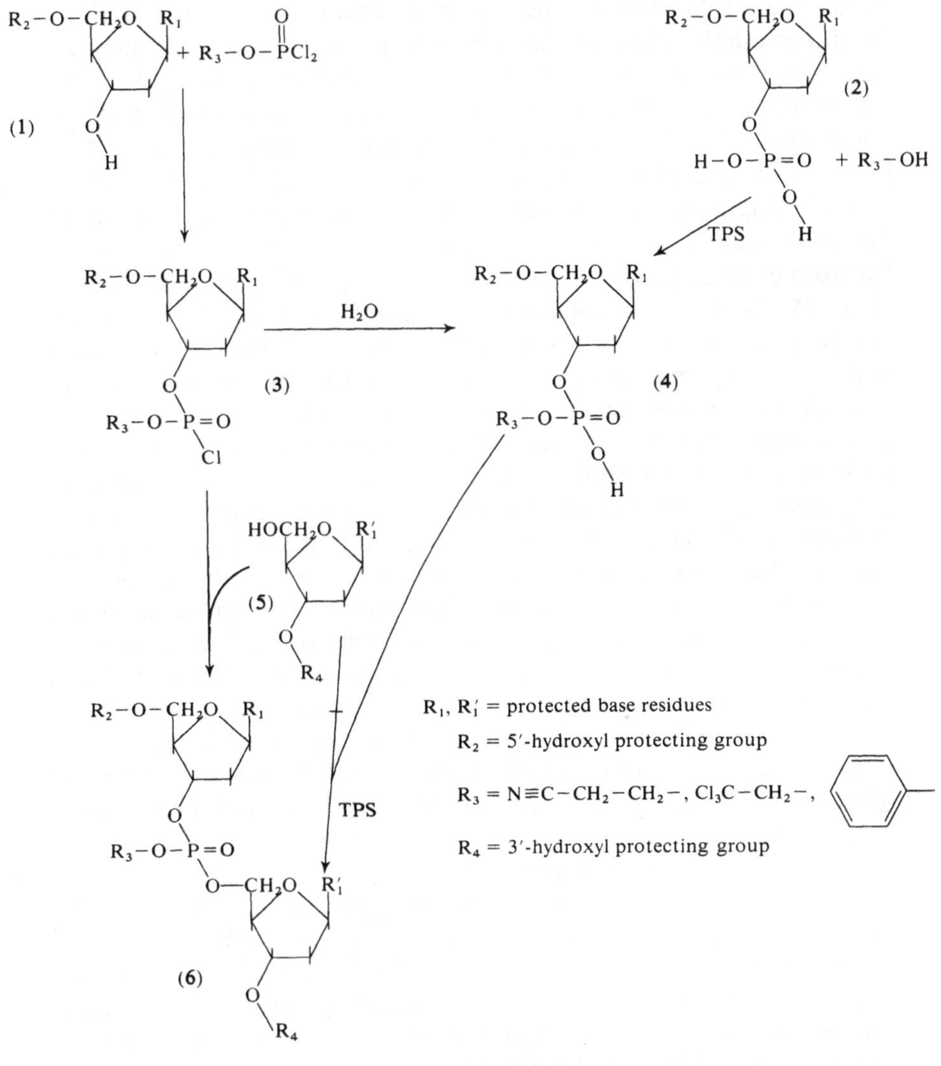

Scheme 4.6. Formation of the internucleotide linkage in the deoxy series *via* phosphotriester intermediates

were protected by further esterification to phosphotriesters and a number of studies have been reported on this so-called *triester approach* to be reviewed in the following chapter.

4.1.1.2. Synthesis *via* Phosphotriester Intermediates

In the deoxy series protection of internucleotidic phosphate by β-cyanoethyl (*87, 235, 236*), trichlorethyl (*42, 75, 76, 78, 94, 418*), phenyl (*73, 352, 358*), and substituted phenyl groups (*352, 358*) has been reported. As outlined in Scheme 4.6 the key intermediate (**4**) is produced either by direct esterification of the monophosphate containing component (**2**) or by concomitant introduction of the protecting group with the phosphate residue. In the latter case a suitably protected 3'-hydroxyl component (**1**) is first reacted with β-cyanoethyl dihydrogen phosphate (*87, 235*) or with phenyl dihydrogen phosphate (*73*) in the presence of a condensing agent as TPS (Scheme 4.6). In other cases the 3'-terminal diester residue is introduced by reaction of a suitably protected 3'-hydroxyl component (**1**) with trichlorethyl or phenyl phosphorodichloridate (*75, 78, 352*). The resulting 3'-phosphorochloridate derivatives (**3**) can be used directly as activated phosphate components for the reaction with a suitably protected 5'-hydroxyl derivative (**5**); alternatively the chloride is removed by hydrolysis, whereby the desired 3'-terminal diester derivatives (**4**) are obtained. The internucleotidic linkage finally is formed by condensation of the latter with suitably protected 5'-hydroxyl bearing components (**5**) in the presence of TPS as condensing agent. Approaches in which esterifications in all steps are achieved by a condensing agent [pathway (**2**) → (**4**) → (**6**) of Scheme 4.6] seem to be more favourable than earlier approaches in which the respective aryl or alkyl phosphorodichloridates were used in stochiometric amount for successive reactions with one 3'-hydroxyl and one 5'-hydroxyl component [Scheme 4.6; pathway (**1**) → (**3**) → (**6**)].

In order to allow repeated condensation steps for the synthesis of longer chains, protecting groups at the 3'- and 5'-termini have to be selected such that selective removal after each condensation step is possible.

Thus, when alkali-labile protecting groups are used at the one end, acid labile protecting groups have to be applied at the other end (*42, 73, 75, 76, 78, 236, 352*). Following this principle oligothymidylic acids could be synthesized by the stepwise (*42, 76, 78, 236, 352*) or blockwise (*42, 73, 75, 78, 236*) approach up to a chain length of eight by using components with protected hydroxyl functions at the respective 3'- and 5'-termini. While this approach has been used for the stepwise synthesis of dinucleoside monophosphates and trinucleoside diphosphate

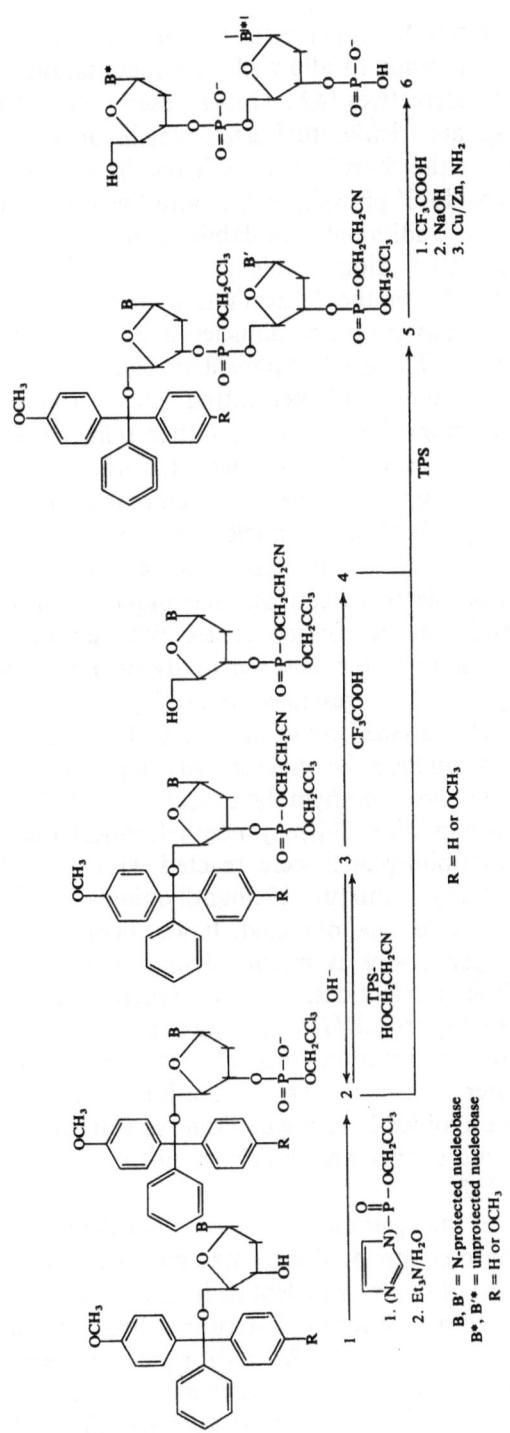

Scheme 4.7. Formation of the internucleotide linkage in the deoxy series *via* phosphotriester intermediates

containing also base residues other than thymine (78, 236), a more recent approach in addition seems to allow block condensation with all four standard nucleotide derivatives (42). This is based on condensation of a nucleoside 3'-phosphate trichlorethyl ester, which carries an acid labile blocking function on the 5'-hydroxyl, with the free 5'-hydroxyl group of a second nucleoside 3'-phosphate trichlorethyl ester, in which the phosphate carries an additional base-labile group (Scheme 4.7). The resulting fully protected dinucleotides can then selectively be deblocked at either the 5'- or the 3'-terminus by the use of acid or base and the resulting partially protected dinucleotides can be used in further condensation reactions. The masked phosphate as 3'-terminal protecting group offers the advantage of permitting the introduction of the phosphate at the mononucleotide stage, rather than before each subsequent condensation. Using this approach a variety of di-, tri-, and tetranucleotides containing all the four common bases could be synthesized in good yields. The tetranucleotides were prepared by block condensations from two dinucleotide units. In view of these encouraging results it seems desirable to extend this technique to the synthesis of longer chains containing specific sequences. Whether in this case no severe limitations arise from the acid sensitivity of the purine glycoside linkages (especially of d bzA residues containing oligonucleotides) or from other side reactions (see below) has yet to be tested.

A systematic study on the polycondensation using triesterintermediates has been reported in the oligothymidylic acid series (87). When 5'-O-monomethoxytrityl-thymidine 3'-[(β-cyanoethyl)phosphate] and thymidine 3'-[(β-cyanoethyl)phosphate] were reacted with aromatic sulfonyl chlorides for 12—14 days a mixture of oligothymidylic acids, the largest being the pentanucleotide, was obtained. It was demonstrated that the failure to yield longer chains is mainly due to the formation of C-pyridinium-thymidine nucleotides, a side reaction previously also observed in the diester approach (166). Owing to the longer reaction times necessary for the triester reactions this side reactions seem to be more severe in the triester approach. The formation of the C-pyridinium nucleotides could be avoided by using collidine as solvent. Production of longer oligonucleotide chains was, however, not observed either when collidine was used (87).

A careful study of the synthesis of the dTpT via the triester approach has shown that deprotection of the triester group by alkali treatment leads to isomerization of the internucleotide linkage to 5'—5'- and 3'—3'-derivatives if the 5'-terminal and/or 3'-terminal hydroxyl functions are free during the alkali treatment (282a, 358). Cyclic triesters seem to be the intermediates of this isomerization (Scheme 4.8) as evidenced from the fact that no isomerization is observed when the hydroxyl functions are blocked by alkali resistant groups (*358*). Protection of the hydroxyl functions, for instance by tetrahydropyranylation (*73*), is therefore necessary before removal of the phosphate protecting aryl groups of oligonucleotides can be carried out by alkali treatment.

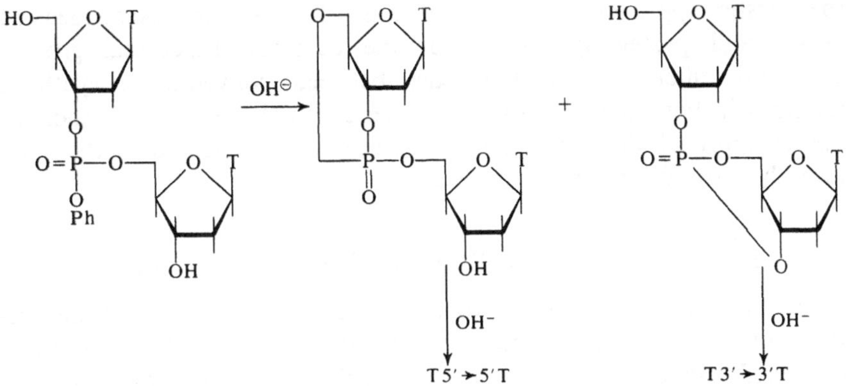

*Scheme 4.8.*Isomerization during alkali treatment of triesters

In view of the progress reviewed here and in view of the reported synthesis of a nonanucleotide in the riboseries (see below) by the triester approach, it seems not unlikely that it finally will allow synthesis of specific gene segments sufficiently long to permit joining reactions which are catalyzed by polynucleotide ligase. The triester approach seems particularly promising as the expectation that rapid and effective purification of intermediates could be effected by silica gel column chromatography (*232*), by preparative thin layer chromatography (see separation techniques) or by short column chromatography (see separation techniques) on a comparatively large scale was fulfilled. As, on the other hand, the products obtained by the triester approach, sometimes do contain considerable amounts of side products (*73, 87*) – even in the oligothymidylic acid series – purification by conventional DEAE cellulose chromatography seems to be necessary, at least for the final products of longer reaction sequences. In the light of the side reactions encountered during both approaches, the question remains open which method – the diester or the triester approach – is superior to the other as far as minimizing such products is concerned.

4.1.2. Synthesis in the Ribo Series

Chemical synthesis of ribopolynucleotides is complicated by the presence of 2'-hydroxyl groups, for which special protecting groups had to be introduced (see chapter of protecting groups). Since the 3'-

phosphomonoesters are more readily available and as the existence of a 3'-phosphate group facilitates the selective protection of the 2'-hydroxyl function of 5'-trityl-N-acyl derivatives, the principle of condensing a protected 3'-phosphate with the free 5'-hydroxyl group of a protected nucleoside component has been used in both the diester and the triester approach (Schemes 4.9 and 4.11). As regards condensing agents and protecting groups for the common functional groups, profound differences do not exist between conventional synthetic methods in both the ribo and deoxy series. In the ribo series, too, the diester approach developed earlier has been complemented by the more recently introduced triester approach.

4.1.2.1. Synthesis *via* Phosphodiester Intermediates

In the classical diester approach utilized, for instance, for the synthesis of all 64 possible ribotrinucleoside diphosphates (249), a suitably protected 3'-phosphate containing the acid labile 5'-monomethoxytrityl protecting group is condensed with the free 5'-hydroxyl group of a protected nucleoside (Scheme 4.9). After selective deblocking of the 5'-hydroxyl function of the resulting protected dinucleoside phosphate, further condensation can be carried out with a new protected 3'-phosphate leading to a protected trinucleoside diphosphate.

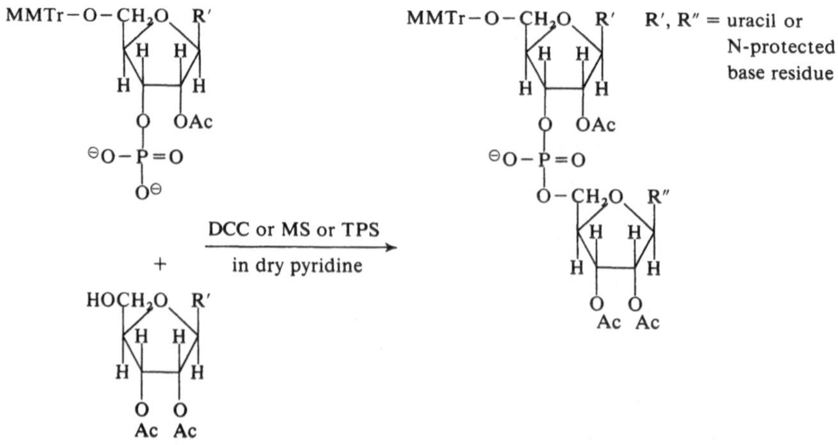

Scheme 4.9. Formation of the internucleotide linkage in the ribo series *via* phosphodiester intermediates

Thus, in contrast to the diester approach in the deoxy series, where removal of an alkali labile 3'-O-acyl group allows further extension at the 3'-end of the growing chain, growth direction in the ribo series occurs towards the 5'-end of the chains at which an acid labile trityl group is removed after each condensation step in order to allow further extension.

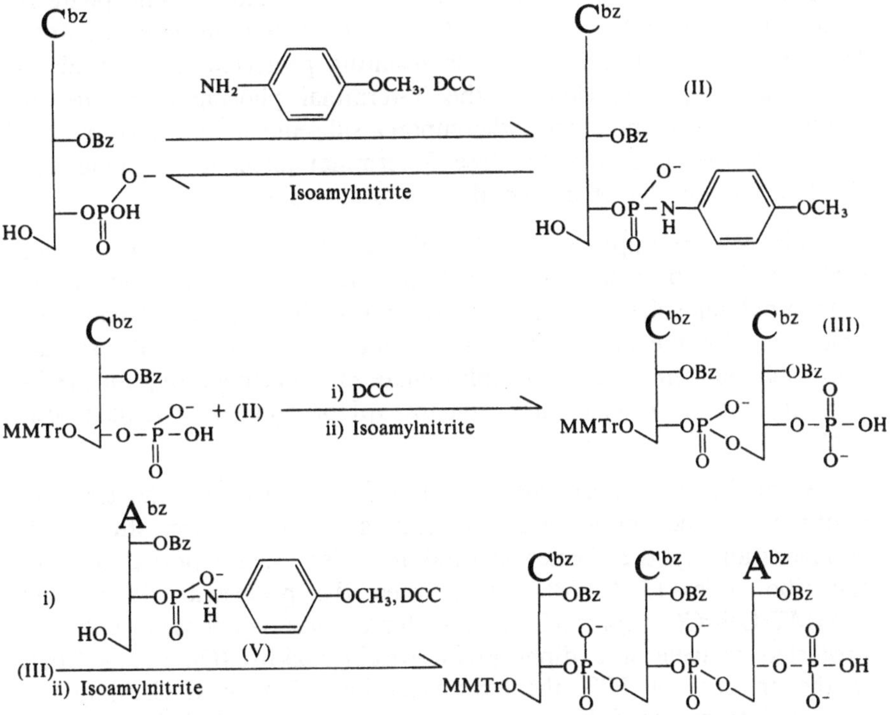

Scheme 4.10. Formation of the internucleotide linkages in the ribo series via phosphodiester intermediates

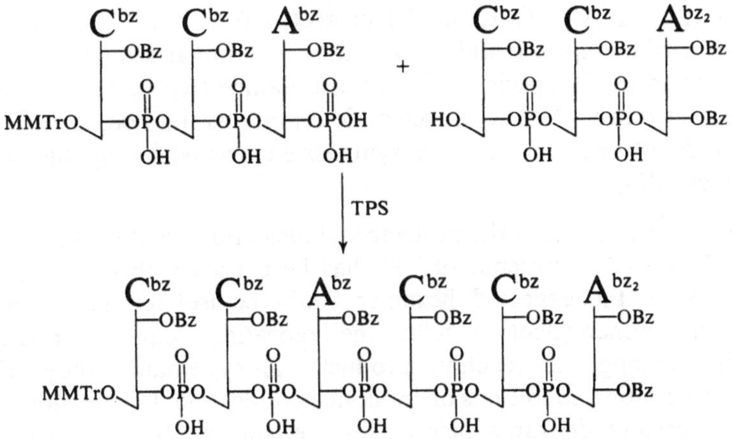

Scheme 4.11

While this principle has been used for trinucleoside diphosphate (*249, 319*) and trinucleotide (*313, 319*) synthesis, more recently a new approach leading to the opposite growth direction could also be developed. According to this principle a suitably protected 3'-phosphate is condensed with the free 5'-hydroxyl of a 3'-phosphoranisidate derivative (*309, 310, 319*, Scheme 4.10). The resulting protected dinucleoside is then selectively unblocked at the 3'-terminal phosphate residue by treatment with isoamylnitrite whereupon a subsequent condensation step can be carried out with the free 3'-terminal phosphate residue and the free 5'-hydroxyl of a second nucleosidic component.

This new principle verified by the synthesis of the trinucleotide CpCpAp in a protected form such that further condensation at the 3'-terminal phosphate is possible (Scheme 4.10), appears of considerable value for the preparation of oligonucleotide blocks. In a similar approach, but using a 2',3'-cyclophosphate as protecting group for the 3'-terminus, the trinucleoside diphosphates GpUpA and CpGpUp have been synthesized (*311, 314, 319*).

While the two approaches mentioned, are based on the stepwise addition of one mononucleotide unit at a time, successful block condensation by the diester method has also been reported now in the ribo series (*313, 318*). Thus, when the protected trinucleotide r[(MMTrO)bzC(Obz)-bzC(Obz)-bzA(Obz)p] was condensed with the protected trinucleoside diphosphate r[bzC(Obz)-bzC(Obz)-bz$_2$A(Obz)] in the presence of TPS, the hexanucleotide r(C–C–A–C–C–A) in the protected form could be isolated in reasonable yield (Scheme 4.11).

This hexanucleotide could be used (after removal of the 5'-O-protecting group) for further block addition to the nonanucleotide r(C–G–U–C–C–A–C–C–A) in the protected form (*318*).

These ribooligonucleotides constitute 3'-terminal sequences derived from certain tRNA species such as yeast alanine tRNA, *E. coli* tyrosine tRNA and others. The nonanucleotide represents the longest ribooligonucleotide of specific sequence synthesized chemically by the diester approach (*319*).

Polycondensation of the protected trinucleotide bzC(OBz)-bzC(OBz)-bzA(OBz)p in the presence of TPS has been tried with limited success (*313*). While the expected hexamer could be isolated in rather low yield, no nonanucleotide with the repeating sequence could be detected among the reaction products. A systematic study of the polycondensation conditions on mononucleotides and dinucleotide blocks seems therefore desirable before this method can be evaluated more thoroughly.

4.1.2.2. Synthesis *via* Triester Intermediates

As in the deoxy series the triester approach was expected 1. to suppress side reactions which occur in the diester synthesis due to the reactivity of the diester internucleotide linkages and 2. to allow rapid isolation and purification of the intermediates by extraction procedures, preparative thin layer and/or short column chromatography. After encouraging results were obtained in the deoxy series a number of attempts were therefore undertaken to extend the triester approach also to the ribo series. Benzyl (*419*), phenyl (*31, 352*), o-chlorphenyl (*31*), trichlorethyl (*296, 298, 299, 300, 418, 419, 462*) and β-cyanoethyl (*113, 417, 418, 419*) groups have been used for the protection of internucleotidic phosphate residues and for the conversion of the 3'-terminal phosphomonoester residues, to the corresponding phosphodiester residues. Thus (Scheme 4.12), after phosphorylation of 5'-O-monomethoxytrityl-2'-O-tetrahydropyranyluridine (**2**) with trichlorethylphosphate in the presence of TPS to give the diester (**3**) further reaction of (**3**) with 2'-O-tetrahydropyranyluridine (**1**) in the presence of TPS leads to the dinucleoside monophosphate derivative (**4**). As observed consistently also in analogous cases with other nucleoside derivatives (*31, 113, 296, 298, 299, 300, 462*) condensation of (**3**) with the 3'-hydroxyl-group of (**1**) to give products containing 3'→3' phosphodiester linkages in the deprotected compounds could not be detected. Apparently the steric hindrance of the neighbouring 2'-O-tetrahydropyranyl group and/or of the bulky arylsulphonic-phosphoric anhydride intermediate, together with the secondary nature of the 3'-hydroxyl group, do not allow reaction at this functional group. Consequently no special 3'-hydroxyl protection is necessary and subsequent reactions can immediately be carried out as outlined in Scheme 4.12 for the synthesis of uridyl-(3'→5')-uridyl-(3'→5')-uridine (**5**). Virtually the same approach seems also feasible when methoxytetrahydropyranyl group and phenyl or o-chlorophenyl groups are chosen for protection of both the 2'- and 5'-hydroxyl functions and the phosphodiester linkages (*31*).

This stepwise approach, in which overall direction of the chain growth towards the 3'-end results, could also be applied to the synthesis of ribooligonucleotides containing all the four common base residues (*298, 299, 300, 462*). More recently this principle could even be extended to the condensation of preformed blocks (*300, 462*) as outlined in Scheme 4.13 for the synthesis of the nonamer GpCmpUpCpApUpApApC (*300*). The latter corresponds to a sequence occurring in the anticodon loop of tRNA$_F^{Met}$ from *E. coli;* this sequence represents the longest ribooligonucleotide of specific sequence synthesized chemically by the triester approach. Protected di- or triribonucleotides (**1, 2, 3, 4** and **5**) were

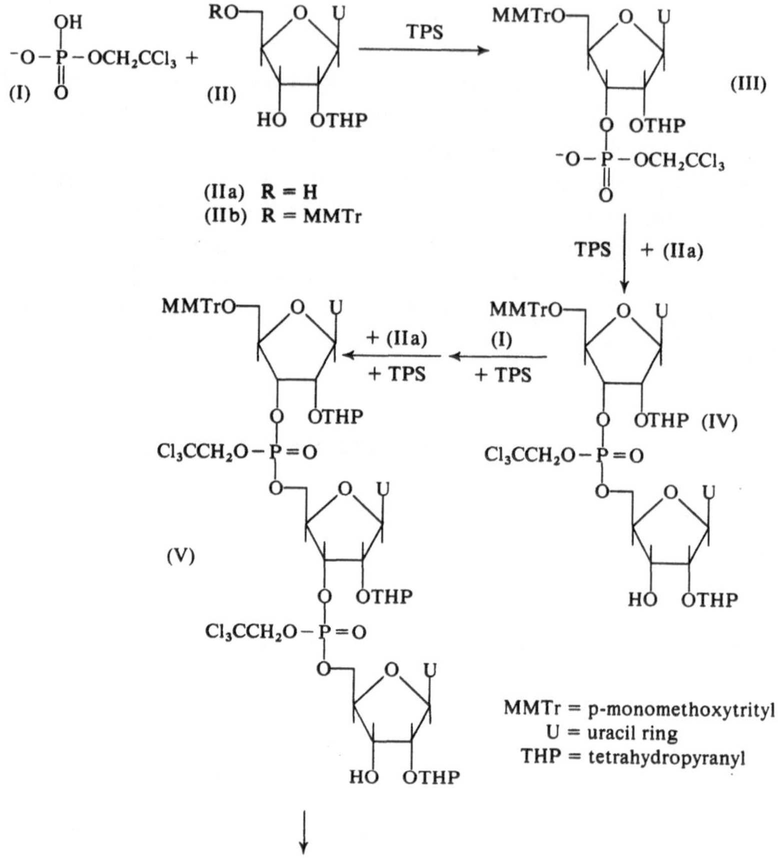

Scheme 4.12. Formation of internucleotide linkages in the ribo series *via* phosphotriester intermediates

assembled stepwise from their 5'-termini starting from 5'-O-trityl-oxyacetyl-2'-tetrahydropyranyl nucleosides by the two-step procedure using trichloroethylphosphate and TPS analogous to the reaction pathway outlined in Scheme 4.12 for the synthesis of UpUpU. Block phosphotriester synthesis of the nonaribonucleotide derivative (**8**) was then accomplished using a similar procedure from protected tetra-nucleotide (**6**) and pentanucleotide (**7**) which had been the coupling products of dinucleotide derivatives (**1** and **2**) and of trinucleotide (**4**) and dinucleotide derivative (**5**) respectively. It is interesting to note that in contrast to the diester approach, in which increasingly large excesses of the incoming nucleotidic components have to be used in order to achiev satisfactory yields, almost equimolar proportions of the reactants could be applied in the triester approach. The yields were satisfactory for coupling of single nucleoside residues (>50%) but dropped to the still respectable range of 20—30% for the condensations involving blocks.

The triester technique has also been used in a complementary strategy leading to growth direction towards the 5'-end (*31, 352, 418, 419*). In

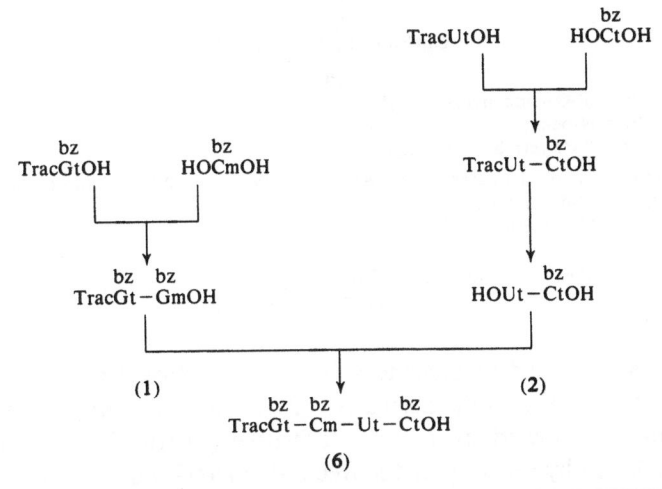

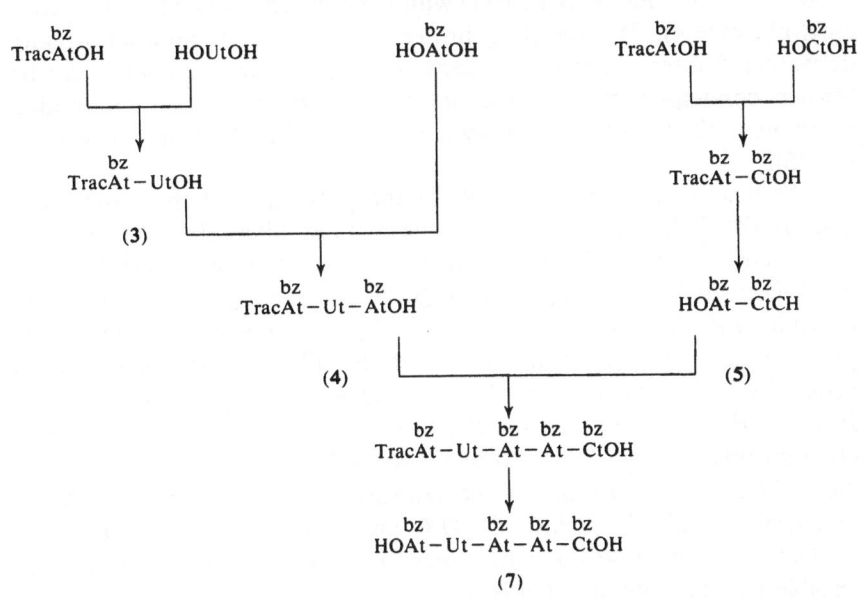

Scheme 4.13. Synthesis of a nonamer of the ribo series by the phosphotriester approach
(continuation p. 426)

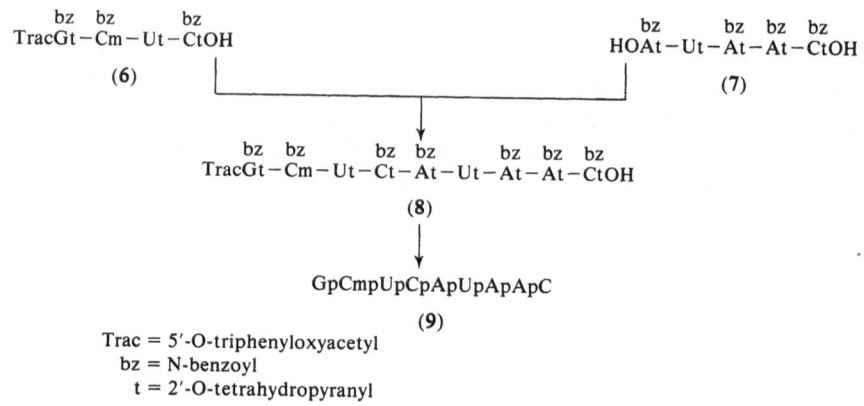

Trac = 5'-O-triphenyloxyacetyl
bz = N-benzoyl
t = 2'-O-tetrahydropyranyl

Hyphen between two characters, e.g. At−Ut, indicates a 2,2,2-trichloroethyl-phospho-triester internucleotide linkage
m = 2'-O-methyl

Scheme 4.13. Synthesis of a nonamer of the ribo series by the phosphotriester approach (continued from page 425)

this case, as outlined in Scheme 4.14, the 5'-hydroxyl group of a partially protected nucleoside derivative (**3**) is reacted with the 3'-diester residue of a protected nucleotide derivative (**2**) (the latter as usual is prepared *in situ* by reaction of the free 3'-hydroxyl function of a partially protected nucleoside derivative (**1**) with phenyltrichloroethyl- or β-cyanoethyl-phosphate). The resulting protected dinucleoside monophosphate derivative (**4**) after selective deprotection of the 5'-hydroxyl function can be further condensed with a second 3'-terminal phosphate containing component (**2**) to give the protected trinucleoside diphosphate derivative (**5**).

The use of acid labile groups for the protection of the 3'-terminus and of the 2'-hydroxyl functions necessitates alkali labile protecting groups such as acetyl, formyl, benzoyl or p-chlorophenoxyacetyl groups for 5'-O-protection. Selective cleavage of the latter groups is, however, difficult, when the alkali labile β-cyanoethyl group is used for phosphate protection. On the other hand, use of the alkali stable trichloroethyl group leaves the protection of the internucleotidic phosphates intact. In contradiction to other reports, however, removal of the trichloro-ethyl group after the final step seems to be far from quantitative (*418*). Nevertheless, synthesis up to the tetrauridine triphosphate in reasonable yields could be achieved (*418*) and the synthesis of ribooligo-nucleotides containing the other three common bases seems also to be feasible by this approach (*419*).

One general drawback of the triester approach as compared to the diester technique consists in the longer reaction times required for

activation and reaction of the diester intermediates. This relative inert-
ness of diesters (as compared to monoesters) is further increased in the
ribooligonucleotide series by the bulky 2′-protecting groups which are

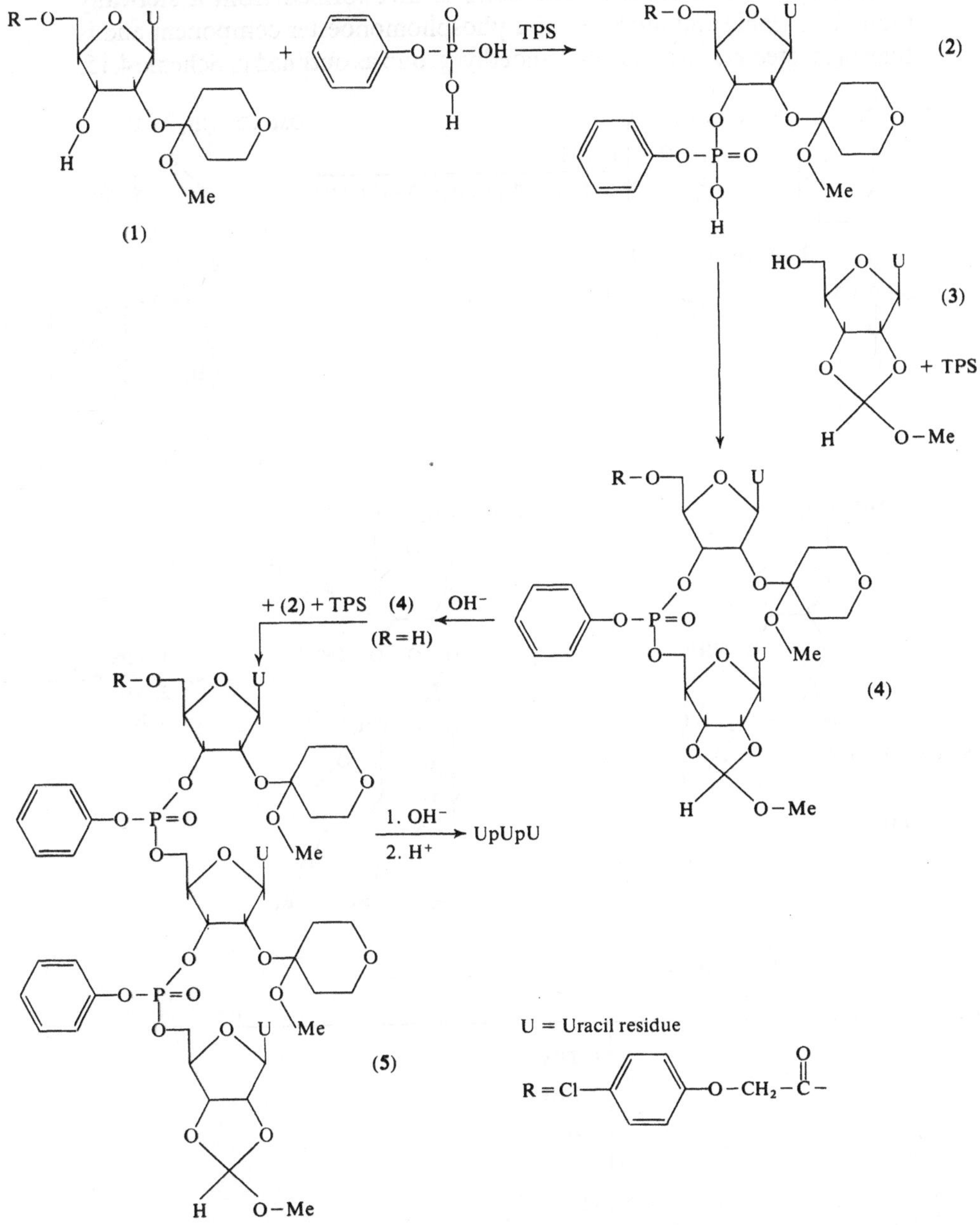

Scheme 4.14. Formation of internucleotide linkages in the ribo series *via* phosphotriester
intermediate

immediately adjacent to the phosphate residues to be activated. An approach has therefore been proposed which would combine the advantageous features of the diester and the triester synthesis (417). Accordingly the internucleotidic bond is first formed from a sterically more favourable and more reactive phosphomonoester component and is then protected *in situ* by the β-cyanoethyl group as outlined in Scheme 4.15.

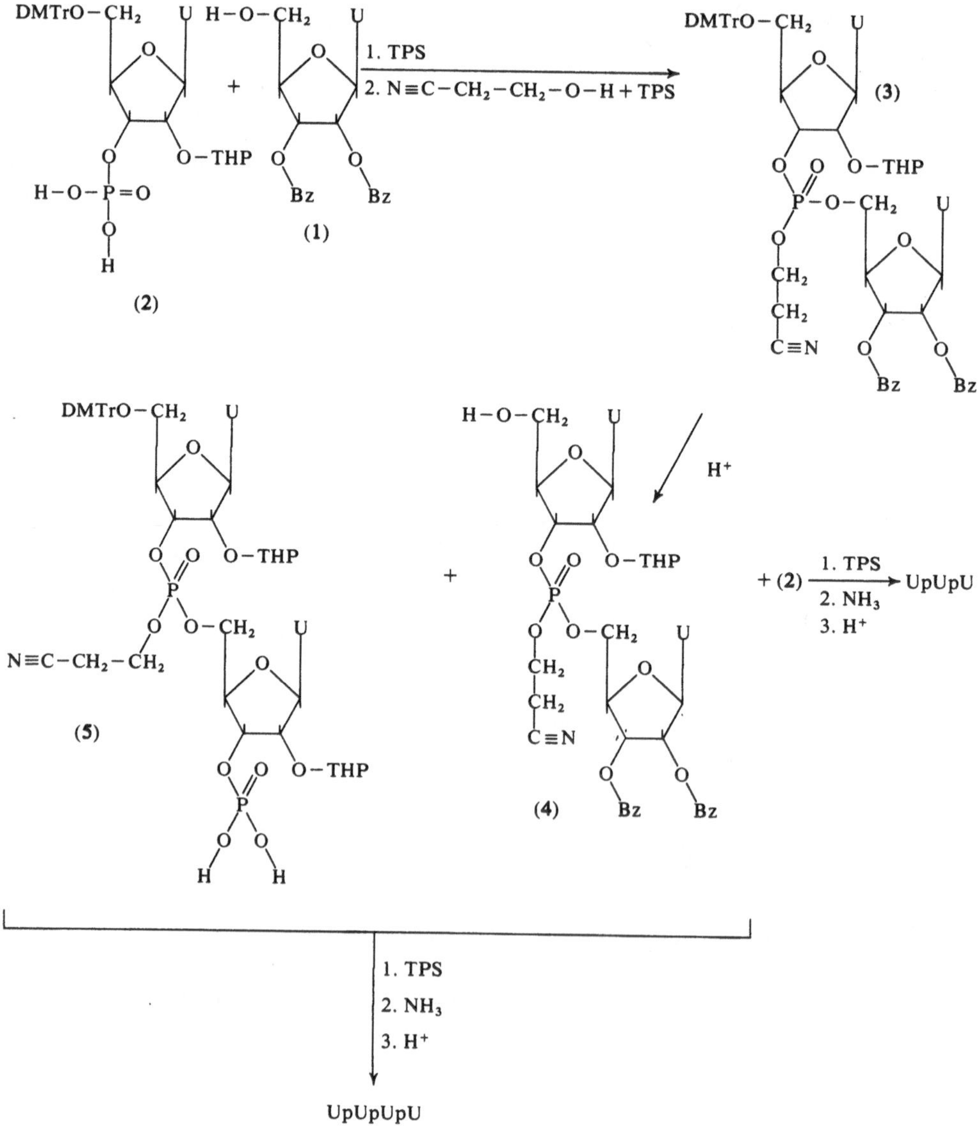

DMTrO—CH₂ U   H—O—CH₂ U

(1)

(2)

1. TPS
2. N≡C—CH₂—CH₂—O—H + TPS

DMTrO—CH₂ U

(3)

(5)

(4)

+ (2)
1. TPS
2. NH₃
3. H⁺
→ UpUpU

H⁺

1. TPS
2. NH₃
3. H⁺

UpUpUpU

Scheme 4.15. Formation of internucleotide linkages in the ribo series by the mixed diester-triester approach

Thus, for the synthesis of oligouridylic acids, 2′,3′-di-O-benzoyluridine
(1) is first condensed with 2′-O-tetrahydropyranyl-5′-O-dimethoxytrityl-
uridine 3′-phosphate (2) in the presence of TPS. The resulting mixture
is subsequently treated with cyanoethyl in the presence of more TPS, to
give the triester (3). After selective removal of the dimethoxytrityl group
with mild acid (whereby the tetrahydropyranyl group is left intact)
the 5′-hydroxyl function of (4) is ready for the second condensation
step with (2) or (5).

The yields obtained in the stepwise synthesis of tri- and tetranucleotides by the use of this mixed diester-triester approach seem to
compare favourable with the yields obtained by application of the pure
diester technique; a more detailed investigation including the synthesis
of oligonucleotides containing also the three other common bases
seems, however, desirable for the final evaluation of this method. An
encouraging yield of 77% could be obtained, when the principle was
applied to the condensation of two dinucleotide blocks (Scheme 4.15;
5 + 4→UpUpUpU; 421, 422).

4.1.3. Modified Oligonucleotides

A variety of modified oligonucleotides has been synthesized to
permit study of their physicochemical, biochemical and/or biological
properties. Comparison of the modified oligomers with the natural
compounds is expected to clarify certain biochemical or biophysical
aspects that deal with the significance of the various functional groups
in polynucleotides. In some cases model compounds have facilitated
the use of physicochemical methods for the investigation of polynucleotides.

Modifications have been introduced at the phosphate groups, at the
sugar moieties and on the base residues, respectively, in the ribo series
as well as in the deoxy series.

Methyl and ethyl phosphotriester derivatives of TpT and of d(ApA)
have been synthesized in order to perform pmr, CD and UV spectroscopic studies in organic solution (261). In order to study the properties of dinucleoside monophosphates containing unnatural internucleotide linkages, thymidylyl-(3′→3′)-, and (5′→5′)-thymidine have
been synthesized (282a, 358). Oligothymidylic acids containing internal
pyrophosphate linkages (384) and oligouridylic acids containing an
internal 5′→5′-linkage (416) have been synthesized in order to test
their primer function for enzymic reactions catalyzed by polymerizing enzymes. As regards altered internucleotide linkages, a number
of ribonucleoside monophosphates and higher oligomers containing

$2' \rightarrow 5'$-internucleotide bonds have been synthesized or isolated as side products (*15, 222, 223, 265, 323, 336, 337, 338, 346, 376, 430, 431, 432, 433, 455, 456, 459*). A "diuridine monophosphate" containing one arabinoside residue instead of a ribose residue has been synthesized as an oligonucleotide in which the sugar moiety is altered (*305*). Phosphor-amidate analogs of oligothymidylic acids have been prepared as a class with a modified phosphate-sugar linkage, which is susceptible to cleavage by mild acid (*237*). In addition to this, synthesis of L-adenylyl-(3'—5')-L-adenosine and of L-adenylyl-(2'—5')-L-adenosine (*438*) has opened a route to oligonucleotides containing modified sugar moieties as well as unnatural internucleotidic linkages. Chemical polymerization of ds^{4}TMP has led to a mixture of oligo-4-thiothymidylic acid (*13*) as an oligomer in which the base residues are modified. In addition, preparation of diribonucleoside monophosphates containing 4-thio-uridine has been reported (*266, 370*). Finally, synthetic dinucleoside monophosphates containing adenine 8-thiocyclonucleosides constitute a type of oligomers in which the modification involves both, base residues as well as the sugar moieties (*453*).

A considerable number of modified oligonucleotides has been derived by chemical modification reactions of natural polymers; as these techniques generally do not involve formation of new internucleotide linkages a survey of these contributions appears to lie beyond the scope of the present review. The numerous modified polymers syn-thesized by enzymic reactions will be surveyed in one of the following chapters.

4.2. Polymer-Support Synthesis of Oligonucleotides

4.2.1. General Reaction Principle

Synthesis of polypeptides and polynucleotides of defined sequence consists basically of a repetition of similar reaction steps with dissimilar monomeric reactants. For multistep reactions of this type R. B. MERRIFIELD (*257*) and R. L. LETSINGER (*226*) have developed the technique of polymer support synthesis. The essential reaction steps are:

1. Attachment of the initial monomer of the projected sequence to a polymeric carrier,

2. Blocking of unreacted functional groups of the carrier and deblocking of the grafted monomer.

3. Chain elongation by a blocked monomer unit.

4. Deblocking of the newly attached monomer.

5. Repetition of steps 3 and 4, until the desired sequence is finished.

6. Cleavage of the product from the support.

The polymer support method has a number of advantages over the conventional technique of condensation in solution: a) Throughout the chain elongation the growing polypeptide or polynucleotide chain is bound to a polymer. The reaction is, therefore, heterogeneous; problems due to solubility differences among the monomers and between them and the growing chain are eliminated. b) The separation and purification procedures during intermediate reaction steps are reduced to simple washing, filtration or precipitation procedures. The time needed for one chain elongation step is greatly reduced. c) The simplicity of separations allows the use of reactants in big excess or repeated treatment, in order to obtain optimum yields of the desired product. d) The repetition of similar or identical reactions allows the automation of the steps of support synthesis.

However, the method also has several drawbacks: a) Unless the yields of all reaction steps are quantitative, a series of homologous truncated sequences are attached to the support along with the desired sequence. The separation of these unwanted sequences after removal from the support can be very difficult, if not impossible in the case of longer chains (468). b) The reactions can be influenced, even controlled by the diffusion of the reactants through the support matrix. Steric hindrance can further inactivate the support bound material and decrease the reaction rates (215, 231). c) The reactions can be influenced by solvation properties of the carrier (395, 396a).

The method of support synthesis has developed very rapidly in the peptide field. Several biologically active polypeptides have been synthesized in this way, such as bradykinin, insulin A and B chains and ribonuclease. However, a great deal of criticism has also arisen due to the fact that long-chain polypeptides synthesized in this way, in spite of their biological activity are not chemically pure compounds.

In oligonucleotide synthesis the success of the support method has been limited due to the complexity of the reactions involved in internucleotide bond formation and the moderate yields which are generally obtained. Difficulties encountered in separating the cleaved product, due to these unsatisfactory yields, has precluded the synthesis of longer polynucleotide chains and no attempts at automation of the process have been made.

4.2.2. Requirements for Supports and Reactants

A) Polymeric Carriers:

The supports are in most cases synthetic macromolecular substances, e.g. polystyrene, but also biopolymers, e.g. polypeptides, or inorganic polymers, e.g. silicagel. As a first approximation these carriers should

be inert substances possessing suitable anchoring groups for an
oligonucleotide chain, which reacts freely as if in solution. This
picture, however, is an inadequate simplification. Carrier syntheses are
complex heterogeneous reactions which have several special features.
One of the reactants is immobilized inside a solvated gel or polymer
coil, an "immobile" or, according to Merrifield, a "solid" phase.
The different molecules of this reactant are in a different environment
as to steric hindrance by the surrounding polymer chains and as to
accessibility to other reactants penetrating inside from the mobile phase.
Once the reaction has taken place unused molecules of mobile reactant
as well as mobile reaction products will have to find their way back
into the mobile phase. Thus chemical reaction and diffusion processes
of various species occur side by side, the overall rate being reaction-
controlled for some and diffusion-controlled for other molecules of the
immobilized reactant. This is evident from the fact that the rate curve
during the initial phase is nearly equal to the one found for the carrier-
free reaction, but subsequently, deviates more and more. The overall
conversion may also be lower than in a parallel reaction in solution,
i.e. some of the reactant molecules may be completely inaccessible.

A basic problem in support synthesis is therefore the optimization of
the properties of the support with regard to steric hindrance and
accessibility of the immobilized reactant. Attempts to solve this in several
ways have been made. The initial approach of R. B. Merrifield (257)
was to use swellable, homogeneously crosslinked gels. The degree of
swelling (mostly about 2%) is chosen so as to ensure good swelling
as well as mechanical stability. Since the degree of swelling can only
be increased within certain limits, an alternative was to try to attach
the immobilized reactant to the surface of pores and cavities of macro-
reticular, i.e. heterogeneously crosslinked gels (68). These gels are rigid,
non-swellable and relatively insensible to differences in the solvating power
of different media. However, here, too, a compromise has to be found
between pore size and mechanical stability. Besides this, the materials
used contain a distribution of pore sizes around an average value. Therefore, part of the reactant will be in small pores, i.e. in less readily
accessible places. This risk is reduced by using microgels, which are very
small rigid gel particles of diameter 0.1—1 µ (214). In these ratio
of outer to inner surface is greatly increased; however, the small particle
size precludes the application of simple filtration processes.

In contrast to this development where the immobilized reactant was
attached to the surface of a rigid lattice, a concurrent line of development
aimed at the use of systems promising maximum chain flexibility.
The simplest solution in this direction was the use of linear, non-cross-

linked polymers as carriers. The first support system of this kind, a chloromethylated polystyrene, was described for peptide synthesis by M. M. Shemyakin and coworkers (404). Due to the internal Brownian movement of the polymer chain, the reactant molecules, which are grafted to it will continually change from positions more to the inside of the coil to others more to the outside and back. Diffusion problems are, therefore, at minimum in this case. Unfortunately, the non-cross-linked polymers do not allow filtration techniques. Removal of mobile reactants and products has to be carried out by precipitation or dialysis. This causes problems as the result of solubility changes, secondary crosslinking and adsorption, which will be discussed later in this section.

A third approach, which combines utmost chain flexibility with an insoluble system, was first studied by R. L. Letsinger and coworkers (226) concurrently for the synthesis of oligopeptides and oligonucleotides. This approach utilizes the so-called popcorn polymers which form on proliferous thermal bulk polymerization of suitable monomers, e.g. styrene. These, too, are heterogeneously crosslinked polymers containing regions of high resp. low coil density. Popcorn polymerization occurs at very low content of divalent monomer ($\sim 1\%$), and crosslinking is effected more by physical than by chemical means. Since the chains are only slightly linked to each other, the chain flexibility in appropriate solvents is nearly as high as in non-crosslinked polymers (34), although the popcorn polymer is insoluble, easily filtrated and mechanically stable. Although more information still has to be collected, it seems that this type of polymer is especially well suited for support purposes in oligonucleotide chemistry.

Another consideration concerns the chain solvation properties of the polymer support. The different steps of oligonucleotide synthesis are carried out almost exclusively in two kinds of media, i.e. pyridine and aqueous or partly aqueous solutions. Especially in the latter case good solvation cannot be expected for supports based on polystyrene. Recent developments in oligonucleotide support synthesis, therefore, aim at the construction of more hydrophilic carriers (33, 44, 392, 396a). Unfortunately, however, all the systems tested so far show a strong tendency to adsorb oligonucleotide chains in aqueous media, thus restricting the utilization of this approach. Cellulose carriers, which show excellent solvation and little adsorption in aqueous media cannot be employed for stepwise oligonucleotide synthesis due to their residual hydroxyl groups. However, they have been very profitably used for enzymatic syntheses (Section 4.2.4) and affinity chromatography (Section 4.2.5).

B) Reactants:

We have distinguished between two types of reactants – the immobilized one, i.e. the growing oligonucleotide chain, and the mobile ones, namely e.g. nucleotide monomers and blocking or deblocking reagents. For the immobilized reactant the polymer can be looked at as a macromolecular blocking group. Since cleavage from the support is the last of all steps leading to the synthesis of a sequence this linkage must be the most stable, and this has to be taken into account on designing a blocking scheme for the intermediates as shown in Section 1.6. From the earlier discussion of the steric influence of the support lattice it is clear that support reactions will be significantly slowed down, when one of the reaction partners is sterically hindered or bulky. This can be seen, for example, on comparing the yields of phosphorylation vs. inter-nucleotide bond formation under comparable conditions. In order to obtain fair yields in a reasonable time it is often necessary to use elevated temperatures and an enormous excess of the mobile reagent (238). This may be a disadvantage, if the reagent is expensive or not readily accessible. On the other hand, the possibility of shifting a reaction equilibrium towards the desired product by using a high excess of mobile reagent or by several repetitions of a reaction cycle may be a reason for using the support approach. It also possesses certain advantages in handling small quantities of reactants.

4.2.3. Chemical Synthesis of Oligonucleotides on Supports

After discussing some general aspects we shall now briefly review the different approaches described for the synthesis of oligonucleotides. Table 4.2 gives a survey of the different methods that have been developed. The methods are classified roughly according to the types of supports used (column 2). The functional groups and the way in which the initial member of the oligonucleotide chain is linked are shown in columns 3 and 4. In the next column the types of intermediates used for oligonucleotide synthesis are described. If necessary, the reaction conditions are indicated. Column 6 gives the conditions for the cleavage of the oligonucleotides from the support. Information on sequences and yields that were obtained is given in column 7. Since it is impossible to list all products, examples are given of a dinucleotide or dinucleoside phosphate, usually the one synthesized in the best yield, and of the longest chain which was obtained by this method. The corresponding references are listed in the last column.

Table 4.2. Approaches to the Synthesis of Oligonucleotides on Polymer Supports

No.	Support	Functional groups	Type of linkage	Conditions and intermediates for oligonucleotide synthesis	Cleavage conditions	Sequences, yields*	References
1	polystyrene, non-crosslinked	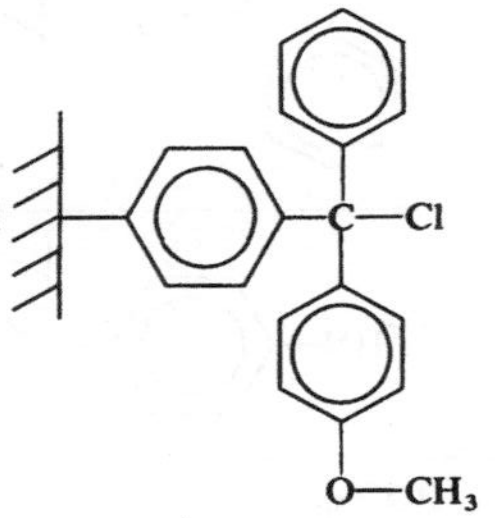	nucleoside-5'-trityl ether	pdT-OAc, pdC^an-OAc, pdG^ac-OAc	trifluoroacetic acid in chloroform	dT(pdT)$_n$ n = 1–2 dinucleotide sequences dTpdT: 96% dT(pdT)$_2$: 83%	(140, 141)
2	polystyrene, non-cross-linked	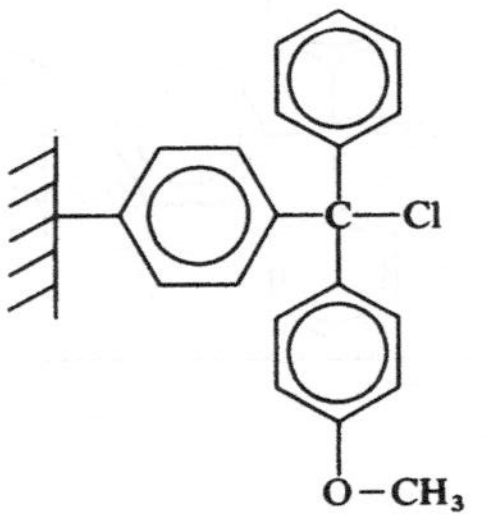	nucleoside-5'-ether	pdT-OAC	50% trifluoro-acetic acid/dioxane 1:100	dT(pdT)$_n$ n = 1–2 dT(pdT)$_2$: 11%	(67)

* Yields given are: best yield of one internucleotide bond formation; overall yield of longest sequence.

Table 4.2 (continued)

No. Support	Functional groups	Type of linkage	Conditions and intermediates for oligonucleotide synthesis	Cleavage conditions	Sequences, yields*	References
3 polystyrene, non-cross-linked	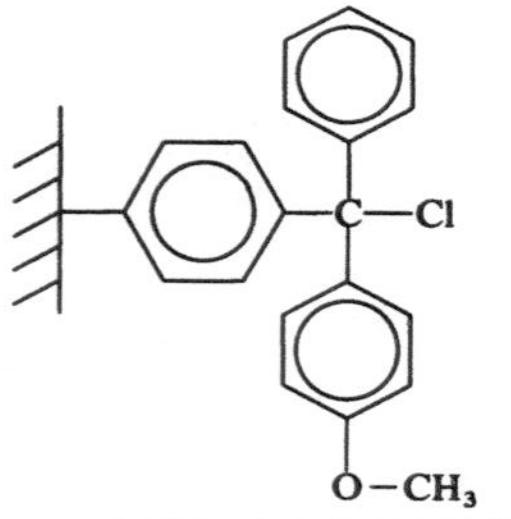	nucleoside-5'-ether	pdT-OAc 1. POCl$_3$/acetonitrile 2. dT-OAc	trifluoroacetic acid in chloroform	dTpdT 67%	(171, 343)
3a polystyrene, non-cross-linked	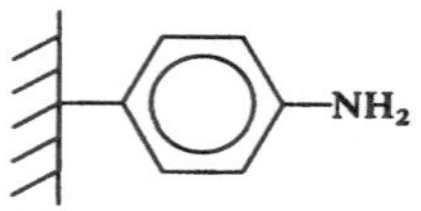	nucleotide-5'-phosphoramidate	pdT-OAc	80% acetic acid		(67, 393)
4 polystyrene, gel. 1% X	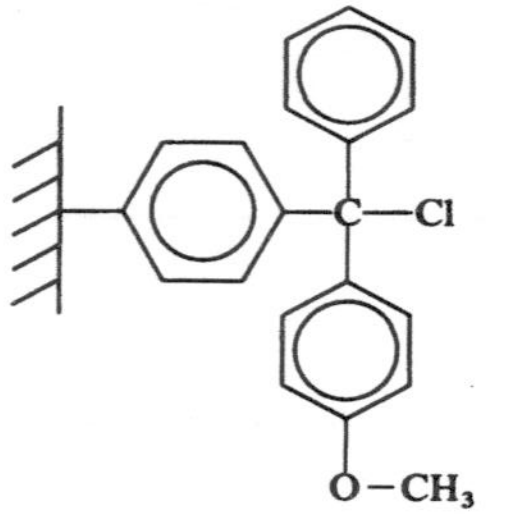	nucleoside-5'-ether	pdTOAC pdA^{bz}, pdA^{ac}OAc pdC^{an}, pdC^{an}OAc pdG^{ac}, pdG^{ac}OAc	acetic acid/water/benzene 32/8/10	dT(pdT)$_n$ n=1-4 di- and tri-nucleotide sequences dTpdT: 75% dT(pdT)$_4$: 6%	(254, 255, 256)

5 polystyrene, gel

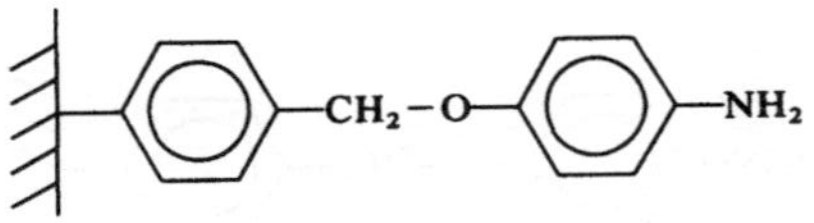

nucleoside-5'-phosphor-amidate

pdTOAc

3×80% acetic acid, 24 h., rt.

$(pdT)_n$ n = 2–3
$(pdT)_2$: 40%
$(pdT)_3$: 5%

(22, 23, 24)

5a polystyrene gel, 2% X

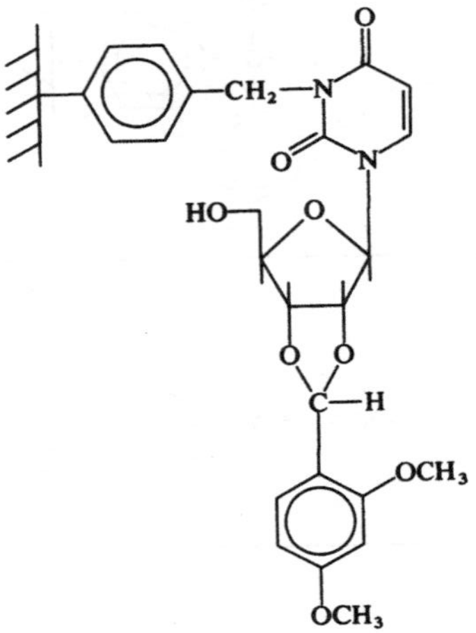

nucleotide-5'-uridinyl-ester

pdT-OAc

1. acid
2. periodate
3. alkali

(390)

Table 4.2 (continued)

No. Support	Functional groups	Type of linkage	Conditions and intermediates for oligonucleotide synthesis	Cleavage conditions	Sequences, yields*	References
6 polystyrene, gel, 2% X	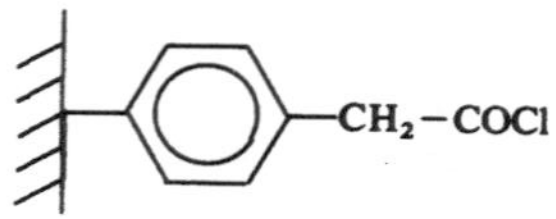 CH_2-COCl	nucleoside-5'-ester	pdT-OAc pdT-OH	dioxane/conc. ammonia 1:1	dT(pdT)$_n$ poly-condensate dTpdT 85%	(220)
7 polystyrene, gel 2% X	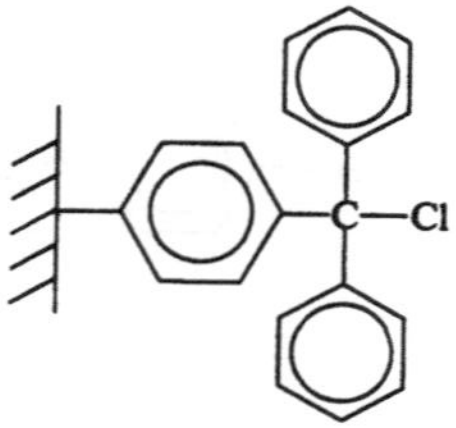 trityl chloride	nucleoside-5'-ether	pdG^{ac}OAc	2% trifluoro-acetic acid in benzene	dT(pdG)$_n$ n = 1–3 dTpdG: 27% dT(pdG)$_3$: ~2%	(476)
8 polystyrene, gel 2% X	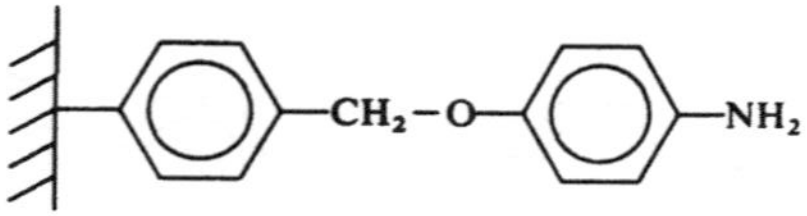 $CH_2-O-\!\!\!-NH_2$	nucleoside-5'-phosphor-amidate	MMTrrU^{bz}-p MMTrC^{bz}-p-OBz	isoamylnitrite in pyridine/acetic acid 1:1	rApUpGp dinucleotides rApUpGp: 10%	(317)

9	polystyrene, macroreticular	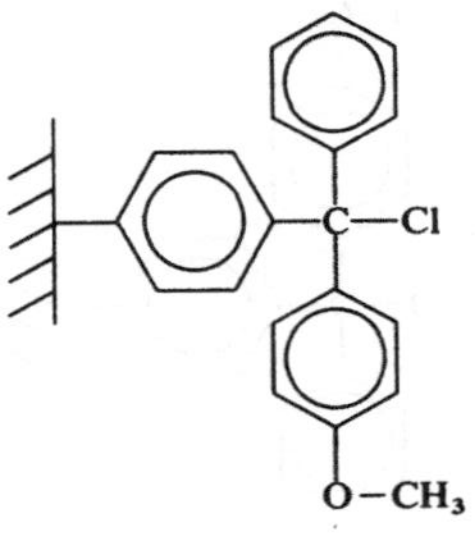	nucleoside-5'-ether	pdT-OAc, pdTpdTOAc, pdC^{an}-OAC, pdA^{bz}-OAc	80% acetic acid	dT(pdT)$_n$ n=1–7 di- and tri-nucleotide sequences d(TTACCTA) dTpdT:50% dT(pdT)$_7$:4% d(TTACCTA): 13%	(68, 215, 217, 218)
10	polystyrene, macroreticular 6%X	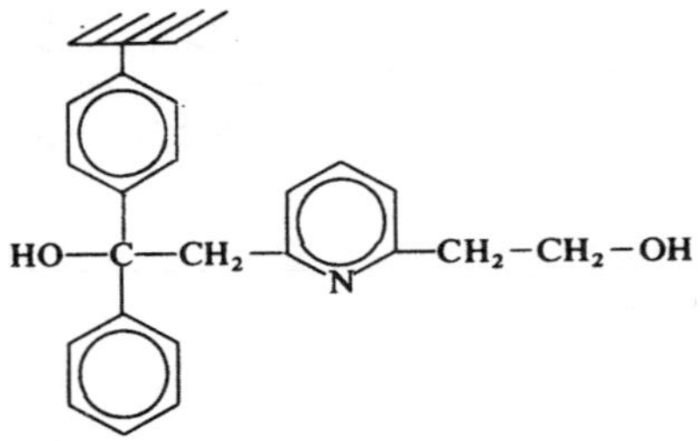	nucleoside-5'-phosphomono-ester	pdT-OAc	2N sodium methylate in methanol/pyridine 1:1	(pdT)$_n$ n=2–6 (pdT)$_2$: 35% (pdT)$_6$~2%	(96)

Table 4.2 (continued)

No.	Support	Functional groups	Type of linkage	Conditions and intermediates for oligonucleotide synthesis	Cleavage conditions	Sequences, yields*	References
11	polystyrene, macroreticular 5% X	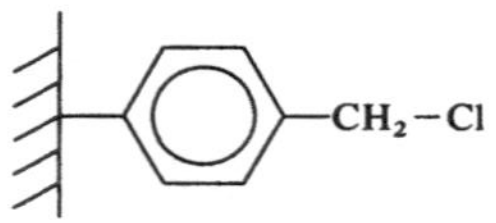	nucleoside-5'-S-benzyl-phosphoro-thioate	pdTOAc, pdC^{an}-OAc, pdA^{bz}-OAc	J_2 in pyridine/water 3:1	(pdT)$_n$ n = 2 – 5 di- and tri-nucleotide sequences (pdT)$_2$: 39% (pdT)$_5$: <1%	(423)
12	polystyrene, macroreticular ca. 3% X	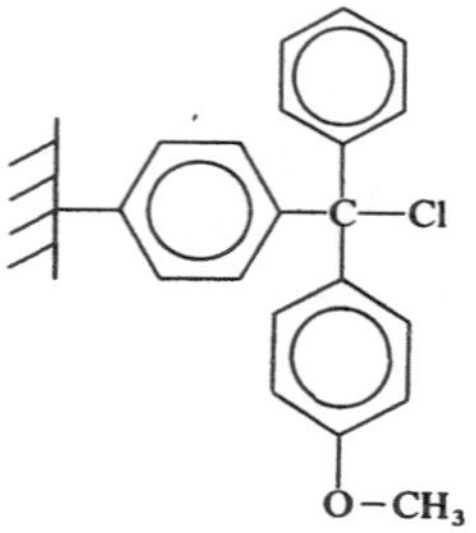	nucleoside-5'-ether	pdTOAc	80% acetic acid	dT(pdT)$_n$ n = 1 – 2 modified dinucleotide dTpdT: 73% dT(pdT)$_2$: 40%	(107)

No.	Support	Structure			Deprotection	Product	Ref.
13	polystyrene, microgel 20% X	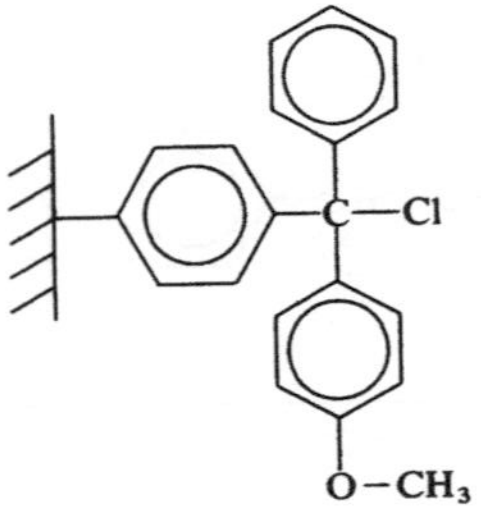	nucleoside-5'-ether	—	80% acetic acid	—	(214)
14	polystyrene, gel 40% X	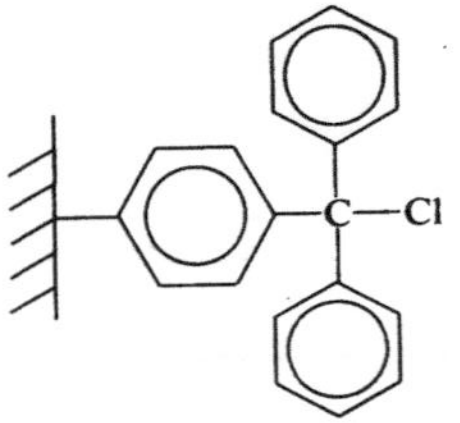	nucleoside-5'-ether	pdA^{bz}-OAc	2% trifluoro-acetic acid in chloroform	dT(pdA)$_n$ n = 1–3 dTpdA: 80% dT(pdA)$_3$: 43%	(174, 239, 342)

Table 4.2 (continued)

No.	Support	Functional groups	Type of linkage	Conditions and intermediates for oligonucleotide synthesis	Cleavage conditions	Sequences, yields*	References
15	polystyrene gel 40% X	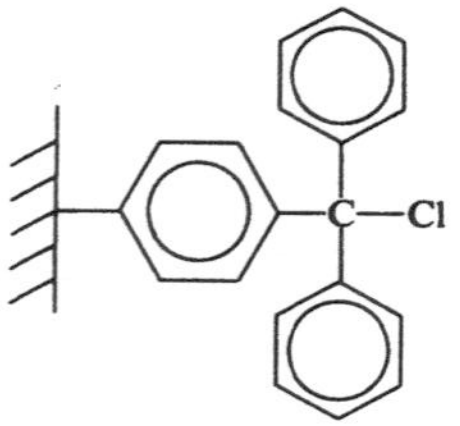	nucleoside-5'-ether	1. PCl_3 2. $HgCl_2$ + nucleotides	2% trifluoro-acetic acid in benzene	$dT(pdT)_n prU$ $n = 1-2$ di- and tri-nucleotide sequences $dTpdT$: 80% $dT(pdT)_2 prU$: 23%	(173)
16	polystyrene, popcorn 0,2% X	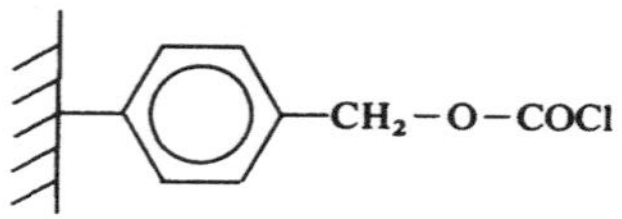	deoxycytidine-N-carbamate	1. β-cyano-ethyl-phosphate, DCC 2. thymidine, MS	0,2 M sodium-hydroxide in dioxane/water 1:1	$dC(pdT)_n$ $n = 1-3$ $dCpdT$: 61% $dC(pdT)_3$: 14%	(226, 227, 228)

| 17 | polystyrene, popcorn 0,1% X | 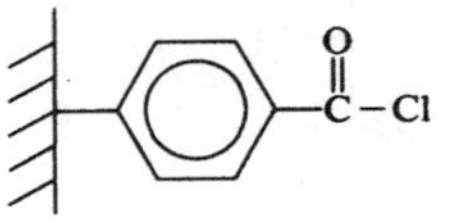 (structure: polymer—C$_6$H$_4$—C(=O)—Cl) | nucleoside-5'-ester | 1. β-cyano-ethyl-phosphate, DCC 2. blocked nucleoside, MS | 0,5 n sodium hydroxide in dioxane/water 1:1 | dT(pdT)$_n$ n = 1–2 dG(pdG)$_n$ n = 1–3 di- and tri-nucleotide sequences d(GGGT) dTpdT: 95% dT(pdT)$_2$: 78% dG(pdG)$_3$: 7% d(GGGT): 18% | (229a, 230, 407, 409) |
| 18 | polystyrene, popcorn | 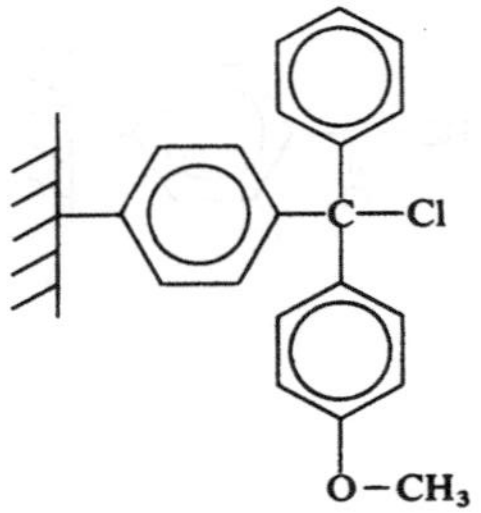 (structure: polymer—C$_6$H$_4$—C(C$_6$H$_5$)(Cl)—C$_6$H$_4$—O—CH$_3$) | nucleoside-5'-ether | pdTOAc, pdTpdTOAc | 80% acetic acid | dT(pdT)$_n$ n = 1–5 dTpdT: 64% dT(pdT)$_5$: 3.5% | (213) |

Table 4.2 (continued)

No.	Support	Functional groups	Type of linkage	Conditions and intermediates for oligonucleotide synthesis	Cleavage conditions	Sequences, yields*	References
19	polystyrene, isotactic	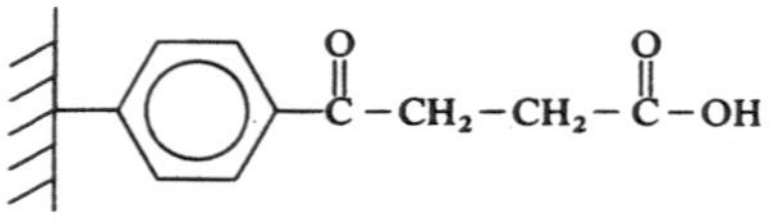	nucleoside-5'-ester	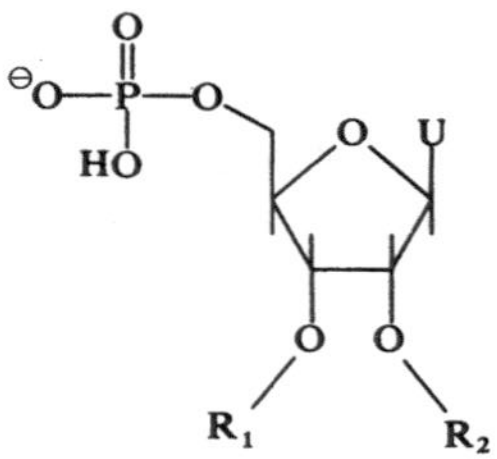 R_1 = methoxy-acetyl R_2 = benzoyl-propionyl	0,5N ammonia	$rU(prU)_n$ n = 1–2 rUprU: 52%	(470)
20	polystyrene, isotactic	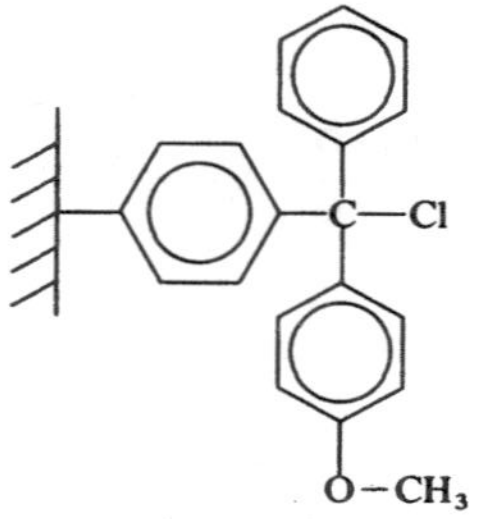	nucleoside-5'-ether	pdTOAc 3'-O-acetyl-5'iododeoxy-uridylic acid	trifluoro-acetic acid in chloroform	$dT(pdT)_n$ n = 1–2 dTpdT: 55% modified dinucleotides	(447)

21	polystyrene, isotactic	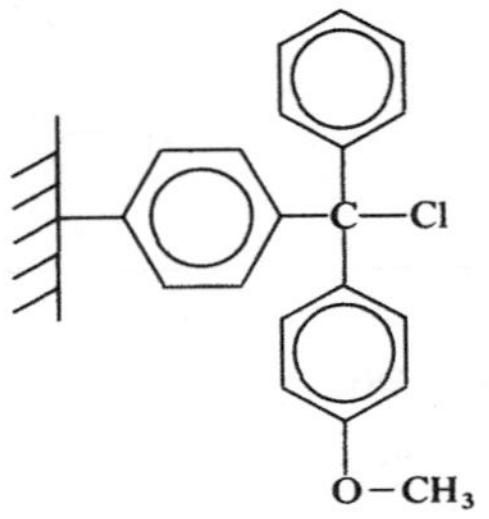	nucleoside-5'-ether	pdA^{bz}-OAc	trifluoroacetic acid in chloroform	dT(pdA)$_n$ n = 1–3 dTpdA: 80% dT(pdA)$_3$: 28%	(342)
22	polyethylene glycol	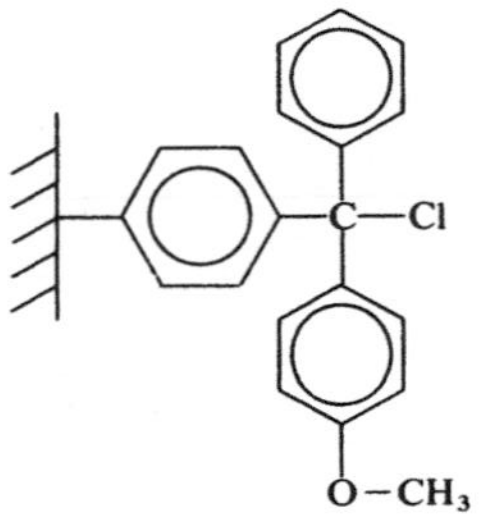	nucleoside-5'-ether	—	—	—	(212)
		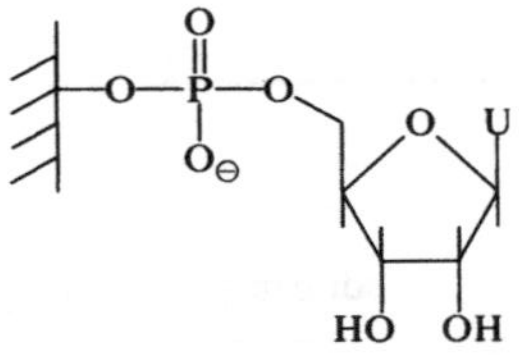	2'(3')-inter-nucleotide linkage	—	—	—	

Table 4.2 (continued)

No.	Support	Functional groups	Type of linkage	Conditions and intermediates for oligonucleotide synthesis	Cleavage conditions	Sequences, yields*	References
23	α,ω-diamino-polyethylene glycol	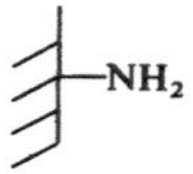 —NH$_2$	phosphor-amidate	pdA^{bz}-OAc	isoamylnitrite in pyridine/acetic acid 1:1	(pdA)$_3$: 14%	(33)
24	polyvinyl-alcohol non-cross-linked	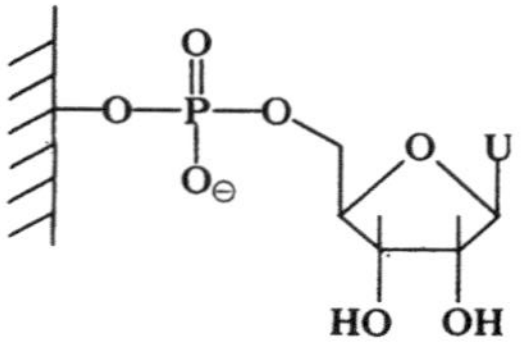	2'(3')-5'-inter-nucleotide linkage	pdTOAc	alkali	(pdT)$_n$ n = 1–5 (pdT)$_2$: 51% (pdT)$_5$: 8%	(33, 378, 382)
25	vinylacetate-N-vinylpyrrol-idone co-polymer non-cross-linked	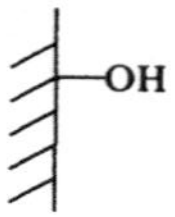 —OH	nucleoside-5'-carbonate	1. pdT, poly-condensation resp. 2. pdT-OMMTr prU(-OAc)$_2$, TPS	conc. ammonia	dT(pdT)$_n$ n = 1–3 dT(pdT)$_m$rU m = 1–2 dTpdT: 56%	(392, 394, 395, 396a)

No.	Support	Structure	Bound group	Reagents	Cleavage conditions	Product/yield	Ref.
26	poly-L-lysine	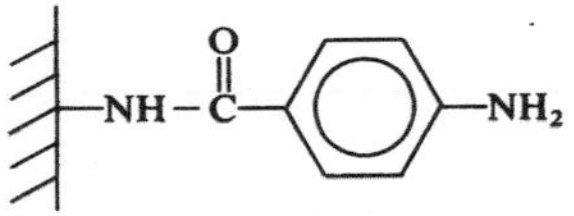	nucleoside-5'-phosphoramidate	pdTOAc	isoamylnitrite in pyridine/acetic acid 1:1	(pdT)$_3$: 14%	*(44)*
27	styrene-acrylic acid copolymer, popcorn	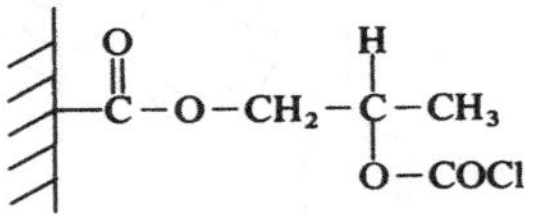	nucleoside-5'-carbonate	1. β-cyano-ethyl-phosphate, MS 2. blocked nucleoside, TPS	0,5 n sodium hydroxide in dioxane/water 1:1	dinucleotide sequences dTpdT: 92%	*(238)*
28	Bio Rex 70 (polyacrylic acid, macro-porous)	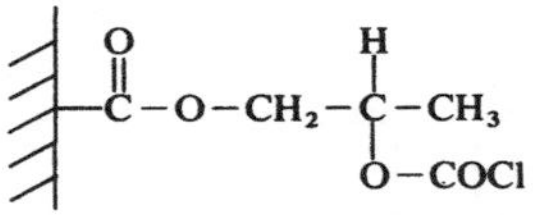	nucleoside-5'-carbonate	1. β-cyano-ethyl-phosphate, MS 2. blocked nucleoside TPS	0,5 n sodium hydroxide in dioxane/ water 1:1	dinucleotide sequences dTpdT: 65%	*(238, 395)*
29	Merckogel-10^6 (polyvinyl-acetate, macroporous)	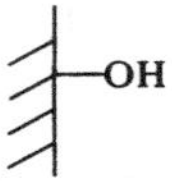	nucleoside-5'-carbonate	1. β-cyano-ethyl-phosphate, MS 2. dT-OβB, TPS	0,5 n sodium hydroxide in dioxane/ water 1:1	dT(pdT)$_n$ n=1–4 dTpdT: 59% dT(pdT)$_4$: 5%	*(238, 395, 397a)*
30	Bio-Beads S-X2 (poly-styrene, macroporous)	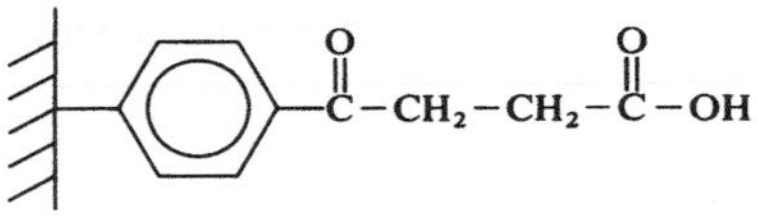	nucleoside-5'-ester	1. MeOPOCl$_2$ 2. dTOAc	0,5 n sodium hydroxide in dioxane/ water 1:1	dT(pdT)$_n$ n=1–2 dTpdT: 38% dT(pdT)$_2$: 10%	*(306)*

Table 4.2 (continued)

No. Support	Functional groups	Type of linkage	Conditions and intermediates for oligonucleotide synthesis	Cleavage conditions	Sequences, yields*	References
31 Sephadex LH 20	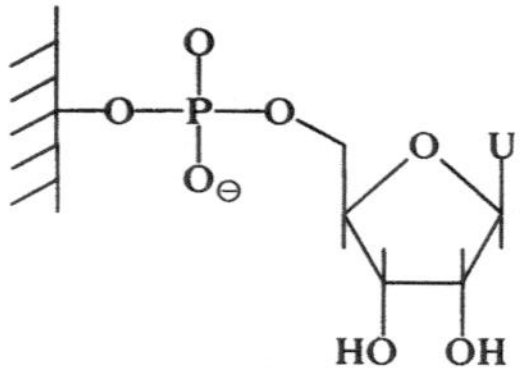	2'(3')-5'-internucleotide linkage	—	0,1 N sodium hydroxide	—	(211)
32 Silicagel	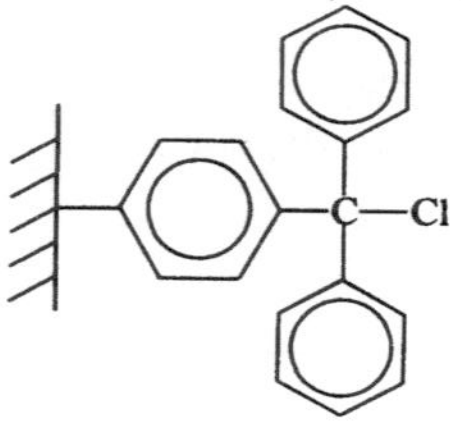	nucleoside-5'-ether	pdTOAc	80% acetic acid	dTpdT 54%	(210)

The majority of systems developed, 21 out of 32, use polystyrene derivatives as carriers. In most of these cases the supports are derivatized so that they possess trityl chloride groups to which nucleosides can be linked as ethers. This acid-labile linkage allows chain extension according to the reaction scheme developed by H. G. KHORANA and coworkers (see Section 1.6). The conditions for acidic removal of the oligonucleotides have to be carefully chosen, since the support reactions are generally slower (see Section 4.1.2) than in the carrier-free case and the risk of decomposition of the oligonucleotides on prolonged exposure to acid is high. Dimethoxytrityl groups have therefore been used in some cases as anchoring groups (147). They allow a more facile acidic cleavage, however, the linkage is so labile that some nucleotide material may be lost during the washing procedures. The yields of internucleotide bond formation have been highest ($>90\%$) with soluble polystyrene derivatives (no. 1—3) (67, 141, 343), where they reach the maximum yields obtained in carrier-free approaches. Lower yields (50—80%) have been reported for all other types of tritylated polystyrene carriers (no. 4, 6, 7, 9, 12, 14, 20, 21) (68, 107, 218, 220, 256, 342, 447, 476). An example of this type of oligonucleotide support synthesis is given in Scheme 4.16.

In this context it should be mentioned that a comparison of the different methods is difficult, even if (as in the above mentioned case), the type of linkage is identical and, in most cases the same product, namely a thymidylic acid dimer, was prepared to test the formation of an internucleotidic linkage. This is due to the complexity of the parameters of support reactions, and, in fact, only few of these support systems seem to have been so extensively studied as to ensure optimization of most factors.

The difficulties encountered in using an acid-labile support linkage have prompted investigations of other types of anchors, which would allow cleavage of the nucleotidic material from the support in alkaline or neutral media. Alkali-labile linkages are formed on conversion of polystyrene resins to macromolecular acids (no. 19 and 30) (306, 470), acid chlorides (6, 17) (220, 229a) and chloroformates (16) (227, 228). This approach, developed first in the laboratory of R. L. LETSINGER, forbids the use of the alkali-labile acetyl groups for the protection of the growing chain end (i.e. in most cases of the 3'-terminus). Development of new blocking groups, such as the β-benzoylpropionyl group, and the application of the triester method made this approach workable and allowed the preparation of small oligonucleotide fragments in excellent, often near quantitative yields (230). The route for the preparation of a trinucleoside diphosphate is shown in Scheme 4.17.

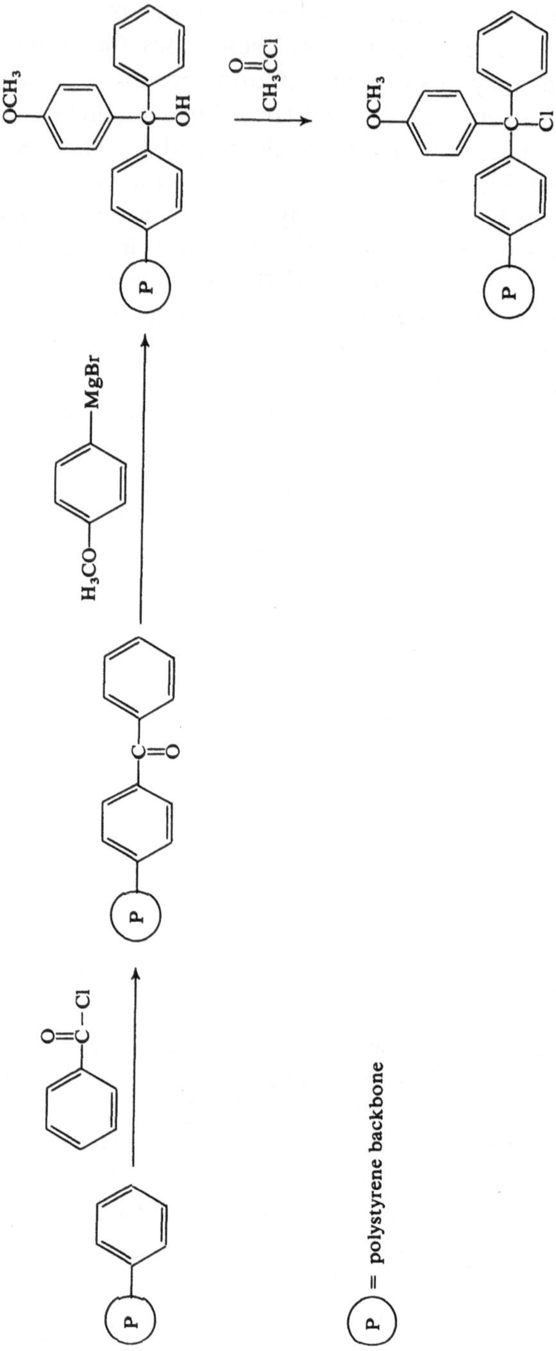

P
O=C–Cl
P
C=O
H₃CO
MgBr
P
OCH₃
C
OH
O
CH₃CCl
OCH₃
C
Cl
P
P = polystyrene backbone

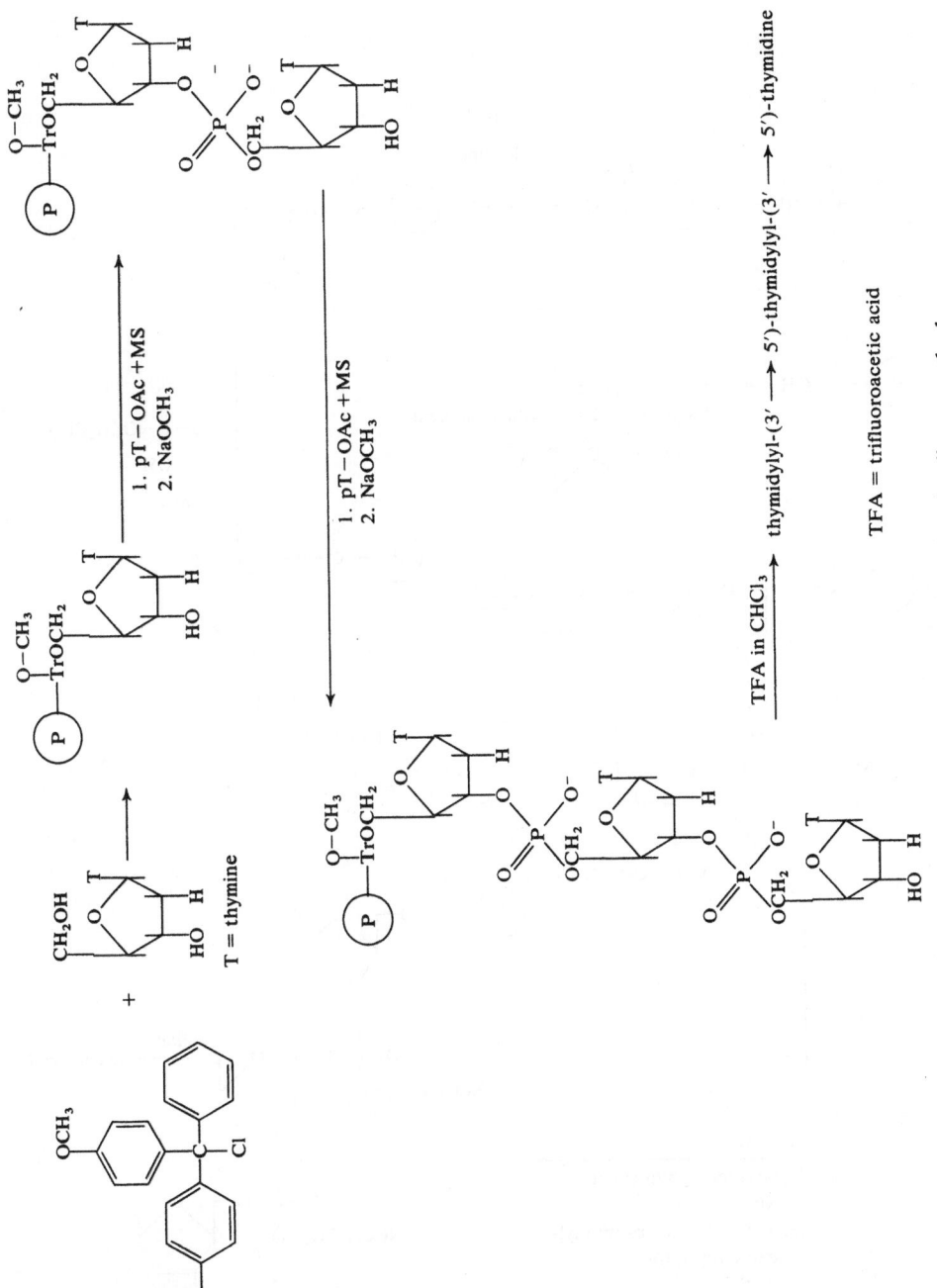

Scheme 4.16. Oligonucleotide support synthesis using the diester method

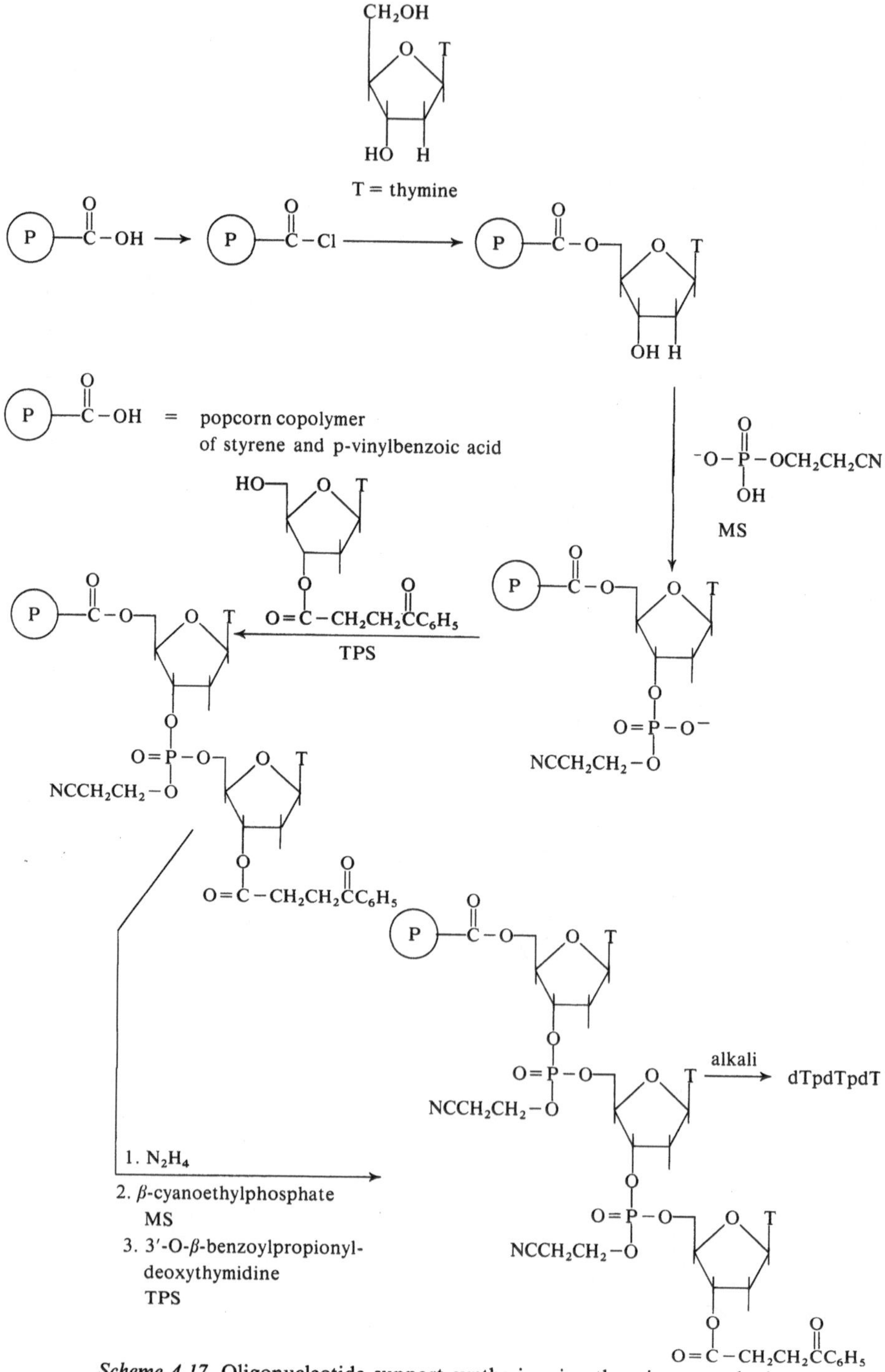

T = thymine

P —C—OH = popcorn copolymer of styrene and p-vinylbenzoic acid

MS

TPS

alkali ⟶ dTpdTpdT

1. N_2H_4
2. β-cyanoethylphosphate MS
3. 3'-O-β-benzoylpropionyl-deoxythymidine TPS

Scheme 4.17. Oligonucleotide support synthesis using the triester method

The alternative possibility, namely the attachment of an oligo-nucleotide through a bond which can be cleaved in neutral medium by a selective reagent, has been developed primarily in the laboratory of F. CRAMER. This alternative is especially attractive, because a selectively labile linkage to the support can be compared to a lock-and-key system and allows the use of a variety of other blocking groups during the elongation of the oligonucleotide chain. Of course, the basic requirement is the availability of suitable selectively labile blocking groups and the possibility to introduce these as anchoring groups into polymers. The introduction of the uridinyl group as anchor into chloro-methylated polystyrene (MERRIFIELD resin) was the first approach of this kind (390) (uridinyl groups as phosphate protecting groups: see Section 1.3.1). Better results for oligonucleotide synthesis were obtained with polystyrene resins using phosphoranilidate (no. 8) (317), α-pyridyl-ethylphosphate (no. 10) (96) and S-benzylphosphorothioate (no. 11) (423) anchors; however, the yields of internucleotide bond formation were only moderate. Phosphoramidate bridges have also been used to attach nucleotides to poly-L-lysine (no. 26) (44) and α,ω-diamino-poly-ethylene glycol (no. 23) (33).

This leads us to another interesting line of development. For some time doubts had arisen as to whether polystyrene resins are really, good carriers for oligonucleotide syntheses at all, as polystyrene is most highly solvated in less polar media, whereas the steps of oligo-nucleotide synthesis are carried out either in water or polar organic media. In order to construct more hydrophilic support resins H. SELIGER and more recently K. K. OGILVIE used either popcorn copolymers of styrene and acrylic acid or commercial hydrophilic macroporous resins as the basis for carriers (238, 306, 395, 397a). Alkali-labile ester or carbonate linkages were used to attach the initial nucleoside and chain elongation was carried out according to Scheme 4.17. Uridylic acid has been used as connection in some instances to procure alkali-labile attachment (no. 22, 24, 31) (211, 212, 378, 382). The yields of internucleotide bond formation were, again, nearly quantitative in the case of the popcorn support (no. 27) (238), whereas the maximum yields obtained with macroporous resins (no. 28, 29, 30) (238, 306, 395) and also with silicagel as porous inorganic support (no. 32) (210) were lower and again in the same range as those obtained previously with macroporous polystyrene. This suggests that the outcome of oligonucleotide reactions is influenced principally by the selection of a certain support structure and only to a lesser extent by better solvation, thus, underlining the importance of steric factors in oligonucleotide support synthesis.

Parallel to the development of more polar insoluble supports non-crosslinked hydrophilic polymers have been modified and tested. These include polyethylene glycol (no. 22, 23) (*33, 212*), polyvinylalcohol or saponified vinylacetate-N-vinylpyrrolidone copolymer (no. 24, 25) (*33, 382, 392*). These approaches differ from the previously described methods using soluble polystyrene (no. 1 and 2) in that the supports are generally soluble in organic media *and* water, and dialysis methods are used for the separation of mobile reactants (for this reason, the approach, has been named "liquid-phase synthesis" (*33*)). An example (*392, 396a*) is illustrated in Scheme 4.18. The yields of oligonucleotides obtained with these systems are still significantly lower than those reported for system no. 1 and also some adsorption difficulties still have to be overcome. Nevertheless, this new technique, which has been successfully applied also in the polypeptide field, should merit further investigation.

Although interest in support synthesis has been focused on the preparation of deoxyoligonucleotides, work in the ribo series has

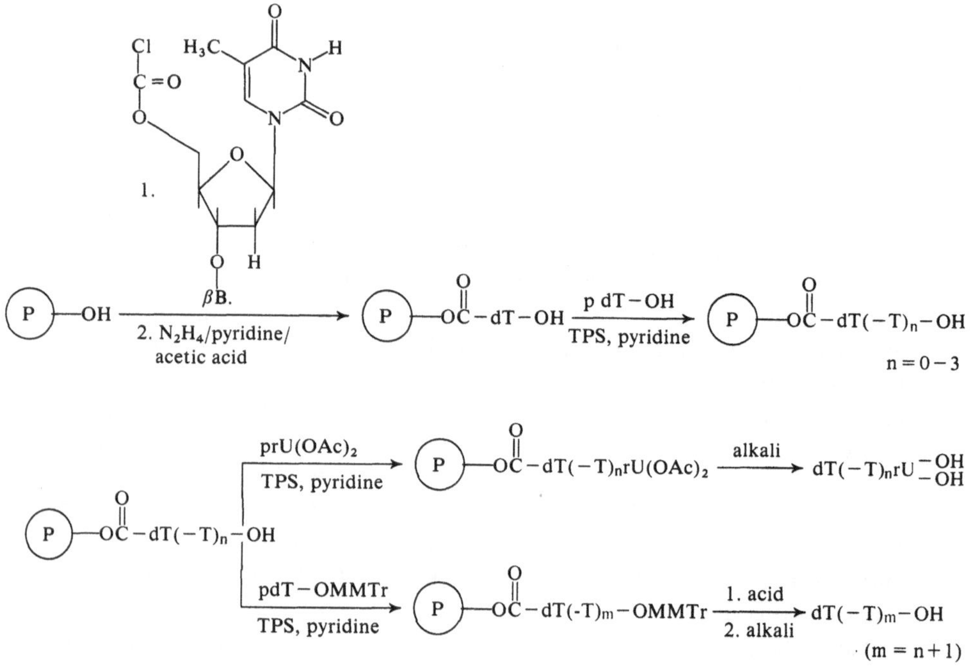

Abbreviations:

$\boxed{P}$ = vinylalcohol−N−vinylpyrrolidone copolymer

βB. = β-benzoylpropionyl-
Ac = acetyl-
MMTr = p-methoxytrityl-

Scheme 4.18. Oligonucleotide synthesis on a soluble support

occasionally been described. Thus K. F. Yɪᴘ and K. C. Tsou (470) have prepared a trimer of uridylic acid, and the sequence rAUG was made by E. Oʜᴛsᴜᴋᴀ et al. (317) using a phosphoranilidate attachment to polystyrene. The carrier method can also serve for the preparation of a modified oligonucleotide, as was demonstrated by the synthesis of two dinucleotides containing 5-fluorouridine (447) and α-pyridone nucleoside (107).

Looking at the results of oligonucleotide support synthesis tabulated in column 7 of Table 4.2, one cannot avoid the statement that, for all the time and effort spent on the development of this technique, the outcome has been relatively meager. Good, even excellent results have been obtained in the synthesis of short oligonucleotide fragments, but here in the meantime the support method has to compete with other effective techniques, such as those using extraction methods or affinity chromatography (see Sections 1.4, 3.1.2, 3.2). For the synthesis of longer oligonucleotide chains, because of the low yields of chain extension, it is not feasible to go beyond the range of good chromatographic separability, if one wants to end up with pure compounds. Few support syntheses of longer oligonucleotide chains containing more than one base have been described; among these the blockwise preparation in good yield of a heptanucleotide sequence by H. Kösᴛᴇʀ, A. Pᴏʟʟᴀᴋ and F. Cʀᴀᴍᴇʀ (218) may show that the preparation of fragments for biologically active polynucleotides can, in fact, be done by support techniques. However, unless this technique is significantly improved, the advantage of time saved by avoiding the column chromatography of intermediates is partly compensated by prolonged reaction times, lower yields and relatively difficult separation of the mixture of products released from the carrier.

4.2.4. Enzymatic Synthesis of Oligo- and Polynucleotides on Supports

The attachment of oligo- and polynucleotides to carriers allows their use as support-bound primers and templates in enzymatic reactions. Generally cellulose is used as a support in these cases, since it is highly polar, insoluble and does not interact strongly with polynucleotides. Residual hydroxyl groups do not disturb as long as there is no chemical internucleotide bond formation. The binding of primers and templates to matrices greatly facilitates the removal of the enzymes and excess substrates from the reaction mixture. In the case of cellulose matrices the bound oligonucleotides react as if free in solution. DNA and RNA polymerase, terminal nucleotidyl transferase and polynucleotide ligase were used for chain extension and joining of polynucleotides. Poly-deoxyadenylate, base paired to poly-dT-cellulose, could be elongated to give support-bound poly dA–dT (*170*). This is schematically represented in equation 4.19.

Cellulose – TTTTTTTTTTTT $\xrightarrow[\text{dTTP, dATP}]{\text{DNA-polymerase}}$ Cellulose – TTTTTTTTTTTTTtttttt
          AAAAAAA   aaaaaAAAAAAA

Scheme 4.19

Similarly, a natural DNA was elongated by a homopolymer piece by terminal deoxynucleotidyl transferase and base-paired to a complementary oligomer bound to cellulose. DNA polymerase then copied the natural DNA to give a support-bound template for further replications. Poly dC could be joined to poly dC, oligo dT-cellulose in the presence of poly dI by polynucleotide ligase (*60*). With equally good results ficoll (*367b*) or polyvinyl alcohol (*361a*) were employed as carriers in the elongation of support-bound oligonucleotides by terminal deoxynucleotidyl transferase; these chains could subsequently be used as templates for polynucleotide synthesis with DNA polymerase I. Guanosine-2′,3′-cyclophosphate, inserted into a polyacrylamide resin, could be condensed with uridine in a reaction catalyzed by guanyl-RNase (*196*). Compared to the alternative approach of enzyme fixation to carriers the use of support-bound primers and templates is still in the beginning; it may, however, become very valuable as soon as the techniques for enzymatic synthesis of specific sequences (see Section 5) are further developed.

4.2.5. Miscellaneous Uses of Supports in Nucleotide Chemistry

In the last part of this section we will discuss some uses of polymer supports pertaining to oligonucleotide synthesis which do not, however, involve internucleotide bond formation with a support-bound nucleic acid constituent.

Phosphorylations of nucleosides bound to supports have already been described as part of oligonucleotide syntheses according to the triester method (see Scheme 4.17). Similarly, conversion of nucleosides to nucleotides in good yield have been described by M. M. Kabachnik *et al.* using tritylchloride carriers and either halides of phosphoric or pyrophosphoric acid or $PCl_3/HgCl_2$ as phosphorylating agents (*171*, *172*).

G. M. Blackburn and coworkers (*22*, *23*, *24*) have used supports for clarifying the mechanism of phosphorylation by dicyclohexyl-carbodiimide and sulfonylchlorides. In both cases polymeric phosphomonoesters were not converted to phosphodiesters with aliphatic alcohols or nucleosides. It was concluded that trimeric or polymeric phosphate esters must be necessary intermediates in phosphorylations or internucleotide bond formation by the above mentioned reagents. Such intermediates would not be formed readily, if the phosphate component were fixed to a polymer matrix. Other findings, e.g. the near quantitative internucleotide bond formation with support bound nucleotides in the triester synthesis (Scheme 4.17) (*230, 238*) are not along this line, although these systems differ from the ones used in the mechanistic investigation in that phosphodiesters were the starting compounds. A more thorough discussion of phosphorylation mechanisms has been given in Section 2.

In all cases discussed so far, a support-bound nucleoside or nucleotide served as a partner in internucleotide bond formation. Alternatively, the condensing agent can be bound to a polymer. Nucleoside and nucleotide component are then incubated with the polymeric condensing reagent and, after a suitable reaction time, the oligonucleotide is filtered off. This principle has been exploited by M. RUBINSTEIN and A. PATCHORNIK (*362*) using poly-3,5-diethylstyrene sulfonylchloride as a macromolecular analog of mesitylenesulfonylchloride. 70—90% of dTpdT could be recovered in a test reaction using alternatively the diester or triester version of internucleotide bond formation. Although separation problems are only partly simplified, the advantage of this approach could lie in an approach to quantitative yields of internucleotide bond formation on further amelioration of the conditions and on use of more sophisticated polymeric reagents.

Other uses of support bound nucleic acids and constituents in nucleic acid chemistry, e.g. in affinity-chromatographic separations, will be discussed in a forthcoming review (*397*).

4.3. Miscellaneous Methods in Chemical Oligonucleotide Synthesis

In a search for other condensation methods, experiments have been undertaken which are aimed at the following goals: first, to activate the hydroxyl residue involved in phosphodiester formation; secondly, to use preactivated phosphomonoester derivatives, thirdly to use unprotected or less protected nucleotide derivatives and finally to achieve non-enzymic synthesis on complementary templates.

4.3.1. Chemical Synthesis via Activation of Hydroxyl Functions

Starting from the earlier observation that nucleoside 3′,5′-cyclophosphates can be synthesized from the corresponding 3′- or 5′-*p*-nitrophenylesters in the presence of anhydrous strong base a new synthetic route for stepwise synthesis of oligodeoxyribonucleotides has been proposed (440). As outlined in Scheme 4.20, the reaction of suitably

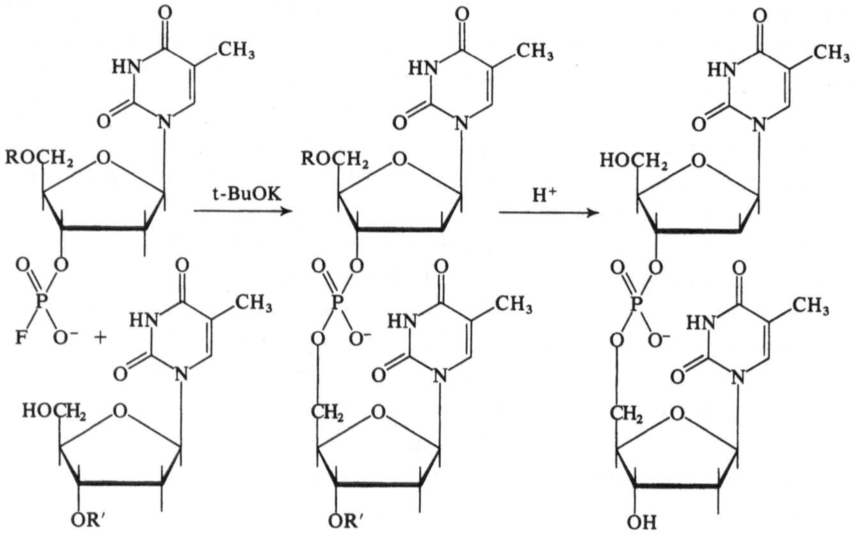

Scheme 4.20. Formation of an internucleotide linkage by activation of hydroxyl groups

protected nucleosides and nucleotides is carried out with anhydrous potassium tertiary butoxide as the base and fluoride as the leaving group on the phosphate. A stepwise synthesis is possible with the use of protecting groups of different stability in acid. The internucleotide bond was also formed when a secondary 3'-hydroxyl group of the one component attacked a 5'-phosphorofluoridate of the second condensation component. In the oligothymidylic acid series this fast reaction (condensation times range between 15 and 30 minutes) has allowed the stepwise synthesis of the corresponding dinucleoside monophosphate and tri-nucleoside diphosphate in excellent yields. As demonstrated by the synthesis of d–ApT in 50% yield this reaction principle can also be extended beyond the oligothymidylic acid series although synthesis of oligonucleotides containing C or G residues has not been reported so far. It is noteworthy that in the case of the d–ApT synthesis protection of the base residues was not necessary and that cleavage of the N-glycosidic linkage in anhydrous base apparently does not constitute a serious problem. Further work seems, however, necessary in order to extend this approach to the synthesis of longer chains as in the synthesis of TpTpTpT only a 19% yield was observed owing to the insolubility of the potassium salt of TpTpT–MMTr in dimethylformamide (440).

4.3.2. Chemical Synthesis via Preactivated Phosphate Derivatives

Preactivation of phosphomonoester residues can be achieved by amidation with p-hydroxyaniline (316). As outlined in Scheme 4.21,

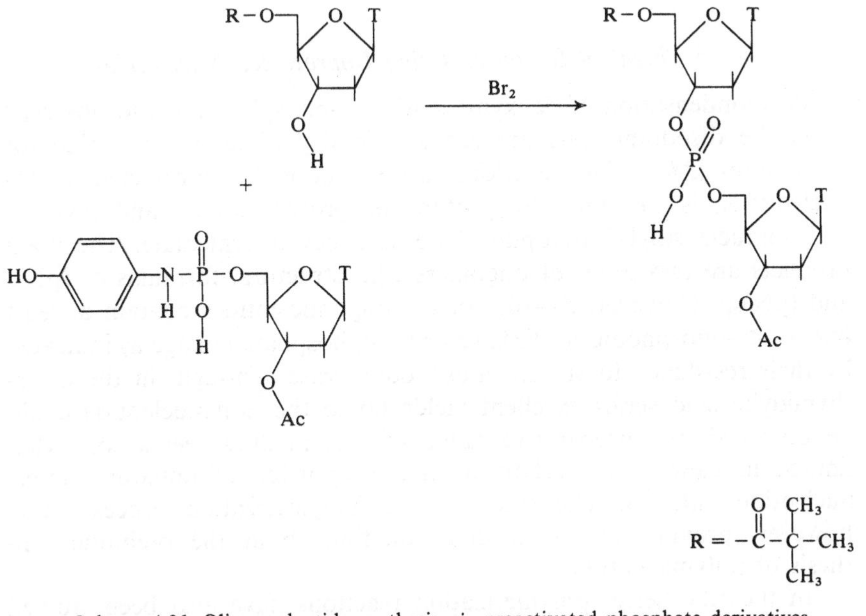

Scheme 4.21. Oligonucleotide synthesis via preactivated phosphate derivatives

in situ activation of 3'-O-acetyl-thymidine 5'-phosphoro-p-hydroxyanilidate by oxidation with bromine leads to the formation of a protected dinucleoside monophosphate when 5'-trimethylacetyl thymidine is offered as hydroxyl component. This principle could be extended to the activation of 3'-monophosphate residues and (in the ribo series) to N-benzoyl-cytosine containing nucleotide derivatives. Since the yields ranged from 24 to 46% for internucleotide bond formation further efforts which would also include attempts at synthesis of longer chains or synthesis of protected adenosine and guanosine derivatives seems desirable.

Polycondensation mediated by preactivated phosphomonoesters has been reported in the oligothymidylic acid series. Thus, activation of thymidine 5'-S-ethyl phosphorothioate (56) with iodine in pyridine in the absence of any external nucleophile produces extensive self-condensation by attack of the 3'-hydroxyl group on the iodine-activated phosphorothioate. After the polycondensation, conversion of the 5'-terminal S-ethyl-phosphorothioate groups to phosphate groups could be achieved by addition of water and the level of pyrophosphate linkages could be reduced by subsequent acetic anhydride treatment. Oligonucleotides up to the nonanucleotide could be isolated in excellent yields. Although this approach has been reported only for the oligothymidylic acid series so far, it seems reasonable to assume that it can be applied successfully also to other series or to block polymerization.

4.3.3. Chemical Synthesis Using Unprotected Nucleotides

Polycondensation of deoxynucleoside 5'-phosphates is also observed when the disodium salts are refluxed in dry dimethylformamide for 30 minutes (340). This reaction, carried out with unprotected mononucleotides, is catalyzed by protons or proton donors and involves P^1,P^2-dinucleosidyl-5'-pyrophosphate as a key intermediate. The main products are two series of oligomers with structural formulas of $(pN)_n$ and $(pN)_np$. However, 5—10% of the oligonucleotides contain at least one 5'—5'-phosphodiester linkage or pyrophosphate linkage as indicated by their resistance to spleen phosphodiesterase. Though, in the oligothymidylic acid series excellent yields up to the nonanucleotide could be observed, the preparative value of this method seems somewhat limited in view of the relatively high proportion of unnatural internucleotide linkages. The study of this polymerization process may, however, provide additional understanding about the prebiotic synthesis of polynucleotides.

In the ribo series polymerization reactions have also been studied under prebiotic conditions. Thus, when excess uridine is heated in the presence of dihydrogen phosphate and urea, di- and oligonucleotides are formed in about 33% yield (323). The majority of the internucleotide bonds formed were 3'—5'-linked; however, large numbers of 2'—5'-bonds and some 5'—5'-bonds were also formed. When adenosine-cyclic 2',3'-phosphate is reacted in the dry state in the presence of aliphatic diamines at moderately elevated temperatures, self-polymerization to give oligonucleotides of chain length up to 6 and higher is observed (455, 456). Here too, the products contain excess of 3'—5'-linkages over 2'—5'-linkages. While the preparative value of these methods seem to be somewhat limited in view of the relatively high frequency of unnatural internucleotide bonds, the study of this polymerization processes may also provide insights into the prebiotic synthesis of polynucleotides.

During conventional diester or triester synthesis the amino functions of the base residues are blocked by acylation. The use of derivatives containing free amino functions seems, however, possible in the case of guanine and adenine mononucleotide blocks, at least for the synthesis of dinucleotides in the deoxy series (*292, 465*) and in the ribo series (*265*). Whether this simplified approach is also feasible for the synthesis of longer chains remains to be shown. As mentioned previously, N-protecting groups are also unnecessary when condensation reactions are carried out in the presence of strong bases by which the hydroxyl functions are activated (Scheme 4.20) (*440*). In the triester approach an activated 3'-phosphodiester of one component seems to react exclusively with the free 5'-hydroxyl of the second component, even if the 3'-hydroxyl function of the latter is also unblocked (Schemes 4.12 and 4.14) (*31, 296, 300*).

4.3.4. Chemical Synthesis on Complementary Templates

This approach tries to mimic enzymic reactions catalyzed by template dependent polymerizing enzymes or by polynucleotide ligases. Accordingly, attempts have been made to combine mononucleotides or oligonucleotides after fixation in mutual vicinity by complex formation on a complementary template. Since the feasibility of this approach was first verified by the synthesis of undecathymidylic acid from two hexathymidylic acid blocks on a poly A template in the presence of a water soluble carbodiimide (*294*), several other groups have tried to improve this technique for synthetic purposes or to investigate its relevance to prebiotic polynucleotide synthesis. Starting with mononucleotide units in the presence of water soluble carbodiimides as activating agents, poly U and poly C have been tested as templates for the synthesis of oligo A, oligo dA and oligo G respectively. Besides some tri- and tetranucleotides, the main products isolated from the rather complex reaction mixtures consisted of the respective dinucleotides or dinucleoside monophosphates (*430, 431, 432*). Similar results were obtained when adenosine cyclic 2',3'-phosphate (*355*) or adenosine-5'-monophosphorimidazolide (*376, 459*) were used as preactivated nucleotide derivatives in the presence of poly U. These reactions seem to be highly specific in respect to base selection by the template, as incorporation of nucleotides not complementary to the template generally is much lower than incorporation of complementary nucleotides. The synthetic value of this monomer approach, however, is severely limited by the observation that – beside other side reactions – the unnatural 2'—5'- and 5'—5'-linkages are frequently formed in preference to the natural 3'—5'-linkages.

When trideoxyadenylic acid is reacted with water soluble carbodiimide in the presence of poly U as template at 20 mM $MgCl_2$ an overall yield of 35% of the polymeric products $d(pApApA)_n$ ($n = 2, 3, 4$) is observed (*12*). It seems that this rather high yield is due to the presence of Mg^{++} which increases the stability of the $[d(pA)_3 \cdot poly U]$ hybrid. When dideoxyadenylic acid or trideoxyadenylic acid, activated by 3'-terminal phosphoamidates, are reacted in the presence of poly U, formation of oligo and poly dA is observed (*295, 403*) in about 10% yield. As evident from degradation with spleen and venom phosphodiesterase, the tetra- and hexaadenylic acids obtained contain natural $5' \rightarrow 3'$-linkages exclusively (*295*).

In order to study the arrangement and reactivity of oligonucleotides on complementary templates, a number of G-containing p-nitrophenyl-oligonucleotide succinates (*408, 410*) have been prepared (Scheme 4.22) and tested for hydrolysis of the p-nitrophenyl residue in the presence of poly C and G-containing oligonucleotide N-acetylhistidates. From the

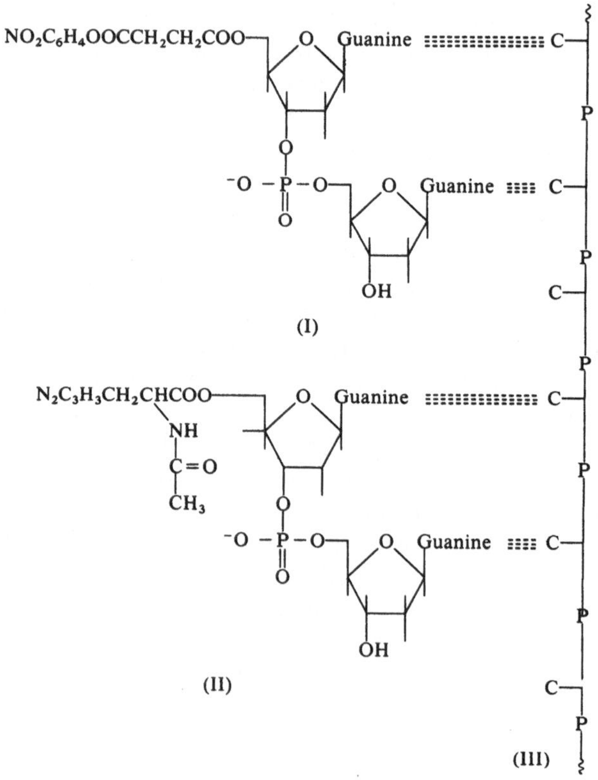

Scheme 4.22. Schematic diagram showing the hydrolysis of I by II on III

hydrolysis rates observed the strength of interaction between the oligo-nucleotide derivatives and poly C was observed to decrease in the order dGpGpG > dGpGp > dGpC > dGpA > dGpA > dGpT. These obser-vations together with the specificities reported in the monomer approach (see above) suggest, that the prerequisite of proper alignement of the nucleotide blocks on the respective template can be fulfilled even with comparatively short chains (or even with monomers). It seems therefore not unlikely that template-directed synthesis especially with oligonucleotide blocks finally will become practical, although further work is necessary in order to improve the yields and/or to guarantee the formation of natural internucleotide bonds. In regard to the synthesis of specific base sequences, one more general drawback of template-directed synthesis (chemical or enzymic), namely the necessity of having proper templates and blocks available, should be kept in mind. Thus, at least for the blocks to be connected and for the chains serving as templates conventional methods of synthesis seem unavoidable.

5. Formation of Internucleotide Linkages by Enzymic Reactions

Three main classes of enzymes have been exploited for synthetic reac-tions in polynucleotide chemistry. The first class, the polymerizing enzymes, can be subdivided into primer dependent and primer-template dependent enzyme species. Polynucleotide phosphorylase appears as the most prominent representative among the primer dependent enzymes, as numerous ribopolymers have been prepared by this enzyme already for the elucidation of the genetic code (193, 194, 197, 439). DNA dependent DNA polymerase I and DNA dependent RNA polymerase from E. coli represent the most frequently used primer-template dependent species and their application was essential, for example, for the pre-paration of polymers containing repeating di-, tri- and tetranucleotide (185, 186, 187, 197, 461). While polymerizing enzymes have long been established as valuable synthetic tools, the second class of enzymes represented by polynucleotide ligases has been introduced only more recently. Its use in connecting chemically synthesized segments has recently culminated in the total synthesis of two tRNA genes (2, 188). Finally ribonucleases, though generally regarded as cleaving agents, have been introduced as a third class of synthetic enzymes, especially for the synthesis of short oligoribonucleotides. This has been made possible by causing reversal of the cleaving reactions by adding a large excess of one of the cleavage products whereby the equilibrium is driven towards the side of internucleotide bond formation.

One general advantage of enzyme-catalyzed synthetic reactions consists in the specificity guaranteed by the respective enzymes. For this reason, blocking groups necessary for the protection of the various functional groups in organic chemical reactions are commonly not needed in enzymic reactions. On the other hand, only comparatively small quantities of synthetic material (often in the range of less than 1 mg) are accessible by enzymic reactions, unless huge amounts of enzymes which usually are costly or time consuming to prepare are applied. Only ribonuclease-catalyzed reactions could be performed on larger scale. While this limitation may be severe for physicochemical measurements, for which several milligrams and more are frequently necessary, studies in the biochemistry of polynucleotides or in molecular genetics quite often have been performed with quantities far below the milligram level of the respective polynucleotides (*39, 130, 206, 366, 467*).

5.1. Reactions Catalyzed by Polymerizing Enzymes

Reactions catalyzed by polymerizing enzymes can be subdivided into two classes. Enzymes of the primer dependent class add activated nucleotide units to the 3'-ends of short oligonucleotide primers to yield homopolymers or random copolymers according to Scheme 5.1. In

$$N_1pN_2pN_3 + \begin{array}{c} ppM \\ or \\ pppM \end{array} \longrightarrow N_1pN_2pN_3(pM)_n + \begin{array}{c} np \\ or \\ npp \end{array}$$
Primer

Scheme 5.1

the ribo series polynucleotide phosphorylase (*175, 193, 194, 197, 439*) (see also references given in Table 5.1) has been used extensively for this type of reaction, in which case ribonucleoside-5'-diphosphates serve as activated nucleotide units together with a dinucleoside monophosphate or longer ribooligonucleotides as primer. Inorganic phosphate is liberated at each step and in the presence of high concentrations of inorganic phosphate the reaction can be reversed towards phosphorolysis of ribopolynucleotides. In the deoxy series the enzyme most frequently used for the primer dependent synthesis of homopolymers or random copolymers is terminal deoxynucleotidyl transferase (*30, 461*), which utilizes deoxyribonucleoside triphosphates as substrates for the polymerization onto the 3'-end of a deoxytrinucleoside diphosphate as minimum primer. One equivalent of pyrophosphate is liberated for each nucleotide added.

In contrast to these primer dependent species in which the addition onto the primer is governed mainly by the availability of the respective di- or triphosphates in the reaction medium, a second class of polymerizing enzymes is dependent on the presence of a template/primer. DNA-dependent DNA polymerases and DNA-dependent RNA polymerases are well-known representatives of this class by which amount *and* sequence of the nucleotide units incorporated are governed by a DNA template. RNA-dependent RNA polymerases (as, for example, replicase induced by the phage Qβ) (*89, 327*) have also long been known and more recently RNA dependent DNA polymerases, so called reverse transcriptases, have been discovered (*102*). Ribo- or deoxyribo-nucleoside triphosphates are the substrates for these enzymes. While a template (DNA or RNA) is compulsory for this class of enzymes, initiation of the polymerizing reaction without a primer is sometimes possible as in the cases of Qβ replicase (*89, 327*) or of DNA-dependent RNA polymerases (*186, 187, 206, 279*). Using Qβ replicase (*89, 327, 377*) or a combination of DNA-dependent DNA polymerase and polynucleotide ligase (*111*) *in vitro* synthesis of the total genoms of phage Qβ or φX174 respectively could be achieved.

Synthetic homopolymers, random copolymers and polymers containing repeating di-, tri- and tetranucleotides obtained by application of the various polymerizing enzymes have been reviewed in detail; these reviews include work published during the past 5 years (*30, 186, 187, 197, 461*). Only a list of polymers containing unusual nucleotides, base pairs or internucleotide linkage will therefore be given in this article (Table 5.1) as a number of polymers has been synthesized containing various modifications in the base, sugar or phosphate moieties.

One general limitation of template dependent polymerizing enzymes arises from the fact that base sequences of the products are entirely governed by the respective templates and that therefore synthesis of specific sequences other than the ones complementary to the template cannot be achieved. Although this limitation converts to a true advantage, where mere copying of already existing templates is desired (as for example in the reported *in vitro* synthesis of the whole genomes of phage Qβ or φX174), enzymic methods which would allow single step additions to a given primer, appear more attractive from a synthetic view point. For this purpose template independent enzymes such as terminal deoxynucleotidyl transferase and polynucleotide phosphorylase represent more favourable candidates, as probably any sequence (not only sequences programmed by templates) could be synthesized once a stepwise procedure is developed.

Table 5.1. *Enzymically Prepared Oligo- and Polynucleotides Containing Unusual Bases, Unusual Base Pairs, Modified Nucleosides or Ribo-Deoxy-Internucleotide Linkages*

Polymer	Polymerizing enzyme	References
$(dI\text{-}dC)_n$	DNA dependent DNA polymerase I (E. coli)	*(114, 115)*
$(dG\text{-}dT)_n$	DNA dependent DNA polymerase I (E. coli)	*(244)*
$(dA\text{-}d4thioT)_n$	DNA dependent DNA polymerase I (E. coli)	*(81, 240, 242, 243)*
$(dA\text{-}d2thioT)_n$	DNA dependent DNA polymerase I (E. coli)	*(241)*
$(dA\text{-}dC)_n \cdot (dT\text{-}d6thioG)_n$	DNA dependent DNA polymerase I (E. coli)	*(17, 242)*
$(dA\text{-}d6thioG)_n \cdot (dT\text{-}dC)_n$	DNA dependent DNA polymerase I (E. coli)	*(17, 242)*
$(dA\text{-}dC)_n \cdot (dT\text{-}d6thioI)_n$	DNA dependent DNA polymerase I (E. coli)	*(17, 242)*
$(dA\text{-}dC)_n \cdot (d4thioT\text{-}dG)_n$	DNA dependent DNA polymerase I (E. coli)	*(17, 242)*
$(dacA)_n$	Terminal deoxynucleotidyl transferase	*(143)*
$(dacG)_n$	Terminal deoxynucleotidyl transferase	*(143)*
$(dacC)_n$	Terminal deoxynucleotidyl transferase	*(127)*
$(dibuG)_n$	Terminal deoxynucleotidyl transferase	*(127)*
$(dU)_n$	Terminal deoxynucleotidyl transferase	*(127)*
$(dN)_{oligo}pA_r$	Terminal deoxynucleotidyl transferase	*(208)*
$(dN)_{oligo}pA_r pA_r$	Terminal deoxynucleotidyl transferase	*(208)*
$(dT)_6 pN_r (N = A, C, G, U)$	Terminal deoxynucleotidyl transferase	*(360, 361)*
$(dT)_6 pN_r (pM_d)_n$ $(N = A, C, G, U; M = A, C, G, T)$	Terminal deoxynucleotidyl transferase	*(360, 361)*
$(dT)_6 pA_r pA_r (pdA)_n$	Terminal deoxynucleotidyl transferase	*(360)*
$(dT)_{30} pU_r$	Terminal deoxynucleotidyl transferase	*(208a)*
$(dT)_{30} pU_r pU_r$	Terminal deoxynucleotidyl transferase	*(208a)*
$(dT)_{30} pA_r$	Terminal deoxynucleotidyl transferase	*(208a)*
$(dT)_{30} pA_r pA_r$	Terminal deoxynucleotidyl transferase	*(208a)*
$(rA)_6 - (dC)_n$	Terminal deoxynucleotidyl transferase	*(90, 91, 92)*
ApApdA	Polynucleotide phosphorylase	*(178a)*
ApApdApdA	Polynucleotide phosphorylase	*(178a)*
ApUpdA	Polynucleotide phosphorylase	*(178a)*
$(rA)_n\text{-}dA$	Polynucleotide phosphorylase	*(50)*
$(rA)_n\text{-}dA\text{-}dA$	Polynucleotide phosphorylase	*(50)*
$(rA)_6\text{-}T_d$ and $(rA)_6\text{-}T_d\text{-}T_d$	Polynucleotide phosphorylase	*(91)*
$(rA)_6\text{-}T_d\text{-}(C_d)_n$	Terminal deoxynucleotidyl transferase	*(91, 92)*
$(rA)_6\text{-}T_d\text{-}T_d\text{-}(C_d)_n$	Terminal deoxynucleotidyl transferase	*(91, 92)*
$(rA, dA)_n$	Polynucleotide phosphorylase	*(49, 51)*
$r(I, 6thioI)_n$	Polynucleotide phosphorylase	*(448)*
$(r4thioU)_n$	Polynucleotide phosphorylase	*(412)*
(r4-thiomethyl U)$_n$	Polynucleotide phosphorylase	*(81, 372)*
(r-phosphothio U)$_n$	Polynucleotide phosphorylase	*(80, 81)*
$(r2thioC)_n$	Polynucleotide phosphorylase	*(373)*
$(rI)_n \cdot (r2thioC)_n$	Polynucleotide phosphorylase	*(86, 373)*

(rU, 4thioU)$_n$	Polynucleotide phosphorylase	*(371)*
(rU, 4thioT)$_n$	Polynucleotide phosphorylase	*(371)*
(rN^3,5-dimethylU)$_n$	Polynucleotide phosphorylase	*(368)*
(r5-hydroxymethyl U)$_n$	Polynucleotide phosphorylase	*(368)*
(r5-methylU)$_n$	Polynucleotide phosphorylase	*(368)*
(r3-methyl T)$_n$	Polynucleotide phosphorylase	*(368)*
(2′-O-methyl A)$_n$	Polynucleotide phosphorylase	*(29)*
(2′-O-methyl C)$_n$	Polynucleotide phosphorylase	*(413, 487)*
(2′-O-methyl U)$_n$	Polynucleotide phosphorylase	*(486)*
(rN^6-etheno A)$_n$	Polynucleotide phosphorylase	*(168, 224)*
(rN^4-etheno C)$_n$	Polynucleotide phosphorylase	*(168)*
(2′-chloro U)$_n$	Polynucleotide phosphorylase	*(148, 150)*
(2′-chloro C)$_n$	Polynucleotide phosphorylase	*(148, 150)*
(2′-fluoro U)$_n$	Polynucleotide phosphorylase	*(167)*
(2′-amino U)$_n$	Polynucleotide phosphorylase	*(149, 151)*
(2′-amino C)$_n$	Polynucleotide phosphorylase	*(151)*
(2′-azido C)$_n$	Polynucleotide phosphorylase	*(151)*
(2′-azido U)$_n$	Polynucleotide phosphorylase	*(442a, 444, 445)*
(rΨ)$_n$	Polynucleotide phosphorylase	*(110)*
(r3-methyl U)$_n$	Polynucleotide phosphorylase	*(445)*
(r 5,6-dihydro U)$_n$	Polynucleotide phosphorylase	*(443, 445)*
(r 5,6-methylene U)$_n$	Polynucleotide phosphorylase	*(443)*
r(A, G)$_n$ · (U)$_n$	Polynucleotide phosphorylase	*(8)*
r(I, U)$_n$ · (rC)$_n$	Polynucleotide phosphorylase	*(9)*
r(U, N^4-hydroxy C)$_n$	Polynucleotide phosphorylase	*(169)*
r(C, N^4-hydroxy C)$_n$	Polynucleotide phosphorylase	*(169)*
(r5-ethyl-U)$_n$	Polynucleotide phosphorylase	*(434)*
(r7-methyl G)$_n$	Polynucleotide phosphorylase	*(334)*
(r7-methyl I)$_n$	Polynucleotide phosphorylase	*(334)*
(rN^6-acetyl C)$_n$	Polynucleotide phosphorylase	*(334)*
(rN^2-methyl G)$_n$	Polynucleotide phosphorylase	*(335)*
(rN^2-dimethyl G)$_n$	Polynucleotide phosphorylase	*(335)*
(r6-chloropurine)$_u$	Polynucleotide phosphorylase	*(449)*
(rC^{1}A)$_n$	Polynucleotide phosphorylase	*(125)*
(rC^{3}A)$_n$	Polynucleotide phosphorylase	*(125)*
(rC^{7}A)$_n$	Polynucleotide phosphorylase	*(125)*
(rH^{6}A)$_n$	Polynucleotide phosphorylase	*(125)*
U-N-U$_n$	Polynucleotide phosphorylase	*(374)*
N-U$_n$	Polynucleotide phosphorylase	*(374)*
(N = 3-deazauridine, 4-deoxy- uridine, 3-deaza-4-deoxyuridine)		
(rA-r4thioU)$_n$	DNA-dependent RNA polymerase (E. coli)	*(71, 81)*
(r-phosphothio A- r-phosphothio U)$_n$	DNA-dependent RNA polymerase (E. coli)	*(77, 80, 81)*

In the ribo series polynucleotide phosphorylase has been successfully used for the stepwise synthesis of oligonucleotides containing specific sequences, when 2′,3′-O-protected ribonucleoside diphosphates were provided as substrates. According to Scheme 5.2 (*179, 252*) addition of such substrate results in a product containing a blocked 3'-terminus, whereby further addition is suppressed. Only after deblocking of the isolated product and addition of new substrate (again in the blocked

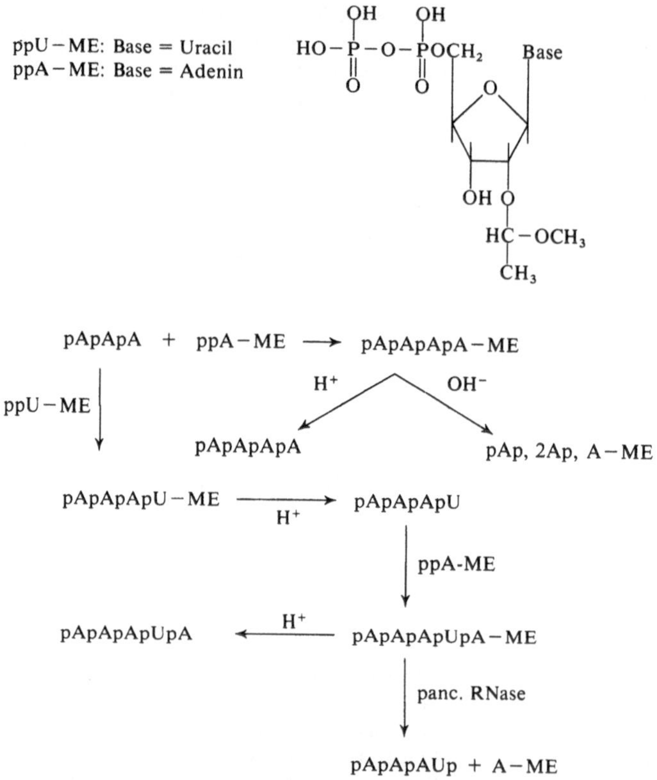

ppU – ME: Base = Uracil
ppA – ME: Base = Adenin

$$HO-\overset{\overset{\displaystyle OH}{|}}{\underset{\underset{\displaystyle O}{\|}}{P}}-O-\overset{\overset{\displaystyle OH}{|}}{\underset{\underset{\displaystyle O}{\|}}{P}}OCH_2 \quad Base$$

OH O
|
HC – OCH₃
|
CH₃

pApApA   +   ppA – ME   ⟶   pApApApA – ME

ppU – ME ↓

H⁺ ╱   ╲ OH⁻

pApApApA   pAp, 2Ap, A – ME

pApApApU – ME ⟶ pApApApU
                  H⁺

↓ ppA-ME

pApApApUpA ⟵ pApApApUpA – ME
            H⁺

↓ panc. RNase

pApApAUp + A – ME

Scheme 5.2. Stepwise oligonucleotide synthesis with polynucleotide phosphorylase

form) can subsequent addition take place. Repetition of this cycle using only one of the four protected ribonucleoside diphosphates as substrate at a time should lead – at least in principle – to any desired ribooligonucleotide sequence if the selection of the ribonucleoside diphosphate used as primer is properly adjusted to the sequence to be synthesized. The selection of the blocking group has to be adjusted to the following conditions: first, there should be no or minimum interference with the substrate binding site of the enzyme; secondly, the blocking group should maximally inhibit the primer function of the single addition product; thirdly the blocking group should be completely stable during the course of the reaction and should be removable under conditions that leave all other functional groups of the product intact. From the various protecting groups tested the 2′(3′)-O (α-methoxyethyl)-group (252) and the 2′(3′)-O-isovaleryl group (179) seem to meet these requirements satisfactorily. In the case where the α-methoxyethyl group is used evidence has been presented that the 2′-O-derivatives of the respective ribonucleoside diphosphates are the substrates accepted by the enzyme leading to a monoaddition product with a blocking group at the 2′-hydroxyl function (18).

Although the 3′-hydroxyl function as such would therefore be free, further addition is apparently suppressed by steric hindrance of the adjacent 2′-O-protecting group which still remains. Extrapolating from this it seems likely that the 2′-O-derivatives represent the substrates also in the cases where the isovaleryl group was used for 2′(3′) protection of ribonucleoside diphosphates (179); in this case, however, rapid equilibration between the 2′- and 3′-isomers will probably allow indirect utilization of the 3′-O-derivatives. Using this stepwise approach ribooligonucleotides of specific sequences up to the size of a pentanucleotide (252) or a heptanucleotide (180) have been prepared by two (252) or three (180) consecutive additions to the respective primers. Nearly quantitative yields seem to be possible in each step using the 2′(3′)-derivatives of all four standard ribonucleoside diphosphates. Problems, may, however, arise 1) from the alkali treatment necessary for removal of the isovaleryl group (179), 2) from primer phosphorolysis induced by the inorganic phosphate liberated during the addition reaction (105), and 3) by transnucleotidation reactions (179). Although the results reported so far seem most encouraging, it remains to be seen whether further improvements are necessary in order to use the enzymic stepwise approach as a routine procedure. Also no reports are so far available on the maximum chain lengths accessible by these approaches.

Analogous attempts in the deoxy series apparently have not been successful. 3′-O-Acetylthymidine triphosphate seems not to be acceptable as a substrate for terminal deoxynucleotidyl transferase (127). This enzyme can, however, utilize ribonucleoside triphosphates (instead of deoxyribonucleoside triphosphates) for the limited addition of one or two nucleotide units to a given primer (208, 360, 361) (see Scheme 5.3).

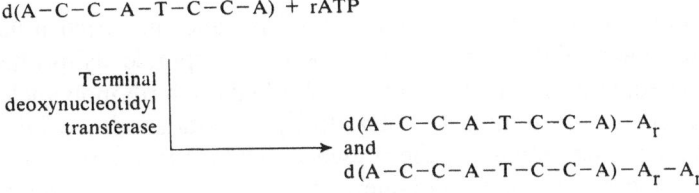

Scheme 5.3

and this reaction offers some synthetic value for the specific extension
of deoxyoligonucleotides. Thus, when the synthetic deoxyoligonucleotide
d(A–C–C–A–T–C–C–A) was enzymically extended to the nonanucleotide
d(A–C–C–A–T–C–C–A)–A$_r$ and decanucleotide d(A–C–C–A–T–C–C–
A)–A$_r$–A$_r$ increased priming efficiency was observed with the latter
in the presence of ØX174 DNA as template and DNA polymerase I
(209). The two additions can also be performed in two successive steps
with two different ribonucleoside triphosphates (R. Roychoudhury and
H. Kössel, unpublished observation). Based on this terminal addition
reaction a method for the 3′-terminal labelling of oligodeoxynucleotides
could be developed (208) which also allows partial sequences determination of oligodeoxynucleotides (379, 467).

The terminal addition of ribonucleotidyl residues to oligodeoxynucleotide primers by terminal deoxynucleotidyl transferase has also
been used in order to make oligodeoxynucleotides more acceptable
as primers for polynucleotide phosphorylase (Feix and Linder, unpublished). As evident from the fact that ribonucleoside triphosphates
are accepted as substrates for a limited terminal addition reaction,
the specificity requirement of the enzyme terminal deoxynucleotidyl
transferase with respect to the sugar moiety of the substrates appears
not to be rigorous. It seems therefore not unlikely that a masked
deoxynucleoside triphosphate which mimics ribonucleoside triphosphate
finally could be used as acceptable substrate for a stepwise terminal
addition approach similar to the one already developed in the ribo
series.

It is interesting to note that a terminal addition reaction reciprocal
to the one observed for terminal transferase has been found for polynucleotide phosphorylase in the terminal addition of deoxynucleotidyl
residues (instead of ribonucleotidyl residues) onto ribooligonucleotide
primers (50, 178a) (Scheme 5.4). This reaction which is also limited

$$\text{ApA + dADP} \xrightarrow[\text{phosphorylase}]{\text{polynucleotide}} \begin{array}{l} \text{ApApA}_d \\ \text{and} \\ \text{ApApA}_d\text{pA}_d \end{array}$$

Scheme 5.4

to the addition of one or two nucleotidyl residues has been utilized in
order to make oligoribonucleotides better acceptable as primers for
terminal deoxynucleotidyl transferase (91, 92) though ribopolynucleotides
and ribonucleotide terminated oligodeoxynucleotides also exhibit considerable priming activity in the presence of this enzyme (90, 360, 361).

The oligomer-initiated polymerization of unprotected deoxyribonucleotide units catalyzed by terminal deoxynucleotidyl transferase can

be limited to the formation of short polymers by using small ratios of deoxyribonucleoside 5′-triphosphates to primer (*142, 349*). The numbers of nucleotide residues added conform reasonably well with a Poisson distribution. When $(T)_4$ is used as primer with one molar equivalent of dCTP followed by another equivalent of dTTP, a product mixture of $T_4C_nT_m(n, m = 0, 1, 2)$ is obtained in which only one member $(n = 0; m = 2)$ is missing (*142*). Thus, successive stepwise addition (even of dinucleotide units) seems feasible by this method although it remains to be seen whether it can be applied to the synthesis of more complex sequences. As under comparable conditions the self-limiting polymerization of dGTP results in a very sharp product distribution, extension or even improvement of this method for the synthesis of G containing deoxyoligonucleotides seems possible. A somewhat similar principle has been applied to the synthesis of a three-section block copolymer of thymidylate, deoxyguanylate and deoxyadenylate by successive terminal additions of oligo dG and oligo dA blocks onto an oligo dT primer catalyzed by terminal deoxynucleotidyl transferase (*349*).

An enzyme capable of catalyzing primer dependent polymerization of deoxyribonucleoside 5′-diphosphates has recently been studied for single step addition onto $d(pA)_4$ (*106*). Although 4 to 8 fold molar excesses of one of the four deoxyribonucleoside 5′-diphosphates in the unprotected form were applied, single terminal addition products were obtained as the main products in reasonable yields. Thus, it seems not unlikely that this enzyme will also be useful in the stepwise synthesis of more complex oligodeoxynucleotides, although no such report has been published so far.

5.2. Reactions Catalyzed by Polynucleotide Ligases

Following the discovery of polynucleotide ligases in the mid sixties a strategy for the total synthesis of the gene coding for an alanine transfer RNA from yeast was immediately envisaged (*2, 6, 185, 188, 189*). The two basic parts of this strategy consist first in synthesizing chemically oligodeoxynucleotide segments according to the known primary structure of the tRNA and second in joining these segments enzymically by polynucleotide ligase. In order to allow the enzymic joining reaction to occur each two segments to be joined must be held in adjacent position by means of base pairing with a third overlapping fragment (the "splint"). The 3′-hydroxyl group of one segment (the "acceptor") is thereby brought into juxtaposition of the 5′-terminal phosphate of the other (the "donor"). The splint thus, provides specific template guidance for the ligation process (Scheme 5.5).

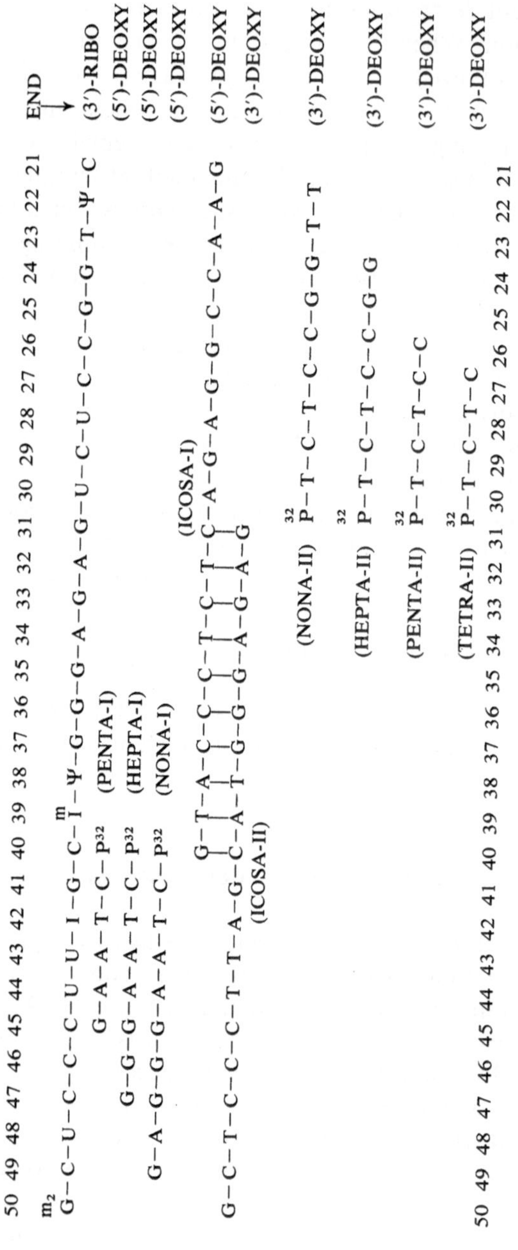

50 49 48 47 46 45 44 43 42 41 40 39 38 37 36 35 34 33 32 31 30 29 28 27 26 25 24 23 22 21   END →

m_2
G—C—U—C—C—C—U—U—I—G—C—I^m—Ψ—G—G—G—A—G—A—G—U—C—U—C—C—G—G—T—Ψ—C   (3')-RIBO

G—A—A—T—C—p^{32} (PENTA-I)   (5')-DEOXY

G—G—G—A—A—T—C—p^{32} (HEPTA-I)   (5')-DEOXY

G—A—G—G—G—A—A—T—C—p^{32} (NONA-I)   (5')-DEOXY

(ICOSA-I)
G—T—A—C—C—C—T—C—T—C—A—G—A—G—G—C—C—A—A—G   (5')-DEOXY

G—C—T—C—C—C—T—T—A—G—C—A—T—G—G—G—A—G—A—G   (3')-DEOXY
(ICOSA-II)

(NONA-II) ^{32}P—T—C—T—C—C—G—G—T—T   (3')-DEOXY

(HEPTA-II) ^{32}P—T—C—T—C—C—G—G   (3')-DEOXY

(PENTA-II) ^{32}P—T—C—T—C—C   (3')-DEOXY

(TETRA-II) ^{32}P—T—C—T—C   (3')-DEOXY

50 49 48 47 46 45 44 43 42 41 40 39 38 37 36 35 34 33 32 31 30 29 28 27 26 25 24 23 22 21

Scheme 5.5. Ligation of oligodeoxynucleotide segments

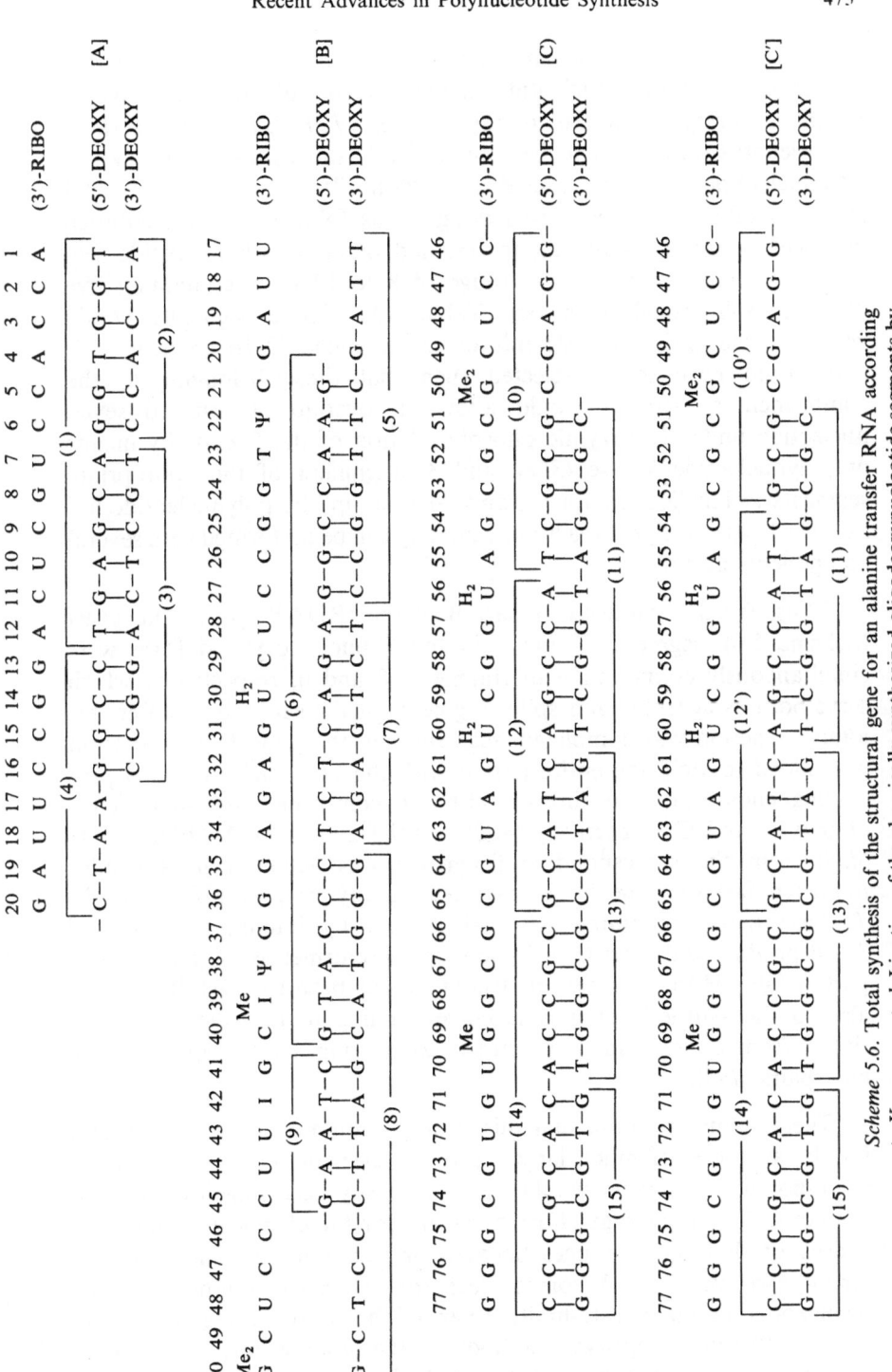

Scheme 5.6. Total synthesis of the structural gene for an alanine transfer RNA according to KHORANA *et al.* Ligation of the chemically synthesized oligodeoxynucleotide segments by polynucleotide ligase

Studies were carried out in order to determine the minimum lengths of the deoxyribooligonucleotide chains which polynucleotide ligases require to bring about the joining reaction (*120, 121, 188*). As these chain lengths turned out to be comparatively small (in one case even a tetranucleotide could be reacted as a "donor") the plan for the total synthesis of a tRNA gene was designed as follows: 1. Conventional chemical synthesis by the diester method of deoxypolynucleotide segments of chain lengths in the range of 8 to 12 units containing free 3′- and 5′-hydroxyl ends (see Table 4.1). These segments would represent the entire two strands of the intended DNA (Scheme 5.6) and would have to be selected such that those belonging to the complementary strands would allow an overlap of four to seven nucleotide units. 2. Enzymic phosphorylation of the 5′-ends by means of polynucleotide kinase (*357*), and 3. alignment of the appropriate segments to bihelical complexes and "sealing up" by polynucleotide ligase. The ligase catalyzed reactions generally can be performed with several components at a time.

Thus, for the synthesis of part B of the tRNA^{Ala} gene from yeast (Scheme 5.6), segments 7 and 9 in the 5′-phosphorylated form were simultaneously connected with fragments 8 and 6, respectively, which serve both as acceptor and splint segments at the same time. After this joining reaction was completed, segment 5 in the phosphorylated form was added to yield the entire part B with the two single stranded protruding ends ready for linkage with the complementary sticky ends of part A and C respectively (*402*). Synthesis of part A (*400*) and C (*454*) from the corresponding fragments was achieved in a similar multistep fashion. The large fragments finally were combined to the 77 base pairs containing bihelical duplex, by joining a preformed A + B product with part C or by joining a preformed B + C adduct with part A (*39*). More recently this technique has been successfully applied also to the synthesis of a gene corresponding to tyrosine suppressor tRNA from *E. coli* and its precursor comprising a total length of 126 base pairs (*190*).

There seems no doubt that this strategy would also prove successful for the synthesis of much larger genes or even of total genoms if the corresponding segments would be more easily accessible by chemical synthesis. As the average effort necessary for the chemical synthesis of a deca or dodecamer ranges between one quarter and one half of a "man-power-year", it becomes clear that chemical synthesis of the segments is the major rate-limiting part of the entire strategy at present and that therefore improved methods for the rapid chemical or enzymic synthesis of oligodeoxynucleotides are highly desirable.

One obvious prerequisite for the application of this strategy is that the base sequence of a gene to be synthesized has to be known in contrast to the approach, whereby genoms have been synthesized *in vitro* by mere copying of input templates in the presence of polymerizing enzymes (*89, 111, 327*). In the case of known RNA sequences the sequences of the respective genes can be deduced unambiguously by the base pairing rules as exemplified by the gene coding for tRNA^{Ala} from yeast (Scheme 5.6). However, due to the degeneracy of the genetic code unequivocal deduction of a nucleotide sequence from a known amino acid sequence can only be achieved to a limited extent (*146, 288*). For instance if the two amino acids methionine and tryptophan, for which only one codon exists, respectively, occur in neighbourhood to each other, oligonucleotide sequences containing no or very few ambiguities can be predicted (*379, 380, 467*). In general, however, more than one third of the nucleotide sequences derived from amino acid sequences are ambiguous and in many of these cases all four possible bases may occur in each ambiguous position. Although this offers the advantage, that the synthetic routes to the gene fragments to be prepared can partly be adjusted to maximum simplicity (*146, 288*), the risk of selecting by chance rare codons has to be kept in mind. As it is hoped that synthetic genes finally will be introduced and transcribed in living cells, selection of rare codons may then forbid effective expression of synthetic genes. The only way of avoiding this problem seems to consist in sequence analysis of the respective mRNAs or DNAs prior to gene synthesis. Sequence analysis will also be a prerequisite to the synthesis of untranslated or untranscribed DNA sequences such as intercistronic regions or regulatory elements such as operator or promotor regions.

In order to obtain maximum yields and/or to avoid certain undesired deviations in the joining reactions catalyzed by polynucleotide ligase, the reaction components and conditions have to be selected carefully. Thus, effective joining of segment 5 in part B of the tRNA^{Ala} gene (Scheme 5.6) requires raising of the temperature to 25° C, whereas a temperature of 15° C is sufficient for the joining of segments 6, 7, 8 and 9 to each other (*188, 402*). This temperature dependency probably reflects internal secondary structures of the oligonucleotide segments, which interfere with the annealing of the segment to be joined.

An instructive deviation from the expected joining reaction was observed when an attempt was made to react the partial duplex obtained from segments 1, 2 and 3 (part A of the tRNA^{Ala} gene, Scheme 5.6) with segment 4 (*188, 400*). When the 5′-terminus of segment (3+2) was phosphorylated, the product formed was a dimer of the starting duplex (comprising two copies of the segments 1, 2 and 3). This dimerization obviously was due to the self-complementary nature of the protruding C–C–G–G single-stranded end of the duplex (1, 2, 3). Joining with the segment 4 was only observed when the 5'-terminal phosphate of the (3+2) segment was absent as this phosphate is required for dimerization.

In order to prevent undesired joining reactions protection of 5'-terminal phosphate groups on the acceptor or splint molecules by alkyl-thio groups has been proposed (129). Interference of such unphysiological groups with the joining reaction apparently is not encountered as they are positioned far enough from the joining center. This modification may therefore indeed prove helpful during further synthetic work in order to prevent alternate wrong joining reactions. Besides self-complementarity of segments to be reacted, infidelity of the joining reaction itself may create additional possible problems. Thus, when pT_{11}–C was reacted (as donor and acceptor molecule) on poly dA as template in the presence of T_4-induced ligase, head to tail joining to the mismatched C-residues was observed (446).

A–C base pairs have also been found acceptable in an oligo-merization reaction observed with segment 4 of the tRNA^{Ala} gene (see Scheme 5.7) (400). Whether other nonclassical base pairs are acceptable or whether more than one mismatched base pair can be tolerated near the joining site remains to be seen. Another interesting type of deviation from the normal joining reactions was observed in the joining of DNA duplexes at completely base paired ends (399, 404). It is difficult to evaluate this type of deviation for synthetic purposes as it may constitute an undesired side reaction in one case or in other cases a synthetic aid.

Ribooligonucleotides can also be reacted as substrates of polynucleo-tide ligase (195, 367a). Thus, head-to-tail joining of ribooligoadenylates in the presence of poly dT was observed. A reciprocal substrate situation (ribo-template, deoxyoligonucleotides to be joined) seems also acceptable as evident from the joining reaction observed with oligo dT on a poly rA template. Sano and Feix have recently demonstrated that – in contrast to earlier results (195) – all-ribo substrates (with the exception of oligo rA on a poly rU template) are also acceptable for polynucleotide ligase catalized reactions (367a). There is no doubt that these "deviations" toward ribo-substrates are widening the possible usefulness of ligases as synthetic tools.

The ligase technique has recently been extended to the joining of a chemically synthesized short oligomer onto a naturally occurring DNA molecule of high molecular weight. Thus, the synthetic dodecamer d(pA–G–G–T–C–G–C–C–G–C–C–C) was annealed and covalently joined to lambda phage DNA in the presence of T_4-ligase (130). This is the

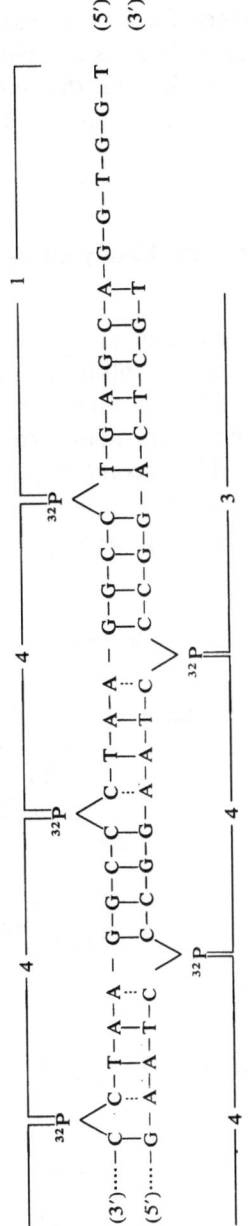

Scheme 5.7. Ligation of partially mismatched oligodeoxynucleotide segments

first time that a chemically synthesized oligonucleotide has been covalently linked to a naturally occurring phage DNA and this approach may be of general importance for future insertion of synthetic genes into living cells by using phage DNA as a "vehicle".

Polynucleotide ligase has allowed the preparation of a circular bihelical DNA containing repeating dinucleotide sequences (328).

5.3. Synthetic Reactions Catalyzed by Ribonucleases

This approach which is restricted to internucleotide bond formation in the ribo series is based on the following observations. During the breakdown of RNA catalyzed by ribonucleases transesterification to the respective 2',3'-cyclophosphates in many cases occurs as the first reaction step (Scheme 5.8). The second step, hydrolysis of the cyclic phosphate, then leads to the final products, the 3'-monophosphate derivatives. Though the latter step is virtually irreversible, in many

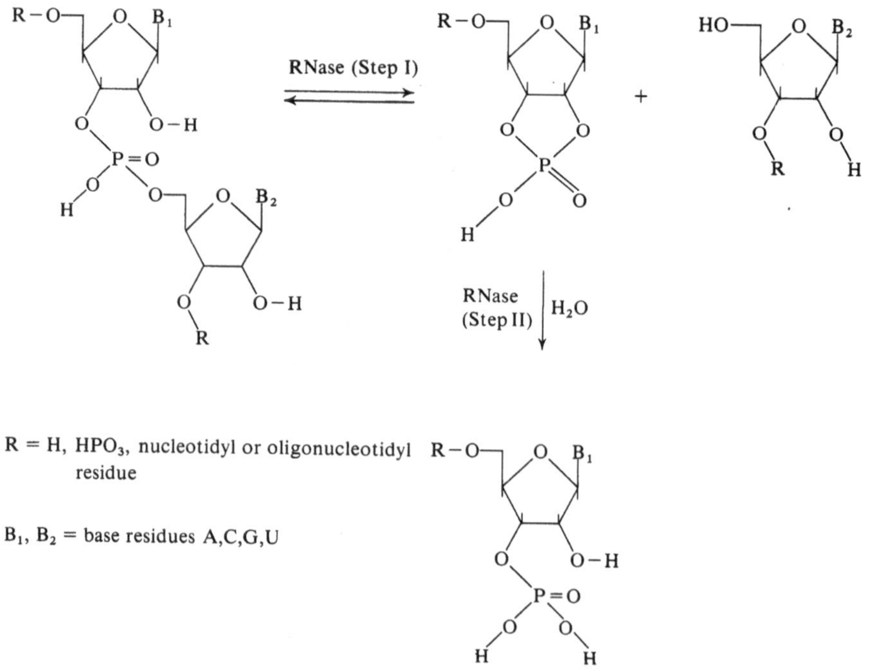

R = H, HPO₃, nucleotidyl or oligonucleotidyl residue

B₁, B₂ = base residues A,C,G,U

Scheme 5.8. Ribonuclease catalyzed formation of internucleotide linkages in the ribo series

References, pp. 483—508

instances conditions could be found under which it proceeds at a much slower rate than the transesterification step. As the transesterification at the same time constitutes a reversible type of reaction, the equilibrium can be shifted from the cyclic phosphate towards the side of internucleotide bond formation merely by application of high concentration of the 5'-hydroxyl carrying component. This approach already used in earlier work with ribonuclease A from bovine pancreas and with ribonuclease T_1 from *Aspergillus oryzae* has led to the stepwise synthesis of dinucleoside monophosphates and of trinucleoside diphosphates (*21, 119, 165, 272, 388*). More recent investigations have demonstrated (a) that several other ribonucleases can serve as synthetic tools (b), that the ribonuclease approach is also feasible for comparatively large scale synthesis (up to gram scale) (c), that coupling of preformed oligonucleotide blocks is possible and (d), that nucleotides containing altered base residues in some cases are also acceptable for the ribonuclease catalyzed reactions.

Besides ribonuclease A and ribonuclease T_1 which specifically catalyze the transesterification reactions from 2',3'-pyrimidine cyclophosphates and from 2',3'-guanosine cyclophosphates, respectively, ribonuclease N_1 from *Neurospora crassa* (*201, 199*), ribonuclease U_2 from *Ustilago sphaerogena* (*165, 200, 450*), and nonspecific ribonucleases from *Bacillus subtilis* (*365*) and *Aspergillus clavatus* (*16, 485*) have now been introduced as synthetic enzymes. Ribonuclease N_1 – like ribonuclease T_1 – specifically catalyzes transesterification reactions from 2',3'-guanosine cyclophosphate containing nucleotides onto a 5'-hydroxyl containing acceptor component (*201, 199*). The yields obtained with both the G specific enzymes (see Table 5.3) are also strongly influenced by the nature of the acceptor molecules with cytidine serving as most efficient acceptor followed by uridine and adenosine. The same order is apparent, where 2',3'-cycloguanosine phosphate has been reacted with dinucleoside monophosphates and where 2',3'-cycloguanosine phosphate terminated dinucleotides have been linked to nucleosides (see Table 5.3). Ribonuclease U_2 exhibits a broader specificity in respect to the 2',3'-cyclophosphate moiety, as it catalyzes transesterification from both the 2',3'-purine cyclophosphates (*165, 200, 450*; Table 5.3). The specificity in respect to the nucleoside acceptors seems to be similar to the one observed with the ribonucleases T_1 and N_1. The two nonspecific ribonucleases from *Bacillus subtilis* (*365*) and from *Aspergillus clavatus* (*16, 485*) have been studied for the synthesis of almost all possible dinucleoside monophosphates, which are obtained in satisfactory yields (Table 5.3). It is noteworthy that the enzyme from *Aspergillus clavatus* has also been used successfully for large scale reactions (*16*) with several grams of each component.

Table 5.3. *Oligoribonucleotides Synthesized by Ribonuclease Catalyzed Reactions*

Reaction type	RNase type	Products (% yield)	References
ApC>p+U	pancr. RNase	ApCpU (12)	*(298)*
ApU>p+Up	pancr. RNase	ApUpUp (11)	*(298)*
Py>p+Pu	pancr. RNase	PypPu (7–15)	*(441)*
ŪpC> +C	pancr. RNase	ŪpCpC (40)*	*(182)*
PupPy>p+N	pancr. RNase	PupPypN (4–12)	*(181)*
U>p+U	pancr. RNase	$(Up)_nU$	*(103)*
	pancr. RNase	$(Up)_n$	*(103)*
	pancr. RNase	$(Up)_n>p$	*(103)*
U>p+N	pancr. RNase	$(Up)_nN$	*(103)*
	pancr. RNase	$(Up)_n$	*(103)*
	pancr. RNase	$(Up)_nU>p$	*(103)*
U>p+PupPu	pancr. RNase polymer bound	UpPupPu (4–18)	*(104)*
G>p+N	T_1-RNase	GpC (20, 66)	*(272, 388)*
	T_1-RNase	GpU (13, 62)	*(272, 388)*
	T_1-RNase	GpI (10, 32)	*(272, 388)*
	T_1-RNase	GpA (5, 4)	*(388)*
	T_1-RNase	GpX (5)	*(374)*
	T_1-RNase	GpthioU (12)	*(165)*
	RNase N_1	GpC (44, 79**)	*(199)*
	RNase N_1	GpU (12, 27)	*(199)*
	RNase N_1	GpA (4)	*(199)*
	RNase N_1	GpthioU (26)	*(165)*
	Actinomycin RNase	GpC (40)	*(442)*
I>p+C	RNase N_1	IpC (22)	*(201)*
Pu>p+N	RNase U_2	ApC (35, 67***)	*(200)*
	RNase U_2	ApU (22, 19, 48***)	*(200, 450)*
	RNase U_2	GpU (18)	*(200)*
	RNase U_2	ApI (10)	*(200)*
	RNase U_2	ApA (6)	*(200)*
	RNase U_2	ApthioU (30)	*(165)*
G>p+Up	T_1-RNase	GpUp (20)	*(272)*
A>p+Np	RNase U_2	ApCp (5)	*(200)*
		ApGp (1)	*(200)*
G>p+NpM	T_1-RNase	GpCpC (20, 27, 40)	*(119, 272, 388)*
	T_1-RNase	GpCpU (14, 23)	*(119, 388)*
	T_1-RNase	GpCpA (12, 14)	*(119, 388)*
	T_1- RNase	GpCpG (3.5, 8)	*(119, 388)*
	T_1-RNase	GpUpC (7, 19, 37)	*(119, 272, 388)*
	T_1-RNase	GpApC (8, 25)	*(24, 119)*
	T_1-RNase	IpCpC (20)	*(119)*

   * Ū symbolyzes a uridine residue modified by addition of water soluble carbodiimide.

   ** Uchida, unpublished.

   *** Uchida and Funayama, unpublished.

	T_1-RNase	IpCpU (19)	*(119)*
	T_1-RNase	IpCpA (15)	*(119)*
	T_1-RNase	IpCpG (7)	*(119)*
G > p + ApApC	RNase N_1	GpApApC (7)	*(201)*
NpG > p + N	T_1-RNase	CpGpC (8)	*(119)*
	T_1-RNase	CpGpU (8, 10)	*(119, 272)*
	T_1-RNase	CpGpA (3)	*(119)*
	T_1-RNase	UpGpC (6)	*(119)*
	T_1-RNase	UpGpU (3, 38)	*(119, 272)*
	T_1-RNase	ApGpU (53)	*(272)*
ApUpG > p + N	T_1-RNase	ApUpGpC (32)	*(272)*
	T_1-RNase	ApUpGpU (37)	*(272)*
	T_1-RNase	ApUpGpA (30)	*(272)*
ApUpG > p + Np	T_1-RNase	ApUpGpCp (15)	*(272)*
	T_1-RNase	ApUpGpUp (15)	*(272)*
	T_1-RNase	ApUpGpAp (5)	*(272)*
ApUpG > p +	T_1-RNase	ApUpGpApAp (11)	*(272)*
oligo(A)$_{2-5}$	T_1-RNase	ApUpGpApApAp (8)	*(272)*
	T_1-RNase	ApUpGpApApApAp (5)	*(272)*
	T_1-RNase	ApUpGpApApApApAp (3)	*(272)*
G > p polymerization	T_1-RNase	(Gp)$_n$ n = 2 (14)	*(139)*
	T_1-RNase	n = 3 (6)	*(139)*
	T_1-RNase	n = 4 (10)	*(139)*
	RNase N_1	n = 2 (22)	*(199)*
	RNase N_1	n = 3 (18)	*(199)*
	RNase N_1	n = 4 (7)	*(199)*
ApG > p polymerization	RNase N_1	(ApGp)$_n$ n = 2 (33)	*(201*
	RNase N_1	n = 3 (7)	*(201)*
A > p polymerization	RNase U_2	(Ap)$_n$ n = 2 (17)	*(200)*
	RNase U_2	n = 3 (5)	*(200)*
	RNase U_2	n = 4 (1.4)	*(200)*
N > p + Py	A. clavatus RNase	NpPy (8–40)	*(16, 485)*
N > p + Py	B. subtilis RNase	NpPy (20–75)	*(365)*

The coupling reaction could be extended to the joining of preformed oligonucleotide blocks (*119, 201, 272, 388*). As summarized in Table 5.3, the cyclophosphate bearing components as well as the acceptor components can constitute longer oligonucleotide chains. The maximum chain length synthesized by the ribonuclease approach seems to be the octanucleotide A–U–G(–A)$_5$ from the blocks A–U–G> and A(pA)$_4$. The coupling of oligomeric components is also evident from the oligomerization reactions observed with 2'-3'-cyclo AMP, 2',3'-cyclo GMP and ApG> in the presence of ribonuclease U_2 and ribonuclease N_1 respectively (*201, 199*).

Oligomerization of 2',3'-cyclo GMP in the presence of ribonuclease T_1 has already been studied earlier (*139*). More recently, however, the product of this reaction has become a matter of controversy as it was demonstrated (*336, 337, 338*) that the products formed at room temperature almost exclusively contain 2'—5'-linkages (as evident from resistance against T_1-ribonuclease and from degradation to 2'-GMP). This is in contradiction to all other reports on ribonuclease T_1 catalyzed synthesis (*119, 139, 165, 272, 388*) where the internucleotide bonds of the products formed at 4° C proved to be entirely susceptible to enzymic degradation, which evidences natural 3'—5'-linkages. Whether this discrepancy is due to a temperature effect or to contaminants in the enzyme preparations used remains open. Even if these controversial findings remain a single case, it nevertheless underlines the necessity of carefully characterizing the products obtained by the ribonuclease approach.

The yields obtained in ribonuclease catalyzed synthetic reactions range between 5 and 30% for short chains (see Table 5.3). In the special case of GpC yields up to 66% and even higher have been reported. Though better yields are generally obtained by chemical methods especially for ribooligonucleotides of medium and higher chain length, the enzymic approach clearly offers the following two advantages: (a) no protecting groups are required either for the 2'-OH functions or for the amino functions on the bases, (b) reactions are carried out entirely in aqueous medium, thereby avoiding the problem of dissolving oligonucleotides in organic solvents such as anhydrous pyridine. On the other hand, the ribonuclease approach is severely limited by the fact that cleavage of the internucleotide bonds already present can only be avoided in the case of base specific ribonucleases (such as T_1) when the respective internucleotide bonds are not present in both the starting components; thus ribonuclease T_1 can only be used for the joining of blocks which do not contain any internal or 5'-terminal guanosine residues.

Synthesis of higher oligonucleotides on larger scale may also be limited because it is difficult to denature or completely remove ribonucleases, which may interfere in subsequent reaction steps or during the workup. In order to eliminate this problem ribonuclease A fixed to solid supports such as CM-cellulose or to maleic anhydride copolymers has been used for the synthesis of the three terminator codons UAA, UAG and UGA from 2',3'-cyclo UMP and the respective dinucleoside monophosphates (*104*). The three codons could be isolated in 4—18% yield. This comparatively low yield is, however, fully counterbalanced (a) by the rapidity of the method, (b) by the nearly quantitative recovery of the unutilized dinucleoside monophosphates and (c) by the possibility to use the CM cellulose bound ribonuclease repeatedly in several reaction steps without loss of activity.

Introduction of nucleotides containing modified base moieties into oligonucleotides by ribonuclease catalyzed reactions has been reported in several cases. Inosine, Xanthosine, and ApI can serve as acceptor molecules in reactions catalyzed by T_1 or U_2 ribonuclease (*119, 272, 388*). 2′,3′-Cyclophosphates derived from inosine, 8-azaguanosine and xanthosine have been used as activated mononucleotides in the presence of ribonuclease T_1 (*119*) and N_1 (*199, 201*). 2′,3′-Cyclo GMP can be reacted with 4-thiouridine in the presence of ribonuclease T_1 or N_1 (*165*) and transfer of 2′,3′-cyclo AMP to 4-thiouridine can also be achieved in the presence of ribonuclease U_2. Codons containing in the wobble position the modified nucleosides 4-deoxyuridine, 3-deazauridine and 3-deaza-4-deoxyuridine have been synthesized by using ribonuclease A (*103*).

The same enzyme has been used for synthetic reactions involving 5′-O-methylphosphoryl-uridine 2′,3′-cyclic phosphate as a class of substrates modified at the 5′-terminal phosphate by an unnatural substituent (*21*).

References

1. AGARWAL, K. L., and M. M. DHAR: The use of 2,3′-anhydronucleosides in the synthesis of the internucleotide bond. Tetrahedron Letters 2451 (1965).
2. AGARWAL, K. L., H. BÜCHI, M. H. CARUTHERS, N. GUPTA, H. G. KHORANA, K. KLEPPE, E. OHTSUKA, U. L. RAJBHANDARY, J. H. VAN DE SANDE, V. SGARAMELLA, H. WEBER, and T. YAMADA: Total synthesis of the gene for an alanine transfer ribonucleic acid from yeast. Nature **227**, 27 (1970).
3. AGARWAL, K. L., A. YAMAZAKI, and H. G. KHORANA: Studies on polynucleotides. XCVIII. A convenient and general method for the preparation of protected dideoxyribonucleotides containing 5′-phosphate end groups. J. Amer. Chem. Soc. **93**, 2754 (1971).
4. AGARWAL, K. L., and H. G. KHORANA: Studies on Polynucleotides. CII. The Use of Aromatic Isocyanates for Selective Blocking of the Terminal 3′-Hydroxyl Group in Protected Deoxyribooligonucleotides. J. Amer. Chem. Soc. **94**, 3578 (1972).
5. AGARWAL, K. L., A. KUMAR, and H. G. KHORANA: Studies on Polynucleotides. CIX. Total synthesis of the structural gene for an alanine transfer ribonucleic acid from yeast. Synthesis of a dodecadeoxynucleotide and a decadeoxynucleotide corresponding to the nucleotide sequence 46 to 65. J. Mol. Biol. **72**, 351 (1972).
6. AGARWAL, K. L., A. YAMAZAKI, P. J. CASHION, and H. G. KHORANA: Chemische Synthese von Polynucleotiden. Angew. Chem. **84**, 489 (1972).
7. AGARWAL, K. L., M. FRIDKIN, E. JAY, and H. G. KHORANA: Deoxynucleotide synthesis using a new phosphate protecting group. J. Amer. Chem. Soc. **95**, 2020 (1973).
8. AKUTSU, H., and M. TSUBOI: Structure of polynucleotide complex with non-complementary nucleosides. I. Poly A, G + poly U. Bull. Chem. Soc. Japan **43**, 3391 (1970).
9. — — Structure of polynucleotide complex with non-complementary nucleosides. II. Poly I, U + poly C. Bull. Chem. Soc. Japan **44**, 20 (1971).
10. AMAGAEVA, A. A., A. M. YURKEVITCH, I. P. RUDAKOVA, L. V. KHRISTENKO, I. M. KUSTANOVICH, and N. A. PREOBRASHENSKII: 5′-O-p-Tolylsulfonyladenosine derivatives. Rhim. Prir. Soedin. **4**, 304 (1968). Chem. Abstr. **70**, 115471 s (1969).
11. ASAI, M., M. MIYAKI, and B. SHIMIZU: Synthetic nucleotides. II. A direct synthetic method for ribonucleotides. Chem. Pharm. Bull. (Tokyo) **15**, 1856 (1967).

12. Badashkeeva, A. G., G. N. Kabasheva, D. G. Knorre, G. G. Shamovskii, and T. N. Shubina: Condensation of trideoxyadenylate by means of a water-soluble carbodiimide in the presence of polyuridylic acid. Dokl. Akad. Nauk SSSR **206**, 870 (1972).

13. Bähr, W., H. Sommer, and K. H. Scheit: Synthesis and properties of oligodeoxy-4-thiothymidylic acid. Biochim. Biophys. Acta **287**, 427 (1972).

14. Baker, B. R., P. M. Tanna, and G. D. F. Jackson: Non-classical antimetabolites. XXII. Simulation of 5'-phosphoribosyl binding. IV. Attempted simulation with nucleoside-5'-carbamates. J. Pharm. Sci. **54**, 987 (1965).

15. Barzilay, I., J. L. Sussman, and Y. Lapidot: Further studies on the chromatograph behaviour of dinucleoside monophosphates. J. Chromatog. **79**, 139 (1973).

16. Bauer, S., R. Lamed, and Y. Lapidot: Large scale synthesis of dinucleoside monophosphates catalyzed by ribonuclease from Aspergillus clavatus. Biotechnol. Bioeng. **XIV**, 861 (1972).

17. Beikirch, H. H., and A. G. Lezius: Double-stranded Polydeoxyribonucleotides containing 6-Thiodeoxyguanosine and 6-Thiodeoxyinosine. $Poly[d(A-C) \cdot d(T-S^6G)]$, $Poly[d(A-S^6G) \cdot d(T-C)]$, and $Poly[d(A-C) \cdot d(T-S^6I)]$. Eur. J. Biochem. **27**, 381 (1972).

18. Bennett, G. N., J. K. Mackey, J. L. Wiebers, and P. T. Gilham: 2'-(O-α-methoxyethyl-)nucleoside-5'-diphosphates as Single Addition Substrates in the Synthesis of Specific Oligoribonucleotides with Polynucleotide Phosphorylase. Biochemistry **12**, 3956 (1973).

19. Berger, H.: Solvolysis and phosphorylating activity of polyphosphate esters. Z. Naturforsch. B **26**, 694 (1971).

20. Berlin, Yu. A., O. G. Chakhmakhcheva, V. A. Efimov, M. N. Kolosov, and V. G. Korobko: Arenesulfonyl imidazolides, new reagents for polynucleotide synthesis. Tetrahedron Letters 1353 (1973).

21. Bernfield, M. R., and F. M. Rottman: Ribonuclease and oligoribonucleotide synthesis. III. Oligonucleotide synthesis with 5'-substituted uridine 2',3'-cyclic phosphates. J. Biol. Chem. **242**, 4134 (1967).

22. Blackburn, G. M., M. J. Brown, and M. R. Harris: Nucleic acid studies on insoluble polymer supports. Chem. Commun. 611 (1966).

23. Blackburn, G. M., M. J. Brown, and M. R. Harris: Synthetic studies of nucleic acids on polymer supports. Part I. Oligodeoxyribonucleotide synthesis on an insoluble polymer support. J. Chem. Soc. 2438 (1967).

24. Blackburn, G. M., M. J. Brown, M. R. Harris, and D. Shire: Synthetic studies of nucleic acids on polymer supports. Part II. Mechanisms of phosphorylation with carbodi-imides and arenesulphonyl chlorides. J. Chem. Soc. 676 (1969).

25. Blank, H. U., D. Frahne, A. Myles, and W. Pfleiderer: Nucleosides. IV. Tritylation and benzylation of adenosine derivatives. Liebigs. Ann. Chem. **742**, 34 (1970).

26. Blank, H. U., and W. Pfleiderer: Nucleosides. I. Syntheses of O'-benzyl derivatives of uridine. Liebigs Ann. Chem. **742**, 1 (1970).

27. — — Nucleosides. II. Tritylations and benzylations of cytidine derivatives. Liebigs Ann. Chem. **742**, 16 (1970).

28. — — Nucleosides. III. New mild method for the selective N^6-acylation of cytidine. Liebigs Ann. Chem. **742**, 29 (1970).

29. Bobst, A. M., P. A. Cerutti, and F. Rottman: The structure of poly 2'-O-methyladenylic acid at acidic and neutral pH. J. Amer. Chem. Soc. **91**, 1246 (1969).

30. Bollum, F. J.: Terminal deoxynucleotidyl transferase. In: The Enzymes, **X** (P. D. Boyer, ed.), 145 (1974).

31. Boom, J. H. van, P. M. J. Burgers, G. P. Owen, C. B. Reese, and R. Saffhill: Approaches to oligoribonucleotide synthesis via phosphotriester intermediates. Chem. Commun. 869 (1971).

32. BOOM, J. H. VAN, P. VAN DEURSEN, J. MEEUWSE, and C. B. REESE: Two sulfur-containing protecting groups for alcoholic hydrogen functions. J. Chem. Soc., Chem. Commun. 766 (1972).

33. BRANDSTETTER, F., H. SCHOTT, und E. BAYER: Liquid-phase-Synthese von Nucleotiden. Tetrahedron Letters 2997 (1973).

34. BREITENBACH, J. W., and O. F. OLAJ: Über die proliferierende Polymerisation in Vernetzungssystemen und dynamisch-kalorimetrische Messungen an Popcorn-Polymeren. Chimia 22, 157 (1968).

35. BROWN, D. M.: Phosphorylation. Advances Org. Chemistry 3, 75 (1963).

36. BRUNNGRABER, E. F., and E. CHARGAFF: Transferase from Escherischia coli effecting low-energy phosphate transfer to nucleosides and nucleotides. Proc. Natl. Acad. Sci. US 67, 107 (1970).

37. BÜCHI, H., and H. G. KHORANA: Studies on Polynucleotides. CV. Total synthesis of the structural gene for an alanine transfer ribonucleic acid from yeast. Chemical synthesis of an icosadeoxyribonucleotide corresponding to the nucleotide sequence 31 to 50. J. Mol. Biol. 72, 251 (1972).

38. CARUTHERS, M. H., and H. G. KHORANA: Studies on Polynucleotides. CXI. Total synthesis of the structural gene for an alanine transfer ribonucleic acid from yeast. Synthesis of a dodecadeoxynucleotide and a heptadeoxynucleotide corresponding to the nucleotide sequence 66 to 77. J. Mol. Biol. 72, 407 (1972).

39. CARUTHERS, M. H., K. KLEPPE, J. H. VAN DE SANDE, V. SGARAMELLA, K. L. AGARWAL, H. BÜCHI, N. K. GUPTA, A. KUMAR, E. OHTSUKA, U. L. RAJ BHANDARY, T. TERAO, H. WEBER, T. YAMADA, and H. G. KHORANA: Studies on Polynucleotides. CXV. Total synthesis of the structural gene for an alanine transfer RNA from yeast. Enzymic joining to form the total DNA duplex. J. Mol. Biol. 72, 475 (1972).

40. CARUTHERS, M. H., J. H. VAN DE SANDE, and H. G. KHORANA: Studies on Polynucleotides. CX. Total synthesis of the structural gene for an alanine transfer ribonucleic acid from yeast. Synthesis of three decadeoxynucleotides corresponding to the nucleotide sequence 51 to 70. J. Mol. Biol. 72, 375 (1972).

41. CASHION, P. J., M. FRIDKIN, K. L. AGARWAL, E. JAY, and H. G. KHORANA: Studies on Polynucleotides. CXXI. The use of trityl- and α-naphthylcarbamoyl cellulose derivatives in oligonucleotide synthesis. Biochemistry 12, 1985 (1973).

42. CATLIN, J. C., and F. CRAMER: Deoxyoligonucleotide synthesis via the triester method. J. Org. Chem. 38, 245 (1973).

43. CEDERGREN, R. J., B. LARUE, and P. LAPORTE: The acylation of ribonucleotides with benzoic and acetic anhydrides in aqueous solutions. Can. J. Biochem. 49, 730 (1971).

44. CHAPMAN, T. M., and D. G. KLEID: Oligonucleotide synthesis on polar polymer supports: The use of a polypeptide support. J. Chem. Soc. D, Chem. Comm. 193 (1973).

45. CHARGAFF, E., and J. N. DAVIDSON: The nucleic acids. Vol. 1, 2 (1955); Vol. 3. New York-London: Academic Press. 1960.

46. CHEN, H.-C., L. C. CRAIG, and E. STONER: On the removal of the residual carboxylic acid groups from cellulosic membranes and Sephadex. Biochem. 11, 3559 (1972).

47. CHLADEK, S., J. ZEMLICKA, and V. GUT: 5-chloro-8-hydroxyquinoline (chloroxine) esters of carboxylic acids — selective reagents for acylation of nucleoside and nucleotide aminoacyl derivatives. Biochem. Biophys. Res. Commun. 35, 306 (1969).

48. CHONG, K. J., and T. HATA: p-Nitrophenyl phosphates as phosphorylating reagents for alcohols. Bull. Chem. Soc. Jap. 44, 2741 (1971).

49. CHOU, J. Y., and M. F. SINGER: Synthesis of a copolymer containing adenylic and deoxyadenylic acid residues with polynucleotide phosphorylase. Biochem. Biophys. Res. Commun. 42, 306 (1971).

50. — — Deoxyadenosine diphosphate as a substrate and inhibitor of polynucleotide phosphorylase of Micrococcus luteus. I. Deoxyadenosine diphosphate as a substrate for

polymerization and the exchange reaction with inorganic ^{32}P. J. Biol. Chem. **246,** 7486 (1971).

51. Chou, J. Y., and M. F. Singer: Deoxyadenosine diphosphate as a substrate and inhibitor of polynucleotide phosphorylase of Micrococcus luteus. III. Copolymerization of adenosine diphosphate and deoxyadenosine diphosphate. J. Biol. Chem. **246,** 7505 (1971).

52. Christensen, L. F., and A. D. Broom: Specific chemical synthesis of ribonucleoside O-Benzyl ethers. J. Org. Chem. **37,** 3398 (1972).

53. Clark, V. M., D. W. Hutchinson, A. J. Kirby, and S. G. Warren: Phosphorylierungsmittel — Bauprinzip und Reaktionsweise. Angew. Chem. **76,** 704 (1964).

54. Cook, A. F.: The use of β,β,β-tribromoethylchloroformate for the protection of nucleoside hydroxyl groups. J. Org. Chemistry **33,** 3589 (1968).

55. Cook, A. F., M. J. Holman, and A. L. Nussbaum: Nucleoside S-alkyl phosphorothioates. II. Preparation and chemical and enzymatic properties. J. Amer. Chem. Soc. **91,** 1522 (1969).

56. — — Nucleoside S-alkyl phosphorothioates. III. Application to oligonucleotide synthesis. J. Amer. Chem. Soc. **91,** 6479 (1969).

57. Cook, A. F., and D. T. Maichuck: Use of chloroacetic anhydride for the protection of nucleoside hydroxyl groups. J. Org. Chem. **35,** 1940 (1970).

58. Cook, A. F., E. P. Heimer, M. J. Holman, D. T. Maichuk, and A. L. Nussbaum: Nucleoside S-alkyl phosphorothioates. V. Synthesis of a tridecadeoxyribonucleotide. J. Amer. Chem. Soc. **95,** 1334 (1972).

59. Cooperman, B. S., G. J. Lloyd, and C.-M. Hsu: Reactivity of phosphorylimidazole, an analog of known phosphorylated enzymes. J. Amer. Chem. Soc. **93,** 4889 (1971).

60. Cozzarelli, N. R., N. E. Melechen, T. M. Jovin, and A. Kornberg: Polynucleotide cellulose as a substrate for polynucleotide ligase induced by phage T4. Biochem. Biophys. Res. Comm. **28,** 578 (1967).

61. Cramer, F.: Probleme der chemischen Polynucleotidsynthese. Angew. Chem. **73,** 49 (1961).

62. Cramer, F., H. Neunhoeffer, K. H. Scheit, G. Schneider, and J. Tennigkeit: Neue Phosphorylierungsreaktionen und Schutzgruppen für Nucleotide. Angew. Chem. **74,** 387 (1962).

63. Cramer, F., H. P. Bär, H. J. Rhaese, W. Saenger, K. H. Scheit, G. Schneider, and J. Tennigkeit: Stabilität von Schutzgruppen für Nucleoside und Nucleotide. Tetrahedron Letters 1039 (1963).

64. Cramer, F., and K. H. Scheit: Über Benzhydrylester von Nucleotiden. Liebigs Ann. Chem. **679,** 150 (1964).

65. Cramer, F.: Die Synthese von Oligo- und Polynucleotiden. Angew. Chem. **78,** 186 (1966).

66. Cramer, F., and T. Hata: Chemie der energiereichen Phosphate. XIX. Phosphorylierung mit der Additionsverbindung aus Bromcyanacetamid und Triphenylphosphin. Liebigs Ann. Chem. **692,** 22 (1966).

67. Cramer, F., R. Helbig, H. Hettler, K. H. Scheit, and H. Seliger: Oligonucleotid-Synthese an einem löslichen Polymeren als Träger. Angew. Chem. **12,** 640 (1966).

68. Cramer, F., and H. Köster: Synthesis of oligonucleotides on a polymeric carrier. Angew. Chem. **80,** 488 (1968). Angew. Chem. Internat. Edit. **7,** 473 (1968).

69. Cramer, F.: Chemical synthesis of oligo- and polynucleotides. Pure Appl. Chem. **18,** 197 (1969).

70. — Recent methods of phosphorylation and the application to nucleotide chemistry. Colloq. Int. Cent. Nat. Rech. Sci. 343 (1970).

71. Cramer, F., E. M. Gottschalk, H. Matzura, K. H. Scheit, and H. Sternbach: The synthesis of the alternating copolymer poly[$r(A-s^4U)$] by RNA polymerase of Escherichia coli. Eur. J. Biochem. **19,** 379 (1971).

72. CUATRECASES, P.: Protein purification by affinity chromatography. Derivatizations of agarose and polyacrylamide beads. J. Biol. Chem. **245**, 3059 (1970).

73. CUSACK, N. J., C. B. REESE, and J. H. VAN BOOM: Block synthesis of oligonucleotides by the phosphotriester approach. Tetrahedron Letters 2209 (1973).

74. DARLIX, J. L., and P. FROMAGEOT: 5'-Monophosphonucleotides. Fr. Pat. 1,539,962 (Cl. C 07f), 20. Sep. 1968, Appl. 11. Aug. 1967. Chem. Abstr. **72**, 32179 t (1970).

75. ECKSTEIN, F., and I. RIZK: Oligonucleotidsynthesen mit Phosphorsäure-β,β,β-trichloräthylester-dichlorid. Angew. Chem. **79**, 939 (1967).

76. — — Synthesis of oligonucleotides by use of phosphoric triesters. Angew. Chem. **6**, 695 (1967).

77. ECKSTEIN, F., and H. GINDL: Polyribonucleotides containing a thiophosphate backbone. FEBS Letters **2**, 262 (1969).

78. ECKSTEIN, F., and I. RIZK: Synthese von Oligodesoxynucleotiden über Phosphorsäuretriester. Chem. Ber. **102**, 2362 (1969).

79. ECKSTEIN, F.: Nucleoside phosphorothioates. J. Amer. Chem. Soc. **92**, 4718 (1970).

80. ECKSTEIN, F., and H. GINDL: Polyribonucleotides containing a phosphorothioate backbone. Eur. J. Biochem. **13**, 558 (1970).

81. ECKSTEIN, F., and K.-H. SCHEIT: Procedures in nucleic acid research. **2**, 665 (G. L. CANTONI and D. R. DAVIES, eds.). New York: Harper and Row. 1971.

82. ECKSTEIN, F.: Protection of phosphoric and related acids. In: Protective groups in organic chemistry. (J. F. W. McOMIE, ed.). London-New York: Plenum Press. 1973.

83. ENDO, T., K. IKEDA, Y. KAWAMURA, and Y. MIZUNO: 1-oxidopyridine-2-yl-diazomethane. A water-soluble alkylating agent for nucleosides and nucleotides. J. Chem. Soc. D, Chem. Comm. 673 (1973).

84. ERMISHKINA, S. A., and A. M. YURKEVICH: 2',3'-O-phenyiboric esters of ribonucleosides in the synthesis of diribonucleoside phosphates. Zh. Obshch. Khim. **40**, 652 (1970).

85. ETO, M., M. SASAKI, M. IIO, M. ETO, and H. OHKAWA: Synthesis of 2-(methylthio)-4H-1,3,2-benzodioxaphosphorine-2-oxide by thiono-thiol conversion and its use as a phosphorylating agent. Tetrahedron Letters 4263 (1971).

86. FAERBER, P., K.-H. SCHEIT, and H. SOMMER: A new poly-nucleotide complex poly-(s^2C) . poly(I). Eur. J. Biochem. **27**, 109 (1972).

87. FALK, W., and C. TAMM: Nucleoside und Nucleotide. Teil 3. Über die Polykondensation von Thymidin-3'-phosphat nach der Triestermethode. Helv. Chim. Acta **55**, 1928 (1972).

88. FARMER, P. B.: personal communication.

89. FEIX, G., R. POLLET, and C. WEISSMANN: Replication of Viral RNA. XVI. Enzymatic Synthesis of Infectious Viral RNA with Noninfectious Q_β Minus Strands as Template. Proc. Natl. Acad. Sci. US **59**, 145 (1968).

90. FEIX, G.: Oligoribonucleotides as primer for terminal deoxynucleotidyl transferase. FEBS Letters **18**, 280 (1971).

91. — Enzymatic synthesis of polydeoxynucleotides covalently linked to an oligoribonucleotide primer. Biochem. Biophys. Res. Commun. **46**, 2141 (1972).

92. — Initiation of DNA synthesis by oligoribonucleotides. In: Gene expression and its regulation (F. T. KENNEY, B. A. HAMKALO, G. FAVELUKES, and J. T. AUGUST, eds.), p. 301. New York: Plenum Press. 1973.

93. FILIP, J., and L. BOHACEK: Preparation of uridine 5'-monophosphate-5-^{3}H and uridine 5'-monophosphate-6-^{3}H with high molar activity. Radioisotopy **12**, 343 (1971).

94. FRANKE, A., F. ECKSTEIN, K. H. SCHEIT, and F. CRAMER: Synthese von Oligo- und Polynucleotiden. XVI. Synthese von Desoxyoligonucleotiden mit der Trichloräthylphosphatschutzgruppe. Chem. Ber. **101**, 944 (1968).

95. FRANKE, A., K. H. SCHEIT, and F. ECKSTEIN: Selektive Phosphorylierung von Nucleosiden. Chem. Ber. **101**, 2998 (1968).

96. Freist, W., and F. Cramer: Synthese von Oligonucleotid-5'-phosphaten an einem polymeren Träger mit 2-(α-Pyridyl)-äthanol als funktioneller Gruppe. Angew. Chem. **82,** 358 (1970).

97. Freist, W., and F. Cramer: Synthese von Oligodesoxynucleotiden mit 2-[α-Pyridyl]-äthanol als Phosphatschutzgruppe. Chem. Ber. **103,** 3122 (1970).

98. Freist, W., R. Helbig, and F. Cramer: 2-[α-Pyridyl]-äthanol als Phosphatschutzgruppe. Chem. Ber. **103,** 1032 (1970).

99. Fromageot, H. P. M., B. E. Griffin, C. B. Reese, and J. E. Sulston: Synthesis of oligoribonucleotides. III. Monoacylation of ribonucleosides and derivatives by orthoester exchange. Tetrahedron **23,** 2315 (1967).

100. Fujimoto, Y. (Kyowa Fermentation Industry Co., Ltd.): 2',3'-O-Substituted ribonucleosides. Jap. Pat. 68 25,496 (Cl. 16 E 362), 4. Nov. 1968, Appl. 10. Nov. 1965. Chem. Abstr. **70,** 68707 m (1969).

101. — 2',3'-O-Substituted ribonucleosides. Jap. Pat. 68 25,498 (Cl. 16 E 431), 4. Nov. 1968, Appl. 10. Nov. 1965. Chem. Abstr. **70,** 68708 n (1969).

102. Gallo, R. C.: Reverse Transcriptase — The DNA Polymerase of Oncogenic RNA Viruses. Nature **234,** 194 (1973).

103. Gassen, H. G.: Synthesis by ribonuclease A of codons containing modified nucleosides in the "wobble" position. FEBS Letters **14,** 225 (1971).

104. Gassen, H. G., and R. Nolte: Synthesis by polymer-bound ribonuclease of the termination codons U-A-A, U-A-G, and U-G-A. Biochem. Biophys. Res. Commun. **44,** 1410 (1971).

105. Gassen, H. G.: personal communication.

106. Gillam, S., and M. Smith: Enzymatic synthesis of deoxyribo-oligonucleotides of defined sequence. Nature **238,** 233 (1972).

107. Glaser, R., U. Sequin, and C. Tamm: Nucleoside und Nucleotide. Festphasensynthese von Oligonucleotiden an einem unlöslichen, makroporösen Träger. Helv. Chim. Acta **56,** 654 (1973).

108. Glinski, R. P., A. B. Ash, C. L. Stevens, M. B. Sporn, and H. M. Lazarus: Nucleotide synthesis. I. Derivatives of thymidine containing p-nitrophenyl phosphate groups. J. Org. Chem. **36,** 245 (1971).

109. Glinski, R. P., C. C. Bacon, and C. L. Stevens: Synthesis of partially protected oligonucleotides. Presented at 161st National Meeting of American Chemical Society, Los Angeles, Calif., No. CARB 29 (1971).

110. Goldberg, I. H.: Preparation and properties of polypseudouridylic acid. In: Methods in enzymology (L. Grossman and K. Moldave, eds.), **XII B,** 519. New York-London: Academic Press. 1968.

111. Goulian, M., A. Kornberg, and R. L. Sinsheimer: Enzymatic Synthesis of DNA. XXIV. Synthesis of Infectious Phage φX 174 - DNA. Proc. Natl. Acad. Sci. **58,** 2321 (1967).

112. Grams, G. W., and R. L. Letsinger: N^6,3'-O-disubstituted deoxyadenosine. J. Org. Chem. **33,** 2589 (1968).

113. — — Synthesis of a diribonucleoside monophosphate by the β-cyanoethyl phosphotriester method. J. Org. Chem. **35,** 868 (1970).

114. Grant, R. C., S. J. Harwood, and R. D. Wells: The synthesis and characterization of poly d(I-C) . poly d(I-C). J. Amer. Chem. Soc. **90,** 4474 (1968).

115. Grant, R. C., M. Kodama, and R. D. Wells: Enzymatic and physical studies on $(dI-dC)_n . (dI-dC)_n$ and $(dG-dC)_n . (dG-dC)_n$. Biochemistry **11,** 805 (1972).

116. Green, D. P. L., T. Ravindranathan, C. B. Reese, and R. Saffhill: Synthesis of oligoribonucleotides. VIII. Preparation of ribonucleoside 2',5'-bisacetals. Tetrahedron Letters 1031 (1970).

117. GRIFFIN, B. E., M. JARMAN, C. B. REESE, and R. E. SULSTON: The synthesis of oligoribonucleotides. II. Methoxymethylidene derivatives of ribonucleosides and 5'-ribonucleotides. Tetrahedron **23**, 2301 (1967).

118. GRINEVA, N. I., V. F. ZARYTOVA, D. G. KNORRE, and E. V. YARMOLINSKAYA: Alkylating derivatives of nucleic acid components. XI. Mechanism of formation of benzylidene derivatives of nucleotides. Izv. Sib. Otd. Akad. Nauk. Chem. Abstr. **76**, 107 (1971).

119. GRÜNBERGER, D., A. HOLÝ, and F. ŠORM: Synthesis of Triribonucleoside Diphosphates with Ribonuclease T₁. Coll. Czech. Chem. Comm. **33**, 286 (1968).

120. GUPTA, N. K., E. OHTSUKA, H. WEBER, S. H. CHANG, and H. G. KHORANA: Studies on Polynucleotides. LXXXVII. The Joining of Short Deoxyribopolynucleotides by DNA Joining Enzymes. Proc. Natl. Acad. Sci. US **60**, 285 (1968).

121. GUPTA, N. K., E. OHTSUKA, V. SGARAMELLA, H. BÜCHI, A. KUMAR, H. WEBER, and H. G. KHORANA: Studies on Polynucleotides. LXXXVIII. Enzymatic Joining of Chemically Synthesized Segments Corresponding to the Gene for Alanine Transfer RNA. Proc. Natl. Acad. Sci. US **60**, 1338 (1968).

122. HAEFFNER, E. W.: Studies on the thermic phosphorylation of activated nucleoside by phosphate anion and nucleotide anion. Biochim. Biophys. Acta **212**, 182 (1970).

123. HACHMANN, J., and H. G. KHORANA: Studies on polynucleotides. XCIII. A further study of the synthesis of deoxyribopolynucleotides using preformed oligonucleotide blocks. J. Amer. Chem. Soc. **91**, 2749 (1969).

124. HAGA, K., M. KAINOSHO, and M. YOSHIKAWA: Phosphorylation. V. Synthesis of inosine-5'-thiophosphates. Bull. Chem. Soc. Jap. **44**, 460 (1973).

125. HAGENBERG, L., H. G. GASSEN, and H. MATTHAEI: Synthesis and coding properties of poly(c¹A), poly(c³A), poly(c⁷A) and poly(h⁶A). Biochem. Biophys. Res. Commun. **50**, 1104 (1973).

126. HALMANN, M., R. A. SANCHEZ, and L. E. ORGEL: Phosphorylation of D-ribose in aqueous solution. J. Org. Chemistry **34**, 3702 (1969).

127. HANSBURY, E., V. N. KERR, V. E. MITCHELL, R. L. RATLIFF, D. A. SMITH, D. L. WILLIAMS, and F. N. HAYES: Synthesis of polydeoxynucleotides using chemically modified subunits. Biochim. Biophys. Acta **199**, 322 (1970).

128. HARVEY, C. L., E. M. CLERICUZIO, and A. L. NUSSBAUM: Small-scale preparation of 5'-nucleotides and analogs by carrot phosphotransferase. Anal. Biochem. **36**, 413 (1970).

129. HARVEY, C. L., R. WRIGHT, A. F. COOK, D. T. MAICHUK, and A. L. NUSSBAUM: Use of phosphate-blocking groups in ligase joining of oligodeoxyribonucleotides. Biochemistry **12**, 208 (1973).

130. HARVEY, C. L., R. WRIGHT, and A. L. NUSSBAUM: Lambda phage DNA: Joining of a chemically synthesized cohesive end. Science **179**, 291 (1973).

131. HARVEY, C. L., A. DE CZEKALA, A. F. COOK, M. J. HOLMAN, T. F. GABRIEL, J. E. MICHALEWSKY, and A. L. NUSSBAUM: High pressure liquid chromatography applied to gene synthesis. Biochim. Biophys. Acta **324**, 433 (1973).

132. HASHIZUME, T.: Synthesis of biochemically significant organic phosphate compounds I. Bis-p-nitrophenyl-phosphorochloridate as a phosphorylating agent. Mem. Coll. Agr., Kyoto Univ., Chem. Ser. **81**, 1 (1959). Chem. Abstr. **57**, 14157 (1962).

133. HATA, T., and J. AZIZIAN: 2-chloroethyl orthoformate as a reagent for protection in nucleotides synthesis. Tetrahedron Letters 4443 (1969).

134. HATA, T., Y. MUSHIKA, and T. MUKAIYAMA: New phosphorylating reagent. I. Preparation of alkyl dihydrogen phosphates by means of 2-chloromethyl-4-nitrophenyl phosphorodichloridate. J. Amer. Chem. Soc. **91**, 4532 (1969).

135. — — — New phosphorylating reagent. II. Preparation of mixed diesters of phosphoric acid by the use of alkyl-2-chloromethyl-4-nitrophenyl hydrogen phosphate. Tetrahedron Letters 3505 (1970).

136. Hata, T., K. Tajima, and T. Mukaiyama: Simple protecting group protection-purification "handle" for polynucleotide synthesis. I. J. Amer. Chem. Soc. **93**, 4928 (1971).

137. Hata, T., and K. J. Chong: p-Nitrophenyl phosphate as a phosphorylating reagent in nucleotide synthesis. Bull. Chem. Soc. Jap. **45**, 654 (1972).

138. Hata, T., I. Nakagawa, and N. Takebayashi: Simple protecting group protection-purification handle for polynucleotide synthesis. III. New method for the synthesis of dinucleotides. Tetrahedron Letters 2931 (1972).

139. Hayashi, H., and F. Egami: Fractionation and Properties of Guanylic acid Polymers synthesized by Ribonuclease T_1. J. Biochem. (Tokyo) **53**, 176 (1963).

140. Hayatsu, H., and H. G. Khorana: Deoxyribooligonucleotide synthesis on a polymer support. J. Amer. Chem. Soc. **88**, 3182 (1966).

141. — — Studies on Polynucleotides. LXXII. Deoxyribooligonucleotide synthesis on a polymer support. J. Amer. Chem. Soc. **89**, 3880 (1967).

142. Hayes, F. N., V. E. Mitchell, R. L. Ratliff, and D. L. Williams: Limited enzymatic addition of deoxyribonucleotide units onto chemically synthesized oligodeoxyribo-5'-nucleotides. Biochemistry **6**, 2488 (1967).

143. Hayes, F. N., E. Hansbury, V. E. Mitchell, R. L. Ratliff, and D. L. Williams: Synthesis of N-acetylated deoxyribonucleoside 5'-triphosphates and their utilization in enzymatic formation of single-stranded polydeoxyribonucleotides. Eur. J. Biochem. **6**, 485 (1968).

144. Hayes, F. N., and V. E. Mitchell: Gel filtration chromatography of polydeoxynucleotides using agarose columns. J. Chromatog. **39**, 139 (1969).

145. Heimer, E. P., M. Ahmad, and A. L. Nussbaum: Chemical synthesis of the "sticky end" of lambda phage DNA r-strand. Biochem. Biophys. Res. Commun. **48**, 348 (1972).

146. Heimer, E., M. Ahmad, S. Roy, A. Ramel, and A. L. Nussbaum: Nucleoside S-alkyl phosphorothioates. VI. Synthesis of deoxyribonucleotide oligomers. J. Amer. Chem. Soc. **94**, 1707 (1972).

147. Helbig, R.: Oligonucleotidsynthesen am polymeren Träger. Dissertation. Techn. Hochschule Braunschweig, 1967.

148. Hobbs, J., H. Sternbach, and F. Eckstein: Poly 2'-deoxy-2'-chlorouridylic and -cytidylic acids. FEBS Letters **15**, 345 (1971).

149. — — — Poly 2'-deoxy-2'-aminouridylic acid. Biochem. Biophys. Res. Commun. **46**, 1509 (1972).

150. Hobbs, J., H. Sternbach, M. Sprinzl, and F. Eckstein: Polynucleotides containing 2'-chloro-2'-deoxyribose. Biochemistry **11**, 4336 (1972).

151. — — — — Polynucleotides containing 2'-amino-2'-deoxyribose and 2'-azido-2'-deoxyribose. Biochemistry **12**, 5138 (1973).

152. Holy, A., and J. Smrt: Oligonucleotidic compounds. XV. A general approach to the stepwise synthesis of ribooligonucleotides. Synthesis of some triribonucleoside diphosphates. Coll. Czech. Chem. Comm. **31**, 3800 (1966).

153. Holy, A., S. Chladek, and J. Zemlicka: Oligonucleotidic compounds. XXIX. Reactions of ribonucleoside 2'(3')-phosphates with dimethylformamide acetals. Collect. Czech. Chem. Commun. **34**, 253 (1969).

154. Holy, A., and J. Zemlicka: Oligonucleotidic compounds. XXXIII. A study on hydrolysis of N-dimethylaminomethylene-cytidine, -adenosine, -guanosine and related 2'-deoxy compounds. Collect. Czech. Chem. Commun. **34**, 2449 (1969).

155. — — Oligonucleotidic compounds. XXXV. Reaction of diribonucleoside phosphates with dimethylformamide acetals. Collect. Czech. Chem. Commun. **34**, 3921 (1969).

156. Holy, A., and G. Kowollik: Nucleic acid components and their analogs. CXXXI. Simple enzymic synthesis of nucleoside-5'-phosphates. Coll. Czech. Chem. Comm. **35**, 1013 (1970).

157. HOLY, A., and M. SOUCEK: Benzoyl cyanide — new benzoylating agent in nucleoside and nucleotide chemistry. Tetrahedron Letters 185 (1971).

158. HOLY, A.: Phosphorylation of nucleosides with trichloromethylphosphonic acid derivatives. Tetrahedron Letters 157 (1972).

159. HONJO, M.,Y. FURUKAWA, K. KOBAYASHI, and R. MARUMOTO (Takeda Chem. Industries, Ltd.): N-Acyl-2',3',5'-tri-O-acyl-cytidines. Ger. Offen. 2,038,807 (Cl. C 07d), 18. Feb. 1971, Appl. 6. Aug. 1969. Chem. Abstr. **74**, 100354 q (1971). Chem. Abstr. **74**, 100354 d (1971).

160. HONJO, M., S. YOSHIKAWA, K. KOBAYASHI, and R. MARUMOTO: $N^4,2',3',5'$-tetra-alkanoylcytidines. Japan Patent 71 37,827 (1971). Chem. Abstr. **76**, 34533k (1972).

161. HUDSON, R. F., and M. GREEN: Die Stereochemie von Substitutionsreaktionen am Phosphor. Angew. Chem. **75**, 47 (1963).

162. IKEHARA, M., and K. MURAO: Nucleosides and Nucleotides. XXXVII. Synthesis of 8-oxoguanosine nucleotides and uric acid-9-D-riboside-5'-phosphate. Chem. Pharm. Bull. (Tokyo) **16**, 1330 (1968).

162a. IKEHARA, M., and S. UESUGI: Selective Tosylation of Adenosine-5'-phosphate. Tetrahedron Letters 713 (1970).

162b. — — Studies on Nucleosides and Nucleotides. LIII. Purine Cyclonucleosides. XVIII. Selective Tosylation of Adenine Nucleotides. Synthesis of 8,2'-Anhydro-8-mercapto-9-β-D-arabinofuranosyl adenine and 5'- and 3'-5'-cyclic phosphate. Tetrahedron **28**, 3687 (1972).

162c. — — Studies on Nucleosides and Nucleotides. LV. Reaction of Cyridine-5'-monophosphate with p-Toluene-sulfonylchloride. Chem. Pharm. Bull. (Tokyo) **21**, 264 (1973).

163. IMURA, N., T. TSURUO, and T. UKITA: On the benzylation of nucleosides. I. Reaction of uridine with benzyl bromide in the presence of sodium hydride. Chem. Pharm. Bull. (Tokyo) **16**, 1105 (1968).

164. IRIE, S.: Selective phosphorylation of thionucleosides. J. Biochem. (Tokyo) **68**, 129 (1970).

165. IRIE, S., T. UCHIDA, and F. EGAMI: Synthesis and ribonuclease degradation of dinucleoside monophosphates containing a thionucleoside. Biochim. Biophys. Acta **209**, 289 (1970).

166. JACOB, T. M., and H. G. KHORANA: Studies on polynucleotides. XXX. A comparative study of reagents for the synthesis of the C_3,-C_5, internucleotidic linkage. J. Amer. Chem. Soc. **86**, 1630 (1964).

167. JANIK, B., M. P. KOTICK, T. H. KREISER, L. F. REVERMAN, R. G. SOMMER, and D. P. WILSON: Synthesis and properties of poly 2'-fluoro-2'-deoxyuridylic acid. Biochem. Biophys. Res. Commun. **46**, 1153 (1972).

168. JANIK, B., R. G. SOMMER, M. P. KOTICK, D. P. WILSON, and R. K. ERICKSON: Synthesis and properties of poly($1,N^6$-ethenoadenylic acid) and poly($3,N^4$-ethenocytidylic acid). Physiol. Chem. Phys. **5**, 27 (1973).

169. JANION, C., and D. SHUGAR: Mechanism of hydroxylamine mutagenesis: Complexing properties of copolymers of hydroxycytidylic acid with cytidylic or uridylic acids. Acta Biochim. Polon. **16**, 219 (1969).

170. JOVIN, T. M., and A. KORNBERG: Oligonucleotide celluloses as solid state primers and templates for polymerases. J. Biol. Chem. **243**, 250 (1968).

171. KABACHNIK, M. M., I. A. POLYAKOVA, V. K. POTAPOV, Z. A. SHABAROVA, and M. A. PROKOF'EV: Phosphorylation of nucleosides on polymeric carriers. Dokl. Akad. Nauk SSSR **195**, 1344 (1970).

172. KABACHNIK, M. M., V. K. POTAPOV, Z. A. SHABAROVA, and M. A. PROKOF'EV: Oxidative phosphorylation of nucleosides. Dokl. Akad. Nauk. SSSR **195**, 1107 (1970).

173. — — — — A new method of synthesis of internucleotide bonds using polymeric supports. Dokl. Akad. Nauk. SSSR **201**, 858 (1971).

174. Kabachnik, M. M., N. G. Timofeeva, M. V. Budanov, V. K. Potapov, Z. A. Shabarova, and M. A. Prokof'ev: Synthesis of oligonucleotides on a polymer carrier. Zhur. Obshch. Khim. **43**, 379 (1973).

175. Kapuler, A. M., C. Monny, and A. M. Michelson: The relationship of mono- and polynucleotide conformation to catalysis by polynucleotide phosphorylase. Biochim. Biophys. Acta **217**, 18 (1970).

176. Katagiri, N., C. P. Bahl, K. Itakura, J. Michniewicz, and S. A. Narang: Use of 9-fluorenylmethanol as phosphate protecting group in the synthesis of deoxyribo-oligonucleotides. J. Chem. Soc. D, Chem. Comm. 803 (1973).

176a. Katagiri, N., K. Itakura, and S. A. Narang: Novel condensing reagents for polynucleotide synthesis. J. Chem. Soc. D, Chem. Comm. 325 (1974).

177. Kathawala, F., and F. Cramer: Synthese von Oligo- und Polynucleotiden. XIII. 2',3'-(2,4-Dimethoxybenzyliden-) als Phosphatschutzgruppe. Liebigs Ann. Chem. **709**, 185 (1967).

178. — — Synthese von Oligo- und Polynucleotiden. XIV. Darstellung von Desoxyoligo-nucleotiden mit 2',3'-(2,4-Dimethoxybenzyliden-)uridin als Phosphatschutzgruppe. Liebigs Ann. Chem. **712**, 195 (1968).

178a. Kaufmann, G., and U. Z. Littauer: Deoxyadenosine diphosphate as substrate for polynucleotide phosphorylase from *Escherichia coli*. FEBS Letters **2**, 79 (1969).

179. Kaufmann, G., M. Fridkin, A. Zutra, and U. Z. Littauer: Monofunctional sub-strates of polynucleotide phosphorylase. Eur. J. Biochem. **24**, 4 (1971).

180. Kaufmann, G., A. Zutra, and U. Z. Littauer: Synthesis of the heptanucleotide U-U-U-G-A-A-G using isovaleryl nucleoside diphosphates and sepharose bound polynucleotide phosphorylase. Israel J. Chem. **9**, 44 BC (1971).

181. Kavunenko, A. P., E. N. Morozova, and N. S. Tikhomirova-Sidorova: Preparation of purine-pyrimidine dinucleotides with terminal 2',3'-cyclophosphate and their use for the synthesis of trinucleoside diphosphates. Zh. Obshch. Khim. **41**, 226 (1971).

182. Kavunenko, A. P., V. P. Sukharevich, and N. S. Tikhomirova-Sidorova: Water-soluble carbodiimide in oligoribonucleotide synthesis catalyzed by pancreatic ribo-nuclease. Zh. Obshch. Khim. **41**, 679 (1971).

183. Kelly, R. C., W. J. Wechter, and D. T. Gish (Upjohn Co.): 5'-O-Derivatives of ara-cytidine. Ger. Offen. 2,025,624 (Cl. C 07d), 3. Dec. 1970, US Appl. 27. May 1969 — 16. Feb. 1970. Chem. Abstr. **74**, 54151 w (1971).

184. Khorana, H. G.: Recent developments in the chemistry of phosphate esters of biological interest. New York: John Wiley & Sons, Inc. 1961.

185. — Synthesis in the study of nucleic acids. Proc. 7th Intern. Congr. Biochem., p. 17, Tokyo (1967).

186. — Polynucleotide synthesis and the genetic code. The Harvey Lectures **62**, 79 (1968).

187. — Nucleinsäuresynthese als Werkzeug für das Studium des Genetischen Codes (Nobel-Vortrag). Angew. Chem. **81**, 1027 (1969).

188. — Total synthesis of the gene for an alanine transfer ribonucleic acid from yeast. Pure Appl. Chem. **25**, 91 (1971).

189. Khorana, H. G., K. L. Agarwal, H. Büchi, M. H. Caruthers, N. K. Gupta, K. Kleppe, A. Kumar, E. Ohtsuka, U. L. Raj Bhandary, J. H. van de Sande, V. Sgaramella, T. Terao, H. Weber, and T. Yamada: Studies on polynucleotides. CIII. Total synthesis of the structural gene for an alanine transfer ribonucleic acid from yeast. J. Mol. Biol, **72**. 209 (1972).

190. Khorana, H. G., K. L. Agarwal, P. Besmer, H. Büchi, M. H. Caruthers, P. J. Cashion, M. Fridkin, E. Jay, D. G. Kleid, A. Kumar, P. C. Loewen, R. Miller, K. Minamoto, R. Rama Moorthy, A. Panet, J. H. van de Sande, T. Sekiya, and N. Sidorova: Synthesis of the gene for the precursor of *E. coli*

tyrosine suppressor tRNA. Abstracts of the 166th Meeting of the Amer. Chem. Soc. (1973).

191. KHWAJA, T. A., and C. B. REESE: Phosphorylation of nucleosides with o-phenylene-phosphorochloridate and o-phenylene phosphate. Tetrahedron **27**, 6189 (1971).

192. KIKUGAWA, K., F. SATO, T. TSURUO, N. IMURA, and T. UKITA: On the benzylation of nucleosides. II. A novel synthesis of 2'-O-benzyl uridine. Chem. Pharm. Bull. (Tokyo) **16**, 1110 (1968).

193. KIMHI, Y., and U. Z. LITTAUER: Polynucleotide phosphorylase from *Escherichia coli*. In: Methods in enzymology **XII B**, 513 (L. GROSSMAN and K. MOLDAVE, eds.). New York-London: Academic Press. 1968.

194. KLEE, C. B.: Procedures in nucleic acid research. **2**, 896 (G. L. CANTONI and D. R. DAVIES, eds.). New York: Harper and Row. 1971.

195. KLEPPE, K., J. H. VAN DE SANDE, and H. G. KHORANA: Polynucleotide ligase-catalyzed joining of deoxyribo-oligonucleotides on ribopolynucleotide templates and of ribo-oligonucleotides on deoxyribopolynucleotide templates. Proc. Nat. Acad. Sci. **67**, 68 (1970).

196. KNORRE, D. G., E. F. MISHENINA, T. I. SHUBINA: Copolymer of acrylamide and 5'-O-acrylylguanosine-2',3'-cyclophosphate as a substrate of guanyl-RNase during the formation of the internucleotide bond. Dokl. Akad. Nauk. SSSR **198**, 1089 (1970).

197. KOCHETKOV, N. K., and E. J. BUDOVSKII: Organic chemistry of nucleic acids. London-New York: Plenum Press. 1971.

198. KOGAN, E. M., E. N. MOROZOVA, N. S. TIKHOMIROVA-SIDOROVA, and G. E. USTYUZHA-NIN: Synthesis of purine-pyrimidine triribonucleotides in the presence of pancreatic ribonuclease. Zh. Obshch. Khim. **39**, 2576 (1969).

199. KOIKE, T., T. UCHIDA, and F. EGAMI: Synthesis of guanylyl-(3',5')-nucleosides and oligoguanylic acids by ribonuclease N_1. Biochim. Biophys. Acta **190**, 257 (1969).

200. KOIKE, T., T. UCHIDA, and F. EGAMI: Synthesis of adenylyl-(3',5')-nucleosides, adenylyl-(3',5')-guanosine 2',3'-cyclic phosphate, and oligoadenylic acids by ribo-nuclease U_2. J. Biochem. (Tokyo) **69**, 111 (1971).

201. KOIKE, T., T. UCHIDA, and F. EGAMI: Synthesis of oligo-ApGp and other oligo-nucleotides by ribonuclease N_1. J. Biochem. (Tokyo) **70**, 55 (1971).

202. KOLODKINA, I. I., A. S. GUSEVA, E. A. IVANOVA, L. S. VARSHAVSKAYA, and A. M. YURKE-VICH: Synthesis and properties of areneboronates of nucleosides and nucleotides. Zh. Obshch. Khim. **40**, 2489 (1970).

203. KOLODKINA, E. E., E. A. IVANOVA, and A. M. YURKEVICH: Ion-exchange chromato-graphy of arylboronic acid complexes of nucleosides and mononucleotides. Khim. Prir. Soedin. **6**, 612 (1970). Chem. Abstr. **74**, 112363 e (1971).

204. KORNBERG, A.: Adenosine phosphokinase. In: Methods in Enzymology, Vol. II. (S. P. COLOWICK and N. O. KAPLAN, eds.), p. 497. New York: Academic Press. 1955.

205. KÖSSEL, H., H. BÜCHI, and H. G. KHORANA: Studies on polynucleotides. LXV. The synthesis of deoxyribopolynucleotides containing repeating tetranucleotide sequences J. Amer. Chem. Soc. **89**, 2185 (1967).

206. KÖSSEL, H., A. R. MORGAN, and H. G. KHORANA: Studies on polynucleotides. LXXIII. Synthesis in vitro of polypeptides containing repeating tetrapeptide sequences dependent upon DNA-like polymers containing repeating tetranucleotide sequences: direction of reading of messenger RNA. J. Mol. Biol. **26**, 449 (1967).

207. KÖSSEL, H., M. W. MOON, and H. G. KHORANA: Studies on Polynucleotides. LX. The Use of Preformed Dinucleotide Blocks in the Stepwise Synthesis of Deoxyribopoly-nucleotides. J. Amer. Chem. Soc. **89**, 2148 (1967).

208. KÖSSEL, H., and R. ROYCHOUDHURY: Synthetic polynucleotides. The terminal addition of riboadenylic acid to deoxyoligonucleotides by terminal deoxynucleotidyl transferase as a tool for the specific labelling of deoxyoligonucleotides at the 3'-ends. Eur. J. Biochem. **22**, 271 (1971).

208a. Kössel, H., and R. Roychoudhury: Proofreading function of DNA polymerase I from *E. coli*. Nature of excision of ribonucleotides from the 3'-termini of oligodeoxynucleotide primers. J. Biol. Chem. **249**, 4094 (1974).

209. Kössel, H., and S. Kühn: unpublished.

210. Köster, H.: Polymer support oligonucleotide synthesis. VI. Inorganic carriers. Tetrahedron Letters 1527 (1972).

211. Köster, H., and K. Heyns: Polymer support oligonucleotide synthesis. VII. Use of Sephadex LH 20. Tetrahedron Letters 1531 (1972).

212. Köster, H.: Polymer support oligonucleotide synthesis. VIII. Use of polyethylene glycol. Tetrahedron Letters 1535 (1972).

213. Köster, H., and F. Cramer: Synthese von Oligonucleotiden an einem Popcorn-Polystyrol als polymerem Träger. Liebigs Ann. Chem. **766**, 6 (1972).

214. Köster, H., and S. Geussenhainer: Ein neuer Träger für die Festphasensynthese von Oligomeren. Angew. Chem. **84**, 712 (1972).

215. Köster, H., and F. Cramer: Reaktionskinetische Untersuchungen an einem makroporösen unquellbaren Polystyrol: Abspaltung von Nucleosiden und Nucleotiden, die über p-Anisyldiphenylmethylätherbindung an das Polymerisat gebunden sind. Makromol. Chem. **167**, 171 (1973).

216. Köster, H., and W. Heidmann: A new approach to the synthesis of oligodeoxyribonucleotides. Angew. Chem. **85**, 871 (1973). Angew. Chem. Internat. Edit. **12**, 859 (1973).

217. Köster, H., and F. Cramer: Synthese von Desoxyoligonucleotiden an einem makroporösen Polystyrol. Liebigs Ann. Chem. 946 (1974).

218. Köster, H., F. Pollack, and F. Cramer: Synthese der Desoxyoligonucleotide dT(pdT)₇ und dTpdTpdApdCpdCpdTpdA an einem makroporösen Polystyrol. Liebigs Ann. Chem. 959 (1974).

218a. Kowollik, G., K. Gaertner, and P. Langen: 2'- and 3'-O-trityluridine. Tetrahedron Letters 3345 (1972).

219. Kumar, A., E. Ohtsuka, and H. G. Khorana: Studies on Polynucleotides. CVI. Total synthesis of the structural gene for an alanine transfer ribonucleic acid from yeast. Synthesis of two nonanucleotides and a heptanucleotide corresponding to nucleotide sequences 22 to 30, 41 to 49 and 28 to 34. J. Mol. Biol. **72**, 289 (1972).

219a. Kumar, A., and H. G. Khorana: Studies on Polynucleotides. CVIII. Total Synthesis of the structural gene for an alanine transfer ribonucleic acid from yeast. Synthesis of an undecadeoxynucleotide, a decadeoxynucleotide and an octadeoxynucleotide corresponding to the nucleotide sequences 7 to 27. J. Mol. Biol. **72**, 329 (1972).

220. Kusama, T., and H. Hayatsu: Use of a derivatized merrifield resin for the polymer-supported synthesis of oligodeoxyribonucleotides. Chem. Pharm. Bull. (Tokyo) **18**, 319 (1970).

221. Lapidot, Y.: Chromatography of diribonucleoside monophosphates on a Bio-Gel P-4 column. J. Chromatog. **56**, 143 (1971).

222. Lapidot, Y., and I. Barzilay: The separation of 2'-5' dinucleoside monophosphates from the corresponding 3'—5'-isomers on a DEAE-sephadex A-25 column. J. Chromatog. **71**, 275 (1972).

223. Lapidot, Y., I. Barzilay, and D. Salomon: Ion-exchange thin-layer chromatography and paper ionophoresis of dinucleoside monophosphates. Anal. Biochem. **49**, 301 (1972).

224. Lehrach, H., and K. H. Scheit: Synthesis and properties of a new fluorescent polynucleotide, poly(1,N⁶-ethenoadenylic acid). Biochim. Biophys. Acta **308**, 28 (1973).

225. Lehrfeld, J.: Silicagel catalyzed detritylation of some carbohydrate derivatives. J. Org. Chemistry **32**, 2544 (1967).

226. Letsinger, R. L., M. J. Kornet, V. Mahadevan, and D. M. Jerina: Reactions on polymer supports. J. Amer. Chem. Soc. **86**, 5163 (1964).

227. LETSINGER, R. L., and V. MAHADEVAN: Oligonucleotide synthesis on a polymer support. J. Amer. Chem. Soc. **87**, 3526 (1965).

228. LETSINGER, R. L., and V. MAHADEVAN: Stepwise synthesis of oligodeoxyribonucleotides on an insoluble polymer support. J. Amer. Chem. Soc. **88**, 5319 (1966).

229. LETSINGER, R. L., and K. K. OGILVIE: Use of *p*-nitrophenyl chloroformate in blocking hydroxyl groups in nucleosides. J. Org. Chem. **32**, 296 (1966).

229a. LETSINGER, R. L., M. H. CARUTHERS, and D. M. JERINA: Reactions of nucleosides on polymer supports. Synthesis of thymidylyl-thymidylyl-thymidine. Biochemistry **6**, 1379 (1967).

230. LETSINGER, R. L., M. H. CARUTHERS, P. S. MILLER, and K. K. OGILVIE: Oligonucleotide syntheses utilizing β-benzoylpropionyl, a blocking group with a trigger for selective cleavage. J. Amer. Chem. Soc. **89**, 7146 (1967).

231. LETSINGER, R. L., and D. M. JERINA: Reactivity of ester groups on insoluble polymer supports. J. Polymer Science **5**, 1977 (1967).

232. LETSINGER, R. L., and K. K. OGILVIE: A convenient method for stepwise synthesis of oligothymidylate derivatives in large-scale quantities. J. Amer. Chem. Soc. **89**, 4801 (1967).

233. LETSINGER, R. L., P. S. MILLER, and G. W. GRAMS: Selective N-debenzoylation of N,O-polybenzoylnucleosides. Tetrahedron Letters 2621 (1968).

234. LETSINGER, R. L., and P. S. MILLER: Protecting groups for nucleosides used in synthesizing oligonucleotides. J. Amer. Chem. Soc. **91**, 3356 (1969).

235. LETSINGER, R. L., and K. K. OGILVIE: Synthesis of oligothymidylates via phosphotriester intermediates. J. Amer. Chem. Soc. **91**, 3350 (1969).

236. LETSINGER, R. L., K. K. OGILVIE, and P. S. MILLER: Developments in syntheses of oligodeoxyribonucleotides and their organic derivatives. J. Amer. Chem. Soc. **91**, 3360 (1969).

237. LETSINGER, R., and W. S. MUNGALL: Phosphoramidate analogs of oligonucleotides. J. Org. Chem. **35**, 3800 (1970).

238. LETSINGER, R. L., and H. H. SELIGER: Polymers with hydroxyl groups as supports for oligonucleotide synthesis. Macromol. Preprints. XXIII rd Internat. Congr. of pure and applied chemistry, Boston, 1261 (1971).

239. LEVINA, A. S., V. K. POTAPOV, D. G. KNORRE, Z. A. SHABAROVA, and T. M. SHUBINA: Individual stages of oligonucleotide synthesis on a highly crosslinked polymer support. Izv. Sib. Otd. Akad. Nauk. 117 (1972).

240. LEZIUS, A. G., and K. H. SCHEIT: Enzymatic synthesis of DNA with 4-thiothymidine triphosphate as substitute for dTTP. Eur. J. Biochem. **3**, 85 (1967).

241. LEZIUS, A. G.: Synthesis and characterization of a copolymer consisting of alternating deoxyadenosine- and 2-thiodeoxythymidine nucleotides. Eur. J. Biochem. **14**, 154 (1970).

242. LEZIUS, A. G., and E. M. GOTTSCHALK: Über eine reversible kooperative Konformationsumwandlung einer synthetischen DNA unter dem Einfluß hoher Salzkonzentrationen. Hoppe Seyler's Zeitschr. Physiol. Chem. **351**, 413 (1970).

243. LEZIUS, A. G., and U. RATH: Synthesis of Poly [d(A-S^{4}T) . d(A-S^{4}T) by Bacillus Subtilis DNA Polymerase. Eur. J. Biochem. **24**, 163 (1971).

244. LEZIUS, A. G., and E. DOMIN: A Wobbly Double-Helix. Nature, New Biol. **254**, 169 (1973).

245. LICHTENTHALER, F. W.: The chemistry and properties of enol phosphates. Chem. Rev. **61**, 607 (1963).

246. LLOYD, G. S., C.-M. HSU, and B. S. COOPERMAN: On the reactivity of phosphorylimidazole, an analog of known phosphorylated enzymes. J. Amer. Chem. Soc. **93**, 4889 (1971).

247. LOEWEN, P. C., and H. G. KHORANA: Studies on Polynucleotides. CXXII. The Dodecanucleotide Sequence Adjoining the CCA-End of the Tyrosine Transfer Ribonucleic Acid Gene. J. Biol. Chem. **248**, 3489 (1973).

248. LOHRMANN, R., and H. G. KHORANA: Studies on Polynucleotides. LII. The use of 2,4,6-triisopropylbenzenesulfonyl chloride for the synthesis of internucleotide bonds. J. Amer. Chem. Soc. **88**, 829 (1966).

249. LOHRMANN, R., D. SÖLL, H. HAYATSU, E. OHTSUKA, and H. G. KHORANA: Studies on polynucleotides. LI. Syntheses of the 64 possible ribotrinucleotides derived from the four major ribomononucleotides. J. Amer. Chem. Soc. **88**, 819 (1966).

250. LOHRMANN, R., and L. E. ORGEL: Urea — inorganic phosphate mixtures as prebiotic phosphorylating agents. Science **171**, 490 (1971).

251. LUK, D. C. M., P. BARTL, and A. L. NUSSBAUM: Chain length characterization of oligo-deoxyribonucleotides by analytical ultracentrifugation. Anal. Biochem. **52**, 118 (1973).

252. MACKEY, J. K., and P. T. GILHAM: New approach to the synthesis of polyribonucleotides of defined sequence. Nature **233**, 551 (1971).

253. MAURER, H. K. (Papierwerke Waldhof-Aschaffenburg A.G.): Stable antileukemic 5'-(adamantanecarbonyl)-N^6-(3-methyl-2-butenyl)-adenosine. Ger. Offen. 2,112 (Cl. C 07 d), 14. Sep. 1972, Appl. P 21 12 263,3, 13. Mar. 1971. Chem. Abstr. **78**, 4491 u (1973).

254. MELBY, L. R., and D. R. STROBACH: Oligonucleotide syntheses on isoluble polymer supports. I. Stepwise synthesis of trithymidine diphosphate. J. Amer. Chem. Soc. **89**, 450 (1967).

255. — — Oligonucleotide syntheses on insoluble polymer supports. II. Pentathymidine tetraphosphate. J. Org. Chem. **34**, 421 (1969).

256. — — Oligonucleotide syntheses on insoluble polymer supports. III. Fifteen di(deoxy-ribonucleoside)monophosphates and several trinucleoside diphosphates. J. Org. Chem. **34**, 427 (1969).

257. MERRIFIELD, R. B.: Automated synthesis of peptides. Science **150**, 178 (1968).

258. MICHELSON, A.: The chemistry of nucleosides and nucleotides. London-New York: Academic Press. 1963.

259. MICHNIEWICZ, J. J., O. S. BHANOT, J. GOODCHILD, S. K. DHEER, R. H. WIGHTMAN, and S. A. NARANG: Benzoylated DEAE-Sephadex. Its preparation and application. Biochim. Biophys. Acta **224**, 626 (1970).

260. MICHNIEWICZ, J. J., C. P. BAHL, K. ITAKURA, N. KATAGIRI, and S. A. NARANG: Fractionation of synthetic deoxyribopolynucleotides on silica-gel-thin-layer plates. J. Chromatog. **85**, 159 (1973).

261. MILLER, P. S., K. N. FANG, N. S. KONDO, and P. O. P. TS'O: Syntheses and properties of adenine and thymine nucleoside alkyl phosphotriesters, the neutral analogs of dinucleoside monophosphates. J. Amer. Chem. Soc. **93**, 6657 (1971).

262. MITSUMO, Y. (Takeda Chem. Industries, Ltd.): Adenosine-2'(3'),5'-diphosphate. Jap. Pat. 70 00,870 (Cl. 16 E 611,2), 12. Jan. 1970, Appl. 8. Feb. 1967. Chem. Abstr. **72**, 111788 u (1970).

263. MITSUNOBU, O., K. KATO, and J. KIMURA: Selective phosphorylation of the 5'-hydroxy groups of thymidine and uridine. J. Amer. Chem. Soc. **91**, 6510 (1969).

264. MITSUNOBU, O., J. KIMURA, and Y. FUJISAWA: Studies on nucleosides and nucleotides. II. Selective acylation of 5'-hydroxyl group of thymidine. Bull. Chem. Soc. Japan **45**, 245 (1972).

265. MIURA, K., and T. UEDA: A convenient synthesis of diribonucleoside monophosphates by the use of unblocked nucleosides. Chem. Pharm. Bull (Tokyo) **19**, 2567 (1971).

266. MIURA, K., M. SHIGA, and T. UEDA: Nucleosides and Nucleotides. VI. Preparation of diribonucleoside monophosphates containing 4-thiouridine. J. Biochem. **73**, 1279 (1973).

267. MIYAUCHI, K., Y. MATSUMOTO, T. FURUYA, and K. UCHIDA: Microbial phosphorylation of inosine and guanosine. Jap. Pat. 70 35,236 (Cl. C 12 d), 11. Nov. 1970, Appl. 31. Mar. 1966 (Yamasa Shoyu Co. Ltd.). Chem. Abstr. **74**, 63175 j (1971).

268. MIZUNO, Y., T. ITOH, and H. TAGAWA: New acetylating agents for nucleosides N-acetyl cyclohydroxamic acids. Chem. and Ind. 1498 (1965).

269. MIZUNO, Y., and T. SASAKI: The synthesis of dinucleoside phosphates of natural linkages by the anhydronucleoside method. Tetrahedron Letters 4579 (1965).

270. MIZUNO, Y., T. SASAKI, T. KAUAI, and H. IGARASHI: Nucleotides I. The reaction of cyclouridines with benzyl hydrogen phosphoric benzoic anhydride. J. Org. Chemistry 30, 1533 (1965).

271. MIZUNO, Y., W. LIMN, K. TSUCHIDA, and K. IKEDA: Novel protecting group for the synthesis of 7α-D-pentofuranosyl-hypoxanthines. J. Org. Chemistry 37, 39 (1972).

272. MOHR, S. C., and R. E. THACH: Application of ribonuclease T_1 to the synthesis of oligoribonucleotides of defined base sequence. J. Biol. Chem. 244, 6566 (1969).

273. MONPARLER, R. L., and G. A. FISCHER: Mammalian deoxynucleoside kinases. I. Deoxycytidine kinase. Purification, properties and kinetic studies with cytosine arabinoside. J. Biol. Chem. 243, 4298 (1968).

274. MOON, M. W., and H. G. KHORANA: Studies on Polynucleotides. LV. The use of mesitoylchloride in the synthesis of internucleotide bonds. J. Amer. Chem. Soc. 88, 1805 (1966).

275. MORAVEK, J.: Formation of oligonucleotides during heating of a mixture of uridine-2',3'-phosphate and uridine. Tetrahedron Letters 1707 (1967).

276. MORAVEK, J., J. KOPECKY, and J. SKODA: Thermal phosphorylations IV. Formation of a natural internucleotide bond in oligonucleotides formed by heating uridine-2',3'-phosphate with uridine. Coll. Czech. Chem. Comm. 33, 960 (1968).

277. — — — Thermal phosphorylations V. Fractionation of products of thermal reaction of uridylic acid with uridine using gel filtration and ion exchange chromatography. Coll. Czech. Chem. Comm. 33, 4407 (1968).

278. — — — Thermal phosphorylations VI. Formation of oligonucleotides from uridine-2',3'-monophosphate. Coll. Czech. Chem. Comm. 33, 4120 (1968).

279. MORGAN, A. R.: Studies on polynucleotides. XCIV. Transcription of DNA's with repeating nucleotide sequences. J. Mol. Biol. 52, 441 (1970).

280. MUKAIYAMA, T., and M. HASHIMOTO: Phosphorylation of alcohols and phosphates by oxidation-reduction condensation. Bull. Chem. Soc. Jap. 44, 106 (1971).

281. MUSHIKA, Y., T. HATA, and T. MUKAIYAMA: New phosphorylating reagent. III. Preparation of mixed diesters of phosphoric acid by the use of an activatable protecting group. Bull. Chem. Soc. Japan 44, 232 (1971).

282. MUSHIKA, Y., and N. YONEDA: New phosphorylating reagent. IV. Preparation of the mixed phosphoric diesters of dl-α-tocopherol and ethylene glycol analogs by means of 2-chlormethyl-4-nitrophenyl phosphorodichloridate. Chem. Pharm. Bull. (Tokyo) 19, 687 (1971).

282a. MYLES, A., W. HUTZENLAUB, G. REITZ, and W. PFLEIDERER: Nucleotide I. Synthese und Eigenschaften von Thymidylyl-(3'→3')-, (3'→5')- und (5'→5')-thymidin; in preparation.

283. NAGYVARY, J., and J. S. ROTH: Studies on the synthesis of the natural internucleotide bond by the use of cyclonucleosides. Tetrahedron Letters 617 (1965).

284. NAKAYAMA, K., and H. TANAKA: Production of nucleic acid related substances. XXXVIII. Production of uridine 5'-monophosphate and orotidine 5'-mono-phosphate by Brevibacterium ammoniagenes. Agr. Biol. Chem. 35, 518 (1971).

285. NARA, T., T. KOMURO, M. MISAWA, and S. KINOSHITA: Production of nucleic acid related substances by fermentative processes. XXIX. Growth responses of Brevibacterium ammoniagenes. Agr. Biol. Chem. 33, 1030 (1969).

286. NARANG, S. A., T. M. JACOB, and H. G. KHORANA: Studies on polynucleotides. LXIII. Deoxyribopolynucleotides containing repeating trinucleotide sequences. The polymerization of protected deoxyribotrinucleotides. J. Amer. Chem. Soc. 89, 2167 (1967).

287. NARANG, S. A., S. K. DHEER, and J. J. MICHNIEWICZ: A new general method for the synthesis of deoxyribopolynucleotides bearing a 5'-phosphomonoester end group. J. Amer. Chem. Soc. 90, 2702 (1968).

288. NARANG, S. A., and S. K. DHEER: Chemical synthesis of three deoxyribododecanucleotide chains of defined sequence. Biochemistry **8**, 3443 (1969).

289. NARANG, S. A., O. S. BHANOT, J. GOODCHILD, and R. WIGHTMAN: Use of substituted phenol as phosphate protecting group in the synthesis of deoxyribo-oligo-nucleotides bearing 5′-phosphomonoester end group. J. Chem. Soc. D Chem. Commun. 91 (1970).

290. NARANG, S. A., O. S. BHANOT, J. GOODCHILD, J. MICHNIEWICZ, R. A. WIGHTMAN, and S. K. DHEER: Use of new protecting groups in the synthesis of deoxyribo-oligonucleotides of defined sequence. J. Chem. Soc. D Chem. Commun. 516 (1970).

290a. NARANG, S. A., O. S. BHANOT, J. GOODCHILD, R. H. WIGHTMAN, and S. K. DHEER: A new general method for the synthesis of phosphate-protected deoxyribo-oligonucleotides. IV. J. Amer. Chem. Soc. **94**, 6183 (1972).

291. NARANG, S. A., K. ITAKURA, C. P. BAHL, and Y. Y. WIGFIELD: Chemical synthesis of two deoxyribopolynucleotide fragments containing the natural sequence of T4 lysozyme gene. Biochem. Biophys. Res. Commun. **49**, 445 (1972).

292. NARANG, S. A., K. ITAKURA, and R. H. WIGHTMAN: A simplification in the synthesis of deoxyribooligonucleotides. Can. J. Chem. **50**, 769 (1972).

293. NARANG, S. A., and J. J. MICHNIEWICZ: Thin-layer chromatography of synthetic polydeoxyribonucleotides. Part III. Anal. Biochem. **49**, 379 (1972).

294. NAYLOR, R., and P. T. GILHAM: Studies on Some Interactions and Reactions of Oligonucleotides in Aqueous Solution. Biochem. **5**, 2722 (1966).

295. NEDRAI, V. K., N. I. SOKOLOVA, Z. A. SHABAROVA, and M. A. PROKOF'EV: Chemical matrix synthesis of oligonucleotides in aqueous solutions. Dokl. Akad. Nauk SSSR **205**, English translation p. 1114 (1972).

296. NEILSON, T.: A novel chemical synthesis for oligoribonucleotides. Chem. Commun. 1139 (1969).

297. NEILSON, T., and E. S. WERSTIUK: Oligoribonucleotide synthesis. II. Preparation of 2′-O-tetrahydropyranyl derivatives of adenosine and cytidine necessary for insertion in stepwise synthesis. Can. J. Chem. **49**, 493 (1971).

298. — — Oligoribonucleotide synthesis III. Synthesis of trinucleotides using a stepwise phosphotriester method. Can. J. Chem. **49**, 3004 (1971).

299. NEILSON, T., E. V. WASTRODOWSKI, and E. S. WERSTIUK: Oligoribonucleotide synthesis. V. Preparation of 2′-O-tetrahydropyranyl derivatives of guanosine and their insertion into a general stepwise synthesis. Can. J. Chem. **51**, 1068 (1973).

300. NELSON, T., and E. S. WERSTINK: Synthesis of the anticodon loop of E. coli methionine transfer ribonucleic acid. J. Amer. Chem. Soc. **96**, 2295 (1974).

301. NEJEDLY, Z., H. SKODOVA, K. HYBS, and J. SKODA: New possibilities of enzyme synthesis of radioactive nucleotides. II. Phosphoribosylation of radioactive bases of nucleic acids by the catalytic effect of unpurified cell-free extract of Brevibacterium ammoniagenes. J. Label. Compounds **6**, 3 (1970).

302. NEUMAN, M. W., W. F. NEUMAN, and K. LANE: Possible role of crystals in the origins of life. IV. The phosphorylation of nucleotides. Curr. Mod. Biol. **3**, 277 (1970).

303. NIKOLENKO, L. N., V. N. NEZAVIBAT'KO, and M. N. SEMENOVA: Selective N-benzoylation of cytidine 5′-monophosphate. Zh. Obshch. Khim. **39**, 223 (1969).

304. OGILVIE, K. K., and R. L. LETSINGER: Use of isobutyloxycarbonyl as a blocking group in preparation of 3′-O-p-monomethoxytritylthymidine. J. Org. Chemistry **32**, 2365 (1967).

305. OGILVIE, K. K., and D. IWACHA: Nucleotide syntheses using O^2, 2′anhydrouridine. Can. J. Chem. **48**, 862 (1970).

306. OGILVIE, K. K., and K. KROEKER: Synthesis of oligothymidylates on an insoluble polymer support. Can. J. Chem. **50**, 1211 (1972).

307. OGILVIE, K. K.: The tert.-butyldimethylsilyl group as a protecting group in deoxynucleosides. Can. J. Chem. **51**, 3799 (1973).

308. OHTSUKA, E., M. W. MOON, and H. G. KHORANA: The synthesis of deoxyribopoly-nucleotides containing repeating dinucleotide sequences. J. Amer. Chem. Soc. **87**, 2954 (1965).

309. OHTSUKA, E., K. MURAO, M. UBASAWA, and M. IKEHARA: A new method for the synthesis of protected ribooligonucleotides with 3′-phosphate end groups. J. Amer. Chem. Soc. **91**, 1537 (1969).

310. — — — — Studies on transfer ribonucleic acids and related compounds. I. Synthesis of ribooligonucleotides using aromatic phosphoramidates as a protecting group. J. Amer. Chem. Soc. **92**, 3441 (1970).

311. OHTSUKA, E., M. UBASAWA, and M. IKEHARA: Studies on transfer ribonucleic acids and related compounds. II. A method for synthesis of protected ribooligonucleotides using a ribonuclease. J. Amer. Chem. Soc. **92**, 3445 (1970).

312. — — — Polynucleotides. VIII. A new method for the synthesis of protected deoxy-ribooligonucleotides with 5′-phosphate. J. Amer. Chem. Soc. **92**, 5507 (1970).

313. — — — Studies on transfer ribonucleic acids and related compounds. III. Synthesis of hexanucleotide having the sequence of the yeast alanine transfer ribonucleic acid 3′ end. J. Amer. Chem. Soc. **93**, 2296 (1971).

314. OHTSUKA, E., H. TAGAWA, and M. IKEHARA: Studies on t-RNA's and related compounds. IV. A simple method for the synthesis of ribotrinucleotides. Chem. Pharm. Bull. **19**, 139 (1971).

315. OHTSUKA, E., A. KUMAR, and H. G. KHORANA: Studies on Polynucleotides. CVII. Total synthesis of the structural gene for an alanine transfer ribonucleic acid from yeast. Synthesis of a dodecadeoxynucleotide and a hexadeoxynucleotide corresponding to the nucleotide sequences 1 to 12. J. Mol. Biol. **72**, 309 (1972).

316. OHTSUKA, E., S. MORIOKA, and M. IKEHARA: Formation of phosphodiester linkages by oxidation of a phosphoramidate. Tetrahedron Letters 2553 (1972).

317. — — — Studies on transfer ribonucleic acids and related compounds. V. Synthesis of ribonucleotides with phosphomonoester end groups on a polymer support. J. Amer. Chem. Soc. **94**, 3229 (1972).

318. OHTSUKA, E., M. UBASAWA, S. MORIOKA, and M. IKEHARA: Studies on Transfer Ribonucleic Acids and Related Compounds. VI. Synthesis of Yeast Alanine Transfer Ribonucleic Acid 3′-terminal Nonanucleotides and 5′-terminal Hexanucleotides. J. Amer. Chem. Soc. **95**, 4725 (1973).

319. OHTSUKA, E.: Chemical synthesis of oligo- and poly-nucleotides. In: Methoden der Organischen Chemie (HOUBEN-WEYL, ed.). In press.

320. OKAZAKI, R., and A. KORNBERG: Deoxythymidine kinase of E. coli I. Purification and some properties of the enzyme. J. Biol. Chem. **239**, 269 (1964).

321. — — Deoxythymidine kinase of E. coli II. Kinetics and feedback control. J. Biol. Chem. **239**, 275 (1964).

322. MCOMIE, J. F. W.: Protective groups. Advances Org. Chemistry **3**, 191 (1963).

323. ÖSTERBERG, R., L. E. ORGEL, and R. LOHRMANN: Further studies of urea-catalyzed phosphorylation reactions. J. Mol. Evol. **2**, 231 (1973).

324. OTT, D. G., V. N. KERR, E. HANSBURY, and F. N. HAYES: Chemical synthesis of nucleoside triphosphates. Anal. Biochem. **21**, 469 (1967).

325. OUCHI, S., T. SOWA, K. TSUNODA, and S. SENOO (Asahi Chem. Industry Co., Ltd.): Direct phosphorylation of nucleosides. Jap. Pat. 7016,708 (Cl. 16 E 461), 10. Jun. 1970, Appl. 14. Jul. 1966. Chem. Abstr. **73**, 66866a (1970).

326. OUCHI, S., T. SOWA, S. KATO, T. OSAWA, and S. SENOH (Asahi Chem. Industry Co., Ltd.): 5′Phosphorylation of unprotected nucleosides. Jap. Pat. 71 08,854 (Cl. C 07 d), 5. Mar. 1971, Appl. 20. Jan. 1967. Chem. Abstr. **75**, 36571z (1971).

327. PACE, N. R., D. H. L. BISHOP, and S. SPIEGELMAN: The Immediate Precursor of Viral RNA in the Q_β-Replicase Reaction. Proc. Natl. Acad. Sci. US **59**, 139 (1968).

328. Paetkau, V. H., and H. G. Khorana: Preparation of a circular bihelical deoxyribo-nucleic acid containing repeating dinucleotide sequences. Biochemistry **10**, 1511 (1971).

329. Paivinen, E., and N. S. Tikhomirova-Sidorova: Selective 4-N-acetylation of 2'-deoxy-cytidine 5'-phosphate. Zh. Obshch. Khim. **41**, 2076 (1971).

330. Paivinen, E., E. N. Morozova, and N. S. Tikhomirova-Sidorova: Acetylation of dinucleotides and synthesis of trinucleotides. Zh. Obshch. Khim. **41**, 219 (1971).

331. Parks, R. E., jr., and R. P. Agarwal: Nucleotide kinases. In: The Enzymes. VI. (P. D. Boyer, ed.), 3rd edition. New York-London: Academic Press. 1972.

332. Philipp, M., and H. Seliger: Formylated deoxynucleotidyl triphosphates as potential substrates of deoxynucleotide polymerizing enzymes. Abstracts. 164th ACS-meeting, New York. CARB 5 (1972).

333. — — Unpublished results.

334. Pochon, F., M. Leng, and A. M. Michelson: Photochimie des polynucléotides. III. Étude de la luminescence de poly-nucléotides à température ordinaire. Biochim. Biophys. Acta **169**, 350 (1968).

335. Pochon, F., and A. M. Michelson: Polynucleotide analogues. XIV. Poly N^2-dimethyl-guanylate. Biochim. Biophys. Acta **182**, 17 (1969).

336. Podder, S. K., and I. Tinoco jr.: Enzymatic synthesis of oligoguanylic acids containing 2'-5' phosphodiester linkages. Biochem. Biophys. Res. Commun. **34**, 569 (1969).

337. Podder, S. K.: Synthetic action of ribonuclease T_1. Biochim. Biophys. Acta **209**, 455 (1970).

338. — On self-interacting oligoribonucleotides. I. absorption and optical rotatory dispersion of 2'-5'- and 3'-5'-oligoguanylic acids. Biochemistry **10**, 2415 (1971).

339. Pongs, O., and P. O. P. Ts'o: Polymerization of 5'-deoxyribonucleotides with β-imidazolyl-4(5)-propanoic acid. Biochem. Biophys. Res. Comm. **36**, 475 (1969).

340. — — Polymerization of unprotected 2'-deoxyribonucleoside-5'-phosphates at elevated temperature. J. Amer. Chem. Soc. **93**, 5241 (1971).

341. Poonian, M. S., E. F. Nowoswiat, and A. L. Nussbaum: Nucleoside S-alkyl phosphoro-thioates. VII. A fragment from the nonsense strand of a modified S-peptide "gene". J. Amer. Chem. Soc. **94**, 3992 (1972).

342. Potapov, V. K., O. G. Chekhmakhcheva, Z. A. Shabarova, and M. A. Prokof'ev: Synthesis of oligonucleotides on polymer carriers. Synthesis of deoxy-[thimidylyl-(3'→5')-adenylyl-(3'→5')-adenylyl-(3'→5')-adenosine]. Dokl. Akad. Nauk SSSR **196**, 360 (1971).

343. Potapov, V. K., S. I. Turkin, and Z. A. Shabarova: Application of Sephadex LH-20 in the synthesis of oligonucleotides on polymeric carriers. Zh. Obshch. Khim. **42**, 2349 (1972).

344. Rabinowitz, J.: 265. Recherche sur la formation et la transformation des esters. LXXXIII. (1). Reaction de condensation et/ou de phosphorylation, en solution aqueuse, de divers composes organiques a fonction —OH, —COOH, —NH₂ ou autre a l'aide de polyphosphates lineares ou cycliques. Helv. Chim. Acta **52**, 2663 (1969).

345. Rabinowitz, J., S. Chang, and C. Ponnamperuma: Phosphorylation by way of inorganic phosphate as a potential prebiotic process. Nature **218**, 442 (1968).

346. Rajabalee, F. J. M.: A convenient synthesis of 2',3',5'-tri-O-acetyladenosine and -uridine. Angew. Chem. **10**, 75 (1971).

347. Ramel, A., E. Heimer, S. Roy, and A. L. Nussbaum: Gel filtration of acylated oligonuc-leotides. Anal. Biochem. **41**, 323 (1971).

348. Randerath, K., and E. Randerath: Thin-layer Separation Methods for Nucleic Acid Derivatives. In: S. P. Colowick and N. O. Kaplan, Methods in Enzymology, Vol. XII (L. Grossmann and K. Moldave, eds.), p. 323. New York-London: Academic Press. 1967.

349. Ratliff, R. L., and F. N. Hayes: Enzymatic synthesis of a three-section block copolymer of thymidylate, deoxyguanylate and deoxyadenylate. Biochim. Biophys. Acta **134**, 203 (1967).

350. REESE, C. B., and R. SAFFHILL: Oligonucleotide synthesis via phosphotriester intermediates: The phenyl protecting group. J. Chem. Soc. D, Chem. Comm. 767 (1968).

351. REESE, C. B., and J. C. M. STEWART: Methoxyacetyl as a protecting group in ribonucleoside chemistry. Tetrahedron Letters 4273 (1968).

352. REESE, C. B.: A systematic approach to oligoribonucleotide synthesis. Chim. Organ. du Phosphore (Colloqu. Intern. du centre national de la recherche scientifique, eds.) **182**, 319 (1969).

353. REESE, C. B., J. H. VAN BOOM, G. R. OWEN, J. PRESTON, and T. RAVINDRANATHAN: Synthesis of oligoribonucleotides. IX. Preparation of ribonucleoside 2'-acetal 5'-esters. J. Chem. Soc. C 3230 (1971).

354. REGEL, W., E. STENGELE, and H. SELIGER: Kinetik der Schutzgruppenabspaltung an Nucleosiden mittels ¹H-NMR-Spektroskopie. Chem. Ber. **107**, 611 (1974).

355. RENZ, M., R. LOHRMANN, and L. E. ORGEL: Catalysts for the polymerization of adenosine cyclic 2',3'-phosphate on a poly (U) template. Biochim. Biophys. Acta **240**, 463 (1971).

356. RICHARDS, G. M., D. J. TUTAS, W. J. WECHTER, and M. LASKOWSKI SR: Hydrolysis of dinucleoside monophosphates containing arabinose in various internucleotide linkages by exonuclease from the venom of crotalus adamanteus. Biochemistry **6**, 2908 (1967).

357. RICHARDSON, C. C.: Phosphorylation of nucleic acid by an enzyme from T4-bacteriophage infected Escherischia coli. Proc. Natl. Acad. Sci. US **54**, 158 (1965).

358. ROKOS, H., W. HUTZENLAUB, A. MYLES, and W. PFLEDERER: Nucleotide, II: Isomerisierung der Internucleotidbindung bei der Abspaltung der Phosphor-Schutzgruppe von Phosphorsäuretriestern. In preparation.

359. RÖSSNER, E.: Synthese von Oligodesoxyribonucleotiden mit 3'-terminalem Ribonucleotid durch Cooligokondensation. Diplomarbeit, Univ. Freiburg, 1972.

360. ROYCHOUDHURY, R., and H. KÖSSEL: Synthetic polynucleotides. Enzymic synthesis of ribonucleotide terminated oligodeoxynucleotides and their use as primers for the enzymic synthesis of polydeoxynucleotides. Eur. J. Biochem. **22**, 310 (1971).

361. ROYCHOUDHURY, R.: Enzymic Synthesis of Polynucleotides. Oligodeoxynucleotides with one 3'-terminal Ribonucleotide as Primers for Polydeoxynucleotide Synthesis. J. Biol. Chem. **247**, 3910 (1972).

361a. ROYCHOUDHURY, R., S. KÜHN, H. SCHOTT, and H. KÖSSEL: Enzymic polynucleotide synthesis primed by polyvinylalcohol linked oligothymidylate. FEBS Letters **50**, 140 (1975).

362. RUBINSTEIN, M., and A. PATCHORNIK: Polymers as chemical reagents. Use of poly-3,5-diethylstyrene sulfonyl chloride for the synthesis of internucleotide bonds. Tetrahedron Letters 2281 (1972).

363. SACHDEV, H. S., and N. A. STARKOVSKY: Enzymatic removal of acyl protecting groups. The use of dihydrocinnamoyl group in oligonucleotide synthesis and its cleavage by α-chymotrypsin. Tetrahedron Letters 733 (1969).

364. SAFFHILL, R.: Selective phosphorylation of the cis-2',3'-diol of unprotected ribonucleosides with trimetaphosphate in aqueous solution. J. Org. Chemistry **35**, 2881 (1970).

365. SAITO, M., Y. FURUICHI, K. TAKEISHI, M. YOSHIDA, M. YAMASAKI, K. ARIMA, H. HAYATSU, and T. UKITA: Synthesis of diribonucleoside monophosphates by use of a nonspecific ribonuclease from Bacillus subtilis. Biochim. Biophys. Acta **195**, 299 (1969).

366. SANGER, F., J. E. DONELSON, A. R. COULSON, H. KÖSSEL, and D. FISCHER: Use of DNA polymerase I primed by a synthetic oligonucleotide to determine a nucleotide sequence in phage f1 DNA. Proc. Nat. Acad. Sci. US **70**, 1209 (1973).

367. SANNO, Y., and A. NOHARA: Phosphorylation of 2',3'-isopropylideneinosine by heating or ultraviolet irradiation in the presence of phosphoric acid and nitriles. Chem. Pharm. Bull. (Tokyo) **16**, 2056 (1968).

367a. Sano, H., and G. Feix: Ribonucleic acid ligase activity of DNA ligase from T$_4$-infected *E. coli.* Biochemistry **13**, 5110 (1974).

367b. Scheffler, I. E., and C. C. Richardson: Chemical and enzymatic studies of DNA covalently linked to ficoll. J. Biol. Chem. **247**, 5736 (1972).

368. Scheit, K. H.: Untersuchungen an poly-5-Hydroxymethyluridylsäure und poly-5-Methyluridylsäure. Biochim. Biophys. Acta **134**, 17 (1967).

369. — Die Benzylester von Desoxydinucleosidphosphaten und 5'-O-(β,β,β,-trichlor-äthylphosphoryl-) thymidylyl-3'-5'-thymidylyl-3'-5'-thymidin. Tetrahedron Letters 3243 (1967).

370. — Über die Synthese und Eigenschaften von 4-Thiouridylyl-(3'-5')-4-thiouridin, 4-Thiouridylyl-(3'-5')-uridin und Uridylyl-(3'-5')-4-thiouridin. Biochim. Biophys. Acta **166**, 285 (1968).

371. Scheit, K. H., und E. Gaertner: Die Polymerisation von 4-Thiouridin-5'-diphosphat und 4-Thiothymidin-5'-diphosphat durch Polynucleotidphosphorylase aus Micrococcus Lysodeikticus. Biochim. Biophys. Acta **182**, 1 (1969).

372. Scheit, K. H.: Enzymatic polymerization of 5-methyl-4-thiouridine-5'-diphosphate by polynucleotide phosphorylase from *Escherichia coli.* Biochim. Biophys. Acta **209**, 445 (1970).

373. Scheit, K.-H., and P. Faerber: Synthesis and properties of poly(s^{2}C), a new Poly(c) analog. Eur. J. Biochem. **24**, 385 (1971).

374. Schetters, H., H. G. Gassen, and H. Matthaei: Codon-anticodon interaction studied with oligonucleotides containing 3-deazauridine, 4-deoxyuridine or 3-deaza-4-deoxyuridine. I. Synthesis by primer-dependent polynucleotide phosphorylase of oligonucleotides containing modified nucleosides. Biochim. Biophys. Acta **272**, 549 (1972).

375. Schneider-Bernloehr, H., R. Lohrmann, J. Sulston, B. J. Weimann, L. E. Orgel, and H. Todd Miles: Non-enzymic synthesis of deoxyadenylate oligonucleotides on a polyuridylate template. J. Mol. Biol. **37**, 151 (1968).

376. Schneider-Bernloehr, H., R. Lohrmann, J. Sulston, L. E. Orgel, and H. Todd Miles: Specificity of template-directed synthesis with adenine nucleosides. J. Mol. Biol. **47**, 257 (1970).

377. Schott, H.: New dihydroxyboryl-substituted polymers for column-chromatographic separation of ribonucleoside-deoxyribonucleoside mixtures. Angew. Chem. **11**, 824 (1972).

378. — Polyvinyl alcohol substituted by nucleotides as carrier for liquid phase oligonucleotide synthesis. Angew. Chem. **12**, 246 (1973).

379. Schott, H., D. Fischer, and H. Kössel: Synthesis of four undecanucleotides complementary to a region of the coat protein cistron of phage fd. Biochemistry **12**, 3447 (1973).

380. Schott, H., and H. Kössel: Synthesis of phage specific deoxyribonucleic acid fragments. I. Synthesis of four undecanucleotides complementary to a mutated region of the coat protein cistron of fd phage deoxyribonucleic acid. J. Amer. Chem. Soc. **95**, 3778 (1973).

381. Schott, H., E. Rudloff, P. Schmidt, R. Roychoudhury, and H. Kössel: A dihydroxyboryl-substituted methacrylic polymer for the column chromatographic separation of mononucleotides, oligonucleotides, and transfer ribonucleic acid. Biochemistry **12**, 932 (1973).

382. Schott, H., F. Brandstetter, and E. Bayer: Liquid-Phase Synthese von Oligothymidylphosphaten. Makromol. Chem. **173**, 247 (1974).

383. Schott, H.: Chemische Synthese eines phagenspezifischen DNA-Fragments. Makrom. Chem. **175**, 1683 (1974).

384. Schwartz, A. W., and F. N. Hayes: Synthesis of a polydeoxyribonucleotide containing an internal pyrophosphate linkage. Biochim. Biophys. Acta **138**, 604 (1967).

385. SCHWARTZ, A. W.: Specific phosphorylation of the 2'- and 3'-position in ribonucleosides. J. Chem. Soc. D, Chem. Comm. 1393 (1969).

386. SCHWARTZ, A., and C. PONNAMPERUMA: Phosphorylation of adenosine with linear polyphosphate salts in aqueous solution. Nature **218,** 449 (1968).

386a. SEDEL'NIKOVA, E. A., and S. M. ZHENODAROVA: Stepwise synthesis of oligonucleotides. I. 5-O-(α-alkoxyalkyl) derivatives of uridine 3'-phosphate. Zh. Obshch. Khim. **38,** 2234 (1968).

387. SEDEL'NIKOVA, E. A., O. A. SMOLYANINOVA, and S. M. ZHENODAROVA: Stepwise synthesis of oligonucleotides. III. Structural analysis of α-alkoxyalkyl groups in O-(α-alkoxylkyl) derivatives of nucleosides and nucleotides. Zh. Obshch. Khim. **38,** 2245 (1968).

388. SEKIYA, T., Y. FURUICHI, M. YOSHIDA, and T. UKITA: Ribonuclease-T₁ catalyzed Synthesis of Triribonucleoside Diphosphates having a Guanosine Residue at the 5'-End. J. Biochem. (Tokyo) **63,** 514 (1968).

389. SELIGER, H.: Versuche zur Einführung und Reaktivierung von Enolgruppen als Schutzgruppen an Phosphorsäuren. Diplom thesis. Technische Hochschule Darmstadt, 1963.

390. SELIGER, H., and F. CRAMER: Nucleophile Substitutionen mit Pyrimidinnucleosid-N³-natriumsalzen. Angew. Chem. **81,** 577 (1969). Angew. Chem. Internat. Edit. **8,** 609 (1969).

391. SELIGER, H.: Chlorameisensäureester von Nucleosiden — neue Zwischenprodukte für Synthesen mit Nucleinsäurebausteinen. Tetrahedron Letters 4043 (1972).

392. SELIGER, H., and G. AUMANN: Oligonucleotide synthesis on a polymer support soluble in water and pyridine. Tetrahedron Letters 2911 (1973).

393. SELIGER, H.: Ein neuer Weg zur Synthese von Polystyrol und Styrol-Copolymeren mit primären aromatischen Aminogruppen. Makromol. Chem. **169,** 83 (1973).

394. SELIGER, H., G. AUMANN, V. GENRICH, M. PHILIPP, and E. RÖSSNER: Improvements in the development of rational methods for oligonucleotide synthesis. Abstracts. XXIVth Internat. Congr. of pure and applied chemistry, Hamburg, p. 136 (1973).

395. SELIGER, H., G. AUMANN, and R. L. LETSINGER: Progress in the synthesis of oligonucleotides on polymer supports. Contributed Papers, Internat. Symposium on Macromolecules, Rio de Janeiro, 221 (1974).

396. SELIGER, H., G. AUMANN, V. GENRICH, M. PHILIPP, E. RÖSSNER, and H. SCHÜTZ: Sequence-specific cooligocondensation of nucleic acid constituents — a new approach to polynucleotide synthesis. Contributed Papers, Internat. Symposium on Macromolecules, Rio de Janeiro, 222 (1974).

396a. SELIGER, H., and G. AUMANN: Oligonukleotidsynthese an unvernetzten Copolymeren des Vinylalkohols und N-Vinylpyrrolidons. Makromol. Chem., in press.

397. SELIGER, H.: Polymers in aid of polynucleotide chemistry. Current topics in chemistry, manuscript in preparation.

397a. SELIGER, H.: Handelsübliche Polymere als Träger in der Oligonukleotidsynthese. I. Synthese eines Pentanucleosidtetraphosphats an Merckogel OR 1 000 000^R. Makromol. Chem., in press.

398. SELIGER, H., H. SCHÜTZ, E. SAUR, and M. PHILIPP: Oligonucleotide synthesis with nucleotide-3'-formyl esters. J. Carbohydr., Nucleosides, Nucleotides, in press.

398a. SELIGER, H., E. RÖSSNER, G. AUMANN, V. GENRICH, M. HOLUPIREK, T. KNÄBLE, and M. PHILIPP: Sequenzspezifische Cooligokondensation von Nukleinsäurebausteinen mit Affinitätsschutzgruppen. I. Desoxyoligonucleotide mit Ribouridin-Terminus. Makromol. Chem., in press.

398b. SELIGER, H., H. SCHÜTZ, and M. PHILIPP: Sequenzspezifische Cooligokondensation von Nukleinsäurebausteinen mit Affinitätsschutzgruppen. II. Cooligomere von 5'-O-(p-Methoxytrityl-)thymidin und Desoxynucleotiden. Makromol. Chem., in press.

399. Sgaramella, V., J. H. van de Sande, and H. G. Khorana: Studies on polynucleotides. C. A novel joining reaction catalyzed by the T4-polynucleotide ligase. Proc. Nat. Acad. Sci. US **67,** 1468 (1970).

400. Sgaramella, V., and H. G. Khorana: Studies on Polynucleotides. CXII. Total synthesis of the structural gene for an alanine transfer RNA from yeast. Enzymic joining of the chemically synthesized polydeoxynucleotides to form the DNA duplex representing nucleotide sequence 1 to 20. J. Mol. Biol. **72,** 427 (1972).

401. — — Studies on polynucleotides. CXVI. A further study of the T4 ligase-catalyzed joining of DNA at base-paired ends. J. Mol. Biol. **72,** 493 (1972).

402. Sgaramella, V., K. Kleppe, T. Terao, N. K. Gupta, and H. G. Khorana: Studies on Polynucleotides. CXIII. Toral synthesis of the structural gene for an alanine transfer RNA from yeast. Enzymic joining of the chemically synthesized segments to form the DNA duplex corresponding to nucleotide sequence 17 to 50. J. Mol. Biol. **72,** 445 (1972).

403. Shabarova, Z. A., and M. A. Prokofiev: A model of enzymatic synthesis of the internucleotide bond between oligodeoxynucleotides. FEBS Letters **11,** 237 (1970).

404. Shemyakin, M. M., Yu. A. Ovchinnikov, A. A. Kiryushkin, and I. V. Kozhevnikova: Synthesis of peptides in solution on a polymeric support. I. Synthesis of glycyl-glycyl-L leucyl-glycin. Tetrahedron Letters 2323 (1965).

405. Shen, T.-Y., and K. H. Boswell (Merck and Co., Inc.): 5'-Adamantoyl-2'-deoxy-5-(methylamino)-uridine. US Pat. 3,676,422 (Cl. 260/211.5R; C 07d), 11. Jul. 1972, Appl. 73,206, 17. Sep. 1970. Chem. Abstr. **77,** 114817s (1972).

406. Shimizu, B., M. Asai, and T. Nishimura: Synthetic nucleotides. I. A convenient synthesis of ribonucleotides. Chem. Pharm. Bull. (Tokyo) **15,** 1847 (1967).   7

407. Shimidzu, T., and R. L. Letsinger: Synthesis of deoxyguanylyldeoxyguanosine on an insoluble polymer support. J. Org. Chem. **33,** 708 (1968).

408. — — Hydrolysis of *p*-nitrophenyl(deoxyguanyl-deoxyguanosine succinate) by deoxyguanyldeoxyguanosine N-acetylhistidate on polycytidylic acid. Bull. Chem. Soc. Japan **44,** 584 (1971).

409. — — The preparation of deoxyguanosine oligomers on an insoluble polymer support. Bull. Chem. Soc. Japan **44,** 1673 (1971).

410. — — Hydrolyses of *p*-Nitrophenyl(oligodeoxyribonucleotide succinate)s by oligo-deoxyribonucleotide N-acetylhistidates on polycytidylic acid. Bull. Chem. Soc. Japan **46,** 3270 (1973).

411. Shinskii, N. G., N. N. Preobrashenskaya, M. G. Ivanovskaya, Z. A. Shabarova, and M. A. Prokof'ev: Phosphorylation of nucleosides by pyrophosphoryl chloride. Dokl. Akad. Nauk SSSR **184,** 622 (1969).

412. Simuth, J., K. H. Scheit, and E. M. Gottschalk: The enzymatic synthesis of poly 4-thiouridylic acid by polynucleotide phosphorylase from *Escherichia coli.* Biochim. Biophys. Acta **204,** 371 (1970).

413. Simuth, J., P. Strehlke, U. Niedballa, H. Vorbrüggen, and K. H. Scheit: 2'-O-methylcytidine 5'-diphosphate as substrate for polynucleotide phosphorylase from *Escherichia coli.* Biochim. Biophys. Acta **228,** 654 (1971).

414. Smirnov, V. D., M. G. Ivanovskaya, E. V. Il'ina, Z. A. Shabarova, and M. A. Prokof'ev: Synthesis of the phenylalanine amide of a pentadeoxynucleotide. Dokl. Akad. Nauk SSSR **206,** English translation p. 1133 (1972).

415. Smrt, J., and J. Catlin: Abnormal course of phosphorylation with methyl phosphate. Tetrahedron Letters 5081 (1970).

416. Smrt, J., and F. Cramer: Oligonucleotidic compounds. XXXVI. Synthesis of uridylyl-(5'-3')-uridylyl-(5'-5')-uridylyl-(3'-5')-uridine and its priming activity for polynucleotide phosphorylase. Collect. Czech. Chem. Commun. **35,** 1456 (1970).

417. Smrt, J.: Protection of the internucleotidic bond after its synthesis. An approach to the synthesis of oligonucleotidic chains. Tetrahedron Letters 3437 (1972).

418. SMRT, J.: Oligonucleotidic compounds. XXXIX. Triester synthesis of oligonucleotides in the ribo series. Collect. Czech. Chem. Commun. **37**, 846 (1972).

419. — Oligonucleotidic compounds. XL. Aspects of the triester synthesis in the ribo series. Collect. Czech. Chem. Commun. **37**, 1870 (1972).

420. — Oligonucleotidic compounds. XLI. On the reaction of ribonucleoside 2'(3')-phosphates with dimethylformamide dimethylacetal. Collect. Czech. Chem. Commun. **37**, 4088 (1972).

421. — A remark on the preparation of protected guanosine-3'-phosphate by means of a mixture of ribonucleases T_1 and T_2. Collect. Czech. Chem. Commun. **39**, 969 (1974).

422. — Combined synthesis of oligonucleotides in the deoxy series. Collect. Czech. Chem. Commun. **39**, 972 (1974).

423. SOMMER, H., and F. CRAMER: Synthese von Oligodesoxynucleotiden mit 5'-terminaler Phosphatgruppe. Angew. Chem. **84**, 710 (1972).

424. SOWA, T., S. OUCHI, and T. OSAWA (Asahi Chem. Industry Co., Ltd.): 5'-Nucleotides by a direct phosphorylation. Jap. Pat. 71 02,025 (Cl. C 07 d), 19. Jan. 1971, Appl. 16. Dec. 1966. Chem. Abstr. **74**, 112405 v (1971).

425. SOWA, T., K. SATO, S. OUCHI, T. OSAWA, and S. SEO (Asahi Chem. Industry Co., Ltd.): Selective phosphorizing to give nucleotides. Jap. Pat. 71 04,986 (Cl. C 07 d), 6. Feb. 1971, Appl. 7. Jan. 1967. Chem. Abstr. **75**, 36578 g (1971).

426. SPORN, M. B., D. M. BERKOWITZ, R. P. GLINSKI, A. B. ASH, and C. L. STEVENS: Irreversible inhibition of nuclear exoribonuclease by thymidine-3'-fluorophosphate and p-haloacetamidophenyl nucleotides. Science **164**, 1408 (1969).

427. SRIVASTAVA, P. C., and M. M. DHAR: The use of a purine cyclonucleoside for the synthesis of a dinucleoside phosphate. Tetrahedron Letters 47 (1968).

428. SRIVASTAVA, P. C., K. L. NAGPAL, and M. M. DHAR: The synthesis of a natural dinucleoside phosphate derivative with the aid of a purine cyclonucleoside. Experientia **24**, 657 (1968).

429. — — — Synthesis of dinucleoside phosphates by reaction of 5'-chloro-5'-deoxynucleosides with nucleotide anions. Experientia **25**, 356 (1969).

429a. STUART, A., and H. G. KHORANA: Studies on polynucleotides. XXXIII. The labelling of end groups in polynucleotide chains: The selective acetylation of terminal hydroxyl groups in deoxyribopolynucleotides. J. Biol. Chem. **239**, 3885 (1964).

430. SULSTON, J., R. LOHRMANN, L. E. ORGEL, and H. TODD MILES: Nonenzymatic synthesis of oligoadenylates on a polyuridylic acid template. Proc. Nat. Acad. Sci. (U.S.) **59**, 726 (1968).

431. — — — — Specificity of oligonucleotide synthesis directed by polyuridylic acid. Proc. Nat. Acad. Sci. (U.S.) **60**, 409 (1968).

432. SULSTON, J., R. LOHRMANN, L. E. ORGEL, H. SCHNEIDER-BERNLOEHR, B. J. WEIMANN, and H. TODD MILES: Non-enzymic oligonucleotide synthesis on a polycytidylate template. J. Mol. Biol. **40**, 227 (1969).

433. SUSSMAN, J. L., I. BARZILAY, M. KEREN-ZUR, and Y. LAPIDOT: Correlation of the differences in conformation between 2'—5' and 3'—5' dinucleoside monophosphates with their behaviour on a Sephadex LH-20 column. Biochim. Biophys. Acta **308**, 189 (1973).

434. SWIERKOWSKI, M., and D. SHUGAR: Poly 5-ethyluridylic acid, a polyuridylic acid analogue. J. Mol. Biol. **47**, 57 (1970).

435. TAJIMA, K., and T. HATA: Simple protecting group protection-purification handle for polynucleotide synthesis. II. Bull. Chem. Soc. Japan **45**, 2608 (1972).

436. TAUNTON-RIGBY, A., Y.-H. KIM, C. J. CROSSCUP, and N. A. STARKOVSKY: Oligonucleotide synthesis. II. The use of substituted trityl groups. J. Org. Chem. **37**, 956 (1972).

437. Taunton-Rigby, A.: Oligonucleotide synthesis. III. Enzymatically removable acyl protecting groups. J. Org. Chem. 38, 977 (1973).

438. Tazawa, I., S. Tazawa, L. M. Stempel, and P. O. P. Ts'o: L-adenylyl-(3'-5')-L-adenosine and L-adenylyl-(2'-5')-L-adenosine. Biochemistry 9, 3499 (1970).

439. Thang, M. N., and M. Grunberg-Manago: Enzymatic synthesis of polyguanylic acid and copolymers containing guanylic acid. In: Methods in enzymology, XII B, 522 (L. Grossman and K. Moldave, eds.). New York-London: Academic Press. 1968.

440. Tigerstrom, R. v., and M. Smith: Oligodeoxyribonucleotides: Chemical synthesis in anhydrous base. Science 167, 1266 (1970).

441. Tikhomirova-Sidorova, N. S., E. M. Kogan, V. A. Sysoev, and G. E. Ustyuzhanin: Alcoholysis of pyrimidine nucleoside 2',3'-cyclophosphates with purine nucleosides in frozen solutions in presence of pancreatic ribonuclease. Zh. Obshch. Khim. 41, 2570 (1971).

442. Tikhomirova-Sidorova, N. S., G. E. Ustyuzhanin, T. N. Kalacheva, and V. I. Kalugina: Hydrolytic and synthetic activity of guanylic ribonuclease of actinomycetes in frozen solutions. Zh. Obshch. Khim. 41, 2108 (1971).

442a. Torrence, P. F., J. A. Waters, and B. Witkop: Unexpected conformational stability of poly(2'-azido-2'-deoxyuridylic acid). J. Amer. Chem. Soc. 94, 3638 (1972).

443. Torrence, P. F., and B. Witkop: Enzymatic synthesis of polynucleotides containing 5,6-methylene- and 5,6-dihydropyrimidines. Biochemistry 11, 1737 (1972).

444. Torrence, P. F., A. M. Bopst, J. A. Waters, and B. Witkop: Synthesis and characterization of potential interferon inducers. Poly(2'-azido-2'-deoxyuridylic acid). Biochemistry 12, 3962 (1973).

445. Torrence, P. F., J. A. Waters, C. E. Buckler, and B. Witkop: Effect of pyrimidine and ribose modifications on the antiviral activity of synthetic polynucleotides. Biochem. Biophys. Res. Commun. 52, 890 (1973).

446. Tsiapalis, C. M., and S. A. Narang: On the fidelity of phage T_4-induced polynucleotide ligase in the joining of chemically synthesized deoxyribooligonucleotides. Biochem. Biophys. Res. Commun. 39, 631 (1970).

447. Tsou, K. C., and K. F. Yip: Synthesis of deoxyoligonucleotides on an isotactic polymer support. J. Macromol. Sci. Chem. A7 (5), 1097 (1973).

448. Uchic, J. T., M. Uchic, and A. D. Broom: Studies on polyribonucleotides. Synthesis of a polyinosinic: 6-thioinosinic acid copolymer. Biochem. Biophys. Res. Commun. 51, 494 (1973).

449. — — — Studies on polyribonucleotides. Polymerization of 6-chloropurine riboside-5'-diphosphate; personal communication.

450. Uchida, T., T. Arima, and F. Egami: Specificity of RNase U_2. J. Biochem. (Tokyo) 67, 91 (1970).

451. Ueda, T., and J. J. Fox: Mononucleotides. Advan. Carbohyd. Chem. Biochem. 22, 307 (1967).

452. Ueda, T., and I. Kawai: A convenient synthesis of ribonucleoside 2',3'-cyclic phosphates from ribonucleosides and ribonucleotides. Chem. Pharm. Bull. (Tokyo) 18, 2303 (1970).

453. Uesugi, S., M. Yasumoto, M. Ikehara, K. N. Fang, and P. O. P. Ts'o: Synthesis and properties of the dinucleoside monophosphate of adenine 8-thiocyclonucleoside. J. Amer. Chem. Soc. 94, 5480 (1972).

454. van de Sande, J. H., M. H. Caruthers, V. Sgaramella, T. Yamada, and H. G. Khorana: Studies on Polynucleotides. CXIV. Total synthesis of the structural gene for an alanine transfer RNA from yeast. Enzymic joining of the chemically synthesized segments to form the DNA duplex corresponding to nucleotide sequence 46 to 77. J. Mol. Biol. 72, 457 (1972).

455. Verlander, M. S., R. Lohrmann, and L. E. Orgel: Catalysts for the selfpolymerization of adenosine cyclic 2',3'-phosphate. J. Mol. Evol. 2, 303 (1973).

456. VERLANDER, M. S., and L. E. ORGEL: Analysis of high molecular weight material from the polymerization of adenosine cyclic 2′,3′-phosphate. J. Mol. Evol. **3**, 115 (1974).

457. WEBER, H., and H. G. KHORANA: Studies on Polynucleotides. CIV. Total synthesis of the structural gene for an alanine transfer ribonucleic acid from yeast. Chemical synthesis of an icosadeoxynucleotide corresponding to the nucleotide sequence 21 to 40. J. Mol. Biol. **72**, 219 (1972).

458. WEIMANN, G., and H. G. KHORANA: Studies on Polynucleotides. XVII. On the mechanism of internucleotide bond synthesis by the carbodiimide method. J. Amer. Chem. Soc. **84**, 4329 (1962).

459. WEIMANN, B. J., R. LOHRMANN, L. E. ORGEL, H. SCHNEIDER-BERNLOEHR, and J. E. SULSTON: Template-directed synthesis with adenosine-5′-phosphorimidazolide. Science **161**, 387 (1968).

460. WEITH, H. L., J. L. WIEBERS, and P. T. GILHAM: Synthesis of cellulose derivatives containing the dihydroxyboryl group and a study of their capacity to form specific complexes with sugars and nucleic acid components. Biochem. **9**, 4396 (1970).

461. WELLS, R. D., and R. M. WARTELL: Review of synthesis of specific DNA polymers. Biochem. Mol. Biol. (ed. K. BURTON). Butterworth and Medical and Technical Publishing Co. (1973).

462. WERSTIUK, E. S., and T. NEILSON: Oligoribonucleotide Synthesis. IV. Approach to block synthesis. Can. J. Chem. **50**, 1283 (1972).

463. WERSTIUK, E. S., and T. NEILSON: Oligoribonucleotide synthesis. VI. Selective deblocking of the 5′-O-triphenylmethoxyacetyl grouping in protected dinucleotides. Can. J. Chem. **51**, 1889 (1973).

464. WESTHEIMER, F. H., C. CLAPP, and J. WISEMAN: Monomeric metaphosphates and metaphosphorimidates. XXIVth IUPAC Congress of Pure and Applied Chemistry, Hamburg, **1973**, Abstracts of Papers, p. 410.

465. WIGHTMAN, R. H., S. A. NARANG, and K. ITAKURA: A novel phosphate protecting group for oligonucleotide synthesis. Can. J. Chem. **50**, 456 (1972).

466. WINDHOLZ, T. B., and D. B. R. JOHNSTON: Trichloroethoxycarbonyl: a generally applicable protecting group. Tetrahedron Letters 2555 (1967).

467. WU, R., C.-P. D. TU, and R. PADMANABHAN: Nucleotide sequence analysis of DNA. XII. The chemical synthesis and sequence analysis of a dodecadeoxynucleotide which binds to the endolysin gene of bacteriophage lambda. Biochem. Biophys. Res. Commun. **55**, 1092 (1973).

467a. WU, R.: personal communication.

468. WÜNSCH, E.: Synthese von Peptidnaturstoffen. Problematik des heutigen Forschungsstandes. Angew. Chem. **83**, 773 (1971).

469. YAMASHITA, T., and T. KATO (Ajinomoto Co, Inc.): 2′,3′-O-Substituted nucleoside 5′-phosphates. Jap. Pat. 69 27,979 (Cl. 16E 611.2), 19. Nov. 1969, Appl. 3. Dec. 1966. Chem. Abstr. **72**, 21913b (1970).

470. YIP, K. F., and K. C. TSOU: A new polymer-support method for the synthesis of ribooligonucleotide. J. Amer. Chem. Soc. **93**, 3272 (1971).

471. YONEI, S., A. KUNINAKA, and H. YOSHINO (Yamasa Shoyu Co., Ltd.): Jap. Pat. 71 31,865 (Cl. C 07 d), 17. Sep. 1971, Appl. 7. Feb. 1969. Chem. Abstr. **75**, 152051 d (1971).

472. YOSHIKAWA, M., T. KATO, and T. TAKENISHI: Selective phosphorylation of unprotected nucleosides. Bull. Chem. Soc. Jap. **42**, 3505 (1969).

473. YOSHIKAWA, M., M. SAKURABA, and K. KUSASHIO: Phosphorylation. IV. Phosphorylation of nucleosides with phosphorus trihalide. Bull. Chem. Soc. Jap. **43**, 456 (1970).

474. YURKEVICH, A. M., L. S. VARSHAVSKAYA, and I. I. KOLODKINA: Reaction of nucleosides with diphenylboric acid esters. Zh. Obshch. Khim. **38**, 2115 (1968).

475. YURKEVICH, A. M., I. I. KOLODKINA, L. S. VARSHAVSKAYA, V. I. BORODULINA-SHVETZ,

I. P. Rudakova, and N. A. Preobrazhenski: The reaction of phenylboronic acid with nucleosides and mononucleosides. Tetrahedron **25**, 477 (1969).

476. Zarytova, V. F., V. K. Potapov, Z. A. Shabarova, and D. G. Knorre: Synthesis of oligonucleotides on polymer supports. Synthesis of oligodeoxynucleotides containing deoxyguanylic acid. Dokl. Akad. Nauk SSSR **199**, 1072 (1970).

477. Zemlicka, J., J. Beranek, and J. Smrt: Preparation and methanolysis of uridine, 6-aza-uridine and 6-aza-cytidine-O-formyl derivatives. Coll. Czech. Chem. Comm. **27**, 2784 (1962).

478. Zemlicka, J., and J. Smrt: The reaction of $O^2,5'$-cyclouridine and $O^2,5'$-cyclocytidine derivatives with nucleotides — A new approach to the synthesis of the 3'-5'-internucleotidic bond. Tetrahedron Letters 2081 (1964).

479. Zemlicka, J.: 2',3'-O-(ethoxymethylene)uridine. Synthetic procedures in nucleic acid chemistry. **1**, 422. New York: Wiley. 1968.

480. Zemlicka, J., and S. Chladek: New method of dephosphorylation of ribonucleoside-2'(3')-phosphates. Tetrahedron Letters 715 (1969).

481. Zemlicka, J.: Nucleic acid components and their analogues. CXXXII. Alkylation of some nucleic acid components and their analogues with dimethylformamide acetals. Collect. Czech. Chem. Commun. **35**, 3572 (1970).

482. — Acetalation and acetylation of pyrimidine nucleosides in dioxane-acetonitrile-hydrogen chloride. J. Org. Chem. **36**, 2383 (1971).

483. Zemlicka, J., and S. Chladek: Synthesis of 2'(3')-O-glycyl derivatives of cytidylyl-(3'-5')-inosine and 2'-deoxycytidylyl-(3'-5')-adenosine. Biochemistry **10**, 1521 (1971).

484. Zemlicka, J., and J. P. Horwitz: Nucleosides. XIII. The concurrent introduction of two different blocking groups into some ribonucleosides. J. Org. Chem. **36**, 2809 (1971).

485. Zhenodarova, S. M., and M. I. Habarova: The enzymic synthesis of adenylyl-(3'-5')-cytidine. Biochim. Biophys. Acta **169**, 559 (1968).

486. Zmudzka, B., and D. Shugar: Role of the 2'-hydroxyl in polynucleotide conformation: poly 2'-O-methyluridylic acid. Acta Biochim. Polon. **18**, 321 (1971).

487. Zmudzka, B., M. Tichy, and D. Shugar: The structure of poly 2'-O-methylcytidylic acid and its complexes with polyinosinic acid. Acta Biochim. Polon. **19**, 149 (1972).

488. Zwierzak, A., and R. Gramze: Organophosphorus esters II. Novel approach to the synthesis of S-alkyl phosphorothioates. Z. Naturforsch. **26b**, 386 (1971).

489. Zwierzak, A., and M. Kluba: Organophosphorus esters — I. *t*-butyl as protecting group in phosphorylation via nucleophilic displacement. Tetrahedron Letters 3163 (1971).

(Received March 1, 1974)

Namenverzeichnis. Author Index

Kursiv gedruckte Seitenzahlen beziehen sich auf Literaturverzeichnisse

Page numbers printed in *italics* refer to References

Sachverzeichnis. Subject Index

Von · By

A. Siegel, Wien

Fortschritte der Chemie organischer Naturstoffe
Progress in the Chemistry of Organic Natural Products

Further volumes see next page

Springer-Verlag Wien · New York

Volume 20: 33 figures. XIII, 509 pages. 1962.
Cloth DM 117,—

Cumulative Index / Generalregister 1—20. 1938—1962. XVI, 369 pages. 1964.
Cloth DM 85,—

Volume 21: 14 figures. VII, 362 pages. 1963.
Cloth DM 88,—

Volume 22: 8 figures. VII, 370 pages. 1964.
Cloth DM 103,—

Volume 23: 58 figures. VIII, 397 pages. 1965.
Cloth DM 109,—

Volume 24: 25 figures. VIII, 475 pages. 1966.
Cloth DM 130,—

Volume 25: 25 figures. VII, 348 pages. 1967.
Cloth DM 92,—

Volume 26: 97 figures. IX, 456 pages. 1968.
Cloth DM 144,—

Volume 27: 47 figures. VIII, 412 pages. 1969.
Cloth DM 130,—

Volume 28: 14 figures. XII, 503 pages. 1970.
Cloth DM 155,—

Volume 29: 18 figures. VIII, 554 pages. 1971.
Cloth DM 181,—

Volume 30: 28 figures. VIII, 666 pages. 1973.
Cloth DM 225,—

Volume 31: 60 figures. IX, 693 pages. 1974.
Cloth DM 231,—

Contents: D. N. McGREGOR, Recent Developments in the Chemistry of Penicillins — CH. TAMM, The Antibiotic Complex of the Verrucarins and Roridins — J. C. ROBERTS, Aflatoxins and Sterigmatocystins — H. WAGNER, Flavonoid-Glykoside — TH. M. HARRIS, C. M. HARRIS, and K. B. HINDLEY, Biogenetic-Type Syntheses of Polyketide Metabolites — J. A. MARSHALL, ST. F. BRADY, and N. H. ANDERSEN, The Chemistry of Spiro [4.5] Decane Sesquiterpenes — E. HECKER and R. SCHMIDT, Phorbolesters — the Irritants and Cocarcinogens of Croton Tiglium L. — E. WINTERFELDT, Stereoselektive Totalsynthese von Indolalkaloiden — G. A. SWAN, Structure, Chemistry, and Biosynthesis of the Melanins — G. N. SCHRAUZER, Mechanisms of Corrin Dependent Enzymatic Reactions — Author Index / Namenverzeichnis — Subject Index / Sachverzeichnis

Further information will be sent on request

Price reduction for subscribers / Preisermäßigung für Subskribenten: 10%.

Special price reduction (20% of the list price) for the Vols. 1—30 plus Cumulative Index. / Vorzugspreis (20% Nachlaß) bei Bezug der Bände 1—30 inklusive Generalregister.